A First Course
in
Fourier
Analysis

David W. Kammler

Department of Mathematics
Southern Illinois University at Carbondale

PRENTICE HALL, Upper Saddle River, New Jersey 07458

Library of Congress Cataloging-in-Publication Data

Kammler, David W.
  A first course in Fourier analysis, David W. Kammler
    p. cm.
    Includes index
    ISBN 0-13-578782-3
    1. Fourier Analysis I. Title: Fourier analysis II. Title

    QA 403.5 K36 2000   99-086439
    $515':2433$–dc21   CIP

Acquisitions Editor: *George Lobell*
Production Editor: *Lynn Savino Wendel*
Assistant Vice President of Production and Manufacturing: *David W. Riccardi*
Executive Managing Editor: *Kathleen Schiaparelli*
Senior Managing Editor: *Linda Mihatov Behrens*
Manufacturing Buyer: *Alan Fischer*
Manufacturing Manager: *Trudy Pisciotti*
Marketing Manager: *Angela Battle*
Marketing Assistant: *Vince Jansen*
Director of Marketing: *John Tweeddale*
Editorial Assistant: *Gale Epps*
Art Director: *Jayne Conte*
Cover Designer: *Bruce Kenselaar*
Cover Illustration: *David W. Kammler*
About the Cover: This image represents the diffraction pattern for colliding gaussian
                  laser beams. *FOURIER* produced the pixels by suitably quantizing
                  samples of the wave function (9.106).

 ©2000 by Prentice-Hall, Inc.
Upper Saddle River, New Jersey 07458

Printed in the United States of America

10 9 8 7 6 5 4 3 2 1

ISBN 0-13-578782-3

Prentice-Hall International (UK) Limited, London
Prentice-Hall of Australia Pty. Limited, Sydney
Prentice-Hall Canada, Inc., Toronto
Prentice-Hall Hispanoamericana, S.A., Mexico
Prentice-Hall of India Private Limited, New Delhi
Prentice-Hall of Japan, Inc. Tokyo
Pearson Education Asia Pte. Ltd.
Editora Prentice-Hall do Brasil, Ltda., Rio de Janeiro

Mathematics: Source and Substance

Profound study of nature is the most fertile source of mathematical discoveries.

Joseph Fourier, *The Analytical Study of Heat*, p. 7

Mathematics is the science of patterns. The mathematician seeks patterns in number, in space, in science, in computers, and in imagination. Mathematical theories explain the relations among patterns; functions and maps, operators and morphisms bind one type of pattern to another to yield lasting mathematical structures. Applications of mathematics use these patterns to *explain* and predict natural phenomena that fit the patterns. Patterns suggest other patterns, often yielding patterns of patterns. In this way mathematics follows its own logic, beginning with patterns from science and completing the portrait by adding all patterns that derive from initial ones.

Lynn A. Steen, The science of patterns, *Science* **240**(1988), 616.

Contents

Preface   ix

The Mathematical Core

Chapter 1   Fourier's Representation for Functions on
$\mathbb{R}$, $\mathbb{T}_p$, $\mathbb{Z}$, and $\mathbb{P}_N$ **1**

1.1	Synthesis and Analysis Equations	1
1.2	Examples of Fourier's Representation	12
1.3	The Parseval Identities and Related Results	23
1.4	The Fourier-Poisson Cube	31
1.5	The Validity of Fourier's Representation	37
	References	59
	Exercise Set 1	61

Chapter 2   Convolution of Functions on $\mathbb{R}$, $\mathbb{T}_p$, $\mathbb{Z}$, and $\mathbb{P}_N$ **89**

2.1	Formal Definitions of $f * g$ and $f \star g$	89
2.2	Computation of $f * g$	91
2.3	Mathematical Properties of the Convolution Product	102
2.4	Examples of Convolution and Correlation	107
	References	115
	Exercise Set 2	116

Chapter 3   The Calculus for Finding Fourier
Transforms of Functions on $\mathbb{R}$ **129**

3.1	Using the Definition to Find Fourier Transforms	131
3.2	Rules for Finding Fourier Transforms	134
3.3	Selected Applications of the Fourier Transform Calculus	147
	References	155
	Exercise Set 3	156

Chapter 4	**The Calculus for Finding Fourier Transforms of Functions on $\mathbb{T}_p$, $\mathbb{Z}$, and $\mathbb{P}_N$**	**173**
4.1	Fourier Series	173
4.2	Selected Applications of Fourier Series	190
4.3	Discrete Fourier Transforms	196
4.4	Selected Applications of the DFT Calculus	212
	References	216
	Exercise Set 4	217
Chapter 5	**Operator Identities Associated with Fourier Analysis**	**239**
5.1	The Concept of an Operator Identity	239
5.2	Operators Generated by Powers of $\mathcal{F}$	243
5.3	Operators Related to Complex Conjugation	251
5.4	Fourier Transforms of Operators	255
5.5	Rules for Hartley Transforms	263
5.6	Hilbert Transforms	266
	References	271
	Exercise Set 5	272
Chapter 6	**The Fast Fourier Transform**	**291**
6.1	Pre-FFT Computation of the DFT	291
6.2	Derivation of the FFT via DFT Rules	296
6.3	The Bit Reversal Permutation	303
6.4	Sparse Matrix Factorization of $\mathcal{F}$ When $N = 2^m$	310
6.5	Sparse Matrix Factorization of $\mathbf{H}$ When $N = 2^m$	323
6.6	Sparse Matrix Factorization of $\mathcal{F}$ When $N = P_1 P_2 \cdots P_m$	327
6.7	Kronecker Product Factorization of $\mathcal{F}$	338
	References	345
	Exercise Set 6	345
Chapter 7	**Generalized Functions on $\mathbb{R}$**	**367**
7.1	The Concept of a Generalized Function	367
7.2	Common Generalized Functions	379
7.3	Manipulation of Generalized Functions	389
7.4	Derivatives and Simple Differential Equations	405
7.5	The Fourier Transform Calculus for Generalized Functions	413
7.6	Limits of Generalized Functions	427
7.7	Periodic Generalized Functions	440
7.8	Alternative Definitions for Generalized Functions	450
	References	452
	Exercise Set 7	453

Selected Applications

Chapter 8   Sampling   **483**

 8.1 Sampling and Interpolation   483
 8.2 Reconstruction of f from Its Samples   487
 8.3 Reconstruction of f from Samples of $a_1 * f$, $a_2 * f$, ... 497
 8.4 Approximation of Almost Bandlimited Functions   505
    References   508
    Exercise Set 8   509

Chapter 9   Partial Differential Equations   **523**

 9.1 Introduction   523
 9.2 The Wave Equation   526
 9.3 The Diffusion Equation   540
 9.4 The Diffraction Equation   553
 9.5 Fast Computation of Frames for Movies   571
    References   573
    Exercise Set 9   574

Chapter 10   Wavelets   **593**

 10.1 The Haar Wavelets   593
 10.2 Support-Limited Wavelets   609
 10.3 Analysis and Synthesis with Daubechies' Wavelets   640
 10.4 Filter Banks   655
    References   673
    Exercise Set 10   674

Chapter 11   Musical Tones   **693**

 11.1 Basic Concepts   693
 11.2 Spectrograms   702
 11.3 Additive Synthesis of Tones   707
 11.4 FM Synthesis of Tones   711
 11.5 Synthesis of Tones from Noise   718
 11.6 Music with Mathematical Structure   723
    References   727
    Exercise Set 11   728

Chapter 12   Probability   **737**

    12.1   Probability Density Functions on $\mathbb{R}$ 737
    12.2   Some Mathematical Tools   741
    12.3   The Characteristic Function   746
    12.4   Random Variables   753
    12.5   The Central Limit Theorem   764
           References   780
           Exercise Set 12   780

Appendices   **A-1**

    Appendix 0   The Impact of Fourier Analysis   A-1
    Appendix 1   Functions and Their Fourier Transforms   A-3
    Appendix 2   The Fourier Transform Calculus   A-13
    Appendix 3   Operators and Their Fourier Transforms   A-18
    Appendix 4   The Whittaker-Robinson Flow Chart
                  for Harmonic Analysis   A-22
    Appendix 5   FORTRAN Code for a Radix 2 FFT   A-26
    Appendix 6   The Standard Normal Probability Distribution   A-32
    Appendix 7   Frequencies of the Piano Keyboard   A-36

Index   **I-1**

Preface

To the Student

This book is about one big idea: You can synthesize a variety of complicated functions from pure sinusoids in much the same way that you produce a major chord by striking nearby C, E, G keys on a piano. A geometric version of this idea forms the basis for the ancient Hipparchus-Ptolemy model of planetary motion (*Almagest*, 2nd century, *cf.* Fig. 1.2). It was Joseph Fourier (*Analytical Theory of Heat*, 1815), however, who developed modern methods for using trigonometric series and integrals as he studied the flow of heat in solids. Today, Fourier analysis is a highly evolved branch of mathematics with an incomparable range of applications and with an impact that is second to none, (*cf.* Appendix 0). If you are a student in one of the mathematical, physical, or engineering sciences, you will almost certainly find it necessary to learn the elements of this subject. My goal in writing this book is to help you acquire a working knowledge of Fourier analysis early in your career.

If you have mastered the usual core courses in calculus and linear algebra, you have the maturity to follow the presentation without undue difficulty. A few of the proofs and more theoretical exercises require concepts (uniform continuity, uniform convergence, ...) from an analysis or advanced calculus course. You may choose to skip over the difficult steps in such arguments and simply accept the stated results. The text has been designed so that you can do this without severely impacting your ability to learn the important ideas in the subsequent chapters. In addition, I will use a *potpourri* of notions from undergraduate courses in differential equations [solve $y'(x) + \alpha y(x) = 0$, $y'(x) = xy(x)$, $y''(x) + \alpha^2 y(x) = 0$, ...], complex analysis (Euler's formula: $e^{i\theta} = \cos\theta + i\sin\theta$, arithmetic for complex numbers, ...), number theory (integer addition and multiplication modulo N, Euclid's gcd algorithm, ...), probability (random variable, mean, variance, ...), physics ($F = ma$, conservation of energy, Huygens' principle, ...), signals and systems (LTI systems, low-pass filters, the Nyquist rate, ...), etc. You will have no trouble picking up these concepts as they are introduced in the text and exercises.

If you wish, you can find additional information about almost any topic in this book by consulting the annotated references at the end of the corresponding chapter. You will often discover that I have abandoned a traditional presentation in favor of one that is in keeping with my goal of making these ideas accessible to undergraduates. For example, the usual presentation of the Schwartz theory of distributions assumes some familiarity with the Lebesgue integral and with a graduate-level functional analysis course. In contrast, my development of δ, III, ... in Chapter 7 uses only notions from elementary calculus. Once you master this theory, you can use generalized functions to study sampling, PDEs, wavelets, probability, diffraction, ...

The exercises (540 of them) are my greatest gift to you! Read each chapter carefully to acquire the basic concepts, and then solve as many problems as you can. You may find it beneficial to organize an interdisciplinary study group, *e.g.*, mathematician + physicist + electrical engineer. Some of the exercises provide routine drill: You must learn to find convolution products, to use the FT calculus, to do routine computations with generalized functions, etc. Some supply historical perspective: You can play Gauss and discover the FFT, analyze Michelson and Stratton's analog supercomputer for summing Fourier series, etc. Some ask for mathematical details: Give a sufficient condition for ... , given an example of ... , show that, ... Some involve your personal harmonic analyzers: Experimentally determine the bandwidth of your eye, describe what would you hear if you replace notes with frequencies $F_1, F_2, \ldots$ by notes with frequencies $C/F_1, C/F_2, \ldots$. Some prepare you for computer projects: Compute π to 1000 digits, prepare a movie for a vibrating string, generate the sound file for Risset's endless glissando, etc. Some will set you up to discover a pattern, formulate a conjecture, and prove a theorem. (It's quite a thrill when you get the hang of it!) I expect you to spend a lot of time working exercises, but I want to help you work efficiently. Complicated results are broken into simple steps so you can do (a), then (b), then (c), ... until you reach the goal. I frequently supply hints that will lead you to a productive line of inquiry. You will sharpen your problem-solving skills as you take this course.

Synopsis

The chapters of the book are arranged as follows:

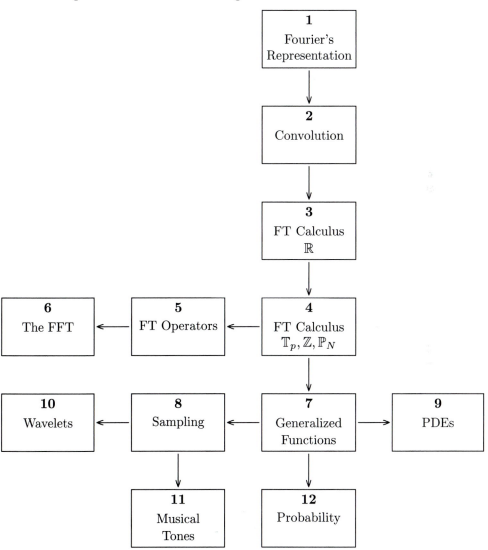

The mathematical core is given in Chapters 1–7 and selected applications are developed in Chapters 8–12.

We present the basic themes of Fourier analysis in the first two chapters. Chapter 1 opens with Fourier's synthesis and analysis equations for functions on the real line $\mathbb{R}$, on the circle $\mathbb{T}_p$, on the integers $\mathbb{Z}$, and on the polygon $\mathbb{P}_N$. We discretize by sampling (obtaining functions on $\mathbb{Z}, \mathbb{P}_N$ from functions on $\mathbb{R}, \mathbb{T}_p$), we periodize by summing translates (obtaining functions on $\mathbb{T}_p, \mathbb{P}_N$ from functions on $\mathbb{R}, \mathbb{Z}$), and we informally derive the corresponding Poisson identities. We combine these mappings to form the Fourier-Poisson cube, a structure that links the Fourier transforms, Fourier series, and discrete Fourier transforms students encounter in their under-graduate classes. We prove that these equations are valid when certain elementary sufficient conditions are satisfied. We complete the presentation of basic themes by describing the convolution product of functions on $\mathbb{R}, \mathbb{T}_p, \mathbb{Z}$, and $\mathbb{P}_N$ in Chapter 2.

Chapters 3 and 4 are devoted to the development of computational skills. We introduce the Fourier transform calculus for functions on $\mathbb{R}$ by finding transforms of the box, $\Pi(x)$, the truncated exponential, $e^{-x} h(x)$, and the unit gaussian $e^{-\pi x^2}$. We present the rules (linearity, translation, dilation, convolution, inversion, ...) and use them to obtain transforms for a large class of functions on $\mathbb{R}$. Various methods are used to find Fourier series. In addition to direct integration (with Kronecker's rule), we present (and emphasize) Poisson's formula, Eagle's method, and the use of elementary Laurent series (from calculus). Corresponding rules facilitate the manipulation of the Fourier representations for functions on $\mathbb{T}_p$ and $\mathbb{Z}$. An understanding of the Fourier transform calculus for functions on $\mathbb{P}_N$ is essential for anyone who wishes to use the FFT. We establish a few well-known DFT pairs and develop the corresponding rules. We illustrate the power of this calculus by deriving the Euler-Maclaurin sum formula from elementary numerical analysis and evaluating the Gauss sums from elementary number theory.

In Chapter 5 we use operators, *i.e.*, function-to-function mappings, to organize the multiplicity of specialized Fourier transform rules. We characterize the basic symmetries of Fourier analysis and develop a deeper understanding of the Fourier transform calculus. We also use the operator notation to facilitate a study of Sine, Cosine, Hartley, and Hilbert transforms.

The subject of Chapter 6 is the FFT (which Gilbert Strang calls the most important algorithm of the 20th century!). After describing the $O(N^2)$ scheme of Horner, we use the DFT calculus to produce an N-point DFT with only $O(N \log_2 N)$ operations. We use an elementary zipper identity to obtain a sparse factorization of the DFT matrix and develop a corresponding algorithm (including the clever enhancements of Bracewell and Buneman) for fast machine computation. We briefly introduce some of the more specialized DFT factorizations that can be obtained by using Kronecker products.

An elementary exposition of generalized functions (the tempered distributions of Schwartz) is given in Chapter 7, the heart of the book. We introduce the Dirac δ [as the second derivative of the ramp $r(x) := \max(x, 0)$], the comb III; the reciprocal "$1/x$", the Fresnel function $e^{i\pi x^2}$, ... and carefully extend the FT calculus rules to this new setting. We introduce generalized (weak) limits so that we can work with infinite series, ordinary derivatives, partial derivatives, ...

Selected applications of Fourier analysis are given in the remaining chapters. (You can find whole textbooks devoted to each of these topics.) Mathematical models based on Fourier synthesis, analysis done with generalized functions, and FFT computations are used to foster insight and understanding. You will experience the enormous "leverage" Fourier analysis can give as you study this material!

Sampling theory, the mathematical basis for digital signal processing, is the focus of Chapter 8. We present weak and strong versions of Shannon's theorem together with the clever generalization of Papoulis. Using these ideas (and characteristics of the human ear) we develop the elements of computer music in Chapter 11. We use additive synthesis and Chowning's FM synthesis to generate samples for musical tones, and we use spectrograms to visualize the structure of the corresponding sound files.

Fourier analysis was invented to solve PDEs, the subject of Chapter 9. We formulate mathematical models for the motion of a vibrating string, for the diffusion of heat (Fourier's work), and for Fresnel diffraction. (The Schrödinger equation from quantum mechanics seems much less intimidating when interpreted within the context of elementary optics!) With minimal effort, we solve these PDEs, establish suitable conservation laws, and examine representative solutions. (The cover illustration was produced by using the FFT to generate slices for the diffraction pattern that results when two gaussian laser beams interfere.)

Chapter 10 is devoted to the study of wavelets, an exciting new branch of mathematics. We introduce the basic ideas using the piecewise constant functions associated with the Haar wavelets. We then use the theory of generalized functions to develop the compactly supported orthogonal wavelets created by I. Daubechies in 1988. Fourier analysis plays an essential role in the study of corresponding filter banks that are used to process audio and image files.

We present the elements of probability theory in Chapter 12 using generalized densities, $e.g.$, $f(x) := (1/2)[\delta(x+1) + \delta(x-1)]$ serves as the probability density for a coin toss. We use Fourier analysis to find moments, convolution products, characteristic functions, and to establish the uncertainty relation (for suitably regular probability densities on $\mathbb{R}$). We then use the theory of generalized functions to prove the central limit theorem, the foundation for modern statistics!

To the Instructor

This book is the result of my efforts to create a modern elementary introduction to Fourier analysis for students from mathematics, science, and engineering. There is more than enough material for a tight one-semester survey or for a leisurely two-semester course that allocates more time to the applications. You can adjust the level and the emphasis of the course to your students by the topics you cover and by your assignment of homework exercises. You can use Chapters 1, 3, 4, 7, and 9 to update a lackluster boundary value problems course. You can use Chapters 1, 3, 4, 7, 8, and 10 to give a serious introduction to sampling theory and wavelets. You can use selected portions of Chapters 2–4, 6, 8, and 11 (with composition exercises!) for a fascinating elementary introduction to the mathematics of computer-generated music. You can use the book for an undergraduate capstone course that emphasizes group learning of the interdisciplinary topics and mastering of some of the more difficult exercises. Finally, you can use Chapters 7–12 to give a graduate-level introduction to generalized functions for scientists and engineers.

This book is not a traditional mathematics text. You will find a minimal amount of jargon and note the absence of a logically complete theorem-proof presentation of elementary harmonic analysis. Basic computational skills are developed for solving real problems, not just for drill. There is a strong emphasis on the visualization of equations, mappings, theorems, ... and on the interpretation of mathematical ideas within the context of some application. In general, the presentation is informal, but there are careful proofs for theorems that have strategic importance, and there are a number of exercises that lead students to develop the implications of ideas introduced in the text.

Be sure to cover one or more of the applications chapters. Students enjoy learning about the essential role Fourier analysis plays in modern mathematics, science, and engineering. You will find that it is much easier to develop and to maintain the market for a course that emphasizes these applications.

When I teach this material I devote 24 lectures to the mathematical core (deleting portions of Chapters 1, 5, and 6) and 18 lectures to the applications (deleting portions of Chapters 10, 11, and 12). I also spend 3–4 hours per week conducting informal problem sessions, giving individualized instruction, etc. I lecture from transparencies and use a PC (with *FOURIER*) for visualization and sonification. This is helpful for the material in Chapters 2, 5, 6, and 12 and essential for the material in Chapters 9, 10, and 11. I use a laser with apertures on 35mm slides to show a variety of diffraction patterns when I introduce the topic of diffraction in Chapter 9. This course is a great place to demonstrate the synergistic roles of experimentation, mathematical modeling, and computer simulation in modern science and engineering.

Course materials related to this book can be downloaded from the author's Web site:

http://www.math.siu.edu/Kammler/

I have one word of caution. As you teach this material you will face the constant temptation to prove too much too soon. My informal use of $\overset{?}{=}$ cries out for the precise statement and proof of some relevant sufficient condition. (In most cases there is a corresponding exercise, with hints, for the student who would really like to see the details.) For every hour that you spend presenting 19th-century advanced calculus arguments, however, you will have one less hour for explaining the 20th-century mathematics of generalized functions, sampling theory, wavelets, ... You must decide which of these alternatives will best serve your students.

Acknowledgments

I wish to thank Southern Illinois University at Carbondale for providing a nurturing environment during the evolution of ideas that led to this book. Sabbatical leaves and teaching fellowships were essential during the early phases of the work. National Science Foundation funding (NSF-USE 89503, NSF-USE 9054179) for four faculty short courses at the Touch of Nature center at SIUC during 1989–1992, and NECUSE funding for a faculty short course at Bates College in 1995 enabled me to share preliminary versions of these ideas with faculty peers. I have profited enormously from their thoughtful comments and suggestions, particularly those of Davis Cope, Bo Green, Carruth McGehee, Dale Mugler, Mark Pinsky, David Snider, Patrick Sullivan, Henry Warchall, and Jo Ward. I deeply appreciate the National Science Foundation course development grant (NSF-USE 9156064) that provided support for the creation of many of the exercise sets as well as for equipment and programming services that were used to develop the second half of the book.

I wish to thank my editors, George Lobell and Lynn Savino Wendel, and the staff at Prentice Hall for investing in this project. I also want to thank the reviewers whose pointed suggestions greatly improved the manuscript. I appreciate the critical role that Pat Van Fleet, David Eubanks, Xinmin Li, Wenbing Zhang, and Jeff McCreight played as graduate students in helping me to learn the details associated with various applications of Fourier analysis. I am particularly indebted to David Eubanks for the innumerable hours he invested in the development of the software package *FOURIER* that I use when I teach this material. I want to acknowledge my debt to Rebecca Parkinson for creating the charming sketch of Joseph Fourier that appears in Fig. 3.4. My heartfelt thanks go to Linda Gibson and to Charles Gibson for preparing the TEX files for the book (and for superhuman patience during the innumerable revisions!). Finally, I express my deep appreciation to my wife, Ruth, for her love and encouragement throughout this project.

I hope that you enjoy this approach for learning Fourier analysis. If you have corrections, ideas for new exercises, suggestions for improving the presentation, etc., I would love to hear from you!

David W. Kammler
Mathematics Department
Southern Illinois University at Carbondale
Carbondale, Illinois 62901-4408
dkammler@math.siu.edu

Chapter 1

Fourier's Representation for Functions on $\mathbb{R}$, $\mathbb{T}_p$, $\mathbb{Z}$, and $\mathbb{P}_N$

1.1 Synthesis and Analysis Equations

Introduction

In mathematics we often try to synthesize a rather arbitrary function f using a suitable linear combination of certain elementary basis functions. For example, the power functions $1, x, x^2, \ldots$ serve as such basis functions when we synthesize f using the power series representation

$$f(x) = a_0 + a_1 x + a_2 x^2 + \cdots . \tag{1}$$

The coefficient a_k that specifies the amount of the basis function x^k needed in the recipe (1) for constructing f is given by the well-known Maclaurin formula

$$a_k = \frac{f^{(k)}(0)}{k!}, \qquad k = 0, 1, 2, \ldots$$

from elementary calculus. Since the equations for $a_0, a_1, a_2, \ldots$ can be used only in cases where $f, f', f'', \ldots$ are defined at $x = 0$, we see that not all functions can be synthesized in this way. The class of analytic functions that do have such power series representations is a large and important one, however, and like Newton [who with justifiable pride referred to the representation (1) as "my method"], you have undoubtedly made use of such power series to evaluate functions, to construct antiderivatives, to compute definite integrals, to solve differential equations, to justify discretization procedures of numerical analysis, etc.

Fourier's representation (developed a century and a half after Newton's) uses as basis functions the complex exponentials

$$e^{2\pi i s x} := \cos(2\pi s x) + i \cdot \sin(2\pi s x) \tag{2}$$

where s is a real frequency parameter that serves to specify the rate of oscillation, and $i^2 = -1$. When we graph this complex exponential, *i.e.*, when we graph

$$u := \operatorname{Re} e^{2\pi i s x} = \cos(2\pi s x)$$
$$v := \operatorname{Im} e^{2\pi i s x} = \sin(2\pi s x)$$

as functions of the real variable x in x, u, v-space, we obtain a helix (a Slinky!) that has the spacing $1/|s|$ between the coils. Projections of this helix on the planes $v = 0$, $u = 0$, $x = 0$ give the sinusoids $u = \cos(2\pi s x)$, $v = \sin(2\pi s x)$, and the circle $u^2 + v^2 = 1$, as shown in Fig. 1.1.

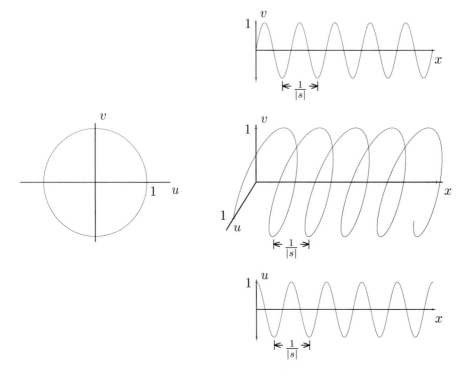

Figure 1.1. The helix $u = \cos(2\pi s x)$, $v = \sin(2\pi s x)$ in x, u, v-space together with projections in the x, u, the x, v, and the u, v planes.

Functions on $\mathbb{R}$

Fourier discovered that any suitably regular complex-valued function f defined on the real line $\mathbb{R}$ can be synthesized by using the integral representation

$$f(x) = \int_{s=-\infty}^{\infty} F(s)e^{2\pi i s x} ds, \qquad -\infty < x < \infty. \tag{3}$$

Here F is also a complex-valued function defined on $\mathbb{R}$, and we think of $F(s)ds$ as being the amount of the exponential $e^{2\pi i s x}$ with frequency s that must be used in the recipe (3) for f. At this point we are purposefully vague as to the exact hypotheses that must be imposed on f to guarantee the existence of such a Fourier representation. Roughly speaking, the Fourier representation (3) is possible in all cases where f does not fluctuate too wildly and where the tails of f at $\pm\infty$ are not too large. It is certainly not *obvious* that such functions can be represented in the form (3) [nor is it *obvious* that $\sin x$, $\cos x$, e^x, and many other functions can be represented using the power series (1)]. At this point we are merely announcing that this is, in fact, the case, and we encourage you to become familiar with the equation (3) along with analogous equations that will be introduced in the next few paragraphs. Later on we will establish the validity of (3) after giving meaning to the intentionally vague term *suitably regular*.

Fourier found that the auxiliary function F from the representation (3) can be constructed by using the integral

$$F(s) = \int_{x=-\infty}^{\infty} f(x)e^{-2\pi i s x} dx, \qquad -\infty < s < \infty. \tag{4}$$

We refer to (3) as the *synthesis equation* and to (4) as the *analysis equation* for f. The function F is said to be the *Fourier transform* of f. We cannot help but notice the symmetry between (3) and (4), *i.e.*, we can interchange f, F provided that we also interchange $+i$ and $-i$. Other less symmetric analysis-synthesis equations are sometimes used for Fourier's representation, *cf.* Ex. 1.4, but we will always use (3)–(4) in this text. We will often display the graphs of f, F side by side, as illustrated in Fig. 1.2. Our sketch corresponds to the case where both f and F are real valued. In general, it is necessary to display the four graphs of Re f, Im f, Re F, and Im F. You will find such displays in Chapter 3, where we develop an efficient calculus for evaluating improper integrals having the form (3) or (4).

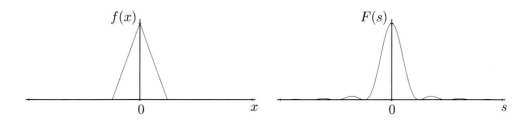

Figure 1.2. The graph of a function f on $\mathbb{R}$ and its Fourier transform F on $\mathbb{R}$.

Functions on $\mathbb{T}_p$

We say that a function f defined on $\mathbb{R}$ is *p-periodic*, $p > 0$, when

$$f(x + p) = f(x), \qquad -\infty < x < \infty.$$

Fourier (like Euler, Lagrange, and D. Bernoulli before him) discovered that a suitably regular p-periodic complex-valued function on $\mathbb{R}$ can be synthesized by using the p-periodic complex exponentials from (2). We will routinely identify any p-periodic function on $\mathbb{R}$ with a corresponding function defined on the circle $\mathbb{T}_p$ having the circumference p as illustrated in Fig. 1.3. [To visualize the process, think of wrapping the graph of $f(x)$ versus x around a right circular cylinder just like the paper label is wrapped around a can of soup!] Of course, separate graphs for Re f and Im f must be given in cases where f is complex valued.

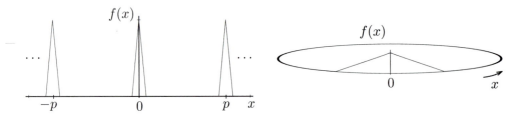

Figure 1.3. Identification of a p-periodic function f on $\mathbb{R}$ with a corresponding function on the circle $\mathbb{T}_p$ having the circumference p.

The complex exponential $e^{2\pi i s x}$ will be p-periodic in the argument x, *i.e.*,

$$e^{2\pi i s (x+p)} = e^{2\pi i s x}, \qquad -\infty < x < \infty,$$

when

$$e^{2\pi i s p} = 1,$$

i.e., when

$$s = k/p \quad \text{for some} \quad k = 0, \pm 1, \pm 2, \dots .$$

In this way we see that the p-periodic exponentials from (2) are given by

$$e^{2\pi i k x / p}, \qquad k = 0, \pm 1, \pm 2, \dots ,$$

cf. Fig. 1.4.

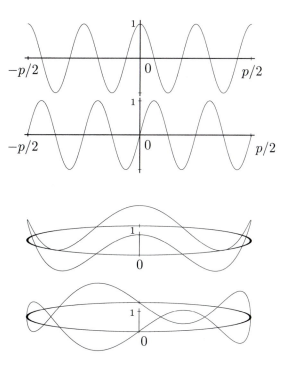

Figure 1.4. Real and imaginary parts of the complex exponen-
tial $e^{8\pi i x/p}$ as functions on $\mathbb{R}$ and as functions on $\mathbb{T}_p$.

Fourier's representation

$$f(x) = \sum_{k=-\infty}^{\infty} F[k]e^{2\pi i k x/p}, \qquad -\infty < x < \infty, \qquad (5)$$

for a p-periodic function f uses all of these complex exponentials. In this case
F is a complex-valued function defined on the integers $\mathbb{Z}$ (from the German word
Zahlen, for *integers*). We use brackets [ ] rather than parentheses ( ) to enclose the
independent variable k in order to remind ourselves that this argument is discrete.
We think of $F[k]$ as being the amount of the exponential $e^{2\pi i k x/p}$ that we must use
in the recipe (5) for f. We refer to (5) as the *Fourier series* for f and we say that
$F[k]$ is the kth *Fourier coefficient* for f. You may be familiar with the alternative
representation

$$f(x) = \frac{a_0}{2} + \sum_{k=1}^{\infty}\{a_k \cos(2\pi k x/p) + b_k \sin(2\pi k x/p)\}$$

for a Fourier series. You can use Euler's identity (2) to see that this representation
is equivalent to (5), *cf.* Ex. 1.16. From time to time we will work with such cos, sin

series, *e.g.*, this form may be preferable when f is real or when f is known to have even or odd symmetry. For general purposes, however, we will use the compact complex form (5).

Fourier found that the coefficients $F[k]$ for the representation (5) can be constructed for a given function f by using the integrals

$$F[k] = \frac{1}{p} \int_{x=0}^{p} f(x) e^{-2\pi i k x/p} dx, \qquad k = 0, \pm 1, \pm 2, \dots . \tag{6}$$

[Before discovering the simple formula (6), Fourier made use of clumsy, mathematically suspect arguments based on power series to find these coefficients.] We refer to (5) as the *synthesis equation* and to (6) as the *analysis equation* for the p-periodic function f, and we say that F is the *Fourier transform* of f within this context. We use small circles on line segments, *i.e.*, *lollipops*, when we graph F (a function on $\mathbb{Z}$), and we often display the graphs of f, F side by side as illustrated in Fig. 1.5. Of course, we must provide separate graphs for Re f, Im f, Re F, Im F in cases where f, F are not real valued. You will find such displays in Chapter 4, where we develop a calculus for evaluating integrals having the form (6).

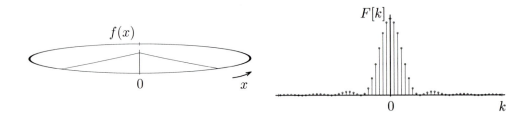

Figure 1.5. The graph of a function f on $\mathbb{T}_p$ and its Fourier transform F on $\mathbb{Z}$.

Functions on $\mathbb{Z}$

There is a Fourier representation for any suitably regular complex-valued function f that is defined on the set of integers, $\mathbb{Z}$. As expected, we synthesize f from the complex exponential functions $e^{2\pi i s n}$ on $\mathbb{Z}$, with s being a real parameter. Now for any real s and any integer m we find

$$e^{2\pi i (s+m) n} = e^{2\pi i s n}, \qquad n = 0, \pm 1, \pm 2, \dots$$

(*i.e.*, the exponentials $e^{2\pi i s n}$, $e^{2\pi i (s\pm 1) n}$, $e^{2\pi i (s\pm 2) n}$, ... are indistinguishable when n is constrained to take integer values). This being the case, we will synthesize f using

$$e^{2\pi i s n}, \qquad 0 \le s < 1$$

or equivalently, using

$$e^{2\pi i s n/p}, \qquad 0 \le s < p$$

where p is some fixed positive number. Figure 1.6 illustrates what happens when we attempt to use some $s > p$. The high-frequency sinusoid takes on the identity or *alias* of some corresponding low-frequency sinusoid. It is easy to see that $e^{2\pi i s n/p}$ oscillates slowly when s is near 0 or when s is near p. The choice $s = p/2$ gives the most rapid oscillation with the complex exponential

$$e^{2\pi i (p/2)n/p} = (-1)^n$$

having the smallest possible period, 2.

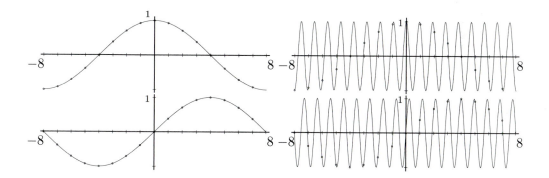

Figure 1.6. The identical samples of $e^{2\pi i x/16}$ and $e^{2\pi i 17x/16}$ at $x = 0, \pm 1, \pm 2, \ldots$.

Fourier's *synthesis equation*,

$$f[n] = \int_{s=0}^{p} F(s)e^{2\pi i s n/p}ds, \tag{7}$$

for a suitably regular function f on $\mathbb{Z}$, uses all of these complex exponentials on $\mathbb{Z}$, and the corresponding *analysis equation* is given by

$$F(s) = \frac{1}{p} \sum_{n=-\infty}^{\infty} f[n]e^{-2\pi i s n/p}. \tag{8}$$

We say that F is the *Fourier transform* of f and observe that this function is p-periodic in s, *i.e.*, that F is a complex-valued function on the circle $\mathbb{T}_p$. Figure 1.7 illustrates such an f, F pair.

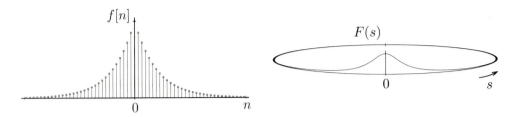

Figure 1.7. The graph of a function f on $\mathbb{Z}$ and its Fourier transform F on $\mathbb{T}_p$.

We have chosen to include the parameter $p > 0$ for the representation (7) (instead of working with the special case $p = 1$) in order to emphasize the *duality* that exists between (5)–(6) and (7)–(8). Indeed, if we replace

$$i, \; x, \; k, \; f, \; F$$

in (5)–(6) by

$$-i, \; s, \; n, \; pF, \; f,$$

respectively, we obtain (7)–(8). Thus every Fourier representation of the form (5)–(6) corresponds to a Fourier representation of the form (7)–(8), and vice versa.

Functions on $\mathbb{P}_N$

Let N be a positive integer, and let $\mathbb{P}_N$ consist of N uniformly spaced points on the circle $\mathbb{T}_N$ as illustrated in Fig. 1.8. We will call this discrete circle a *polygon* even in the degenerate cases where $N = 1, 2$.

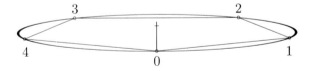

Figure 1.8. The polygon $\mathbb{P}_5$.

The simplest Fourier representation [found by Gauss in the course of his study of interpolation by trigonometric polynomials a few years before Fourier discovered either (3)–(4) or (5)–(6)] occurs when f is a complex-valued N-periodic function defined on $\mathbb{Z}$. We will routinely identify such an N-periodic f with a corresponding function that is defined on $\mathbb{P}_N$ as illustrated in Fig. 1.9. Of course, we must provide separate graphs for Re f, Im f when f complex valued. Since f is completely specified by the N function values $f[n]$, $n = 0, 1, \ldots, N - 1$, we will sometimes find that it is convenient to use a complex N-vector

$$f = (f[0], f[1], \ldots, f[N - 1])$$

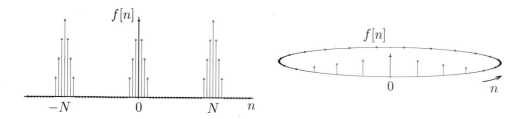

Figure 1.9. Identification of an N-periodic discrete function on $\mathbb{Z}$ with a corresponding function on the polygon $\mathbb{P}_N$.

to represent this function. This is particularly useful when we wish to process f numerically. You will observe that we always use $n = 0, 1, \ldots, N - 1$ (not $n = 1, 2, \ldots, N$) to index the components of f.

The complex exponential $e^{2\pi i s n}$ (with s being a fixed real parameter) will be N-periodic in the integer argument n, *i.e.*,

$$e^{2\pi i s(n+N)} = e^{2\pi i s n} \quad \text{for all} \quad n = 0, \pm 1, \pm 2, \ldots$$

when

$$e^{2\pi i s N} = 1,$$

i.e., when $s = k/N$ for some integer k. On the other hand, when m is an integer we find

$$e^{2\pi i k n/N} = e^{2\pi i (k+mN)n/N} \quad \text{for all} \quad n = 0, \pm 1, \pm 2, \ldots,$$

so the parameters

$$s = \frac{k}{N}, \quad s = \frac{k \pm N}{N}, \quad s = \frac{k \pm 2N}{N}, \quad \ldots$$

all give the same function. Thus we are left with precisely N distinct discrete N-periodic complex exponentials

$$e^{2\pi i k n/N}, \qquad k = 0, 1, \ldots, N - 1.$$

The complex exponentials with $k = 1$ or $k = N - 1$ make one complete oscillation on $\mathbb{P}_N$, those with $k = 2$ or $k = N - 2$ make two complete oscillations, etc., as illustrated in Fig. 1.10. The most rapid oscillation occurs when N is even and $k = N/2$ with the corresponding complex exponential

$$e^{2\pi i (N/2)n/N} = (-1)^n$$

having the smallest possible period, 2.

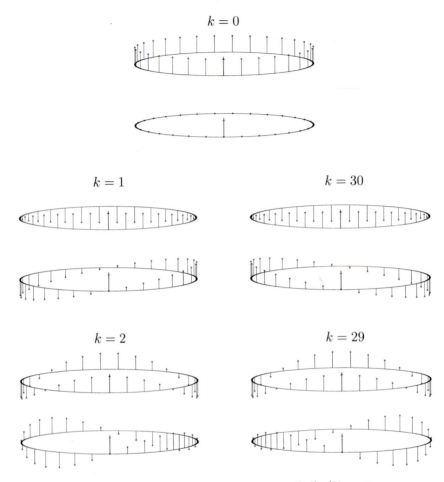

Figure 1.10. Complex exponentials $e^{2\pi i k n/31}$ on $\mathbb{P}_{31}$.

Fourier's *synthesis equation* takes the form

$$f[n] = \sum_{k=0}^{N-1} F[k]e^{2\pi i k n/N}, \qquad n = 0, \pm 1, \pm 2, \ldots \tag{9}$$

within this setting. Again we regard $F[k]$ as the amount of the discrete exponential $e^{2\pi i k n/N}$ that must be used in the recipe for f, we refer to (9) as the *discrete Fourier series* for f, and we say that $F[k]$ is the kth *Fourier coefficient* for f. The corresponding *analysis equation*

$$F[k] = \frac{1}{N} \sum_{n=0}^{N-1} f[n]e^{-2\pi i k n/N}, \qquad k = 0, 1, \ldots, N-1 \tag{10}$$

enables us to find the coefficients $F[0], F[1], \ldots, F[N-1]$ for the representation
(9) from the known function values of $f[0], f[1], \ldots, f[N-1]$. We refer to F as
the *discrete Fourier transform* (DFT) or more simply as the *Fourier transform* of
f within this context. The formula (10) gives an N-periodic discrete function on
$\mathbb{Z}$ when we allow k to take all integer values, so we will say that F is a function
on $\mathbb{P}_N$. Again, we plot graphs of f, F side by side, as illustrated in Fig. 1.11. You
will find such displays in Chapter 4, where we develop a calculus for evaluating the
finite sums (9), (10). Later on, in Chapter 6, you will learn an efficient way to do
such calculations on a computer.

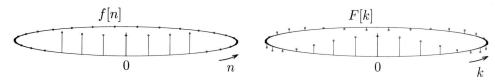

Figure 1.11. The graph of a function f on $\mathbb{P}_{31}$ and its Fourier
transform F on $\mathbb{P}_{31}$.

Summary

The following observations will help you remember Fourier's synthesis and analysis equations.

Functions on $\mathbb{R}$

(3) $f(x) = \int_{s=-\infty}^{\infty} F(s)e^{2\pi i s x} ds$

(4) $F(s) = \int_{x=-\infty}^{\infty} f(x)e^{-2\pi i s x} dx$

Functions on $\mathbb{T}_p$

(5) $f(x) = \sum_{k=-\infty}^{\infty} F[k]e^{2\pi i k x/p}$

(6) $F[k] = \dfrac{1}{p}\int_{x=0}^{p} f(x)e^{-2\pi i k x/p} dx$

Functions on $\mathbb{Z}$

(7) $f[n] = \int_{s=0}^{p} F(s)e^{2\pi i s n/p} ds$

(8) $F(s) = \dfrac{1}{p}\sum_{n=-\infty}^{\infty} f[n]e^{-2\pi i s n/p}$

Functions on $\mathbb{P}_N$

(9) $f[n] = \sum_{k=0}^{N-1} F[k]e^{2\pi i k n/N}$

(10) $F[k] = \dfrac{1}{N}\sum_{n=0}^{N-1} f[n]e^{-2\pi i k n/N}$

- The Fourier transform F has the real argument s when f is aperiodic and the
  integer argument k when f is periodic. The function f has the real argument
  x when F is aperiodic and the integer argument n when F is periodic.

- The argument of the exponentials that appear in the synthesis-analysis equations is the product of $\pm 2\pi i$, the argument s or k of F, the argument x or n of f, and the reciprocal $1/p$ or $1/N$ of the period if either f or F is periodic.
- The synthesis equation uses the $+i$ exponential and *all* values of F to form f. The analysis equation uses the $-i$ exponential and *all* values of f to form F.
- The reciprocal $1/p$ or $1/N$ of the period serves as a scale factor on the analysis equation in cases where either f or F is periodic. No such factor is used on the synthesis equation.

During the opening scene of an opera you catch glimpses of the main characters, but you have not yet learned the subtle personality traits or relationships that will unfold during the rest of the performance. In much the same way, you have been briefly introduced to the eight remarkable identities (3)–(10) that will appear throughout this course. (You can verify this by skimming through the text!) At this point, it would be very beneficial for you to spend a bit of time getting acquainted with these identities. Begin with a function f from Exs. 1.1, 1.8–1.10, 1.13, 1.14, evaluate the sum or integral from the analysis equation to find F, and then evaluate the sum or integral from the synthesis equation to establish the validity of Fourier's representation for this f. (It's hard to find examples where both of these sums, integrals can be found by using the tools from calculus!) See if you can determine how certain symmetries possessed by f are made manifest in F by doing Exs. 1.2, 1.11, 1.15. Explore alternative ways for writing the synthesis-analysis equations as given in Exs. 1.3, 1.4, 1.12, 1.16. And if you are interested in associating some physical meaning with the synthesis and analysis equations, then do try Ex. 1.17!

1.2 Examples of Fourier's Representation

Introduction

What can you do with Fourier's representation? In this section we will briefly describe six diverse settings for these ideas that will help you learn to recognize the patterns (3)–(10). Other applications will be developed with much more detail in Chapters 8–12. (You may want to read the first few pages of some of these chapters at this time!)

The Hipparchus-Ptolemy model of planetary motion

One of the most difficult problems faced by the ancient Greek astronomers was that of predicting the position of the planets. A remarkably successful model of planetary motion that is described in Ptolemy's *Almagest* leads to an interesting geometric interpretation for truncated Fourier series. Using modern notation we write

$$z_1(t) = a_1 e^{2\pi i t / T_1}, \qquad -\infty < t < \infty$$

with

$$a_1 = |a_1| e^{i\phi_1}, \qquad 0 \le \phi_1 < 2\pi$$

to describe the uniform circular motion of a planet P around the Earth E at the origin. Here $|a_1|$ is the radius of the orbit, T_1 is the period, and the phase parameter ϕ_1 serves to specify the location of the planet at time $t = 0$. Such a one-circle model cannot account for the occasional retrograde motion of the outer planets Mars, Jupiter, and Saturn. We build a more sophisticated two-circle model by writing

$$z_2(t) = z_1(t) + a_2 e^{2\pi i t / T_2}$$

with

$$a_2 = |a_2| e^{i\phi_2}, \qquad 0 \le \phi_2 < 2\pi.$$

The planet P now undergoes uniform circular motion about a point that undergoes uniform circular motion around the Earth E at the origin, *cf.* Fig. 1.12. This two-circle model can produce the observed retrograde motion (try a computer simulation using the data from Ex. 1.18!), but it cannot fit the motion of the planets to observational accuracy.

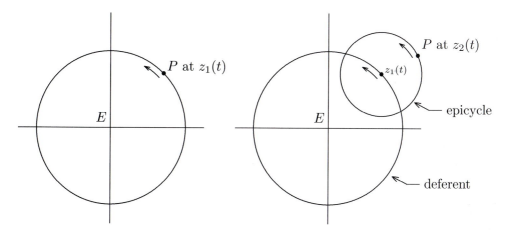

Figure 1.12. The addition of uniform circular motions.

Proceeding in this way we obtain a geometric interpretation of the motion described by the exponential sum

$$z_n(t) = a_1 e^{2\pi i t / T_1} + a_2 e^{2\pi i t / T_2} + \cdots + a_n e^{2\pi i t / T_n}$$

using a fixed circle (called the *deferent*) and $n-1$ moving circles (called *epicycles*). Such a motion is periodic when $T_1, T_2, \ldots, T_n$ are integral multiples of some $T > 0$, in which case the sum is a Fourier series with finitely many terms. Hipparchus and Ptolemy used a shifted four-circle construction of this type (with the Earth near but not at the origin) to fit the motion of each planet. These models were used for predicting the positions of the five planets of antiquity until Kepler and Newton discovered the laws of planetary motion some 1300 years later.

Gauss and the orbits of the asteroids

On the first day of the 19th century the asteroid Ceres was discovered, and in rapid succession the asteroids Pallas, Vesta, and Juno were also found. Gauss became interested in the problem of determining the orbits of such planetoids from observational data. In 1802, Baron von Zach published the 12 data points for the orbit of the asteroid Pallas that are plotted in Fig. 1.13. Gauss decided to interpolate this data by using a 360°-periodic trigonometric polynomial

$$y(x) = \sum_{k=0}^{11} c_k \, e^{2\pi i k x/360}$$

with the 12 coefficients $c_0, c_1, \ldots, c_{11}$ being chosen to force the graph of y to pass through the 12 known points $(n \cdot 30°, y_n)$, $n = 0, 1, \ldots, 11$, *i.e.*, so as to make

$$y_n = \sum_{k=0}^{11} c_k \, e^{2\pi i k n/12}, \qquad n = 0, 1, \ldots, 11.$$

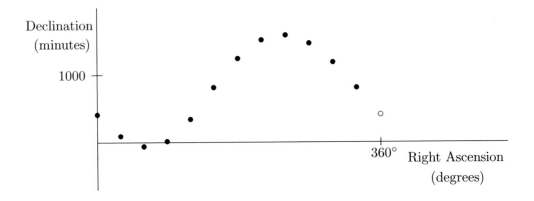

Figure 1.13. Declination of the asteriod Pallas as a function of right ascension as published by Baron von Zach. (Declination and right ascension are measures of latitude and longitude on the celestial sphere.)

We recognize this as the synthesis equation (9) (with $N = 12$, $F[k] = c_k$), and use the corresponding analysis equation (10) to obtain the coefficients

$$c_k = \frac{1}{12} \sum_{n=0}^{11} y_n \, e^{-2\pi ikn/12}, \qquad k = 0, 1, \dots, 11.$$

Of course, it is one thing to write down such a formula and quite another to obtain a numerical value for each of the c_k's. (Remember that Gauss did all of the arithmetic by hand.) You will find Baron von Zach's data in Ex. 1.19. Perhaps as you analyze this data (with a computer!) you will share in Gauss's discovery of a very clever way to expedite such calculations.

Fourier and the flow of heat

Shortly after the above work of Gauss was completed, Fourier invented the representations (5)–(6) and (3)–(4) (*i.e.*, Fourier series and Fourier integrals) to use for solving problems involving the flow of heat in solids. He first showed that the temperature $u(x, t)$ at time $t \geq 0$ and coordinate x along a thin insulated rod of uniform cross section is a solution of the partial differential equation

$$\frac{\partial u}{\partial t}(x, t) = a^2 \frac{\partial^2 u}{\partial x^2}(x, t),$$

with the *thermal diffusivity* parameter a^2 depending on the material of which the rod is made. (You will find an elementary derivation in Section 9.3.) Fourier observed that the function

$$e^{2\pi isx} \cdot e^{-4\pi^2 a^2 s^2 t}$$

satisfies the partial differential equation for every choice of the real parameter s. He conceived the idea of combining such elementary solutions to produce a temperature function $u(x, t)$ that agrees with some prescribed initial temperature when $t = 0$.

For the temperature in a *rod* (that extends from $x = -\infty$ to $x = +\infty$) Fourier wrote

$$u(x, t) = \int_{s=-\infty}^{\infty} A(s) e^{2\pi isx} e^{-4\pi^2 a^2 s^2 t} ds$$

with the intention of choosing the amplitude function $A(s)$, $-\infty < s < \infty$, to make his formula to agree with the known initial temperature $u(x, 0)$ at time $t = 0$, *i.e.*, to make

$$u(x, 0) = \int_{s=-\infty}^{\infty} A(s) e^{2\pi isx} ds.$$

We recognize this identity as the synthesis equation (3) for the function $u(x, 0)$ and use the corresponding analysis equation (4) to write

$$A(s) = \int_{x=-\infty}^{\infty} u(x, 0) e^{-2\pi isx} dx, \qquad -\infty < s < \infty,$$

thereby expressing A in terms of the initial temperature. In this way Fourier solved the heat flow problem for a doubly infinite rod. You can work out the details for a rod with an initial hot spot by solving Ex. 1.20.

For the temperature in a *ring* of circumference $p > 0$, Fourier used the p-periodic solutions

$$e^{2\pi ikx/p} \cdot e^{-4\pi^2 a^2 (k/p)^2 t}, \qquad k = 0, \pm 1, \pm 2, \ldots$$

of the diffusion equations (with $s = k/p$) to write

$$u(x, t) = \sum_{k=-\infty}^{\infty} c_k \, e^{2\pi ikx/p} \cdot e^{-4\pi^2 a^2 (k/p)^2 t}$$

with the intention of choosing the coefficients c_k, $k = 0, \pm 1, \pm 2, \ldots$ to make

$$u(x, 0) = \sum_{k=-\infty}^{\infty} c_k \, e^{2\pi ikx/p}.$$

We recognize this as the synthesis equation (5) for the initial temperature $u(x, 0)$ and use the corresponding analysis equation (6) to express the coefficients

$$c_k = \frac{1}{p} \int_{x=0}^{p} u(x, 0) e^{-2\pi ikx/p} dx, \qquad k = 0, \pm 1, \pm 2, \ldots$$

in terms of the known initial temperature. In this way Fourier solved the heat flow problem for a ring.

Today, such procedures are used to solve a number of partial differential equations that arise in science and engineering, and we will develop these ideas in Chapter 9. It is somewhat astonishing, however, to realize that Fourier chose periodic functions to study the flow of heat, a physical phenomenon that is as intrinsically aperiodic as any that we can imagine!

Fourier's representation and LTI systems

Function-to-function mappings are commonly studied in many areas of science and engineering. Within the context of engineering we focus attention on the device that effects the input-to-output transformation, and we represent such a *system* using a diagram of the sort shown in Fig. 1.14. In mathematics, such function-to-function mappings are called *operators*, and we use the notation

$$f_o = \mathbf{A} f_i$$

or

$$\mathbf{A} : f_i \to f_o \quad \text{by} \quad f_o(t) = (\mathbf{A} f_i)(t)$$

(where $\mathbf{A}$ is the name of the operator) to convey the same idea.

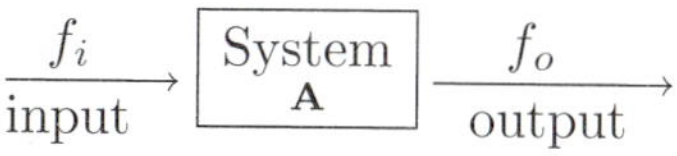

$$\underset{\text{input}}{\xrightarrow{\quad f_i \quad}} \boxed{\begin{array}{c} \text{System} \\ \textbf{A} \end{array}} \underset{\text{output}}{\xrightarrow{\quad f_o \quad}}$$

Figure 1.14. Schematic representation of a system **A**.

In practice we often deal with systems that are *homogeneous* and *additive, i.e.*,

$$\mathbf{A}(cf) = c(\mathbf{A}f)$$
$$\mathbf{A}(f + g) = (\mathbf{A}f) + (\mathbf{A}g)$$

when f, g are arbitrary inputs and c is an arbitrary scalar. Such systems are said to be *linear*. Many common systems also have the property of translation invariance. We say that a system is *translation invariant* if the output

$$g_o = \mathbf{A}g_i$$

of an arbitrary τ-translate

$$g_i(t) := f_i(t + \tau), \qquad -\infty < t < \infty,$$

of an arbitrary input function f_i is the corresponding τ-translate

$$g_o(t) = f_o(t + \tau), \qquad -\infty < t < \infty$$

of the output

$$f_o = \mathbf{A}f_i$$

to f_i, *i.e.*, when we translate f_i by τ the system responds by shifting f_o by τ, $-\infty < \tau < \infty$. Systems that are both linear and translation invariant are said to be LTI.

A variety of signal processing devices can be modeled by using LTI systems. For example, the speaker for an audio system maps an electrical input signal from an amplifier to an acoustical output signal, with time being the independent variable. A well-designed speaker is more-or-less linear. If we simultaneously input signals from two amplifiers, the speaker responds with the sum of the corresponding outputs, and if we scale the input signal, *e.g.*, by adjusting the volume control, the acoustical response is scaled in a corresponding manner (provided that we do not exceed the power limitations of the speaker!) Of course, when we play a familiar CD or tape on different occasions, *i.e.*, when we time shift the input signal, we expect to hear an acoustical response that is time shifted in exactly the same fashion (provided that the time shift amounts to a few hours or days and not to a few million years!)

A major reason for the importance of Fourier analysis in electrical engineering is that every complex exponential

$$e_s(t) := e^{2\pi i s t}, \qquad -\infty < t < \infty$$

(with s being a fixed real parameter) is an eigenfunction of every LTI system. We summarize this by saying, *An LTI system responds sinusoidally when it is shaken sinusoidally.* The proof is based on the familiar multiplicative property

$$e_s(t + \tau) = e_s(\tau) \cdot e_s(t)$$

of the complex exponential. After applying the LTI operator $\mathbf{A}$ to both sides of this equation, we use the translation invariance to simplify the left side, we use the linearity to simplify the right side, and thereby write

$$(\mathbf{A}e_s)(t + \tau) = e_s(\tau) \cdot (\mathbf{A}e_s)(t), \qquad -\infty < t < \infty, \ -\infty < \tau < \infty.$$

We now set $t = 0$ to obtain the eigenfunction relation

$$(\mathbf{A}e_s)(\tau) = \alpha(s) \cdot e_s(\tau), \qquad -\infty < \tau < \infty$$

with the system function

$$\alpha(s) := (\mathbf{A}e_s)(0), \qquad -\infty < s < \infty$$

being the corresponding eigenvalue.

If we know the system function $\alpha(s)$, $-\infty < s < \infty$, we can find the system response to any suitably regular input function f_i. Indeed, using Fourier's representation (3) we write

$$f_i(t) = \int_{s=-\infty}^{\infty} F_i(s)e^{2\pi i s t} ds$$

and approximate the integral of this synthesis equation with a Riemann sum of the form

$$f_i(t) \approx \sum_{k=1}^{N} F_i(s_k)e^{2\pi i s_k t} \, \Delta s_k.$$

Since the linear operator $\mathbf{A}$ maps

$$e^{2\pi i s t} \quad \text{to} \quad \alpha(s)e^{2\pi i s t}$$

for every choice of the frequency parameter s, it must map the Riemann sum

$$\sum_{k=1}^{N} F_i(s_k)e^{2\pi i s_k t}\Delta s_k \quad \text{to} \quad \sum_{k=1}^{N} F_i(s_k)\alpha(s_k)e^{2\pi i s_k t}\Delta s_k,$$

with the sum on the right being an approximation to the integral

$$\int_{s=-\infty}^{\infty} F_i(s)\alpha(s)e^{2\pi i s t} ds.$$

We conclude that **A** maps

$$f_i(t) = \int_{-\infty}^{\infty} F_i(s)e^{2\pi i s t}ds \quad \text{to} \quad f_o(t) = \int_{-\infty}^{\infty} F_i(s)\alpha(s)e^{2\pi i s t}ds$$

(provided that the system possesses a continuity property that enables us to justify the limiting process involved in passing from an approximating Riemann sum to the corresponding integral). In this way we see that the Fourier transform of the output is obtained by multiplying the Fourier transform of the input by the LTI system function α.

The above discussion deals with systems that map functions on $\mathbb{R}$ to functions on $\mathbb{R}$. Analogous considerations can be used for LTI systems that map functions on $\mathbb{T}_p$, $\mathbb{Z}$, $\mathbb{P}_N$ to functions on $\mathbb{T}_p$, $\mathbb{Z}$, $\mathbb{P}_N$, respectively, *cf.* Ex. 1.21.

Schoenberg's derivation of the Tartaglia-Cardan formulas

The discrete Fourier representation of (9)–(10) can be used to find formulas for the roots of polynomials of degree 2,3,4, *cf.* I. Schoenberg, pp. 79–81. To illustrate the idea, we will derive the familiar quadratic formula for the roots x_0, x_1 of the quadratic polynomial

$$x^2 + bx + c = (x - x_0)(x - x_1)$$

as functions of the coefficients b, c. In view of the synthesis equation (9) we can write

$$x_0 = X_0 + X_1, \qquad x_1 = X_0 - X_1$$

(where we take $N = 2$ and use x_0, x_1, X_0, X_1 instead of the more cumbersome $x[0]$, $x[1]$, $X[0]$, $X[1]$). It follows that

$$x^2 + bx + c = \{x - (X_0 + X_1)\}\{x - (X_0 - X_1)\}$$
$$= x^2 - 2X_0 x + (X_0^2 - X_1^2),$$

and upon equating coefficients of like powers of x we find

$$b = -2X_0, \qquad c = X_0^2 - X_1^2.$$

We solve for X_0, X_1 in turn and write

$$X_0 = -\frac{1}{2}b, \qquad X_1 = \frac{1}{2}(b^2 - 4c)^{1/2}.$$

Knowing X_0, X_1 we use the synthesis equation to obtain the familiar expressions

$$x_0 = \frac{1}{2}\{-b + (b^2 - 4c)^{1/2}\}, \qquad x_1 = \frac{1}{2}\{-b - (b^2 - 4c)^{1/2}\}.$$

The same procedure enables us to derive the Tartaglia-Cardan formulas for the roots x_0, x_1, x_2 of the cubic polynomial

$$x^3 + bx^2 + cx + d = (x - x_0)(x - x_1)(x - x_2)$$

in terms of the coefficients b, c, d. We define

$$\omega := e^{2\pi i/3} = \frac{-1 + \sqrt{3}i}{2}$$

so that we can use the compact form of the synthesis equations

$$x_0 = X_0 + X_1 + X_2, \qquad x_1 = X_0 + \omega X_1 + \omega^2 X_2, \qquad x_2 = X_0 + \omega^2 X_1 + \omega X_2$$

to express x_0, x_1, x_2 in terms of the discrete Fourier transform X_0, X_1, X_2. After a bit of nasty algebra (*cf*. Ex. 1.22) we find

$$x^3 + bx^2 + cx + d = (x - X_0)^3 - 3X_1 X_2 (x - X_0) - X_1^3 - X_2^3$$

so that

$$X_0 = -\frac{b}{3}, \qquad X_1 X_2 = \frac{b^2 - 3c}{9}, \qquad X_1^3 + X_2^3 = \frac{-27d + 9bc - 2b^3}{27}.$$

From the last pair of equations we see that $Y = X_1^3, X_2^3$ are the roots of the quadratic polynomial

$$(Y - X_1^3)(Y - X_2^3) = Y^2 - (X_1^3 + X_2^3)Y + (X_1 X_2)^3$$

$$= Y^2 + \left(\frac{27d - 9bc + 2b^3}{27}\right)Y + \left(\frac{b^2 - 3c}{9}\right)^3,$$

i.e.,

$$X_1 = \left\{-\left(\frac{27d - 9bc + 2b^3}{54}\right) + \left[\left(\frac{27d - 9bc + 2b^3}{54}\right)^2 - \left(\frac{b^2 - 3c}{9}\right)^3\right]^{1/2}\right\}^{1/3}$$

$$X_2 = \left\{-\left(\frac{27d - 9bc + 2b^3}{54}\right) - \left[\left(\frac{27d - 9bc + 2b^3}{54}\right)^2 - \left(\frac{b^2 - 3c}{9}\right)^3\right]^{1/2}\right\}^{1/3}.$$

Knowing X_0, X_1, X_2 we use the synthesis equation to write

$$x_0 = X_0 + X_1 + X_2$$

$$x_1 = X_0 - \frac{X_1 + X_2}{2} + \frac{i\sqrt{3}(X_1 - X_2)}{2}$$

$$x_2 = X_0 - \frac{X_1 + X_2}{2} - \frac{i\sqrt{3}(X_1 - X_2)}{2}.$$

The roots of a quartic polynomial can be found in a similar manner (but it takes a lot of very nasty algebra to do the job!).

Fourier transforms and spectroscopy

We can produce an exponentially damped complex exponential

$$y_0(t) := \begin{cases} e^{-\alpha t} e^{2\pi i s_0 t} & \text{if } t > 0 \\ 0 & \text{if } t < 0 \end{cases}$$

by subjecting a damped harmonic oscillator (*e.g.*, a mass on a spring with damping) to a suitable initial excitation. Here $\alpha > 0$ and $-\infty < s_0 < \infty$. Graphs of y_0 and the Fourier transform

$$Y_0(s) = \int_0^\infty e^{-2\pi i s t} e^{-\alpha t} e^{2\pi i s_0 t} dt = \frac{1}{\alpha + 2\pi i (s - s_0)}$$

are shown in Fig. 1.15. The function Y_0, which is called a *Lorenzian*, is concentrated near $s = s_0$ with an approximate width $\alpha/2\pi$. (You can learn more about such functions by doing Ex. 3.34 a little later in the course.)

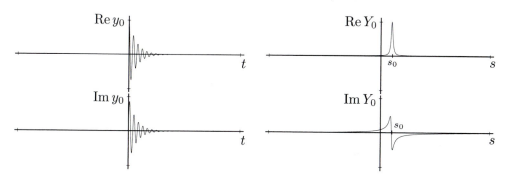

Figure 1.15. The function $y_0(t) = e^{-\alpha t} e^{2\pi i s_0 t}$ and its Fourier transform $Y_0(s)$.

When we subject molecules to a burst of electromagnetic radiation (radio frequency, microwave, infrared, ...) we induce various damped oscillations. The resulting transient has the form

$$y(t) = \begin{cases} \sum_k A_k e^{-\alpha_k t} e^{2\pi i s_k t} & \text{if } t > 0 \\ 0 & \text{if } t < 0 \end{cases}$$

with parameters $\alpha_k > 0$, $-\infty < s_k < \infty$ that are determined by the arrangement of the atoms that form the molecules. We can observe these parameters when we graph the Fourier transform

$$Y(s) = \sum_k \frac{A_k}{\alpha_k + 2\pi i (s - s_k)}.$$

Within this context Y is said to be a *spectrum*.

For example, when a sample of the amino acid *arginine*

$$O = C - C - C - C - C - N - C = NH_2^+$$

(with H above the four central carbons, and O^-, NH_3^+, H, H, H, H, NH_2 below)

is placed in a strong magnetic field and subjected to a 500-MHz pulse, the individual protons precess. The resulting free induction decay voltage, $y(t)$, and corresponding spectrum, $Y(s)$, are shown in Fig. 1.16. (You can see the individual Lorenzians!) Richard Earnst won the 1991 Nobel prize in chemistry for developing this idea into a powerful tool for determining the structure of organic molecules.

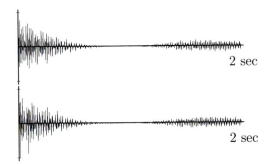

Free induction decay samples

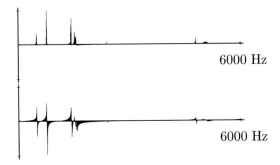

DFT of the FID samples

Figure 1.16. FT-NMR analysis of arginine.

1.3 The Parseval Identities and Related Results

The Parseval identities

Let f, g be suitably regular functions on $\mathbb{R}$ with Fourier transforms F, G, respectively. Using the synthesis equation for g and the analysis equation for f (and using a bar to denote the complex conjugate), we formally write

$$\int_{x=-\infty}^{\infty} f(x)\overline{g(x)}dx = \int_{x=-\infty}^{\infty} f(x)\left\{ \overline{\int_{s=-\infty}^{\infty} G(s)e^{2\pi isx}ds} \right\} dx$$

$$\overset{?}{=} \int_{s=-\infty}^{\infty} \int_{x=-\infty}^{\infty} f(x)e^{-2\pi isx}\overline{G(s)}dx \, ds$$

$$= \int_{s=-\infty}^{\infty} F(s)\overline{G(s)}ds,$$

assuming that we can somehow justify the exchange in the order of the integration processes in the step marked with the question mark (*e.g.*, by imposing restrictive hypotheses on f, g and using a suitable Fubini theorem from advanced calculus). We refer to the resulting equation

$$\int_{x=-\infty}^{\infty} f(x)\overline{g(x)}dx = \int_{s=-\infty}^{\infty} F(s)\overline{G(s)}ds \tag{11}$$

as the *Parseval identity* for functions on $\mathbb{R}$. Analogous arguments lead to the corresponding Parseval identities

$$\int_{x=0}^{p} f(x)\overline{g(x)}dx = p \sum_{k=-\infty}^{\infty} F[k]\overline{G[k]} \tag{12}$$

$$\sum_{n=-\infty}^{\infty} f[n]\overline{g[n]} = p \int_{s=0}^{p} F(s)\overline{G(s)}ds \tag{13}$$

$$\sum_{n=0}^{N-1} f[n]\overline{g[n]} = N \sum_{k=0}^{N-1} F[k]\overline{G[k]} \tag{14}$$

for functions on $\mathbb{T}_p$, $\mathbb{Z}$, $\mathbb{P}_N$, respectively. The period p or N appears as a factor on the transform side of these equations. An exchange of infinite summation and integration processes is involved in this heuristic derivation of (12)–(13). In contrast, only finite sums are used in the derivation of (14) from the synthesis and analysis equations (9)–(10) that we will establish in a subsequent discussion. You will find alternative forms for the Parseval identities (11)–(14) in Ex. 1.24.

The Plancherel identities

When we set $g = f$ in (11)–(14) we obtain the equations

$$\int_{x=-\infty}^{\infty} |f(x)|^2 dx = \int_{s=-\infty}^{\infty} |F(s)|^2 ds \tag{15}$$

$$\int_{x=0}^{p} |f(x)|^2 dx = p \sum_{k=-\infty}^{\infty} |F[k]|^2 \tag{16}$$

$$\sum_{n=-\infty}^{\infty} |f[n]|^2 = p \int_{0}^{p} |F(s)|^2 ds \tag{17}$$

$$\sum_{n=0}^{N-1} |f[n]|^2 = N \sum_{k=0}^{N-1} |F[k]|^2 \tag{18}$$

that link the aggregate squared size (or *energy*) of a function f on $\mathbb{R}$, $\mathbb{T}_p$, $\mathbb{Z}$, $\mathbb{P}_N$, respectively, to that of its Fourier transform F. We will refer to (15)–(18) as the *Plancherel identities* (although the names of Bessel, Lyapunov, Parseval, and Rayleigh are also properly associated with these equations).

As we have noted, (15)–(18) can be obtained from (11)–(14) simply by setting $g = f$. The corresponding identities are really equivalent, however, since we can obtain a Parseval identity from the corresponding (seemingly less general) Plancherel identity by using the *polarization identities*

$$f\bar{g} = \frac{1}{4}\left\{|f + g|^2 + i|f + ig|^2 + i^2|f + i^2 g|^2 + i^3|f + i^3 g|^2\right\}$$

$$F\bar{G} = \frac{1}{4}\left\{|F + G|^2 + i|F + iG|^2 + i^2|F + i^2 G|^2 + i^3|F + i^3 G|^2\right\}$$

together with the linearity of the Fourier transform process, *cf.* Ex. 1.25.

Orthogonality relations for the periodic complex exponentials

It is a simple matter to verify the *orthogonality relations*

$$\int_{x=0}^{p} e^{2\pi i k x/p} e^{-2\pi i \ell x/p} dx = \begin{cases} p & \text{if } k = \ell \\ 0 & \text{otherwise,} \end{cases} \qquad k, \ell = 0, \pm 1, \pm 2, \ldots \tag{19}$$

for the p-periodic complex exponentials on $\mathbb{R}$. The corresponding *discrete orthogonality relations*

$$\sum_{n=0}^{N-1} e^{2\pi i k n/N} e^{-2\pi i \ell n/N} = \begin{cases} N & \text{if } k = \ell, \ell \pm N, \ell \pm 2N, \dots \\ 0 & \text{otherwise,} \end{cases} \tag{20}$$

$$k, \ell = 0, \pm 1, \pm 2, \dots$$

can be proved by using the formula

$$1 + z + z^2 + \cdots + z^{N-1} = \begin{cases} N & \text{if } z = 1 \\ (z^N - 1)/(z - 1) & \text{otherwise} \end{cases}$$

for the sum of a geometric progression with

$$z := e^{2\pi i (k - \ell)/N}.$$

We easily verify that

$$z = 1 \quad \text{if} \quad k - \ell = 0, \pm N, \pm 2N, \dots$$

while

$$z^N = 1 \quad \text{for all} \quad k, \ell = 0, \pm 1, \pm 2, \dots$$

and thereby complete the argument. An alternative geometric proof of (20) is the object of Ex. 1.26. Real versions of (19)–(20) are developed in Ex. 1.27.

The orthogonality relations (19),(20) are the special cases of the Parseval identities (12),(14) that result when the discrete functions F, G vanish at all but one of the points of $\mathbb{Z}, \mathbb{P}_N$, respectively, where the value 1 is taken.

Bessel's inequality

Let f be a function on $\mathbb{T}_p$ and let

$$\tau_n(x) := \sum_{k=-n}^{n} c_k e^{2\pi i k x/p} \tag{21}$$

be any p-periodic trigonometric polynomial of degree n or less with complex coefficients c_k, $k = 0, \pm 1, \pm 2, \dots, \pm n$. By using the analysis equation (6) for the Fourier coefficients of f and the orthogonality relations (19), we find

$$
\int_{x=0}^{p} |f(x) - \tau_n(x)|^2 dx
$$

$$
= \int_{x=0}^{p} \left\{ f(x) - \sum_{k=-n}^{n} c_k\, e^{2\pi i k x/p} \right\} \left\{ \overline{f(x)} - \sum_{\ell=-n}^{n} \overline{c_\ell}\, e^{-2\pi i \ell x/p} \right\} dx
$$

$$
= \int_{x=0}^{p} |f(x)|^2 dx - \sum_{\ell=-n}^{n} \overline{c_\ell} \int_{x=0}^{p} f(x) e^{-2\pi i \ell x/p} dx
$$

$$
- \sum_{k=-n}^{n} c_k \int_{x=0}^{p} \overline{f(x)} e^{2\pi i k x/p} dx \tag{22}
$$

$$
+ \sum_{k=-n}^{n} \sum_{\ell=-n}^{n} c_k \overline{c_\ell} \int_{x=0}^{p} e^{2\pi i k x/p} \cdot e^{-2\pi i \ell x/p} dx
$$

$$
= \int_{x=0}^{p} |f(x)|^2 dx - p \sum_{\ell=-n}^{n} \overline{c_\ell}\, F[\ell] - p \sum_{k=-n}^{n} c_k\, \overline{F[k]} + p \sum_{k=-n}^{n} c_k \overline{c_k}
$$

$$
= \int_{x=0}^{p} |f(x)|^2 dx - p \sum_{k=-n}^{n} |F[k]|^2 + p \sum_{k=-n}^{n} |F[k] - c_k|^2
$$

when all of the integrals exist and are finite, *e.g.*, as is certainly the case when f is bounded and continuous at all but finitely many points of $\mathbb{T}_p$.

If we specialize (22) by taking $c_k = F[k]$ for $k = 0, \pm 1, \pm 2, \ldots, \pm n$, the right-most sum vanishes and we find

$$
\int_{x=0}^{p} |f(x)|^2 dx - p \sum_{k=-n}^{n} |F[k]|^2 = \int_{0}^{p} |f(x) - \sum_{k=-n}^{n} F[k] e^{2\pi i k x/p}|^2 dx \geq 0
$$

for every choice of $n = 1, 2, \ldots$. In this way we prove *Bessel's inequality*,

$$
\int_{x=0}^{p} |f(x)|^2 dx \geq p \sum_{k=-\infty}^{\infty} |F[k]|^2, \tag{23}
$$

a one-sided version of (16).

The Weierstrass approximation theorem

Let f be a continuous function on $\mathbb{T}_p$. We will show that we can uniformly approximate f as closely as we please with a p-periodic trigonometric polynomial (21). More specifically, we will construct trigonometric polynomials $\tau_1, \tau_2, \ldots$ such that

$$
\lim_{n \to \infty} \max_{0 \leq x \leq p} |f(x) - \tau_n(x)| = 0.
$$

This result is known as the *Weierstrass approximation theorem.* We need this result to establish the validity of (5)–(6). The proof will use a few ideas from intermediate analysis. (You may wish to jump to Section 1.4 and come back later to sort out the details.)

For each $n = 1, 2, \ldots$ we define the de la Vallée-Poussin power kernel

$$\delta_n(x) := p^{-1} \, 4^n \binom{2n}{n}^{-1} \cos^{2n}(\pi x/p), \tag{24}$$

cf. Fig. 1.17. We will show that this nonnegative function has a unit area concentrated at the origin of $\mathbb{T}_p$. By using the Euler identity for cos and the binomial formula, we write

$$
\begin{aligned}
\delta_n(x) = {}& p^{-1} \binom{2n}{n}^{-1} \left\{ e^{\pi i x/p} + e^{-\pi i x/p} \right\}^{2n} \\[2mm]
= {}& p^{-1} \binom{2n}{n}^{-1} \left\{ \binom{2n}{0} e^{2\pi i n x/p} + \binom{2n}{1} e^{2\pi i (n-1) x/p} \right. \\[2mm]
& + \binom{2n}{2} e^{2\pi i (n-2) x/p} + \cdots \\[2mm]
& \left. + \binom{2n}{n} 1 + \cdots + \binom{2n}{2n} e^{-2\pi i n x/p} \right\}.
\end{aligned}
\tag{25}
$$

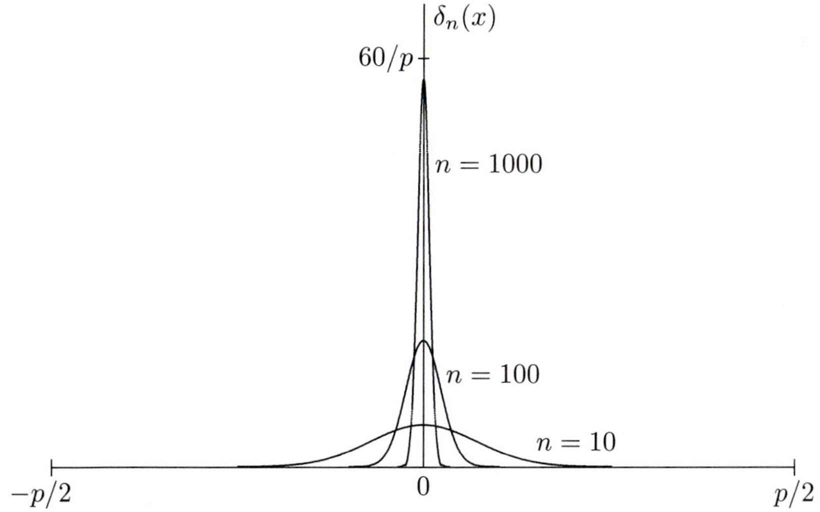

Figure 1.17. The de la Vallée-Poussin power kernel (24) for $n = 10^1, 10^2, 10^3$.

Moreover, after noting that

$$\delta_n(0) = p^{-1}4^n \binom{2n}{n}^{-1} = \frac{4^n n! n!}{p(2n)!}$$

$$= \frac{2n}{p} \frac{2n-2}{2n-1} \frac{2n-4}{2n-3} \cdots \frac{2}{3} < \frac{2n}{p}$$

and observing that $\delta_n(x)$ is monotonic on $(-p/2, 0)$ and on $(0, p/2)$, we see that

$$\delta_n(x) \leq \frac{2n}{p} \cos^{2n}(\pi\alpha/p) \quad \text{when} \quad \alpha \leq |x| \leq p/2.$$

In this way we verify that the kernel δ_n has the following properties:

$$\delta_n(x) \geq 0 \text{ for } -p/2 \leq x < p/2 \qquad\qquad \text{(positivity)};$$

$$\int_{-p/2}^{p/2} \delta_n(x)dx = 1 \qquad\qquad\qquad \text{(unit area); and} \qquad (26)$$

$$\max_{\alpha \leq |x| \leq p/2} \delta_n(x) \to 0 \text{ as } n \to \infty \text{ when } 0 < \alpha < p/2 \quad \text{(small tail).}$$

We now define

$$\tau_n(x) := \int_{u=0}^{p} f(u)\delta_n(x-u)du, \quad n = 1, 2, \ldots, \qquad (27)$$

cf. Fig. 1.18. [After studying Chapter 2 you will recognize (27) as a convolution product, and after studying Chapter 7 you will recognize (27) as an approximation for the sifting relation (7.64) for Dirac's delta.] We use (25), (27), and (6) to write

$$\tau_n(x) = \binom{2n}{n}^{-1} \left\{ \binom{2n}{0} F[n]e^{2\pi inx/p} + \binom{2n}{1} F[n-1]e^{2\pi i(n-1)x/p} \right.$$

$$\left. + \cdots + \binom{2n}{2n} F[-n]e^{-2\pi inx/p} \right\}, \qquad (28)$$

and thereby see that τ_n is a p-periodic trigonometric polynomial of degree n or less that is easily constructed from the Fourier coefficients (6) of f.

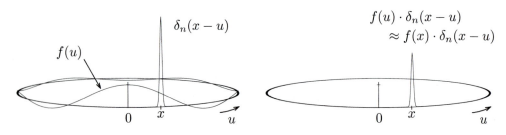

Figure 1.18. Construction of the approximation (27) to f on $\mathbb{T}_p$.

Suppose now that $\epsilon > 0$ is given. We first choose $0 < \alpha < p/2$ so small that

$$|f(x) - f(u)| < \epsilon/2 \quad \text{when} \quad |x - u| < \alpha.$$

With α thus chosen, we use the tail hypothesis of (26) and select n so large that

$$\max_{\alpha \leq |x| \leq p/2} \delta_n(x) < \frac{\epsilon}{4Mp}$$

where $M > 0$ is some uniform bound for $|f(x)|$. By using the unit area hypothesis of (26) with the p-periodicity of δ_n we see that

$$f(x) = \int_{x=0}^{p} f(x)\delta_n(x - u)du$$

[since $f(x)$ is a constant with respect to the u-integration]. It follows that

$$f(x) - \tau_n(x) = \int_{u=0}^{p} \{f(x) - f(u)\}\delta_n(x - u)du$$

$$= \int_{u=x-p/2}^{x+p/2} \{f(x) - f(u)\}\delta_n(x - u)du.$$

In conjunction with the positivity hypothesis of (26) and our choices for α, n this leads to the uniform bound

$$|f(x) - \tau_n(x)| \leq \int_{|x-u|\leq\alpha} |f(x) - f(u)|\delta_n(x - u)du$$

$$+ \int_{\alpha \leq |x-u| \leq p/2} |f(x) - f(u)|\delta_n(x - u)du$$

$$\leq \max_{|x-u|\leq\alpha} |f(x) - f(u)| \cdot \int_{u=0}^{p} \delta_n(x - u)du$$

$$+ \max_{\alpha \leq |x-u| \leq p/2} |f(x) - f(u)| \cdot \delta_n(x - u) \cdot p$$

$$\leq \frac{\epsilon}{2} \cdot 1 + 2M \cdot \frac{\epsilon}{4Mp} \cdot p = \epsilon,$$

thus completing the proof.

You can use variations of this argument to study the pointwise convergence of Fourier series, *cf.* Exs. 1.31, 1.32.

There is a second (mean square) form of the Weierstrass approximation theorem that can be used when f is bounded on $\mathbb{T}_p$ and continuous at all points of $\mathbb{T}_p$ but $x_1, x_2, \ldots, x_m$ where jumps occur. In this case we can show that

$$\lim_{n\to\infty} \int_0^p |f(x) - \tau_n(x)|^2 dx = 0$$

for suitably chosen trigonometric polynomials $\tau_1, \tau_2, \dots$. We will form τ_n as before, noting that $|\tau_n(x)| \leq M$ when M is a uniform bound for f. We let $J(\delta)$ be the portion of $\mathbb{T}_p$ that remains after we remove small open intervals $I_1(\delta), I_2(\delta), \dots, I_m(\delta)$ of length δ centered at $x_1, x_2, \dots, x_m$. We can then write

$$\int_0^p |f(x) - \tau_n(x)|^2 dx \leq \sum_{\ell=1}^m \int_{I_\ell(\delta)} |f(x) - \tau_n(x)|^2 dx + \int_{J(\delta)} |f(x) - \tau_n(x)|^2 dx$$

$$\leq m \cdot (2M)^2 \cdot \delta + p \cdot \max_{x \in J(\delta)} |f(x) - \tau_n(x)|^2.$$

Given $\epsilon > 0$ we can make

$$m \cdot (2M)^2 \cdot \delta < \frac{\epsilon}{2}$$

by choosing a sufficiently small δ. Since f is continuous on $J(\delta)$, the above argument shows that

$$\lim_{n \to \infty} \max_{x \in J(\delta)} |f(x) - \tau_n(x)| = 0,$$

so we will have

$$p \cdot \max_{x \in J(\delta)} |f(x) - \tau_n(x)|^2 < \frac{\epsilon}{2}$$

for all sufficiently large n.

A proof of Plancherel's identity for functions on $\mathbb{T}_p$

Let f be a piecewise continuous function on $\mathbb{T}_p$. We drop the nonnegative rightmost sum from (22) to obtain the inequality

$$\int_{x=0}^p |f(x)|^2 dx - p \sum_{k=-n}^n |F[k]|^2 \leq \int_{x=0}^p |f(x) - \tau_n(x)|^2 dx$$

whenever τ_n is any p-periodic trigonometric polynomial (21) of degree n or less. We have shown that the right-hand side vanishes in the limit as $n \to \infty$ when we use the construction (27), (29) of de la Vallée-Poussin to produce $\tau_1, \tau_2, \dots$, and in this way we see that

$$\int_{x=0}^p |f(x)|^2 dx - p \sum_{k=-\infty}^\infty |F[k]|^2 \leq 0.$$

In conjunction with the Bessel inequality (22), this proves the Plancherel identity (16) for all piecewise continuous functions f on $\mathbb{T}_p$. A proof of the Plancherel identity (15) for suitably restricted functions f on $\mathbb{R}$ is given in Ex. 1.40.

Two essentially different piecewise continuous functions f, g on $\mathbb{T}_p$ cannot have the same Fourier coefficients. Indeed, if $F[k] = G[k]$ for all $k = 0, \pm 1, \pm 2, \dots$, then we can use Plancherel's identity to write

$$\int_0^p |f(x) - g(x)|^2 dx = p \sum_{k=-\infty}^{\infty} |F[k] - G[k]|^2 = 0.$$

It follows that $f(x) = g(x)$ at all points x where f and g are continuous.

1.4 The Fourier-Poisson Cube

Introduction

Classical applications of Fourier analysis use the integral (3) or the infinite series (5). Digital computers can be programmed to evaluate the finite sums (9)–(10) with great efficiency, *cf.* Chapter 6. We are now going to derive some identities that connect these seemingly unrelated forms of Fourier analysis. This will make it possible for us to use discrete Fourier analysis to prepare computer simulations for vibrating strings, diffusing heat, diffracting light, etc. (as described in Section 9.5).

The synthesis-analysis equations (3)–(4), (5)–(6), (7)–(8), (9)–(10) establish bidirectional mappings $f \leftrightarrow F$, $g \leftrightarrow G$, $\phi \leftrightarrow \Phi$, $\gamma \leftrightarrow \Gamma$ that link suitably regular functions f, g, ϕ, γ defined on $\mathbb{R}$, $\mathbb{T}_p$, $\mathbb{Z}$, $\mathbb{P}_N$ and their corresponding Fourier transforms F, G, Φ, Γ. We will formally establish certain connections between these four kinds of univariate Fourier analysis. In so doing, we introduce eight unidirectional mappings $f \rightarrow g$, $f \rightarrow \phi$, $g \rightarrow \gamma$, $\phi \rightarrow \gamma$, $F \rightarrow G$, $F \rightarrow \Phi$, $G \rightarrow \Gamma$, $\Phi \rightarrow \Gamma$ that serve to link the unconnected adjacent corners of the incomplete cube of Fig. 1.19. In this way we begin the process of unifying the various Fourier representations, and we prepare some very useful computational tools.

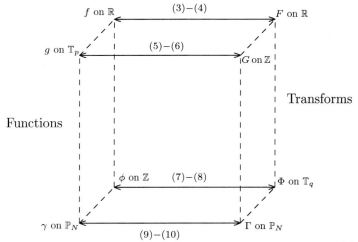

Figure 1.19. Functions from the four Fourier transform pairs (3)–(4), (5)–(6), (7)–(8), and (9)–(10) arranged on the corners of a cube.

Discretization by h-sampling

Given a function f on the continuum $\mathbb{R}$ and a spacing parameter $h > 0$, we can construct a corresponding discrete function ϕ on $\mathbb{Z}$ by defining

$$\phi[n] := f(nh), \qquad n = 0, \pm 1, \pm 2, \ldots .$$

We say that ϕ is constructed from f by h-*sampling*. The same process can be used to construct a discrete function γ on $\mathbb{P}_N$ from a function g on the continuum $\mathbb{T}_p$, i.e., to construct an N-periodic function on $\mathbb{Z}$ from a p-periodic function on $\mathbb{R}$, but in this case we must take $h := p/N$ (so that N steps of size h will equal the period p). With this in mind we define

$$\gamma[n] := g\left(\frac{np}{N}\right), \qquad n = 0, \pm 1, \pm 2, \ldots .$$

These discretization mappings $f \to \phi$ and $g \to \gamma$ are illustrated in Fig. 1.20. The discrete functions ϕ, γ provide good representations for f, g in cases where f, g do not vary appreciably over any interval of length h.

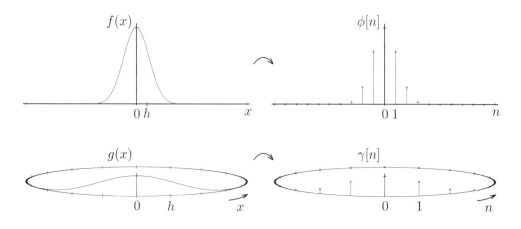

Figure 1.20. Construction of functions ϕ, γ on $\mathbb{Z}, \mathbb{P}_N$ from functions f, g on $\mathbb{R}, \mathbb{T}_p$ by h-sampling.

Periodization by p-summation

Let f be a function on $\mathbb{R}$ and assume that $f(x)$ rapidly approaches 0 as $x \to \pm\infty$. We can sum the translates

$$\ldots, f(x + 2p), f(x + p), f(x), f(x - p), f(x - 2p), \ldots$$

to produce the p-periodic function

$$g(x) := \sum_{m=-\infty}^{\infty} f(x - mp), \qquad -\infty < x < \infty$$

when $p > 0$. We say that the function g on $\mathbb{T}_p$ is produced from f by *p-summation*. Analogously, when ϕ is a function on $\mathbb{Z}$ and $\phi[n]$ rapidly approaches 0 as $n \to \pm\infty$ we can construct a function γ on $\mathbb{P}_N$, $N = 1, 2, \ldots$ by writing

$$\gamma[n] := \sum_{m=-\infty}^{\infty} \phi[n - mN], \qquad n = 0, \pm 1, \pm 2, \ldots .$$

These periodization mappings $f \to g$ and $\phi \to \gamma$ are illustrated in Fig. 1.21. The periodic functions g, γ provide good representations for f, ϕ when the graphs of f, ϕ are concentrated in intervals of length p, N, respectively.

The Poisson relations

Let ϕ be a function on $\mathbb{Z}$. We will assume that ϕ is *absolutely summable, i.e.,*

$$\sum_{m=-\infty}^{\infty} |\phi[m]| < \infty,$$

(to ensure that the above sum for $\gamma[n]$ is convergent) and use the analysis equation (10) to obtain the discrete Fourier transform

$$\Gamma[k] = \frac{1}{N} \sum_{n=0}^{N-1} \gamma[n] e^{-2\pi i k n / N}$$

$$= \frac{1}{N} \sum_{n=0}^{N-1} \sum_{m=-\infty}^{\infty} \phi[n - mN] e^{-2\pi i k n / N}.$$

Now since $e^{-2\pi i k n / N}$ is N-periodic in n and since every integer ν has a unique representation

$$\nu = n - mN \text{ with } n = 0, 1, \ldots, N-1 \text{ and } m = 0, \pm 1, \pm 2, \ldots,$$

we can write

$$\Gamma[k] = \frac{1}{N} \sum_{m=-\infty}^{\infty} \sum_{n=0}^{N-1} \phi[n - mN] e^{-2\pi i k (n - mN)/N}$$

$$= \frac{1}{N} \sum_{\nu=-\infty}^{\infty} \phi[\nu] e^{-2\pi i k \nu / N}.$$

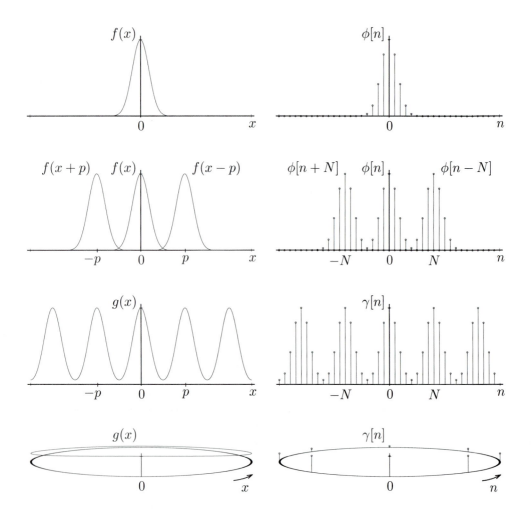

Figure 1.21. Construction of functions g, γ on $\mathbb{T}_p, \mathbb{P}_N$ from functions f, ϕ on $\mathbb{R}, \mathbb{Z}$ by p-summation, N-summation, respectively.

We now use the analysis equation (8) (with p replaced by q to avoid confusion at a later point in the presentation) to obtain

$$\Gamma[k] = \frac{q}{N} \cdot \frac{1}{q} \sum_{\nu=-\infty}^{\infty} \phi[\nu] e^{-2\pi i (kq/N)\nu/q}$$

$$= \frac{q}{N} \Phi\left(\frac{kq}{N}\right), \qquad k = 0, \pm 1, \pm 2, \dots .$$

If we construct γ from ϕ by N-summation, then we can obtain Γ from Φ by q/N-sampling and q/N-scaling. (The Fourier transform Φ of ϕ is assumed to be a function on $\mathbb{T}_q$, $q > 0$.)

Analogously, when f is a suitably regular function on $\mathbb{R}$ we can find the Fourier coefficients of the p-periodic function

$$g(x) := \sum_{m=-\infty}^{\infty} f(x - mp)$$

by writing

$$\begin{aligned}
G[k] &= \frac{1}{p} \int_{x=0}^{p} g(x)e^{-2\pi ikx/p}dx \\
&= \frac{1}{p} \int_{x=0}^{p} \sum_{m=-\infty}^{\infty} f(x - mp)e^{-2\pi ikx/p}dx \\
&\overset{?}{=} \frac{1}{p} \sum_{m=-\infty}^{\infty} \int_{0}^{p} f(x - mp)e^{-2\pi ik(x-mp)/p}dx \\
&= \frac{1}{p} \int_{\xi=-\infty}^{\infty} f(\xi)e^{-2\pi ik\xi/p}d\xi \\
&= \frac{1}{p} F\left(\frac{k}{p}\right), \qquad k = 0, \pm 1, \pm 2, \dots .
\end{aligned}$$

Of course, we must impose a mild regularity condition on f to ensure that the functions g, G are well defined and to ensure that the exchange of the summation and integration processes is permissible. In this way we see that if g is formed from f by p-summation, then G is formed from F by $1/p$-sampling and $1/p$-scaling.

We have used the analysis equations (4) and (6), (8) and (10) to obtain the Fourier transform pairs

$$g(x) := \sum_{m=-\infty}^{\infty} f(x - mp), \qquad G[k] = \frac{1}{p}F\left(\frac{k}{p}\right), \tag{29}$$

$$\gamma[n] := \sum_{m=-\infty}^{\infty} \phi[n - mN], \qquad \Gamma[k] = \frac{q}{N}\Phi\left(\frac{kq}{N}\right), \tag{30}$$

when f, ϕ are suitably regular functions on $\mathbb{R}, \mathbb{Z}$ with Fourier transforms F, Φ on $\mathbb{R}, \mathbb{T}_q$, respectively. Analogous arguments [using the synthesis equations (3) and (7), (5) and (9)] can be used to obtain the Fourier transform pairs

$$\phi[n] := f\left(\frac{np}{N}\right), \qquad \Phi(s) = \sum_{m=-\infty}^{\infty} F\left(s - \frac{mN}{p}\right), \tag{31}$$

$$\gamma[n] := g\left(\frac{np}{N}\right), \qquad \Gamma[k] = \sum_{m=-\infty}^{\infty} G[k - mN], \tag{32}$$

when f, g are suitably regular functions on $\mathbb{R}, \mathbb{T}_p$ with Fourier transforms F, G on $\mathbb{R}, \mathbb{Z}$, respectively, *cf.* Ex. 1.34. We will refer to (29)–(32) as the *Poisson relations*. You will observe the dual roles played by sampling and summation in these equations.

The Fourier-Poisson cube

We use the Poisson relations together with the analysis and synthesis equations of Fourier (as arranged in Fig. 1.19) to produce the *Fourier-Poisson cube* of Fig. 1.22. Suitably regular functions f, g, ϕ, γ that are defined on $\mathbb{R}, \mathbb{T}_p, \mathbb{Z}, \mathbb{P}_N$ lie at the corners of the left face of this cube, and the corresponding Fourier transforms F, G, Φ, Γ defined on $\mathbb{R}, \mathbb{Z}, \mathbb{T}_{N/p}, \mathbb{P}_N$, respectively, lie on the corners of the right face. Of necessity we must work with both p-periodic and q-periodic functions with $q = N/p$ in this diagram (and this is why we introduced the parameter q in the previous section). The synthesis-analysis equations (3)–(10) allow us to pass back and forth from function to transform. The process of h-sampling and p-summation provide us with one-way mappings that connect adjacent corners of the left (function) face of the cube, and Poisson's formulas (29)–(32) induce corresponding one-way mappings that connect adjacent corners of the right (transform) face of the cube.

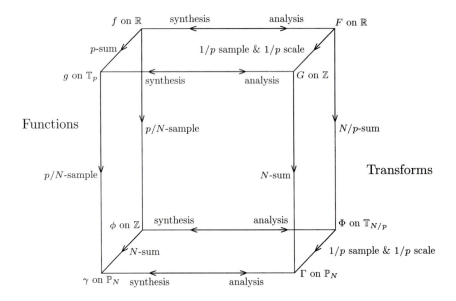

Figure 1.22. The Fourier-Poisson cube is a commuting diagram formed from the 8 mappings of (3)–(10) and the 12 mappings of (29)–(32).

You will observe that it is possible to move from the f corner to the γ corner along the alternative routes $f \to g \to \gamma$ or $f \to \phi \to \gamma$. We use the Poisson relations to verify that these mappings produce the same function γ:

$$\gamma[n] = g\left(\frac{np}{N}\right) = \sum_{m=-\infty}^{\infty} f\left(\frac{np}{N} - mp\right),$$

$$\gamma[n] = \sum_{m=-\infty}^{\infty} \phi[n - mN] = \sum_{m=-\infty}^{\infty} f\left([n - mN]\frac{p}{N}\right).$$

You can use similar arguments to verify that any two paths joining one corner of the cube to another (in a way that is consistent with the arrows) correspond to the same composite mapping. We summarize this by saying that the Fourier-Poisson cube is a *commuting diagram*.

The Fourier-Poisson cube is a helpful way to visualize the connections between (3)–(10) and (29)–(32). You will learn to work with all of these mappings as the course progresses. Practical methods for finding Fourier transforms of functions on $\mathbb{R}$, *i.e.*, for using the mappings $f \leftrightarrow F$, will be developed in Chapter 3. The Fourier series mappings $g \leftrightarrow G$, $\phi \leftrightarrow \Phi$ and the DFT mappings $\gamma \leftrightarrow \Gamma$ will be studied in Chapter 4. You will learn to use the equivalence of $f \to g \to G$ and $f \to F \to G$ to find many Fourier series with minimal effort! The fast Fourier transform (FFT), an efficient algorithm for effecting the mappings $\gamma \leftrightarrow \Gamma$ on a computer, will be the focus of Chapter 6. You will even learn to invert the one-way discretization maps $f \to \phi$, $g \to \gamma$ (when F, G are suitably localized) as you study the sampling theorem in Chapter 8. At this point, however, you will find it most helpful to work through Exs. 1.35,1.36 so that you will see how Poisson's relations can be used to analyze the error associated with certain discrete approximations to the integrals (6),(4) for Fourier transforms on $\mathbb{T}_p, \mathbb{R}$, respectively.

1.5 *The Validity of Fourier's Representation*

Introduction

In this section we will establish the validity of Fourier's representation for suitably regular functions on $\mathbb{P}_N, \mathbb{Z}, \mathbb{T}_p, \mathbb{R}$ and some of the arguments use ideas from intermediate analysis. Focus on the flow of the argument as you read the proof for the first time, skipping over the steps that you do not understand. You can come back and sort out the troublesome details after you have studied the more concrete material in Chapters 2–4.

We will continue to use the letter pairs f, F, g, G, ϕ, Φ, and γ, Γ (instead of the generic f, F of (3)–(4), (5)–(6), (7)–(8), and (9)–(10)) to help you follow the course of the argument as we move around the Fourier-Poisson cube, establishing in turn the links $\gamma \leftrightarrow \Gamma$, $\phi \leftrightarrow \Phi$, $g \leftrightarrow G$, and finally, $f \leftrightarrow F$.

Functions on $\mathbb{P}_N$

Let γ be any function on $\mathbb{P}_N$, *i.e.*, let the complex numbers $\gamma[0], \gamma[1], \ldots, \gamma[N-1]$ be given. Using the analysis equation (10) we define

$$\Gamma[k] := \frac{1}{N} \sum_{m=0}^{N-1} \gamma[m] e^{-2\pi ikm/N}, \qquad k = 0, 1, \ldots, N-1.$$

By using this expression together with the orthogonality relations (20) we find

$$\sum_{k=0}^{N-1} \Gamma[k] e^{2\pi ikn/N} = \sum_{k=0}^{N-1} \left\{ \frac{1}{N} \sum_{m=0}^{N-1} \gamma[m] e^{-2\pi ikm/N} \right\} e^{2\pi ikn/N}$$

$$= \sum_{m=0}^{N-1} \gamma[m] \left\{ \frac{1}{N} \sum_{k=0}^{N-1} e^{2\pi ikn/N} \cdot e^{-2\pi ikm/N} \right\}$$

$$= \gamma[n], \quad n = 0, 1, \ldots, N-1,$$

i.e., the synthesis equation (9) holds. Thus we see that Fourier's representation can be used for any function on $\mathbb{P}_N$, so the bottom front link from the Fourier-Poisson cube of Fig. 1.22 is secure.

Absolutely summable functions on $\mathbb{Z}$

Let ϕ be an absolutely summable function on $\mathbb{Z}$, *i.e.*, $\phi[n] \to 0$ as $n \to \pm\infty$ so rapidly that

$$\sum_{n=-\infty}^{\infty} |\phi[n]| < \infty.$$

This hypothesis of absolute summability ensures that the Fourier transform

$$\Phi(s) := \frac{1}{q} \sum_{n=-\infty}^{\infty} \phi[n] e^{-2\pi isn/q}$$

is well defined, with the series converging absolutely and uniformly on $\mathbb{R}$ to the continuous q-periodic function Φ. Moreover, the same hypothesis guarantees that the N-periodic discrete function

$$\gamma[n] := \sum_{m=-\infty}^{\infty} \phi[n - mN], \qquad n = 0, \pm 1, \pm 2, \ldots$$

is well defined by N-summation with the corresponding discrete Fourier transform being given by the Poisson relation

$$\Gamma[k] = \frac{q}{N} \Phi\left(\frac{kq}{N}\right), \qquad k = 0, \pm 1, \pm 2, \ldots$$

of (30). We use these expressions for γ, Γ in the synthesis equation

$$\gamma[n] = \sum_{k=0}^{N-1} \Gamma[k] e^{2\pi i n k/N}, \qquad n = 0, \pm 1, \pm 2, \ldots$$

(which we have just established) to obtain the discrete *Poisson sum formula*

$$\sum_{m=-\infty}^{\infty} \phi[n - mN] = \frac{q}{N} \sum_{k=0}^{N-1} \Phi\left(\frac{kq}{N}\right) e^{2\pi i (kq/N)(n/q)}, \qquad n = 0, \pm 1, \pm 2, \ldots . \quad (33)$$

As $N \to \infty$, the translates $\phi[n - mN]$, $m = \pm 1, \pm 2, \ldots$ from the sum on the left of (33) move off to $\pm \infty$, while the Riemann sums on the right converge to a corresponding integral. Thus in the limit as $N \to \infty$ (33) yields the Fourier synthesis equation

$$\phi[n] = \int_{s=0}^{q} \Phi(s) e^{2\pi i s n/q} ds, \qquad n = 0, \pm 1, \pm 2, \ldots .$$

In this way we prove that Fourier's representation (7)–(8) is valid for any absolutely summable function on $\mathbb{Z}$. The four links at the bottom of the Fourier-Poisson cube are secure when ϕ is such a function.

Continuous piecewise smooth functions on $\mathbb{T}_p$

Let g be a continuous piecewise smooth function on $\mathbb{T}_p$, *i.e.*, g is continuous on $\mathbb{T}_p$ and g' is defined and continuous at all but a finite number of points of $\mathbb{T}_p$ where finite jump discontinuities can occur. The graph of g on $\mathbb{T}_p$ is thus formed from finitely many smooth curves joined end-to-end with corners being allowed at the points of connection, *e.g.*, as illustrated in Fig. 1.3. The Fourier coefficients

$$G[k] := \frac{1}{p} \int_0^p g(x) e^{-2\pi i k x/p} dx, \qquad k = 0, \pm 1, \pm 2, \ldots ,$$

$$G_1[k] := \frac{1}{p} \int_0^p g'(x) e^{-2\pi i k x/p} dx, \qquad k = 0, \pm 1, \pm 2, \ldots$$

of g, g' are then well defined. Since $g(0+) = g(0-) = g(p-)$, we can use an integration by parts argument to verify that

$$G_1[k] = (2\pi i k/p) G[k], \qquad k = 0, \pm 1, \pm 2, \ldots .$$

We use this identity with the real inequality

$$|ab| \leq \frac{1}{2}(a^2 + b^2)$$

and Bessel's inequality (23) (for g') to see that

$$\sum_{k=-\infty}^{\infty} |G[k]| = |G[0]| + \frac{p}{2\pi} \sum_{k \neq 0} \left| \frac{1}{k} G_1[k] \right|$$

$$\leq |G[0]| + \frac{p}{4\pi} \sum_{k \neq 0} \left\{ \frac{1}{k^2} + |G_1[k]|^2 \right\}$$

$$\leq |G[0]| + \frac{p}{2\pi} \sum_{k=1}^{\infty} \frac{1}{k^2} + \frac{1}{4\pi} \int_0^p |g'(x)|^2 dx$$

$$< \infty.$$

Thus the function

$$\phi[k] := p \, G[-k], \qquad k = 0, \pm 1, \pm 2, \dots$$

is absolutely summable on $\mathbb{Z}$. We have shown that any such function has the Fourier representation

$$\phi[k] = \int_{x=0}^p \Phi(x) e^{2\pi i k x/p} dx, \qquad k = 0, \pm 1, \pm 2, \dots$$

where

$$\Phi(x) := \frac{1}{p} \sum_{k=-\infty}^{\infty} \phi[k] e^{-2\pi i k x/p}, \qquad -\infty < x < \infty.$$

After expressing ϕ in terms of G this synthesis-analysis pair takes the form

$$G[k] = \frac{1}{p} \int_{x=0}^p \Phi(x) e^{-2\pi i k x/p} dx, \qquad k = 0, \pm 1, \pm 2, \dots$$

$$\Phi(x) = \sum_{k=-\infty}^{\infty} G[k] e^{2\pi i k x/p}, \qquad -\infty < x < \infty.$$

In this way we see that the original function g and the auxiliary function Φ have the same Fourier coefficients. We apply the Plancherel identity (16) to the continuous p-periodic function $g - \Phi$ and thereby conclude that $g = \Phi$. In this way we establish the desired synthesis equation

$$g(x) = \sum_{k=-\infty}^{\infty} G[k] e^{2\pi i k x/p}, \qquad -\infty < x < \infty,$$

and prove that Fourier's representation (5)–(6) is valid for any continuous piecewise smooth function on $\mathbb{T}_p$.

Since the Fourier coefficients are absolutely summable, the sequence of partial sums

$$s_n(x) := \sum_{k=-n}^{n} G[k]e^{2\pi i k x/p}, \qquad n = 0, 1, 2, \ldots$$

of the Fourier series converges absolutely and uniformly on $\mathbb{T}_p$ to g. In particular, all but finitely many of the trigonometric polynomials $s_0, s_1, s_2, \ldots$ have graphs that lie within an arbitrarily small ϵ-tube

$$\{(x, z) : x \in \mathbb{T}_p, \ z \in \mathbb{C}, \ |g(x) - z| < \epsilon\}, \quad \epsilon > 0$$

drawn about the graph of g on $\mathbb{T}_p$, cf. Fig. 1.23.

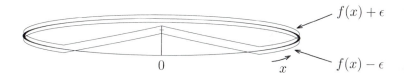

Figure 1.23. Any real ϵ-tube drawn about the graph of the continuous piecewise smooth function f from Fig. 1.3 contains the graphs of all but finitely many of the partial sums $s_0, s_1, s_2, \ldots$ of the corresponding Fourier series.

The sawtooth singularity function on $\mathbb{T}_1$

In this section we study the convergence of the Fourier series of the 1-periodic sawtooth function

$$w_0(x) := \begin{cases} 0 & \text{if } x = 0 \\ \dfrac{1}{2} - x & \text{if } 0 < x < 1 \end{cases} \tag{34}$$

that is continuously differentiable at all points of $\mathbb{T}_1$ except the origin, where a unit jump occurs, cf. Fig. 1.24. Using integration by parts we compute the Fourier coefficients

$$W_0[k] := \int_0^1 \left(\frac{1}{2} - x\right) e^{-2\pi i k x} dx$$

$$= -\frac{1}{2\pi i k}\left\{\left(\frac{1}{2} - x\right) e^{-2\pi i k x}\Big|_{x=0}^{1} + \int_0^1 e^{-2\pi i k x} dx\right\}$$

$$= \frac{1}{2\pi i k}, \qquad k = \pm 1, \pm 2, \ldots$$

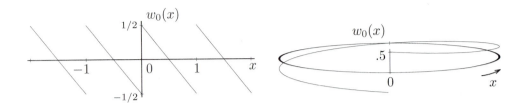

Figure 1.24. Graphs of the sawtooth singularity function (34) as a 1-periodic function on $\mathbb{R}$ and as a function on $\mathbb{T}_1$.

with

$$W_0[0] = \int_0^1 \left(\frac{1}{2} - x\right) dx = 0.$$

We will show that the slightly modified Fourier representation

$$w_0(x) = \lim_{L \to \infty} \sum_{k=-L}^{L} W_0[k]e^{2\pi i k x} = \sum_{k=1}^{\infty} \frac{\sin(2\pi k x)}{\pi k} \tag{35}$$

(with the limits at $\pm\infty$ taken symmetrically) is valid at each point x. Since (35) holds trivially when $x = 0$, we need only give a proof for points $0 < x < 1$.

We construct a continuous piecewise smooth 1-periodic function

$$w_1(x) := -\frac{1}{12} + \int_{u=0}^{x} w_0(u)du, \qquad -\infty < x < \infty,$$

having the derivative $w_0(x)$ (at points $x \neq 0, \pm 1, \pm 2, \dots$), noting that

$$w_1(x) = -\frac{1}{12} + \frac{x}{2} - \frac{x^2}{2} \quad \text{when} \quad 0 \leq x \leq 1.$$

The constant $-1/12$ has been chosen to make

$$W_1[0] := \int_0^1 w_1(x)dx = 0,$$

and an integration by parts argument can be used to verify that

$$W_1[k] = \frac{1}{(2\pi i k)^2}, \qquad k = \pm 1, \pm 2, \dots ,$$

cf. (4.19)–(4.24). We have already shown that such a function w_1 has the Fourier representation

$$w_1(x) = \sum_{k=-\infty}^{\infty} W_1[k]e^{2\pi i k x} = \sum_{k=1}^{\infty} -\frac{\cos(2\pi k x)}{\pi k(2\pi k)}, \qquad -\infty < x < \infty$$

with this Fourier series converging absolutely and uniformly. This being the case, we can establish (35), *i.e.*, we can justify the term-by-term differentiation of the Fourier series for w_1, by showing that the series (35) converges uniformly on every closed interval $\alpha \leq x \leq 1 - \alpha$ with $0 < \alpha < 1/2$.

We will use a classic argument of Abel and Dedekind to verify that the sequence of partial sums

$$s_n(x) := \sum_{k=-n}^{n} W_0[k]e^{2\pi ikx} = \sum_{k=1}^{n} \frac{\sin(2\pi kx)}{\pi k}, \qquad n = 1, 2, \ldots$$

of the Fourier series (35) converges uniformly on $0 < \alpha \leq x \leq 1 - \alpha$. We introduce the auxiliary functions

$$p_k(x) := -\frac{\cos(2\pi kx + \pi x)}{2 \sin(\pi x)}, \qquad 0 < x < 1, \quad k = 0, 1, 2, \ldots$$

that have the two properties

$$p_k(x) - p_{k-1}(x) = \frac{\cos(2\pi kx - \pi x) - \cos(2\pi kx + \pi x)}{2 \sin(\pi x)}$$

$$= \sin(2\pi kx), \qquad 0 < x < 1, \qquad k = 1, 2, 3, \ldots$$

and

$$|p_k(x)| \leq \frac{1}{2 \sin(\pi \alpha)} \quad \text{for } \alpha \leq x \leq 1 - \alpha, \quad k = 1, 2, \ldots .$$

When $m > n > 0$ and $\alpha \leq x \leq 1 - \alpha$ we use the first of these to write

$$s_m(x) - s_n(x) = \frac{\sin\{2\pi(n+1)x\}}{\pi(n+1)} + \frac{\sin\{2\pi(n+2)x\}}{\pi(n+2)} + \cdots + \frac{\sin\{2\pi mx\}}{\pi m}$$

$$= \frac{p_{n+1}(x) - p_n(x)}{\pi(n+1)} + \frac{p_{n+2}(x) - p_{n+1}(x)}{\pi(n+2)} + \cdots + \frac{p_m(x) - p_{m-1}(x)}{\pi m}$$

$$= \frac{1}{\pi} \left\{ \left(\frac{-1}{n+1} \right) p_n(x) + \left(\frac{1}{n+1} - \frac{1}{n+2} \right) p_{n+1}(x) + \cdots \right.$$

$$\left. + \left(\frac{1}{m-1} - \frac{1}{m} \right) p_{m-1}(x) + \left(\frac{1}{m} \right) p_m(x) \right\}.$$

We then use the second to obtain the uniform bound

$$|s_m(x) - s_n(x)| \leq \frac{1}{2\pi \sin(\pi\alpha)} \left\{ \frac{1}{n+1} + \left(\frac{1}{n+1} - \frac{1}{n+2} \right) + \cdots \right.$$

$$\left. + \left(\frac{1}{m-1} - \frac{1}{m} \right) + \frac{1}{m} \right\}$$

$$= \frac{1}{\pi(n+1) \sin(\pi\alpha)} \quad \text{when } \alpha \leq x \leq 1 - \alpha, \quad m > n > 0,$$

and thereby establish the validity of (35).

The Gibbs phenomenon for w_0

Before leaving this discussion, we will show *how* $s_n(x)$ converges to $w_0(x)$ in a neighborhood of the point of discontinuity $x = 0$. The analysis will help you understand the annoying ripples you will observe every time you approximate a discontinuous function on $\mathbb{T}_p$ with a partial sum from its Fourier series. We find it convenient to define

$$\xi := 2nx = \frac{x}{1/2n}$$

so that ξ provides us with a measure of x in units of $1/2n$. We can then write

$$s_n(x) = s_n(\xi/2n)$$
$$= \sum_{k=1}^{n} \frac{\sin\{2\pi k(\xi/2n)\}}{\pi k}$$
$$= \sum_{k=1}^{n} \frac{\sin\{\pi(k\,\xi/n)\}}{\pi(k\,\xi/n)} \cdot (\xi/n).$$

We regard this as a very good Riemann sum approximation

$$s_n(x) \approx \mathcal{G}(\xi) \tag{36}$$

to the Gibbs function

$$\mathcal{G}(\xi) := \int_0^{\xi} \frac{\sin \pi u}{\pi u} du \tag{37}$$

when n is large and $2nx$ is of modest size. (A large x analysis is given in Ex. 1.37.) The odd function $\mathcal{G}$, shown in Fig. 1.25, takes the extreme values

$$\mathcal{G} = .5894\ldots,\ .4514\ldots,\ .5330\ldots,\ .4749\ldots,\ .5201\ldots,\ \ldots$$

that oscillate about the line $\mathcal{G} = 1/2$ with decreasing amplitude at the points $\xi = 1, 2, 3, 4, 5, \ldots$, corresponding to the abscissas $x = 1/2n, 2/2n, 3/2n, \ldots$.

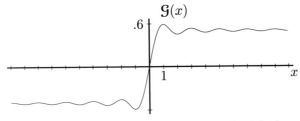

Figure 1.25. The Gibbs function $\mathcal{G}$ of (37).

Using the approximation (36) and the graph of Fig. 1.25, we see that when n is large the graph of $s_n(x)$ will overshoot the value $w_0(0+) = 1/2$ by about 9% of the total jump $J := w_0(0+) - w_0(0-) = 1$ in w_0, taking an extreme value close to .59 near the abscissa $x = 1/2n$. This behavior, well illustrated in the plots of s_n in Fig. 1.26, is known as *Gibbs phenomenon*. It was observed by Michelson when he plotted partial sums for dozens of Fourier series with his mechanical harmonic analyzer, *cf.* Ex. 1.45, and subsequently described by J.W. Gibbs, *cf. Nature* **58**(1898), 544–545, **59**(1898), 200, and **59**(1899), 606. (An earlier exposition was given by Wilbraham, in 1841, *cf.* E. Hewitt and R.E. Hewitt, The Gibbs-Wilbraham phenomenon: An episode in Fourier analysis, *Arch. History Exact Sci.* **21**(1979), 129–160 for additional details.)

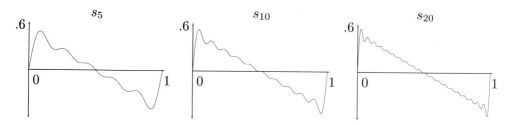

Figure 1.26. The partial sums s_5, s_{10}, s_{20} of the Fourier series (35) for the sawtooth singularity function w_0 of (34).

Piecewise smooth functions on $\mathbb{T}_p$

Let g be a piecewise smooth function on $\mathbb{T}_p$, *i.e.*, g, g' are continuous at all but finitely many points of $\mathbb{T}_p$ where finite jump discontinuities can occur, and assume that g has the *midpoint regularization*

$$g(x) = \frac{1}{2}\{g(x+) + g(x-)\}$$

at every point x. The graph of g on $\mathbb{T}_p$ is thus formed from finitely many smooth curves and isolated midpoints, as illustrated in Fig. 1.27.

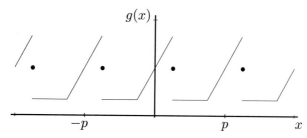

Figure 1.27. A piecewise smooth p-periodic function g on $\mathbb{R}$ with midpoint regularization at points of discontinuity.

Any such function g has the decomposition

$$g(x) = J_1 w_0 \left(\frac{x - x_1}{p} \right) + J_2 w_0 \left(\frac{x - x_2}{p} \right) + \cdots + J_m w_0 \left(\frac{x - x_m}{p} \right) + g_r(x) \quad (38)$$

where $0 \leq x_1 < x_2 < \cdots < x_m < p$ are the points where g has the nonzero jumps

$$J_\ell := g(x_\ell+) - g(x_\ell-), \quad \ell = 1, 2, \ldots, m, \tag{39}$$

where w_0 is the sawtooth singularity function (34), and where the function $g_r(x)$ is continuous and piecewise smooth. The decomposition (38) for the function of Fig. 1.27 is shown in Fig. 1.28.

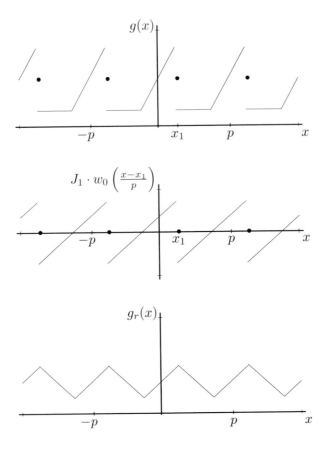

Figure 1.28. The decomposition (38) for the piecewise smooth function of Fig. 1.27.

We know that the Fourier series for the continuous piecewise smooth function g_r converges uniformly on $\mathbb{T}_p$ to g_r. Fourier's representation is also valid for the shifted and dilated sawtooth function

$$w_0\left(\frac{x-x_\ell}{p}\right) = \lim_{L\to\infty}\sum_{k=-L}^{L} W_0[k]e^{2\pi ik(x-x_\ell)/p}$$

$$= \lim_{L\to\infty}\sum_{k=-L}^{L}\left\{W_0[k]e^{-2\pi ikx_\ell/p}\right\}e^{2\pi iks/p}$$

with the convergence being uniform on the portion of $\mathbb{T}_p$ that remains after a small interval centered at the point of discontinuity $x = x_\ell$ has been removed, $\ell = 1, 2, \ldots, m$. Since each term from the decomposition (38) has a valid Fourier representation, we can write

$$g(x) = \lim_{L\to\infty}\sum_{k=-L}^{L} G[k]e^{2\pi ikx/p}, \qquad -\infty < x < \infty.$$

Once again we must sum the terms in a symmetric fashion.

From the above analysis we see that the partial sums of the Fourier series for g converge uniformly on that portion of $\mathbb{T}_p$ that remains after m arbitrarily small intervals centered at the points of discontinuity $x_1, x_2, \ldots, x_m$ have been removed. The term $J_\ell w_0((x-x_\ell)/p)$ introduces a Gibbs overshoot near the point $x = x_\ell$, as illustrated in Fig. 1.29 for the function of Fig. 1.27. The graphs of all but finitely many of these partial sums will be contained in the region obtained by adding to the ϵ-tube about the graph of g the Gibbs' ϵ-tubes

$$\{(x, z) : x \in \mathbb{T}_p, \ z \in \mathbb{C}, \ |x - x_\ell| < \epsilon, \ |z - g(x_\ell) - tJ_\ell| < \epsilon, \ -.59 < t < .59\},$$

$$\ell = 1, 2, \ldots, m$$

containing the singularities (together with their $\pm p, \pm 2p, \ldots$ translates), $cf.$ Fig. 1.30.

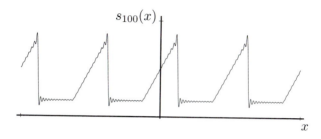

$s_{100}(x)$

x

Figure 1.29. The partial sum s_{100} of the Fourier series for the function of Fig. 1.27 exhibits the 9% Gibbs overshoot at each point of discontinuity.

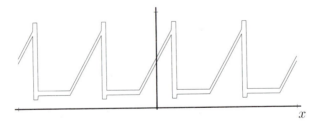

Figure 1.30. All but finitely many of the partial sums of the Fourier series for the function g of Fig. 1.27 lie in the region formed by combining the ϵ-tube about g and the Gibbs ϵ-tubes at the jumps of g.

Smooth functions on $\mathbb{R}$ with small regular tails

Let f and its derivative f' be defined and continuous on $\mathbb{R}$, and let f'' be defined and continuous at all but finitely many points of $\mathbb{R}$ where finite jump discontinuities can occur. Assume further that the tails of f at $\pm\infty$ are small and regular in the sense that

$$|f(x)| + |f'(x)| \leq T(|x|), \qquad -\infty < x < \infty \tag{40}$$

for some choice of the continuous, nonincreasing, integrable function $T(x)$ on the half line $0 \leq x < \infty$, and that

$$\int_{-\infty}^{\infty} |f''(x)| dx < \infty. \tag{41}$$

These hypotheses ensure that the Fourier transforms

$$F(s) := \int_{-\infty}^{\infty} f(x) e^{-2\pi i s x} dx,$$

$$F_1(s) := \int_{-\infty}^{\infty} f'(x) e^{-2\pi i s x} dx, \quad F_2(s) := \int_{-\infty}^{\infty} f''(x) e^{-2\pi i s x} dx$$

are well-defined continuous functions on $\mathbb{R}$.

Using the fundamental theorem of calculus, we write

$$f(x) = f(0) + \int_{0}^{x} f'(\xi) d\xi, \qquad -\infty < x < \infty, \tag{42}$$

and since f' is absolutely integrable, the limits

$$f(\pm\infty) = f(0) + \int_{0}^{\pm\infty} f'(\xi) d\xi \tag{43}$$

must exist. Since f itself is absolutely integrable, it follows that $f(\pm\infty) = 0$. Analogously, $f'(\pm\infty) = 0$ as well. Knowing that f, f' vanish at $\pm\infty$ we integrate by parts to verify that

$$F_2(s) = 2\pi is \, F_1(s) = (2\pi is)^2 F(s), \qquad -\infty < s < \infty$$

with

$$s^2|F(s)| = \frac{1}{4\pi^2}|F_2(s)| = \frac{1}{4\pi^2}\left|\int_{-\infty}^{\infty} f''(x)e^{-2\pi isx}dx\right| \le \frac{1}{4\pi^2}\int_{-\infty}^{\infty} |f''(x)|dx < \infty.$$

Thus the tails of F at $\pm\infty$ go to zero so rapidly that

$$|F(s)| < M/s^2, \qquad s \ne 0 \qquad (44)$$

for some constant M.

We now show that the function

$$g(x) := \sum_{m=-\infty}^{\infty} f(x - mp), \qquad p > 0,$$

produced by the p-summation process is continuously differentiable on $\mathbb{T}_p$. Indeed, by using the hypothesis (40) together with the integral test for convergence of an infinite series, we find

$$\sum_{m=-\infty}^{\infty} |f(x - mp)| \le \sum_{m=-\infty}^{\infty} T(|x - mp|) \le 2\left\{T(0) + \frac{1}{p}\int_0^{\infty} T(\xi)d\xi\right\} < \infty,$$

and analogously,

$$\sum_{m=-\infty}^{\infty} |f'(x - mp)| \le 2\left\{T(0) + \frac{1}{p}\int_0^{\infty} T(\xi)d\xi\right\} < \infty.$$

Knowing that these series converge absolutely and uniformly on $\mathbb{T}_p$, we conclude that the function g is well defined and continuously differentiable with

$$g'(x) = \sum_{m=-\infty}^{\infty} f'(x - mp).$$

Since f is absolutely integrable, the Fourier coefficients of the p-periodic function g are given by the Poisson relation (29), *i.e.*,

$$G[k] = \frac{1}{p}F\left(\frac{k}{p}\right), \qquad k = 0, \pm 1, \pm 2, \dots .$$

We have already proved that Fourier's representation is valid for a continuous piecewise smooth function on $\mathbb{T}_p$, so

$$g(x) = \sum_{k=-\infty}^{\infty} G[k]e^{2\pi ikx/p}, \qquad -\infty < x < \infty.$$

After expressing g, G in terms of f, F, we obtain the continuous version of the *Poisson sum formula*

$$\sum_{m=-\infty}^{\infty} f(x-mp) = \sum_{k=-\infty}^{\infty} \frac{1}{p} F\left(\frac{k}{p}\right) e^{2\pi ikx/p}, \qquad -\infty < x < \infty. \qquad (45)$$

[The discrete version appears in (33).]

The desired Fourier synthesis equation (3) is now obtained by using (45) in conjunction with a suitable limiting argument. Indeed, for any choice of x, $p > |x|$ and $L > 1/p$, we use (45), (40), and (44) in turn to write

$$\left| f(x) - \int_{-\infty}^{\infty} F(s)e^{2\pi isx}ds \right|$$

$$= \left| f(x) - \sum_{m=-\infty}^{\infty} f(x-mp) + \sum_{k=-\infty}^{\infty} \frac{1}{p} F\left(\frac{k}{p}\right) e^{2\pi ikx/p} - \int_{-\infty}^{\infty} F(s)e^{2\pi isx}ds \right|$$

$$\leq \sum_{m \neq 0} |f(x-mp)| + \left| \sum_{|k| \leq Lp} \frac{1}{p} F\left(\frac{k}{p}\right) e^{2\pi ikx/p} - \int_{-L}^{L} F(s)e^{2\pi isx}ds \right|$$

$$+ \sum_{|k|>Lp} \left| \frac{1}{p} F\left(\frac{k}{p}\right) \right| + \int_{|s|>L} |F(s)|ds$$

$$\leq 2\sum_{m=1}^{\infty} T(mp - |x|) + \left| \sum_{|k| \leq Lp} \frac{1}{p} F\left(\frac{k}{p}\right) e^{2\pi ikx/p} - \int_{-L}^{L} F(s)e^{2\pi isx}ds \right|$$

$$+ \frac{2}{p}\sum_{k>Lp} \frac{M}{(k/p)^2} + 2\int_{L}^{\infty} \frac{M}{s^2}ds.$$

Suppose now that $\epsilon > 0$ is given. We will have

$$2\int_{L}^{\infty} \frac{M}{s^2}ds = \frac{2M}{L} < \frac{\epsilon}{4}$$

and

$$\frac{2}{p} \sum_{k>Lp} \frac{M}{(k/p)^2} \leq 2Mp\left\{ \frac{1}{(Lp)^2} + \frac{1}{(Lp+1)^2} + \frac{1}{(Lp+2)^2} + \cdots \right\}$$

$$\leq 2Mp \int_{Lp-1}^{\infty} \frac{ds}{s^2} = \frac{2Mp}{Lp-1} < \frac{\epsilon}{4}$$

provided that $L > 16M/\epsilon$ and $Lp > 2$. With L so chosen, we force

$$\left| \sum_{|k| \leq Lp} \frac{1}{p} F\left(\frac{k}{p}\right) e^{2\pi i kx/p} - \int_{-L}^{L} F(s)e^{2\pi isx}ds \right| \leq \frac{\epsilon}{4}$$

by making the mesh $1/p$ in this approximating sum to the integral sufficiently small, *i.e.*, by choosing p sufficiently large. Finally, by using the monotonicity and integrability of T, we see that

$$2 \sum_{m=1}^{\infty} T(mp - |x|) < 2\left\{ T(p - |x|) + \frac{1}{p} \int_{p-|x|}^{\infty} T(u)du \right\} < \frac{\epsilon}{4}$$

when p is sufficiently large. In this way we prove that

$$\left| f(x) - \int_{s=-\infty}^{\infty} F(s)e^{2\pi isx}ds \right| < \epsilon$$

for every choice of $\epsilon > 0$, thereby establishing the validity of Fourier's representation

$$f(x) = \int_{-\infty}^{\infty} F(s)e^{2\pi isx}ds, \qquad -\infty < x < \infty.$$

Indeed, all 12 links of the Fourier-Poisson cube of Fig. 1.22 are secure for such a function f.

Singularity functions on $\mathbb{R}$

We define the singularity functions

$$y_0(x) := -\frac{1}{4}\begin{cases} (x+2)e^x & \text{if } x < 0 \\ 0 & \text{if } x = 0 \\ (x-2)e^{-x} & \text{if } x > 0, \end{cases}$$

$$y_1(x) := -\frac{1}{4}\begin{cases} (x+1)e^x & \text{if } x \leq 0 \\ (-x+1)e^{-x} & \text{if } x \geq 0, \end{cases} \tag{46}$$

$$y_2(x) := -\frac{1}{4}\begin{cases} xe^x & \text{if } x \leq 0 \\ xe^{-x} & \text{if } x \geq 0 \end{cases}$$

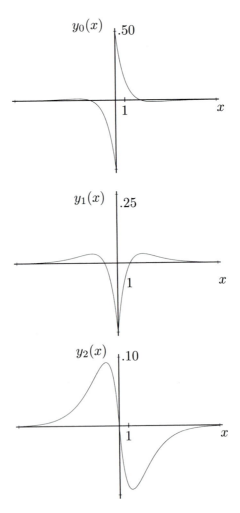

Figure 1.31. The singularity functions y_0, y_1, y_2 of (46).

shown in Fig. 1.31 to use for the purpose of removing jump discontinuities from a function on $\mathbb{R}$ and from its first two derivatives. By construction

$$\begin{aligned} y_2'(x) &= y_1(x), & -\infty < x < \infty \\ y_2''(x) &= y_1'(x) = y_0(x), & -\infty < x < 0 \text{ or } 0 < x < \infty \end{aligned} \tag{47}$$

with y_1, y_2 being continuous on $\mathbb{R}$ and with y_0 being continuous at all points of $\mathbb{R}$ except the origin, where we find

$$\begin{aligned} y_0(0+) - y_0(0-) &= 1 \\ y_0'(0+) - y_0'(0-) &= 0. \end{aligned} \tag{48}$$

We will show that Fourier's representation is valid for y_0, y_1, y_2 and then use this fact to show that Fourier's representation can also be used for all piecewise smooth functions having small regular tails. As a first step, we integrate by parts to verify that y_2 has the Fourier transform

$$
\begin{aligned}
Y_2(s) : &= \int_{x=-\infty}^{\infty} y_2(x) e^{-2\pi i s x} dx \\
&= -\frac{1}{4} \int_{-\infty}^{0} x \, e^{x(1-2\pi i s)} dx - \frac{1}{4} \int_0^{\infty} x \, e^{-x(1+2\pi i s)} dx \\
&= -\frac{1}{4} \left\{ \frac{x}{1-2\pi i s} - \frac{1}{(1-2\pi i s)^2} \right\} e^{x(1-2\pi i s)} \Big|_{-\infty}^{0} \\
&\quad + \frac{1}{4} \left\{ \frac{x}{1+2\pi i s} + \frac{1}{(1+2\pi i s)^2} \right\} e^{-x(1+2\pi i s)} \Big|_0^{\infty} \\
&= \frac{2\pi i s}{(1+4\pi^2 s^2)^2}, \qquad -\infty < s < \infty.
\end{aligned}
$$

Using (47) and integration by parts, we then find in turn the Fourier transforms

$$
Y_1(s) : = \int_{x=-\infty}^{\infty} y_2'(x) e^{-2\pi i s x} dx = 2\pi i s \, Y_2(s) = \frac{(2\pi i s)^2}{(1+4\pi^2 s^2)^2}, \quad -\infty < s < \infty,
$$

$$
Y_0(s) : = \int_{x=-\infty}^{\infty} y_1'(x) e^{-2\pi i s x} dx = 2\pi i s \, Y_1(s) = \frac{(2\pi i s)^3}{(1+4\pi^2 s^2)^2}, \quad -\infty < s < \infty
$$

of y_1, y_0, respectively.

The smooth odd function y_2 has small regular tails, so from the analysis of the preceding section we know that y_2 has the Fourier representation

$$
\begin{aligned}
y_2(x) &= \int_{s=-\infty}^{\infty} Y_2(s) e^{2\pi i s x} dx \\
&= -2 \int_{s=0}^{\infty} \frac{(2\pi s) \sin(2\pi s x) ds}{(1+4\pi^2 s^2)^2}, \qquad -\infty < x < \infty.
\end{aligned}
\tag{49}
$$

Moreover, since the integral of (49) and the integral obtained by formally differentiating (49) with respect to x both converge uniformly, we can write

$$
y_1(x) = y_2'(x) = -2 \int_{s=0}^{\infty} \frac{(2\pi s)^2 \cos(2\pi s x) ds}{(1+4\pi^2 s^2)^2}, \qquad -\infty < x < \infty.
\tag{50}
$$

Thus, y_1 has the Fourier representation

$$y_1(x) = \int_{s=-\infty}^{\infty} Y_1(s)e^{2\pi i s x} ds. \tag{51}$$

As a final step we will show that y_0 has the Fourier representation

$$y_0(x) = \lim_{L \to \infty} \int_{s=-L}^{L} Y_0(s)e^{2\pi i s x} ds \tag{52}$$

(with symmetric limits on the integral), or equivalently, that

$$y_0(x) = 2 \int_{s=0}^{\infty} \frac{(2\pi s)^3 \sin(2\pi s x) ds}{(1 + 4\pi^2 s^2)^2}, \qquad -\infty < x < \infty. \tag{53}$$

Since (53) holds trivially at the point of discontinuity, $x = 0$, and since y_0 is odd, it is enough to verify that (53) holds at each point $x > 0$. We will show that the integral of (53) converges uniformly on the interval $\alpha \le x < \infty$ for every $\alpha > 0$. Since the same is true of the integral of (50), we thereby justify the process of differentiating under the integral sign of (50) to produce (53). With this in mind, we let $\epsilon > 0$ be selected and let $M > L > \alpha > 0$ be chosen with L lying to the right of all of the local extreme points of the kernel

$$G(s) := \frac{2(2\pi s)^3}{(1 + 4\pi^2 s^2)^2} .$$

That portion of the tail of the integral in (53) between L and M is then uniformly bounded by

$$\left| \int_{s=L}^{M} G(s) \sin(2\pi s x) ds \right|$$

$$= \frac{1}{2\pi x} \left| \int_{L}^{M} G(s)\{-2\pi x \sin(2\pi s x)\} ds \right|$$

$$= \frac{1}{2\pi x} \left| G(s) \cos(2\pi s x) \big|_{s=L}^{M} - \int_{s=L}^{M} G'(s) \cos(2\pi s x) ds \right|$$

$$\le \frac{1}{2\pi \alpha} \left\{ G(M) + G(L) + \left| \int_{s=L}^{M} G'(s) ds \right| \right\}$$

$$\le \frac{2G(L)}{2\pi \alpha}$$

$$< \epsilon$$

if L is sufficiently large. Thus the integral of (53) converges uniformly on $\alpha \le x < \infty$ and the validity of Fourier's representation of y_0 is established.

Piecewise smooth functions on $\mathbb{R}$ with small regular tails

Let f be defined on $\mathbb{R}$ and assume that f is continuous except for finitely many points $x_1 < x_2 < \cdots < x_m$ where finite jumps

$$J_\mu := f(x_\mu+) - f(x_\mu-), \qquad \mu = 1, 2, \ldots, m$$

occur and where f has the midpoint regularization

$$f(x_\mu) = \frac{1}{2}\{f(x_\mu+) + f(x_\mu-)\}, \qquad \mu = 1, 2, \ldots, m,$$

e.g., as illustrated by Fig. 1.32. Let f' be defined and continuous except for finitely many points $x'_1 < x'_2 < \ldots < x'_{m'}$ where finite jumps

$$J'_\mu := f'(x'_\mu+) - f'(x'_\mu-), \qquad \mu = 1, 2, \ldots, m'$$

occur. Let f'' be defined and continuous except for finitely many points where finite jump discontinuities can occur, and assume that

$$\int_{-\infty}^{\infty} |f''(x)|dx < \infty.$$

Finally, assume that the tails of f at $\pm\infty$ are small and regular in the sense that

$$|f(x)| + |f'(x)| < T(|x|), \qquad x \neq x'_1, \ldots, x'_{m'}$$

where T is continuous, nonincreasing, and integrable on the half line $0 \leq x < \infty$. Using (46)–(48) we see that such a function f has the decomposition

$$f(x) = \sum_{\mu=1}^{m} J_\mu\, y_0(x - x_\mu) + \sum_{\mu=1}^{m'} J'_\mu\, y_1(x - x'_\mu) + f_r(x) \qquad (54)$$

where f_r, f'_r are continuous on $\mathbb{R}$, where

$$\int_{-\infty}^{\infty} |f''_r(x)|dx < \infty,$$

and where f_r (like f, y_0, y_1) has small regular tails at $\pm\infty$. By using the Fourier representations of f_r, y_0, y_1 developed in the two preceding sections, we conclude that

$$f(x) = \lim_{L \to +\infty} \int_{-L}^{L} F(s)e^{2\pi i s x}ds, \qquad -\infty < x < \infty \qquad (55)$$

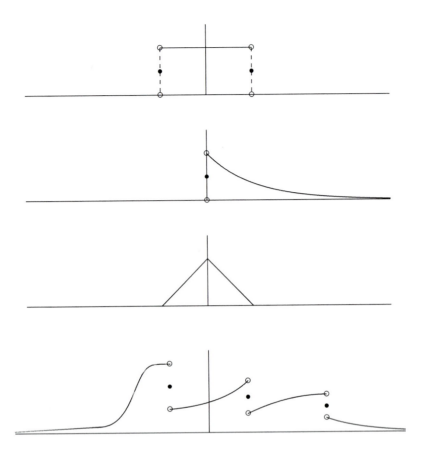

Figure 1.32. Piecewise smooth functions on $\mathbb{R}$ with small regular tails and midpoint regularization.

[*i.e.*, (3) holds when the integration limits are taken in a symmetric fashion]. Here

$$
F(s) := \int_{-\infty}^{\infty} f(x)e^{-2\pi isx}dx
$$

$$
= \sum_{\mu=1}^{m} J_\mu \, e^{-2\pi i x_\mu s} Y_0(s) + \sum_{\mu=1}^{m'} J'_\mu \, e^{-2\pi i x'_\mu s} Y_1(s) + F_r(s),
\tag{56}
$$

where F_r is the Fourier transform of f_r. Thus Fourier's representation is valid for such a function f.

We now know that if f satisfies the above hypotheses, then the Fourier transform F is well defined by the analysis equation (4), and the synthesis equation (3) can be used to represent f provided that we rewrite (3) in the form (55). In view of the symmetry present in the synthesis-analysis equations (3)–(4), the same argument shows that if a function F satisfies the above hypotheses, then we can use the synthesis equation (3) to construct a function f that will have F as its Fourier transform provided that we rewrite (4) in the form

$$F(s) = \lim_{L \to +\infty} \int_{-L}^{L} f(x)e^{-2\pi isx} dx.$$

Thus if f, F are piecewise smooth functions having midpoint regularization at any points of discontinuity, if at least one of the functions f, F has small regular tails, and if f, F are linked by either (3) or (4), then f, F are linked by both (3) and (4). This observation can be used to justify the use of Fourier's representation (3)–(4) for almost all of the common transform pairs of ordinary functions f, F that we meet in the elementary applications of Fourier analysis.

Extending the domain of validity

We have shown that Fourier's representation is valid for all functions defined on $\mathbb{P}_N$ and for certain large but not universal classes of functions defined on $\mathbb{Z}$, $\mathbb{T}_p$, and $\mathbb{R}$. The restrictive hypotheses that we have imposed can be weakened considerably, e.g., a function on $\mathbb{T}_p$ or $\mathbb{R}$ has a valid Fourier representation if the following four *Dirichlet conditions* are satisfied:

- $\int |f(x)|dx < \infty$, with the integral taken over the domain $\mathbb{T}_p$ or $\mathbb{R}$ of f;
- At each point x in the domain of f, finite limits $f(x+), f(x-)$ exist and

$$f(x) = \tfrac{1}{2}\{f(x+) + f(x-)\};$$

- The points where f is discontinuous, if any, are isolated, *i.e.*, there are only finitely many such points in any bounded portion of the domain; and
- Any open interval (a, b) from the domain of f can be broken into subintervals $(a, x_1), (x_1, x_2), \ldots, (x_n, b)$ on each of which f is monotonic by the deletion of finitely many points $x_1, x_2, \ldots, x_n$ [which depend on the choice of (a, b)].

You will find many other sufficient conditions in the literature, but there is still no known necessary and sufficient condition for the validity of Fourier's representation of functions on $\mathbb{Z}$, $\mathbb{T}_p$, or $\mathbb{R}$. Fourier believed (5)–(6) to be valid at all points when f is continuous, but a half century after his death DuBois-Reymond constructed a rather bizarre continuous function with a Fourier series that diverged at some points. A simpler example of Fejér is developed in Ex. 1.44. Such points of divergence are relatively rare, however, for in a remarkable theorem published in 1966, Carlson proved that the set of points where the Fourier series of a given continuous function on $\mathbb{T}_p$ fails to represent the function can be covered with a sequence of intervals $(a_1, b_1), (a_2, b_2), \ldots$ having total length $\sum(b_n - a_n)$ less than a preassigned $\epsilon > 0$.

The search for a deeper understanding of Fourier's representations (3)–(4), (5)–(6), (7)–(8) has been most fruitful in spite of the fact that these basic validity questions remain unresolved, *cf.* Appendix 0. You have undoubtedly observed that we must answer the questions

What do I mean by function?

What do I mean by convergence?

as we formulate our theorems, *e.g.*, we can allow our piecewise smooth functions on $\mathbb{T}_p$ to have finite jumps if we use symmetric limits on the sum (5). Our simple sufficient conditions can be relaxed considerably, but the proofs become more difficult. (You can find a remarkable 150-year summary of such work in the books by Bochner, Titchmarsh, and Zygmund that are cited in the following references, but you will need some understanding of Lebesgue's theory of integration to follow many of the arguments!)

We will briefly return to the validity question again in Chapter 7 after introducing you to a new concept of function and to a new definition for convergence, *cf.* Fig. 1.33. We will even show that Fourier was right after all: Every continuous p-periodic function *is* represented by its Fourier series ... when we have the right understanding of convergence!

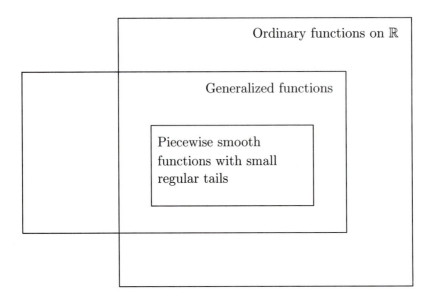

Figure 1.33. Functions on $\mathbb{R}$ that are known to have a valid Fourier representation: Piecewise smooth functions with small regular tails and generalized functions (as defined in Chapter 7).

References

J.J. Benedetto, *Harmonic Analysis and Applications*, CRC Press, Boca Raton, FL, 1997.

A modern introduction to Fourier analysis.

S. Bochner, *Lectures on Fourier Integrals* (English translation by M. Tenenbaum and H. Pollard), Princeton University Press, Princeton, NJ, 1959.

An exceptionally well written mid-20th-century mathematical treatise for Fourier transforms of functions on $\mathbb{R}$.

W.L. Briggs and V.E. Henson, *The DFT*, SIAM, Philadelphia, 1995.

A comprehensive introduction to discrete Fourier analysis.

J.W. Brown and R.V. Churchill, *Complex Variables and Applications, 6th ed.*, McGraw-Hill, New York, 1996.

You will find the elements of complex arithmetic and a discussion of complex-valued functions in the first few sections of this highly evolved undergraduate mathematics text.

H.S. Carslaw, *An Introduction to the Theory of Fourier's Series and Integrals, 3rd ed.*, Macmillan and Company, New York, 1930; reprinted by Dover Publications, New York, 1950.

Fourier analysis as known in the early 20th century.

D.C. Champeney, *A Handbook of Fourier Theorems*, Cambridge University Press, Cambridge, 1987.

A descriptive survey of basic Fourier analysis theorems that is written for scientists and engineers.

R. Courant and F. John, *Introduction to Calculus and Analysis, Vol. I*, John Wiley & Sons, New York, 1965.

Chapters 7, 8 of this classic intermediate-level mathematics text have exceptionally nice expositions of uniform convergence, Fourier series, respectively.

H. Dym and H.P. McKean, *Fourier Series and Integrals*, Academic Press, New York, 1972.

A mathematical account of Fourier analysis with numerous applications.

J. Fourier, *The Analytical Theory of Heat* (English translation by A. Freeman), Cambridge University Press, Cambridge, 1878; reprinted by Dover Publications, New York, 1955.

Fourier's early 19th-century account of the series and transforms he created to study the conduction of heat in solids.

J.D. Gaskill, *Linear Systems, Fourier Transforms, and Optics*, John Wiley & Sons, New York, 1978.

Chapters 1–7 provide elementary introduction to Fourier analysis for physicists.

R.R. Goldberg, *Fourier Transforms*, Cambridge University Press, Cambridge, 1965.

A tightly written classical introduction to the Fourier transform of functions on $\mathbb{R}$ (that uses Lebesgue and Riemann-Stieltjes integration).

E.A. González-Velasco, *Fourier Analysis and Boundary Value Problems*, Academic Press, San Diego, CA, 1996.

Formal mathematics for solving partial differential equations interspersed with historical accounts (and photographs) of Fourier, Poisson, Dirichlet, Gauss, ...

Y. Katznelson, *An Introduction to Harmonic Analysis*, John Wiley & Sons, New York, 1968.

An elegant first course in Fourier analysis for mathematics graduate students.

T.W. Körner, *Fourier Analysis*, Cambridge University Press, Cambridge, 1988.

A delightful intermediate-level introduction to Fourier analysis and selected applications to probability, differential equations, etc.

C. Lanczos, *Discourse on Fourier Series*, Oliver & Boyd, Edinburgh, 1966.

A mid-20th-century introduction to Fourier analysis for scientists and engineers by a distinguished mathematician with a gift for exposition.

A.G. Marshall and F.R. Verdun, *Fourier Transforms in NMR, Optical, and Mass Spectrometry*, Elsevier, Amsterdam, 1990.

Chapters 1–3 provide an introduction to the Fourier transform for chemists.

I.P. Natanson, *Constructive Function Theory, Vol. I* (English translation by A.N. Obolensky), Frederick Unger Publishing, New York, 1964.

A classic intermediate-level exposition of uniform approximation by truncated Fourier series, Fejér sums, etc.

A.V. Oppenheim, A.S. Willsky, and I.T. Young, *Signals and Systems*, Prentice Hall, Englewood Cliffs, NJ, 1988.

Chapters 4–5 contain exceptionally well written elementary accounts of discrete and continuous Fourier analysis as used in electrical engineering.

I.J. Schoenberg, The finite Fourier series I, II, III, and IV, *Mathematical Time Exposures*, Mathematical Association of America, Washington, DC, 1982.

A charming introduction to Fourier analysis of functions on $\mathbb{P}_N$ with applications to geometry.

E.C. Titchmarsh, *Introduction to the Theory of Fourier Integrals, 2nd ed.*, Clarendon Press, Oxford, 1948; reprinted by Chelsea, New York, 1986.

A mid-20th-century mathematical treatise for Fourier transforms of functions
on $\mathbb{R}$.

G.P. Tolstov, *Fourier Series* (English translation by R.A. Silverman), Prentice Hall,
Englewood Cliffs, NJ, 1962; reprinted by Dover Publications, New York, 1976.

An exceptionally well written elementary exposition of Fourier series (with a
brief introduction to the Fourier transform of functions on $\mathbb{R}$).

J.S. Walker, *Fourier Analysis*, Oxford University Press, New York, 1988.

An intermediate-level introduction to the basic mathematical theory as well as
some of the principal applications of Fourier analysis.

A. Zygmund, *Trigonometric Series, 2nd ed.*, Cambridge University Press, Cambridge, 1959.

The definitive mathematical treatise on Fourier series.

Exercise Set 1

▶ **EXERCISE 1.1:** In this exercise you will verify that Fourier's representation
(3)–(4) is valid for the box function

$$f(x) := \begin{cases} 1 & \text{if } -\frac{1}{2} < x < \frac{1}{2} \\ \frac{1}{2} & \text{if } x = \pm\frac{1}{2} \\ 0 & \text{if } x < -\frac{1}{2} \text{ or } x > \frac{1}{2}. \end{cases}$$

(a) Evaluate the integral (4) in this particular case and thereby show that

$$F(s) = \begin{cases} 1 & \text{if } s = 0 \\ \dfrac{\sin(\pi s)}{\pi s} & \text{if } s \neq 0. \end{cases}$$

(b) By using the fact that F is even, show that the synthesis equation (3) for f reduces
to the identity

$$f(x) \stackrel{?}{=} \int_{s=0}^{\infty} 2\,\frac{\sin(\pi s)}{\pi s}\,\cos\left(2\pi s x\right) ds.$$

The integral is not easily evaluated with the techniques of elementary calculus. The
remaining steps will show you how this can be done.

(c) Use integration by parts to verify that

$$\int_{s=0}^{\infty} e^{-ps} \cos(\pi qs)ds = \frac{p}{p^2 + (\pi q)^2}, \qquad p > 0.$$

(d) Integrate the identity of (c) with respect to q from $q = 0$ to $q = a$ to obtain

$$\int_{s=0}^{\infty} e^{-ps} \frac{\sin(\pi as)}{\pi s} ds = \frac{1}{\pi} \arctan\left(\frac{\pi a}{p}\right).$$

(e) Let $p \to 0+$ in the identity of (d), and thereby show that

$$\int_{s=0}^{\infty} \frac{\sin(\pi as)}{\pi s} ds = \begin{cases} -\frac{1}{2} & \text{if } a < 0 \\ 0 & \text{if } a = 0 \\ \frac{1}{2} & \text{if } a > 0. \end{cases}$$

(f) Use a trigonometric identity to write the integral from the synthesis equation of (b) in the form

$$\int_{s=0}^{\infty} \frac{2\sin(\pi s)\cos(2\pi sx)ds}{\pi s} = \int_{s=0}^{\infty} \frac{\sin[\pi(1+2x)s]}{\pi s} ds + \int_{s=0}^{\infty} \frac{\sin[\pi(1-2x)s]}{\pi s} ds.$$

(g) Finally, use the result of (e) (with $a = 1 \pm 2x$) to evaluate the integrals of (f) and thereby verify the synthesis identity from (b).

Note. As you study Chapter 3 you will learn that we almost never perform such tedious calculations. We use the analysis of Section 1.5 to infer that the identity from (b) is valid!

▶ **EXERCISE 1.2:**   Let f be a suitably regular function on $\mathbb{R}$. We use (2) with (4) to write

$$F(s) = \int_{x=-\infty}^{\infty} \cos(2\pi sx) f(x)dx - i \int_{x=-\infty}^{\infty} \sin(2\pi sx) f(x)dx.$$

The first integral is an even function of s and the second integral is an odd function of s. What can you say about the Fourier transform F if you know that f is

(a) even?

(b) odd?

(c) real-valued and even?

(d) real-valued and odd?

(e) pure imaginary and even?

(f) pure imaginary and odd?

(g) hermitian, *i.e.*, $\overline{f(x)} = f(-x)$?

(h) antihermitian, *i.e.*, $\overline{f(x)} = -f(-x)$?

(i) real?

(j) pure imaginary?

Hint. You should discover some connection between (g), (h) and (i), (j).

▶ **EXERCISE 1.3:**   Let g be a suitably regular function defined on the half line $x \geq 0$.

(a) Use (3)–(4) to derive the *cosine transform* pair

$$g(x) = 2 \int_{s=0}^{\infty} G(s) \cos(2\pi sx) ds, \quad G(s) = 2 \int_{x=0}^{\infty} g(x) \cos(2\pi sx) dx$$

that shows how to synthesize g from cosine functions.

Hint. Specialize (3)–(4) to the case where f is the even function

$$f(x) := \begin{cases} g(x) & \text{for } x \geq 0 \\ g(-x) & \text{for } x < 0. \end{cases}$$

(b) Derive the analogous *sine transform* pair

$$g(x) = 2 \int_{s=0}^{\infty} G(s) \sin(2\pi sx) ds, \quad x > 0, \qquad G(s) = 2 \int_{x=0}^{\infty} g(x) \sin(2\pi sx) dx, \quad s > 0$$

that shows how to synthsize g from sine functions.

Note. The cosine and sine transforms of a real-valued function are real-valued.

▶ **EXERCISE 1.4:**   A factor of 2π is included in the arguments of the exponential functions that appear in (3) and (4). In this exercise you will determine what happens when we remove this factor.

(a) Some authors define the Fourier transform by replacing (4) with the integral

$$F_1(\xi) := \int_{x=-\infty}^{\infty} f(x) e^{-i\xi x} dx, \qquad -\infty < \xi < \infty.$$

Use $F(s) = F_1(2\pi s)$ in (3) and thereby show that

$$f(x) = \frac{1}{2\pi} \int_{\xi=-\infty}^{\infty} F_1(\xi) e^{i\xi x} d\xi, \qquad -\infty < x < \infty.$$

The 2π reappears and destroys the symmetry!

(b) Likewise, show that when we replace (4) by

$$F_2(\xi) := \frac{1}{\sqrt{2\pi}} \int_{x=-\infty}^{\infty} f(x) e^{-i\xi x} dx, \qquad -\infty < \xi < \infty,$$

the synthesis equation takes the form

$$f(x) = \frac{1}{\sqrt{2\pi}} \int_{-\infty}^{\infty} F_2(\xi) e^{i\xi x} d\xi, \qquad -\infty < x < \infty.$$

Symmetry is restored, but we must still contend with a pair of 2π's in our equations.

(c) More generally, show that if we replace (4) by the analysis equation

$$F_3(\xi) := A \int_{x=-\infty}^{\infty} f(x) e^{-i\alpha\xi x} dx, \quad -\infty < \xi < \infty$$

where α, A are any nonzero real constants, then we must use the synthesis equation

$$f(x) = \frac{|\alpha|}{2\pi A} \int_{\xi=-\infty}^{\infty} F_3(\xi) e^{i\alpha\xi x} d\xi, \quad \infty < x < \infty.$$

Note. The 2π's *must* appear in the exponentials from (9)–(10). We have chosen to include the optional 2π's in the exponentials from (3)–(8) to make it easier for you to remember the analysis and synthesis equations in all four settings.

▶ **EXERCISE 1.5:**   Let f be a suitably regular function on $\mathbb{R}$.

(a) What function do we obtain when we take the Fourier transform of the Fourier transform of f?

   Hint. Change the sign of x in the synthesis equation (3)!

(b) What function do we obtain when we take the Fourier transform of the Fourier transform of the Fourier transform of the Fourier transform of f?

▶ **EXERCISE 1.6:**   Let the complex-valued function f on $\mathbb{T}_p$, $p > 0$, be defined by specifying $f(x)$ for $0 \le x < p$. Explain what we mean when we say that f is

(a) even on $\mathbb{T}_p$;

(b) odd on $\mathbb{T}_p$;

(c) continuous on $\mathbb{T}_p$;

(d) continuously differentiable on $\mathbb{T}_p$;

(e) absolutely integrable on $\mathbb{T}_p$;

(f) $p/2$-periodic on $\mathbb{T}_p$.

▶ **EXERCISE 1.7:**   Sketch the graphs of $f(x)$, $f(-x)$, $f(x-1)$, $f(1-x)$ when

$$f(x) := \begin{cases} 1-x & 0 \le x \le 1 \\ 0 & \text{otherwise} \end{cases}$$

is a function on:

(a) $\mathbb{R}$;   (b) $\mathbb{T}_1$;   (c) $\mathbb{T}_2$.

▶ **EXERCISE 1.8:**   Let f be a p-periodic trigonometric polynomial of degree N or less, *i.e.*,

$$f(x) := \sum_{k=-N}^{N} c_k e^{2\pi i k x/p},$$

for some choice of $p > 0$, $N = 0, 1, 2, \ldots$, and complex coefficients c_k, $k = 0, \pm 1, \ldots, \pm N$. Show that f is given by (5) when (6) is used to define F.

Hint. Multiply both sides of the defining equation for f by $e^{-2\pi i \ell x/p}$, integrate from $x = 0$ to $x = p$, and use the orthogonality relations (20) to show that

$$F[k] = \begin{cases} c_k & \text{if } k = 0, \pm 1, \dots, \pm N \\ 0 & \text{otherwise.} \end{cases}$$

Note. The class of p-periodic functions for which the Fourier representation (5)–(6) is valid includes the set of all trigonometric polynomials. This is analogous to the fact that the set of functions for which the Newton representation (1) is valid includes the set of all algebraic polynomials.

▶ **EXERCISE 1.9:** In this exercise you will verify that Fourier's representation (5)–(6) is valid for the Poisson kernel

$$f(x) := \frac{1 - r^2}{1 - 2r\cos(2\pi x) + r^2}, \qquad -\infty < x < \infty.$$

Here r is a real parameter with $0 \le r < 1$.

(a) Sketch the graph of f when $r = 0, .5, .9, .99$.

(b) Use the identity $1 + z + z^2 + z^3 + \cdots = 1/(1 - z)$, $|z| < 1$, with $z = re^{2\pi i x}$ to show that

$$\sum_{k=0}^{\infty} r^k e^{2\pi i k x} = \frac{\{1 - r\cos(2\pi x)\} + i\,r\sin(2\pi x)}{1 - 2r\cos(2\pi x) + r^2}.$$

(c) Use the result of (b) to show that

$$f(x) = \sum_{k=-\infty}^{\infty} r^{|k|} e^{2\pi i k x}, \qquad -\infty < x < \infty.$$

In this way you obtain (5) with $F[k] := r^{|k|}$ for each $k = 0, \pm 1, \pm 2, \dots$.

(d) A direct computation of the integrals

$$F[k] = \int_{x=0}^{1} \frac{(1 - r^2)e^{-2\pi i k x}}{1 - 2r\cos(2\pi x) + r^2}\,dx, \qquad k = 0, \pm 1, \pm 2, \dots$$

from the analysis equation (6) is fairly difficult. In this case, however, we can use the rapidly converging series from (c) to write

$$F[k] = \int_{x=0}^{1} \left(\sum_{\ell=-\infty}^{\infty} r^{|\ell|} e^{2\pi i \ell x} \right) e^{-2\pi i k x}\,dx = \sum_{\ell=-\infty}^{\infty} r^{|\ell|} \left(\int_{x=0}^{1} e^{2\pi i \ell x} e^{-2\pi i k x}\,dx \right).$$

Evaluate the remaining integrals and verify that (6) does yield $F[k] = r^{|k|}$.

Note. It is not so easy to find a function (other than a trigonometric polynomial) where we can evaluate the integrals (6) and sum the series (5) using the tools from calculus. We need the analysis of Section 1.5 to show that a suitably regular p-periodic function can be represented by its Fourier series!

▶ **EXERCISE 1.10:**   Let N be a positive integer and let f be defined on $\mathbb{Z}$ by

$$f[n] := \begin{cases} 1 & \text{if } n = 0, \pm 1, \pm 2, \dots, \pm N \\ 0 & \text{otherwise.} \end{cases}$$

(a) Show that f has the p-periodic Fourier transform

$$F(s) = \frac{1}{p} \begin{cases} 2N + 1 & \text{if } s = 0 \\ \dfrac{\sin\{(2N+1)\pi s/p\}}{\sin(\pi s/p)} & \text{if } 0 < s < p. \end{cases}$$

Hint. Set $z = e^{-i\pi s/p}$ in the formula

$$\sum_{n=-N}^{N} z^{2n} = z^{-2N} \sum_{n=0}^{2N} z^{2n} = \frac{z^{2N+1} - z^{-2N-1}}{z - z^{-1}}, \qquad z \neq \pm 1.$$

(b) Verify that f has the Fourier representation

$$f[n] = \int_0^p \frac{\sin\{(2N+1)\pi s/p\}}{p\sin(\pi s/p)} e^{2\pi ins/p} ds, \quad n = 0, \pm 1, \pm 2, \dots$$

of (7)–(8).

(c) Without performing any additional computation, write down the Fourier series (5) for the p-periodic function

$$g(x) = \begin{cases} 2N + 1 & \text{if } x = 0, \pm p, \pm 2p, \dots \\ \dfrac{\sin\{(2N+1)\pi x/p\}}{\sin(\pi x/p)} & \text{otherwise.} \end{cases}$$

Hint. Use (a) and the duality between (5)–(6) and (7)–(8).

▶ **EXERCISE 1.11:**   Answer the questions from Ex. 1.2 when f is a suitably regular function on $\mathbb{T}_p$.

Hint. $F[k] = \dfrac{1}{p} \displaystyle\int_{x=-p/2}^{p/2} \cos\left(\frac{2\pi kx}{p}\right) f(x)dx - \frac{i}{p} \int_{x=-p/2}^{p/2} \sin\left(\frac{2\pi kx}{p}\right) f(x)dx.$

▶ **EXERCISE 1.12:**   Let g be a suitably regular real-valued function defined on the "right" half of $\mathbb{T}_p$, i.e., for $0 \le x < p/2$.

(a) Use (5)–(6) to derive the *cosine transform* pair

$$g(x) = G[0] + 2\sum_{k=1}^{\infty} G[k] \cos\left(\frac{2\pi kx}{p}\right), \quad 0 \le x < \frac{p}{2}, \quad G[k] = \frac{2}{p} \int_{x=0}^{p/2} g(x) \cos\left(\frac{2\pi kx}{p}\right) dx$$

that shows how to synthsize g from cosine functions.

(b) Derive the analogous *sine transform* pair

$$g(x) = 2 \sum_{k=1}^{\infty} G[k] \sin\left(\frac{2\pi k x}{p}\right), \quad 0 < x < \frac{p}{2}, \qquad G[k] = \frac{2}{p} \int_0^{p/2} g(x) \sin\left(\frac{2\pi k x}{p}\right) dx.$$

Hint. Adapt the analysis of Ex. 1.3 to this setting.

▶ **EXERCISE 1.13:**   In this exercise you will establish (9)–(10) when $N = 2, 4$.

(a) Let $N = 2$ so that $e^{-2\pi i/N} = -1$. Show that the function f on $\mathbb{P}_2$ with components $f[0] := a$, $f[1] := b$ has the discrete Fourier transform F with components $F[0] = (a+b)/2$, $F[1] = (a-b)/2$, and then verify that f has the representation (9).

(b) Let $N = 4$ so that $e^{-2\pi i/N} = -i$. Find the components of the discrete Fourier transform F of the function f on $\mathbb{P}_4$ with components $f[0] := a$, $f[1] := b$, $f[2] := c$, $f[3] := d$, and verify that f has the representation (9).

▶ **EXERCISE 1.14:**   Find complex coefficients $c_0, c_1, \ldots, c_5$ such that

$$\sum_{k=0}^{5} c_k e^{2\pi i k n/6} = n, \qquad n = 0, 1, \ldots, 5.$$

Hint. The coefficients $c_0, c_1, \ldots, c_5$ play the role of $F[0], F[1], \ldots, F[5]$ in the synthesis equation (9). Use the corresponding analysis equation (10) to obtain $c_0 = 5/2$, $c_1 = -1/2 + (\sqrt{3}/2)i, \ldots, c_5 = -1/2 - (\sqrt{3}/2)i$.

Note. A generalization of this result is given in Ex. 4.26.

▶ **EXERCISE 1.15:**   Answer the questions from Ex. 1.2 when f is a function on $\mathbb{P}_N$.

Hint. $F[k] = \dfrac{1}{N} \displaystyle\sum_{-N/2 < n \le N/2} \cos\left(\frac{2\pi k n}{N}\right) f[n] - \frac{i}{N} \sum_{-N/2 < n < N/2} \sin\left(\frac{2\pi k n}{N}\right) f[n].$

▶ **EXERCISE 1.16:**   In this exercise you will rewrite the synthesis and analysis equations (3)–(10) using the familiar sine and cosine functions.

(a) Show that if (3)–(4) hold, then

$$f(x) = \int_{s=0}^{\infty} a(s) \cos(2\pi s x) ds + \int_{s=0}^{\infty} b(s) \sin(2\pi s x) ds$$

where

$$a(s) := 2 \int_{x=-\infty}^{\infty} f(x) \cos(2\pi s x) dx, \qquad b(s) := 2 \int_{x=-\infty}^{\infty} f(x) \sin(2\pi s x) dx.$$

Hint. Use (3)–(4) and Euler's identity (2) with $a(s) := F(s) + F(-s)$, $b(s) := i[F(s) - F(-s)]$.

(b) Show that if (5)–(6) hold, then

$$f(x) = \frac{a_0}{2} + \sum_{k=1}^{\infty} \left\{ a_k \cos\left(\frac{2\pi kx}{p}\right) + b_k \sin\left(\frac{2\pi kx}{p}\right) \right\}$$

where

$$a_k := \frac{2}{p} \int_{x=0}^{p} f(x) \cos\left(\frac{2\pi kx}{p}\right), \qquad b_k := \frac{2}{p} \int_{x=0}^{p} f(x) \sin\left(\frac{2\pi kx}{p}\right).$$

(c) Show that if (7)–(8) hold, then

$$f[n] = \int_{s=0}^{p/2} a(s) \cos\left(\frac{2\pi sn}{p}\right) ds + \int_{s=0}^{p/2} b(s) \sin\left(\frac{2\pi sn}{p}\right) ds$$

where

$$a(s) := \frac{2}{p} \sum_{n=-\infty}^{\infty} f[n] \cos\left(\frac{2\pi sn}{p}\right), \qquad b(s) := \frac{2}{p} \sum_{n=-\infty}^{\infty} f[n] \sin\left(\frac{2\pi sn}{p}\right).$$

(d) Show that if (9)–(10) hold, then

$$f[n] = \frac{a_0}{2} + \sum_{k=1}^{(N-2)/2} \left\{ a_k \cos\left(\frac{2\pi kn}{N}\right) + b_k \sin\left(\frac{2\pi kn}{N}\right) \right\} + \frac{a_{N/2}}{2} \cos(n\pi)$$

$$\text{if } N = 2, 4, 6, \dots , \text{ and}$$

$$f[n] = \frac{a_0}{2} + \sum_{k=1}^{(N-1)/2} \left\{ a_k \cos\left(\frac{2\pi kn}{N}\right) + b_k \sin\left(\frac{2\pi kn}{N}\right) \right\}$$

$$\text{if } N = 1, 3, 5, \dots ,$$

where

$$a_k := \frac{2}{N} \sum_{n=0}^{N-1} f[n] \cos\left(\frac{2\pi kn}{N}\right), \qquad b_k := \frac{2}{N} \sum_{n=0}^{N-1} f[n] \sin\left(\frac{2\pi kn}{N}\right).$$

▶ **EXERCISE 1.17:**   This exercise will show you how to attach units to the variables s, x, k, n that appear in the dimensionless expressions

$$sx, \quad \frac{kx}{p} \quad \frac{sn}{p}, \quad \frac{kn}{N}$$

from the complex exponentials in (3)–(4), (5)–(6), (7)–(8), and (9)–(10). Let $\alpha > 0$ be a unit of time, *e.g.*, we might use $\alpha = 10^{-15}$ sec, 10^{-3} sec, 1 sec when we work with light waves, sound waves, water waves, respectively. We rewrite the above expressions in the form

$$\left(\frac{s}{\alpha}\right)(x\alpha), \quad \left(\frac{k}{\alpha p}\right)(x\alpha), \quad \left(\frac{s}{\alpha p}\right)(n\alpha), \quad \left(\frac{k}{\alpha N}\right)(n\alpha),$$

identifying the first factor with frequency, F, and the second with time, t.

(a) Within this context, x measures time in units of α and s measures frequency in units of $1/\alpha$ when f is a function on $\mathbb{R}$. What is the analogous meaning of x, s, n, k when f is a function on $\mathbb{T}_p$? $\mathbb{Z}$? $\mathbb{P}_N$?

(b) When f is a function on $\mathbb{T}_p$, $x\alpha$ ranges over a time interval of length $T = \alpha p$ and frequency is *quantized* in units of $\Delta F = 1/T$. The product gives the *reciprocity relation*

$$T\,\Delta F = 1.$$

Find the analogous reciprocity relations for functions on $\mathbb{Z}$ and $\mathbb{P}_N$.

Hint. There are two such relations when f is a function on $\mathbb{P}_N$.

(c) The accoustical waveform

$$w(t) := e^{-(t/1\,\text{sec})^2} \cdot \cos\{2\pi \cdot 200\,\text{Hz} \cdot t\}$$

corresponds to a flute-like tone with a pitch of 200 Hz that sounds from $t \approx -2$ sec to $t \approx 2$ sec. What values of the frequency parameter s or k correspond to the frequencies F = ± 200 Hz when we set $\alpha = 10^{-3}$ sec and examine the Fourier transform of the function

$$f_{\mathbb{R}}(x) := w(x\alpha) \text{ on } \mathbb{R}? \qquad f_{\mathbb{T}}(x) := \sum_{m=-\infty}^{\infty} w([x + 100m]\alpha) \text{ on } \mathbb{T}_{100}?$$

$$f_{\mathbb{Z}}[n] := w(n\alpha) \text{ on } \mathbb{Z}? \qquad f_{\mathbb{P}}[n] := \sum_{m=-\infty}^{\infty} w([n + 1000m]\alpha) \text{ on } \mathbb{P}_{1000}?$$

(d) If α is "inappropriately chosen," we cannot locate the frequencies F > 0 and $-$F in the Fourier transforms of the processed functions $f_{\mathbb{T}}$, $f_{\mathbb{Z}}$, $f_{\mathbb{P}}$ as given in (c). Explain why we might wish to impose the respective constraints

$$\frac{1}{2p} < \text{F}\alpha, \quad \text{F}\alpha < \frac{1}{2}, \quad \frac{1}{2N} < \text{F}\alpha < \frac{1}{2}\,.$$

Hint. When we work on $\mathbb{T}_p$, frequency is quantized in units of $1/\alpha p$ so we cannot expect to detect a frequency less than $1/2\alpha p$.

Note. Newton and Fourier discovered the first principles of *dimensional analysis* that are used in this exercise, *cf.* H.E. Huntley, *Dimensional Analysis*, McDonald & Co., London, 1952; reprinted by Dover Publications, New York, 1967, pp. 33–37.

▶ **EXERCISE 1.18:**   The planets Earth, Mars orbit the sun with periods $T_E = 1$ yr, $T_M = 1.88$ yr at a mean distance 1 au := $150 \cdot 10^6$ km, 1.52 au = $228 \cdot 10^6$ km, respectively. In this exercise you will use the simple approximations

$$Z_E(t) := e^{2\pi i t/1 \text{ yr}}, \qquad Z_m(t) := 1.5 e^{2\pi i t/2 \text{ yr}}$$

to study the motion of Mars as seen from Earth.

(a) Draw concentric circles with radii 1, 1.5 and label points $A, B, \ldots, I$ on each that locate Earth, Mars at times $t = 0, 1/4, \ldots, 8/4$ yr.

(b) Draw the orbit $Z(t) = Z_M(t) - Z_E(t)$, $0 \leq t \leq 2$ yr, that shows the position of Mars as seen from Earth. This orbit corresponds to one of the two circle approximations of Hipparchus and Ptolemy as described in the text.

(c) Normally Mars moves across the night sky in the same direction as the moon. There is a three month period every other year, however, when this planet moves in the opposite direction. Use your analysis from (b) to explain this retrograde motion.

Note. Lagrange was the first to recognize the connection between Fourier analysis and the ancient Hipparchus-Ptolemy model for planetary motion, *cf.* H. Goldstein, *A History of Numerical Analysis from the 16th through the 19th Century*, Springer-Verlag, New York, 1977, p. 171.

▶ **EXERCISE 1.19:**   The direction a telescope must be pointed in order to see a given star, planet, asteroid, ... is specified by giving the right ascension and declination, *i.e.*, the longitude and lattitude of the corresponding point on the celestial sphere. In 1802, Baron Von Zach published the observed declination

$$y = 408, \ 89, \ -66, \ 10, \ 338, \ 807, \ 1238, \ 1511, \ 1583, \ 1462, \ 1183, \ 804$$

(in minutes) for the orbit of the asteroid Pallas at the right ascension $x = 0°$, $30°$, $60°$, ..., $330°$, respectively, *cf.* Fig. 1.13. As you analyze this data you will share in a very important discovery of Gauss.

(a) Use the analysis equation (10) to find the coefficients $c_0, c_1, \ldots, c_{11}$ for the trigonometric polynomial

$$y(x) = \sum_{k=0}^{11} c_k e^{2\pi i k x/360}$$

that fits the data at the 12 points (0,408), (30,89), ... ,(330,804).

Hint. Use a computer. You should find $c_0 = 780.5833$, $c_1 = -205.5072 + 360.1139i$, ...

Note. Gauss used real arithmetic and did such computations by hand!

(b) Explain how to use symmetry to reduce the amount of computation in (a).

Hint. Compare c_k with c_{12-k} and use Ex. 1.15(i).

(c) The above form for $y(x)$ is equivalent to

$$y(x) = \text{Re}\left\{ \sum_{k=-5}^{6} c_k e^{2\pi i k x/360} \right\}$$

when x is a multiple of $30°$ (and we set $c_{-k} := c_{12-k}$, $k = 1, 2, \ldots, 5$), but the latter is preferable when this is not the case. Explain why.

(d) The twelve-term trigonometric polynomial found in (a) exactly interpolates the twelve data points, but only half of the coefficients seem to contribute to the sum in a significant way. For this reason Gauss was undoubtedly motivated to fit the data approximately using a smaller number of terms. For example, he could exactly fit the six *even* data points (with $x = 0, 60, 120, \ldots, 300$) using

$$y_e(x) = \sum_{k=0}^{5} c_k^e e^{2\pi i k x / 360}$$

and he could equally well fit the six *odd* data points (with $x = 30, 90, 150, \ldots, 330$) using

$$y_o(x) = \sum_{k=0}^{5} c_k^o e^{2\pi i k x / 360},$$

and then use either of these trigonometric polynomials to generate a curve analogous to the one sketched in Fig. 1.13. Compute the six coefficients c_k^e and the six coefficients c_k^o.

Hint. Compute $d_k := c_k^o \cdot e^{2\pi i k / 12}$ before you compute c_k^o. You will find

$$c_0^e = 780.6667, \quad c_1^e = -205.3333 + 359.9779i, \ldots,$$
$$c_0^o = 780.5000, \quad c_1^o = -205.6810 + 360.2500i, \ldots.$$

(e) What symmetry is possessed by the coefficients you have computed in (d)? Is it necessary to compute all 12 of these coefficients directly? Explain.

Hint. Compare c_k^e, c_k^o with c_{6-k}^e, c_{6-k}^o, respectively.

(f) You now have two equally valid choices c_k^e (based on the six even data points) and c_k^o (based on the six odd data points) for the kth coefficient of a six-term trigonometric sum to use for fitting the given data. How would *you* combine these two estimates to produce a better coefficient c_k^* that depends on all twelve data points? Compare the c_k^*'s produced by your "natural" choice with the coefficients you obtained in (a).

(g) Show that the observations of (b), (e), (f) lead to the relations

$$c_k = \tfrac{1}{2}(c_k^e + c_k^o), \qquad c_{k+6} = \tfrac{1}{2}(c_k^e - c_k^o), \qquad k = 0, 1, \ldots, 5.$$

In this way you see that all twelve of the twelve-term coefficients $c_0, c_1, \ldots, c_{11}$ can be obtained from the eight 6-term coefficients c_k^e, c_k^o, $k = 0, 1, 2, 3$, with about one third of the effort!

Note. The recursive use of this observation leads to an FFT algorithm. For additional details, *cf.* Ex. 6.24 and M.T. Heideman, D.H. Johnson, and C.S. Burrus, Gauss and the history of the fast Fourier transform, *Arch. Hist. Exact Sci.* **34**(1985), 265–277.

▶ **EXERCISE 1.20:**   Fourier derived the formula

$$u(x,t) = \int_{s=-\infty}^{\infty} A(s)e^{2\pi i s x}e^{-4\pi^2 a^2 s^2 t}\,ds$$

for the temperature u at the point x, $-\infty < x < \infty$, at time $t \geq 0$ along an infinite one-dimensional rod with thermal diffusivity a^2. Suppose that when $t < 0$ the rod is held at the uniform temperature $u = 0$. At time $t = 0$ that portion of the rod from $x = -1/2$ to $x = +1/2$ is instantaneously heated to the temperature $u = 100$ (e.g., by using a laser), thereby producing the initial temperature

$$\int_{s=-\infty}^{\infty} A(s)e^{2\pi i s x}\,ds = u(x,0+) := \begin{cases} 100 & \text{if } |x| < \frac{1}{2} \\ 0 & \text{if } |x| > \frac{1}{2} . \end{cases}$$

Use the analysis equation (4) together with the Fourier transform pair of Ex. 1.1 to find A, and thereby produce a formula for $u(x,t)$.

Note. An extensive discussion of such problems is given in Section 9.3, and plots of $u(x,t)$ vs. x at times $t = 0+$, $.001/a^2$, $.01/a^2$, $.1/a^2$ are shown in Fig. 9.8. Fourier's formula predicts temperature distributions that match our physical intuition of how an initial hot spot in a conducting rod dissipates over time!

▶ **EXERCISE 1.21:**   Let $h > 0$ and let the averaging operator

$$(\mathbf{A}_h f)(x) := \frac{1}{2h}\int_{x-h}^{x+h} f(u)\,du, \qquad -\infty < x < \infty$$

be applied to any suitably regular function f on $\mathbb{R}$.

(a) Show that $\mathbf{A}_h$ is linear and translation invariant.

(b) Let $e_s(x) := e^{2\pi i s x}$ for $-\infty < s < \infty$. Show that

$$\mathbf{A}_h e_s = \alpha(s) \cdot e_s \quad \text{where } \alpha(s) := \begin{cases} 1 & \text{if } s = 0 \\ \sin(2\pi sh)/2\pi sh & \text{otherwise.} \end{cases}$$

(c) Assume that f has the Fourier representation (3). What is the corresponding Fourier representation for $\mathbf{A}_h f$?

(d) Assume that f has the Fourier representation (5). What is the corresponding Fourier representation for $\mathbf{A}_h f$?

▶ **EXERCISE 1.22:**   In this exercise you will verify certain algebraic identities that Schoenberg used to derive the Tartaglia-Cardan formula for the roots x_0, x_1, x_2 of a cubic equation $x^3 + bx^2 + cx + d = 0$.

(a) Let $x_n = \sum_{k=0}^{2} X_k \omega^{kn}$, $n = 0, 1, 2$, where $\omega := e^{2\pi i/3}$. Show that

$$(x - x_0)(x - x_1)(x - x_2) = (x - X_0)^3 - 3X_1 X_2(x - X_0) - (X_1^3 + X_2^3).$$

Hint. $\omega^3 = 1$ and $1 + \omega + \omega^2 = 0$.

(b) Using (a), show that

$$X_0 = -\frac{b}{3}, \qquad X_1 X_2 = \frac{b^2 - 3c}{9}, \qquad X_1^3 + X_2^3 = -\frac{27d - 9bc + 2b^3}{27}.$$

▶ **EXERCISE 1.23:**   In this (experimental) exercise you are to use a piano to do harmonic analysis and harmonic synthesis of simple vocal *tones*, *cf.*, Section 11.1. Depress the damper pedal (the rightmost one) of a piano, thereby allowing the strings to vibrate freely. Shout sustained vowels aaaa ... , eeee ... , iiii ... , oooo ... , uuuu ... , semivowels rrrr ... , *ℓℓℓℓ* ... , wwww ... , jjjj ... , fricative consonants zzzz ... , ssss ... , vvvv ... , ffff ... , *θθθθ* ... , shhh ... , or nasal consonants nnnn ... , mmmm ... of various pitches toward the sounding board and then listen as the strings return the waveform you have generated. During the vocalization process you produce a more or less periodic waveform that induces some of the piano strings to vibrate, with shorter, longer strings being stimulated by higher, lower pitches and with stronger, weaker vibrations being induced by louder, softer components of the given frequency in the tone. In this way each tone induces a characteristic pattern of vibration of the piano's strings, *i.e.*, these strings serve to analyze the original waveform. After you stop producing sound, the piano strings continue to vibrate, and at this point they serve to synthesize the periodic waveform associated with your vocalization.

▶ **EXERCISE 1.24:**   In this exercise you will informally derive alternative forms of the Parseval identities (11), (12), (13), (14) by suitably using the corresponding synthesis-analysis equations and freely interchanging the limiting processes associated with integration and summation.

(a) Use (3)–(4) to show that

$$\int_{x=-\infty}^{\infty} f(x)g(x)dx = \int_{s=-\infty}^{\infty} F(s)G(-s)ds,$$

$$\int_{x=-\infty}^{\infty} f(x)G(x)dx = \int_{s=-\infty}^{\infty} F(s)g(s)ds$$

when F, G are the Fourier transforms of the suitably regular functions f, g on $\mathbb{R}$.

(b) Use (9)–(10) to show that

$$\sum_{n=0}^{N-1} f[n]g[n] = N \sum_{k=0}^{N-1} F[k]G[-k],$$

$$\sum_{n=0}^{N-1} f[n]G[n] = \sum_{k=0}^{N-1} F[k]g[k]$$

when F, G are the Fourier transforms of the functions f, g on $\mathbb{P}_N$.

(c) Use (5)–(6) to show that

$$\int_{x=0}^{p} f(x)g(x)dx = p \sum_{k=-\infty}^{\infty} F[k]G[-k]$$

when F, G are the Fourier transforms of the suitably regular functions f, g on $\mathbb{T}_p$.

(d) Use (7)–(8) to show that

$$\sum_{n=-\infty}^{\infty} f[n]g[n] = p \int_{s=0}^{p} F(s)G(-s)ds$$

when F, G are the Fourier transforms of the suitably regular functions f, g on $\mathbb{Z}$.

(e) Use (5)–(6) and (7)–(8) to show that

$$\int_{x=0}^{p} f(x)G(x)dx = \sum_{k=-\infty}^{\infty} F[k]g[k]$$

when F, G are the Fourier transforms of the suitably regular functions f, g on $\mathbb{T}_p$, $\mathbb{Z}$, respectively.

▶ **EXERCISE 1.25:**   In this exercise you will derive the Parseval relations from the seemingly less general Plancherel identities.

(a) Let a, b be complex numbers. Show that

$$a\bar{b} = \tfrac{1}{4}\{|a + b|^2 + i|a + ib|^2 + i^2|a + i^2 b|^2 + i^3|a + i^3 b|^2\}.$$

(b) Use the polarization identity of (a) (on a point-by-point basis) to derive the Parseval identities (11), (12), (13), (14) from the corresponding Plancherel identities (15), (16), (17), (18).

▶ **EXERCISE 1.26:**   In this exercise you will use properties of the centroid (from calculus) to prove the discrete orthogonality relation (20).

(a) Let $N = 6$ and let $\omega := e^{2\pi i/6} = (1 + \sqrt{3}i)/2$. Given some $k = 0, 1, \ldots, 5$ we will place a unit mass at each point $z_0 := 1$, $z_1 := \omega^k$, $z_2 := \omega^{2k}$, $\ldots$, $z_5 := \omega^{5k}$ in the complex plane. Sketch the six mass distributions that are produced in this way, and use symmetry to explain why the centroid is at the origin when $k = 1, 2, 3, 4, 5$.

Hint. When $k = 3$ you will end up with a mass of size 3 at $z = \pm 1$.

(b) Generalize (a) and thereby prove (20) for each $N = 2, 3, \ldots$.

▶ **EXERCISE 1.27:**   In this exercise you will derive real versions of the orthogonality relations (19), (20).

(a) Let k, ℓ be nonnegative integers. Use suitable trigonometric identities to show that

$$\int_{x=0}^{p} \cos\left(\frac{2\pi kx}{p}\right) \cos\left(\frac{2\pi \ell x}{p}\right) dx = \begin{cases} p & \text{if } k = \ell = 0 \\ p/2 & \text{if } k = \ell \neq 0 \\ 0 & \text{otherwise,} \end{cases}$$

$$\int_{x=0}^{p} \cos\left(\frac{2\pi kx}{p}\right) \sin\left(\frac{2\pi \ell x}{p}\right) dx = 0,$$

$$\int_{x=0}^{p} \sin\left(\frac{2\pi kx}{p}\right) \sin\left(\frac{2\pi \ell x}{p}\right) dx = \begin{cases} p/2 & \text{if } k = \ell \neq 0 \\ 0 & \text{otherwise.} \end{cases}$$

(b) Let k, ℓ take the values $0, 1, \ldots, N - 1$. Show that

$$\sum_{n=0}^{N-1} \cos\left(\frac{2\pi kn}{N}\right) \cos\left(\frac{2\pi \ell n}{N}\right) = \begin{cases} N & \text{if } k = \ell = 0, \ N/2 \\ N/2 & \text{if } k = \ell \text{ and } k \neq 0, \ N/2 \\ N/2 & \text{if } k = N - \ell \text{ and } k \neq 0, \ N/2 \\ 0 & \text{otherwise,} \end{cases}$$

$$\sum_{n=0}^{N-1} \cos\left(\frac{2\pi kn}{N}\right) \sin\left(\frac{2\pi \ell n}{N}\right) = 0,$$

$$\sum_{n=0}^{N-1} \sin\left(\frac{2\pi kn}{N}\right) \sin\left(\frac{2\pi \ell n}{N}\right) = \begin{cases} N/2 & \text{if } k = \ell \text{ and } k \neq 0, N/2 \\ -N/2 & \text{if } k = N - \ell \text{ and } k \neq 0, N/2 \\ 0 & \text{otherwise.} \end{cases}$$

▶ **EXERCISE 1.28:**   Let f be a piecewise smooth function on $\mathbb{T}_p$.

(a) Let τ_n be an arbitrary p-periodic trigonometric polynomial (21) of degree n or less, and let s_n be the corresponding nth partial sum of the Fourier series for f. Use (22) to show that

$$\int_0^p |f(x) - s_n(x)|^2 dx \leq \int_0^p |f(x) - \tau(x)|^2 dx$$

with equality if and only if $\tau = s_n$. The truncated Fourier series gives the best *least squares approximation* to f by a p-periodic trigonometric polynomial of degree n or less!

(b) Derive the identity

$$\int_0^p |f(x) - s_n(x)|^2 dx = p \sum_{k=n+1}^{\infty} \left\{ |F[k]|^2 + |F[-k]|^2 \right\}.$$

The *quality* of the optimal least squares approximation depends on how rapidly the Fourier coefficients $F[k]$ go to zero as $k \to \pm\infty$!

▶ **EXERCISE 1.29:**   Let f be a function on $\mathbb{P}_N$ with Fourier transform F. Let m be a nonnegative integer with $2m + 1 \leq N$, and let

$$\tau_m[n] := \sum_{k=-m}^{m} c_k e^{2\pi ikn/N}, \qquad s_m[n] := \sum_{k=-m}^{m} F[k] e^{2\pi ikn/N}.$$

(a) Show that

$$\sum_{n=0}^{N-1} \left| f[n] - s_m[n] \right|^2 \leq \sum_{n=0}^{N-1} \left| f[n] - \tau_m[n] \right|^2$$

with equality if and only $\tau_m = s_m$. The truncated discrete Fourier series gives the best least squares approximation, *cp.* Ex. 1.28(a).

(b) Derive the identity

$$\sum_{n=0}^{N-1} \left| f[n] - s_m[n] \right|^2 = N \sum_{m < k < N-m} \left| F[k] \right|^2.$$

The quality of the optimal least squares approximation depends on the size of the "high-frequency" Fourier coefficients, *cp.* Ex. 1.28(b).

▶ **EXERCISE 1.30:**   In the book we use the Weierstrass theorem to show that

$$\int_0^p |f(x)|^2 dx \leq p \sum_{k=-\infty}^{\infty} \left| F[k] \right|^2$$

when f is any piecewise continuous function on $\mathbb{T}_p$. (In conjunction with (23), this gives (16).) Show that this inequality also holds when f is any square integrable function on $\mathbb{T}_p$ that is continuous except for finitely many points where jumps can occur.

Hint. Use (22) and the inequality $|a + b|^2 \leq 2|a|^2 + 2|b|^2$ to write

$$\int_0^p |f(x)|^2 dx - p \sum_{k=-\infty}^{\infty} \left| F[k] \right|^2 \leq \int_{x=0}^p |f(x) - \tau(x)|^2 dx$$

$$\leq 2 \int_{x=0}^p |f(x) - f_c(x)|^2 dx + 2 \int_{x=0}^p |f_c(x) - \tau(x)|^2 dx.$$

Here f_c is any continuous function on $\mathbb{T}_p$ and τ is a trigonometric polynomial. When $\epsilon > 0$ is given, you can make the first integral on the right less than $\epsilon/2$ by properly choosing f_c, and you can then make the second integral less than $\epsilon/2$ by properly choosing τ. Fill in the details.

▶ **EXERCISE 1.31:**   In this exercise you will examine a proof of the validity of Fourier's representation (5)–(6) that uses the real, even, p-periodic *Dirichlet kernel*

$$\delta_n(x) := \frac{1}{p} \sum_{k=-n}^{n} e^{2\pi i k x p} = \begin{cases} (2n+1)/p & \text{if } x = 0, \pm p, \pm 2p, \ldots \\ \dfrac{\sin\{\pi(2n+1)x/p\}}{p\sin(\pi x/p)} & \text{otherwise.} \end{cases}$$

(a) Verify the above formula for the sum that defines δ_n.

(b) Let f be a square integrable function on $\mathbb{T}_p$ that is continuous except for finitely many points where jumps can occur. Show that the symmetrically truncated Fourier series (5) has the integral representation

$$s_n(x) := \sum_{k=-n}^{n} F[k] e^{2\pi i k x/p} = \int_{u=0}^{p} f(u)\delta_n(x-u)du, \qquad n = 0, 1, 2, \ldots .$$

Thus, we can verify that Fourier's representation (5) holds at the point x (when the limits are taken symmetrically) by showing that

$$f(x) = \lim_{n\to\infty} \int_0^p f(u)\frac{\sin\{\pi(2n+1)(x-u)/p\}}{p\sin\{\pi(x-u)/p\}}\,du.$$

(c) Try to establish the validity of Fourier's representation at some point x where f is continuous by replacing the de la Vallée-Poussin power kernel with the above Dirichlet kernel in the proof of the Weierstrass theorem as given in the text. Why does this attempt fail?

(d) Show that

$$s_n(x) - f(x) = \int_{u=0}^{p} \{f(u)-f(x)\}\delta_n(x-u)du = \Phi_+[-n] - \Phi_-[n], \qquad n = 0, 1, 2, \ldots$$

where (for fixed x) $\Phi_\pm$ are the Fourier transforms of the p-periodic functions

$$\phi_\pm(v) := \frac{\{f(x+v)-f(x)\}e^{\pm\pi i v/p}}{2i\sin(\pi v/p)}.$$

(e) Show that if f has a derivative at x, then the functions $\phi_\pm$ have removable singularities at $v = 0, \pm p, \pm 2p, \ldots$ with

$$\lim_{v\to mp} \phi_\pm(v) = \frac{f'(x)}{2\pi i/p}, \qquad m = 0, \pm 1, \pm 2, \ldots .$$

(f) Using (d)–(e) show that if f is any square integrable piecewise continuous function on $\mathbb{T}_p$ with a derivative at the point x, then

$$f(x) = \lim_{n\to\infty} \sum_{k=-n}^{n} F[k] e^{2\pi i k x/p}.$$

Hint. Use Bessel's inequality to show that $\Phi_\pm[n] \to 0$ as $n \to \pm\infty$.

▶ **EXERCISE 1.32:**   In this exercise you will prove the Weierstrass theorem by using the real, even p-periodic *Fejér kernel*

$$\delta_n(x) := \frac{1}{(n+1)p} \begin{cases} (n+1)^2 & \text{if } x = 0, \pm p, \pm 2p, \dots \\[2mm] \dfrac{\sin^2\{\pi(n+1)x/p\}}{\sin^2(\pi x/p)} & \text{otherwise.} \end{cases}$$

(a) Show that

$$\delta_n(x) = \frac{1}{(n+1)p} \sum_{k=-n}^{n} (n+1-|k|)e^{2\pi i k x/p}.$$

Hint. $\displaystyle \sum_{k=-n}^{n} (n+1-|k|)z^k = z^{-n}\left\{ \sum_{\ell=0}^{n} z^\ell \right\}^2 = \left\{ \frac{z^{(n+1)/2} - z^{-(n+1)/2}}{z^{1/2} - z^{-1/2}} \right\}^2, \quad z \neq 1.$

(b) Let f be a continuous function on $\mathbb{T}_p$ and for $n = 0, 1, 2, \dots$ let

$$\sigma_n(x) := \int_0^p f(u)\delta_n(x-u)du.$$

Show that $\sigma_0, \sigma_1, \sigma_2, \dots$ is a sequence of p-periodic trigonometric polynomials that converges uniformly to f.

Hint. Does δ_n have the properties (26)?

▶ **EXERCISE 1.33:**   Let f be a continuous function on $\mathbb{T}_p$ and let s_n, σ_n be as in Exs. 1.31 and 1.32.

(a) Verify that

$$\sigma_n(x) = \frac{1}{n+1}\{s_0(x) + s_1(x) + s_2(x) + \dots + s_n(x)\}, \quad n = 0, 1, 2, \dots \ .$$

(b) Show that if $\lim s_n(x) = L$ at some point x, then $\lim \sigma_n(x) = L$.

Hint. First use (a) to show that

$$|\sigma_{n+m} - L| \leq \frac{n \max\limits_{k \leq n} |s_k - L| + m \max\limits_{k > n} |s_k - L|}{n + m + 1}.$$

(c) Show that if the Fourier series converges at some point x, then it must converge to $f(x)$.

Hint. Use (b) and Ex. 1.32(b).

▶ **EXERCISE 1.34:**   Establish the Fourier transform pairs (31), (32) when f, g are suitably regular functions on $\mathbb{R}$, $\mathbb{T}_p$.

▶ **EXERCISE 1.35:**   In this exercise you will analyze a simple numerical procedure for computing Fourier coefficients of a suitably regular function g on $\mathbb{T}_p$.

(a) Let $N = 2M + 1$ where $M = 1, 2, \ldots$, and let the vector

$$\gamma := \left[g(0), g\left(\frac{p}{N}\right), g\left(\frac{2p}{N}\right), \ldots, g\left(\frac{(N-1)p}{N}\right) \right]$$

be generated from samples of g. Explain why the discrete Fourier transform Γ of γ is close to the vector

$$(G[0], G[1], G[2], \ldots, G[M], G[-M], \ldots, G[-2], G[-1])$$

of Fourier coefficients of g when the tails of G at $\pm\infty$ are small and when M is large.

Hint. Use (32) to show that

$$\left| \Gamma[k] - G[k] \right| \leq \sum_{\substack{m=-\infty \\ m \neq 0}}^{\infty} \left| G[k - mN] \right|.$$

(b) Verify that $\Gamma[k]$ is just the Riemann sum approximation

$$G[k] \approx \frac{1}{p} \sum_{n=0}^{N-1} g\left(n\frac{p}{N}\right) e^{-2\pi i(k/p)(np/N)} \frac{p}{N}, \qquad k = 0, \pm 1, \ldots, \pm M$$

for the corresponding integral of the analysis equation (6).

Note. You may wish to try this numerical scheme with the function g from Ex. 1.9.

▶ **EXERCISE 1.36:**   In this exercise you will analyze a numerical procedure for computing samples of the Fourier transform F of a suitably regular function f on $\mathbb{R}$.

(a) Let $N = 2M + 1$ where $M = 1, 2, \ldots$. Show that the discrete Fourier transform of the vector

$$\left[pf(0), pf\left(\frac{p}{N}\right), pf\left(\frac{2p}{N}\right), \ldots, pf\left(\frac{Mp}{N}\right), pf\left(\frac{-Mp}{N}\right), \ldots, pf\left(\frac{-2p}{N}\right), pf\left(\frac{-p}{N}\right) \right]$$

of p-scaled samples of f is close to the N-vector

$$\left[F(0), F\left(\frac{1}{p}\right), F\left(\frac{2}{p}\right), \ldots, F\left(\frac{M}{p}\right), F\left(\frac{-M}{p}\right), \ldots, F\left(\frac{-2}{p}\right), F\left(\frac{-1}{p}\right) \right]$$

of samples of F when the tails of both f and F are small and when p, N are chosen so that both p and N/p are large.

Hint. Use the discrete Fourier transform pair

$$\gamma[n] := \sum_{m=-\infty}^{\infty} f\left(\frac{np}{N} - mp\right), \qquad \Gamma[k] = \frac{1}{p} \sum_{m=-\infty}^{\infty} F\left(\frac{k}{p} - m\frac{N}{p}\right)$$

from (29) and (32) to show that the error in component $k = 0, \pm 1, \ldots, \pm M$ is bounded by

$$\max_{-M \leq n \leq M} \quad p \sum_{\substack{m=-\infty \\ m \neq 0}}^{\infty} \left| f\left(\frac{np}{N} - mp\right) \right| + \sum_{\substack{m=-\infty \\ m \neq 0}}^{\infty} \left| F\left(\frac{k}{p} - m\frac{N}{p}\right) \right|.$$

(b) Verify that $\Gamma[k]$ is just the truncated Riemann sum approximation

$$F\left(\frac{k}{p}\right) \approx \sum_{m=-M}^{M} f\left(\frac{mp}{N}\right) e^{-2\pi i \frac{k}{p} \frac{mp}{N}} \frac{p}{N}, \qquad k = 0, \pm 1, \ldots, \pm M$$

for the corresponding integral of the analysis equation (4).

Note. You may wish to try this numerical scheme with the Fourier transform pair

$$f(x) = e^{-\pi x^2}, \qquad F(s) = e^{-\pi s^2}$$

from Appendix 1. If you use $p = 8$ and $N = 64$, you can compute approximations to $F(k/p)$ that are accurate to 16 decimal places!

▶ **EXERCISE 1.37:**   Let w_0 be the 1-periodic sawtooth singularity function of (34)–(35). When x is small, we can study the Gibbs phenomenon by using the approximation (36)–(37) from the book. In this exercise you will develop the large x approximation of Bochner that reveals the structure shown in Fig. 1.26.

(a) Let s_n be the nth partial sum of (35). Show in turn that

$$s_n'(x) = \frac{\sin\{(2n+1)\pi x\}}{\sin(\pi x)} - 1, \qquad s_n(x) = \int_0^x \frac{\sin\{(2n+1)\pi u\}}{\sin(\pi u)} du - x \quad \text{for} \quad |x| < 1.$$

Hint. Use Ex. 1.31(a).

(b) Using (a), show that when $|x| < 1$,

$$s_n(x) = w_0(x) + \mathcal{G}\{(2n+1)x)\} - \tfrac{1}{2}\operatorname{sgn}(x) + R_n(x)$$

where $\mathcal{G}$ is the Gibbs function (37) and

$$R_n(x) := \frac{1}{\pi} \int_0^{\pi x} \left(\frac{1}{\sin u} - \frac{1}{u}\right) \sin\{(2n+1)u\} du.$$

(c) Use an integration by parts argument to show that

$$|R_n(x)| \leq \frac{C}{2n+1}, \qquad |x| \leq \tfrac{1}{2}$$

for a suitably chosen constant C.

Note. I am indebted to Henry Warchall for bringing this analysis to my attention and for pointing out that we can choose $C = 1/\pi$ in (c).

▶ **EXERCISE 1.38:** Let f be an absolutely integrable function on $\mathbb{R}$ that is continuous except for finitely many points where jumps can occur.

(a) Show that the Fourier transform F is well defined by (4).

(b) Show that F is uniformly continuous.

> *Hint.* Let A be that portion of $\mathbb{R}$ within the union of tiny intervals (b_1, a_2), $(b_2, a_3), \ldots, (b_{N-1}, a_N)$ containing the points of discontinuity of f and the semi-infinite intervals $(-\infty, a_1)$, $(b_N, +\infty)$. You can then write
>
> $$|F(s+h) - F(s)| = \left| \int_{x=-\infty}^{\infty} f(x)e^{-2\pi isx}\{e^{-2\pi ihx} - 1\}dx \right|$$
>
> $$\leq \int_A 2|f(x)|dx + \sum_{n=1}^{N} \int_{a_n}^{b_n} |f(x)||e^{-2\pi ihx} - 1|dx.$$
>
> How must $a_1 < b_1 < a_2 < b_2 < \cdots < a_N < b_N$ and h be chosen to make this less than some preassigned $\epsilon > 0$?

(c) Show that $F(s) \to 0$ as $s \to \pm\infty$. This is known as the *Riemann-Lebesgue lemma.*

> *Hint.* Let A, $a_1, b_1, \ldots, a_N, b_N$ be as in (b) and write
>
> $$|F(s)| = \left| \int_A f(x)e^{-2\pi isx}dx + \sum_{n=1}^{N} \int_{a_n}^{b_n}\{f(x) - y_n(x)\}e^{-2\pi isx}dx + \sum_{n=1}^{N} \int_{a_n}^{b_n} y_n(x)e^{-2\pi isx}ds \right|$$
>
> $$\leq \int_A |f(x)|dx + \sum_{n=1}^{N} \int_{b_n}^{a_n} |f(x) - y_n(x)|dx + \sum_{n=1}^{N} \left| \int_{a_n}^{b_n} y_n(x)e^{-2\pi isx}dx \right|$$
>
> where y_n is a step function (*i.e.*, a piecewise constant function) on the interval $a_n \leq x \leq b_n$. You can make the first and second expressions small by suitably choosing A and $y_1, y_2, \ldots, y_n$. When $s \neq 0$, the third expression can be majorized by a finite sum of terms of the form
>
> $$\left| \int_\alpha^\beta C e^{-2\pi isx}dx \right| \leq \left| \frac{C}{\pi s} \right|.$$

▶ **EXERCISE 1.39:** Let f be an absolutely integrable function on $\mathbb{R}$ that is continuous except for finitely many points where jumps can occur. In this exercise you will show that Fourier's representation (55) is valid at any point x where f is differentiable. (You may wish to refer to Ex. 1.31 as you sort out the details.)

(a) Show that

$$f_L(x) := \int_{-L}^{L} \left\{ \int_{-\infty}^{\infty} f(u)e^{-2\pi i su} \right\} e^{2\pi i sx} du\, ds = \int_{-\infty}^{\infty} f(u)\delta_L(x-u)du$$

where

$$\delta_L(x) := \int_{-L}^{L} e^{2\pi i sx} ds = \begin{cases} 2L & \text{if } x = 0 \\ \sin(2\pi Lx)/\pi x & \text{otherwise.} \end{cases}$$

Thus we can show that (55) holds at some point x by showing that $f_L(x) \to f(x)$ as $L \to \infty$.

(b) Use the result of Ex. 1.1(e) to show that

$$f_L(x) - f(x) = \int_{u=-\infty}^{\infty} \{f(u) - f(x)\}\delta_L(x-u)du = \Phi(-L) - \Phi(L)$$

where (for fixed x) Φ is the Fourier transform of

$$\phi(v) := \frac{f(x+v) - f(x)}{2\pi i v}, \qquad v \neq 0.$$

(c) Show that f has the representation (55) at any point x where f is differentiable.

Hint. Verify that the Fourier integrals for Φ are well defined at $\pm L$ and then use the Riemann-Lebesgue lemma of Ex. 1.38(c) to see that $\Phi(\pm L) \to 0$ as $L \to \infty$.

▶ **EXERCISE 1.40:**   In this exercise you will prove that the Plancherel identity (15) is valid for any function f having the representation (54). In conjunction with Ex. 1.25, this shows that the Parseval identity (11) is valid for all such functions f, g.

(a) Let f have the representation (54), let

$$f_0(x) := \sum_{\mu=1}^{m} J_\mu z(x - x_\mu) \quad \text{where} \quad z(x) := \frac{1}{2} \begin{cases} -e^x & \text{if } x < 0 \\ 0 & \text{if } x = 0 \\ e^{-x} & \text{if } x > 0, \end{cases}$$

and let $f_1 := f - f_0$. Explain why the functions f_0, f_1 and the Fourier transform F_1 of f_1 are all piecewise continuous absolutely integrable functions on $\mathbb{R}$.

(b) Verify that

$$\int_{-\infty}^{\infty} f_1(x)\overline{f_1(x)}dx = \int_{-\infty}^{\infty} F_1(s)\overline{F_1(s)}ds, \quad \int_{-\infty}^{\infty} f_0(x)\overline{f_1(x)}dx = \int_{-\infty}^{\infty} F_0(s)\overline{F_1(s)}ds.$$

Hint. The functions $f_1(x)\overline{F_1(s)}$ and $f_0(x)\overline{F_1(s)}$ are absolutely integrable on the plane $-\infty < x < \infty$, $-\infty < s < \infty$, so it is possible to interchange the order of the x and s integrations.

(c) Using (b), verify that the proof of Plancherel's identity can be reduced to showing that

$$\int_{x=-\infty}^{\infty} |f_0(x)|^2 dx = \int_{s=-\infty}^{\infty} |F_0(s)|^2 ds$$

whenever f_0 is a function of the form given in (a).

(d) Using (c), verify that the proof of Plancherel's identity can be reduced to showing that

$$\int_{x=-\infty}^{\infty} z(x-x_\mu)z(x-x_\nu)dx = \int_{s=-\infty}^{\infty} Z(s)e^{-2\pi i x_\mu s}\overline{Z(s)}e^{2\pi i x_\nu s} ds$$

whenever $-\infty < x_\mu \leq x_\nu < \infty$.

(e) Using (d), verify that the proof of Plancherel's identity can be reduced to showing that

$$\int_{x=-\infty}^{\infty} z\left(x+\frac{h}{2}\right) z\left(x-\frac{h}{2}\right) dx = \int_{s=-\infty}^{\infty} |Z(s)|^2 e^{-2\pi i h s} ds$$

whenever $h \geq 0$.

(f) Show that

$$Z(s) = -\frac{2\pi i s}{1+4\pi^2 s^2}, \quad \int_{-\infty}^{\infty} z\left(x+\frac{h}{2}\right) z\left(x-\frac{h}{2}\right) dx = \frac{1}{4}(1-h)e^{-h}, \quad h \geq 0,$$

and thereby establish the identity (e).

Hint. The singularity function y_1 has the Fourier representation (51).

▶ **EXERCISE 1.41:**   Let $a_1 < b_1 \leq a_2 < b_2 \leq \cdots \leq a_N < b_N$, and for each $n = 1, 2, \ldots, N$, let

$$f_n(x) := \int_{x=a_n}^{b_n} F_n(s)e^{2\pi i s x} ds$$

where F_n is a piecewise smooth function on $[a_n, b_n]$ with

$$\int_{s=a_n}^{b_n} |F_n(s)|^2 ds = 1.$$

Show that $f_1, f_2, \ldots, f_n$ satisfy the orthogonality relations

$$\int_{x=-\infty}^{\infty} f_n(x)\overline{f_m(x)}dx = \begin{cases} 1 & \text{if } n = m = 1, 2, \ldots, N \\ 0 & \text{if } n \neq m. \end{cases}$$

Hint. In view of Ex. 1.40, you can use the Parseval identity (11).

▶ **EXERCISE 1.42:**   Let F be a piecewise smooth function on $\mathbb{R}$ with small regular tails at $\pm\infty$, let Y be any piecewise smooth function on the finite interval $a \leq x \leq b$, and let

$$f(x) := \int_{-\infty}^{\infty} F(s)e^{2\pi isx}\,ds, \quad f_{a,b}(x) := \int_{a}^{b} F(s)e^{2\pi isx}\,ds, \quad y(x) := \int_{a}^{b} Y(s)e^{2\pi isx}\,ds.$$

(a) Show that

$$\int_{-\infty}^{\infty} |f(x) - y(x)|^2\,dx = \int_{-\infty}^{\infty} |f(x)|^2\,dx - \int_{a}^{b} |F(s)|^2\,ds + \int_{a}^{b} |F(s) - Y(s)|^2\,ds,$$

and thereby obtain a version of (22) that is appropriate for functions on $\mathbb{R}$.

Hint. Use the Plancherel identity from Ex. 1.40.

(b) Show that

$$\int_{-\infty}^{\infty} |f(x) - f_{a,b}(x)|^2\,dx \leq \int_{-\infty}^{\infty} |f(x) - y(x)|^2\,dx$$

with equality if and only if $y = f_{a,b}$. The truncated Fourier transform gives the best least squares approximation, *cp.* Ex. 1.28 and Ex. 1.29.

(c) Derive the identity

$$\int_{-\infty}^{\infty} |f(x) - f_{a,b}(x)|^2\,dx = \int_{-\infty}^{a} |F(s)|^2\,ds + \int_{b}^{\infty} |F(s)|^2\,ds.$$

The quality of the optimal least squares approximation depends on the size of the "high-frequency" portion of the Fourier transform.

▶ **EXERCISE 1.43:**   In this exercise you will study the Gibbs phenomenon associated with Fourier's representation of piecewise smooth functions on $\mathbb{R}$ with small regular tails. The analysis parallels that of Ex. 1.37.

(a) Show that the glitch function from Ex. 1.40 has the Fourier representation

$$z(x) := \frac{1}{2}\begin{cases} -e^x & \text{if } x < 0 \\ 0 & \text{if } x = 0 \\ e^{-x} & \text{if } x > 0 \end{cases} = \int_{0}^{\infty} \frac{4\pi s \sin(2\pi sx)}{1 + 4\pi^2 s^2}\,ds.$$

(b) Let

$$z_L(x) := \int_{0}^{L} \frac{4\pi s \sin(2\pi sx)}{1 + 4\pi^2 s^2}\,ds$$

be the approximation to z that uses only the complex exponentials having frequencies in the band $-L \leq s \leq L$, *cf.* Ex. 1.42. Show that

$$z_L(x) = z(x) + \mathcal{G}(2Lx) - \tfrac{1}{2}\operatorname{sgn}(x) + R_L(x)$$

where $\mathcal{G}$ is the Gibbs function (37) and

$$R_L(x) := \int_L^\infty \frac{\sin(2\pi sx)}{\pi s(1 + 4\pi^2 s^2)}\, ds.$$

Hint. Use the integral from Ex. 1.1(e).

(c) Let f be a piecewise smooth function with small regular tails, and let F be the Fourier transform. Describe the appearance of the approximation

$$f_L(x) := \int_{-L}^{L} F(s)e^{2\pi isx}\, ds$$

to the function in a neighborhood of some point where f has a jump discontinuity.

▶ **EXERCISE 1.44:**   This exercise will introduce you to Fejér's example of a continuous 1-periodic function having a Fourier series that diverges at the point $x = 0$.

(a) For $n = 1, 2, \ldots$ we define

$$f_n(x) := 2\pi \sin\{2\pi(n+1)x\} \cdot \sum_{k=1}^{n} \frac{\sin(2\pi kx)}{\pi k}.$$

Use (35) and your knowledge of the Gibbs phenomenon (as illustrated in Fig. 1.26 and analyzed in Ex. 1.37) to explain why f_n is bounded.

(b) Use a suitable trigonometric identity to show that

$$f_n(x) = \frac{\cos(2\pi x)}{n} + \frac{\cos(2 \cdot 2\pi x)}{n-1} + \cdots + \frac{\cos(n \cdot 2\pi x)}{1}$$
$$- \frac{\cos\{(n+2) \cdot 2\pi x\}}{1} - \frac{\cos\{(n+3) \cdot 2\pi x\}}{2} - \cdots - \frac{\cos\{(2n+1) \cdot 2\pi x\}}{n},$$

and thereby determine the values of the partial sums

$$S_{n,m}(x) := \sum_{k=-m}^{m} F_m[k]e^{2\pi ikx}, \qquad m = 0, 1, 2, \ldots$$

of the Fourier series for f_n at $x = 0$.

(c) Show that we can define a continuous 1-periodic function by writing

$$f(x) := \sum_{n=1}^{\infty} \frac{1}{n^2} f_{2^{n^3}}(x).$$

(d) Show that the mth symmetric partial sum of the Fourier series for f takes a value in excess of

$$\frac{1}{n^2}\left\{\frac{1}{1}+\frac{1}{2}+\frac{1}{3}+\cdots+\frac{1}{2^{n^3}}\right\} > n\ln 2$$

when $x = 0$, $m = 2^{n^3}$, and thereby prove that the Fourier series diverges at this point.

Note. The Fourier series *does* converge to f when we use the weak limit concept that will be introduced in Section 7.6.

▶ **EXERCISE 1.45:** This exercise will introduce you to a cleverly designed mechanical device for harmonic synthesis that was invented at the end of the 19th century.

(a) Study the mechanical linkage of Fig. 1.34 to see how uniform circular motion can be used to produce the displacement function $y(t) = y_0 + c\cos(2\pi st + \alpha)$.

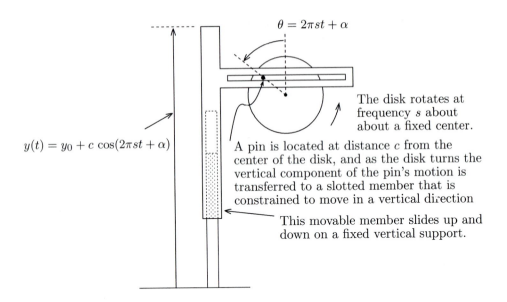

$\theta = 2\pi st + \alpha$

$y(t) = y_0 + c\,\cos(2\pi st + \alpha)$

The disk rotates at frequency s about about a fixed center.

A pin is located at distance c from the center of the disk, and as the disk turns the vertical component of the pin's motion is transferred to a slotted member that is constrained to move in a vertical direction

This movable member slides up and down on a fixed vertical support.

Figure 1.34. A mechanical linkage that produces a sinusoidal displacement.

(b) One large spring with force constant K, and N smaller springs each with force constant k are configured as shown in Fig. 1.35. Let y be the displacement of the lower end of the large spring when the upper ends of the small springs are given the small displacements $y_1, y_2, \ldots, y_N$. Explain why the large spring has the elongation $L + y$ and the nth small spring has the elongation $\ell + y_n - (a/A)y$, $n = 1, 2, \ldots, N$. Here L is the elongation of the large spring and ℓ is the common elongation of the small springs when $y = y_1 = y_2 = \cdots = y_N = 0$.

(c) Show that the apparatus from Fig. 1.35 mechanically sums the displacements $y_1, y_2, \ldots, y_N$ in the sense that

$$y = C \sum_{n=1}^{N} y_n \quad \text{where} \quad C := \left\{ N \left[\frac{a}{A} + \frac{\ell}{L} \right] \right\}^{-1}.$$

Hint. The moments balance when

$$\sum_{n=1}^{N} \left[\ell + y_n - \left(\frac{a}{A} \right) y \right] ka = [L + y]KA.$$

Note. Michelson and Stratton built an 80 term harmonic synthesizer using springs (as shown in Fig. 1.35) to sum displacements $y_n(t) = c_n \cos(2\pi nt + \alpha_n)$, $n = 1, 2, \ldots, 80$, produced with a somewhat more sophisticated linkage than the one shown in Fig. 1.34. A graph of the sum was drawn by a pen driven by $y(t)$. A fascinating collection of such graphs and a photograph of this old "supercomputer" can be found in A. Michelson and S. Stratton, A new harmonic analyzer, *Am. Jour. Sci-Fourth Ser.* **V**(1898), 1–13.

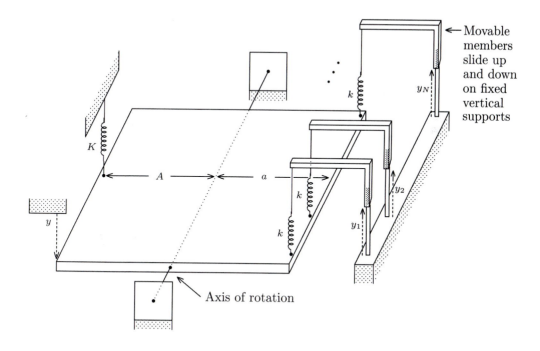

Figure 1.35. A mechanical device that uses springs to add small displacements.

▶ **EXERCISE 1.46:**   Let the function y have the representation

$$y(t) := \sum_{k=1}^{N} c_k e^{2\pi i s_k t}, \qquad -\infty < t < \infty$$

where $N = 1, 2, \ldots$ and the (not necessarily uniformly spaced) frequencies $s_1 < s_2 < \cdots < s_N$ are known but where $c_1, c_2, \ldots, c_N$ are unknown complex parameters. In this exercise you will develop an analysis equation that can be used to find these coefficients.

Such problems occasionally arise in the natural sciences. For example, the height of sea water in a given harbor is well modeled by a sum of sinusoids that correspond to the earth's rotation (with the sun, moon giving rise to terms with frequencies 2/day, 1.9323/day), to the moon's revolution about the earth, to the earth's revolution about the sun, to the moon's motion out of the plane of the earth's equator, etc., *cf.* R.A.R. Tricker, *Bores, Breakers, Waves, and Wakes*, American Elsevier, New York, 1965, pp. 1–22.

(a) Show that for each $k = 1, 2, \ldots, N$

$$c_k = \lim_{t_1 \to -\infty} \frac{1}{t_2 - t_1} \int_{t=t_1}^{t_2} e^{-2\pi i s_k t} y(t)dt = \lim_{t_2 \to +\infty} \frac{1}{t_2 - t_1} \int_{t=t_1}^{t_2} e^{-2\pi i s_k t} y(t)dt \, .$$

   In particular, if we have observed $y(t)$ at all times $t \leq t_0$, we can use the above analysis equation to find $c_1, c_2, \ldots, c_N$ and then predict $y(t)$ at all times $t > t_0$.

(b) Show that the above trigonometric sum vanishes for $-\infty < x < \infty$ if and only if $c_1 = c_2 = \cdots = c_N = 0$.

(c) Assume that $c_1, c_2, \ldots, c_N$ are all nonzero. Show that y is p-periodic for some $p > 0$, if and only if the products $ps_1, ps_2, \ldots, ps_N$ are all integers. (This will be the case when $s_1, s_2, \ldots, s_N$ are *commensurate*, i.e., s_k/s_ℓ is a rational number for each choice of $k, \ell = 1, 2, \ldots, N$ with $s_k \neq 0$.)

Note. In cases where the frequencies $s_1 < s_2 < \cdots < s_N$ are not commensurate, the trigonometric sum y is not periodic. Such a function is *almost periodic*, however, in the sense that for every choice of $\epsilon > 0$ there are infinitely many ϵ-approximate periods p_n, with

$$|y(x + p_n) - y(x)| < \epsilon, \qquad -\infty < x < \infty, \qquad n = 0, \pm 1, \pm 2, \ldots \, .$$

These p_n's are more or less uniformly distributed on the real line in the sense that every interval of length B contains at least one of them when $B > 0$ is sufficiently large, *cf.* H. Bohr, *Almost Periodic Functions*, Julius Springer, Berlin, 1933; English translation by H. Cohn, Chelsea, New York, 1947, pp. 32, 80.

Chapter 2

Convolution of Functions on $\mathbb{R}$, $\mathbb{T}_p$, $\mathbb{Z}$, and $\mathbb{P}_N$

2.1 Formal Definitions of $f * g$, $f \star g$

In elementary algebra you learned to combine functions f, g by using the binary operations of pointwise addition, subtraction, multiplication, and division, *i.e.*,

$$f + g, \quad f - g, \quad f \cdot g, \quad f/g.$$

For example, when f, g are functions on $\mathbb{R}$ or $\mathbb{Z}$ we define

$$(f \cdot g)(x) := f(x) \cdot g(x), \qquad x \in \mathbb{R}$$

or

$$(f \cdot g)[n] := f[n] \cdot g[n], \qquad n \in \mathbb{Z}.$$

We will use the symbols $*, \star$ for two closely related binary operations, convolution and correlation, that will appear from time to time in the remainder of the book. The purpose of this short chapter is to introduce you to these two new operations that result from the accumulation of certain pointwise arithmetic products.

We define the *convolution product* $f * g$ of two suitably regular functions f, g by writing

$$(f * g)(x) := \int_{u=-\infty}^{\infty} f(u)g(x - u)du \tag{1}$$

when f, g, (and $f * g$) are functions on $\mathbb{R}$,

$$(f * g)(x) := \int_{u=0}^{p} f(u)g(x - u)du \tag{2}$$

when f, g, (and $f * g$) are functions on $\mathbb{T}_p$,

$$(f * g)[n] := \sum_{m=-\infty}^{\infty} f[m]g[n-m] \tag{3}$$

$$\text{when } f, g, \text{ (and } f * g) \text{ are functions on } \mathbb{Z},$$

$$(f * g)[n] := \sum_{m=0}^{N-1} f[m]g[n-m] \tag{4}$$

$$\text{when } f, g, \text{ (and } f * g) \text{ are functions on } \mathbb{P}_N.$$

The integral, sum for computing $(f * g)(x)$, $(f * g)[n]$ gives the aggregate of all possible products $f(u)g(x - u)$, $f[m]g[n - m]$ with arguments that sum to x, n, respectively. We must impose conditions on f, g to ensure that the integral or sum for $f * g$ is well defined. For example, when f, g are piecewise continuous functions on $\mathbb{R}$ we can form $f * g$ if one of the functions is bounded and the other is absolutely integrable.

You will observe that (1)–(4) give four distinct ways to combine functions f, g, and it would not be inappropriate for us to introduce four distinct symbols *e.g.*, $\circledast_{\mathbb{R}}$, $\circledast_{\mathbb{T}_p}$, $\circledast_{\mathbb{Z}}$, $\circledast_{\mathbb{P}_N}$, for the corresponding binary operations. In practice, this proves to be unnecessarily cumbersome, and we will use the same symbol $*$ in (1)–(4). You must determine the context (*i.e.*, ask the question, "Are f and g functions on $\mathbb{R}$, $\mathbb{T}_p$, $\mathbb{Z}$, or $\mathbb{P}_N$?") when you assign meaning to $f * g$.

We define the *correlation product* $f \star g$ of two suitably regular functions f, g by writing

$$(f \star g)(x) := \int_{u=-\infty}^{\infty} \overline{f(u)}g(u+x)du \tag{5}$$

$$\text{when } f, g, \text{ (and } f \star g) \text{ are functions on } \mathbb{R},$$

$$(f \star g)(x) := \int_{u=0}^{p} \overline{f(u)}g(u+x)du \tag{6}$$

$$\text{when } f, g, \text{ (and } f \star g) \text{ are functions on } \mathbb{T}_p,$$

$$(f \star g)[n] := \sum_{m=-\infty}^{\infty} \overline{f[m]}g[m+n] \tag{7}$$

$$\text{when } f, g, \text{ (and } f \star g) \text{ are functions on } \mathbb{Z},$$

$$(f \star g)[n] := \sum_{m=0}^{N-1} \overline{f[m]}g[m+n] \tag{8}$$

$$\text{when } f, g, \text{ (and } f \star g) \text{ are functions on } \mathbb{P}_N.$$

The overbar denotes the complex conjugate. We again use the context to determine which meaning (5)–(8) is intended for $f \star g$.

Correlation and conjugation are closely related

When f, g are suitably regular functions on $\mathbb{R}$ we can use the change of variable $u := -v$ in (5) to write

$$(f \star g)(x) = \int_{v=-\infty}^{\infty} \overline{f(-v)} g(x - v) dv$$

in the form of a convolution product. Similar arguments can be used with (6)–(8), so we can always express $\star$ in terms of $*$ by writing

$$f \star g = f^{\dagger} * g. \tag{9}$$

Here $f^{\dagger}$ is the *hermitian conjugate* of f, *i.e.*,

$$f^{\dagger}(x) := \overline{f(-x)} \quad \text{when } f \text{ is a function on } \mathbb{R} \text{ or } \mathbb{T}_p,$$
$$f^{\dagger}[n] := \overline{f[-n]} \quad \text{when } f \text{ is a function on } \mathbb{Z}, \ \mathbb{P}_N,$$

cf. Exs. 1.2, 1.11, 1.15. Since

$$f^{\dagger\dagger} = f$$

we can use (9) to write

$$f * g = (f^{\dagger})^{\dagger} * g = f^{\dagger} \star g \tag{10}$$

and thereby express $*$ in terms of $\star$. From now on we will focus on the convolution product. You can always use (9) to convert a statement about $f * g$ to an equivalent statement about $f \star g$.

2.2 Computation of $f * g$

Direct evaluation

When f and g have a particularly simple structure, we can use the defining integral or sum to obtain $f * g$.

Example: Let $h > 0$. Find a simple expression for the convolution product of the piecewise continuous function f on $\mathbb{R}$ and

$$a_h(x) := \begin{cases} 1/2h & \text{if } -h < x < h \\ 0 & \text{otherwise.} \end{cases}$$

Solution: We use (1) to write

$$(f * a_h)(x) = \int_{-\infty}^{\infty} f(u)a_h(x-u)du$$

$$= \frac{1}{2h} \int_{-h<x-u<h} f(u)du \tag{11}$$

$$= \frac{1}{2h} \int_{x-h}^{x+h} f(u)du,$$

i.e., $(f * a_h)(x)$ is the average value of f on the interval $[x-h, x+h]$.   ∎

Example: Find a simple expression for the convolution product of the piece-wise continuous function f on $\mathbb{T}_p$ and

$$e_k(x) := e^{2\pi i k x/p}, \qquad k = 0, \pm 1, \pm 2, \dots .$$

Solution: We use (2) with the analysis equation for f to write

$$(f * e_k)(x) = \int_{u=0}^{p} f(u)e^{2\pi i k(x-u)/p}du$$

$$= e^{2\pi i k x/p} \cdot p \cdot \frac{1}{p} \int_{0}^{p} f(u)e^{-2\pi i k u/p}du \tag{12}$$

$$= p\,F[k]e_k(x), \qquad k = 0, \pm 1, \pm 2, \dots .$$

In particular,

$$e_\ell * e_k = \begin{cases} p\,e_k & \text{if } k = \ell \\ 0 & \text{if } k \neq \ell. \end{cases}$$ ∎

The sum of scaled translates

When f, g are functions on $\mathbb{P}_N$, we can use (4) to write

$$(f * g)[n] = f[0] \cdot g[n] + f[1] \cdot g[n-1] + f[2] \cdot g[n-2] + \cdots + f[N-1] \cdot g[n-(N-1)].$$

We regard $f[0], f[1], f[2], \dots, f[N-1]$ as scalars that are applied to the translated "vectors" $g[n], g[n-1], g[n-2], \dots, g[n-(N-1)]$, respectively. The same idea can be used when f, g are functions on $\mathbb{Z}$, but the sum may have infinitely many terms.

Example: Find the convolution product of

$$f = (f[0], f[1], f[2], f[3]) := (3, 1, 4, 1),$$
$$g = (g[0], g[1], g[2], g[3]) := (5, 9, 2, 6).$$

Solution: We compute $f * g$ on $\mathbb{P}_4$ by summing translates as follows.

$$(f * g)[0] = 3 \cdot 5 + 1 \cdot 6 + 4 \cdot 2 + 1 \cdot 9 = 38$$
$$(f * g)[1] = 3 \cdot 9 + 1 \cdot 5 + 4 \cdot 6 + 1 \cdot 2 = 58$$
$$(f * g)[2] = 3 \cdot 2 + 1 \cdot 9 + 4 \cdot 5 + 1 \cdot 6 = 41$$
$$(f * g)[3] = 3 \cdot 6 + 1 \cdot 2 + 4 \cdot 9 + 1 \cdot 5 = 61$$

We check our work using the identity $\left(\sum f[n]\right)\left(\sum g[n]\right) = \sum (f * g)[n]$ of Ex. 2.26, i.e.,

$$(3 + 1 + 4 + 1) \cdot (5 + 9 + 2 + 6) = 9 \cdot 22 = 198 = 38 + 58 + 41 + 61. \qquad \blacksquare$$

Example: Let f be a function on $\mathbb{Z}$ with

$$f[n] := \begin{cases} 1 & \text{if } n = 0 \text{ or } n = 1 \\ 0 & \text{otherwise.} \end{cases}$$

Find a formula for the components of $f_1 := f$, $f_2 := f * f$, $f_3 := f * (f * f)$, $\dots$.

Solution: We use a sum of translates to write

$$(f * g)[n] = g[n] + g[n - 1], \qquad n = 0, \pm 1, \pm 2, \dots ,$$

and thereby produce f_1, f_2, f_3 as shown in Fig. 2.1. In conjunction with the Pascal triangle relation for the binomial coefficients this gives the formula

$$f_m[n] = \binom{m}{n}. \qquad \blacksquare$$

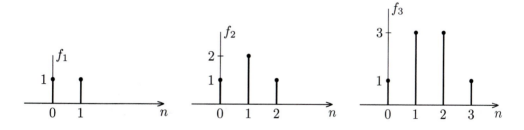

Figure 2.1. Convolution products $f_1 = f$, $f_2 = f * f$, $f_3 = f * f * f$.

Example: Let $f := (1,1,0,0,0,0)$ be a function on $\mathbb{P}_6$. Find a formula for the components of

$$f_1 := f, \quad f_2 := f * f, \quad f_3 := f * (f * f), \ \dots \ .$$

Solution: We use a sum of translates to write

$$(1,1,0,0,0,0) * (a_0, a_1, a_2, a_3, a_4, a_5)$$
$$= (a_0, a_1, a_2, a_3, a_4, a_5) + (a_5, a_0, a_1, a_2, a_3, a_4)$$
$$= (a_5 + a_0, a_0 + a_1, a_1 + a_2, a_2 + a_3, a_3 + a_4, a_4 + a_5).$$

We then use this cyclic version of the Pascal triangle relation to write

$$f_1 = (1,1,0,0,0,0), \quad f_2 = (1,2,1,0,0,0), \quad f_3 = (1,3,3,1,0,0)$$
$$f_4 = (1,4,6,4,1,0), \quad f_5 = (1,5,10,10,5,1), \quad f_6 = (2,6,15,20,15,6), \ \dots$$

and thereby see that

$$f_m[n] = \binom{m}{n} + \binom{m}{n+6} + \binom{m}{n+12} + \cdots . \qquad \blacksquare$$

The sliding strip method

We often find it necessary to compute the convolution product of functions f, g that are defined piecewise on $\mathbb{R}$, *e.g.*,

$$f(x) := \begin{cases} 1 & \text{if } 0 < x < 1 \\ 0 & \text{otherwise,} \end{cases} \qquad g(x) := \begin{cases} x & \text{if } 0 < x < 2 \\ 0 & \text{otherwise.} \end{cases} \qquad (13)$$

In such cases it is sometimes possible to split the integral of (1) into a sum of subintegrals that we can evaluate by using the fundamental theorem of calculus. In practice, however, we are usually overwhelmed with the task of determining how the various limits of integration depend on the argument x. For example, when f, g are the simple functions of (13) we find

$$(f * g)(x) = \int_{\substack{0 < u < 1 \\ \text{and} \\ 0 < x - u < 2}} f(u)g(x - u)du = \int_{\min\{1, \max\{0, x-2\}\}}^{\max\{0, \min\{1, x\}\}} (x - u)du. \qquad (14)$$

There is a much better way to organize such a calculation, and we will use the functions (13) to illustrate the procedure. For representative choices of x we will sketch the graphs of $f(u)$ and $g(x - u)$ as functions of u, form the integrand $f(u)g(x - u)$, and evaluate the integral (1).

We can obtain the graph of $g(x - u)$ as a function of u by reflecting $g(u)$ to get $g(-u)$ and then translating $g(-u)$ by x to get $g(-(u - x)) = g(x - u)$. As an alternative, we simply reflect the graph of $g(u)$ about the line $u = x/2$ to obtain $g(u - 2(u - x/2)) = g(x - u)$, cf. Fig. 2.2. If you observe that the point $u = x$ on the graph of $g(x - u)$ corresponds to the point $u = 0$ on the graph of $g(u)$, you will find it easy to visualize how $g(x - u)$ slides along the u-axis as the parameter x increases from $-\infty$ to $+\infty$.

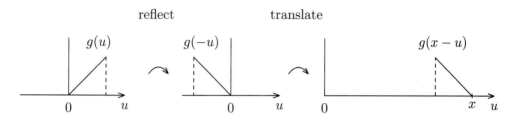

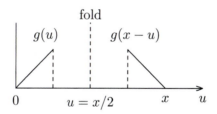

Figure 2.2. Construction of $g(x - u)$ from $g(u)$ by reflection followed by translation and by folding.

For each real argument x we evaluate the integral of $f(u) \cdot g(x - u)$ (by using the formula for the area of a right triangle!) as shown in Fig. 2.3. In this way we obtain the convolution product

$$(f * g)(x) = \frac{1}{2} \begin{cases} x^2 & \text{if } 0 \le x \le 1 \\ 2x - 1 & \text{if } 1 \le x \le 2 \\ -x^2 + 2x + 3 & \text{if } 2 \le x \le 3 \\ 0 & \text{otherwise} \end{cases}$$

shown in Fig. 2.4. Of course, you could also obtain this result by evaluating the integral in (14), but by now you should be convinced that the analysis of Fig. 2.3 gives the desired result with much less effort.

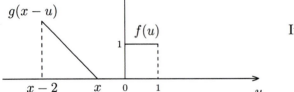

If $x \leq 0$

$$(f * g)(x) = 0.$$

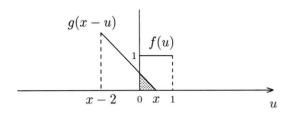

If $0 \leq x \leq 1$

$$(f * g)(x) = \frac{x^2}{2}.$$

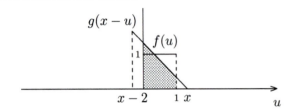

If $1 \leq x \leq 2$

$$(f * g)(x) = \frac{x^2}{2} - \frac{(x-1)^2}{2}.$$

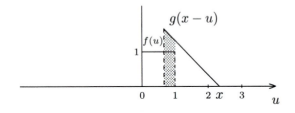

If $2 \leq x \leq 3$

$$(f * g)(x) = \frac{2^2}{2} - \frac{(x-1)^2}{2}.$$

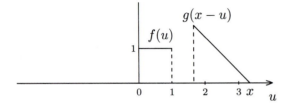

If $3 \leq x$

$$(f * g)(x) = 0.$$

Figure 2.3. Sliding strip computation of the convolution product of the functions (13).

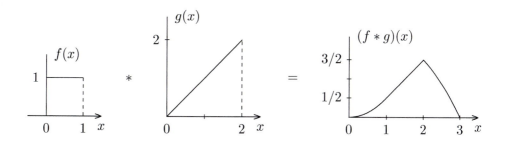

Figure 2.4. The convolution product of the functions (13).

Example: Find the convolution product $f * g$ when

$$f(x) := g(x) := \begin{cases} 1 & \text{if } -\dfrac{1}{2} < x < \dfrac{1}{2} \\ 0 & \text{otherwise.} \end{cases}$$

Solution: Visualize $f(u)$ as a fixed box with edges at $u = \pm 1/2$ and visualize $g(x - u)$ as a box with edges at $u = x \pm 1/2$. As x increases from $-\infty$ to $+\infty$, $g(x - u)$ slides along the x-axis from $u = -\infty$ to $u = +\infty$. The boxes first make contact when $x = -1$ [i.e., when the right edge of $g(x - u)$ at $u = x + 1/2$ coincides with the left edge of $f(u)$ at $u = -1/2$]. The area under the product $f(u)g(x - u)$ linearly increases from 0 to 1 as x increases from -1 to 0 (where the boxes coincide). The overlap area linearly decreases from 1 to 0 as x increases from 0 to 1 [where the left edge of $g(x - u)$ at $u = x - 1/2$ coincides with the right edge of $f(u)$ at $u = 1/2$]. There is no overlap when $x > 1$. In this way we find

$$(f * g)(x) = \begin{cases} 0 & \text{if } x \le -1 \\ 1 + x & \text{if } -1 \le x \le 0 \\ 1 - x & \text{if } 0 \le x \le 1 \\ 0 & \text{if } 1 \le x \end{cases} = \begin{cases} 1 - |x| & \text{if } |x| \le 1 \\ 0 & \text{if } |x| \ge 1, \end{cases}$$

as shown in Fig. 2.5.   ∎

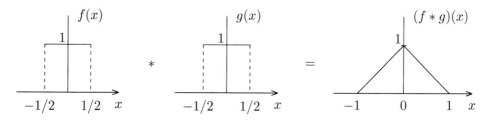

Figure 2.5. The convolution product of two boxes.

Example: Let $\alpha > 0, \beta > 0$. Find the convolution product of

$$f(x) := \begin{cases} e^{-\alpha x} & \text{if } x \geq 0 \\ 0 & \text{if } x < 0, \end{cases} \qquad g(x) := \begin{cases} e^{-\beta x} & \text{if } x \geq 0 \\ 0 & \text{if } x < 0. \end{cases}$$

Solution: After noting that $f(u)$ is nonzero on $[0, +\infty)$ and that $g(x - u)$ is nonzero on $(-\infty, x]$ we write

$$(f * g)(x) = \begin{cases} 0 & \text{if } x \leq 0 \\ \displaystyle\int_{u=0}^{x} e^{-\alpha u} e^{-\beta(x-u)} du & \text{if } x \geq 0 \end{cases}$$

$$= \begin{cases} 0 & \text{if } x \leq 0 \\ x e^{-\alpha x} & \text{if } x \geq 0 \text{ and } \beta = \alpha \\ \dfrac{e^{-\beta x} - e^{-\alpha x}}{\alpha - \beta} & \text{if } x \geq 0 \text{ and } \beta \neq \alpha. \end{cases} \tag{15}$$

Once we visualize $f(u)$ and $g(x-u)$ it is easy to determine the limits of integration and perform the calculation that gives $f * g$, cf. Fig. 2.6. ∎

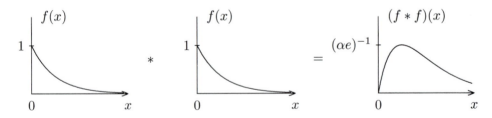

Figure 2.6. The convolution product (15) of truncated exponentials with $\beta = \alpha$.

Example: Find the convolution product $f * g$ for the functions of Fig. 2.7.

Solution: The step functions f, g are constant on every interval $(n, n + 1)$, $n = 0, \pm 1, \pm 2, \ldots$, so we can easily obtain

$$(f * g)(1) = 3, \quad (f * g)(2) = 7, \quad (f * g)(3) = 3, \quad (f * g)(4) = 2,$$
$$(f * g)(n) = 0 \quad \text{for} \quad n = 0, -1, -2, \ldots \text{ and } n = 5, 6, 7, \ldots$$

by adding areas of rectangles. Since $(f * g)(x)$ is linear on every interval $n < x < n+1$, $n = 0, \pm 1, \pm 2, \ldots$, we can produce the graph of $f * g$ as shown in Fig. 2.7 by connecting the dots with line segments. You can find each of the convolution products from Ex. 2.1 and Ex. 2.2 by using this procedure. ∎

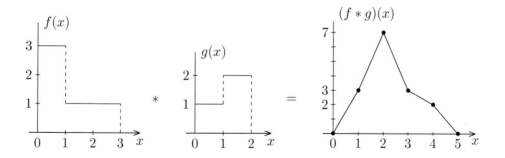

Figure 2.7. The piecewise linear convolution product $f * g$ of step functions f, g.

Variations of the sliding strip method can be used to find the convolution product of functions f, g on $\mathbb{T}_p$. Of course, you can move from $g(u)$ to $g(-u)$ to $g(x - u)$ by using reflection on $\mathbb{T}_p$ followed by translation on (*i.e.*, rotation around) $\mathbb{T}_p$. The integral of $f(u) \cdot g(x - u)$ over $\mathbb{T}_p$ then gives $(f * g)(x)$. If you wish, you can regard f as a p-periodic function on $\mathbb{R}$, define

$$g_0(x) := \begin{cases} g(x) & \text{if } 0 \le x < p \\ 0 & \text{otherwise} \end{cases}$$

on $\mathbb{R}$, compute $f * g_0$ on $\mathbb{R}$ (as described above), and set

$$(f * g)(x) = (f * g_0)(x) \quad \text{for} \quad 0 \le x < p.$$

You can also define

$$f_0(x) := \begin{cases} f(x) & \text{for } 0 \le x < p \\ 0 & \text{otherwise,} \end{cases} \qquad g_0(x) := \begin{cases} g(x) & \text{for } 0 \le x < p \\ 0 & \text{otherwise} \end{cases}$$

as functions on $\mathbb{R}$, compute $f_0 * g_0$ on $\mathbb{R}$ (as described above), and take

$$(f * g)(x) = (f_0 * g_0)(x) + (f_0 * g_0)(x + p) \qquad \text{for} \quad 0 \le x < p.$$

This third alternative clearly displays the wraparound effect that can occur when f, g are not suitably localized on $\mathbb{T}_p$.

Example: Let $0 < \alpha \le 1$. Find $f * f$ when

$$f(x) := \begin{cases} 1 & \text{if } 0 \le x < \alpha \\ 0 & \text{if } \alpha < x < 1 \end{cases}$$

is a function on $\mathbb{T}_1$.

Solution: If we define

$$f_0(x) := \begin{cases} 1 & \text{if } 0 \le x < \alpha \\ 0 & \text{if } x < 0 \text{ or } x \ge \alpha \end{cases}$$

as a function on $\mathbb{R}$, we find

$$(f_0 * f_0)(x) = \begin{cases} x & \text{if } 0 \le x \le \alpha \\ 2\alpha - x & \text{if } \alpha \le x \le 2\alpha \\ 0 & \text{otherwise.} \end{cases}$$

When $0 < \alpha \le 1/2$ there is no wraparound effect and we write

$$(f * f)(x) = (f_0 * f_0)(x) = \begin{cases} x & \text{if } 0 \le x \le \alpha \\ 2\alpha - x & \text{if } \alpha \le x \le 2\alpha \\ 0 & \text{if } 2\alpha \le x \le 1. \end{cases}$$

When $1/2 < \alpha < 1$ there is a wraparound effect, *cf.* Fig. 2.8, and we have

$$(f * f)(x) = (f_0 * f_0)(x) + (f_0 * f_0)(x+1) = \begin{cases} 2\alpha - 1 & \text{if } 0 \le x \le 2\alpha - 1 \\ x & \text{if } 2\alpha - 1 \le x \le \alpha \\ 2\alpha - x & \text{if } \alpha \le x \le 1. \end{cases} \quad \blacksquare$$

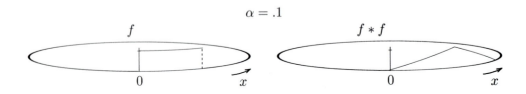

$$\alpha = .1$$

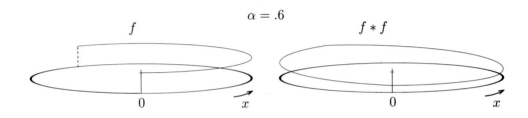

$$\alpha = .6$$

Figure 2.8. The functions f and $f * f$ on $\mathbb{T}_1$ when $\alpha = .1$ and when $\alpha = .6$.

The sliding strip method can be used to obtain the convolution product of functions f, g on $\mathbb{Z}$ or $\mathbb{P}_N$. For such discrete functions we compute $(f * g)[n]$ by summing $f[m]g[n - m]$ with respect to m [instead of integrating $f(u)g(x - u)$ with respect to u].

Generating functions

When a, b are suitably regular functions on $\mathbb{Z}$ we can form the corresponding *generating functions*

$$A(z) := \sum_{n=-\infty}^{\infty} a[n]z^n, \qquad B(z) := \sum_{n=-\infty}^{\infty} b[n]z^n.$$

The coefficients of the product

$$A(z) \cdot B(z) = \sum_{k=-\infty}^{\infty} \sum_{m=-\infty}^{\infty} a[m]b[k]z^{m+k} = \sum_{n=-\infty}^{\infty} \left\{ \sum_{m=-\infty}^{\infty} a[m]b[n - m] \right\} z^n$$

are the components of $a * b$. If you can find the power series for $A(z) \cdot B(z)$, you can compute the convolution product $a * b$. For example, the identity

$$(3 + z + z^2) \cdot (1 + 2z) = 3 + 7z + 3z^2 + 2z^3$$

corresponds to

$$(\ldots, 3, 1, 1, 0, \ldots) * (\ldots, 1, 2, 0, \ldots) = (\ldots, 3, 7, 3, 2, 0, \ldots).$$

You can use various techniques from algebra, calculus, ... to find the power series for $A(z) \cdot B(z)$ and thereby find the convolution product $a * b$.

Example: Let α, β be complex numbers with $0 < |\alpha| < 1$, $0 < |\beta| < 1$, and let

$$a[n] := \begin{cases} \alpha^n & \text{if } n = 0, 1, \ldots \\ 0 & \text{otherwise,} \end{cases} \qquad b[n] := \begin{cases} \beta^n & \text{if } n = 0, 1, \ldots \\ 0 & \text{otherwise.} \end{cases}$$

Find the convolution product $a * b$ on $\mathbb{Z}$.

Solution: When $|z|$ is sufficiently small, the formula for the sum of a geometric progression gives the generating functions

$$A(z) := \sum_{n=0}^{\infty} \alpha^n z^n = \frac{1}{1 - \alpha z}, \qquad B(z) = \sum_{n=0}^{\infty} \beta^n z^n = \frac{1}{1 - \beta z}.$$

When $\alpha \neq \beta$ we write

$$A(z) \cdot B(z) = \frac{1}{1 - \alpha z} \cdot \frac{1}{1 - \beta z} = \frac{1}{(\alpha - \beta)z} \cdot \left\{ \frac{1}{1 - \alpha z} - \frac{1}{1 - \beta z} \right\}$$

$$= \frac{1}{(\alpha - \beta)z} \sum_{m=0}^{\infty} (\alpha^m - \beta^m) z^m = \sum_{n=0}^{\infty} \frac{\alpha^{n+1} - \beta^{n+1}}{\alpha - \beta} z^n$$

and thereby show that

$$(a * b)[n] = \begin{cases} \dfrac{\alpha^{n+1} - \beta^{n+1}}{\alpha - \beta} & \text{if } n = 0, 1, \ldots \\ 0 & \text{otherwise.} \end{cases}$$

When $\alpha = \beta$ we use a similar argument to obtain

$$(a * b)[n] = \begin{cases} (n+1)\alpha^n & \text{if } n = 0, 1, \ldots \\ 0 & \text{otherwise.} \end{cases}$$

You may wish to compare these results to those of (15).   ∎

2.3 *Mathematical Properties of the Convolution Product*

Introduction

There are a number of basic properties of the convolution product that follow from the defining relations (1)–(4).

The Fourier transform of $f * g$

Let f, g be suitably regular functions on $\mathbb{R}$ and let $q := f * g$. We use the definition of the Fourier transform together with (1) to write

$$Q(s) := \int_{x=-\infty}^{\infty} (f * g)(x) e^{-2\pi i s x} dx$$

$$= \int_{x=-\infty}^{\infty} \int_{u=-\infty}^{\infty} f(u) g(x - u) e^{-2\pi i s x} du \, dx$$

$$\overset{?}{=} \int_{u=-\infty}^{\infty} \int_{x=-\infty}^{\infty} f(u) g(x - u) e^{-2\pi i s x} dx \, du$$

$$= \int_{u=-\infty}^{\infty} f(u)e^{-2\pi isu} \cdot \int_{x=-\infty}^{\infty} g(x-u)e^{-2\pi is(x-u)}dx\, du$$

$$= \int_{u=-\infty}^{\infty} f(u)e^{-2\pi isu} \cdot G(s)du$$

$$= F(s) \cdot G(s),$$

assuming that we can justify the exchange of limits in the $\overset{?}{=}$ step. In this way we see that the Fourier transform converts convolution into ordinary multiplication. With such informal arguments we find

$$q(x) := (f * g)(x) \quad \text{on } \mathbb{R} \quad \text{has the FT } Q(s) = \quad F(s) \cdot G(s) \quad \text{on } \mathbb{R}, \quad (16)$$

$$q(x) := (f * g)(x) \quad \text{on } \mathbb{T}_p \quad \text{has the FT } Q[k] = p \cdot F[k] \cdot G[k] \quad \text{on } \mathbb{Z}, \quad (17)$$

$$q[n] := (f * g)[n] \quad \text{on } \mathbb{Z} \quad \text{has the FT } Q(s) = p \cdot F(s) \cdot G(s) \quad \text{on } \mathbb{T}_p, \quad (18)$$

$$q[n] := (f * g)[n] \quad \text{on } \mathbb{P}_N \text{ has the FT } Q[k] = N \cdot F[k] \cdot G[k] \quad \text{on } \mathbb{P}_N. \quad (19)$$

These four equations, Fourier's synthesis-analysis equations, Parseval's identities, and Poisson's relations are the most basic identities of elementary Fourier analysis!

    An *indirect* scheme for finding convolution products is suggested by (16)–(19), *e.g.*,

$$f \overset{\text{FT}}{\leadsto} F, \quad g \overset{\text{FT}}{\leadsto} G, \quad F \cdot G \overset{\text{IFT}}{\leadsto} f * g \tag{20}$$
$$\text{when } f, g, (\text{and } f * g) \text{ are functions on } \mathbb{R},$$

$$f \overset{\text{DFT}}{\leadsto} F, \quad g \overset{\text{DFT}}{\leadsto} G, \quad N \cdot F \cdot G \overset{\text{IDFT}}{\leadsto} f * g \tag{21}$$
$$\text{when } f, g, (\text{and } f * g) \text{ are functions on } \mathbb{P}_N.$$

You will learn to use this very powerful method for computing $f * g$ as you study Chapters 3, 4.

Algebraic structure

    The convolution product that we have defined by (1), (2), (3), or (4) has many of the familiar properties of pointwise multiplication. We easily verify that $*$ is *homogeneous*,

$$(\alpha f) * g = \alpha(f * g) = f * (\alpha g),$$

and that $*$ *distributes* over addition,

$$f * (g_1 + g_2) = (f * g_1) + (f * g_2)$$
$$(f_1 + f_2) * g = (f_1 * g) + (f_2 * g),$$

by using the corresponding properties of the integrals or sums in (1)–(4). These give the *linearity* relation

$$\left(\sum_{m=1}^{M} \alpha_m f_m \right) * \left(\sum_{n=1}^{N} \beta_n g_n \right) = \sum_{m=1}^{M} \sum_{n=1}^{N} \alpha_m \beta_n (f_m * g_n) \tag{22}$$

when $f_1, \ldots, f_M$, $g_1, \ldots, g_N$ are suitably regular functions and $\alpha_1, \ldots, \alpha_M$, $\beta_1, \ldots, \beta_N$ are scalars. A simple change of variables ($u' := x - u$ or $m' := n - m$) in (1), (2), (3), or (4) shows that $*$ is *commutative*,

$$f * g = g * f, \tag{23}$$

whenever these products are defined, *cf.* Ex. 2.19.

The *associativity* property,

$$f_1 * (f_2 * f_3) = (f_1 * f_2) * f_3, \tag{24}$$

that allows us to write

$$f_1 * f_2 * \cdots * f_N$$

(without inserting parentheses to specify the order in which these products are formed) is true — most of the time, *cf.* Ex. 2.20. Since the Fourier transform of $f * g$ is a scalar multiple of $F \cdot G$ you might expect the associativity of $*$ to follow from that of $\cdot$, and this is the case when f_1, f_2, f_3, $f_1 * f_2$, $f_2 * f_3$, $(f_1 * f_2) * f_3$, $f_1 * (f_2 * f_3)$ are all suitably regular or equivalently when the Fourier transforms F_1, F_2, F_3, $F_1 \cdot F_2$, $F_2 \cdot F_3$, $(F_1 \cdot F_2) \cdot F_3$, $F_1 \cdot (F_2 \cdot F_3)$ are all suitably regular. It is not so easy to convert such vague statements into useful theorems, however. We will study such products and convolution products more carefully in Chapters 7 and 12. For now, you might like to see what can go wrong and explore a possible fix by working through Exs. 2.35 and 2.36.

You will recall that the function 1 serves as a multiplicative identity for $\cdot$, *i.e.*,

$$1 \cdot f = f \cdot 1 = f.$$

The functions

$$\delta[n] := \begin{cases} 1 & \text{if } n = 0 \\ 0 & \text{if } n = \pm 1, \pm 2, \ldots, \end{cases} \qquad \delta[n] := \begin{cases} 1 & \text{if } n = 0 \\ 0 & \text{if } n = 1, 2, \ldots, N - 1 \end{cases}$$

serve as *identities* for the convolution product of functions on $\mathbb{Z}$, $\mathbb{P}_N$, *i.e.*,

$$\delta * f = f * \delta = f. \tag{25}$$

No ordinary function on $\mathbb{R}$ or $\mathbb{T}_p$ serves as an identity. (We will create generalized functions δ and III for this purpose in Chapter 7.)

A number of interesting new mathematical structures can be obtained by re-placing $\cdot$ by $*$ within some familiar context. You can find the convolution square root s of a function a (on $\mathbb{R}$, $\mathbb{T}_p$, $\mathbb{Z}$, or $\mathbb{P}_N$) by solving

$$s * s = a,$$

and use it to find roots of the quadratic equation

$$a * x * x + b * x + c = 0.$$

You can formulate conditions for the convergence of the infinite series

$$a_0 + a_1 * x + a_2 * x * x + \cdots .$$

You can devise procedures for solving a linear equation

$$a * x = b$$

or a system of linear equations

$$a_{11} * x_1 + a_{12} * x_2 = b_1$$
$$a_{21} * x_1 + a_{22} * x_2 = b_2.$$

Related exercises can be found in Chapters 3, 4. (You will work more efficiently after you learn a few basic skills for taking Fourier transforms.)

Translation invariance

Let f, g be suitably regular functions on $\mathbb{R}$, and let $-\infty < a < \infty$. The function $g(x + a)$ results when we a-translate g, and with a slight abuse of notation we write

$$f(x) * g(x + a) = \int_{u=-\infty}^{\infty} f(u)g\{(x - u) + a\}du$$
$$= \int_{u=-\infty}^{\infty} f(u)g\{(x + a) - u\}du$$
$$= (f * g)(x + a).$$

In this way we show that the convolution product is *translation invariant*: If we convolve f with the a-translate of g, we obtain the a-translate of $f * g$, cf. Ex. 2.21. Analogous arguments show that the convolution product of suitably regular functions on $\mathbb{T}_p$, $\mathbb{Z}$, or $\mathbb{P}_N$ are translation invariant. Since the convolution product $f * g$ is linear in g, the corresponding convolution operator

$$\mathbf{A}g := f * g \tag{26}$$

is always LTI.

A function g on $\mathbb{P}_N$ can be written as a sum of translates of the identity δ for the convolution product, *i.e.*,

$$
\begin{aligned}
g[n] = g[0] \cdot \delta[n] &+ g[1] \cdot \delta[n-1] + g[2] \cdot \delta[n-2] \\
&+ \cdots + g[N-1] \cdot \delta[n-(N-1)].
\end{aligned}
$$

If we pass g through an LTI system $\mathbf{A}$, we obtain the output

$$
\begin{aligned}
(\mathbf{A}g)[n] = g[0] \cdot (\mathbf{A}\delta)[n] &+ g[1] \cdot (\mathbf{A}\delta)[n-1] + g[2] \cdot (\mathbf{A}\delta)[n-2] \\
&+ \cdots + g[N-1] \cdot (\mathbf{A}\delta)[n-(N-1)].
\end{aligned}
$$

In this way we see that $\mathbf{A}$ has the representation (26) with

$$
f := \mathbf{A}\delta \tag{27}
$$

being known as the *impulse response* of the system $\mathbf{A}$. The same argument can be used when f is a function on $\mathbb{Z}$ provided that we can apply $\mathbf{A}$ to an infinite series on a term-by-term basis. After introducing generalized functions in Chapter 7 we will see that the representation (26) or (27) can be used for many LTI operators that are applied to functions on $\mathbb{R}$ or $\mathbb{T}_p$, *cf.* Ex. 2.28.

Differentiation of $f * g$

When f, g are suitably regular functions on $\mathbb{R}$ we can write

$$
\begin{aligned}
(f * g)'(x) &= \frac{d}{dx} \int_{u=-\infty}^{\infty} f(u)g(x-u)\,du \\
&\overset{?}{=} \int_{u=-\infty}^{\infty} f(u)g'(x-u)\,du \\
&= (f * g')(x).
\end{aligned}
$$

Of course, we must impose hypotheses on f, g that allow us to exchange the order of differentiation and integration, *cf.* Ex. 2.34. Since $f * g = g * f$, it follows that

$$
\begin{aligned}
(f * g)' &= f' * g = f * g' \\
(f * g)'' &= f'' * g = f' * g' = f * g''
\end{aligned}
$$

$$
\vdots
$$

with

$$
(f * g)^{(n)} = f^{(m)} * g^{(n-m)}, \quad m = 0, 1, \ldots, n \tag{28}
$$

when f has m derivatives and g has $n - m$ derivatives for some $m = 0, 1, \ldots, n$. (The corresponding *Leibnitz rule*

$$(f \cdot g)^{(n)} = \sum_{m=0}^{n} \binom{n}{m} f^{(m)} \cdot g^{(n-m)} \tag{29}$$

from calculus requires both f and g to have n derivatives.)

We can use (28) and the linearity (22) to write

$$p(\mathbf{D})\{f * g\} = f * \{p(\mathbf{D})g\} \tag{30}$$

when

$$p(\mathbf{D}) := c_0 + c_1 \mathbf{D} + c_2 \mathbf{D}^2 + \cdots + c_n \mathbf{D}^n, \quad \mathbf{D} := \frac{d}{dx}$$

and $c_0, c_1, \ldots, c_n$ are constants. This identity will prove to be useful when we study ordinary and partial differential equations.

As you compute convolution products of functions on $\mathbb{R}$, you will observe that $f * g$ is always smoother than either f or g, cf. Figs. 2.4–2.7 and 2.9 (which corresponds to Ex. 2.7). Of course, an integration process such as (1) will always produce a function that is smoother than the integrand. The identity (28) shows that $f * g$ also inherits all of the smoothness of f plus all of the smoothness of g. Exercises 2.29 and 2.37 will help you sort out the details.

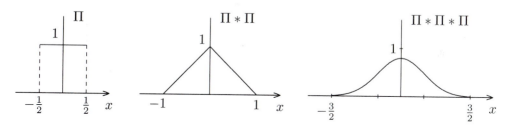

Figure 2.9. The functions Π, $\Pi * \Pi$, and $\Pi * \Pi * \Pi$.

2.4 Examples of Convolution and Correlation

Convolution as smearing

A discrete representation of Jefferson's Monticello is obtained by specifying real numbers $s[n]$, $n = 0, 1, \ldots, 319$ as shown in Fig. 2.10. A corresponding blurred image

$$b[n] := \frac{1}{17}\{s[n - 8] + s[n - 7] + \cdots + s[n + 8]\}$$

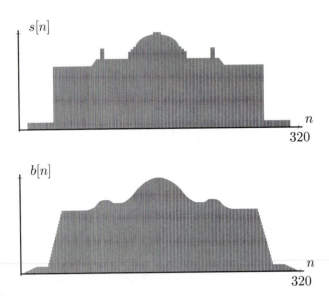

Figure 2.10. A discrete representation s, of Jefferson's Monti-cello together with a blurred image $b = a * s$.

(with indices taken mod 320) is also shown in Fig. 2.9. The blurred image b does not exhibit any of the fine detail (sharp corners, chimneys, dome rings, ...) that is present in the original scene function s.

You will observe that we can write

$$a * s = b \tag{31}$$

where

$$a[n] := \begin{cases} 1/17 & \text{if } n = 0, \pm 1, \pm 2, \ldots, \pm 8 \\ 0 & \text{if } n = \pm 9, \pm 10, \ldots, \pm 159, 160 \end{cases} \tag{32}$$

is a discrete box, *cf.* (11). Knowing a (*i.e.*, knowing the characteristics of the defective "camera" that produced the blurred image b) we can attempt to reconstruct s from b by solving the convolution equation (31). In view of (19), the DFTs of a, s, b satisfy

$$A[k] \cdot S[k] = 320 B[k], \qquad k = 0, 1, \ldots, 319$$

so we can synthesize s from

$$S[k] = 320 \frac{B[k]}{A[k]}.$$

[After you learn a bit more about DFTs you will find it easy to verify that the $A[k]$ for (32) is never 0.] Such deconvolution techniques can be used to sharpen blurred images, enhance old audio recordings, etc.

Echo location

Geologists locate a layer of oil bearing sedimentary rock by sending a real signal $x(t)$ into the earth, listening for an attenuated echo $Ax(t-T)$ from that layer, and using the time shift T to determine the depth of the oil. The scheme is complicated by the fact that the attenuation parameter A is normally so small that the reflected signal is buried in the noise from mini earthquakes, thermal creaking, freeway and runway vibrations, etc. Figure 2.11 shows a chirp x and a noise-contaminated echo

$$e(t) := A\,x(t-T) + n(t).$$

It does not seem possible to determine the time shift T from such data.

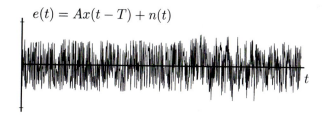

Figure 2.11. An FM chirp, $x(t)$, and a noise-contaminated echo $e(t) = Ax(t-T) + n(t)$.

As you may have guessed, however, the transmitted signal x has been designed with some care. (Bats, dolphins, and whales use similar frequency-modulated chirps for echo location!) The autocorrelation function

$$(x \star x)(t-T) = \int_{u=-\infty}^{\infty} x(u)x(u+t-T)du$$

has a sharp peak at $t = T$, *cf.* Fig. 2.12. If A is not too small, the correlation product

$$(x \star e)(t) = \int_{u=-\infty}^{\infty} x(u)e(u+t)du$$

$$= A\,(x \star x)(t-T) + \int_{u=-\infty}^{\infty} x(u)n(u+t)du$$

will have a corresponding peak that reveals the precise location of T, *cf.* Fig. 2.12. We do the impossible by computing $x \star e$!

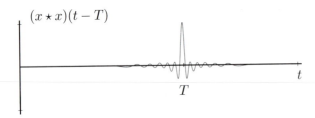

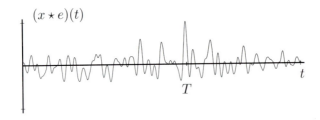

Figure 2.12. The autocorrelation $(x \star x)(t-T)$ and the correlation product $(x \star e)(t)$ for the functions x, e of Fig. 2.11.

Convolution and probability

The probability of throwing the integer n with a fair die is given by the discrete density

$$p[n] := \begin{cases} 1/6 & \text{if } n = 1, 2, \ldots, 6 \\ 0 & \text{otherwise.} \end{cases}$$

We can throw the integer 6 with a pair of dice if the first and second come up 1 and 5, 2 and 4, 3 and 3, 4 and 2, or 5 and 1. Assuming that the dice are thrown independently, the probability for throwing a 6 is obtained by writing

$$p[1] \cdot p[5] + p[2] \cdot p[4] + p[3] \cdot p[3] + p[4] \cdot p[2] + p[5] \cdot p[1].$$

In this way we see that

$$p_2[n] := \sum_{m=-\infty}^{\infty} p[m]p[n-m] = (p * p)[n], \quad n = 0, \pm 1, \pm 2, \dots$$

gives the probability of throwing the integer n with a pair of dice.

If we have 3 dice we will obtain the integer n if we throw 1 with the first die and $n - 1$ with the remaining two, 2 with the first and $n - 2$ with the remaining two, etc. This analysis gives the formula

$$p_3[n] = \sum_{m=-\infty}^{\infty} p[m]p_2[n-m] = (p * p * p)[n]$$

for the probability of throwing the integer n with three dice. Analogously,

$$p_4[n] = (p * p * p * p)[n], \quad p_5[n] = (p * p * p * p * p)[n], \dots$$

give the probabilities for throwing the integer n with $4, 5, \dots$ dice.

You can generate the probability densities $p_2, p_3, \dots$ by using the sliding strip method to convolve p with $p, p_2, \dots$ as follows.

n	1	2	3	4	5	6
$6p[n]$	1	1	1	1	1	1

n	2	3	4	5	6	7	8	9	10	11	12
$36(p*p)[n]$	1	2	3	4	5	6	5	4	3	2	1

n	3	4	5	6	7	8	9	10	11	12	13	14	15	16	17	18
$216(p*p*p)[n]$	1	3	6	10	15	21	25	27	27	25	21	15	10	6	3	1

You can also compute p, $p * p$, $p * p * p$, $\dots$ in cases where the face numbers 1,2,3,4,5,6 on the die are replaced by other integers, *e.g.*, when the six faces are numbered 1,1,1,2,2,3 we begin with the density

$$p[n] := \begin{cases} 3/6 & \text{if } n = 1 \\ 2/6 & \text{if } n = 2 \\ 1/6 & \text{if } n = 3 \\ 0 & \text{otherwise,} \end{cases}$$

cf. Ex. 2.30.

Convolution and arithmetic

Let $\mathcal{A}, \mathcal{B}$ be positive integers with the base 10 representations

$$\mathcal{A} = a_0 + a_1 10 + a_2 10^2 + \cdots, \quad \mathcal{B} = b_0 + b_1 10 + b_2 10^2 + \cdots .$$

The digits a_n, b_n take values $0, 1, \ldots, 9$, and we will assume that a_n, b_n are defined for all $n = 0, \pm 1, \pm 2, \ldots$. We can form the product $\mathcal{C} := \mathcal{A} \cdot \mathcal{B}$ by computing the convolution product $c = a * b$ of the digit strings and writing

$$\mathcal{C} = c_0 + c_1 10 + c_2 10^2 + \cdots .$$

The "digits" $c_0, c_1, c_2, \ldots$ that we obtain in this way may take values greater than 9, so we must use a suitable carrying process to obtain the canonical base 10 form

$$\mathcal{C} = c_0' + c_1' \, 10 + c_2' \, 10^2 + \cdots$$

with digits $c_0', c_1', c_2', \ldots$ that take the values $0, 1, \ldots, 9$.

You will recognize these steps in the following computation of the product $3141 \cdot 5926 = 18613566$.

10^7	10^6	10^5	10^4	10^3	10^2	10^1	10^0
				3	1	4	1
			$\times$	5	9	2	6
				3×6	1×6	4×6	1×6
			3×2	1×2	4×2	1×2	
		3×9	1×9	4×9	1×9		
	3×5	1×5	4×5	1×5			
	15	32	35	61	23	26	6
1	8	6	1	3	5	6	6

When we use the familiar elementary school algorithm for multiplication, we do a portion of the carrying process as we write each line. In the above computation we allow "digits" greater than 9 in the intermediate steps but we reduce the "digits" to canonical form in the last line.

We can also regard the digit strings a, b for $\mathcal{A}, \mathcal{B}$ as functions on $\mathbb{P}_N$ and generate $a * b$ within this context provided that N is large enough to avoid wraparound effects. For example, we can use the function d on $\mathbb{P}_{64}$ with

$$d[0] = 6, d[1] = 7, \ldots, d[30] = 1, d[31] = \cdots = d[63] = 0$$

to represent the digits of

$$2^{100} = 1267650600228229401496703205376 .$$

Figure 2.13 shows d, $d * d$, and the digits of $2^{100} \cdot 2^{100}$ on $\mathbb{P}_{64}$.

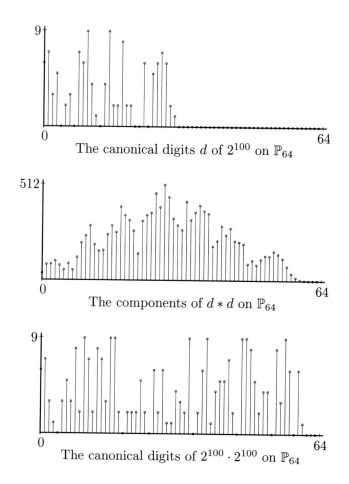

The canonical digits d of 2^{100} on $\mathbb{P}_{64}$

The components of $d * d$ on $\mathbb{P}_{64}$

The canonical digits of $2^{100} \cdot 2^{100}$ on $\mathbb{P}_{64}$

Figure 2.13. Computation of $2^{100} \cdot 2^{100}$ by convolving digit strings on $\mathbb{P}_{64}$.

Suppose that we wish to find the product of two M-digit numbers. If we use the usual elementary school algorithm, we must multiply each digit in the multiplicand by each digit in the multiplier, thus forming M^2 digit-digit products in the course of the calculation. You can easily write down 4 or 9 such products when $M = 2$ or 3, but not even the fastest computer can form the $10^9 \cdot 10^9$ digit-digit products that would be required for the multiplication of billion digit integers. How then is it possible to compute a billion digits of $\pi = 3.14159\ldots$?

The fast Fourier transform (FFT), an algorithm we will describe in Chapter 6, uses a small multiple of $N \log_2 N$ arithmetic operations to produce the discrete Fourier transform of an N-component vector. If we use the FFT with (21), we can compute the convolution product of N-component vectors by expending a small multiple of $N \log_2 N$ arithmetic operations. This makes it possible to multiply two M-digit integers using a small multiple of $2M \log_2(2M)$ arithmetic operations. (When you compare $M^2 = 10^{18}$ with $2M \log_2 2M = 62 \cdot 10^9$ for $M = 10^9$, you will see that the familiar elementary school algorithm is hopelessly inefficient!) You will find additional details for this multiplication scheme in Ex. 2.31.

You can add, subtract, multiply large integers by adding, subtracting, convolving the digit strings and then reducing the resulting "digits" to canonical form. If you keep track of exponents, *e.g.*,

$$(123 \cdot 10^{-30}) \cdot (456 \cdot 10^{22}) = (123 \cdot 456) \cdot 10^{-8}$$

you can add, subtract, multiply floating point numbers. A trick from calculus provides an efficient scheme for doing division. (Don't even think of coding the wretched long-division algorithm from elementary school!) Newton's iteration

$$x_{\nu+1} := x_\nu \cdot (2 - a \cdot x_\nu), \quad \nu = 0, 1, \ldots$$

for computing the root, $1/a$, of $f(x) := a - 1/x$, uses only multiplication and subtraction. The iterates converge quadratically, so if x_0 is good to 10 digits, then $x_1, x_2, x_3, \ldots$ will be good to $20, 40, 80, \ldots$ digits. Once you have computed the reciprocal $1/a$ to the desired precision, you can find the quotient $b/a = b \cdot (1/a)$ by using fast multiplication. The familiar Newton iteration

$$x_{\nu+1} = \frac{x_\nu + a}{2x_\nu}, \quad \nu = 0, 1, \ldots$$

then leads to a fast algorithm for computing $\sqrt{a}$, $a > 0$.

If you like to code, you can use these ideas to develop software for doing "high-precision" calculations using $+, -, \times, \div, \sqrt{\ }$. You may not reach a billion digits on your PC, but you can use the AGM iteration

$$a_0 := 1, \quad b_0 := \frac{1}{\sqrt{2}}, \quad t_0 := \frac{1}{4}, \quad x_0 := 1$$

$$a_{\nu+1} := (a_\nu + b_\nu)/2, \quad b_{\nu+1} := \sqrt{(a_\nu \cdot b_\nu)}, \quad t_{\nu+1} := t_\nu - x_\nu \cdot (a_{\nu+1} - a_\nu)^2,$$

$$x_{\nu+1} := 2x_\nu, \quad \nu = 0, 1, \ldots$$

to compute a few thousand digits of

$$\pi = \frac{a^2}{t} = \frac{(a+b)^2}{4t},$$

where $a := \lim a_\nu$, $b := \lim b_\nu$, $t := \lim t_\nu$, *cf.* J.M. Borwein and P.B. Borwein, *Pi and the AGM*, John Wiley & Sons, New York, 1987, p. 48.

References

R.N. Bracewell, *The Fourier Transform and Its Applications, 2nd ed.*, McGraw-Hill, New York, 1986.

A nice introduction to convolution is given in Chapter 3 of this popular text for scientists and engineers.

W.L. Briggs and V.E. Henson, *The DFT*, SIAM, Philadelphia, 1995.

Properties of the discrete convolution product and numerous applications are described in this book.

W. Feller, *An Introduction to Probability Theory and Its Applications, Vol. 1, 2nd ed.*, John Wiley & Sons, New York, 1957.

Chapter 9 describes the convolution of functions on $\mathbb{Z}$ within the context of probability theory.

W. Feller, *An Introduction to Probability Theory and Its Applications, Vol. 2*, John Wiley & Sons, New York, 1966.

Chapter 5 describes the convolution of functions on $\mathbb{T}_p$, $\mathbb{R}$ within the context of probability theory.

J.D. Gaskill, *Linear Systems, Fourier Transforms, and Optics*, John Wiley & Sons, New York, 1978.

An exposition of convolution as used in optics is given in Chapters 6 and 9.

P. Henrici, *Applied and Computational Complex Analysis, Vol. 3*, John Wiley & Sons, New York, 1986.

Discrete convolution (with corresponding fast algorithms for arithmetic and for manipulating power series) is discussed in Chapter 13.

A.V. Oppenheim, A.S. Willsky, and I.T. Young, *Signals and Systems*, Prentice Hall, Englewood Cliffs, NJ, 1983.

An exposition of convolution as used in systems theory is given in Chapter 3.

I.J. Schoenberg, *Mathematical Time Exposures*, Mathematical Association of America, Washington, D.C., 1982.

Convolution within the context of geometry is discussed in Chapter 6.

J.S. Walker, *Fast Fourier Transforms, 2nd ed.*, CRC Press, Boca Raton, FL, 1996.

Many applications of convolution can be found in this elementary text.

Exercise Set 2

▶ **EXERCISE 2.1:**   Find the following convolution products (of functions on $\mathbb{R}$).

(a)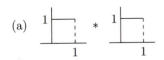

(d)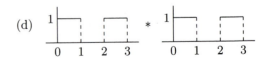

(b)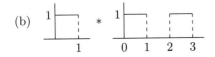

(e)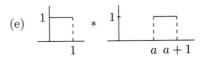

(c)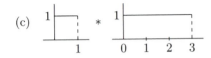

(f)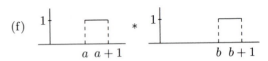

▶ **EXERCISE 2.2:**   Functions f, g on $\mathbb{R}$ are given by the following graphs.

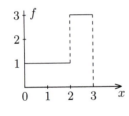

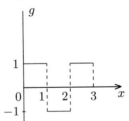

Find the following convolution and correlation products.

(a) $f * f$ (b) $f * g$ (c) $g * f$ (d) $g * g$

(e) $f \star f$ (f) $f \star g$ (g) $g \star f$ (h) $g \star g$

▶ **EXERCISE 2.3:**   Let h be the *Heaviside function,*

$$h(x) := \begin{cases} 1 & \text{if } x > 0 \\ 0 & \text{if } x < 0. \end{cases}$$

Find the following convolution products.

(a) $h_2 := h * h$

(b) $h_3 := h * h * h$

(c) $h_n := h * h * \cdots * h$ (with n factors of h)

(d) $h(x - x_1) * h(x - x_2) * \cdots * h(x - x_n)$ where $x_1, x_2, \ldots, x_n$ are any real numbers.

(e) $\{x^m h(x)\} * \{x^n h(x)\}$ where m, n are nonnegative integers.

   Hint: This convolution product is a scalar multiple of h_{m+n+2}.

▶ **EXERCISE 2.4:**   Let $f(x) := h(x)e^{-x}$ where h is the Heaviside function of Ex. 2.3. Find the following convolution products.

(a) $f_2 := f * f$

(b) $f_3 := f * f * f$

(c) $f_n := f * f \cdots * f$ (with n factors of f)

(d) $\{x^n f(x)\} * \{x^m f(x)\}$ where m, n are nonnegative integers.

▶ **EXERCISE 2.5:**   Let $f(x) := e^{-|x|}$. Show that $(f * f)(x) = (1 + |x|)e^{-|x|}$.

Hint. When $x < 0$ you can split the integral over $\mathbb{R}$ into easily evaluated integrals over $(-\infty, x)$, $(x, 0)$, $(0, \infty)$. Do this and then make clever use of symmetry!

▶ **EXERCISE 2.6:**   Let $f(x) := e^{-\pi x^2}$, $-\infty < x < \infty$.

(a) Verify that $f(u)f(x - u) = e^{-2\pi(u - x/2)^2} e^{-\pi x^2/2}$.

(b) Using (a), show that $(f * f)(x) = Ie^{-\pi x^2/2}$ where $I := \int\limits_{-\infty}^{\infty} e^{-2\pi y^2} dy$.

Note. In Chapter 3 you will learn an easy way to show that $I = 1/\sqrt{2}$.

▶ **EXERCISE 2.7:**   For $n = 0, 1, 2, \ldots$ we define the *B-spline* B_n on $\mathbb{R}$ by writing $B_0 := \Pi$, $B_1 := \Pi * \Pi$, $B_2 := \Pi * \Pi * \Pi$, $\ldots$ where

$$\Pi(x) := \begin{cases} 1 & \text{for } |x| < \frac{1}{2} \\ 0 & \text{for } |x| > \frac{1}{2}. \end{cases}$$

Graphs of B_0, B_1, B_2 are shown in Fig. 2.9.

(a) For which values of x is $B_n(x) > 0$?

(b) Show that B_n is differentiable with

$$B_n'(x) = B_{n-1}(x + \tfrac{1}{2}) - B_{n-1}(x + \tfrac{1}{2}), \quad n = 1, 2, \ldots .$$

Hint. Begin with the defining integral for $B_n = \Pi * B_{n-1}$.

(c) For which values of m is $B_n^{(m)}$ continuous?

(d) Express $B_n^{(n)}$ as a sum of scaled translates of $B_0 = \Pi$.

Hint. Use (b) to express B_n'' in terms of B_{n-2}, B_n''' in terms of B_{n-3}, etc.

Note. The B-splines have many applications in numerical analysis. We use the formula

$$B_n(x) = \frac{1}{n} \left\{ [\tfrac{1}{2}(n + 1) + x]B_{n-1}(x + \tfrac{1}{2}) + [\tfrac{1}{2}(n + 1) - x]B_{n-1}(x - \tfrac{1}{2}) \right\}, \quad n = 1, 2, \ldots$$

recursively [with $B_0(x) = \Pi(x)$] when we want to evaluate B_n at some point x, *cf.* Ex. 3.31.

▶ **EXERCISE 2.8:**   Let

$$f(x) := \begin{cases} 1 & \text{for } 0 \leq x < 1 \text{ or } 3 \leq x < 4 \\ 0 & \text{for } 1 \leq x < 3, \end{cases} \qquad g(x) := x \text{ for } 0 \leq x < 4$$

be regarded as functions on the circle $\mathbb{T}_4$. Find the following convolution products.

(a) $f * f$ (b) $f * g$ (c) $g * g$

▶ **EXERCISE 2.9:**   Let

$$e_k(x) := e^{2\pi i k x/p}, \qquad k = 0, \pm 1, \pm 2, \ldots$$
$$c_k(x) := \cos(2\pi k x/p), \qquad k = 0, 1, 2, \ldots$$
$$s_k(x) := \sin(2\pi k x/p), \qquad k = 1, 2, 3, \ldots$$

be regarded as functions on the circle $\mathbb{T}_p$. Find the following convolution products.

(a) $e_k * e_\ell$ (b) $e_k * c_\ell$ (c) $e_k * s_\ell$ (d) $c_k * c_\ell$ (e) $c_k * s_\ell$ (f) $s_k * s_\ell$

▶ **EXERCISE 2.10:**   Let f be a function on $\mathbb{Z}$ with $f[n] := 1/n!$ when $n = 0, 1, \ldots$ and $f[n] = 0$ when $n = -1, -2, \ldots$. Let $f_1 := f$, $f_2 := f * f$, $f_3 := f * f * f$, $\ldots$. Find a simple formula for the components of f_m, $m = 1, 2, \ldots$.

Hint. Make use of the Maclaurin series for e^x.

▶ **EXERCISE 2.11:**   Let f, δ be functions on $\mathbb{Z}$ with $\delta[0] := 1$ and $\delta[n] := 0$ if $n = \pm 1, \pm 2, \ldots$. Let the translation parameters m, m_1, m_2 be integers. Find simple expressions for the following convolution products.

(a) $\delta[n - m_1] * \delta[n - m_2]$ (b) $\delta[n - m] * f[n]$ (c) $\delta[n - m_1] * f[n - m_2]$

▶ **EXERCISE 2.12:**   Let δ, Δ, h be functions on $\mathbb{Z}$ with

$$\delta[n] := \begin{cases} 1 & \text{if } n = 0 \\ 0 & \text{otherwise,} \end{cases} \quad \Delta[n] := \begin{cases} -1 & \text{if } n = 1 \\ 1 & \text{if } n = 0 \\ 0 & \text{otherwise,} \end{cases} \quad h[n] := \begin{cases} 1 & \text{if } n = 0, 1, 2, \ldots \\ 0 & \text{otherwise.} \end{cases}$$

(a) Show that $\Delta * h = h * \Delta = \delta$.

(b) Let $\Delta_1 := \Delta$, $\Delta_2 := \Delta * \Delta$, $\Delta_3 := \Delta * \Delta * \Delta, \ldots$. Find a simple formula for $\Delta_p[n]$, $p = 1, 2, \ldots$.

(c) Let $h_1 := h$, $h_2 := h * h$, $h_3 := h * h * h, \ldots$. Find a simple formula for $h_p[n]$, $p = 1, 2, \ldots$.

    Hint. When $n \geq 0$, $h_2[n] = (n+1)/1!$, $h_3[n] = (n+1)(n+2)/2!$, $\ldots$.

(d) Using (a), show that if p, q are nonnegative integers, then

$$\Delta_p * h_q = \begin{cases} \Delta_{p-q} & \text{if } p > q \\ \delta & \text{if } p = q \\ h_{q-p} & \text{if } q > p. \end{cases}$$

Note. You may wish to compare this result to that of Ex. 2.24(c).

▶ **EXERCISE 2.13:**   Let f, g be suitably regular p-periodic functions on $\mathbb{R}$ having the Fourier series

$$f(x) = \sum_{k=-\infty}^{\infty} F[k]\, e^{2\pi i k x/p}, \qquad g(x) = \sum_{k=-\infty}^{\infty} G[k]\, e^{2\pi i k x/p}.$$

(a) Formally multiply these Fourier series and combine like terms to obtain the Fourier series for the product $f \cdot g$. In so doing, you should make suitable use of the convolution product $F * G$.

(b) Formally convolve these Fourier series and simplify to obtain the Fourier series for the convolution product $f * g$. In so doing, freely make use of the result from Ex. 2.9(a).

▶ **EXERCISE 2.14:**   Find the convolution product $f * g$ when the functions f, g on $\mathbb{P}_4$ are given by:

(a) $f := (1, 2, 3, 4), \quad g := (1, 0, 0, 0);$ (b) $f := (1, 2, 3, 4), \quad g := (0, 0, 1, 0);$

(c) $f := (1, 2, 3, 4), \quad g := (1, -1, 0, 0);$ (d) $f := (1, 2, 3, 4), \quad g := (1, 1, 1, 1).$

▶ **EXERCISE 2.15:**   Find the convolution products $f * f$, $f * f * f$, $f * f * f * f$ when the function f on $\mathbb{P}_4$ is given by:

(a) $f := (1, 1, 0, 0);$ (b) $f := (1, -1, 0, 0);$ (c) $f := (1, 0, 0, 0);$

(d) $f := (0, 1, 0, 0);$ (e) $f := (0, 1, 0, 1);$ (f) $f := (1, 1, 1, 1).$

▶ **EXERCISE 2.16:**   A function f on $\mathbb{P}_4$ has the components $f[0] = -1$, $f[1] = f[2] = f[3] = 1$. Find the convolution product $f * f$:

(a) by using a direct computation;

(b) by writing $f[n] = u[n] - 2\delta[n]$ where

$$u[n] := 1 \quad n = 0, 1, 2, 3, \qquad \delta[n] := \begin{cases} 1 & \text{if } n = 0 \\ 0 & \text{if } n = 1, 2, 3, \end{cases}$$

and using algebraic properties of the convolution product with the identities $u*u = 4u$, $u * \delta = u$, $\delta * \delta = \delta$.

▶ **EXERCISE 2.17:**   Find the convolution product $f * f$ when:

(a) f is a function on $\mathbb{P}_5$ having components $f[0] = f[1] = 0$, $f[2] = 1$, $f[3] = 2$, $f[4] = 3$;

(b) f is a function on $\mathbb{Z}$ having the components of (a) with $f[n] = 0$ if $n < 0$ or $n > 4$;

(c) f is a function on $\mathbb{T}_5$ given by the following graph;

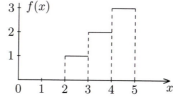

(d) f is a function on $\mathbb{R}$ given by the graph of (c) with $f(x) = 0$ if $x < 0$ or $x > 5$.

▶ **EXERCISE 2.18:**   Find all functions f on $\mathbb{P}_2$ such that:

(a) $f * f = 1 := (1,1)$;

(b) $f * f = \delta := (1,0)$.

▶ **EXERCISE 2.19:**   Show that the convolution product is *commutative, i.e.,*
$f_1 * f_2 = f_2 * f_1$ when f_1, f_2 are suitably regular functions on:

(a) $\mathbb{R}$;   (b) $\mathbb{T}_p$;   (c) $\mathbb{Z}$;   (d) $\mathbb{P}_N$.

Hint. Set $u' := x - u$ or $m' := n - m$ in the defining integral or sum.

▶ **EXERCISE 2.20:**   Show that the convolution product is *associative, i.e.,*
$f_1 * (f_2 * f_3) = (f_1 * f_2) * f_3$ when f_1, f_2, f_3 are suitably regular functions on:

(a) $\mathbb{R}$;   (b) $\mathbb{T}_p$;   (c) $\mathbb{Z}$;   (d) $\mathbb{P}_N$.

Hint. Give an informal argument in which you freely interchange the order of integration or summation. (Conditions that justify such an interchange of limits for functions on $\mathbb{Z}$, $\mathbb{R}$ can be found in Exs. 2.35, 2.36.)

▶ **EXERCISE 2.21:**   Let f_1, f_2 be functions on $\mathbb{R}$ and assume that $f_1 * f_2$ is well defined.

(a) Show that if we translate f_1 or f_2, then $f_1 * f_2$ is translated in the same direction by the same amount, *i.e.,*

$$f_1(x + a) * f_2(x) = (f_1 * f_2)(x + a) = f_1(x) * f_2(x + a), \quad -\infty < a < \infty.$$

(b) Draw a sketch to illustrate the result of (a) in the case where f_1, f_2 are rectangular pulses, *cp.* Ex. 2.1.

(c) Formulate a rule for computing $f_1(x + a_1) * f_2(x + a_2) * \cdots * f_n(x + a_n)$.

▶ **EXERCISE 2.22:**   Let the functions f, g on $\mathbb{R}$ be given by

$$f(x) := \begin{cases} 1 & \text{if } 0 \leq x \leq 2 \\ 0 & \text{otherwise,} \end{cases} \qquad g(x) := \begin{cases} 1 & \text{if } 0 \leq x \leq 3 \\ 0 & \text{otherwise.} \end{cases}$$

(a) Find and sketch the convolution product $f * g$.

(b) Use (a) to find and sketch the convolution product of the functions

$$f_4(x) := \sum_{m=-\infty}^{\infty} f(x - 4m), \qquad g_4(x) := \sum_{m=-\infty}^{\infty} g(x - 4m)$$

on the circle $\mathbb{T}_4$.

▶ **EXERCISE 2.23:**   This exercise will help you determine where certain convolution products must vanish.

(a) Let f_1, f_2 be piecewise continuous functions on $\mathbb{R}$ that vanish when $x < a_1$, $x < a_2$, respectively. Such functions are said to have a *finite past*. Show that $f_1 * f_2$ is a well-defined continuous function that vanishes when $x \leq a_1 + a_2$.

(b) Let $f_1, \ldots, f_n$ be piecewise continuous functions on $\mathbb{R}$ that vanish when $x < a_1$, $\ldots, x < a_n$, respectively. What can you infer about $f_1 * \cdots * f_n$?

(c) Formulate an analogous result for the convolution product $f_1 * \cdots * f_n$ of piecewise continuous functions on $\mathbb{R}$ that have *finite futures*.

(d) Let $f_1, \ldots, f_n$ be piecewise continuous functions that vanish outside of the finite intervals $[a_1, b_1], \ldots, [a_n, b_n]$, respectively. What can you infer about the convolution product $f_1 * \cdots * f_n$?

Note. You can use Exs. 2.1, 2.7, etc. to illustrate this result.

▶ **EXERCISE 2.24:**   Let **D** be the *derivative operator*, i.e., $(\mathbf{D}f)(x) := f'(x)$, and let $h_1 = h$, $h_2 = h * h$, $h_3 = h * h * h$, $\ldots$ where h is the Heaviside function, of Ex. 2.3.

(a) Show that the convolution integrals

$$(h * f)(x) = \int_{-\infty}^{x} f(u)\,du, \qquad (h * f')(x) = \int_{-\infty}^{x} f'(u)\,du$$

are well defined when f is continuous, f' is piecewise continuous, and both f and f' are absolutely integrable on $\mathbb{R}$.

(b) Using (a), show that $\mathbf{D}(h * f) = h * (\mathbf{D}f) = f$ when f is suitably regular.

(c) Let p, q be nonnegative integers. Show that if $f, f', \ldots, f^{(p-1)}$ are continuous and absolutely integrable, and if $f^{(p)}$ is piecewise continuous and absolutely integrable then

$$\mathbf{D}^p(h_q * f) = \begin{cases} \mathbf{D}^{p-q} f & \text{if } p > q \\ f & \text{if } p = q \\ h_{q-p} * f & \text{if } q > p. \end{cases}$$

Note. You may wish to compare this result to that of Ex. 2.12(d).

▶ **EXERCISE 2.25:**   Let $f_n(x) := x^n e^{-\pi x^2}$, $n = 0, 1, 2$. Use the differentiation rule (28) with the known convolution product $f_0 * f_0$ from Ex. 2.6 to find:

(a) $f_0 * f_1$;   (b) $f_1 * f_1$;   (c) $f_0 * f_2$.

Hint. Observe that $f_0' = -2\pi f_1$ so that $f_0 * f_1 = (-1/2\pi)(f_0 * f_0)'$.

▶ **EXERCISE 2.26:**   In this exercise you will establish a multiplicative relation for the convolution product.

(a) Let f, g be piecewise continuous, absolutely integrable functions on $\mathbb{R}$. Show that

$$\int_{x=-\infty}^{\infty} (f * g)(x)dx = \left\{ \int_{x=-\infty}^{\infty} f(x)dx \right\} \left\{ \int_{x=-\infty}^{\infty} g(x)dx \right\}.$$

Hint. Since $f(u)g(x - u)$ is absolutely integrable on $\mathbb{R}^2$ you can exchange the limits of integration.

(b) Formulate an analogous result for functions f, g on $\mathbb{T}_p$.

(c) Let f, g be functions on $\mathbb{P}_N$. Show that

$$\sum_{n=0}^{N-1} (f * g)[n] = \left\{ \sum_{n=0}^{N-1} f[n] \right\} \left\{ \sum_{n=0}^{N-1} g[n] \right\}.$$

(d) Formulate an analogous result for functions f, g on $\mathbb{Z}$.

Note. A nonnegative function f on $\mathbb{R}$, $\mathbb{T}_p$ or $\mathbb{Z}$, $\mathbb{P}_N$ is said to be a probability density if its global integral or sum is 1. This exercise shows that the convolution product of probability densities is a probability density.

▶ **EXERCISE 2.27:**   Let $\mathbf{c} := (c_0, c_1, \dots, c_{N-1})$, $\mathbf{d} := (d_0, d_1, \dots, d_{N-1})$ and let $\mathbf{C}, \mathbf{D}$ be the corresponding $N \times N$ *circulant* matrices

$$\mathbf{C} = \begin{bmatrix} c_0 & c_{N-1} & c_{N-2} & \cdots & c_1 \\ c_1 & c_0 & c_{N-1} & \cdots & c_2 \\ c_2 & c_1 & c_0 & \cdots & c_3 \\ \vdots & \vdots & \vdots & & \vdots \\ c_{N-1} & c_{N-2} & c_{N-3} & \cdots & c_0 \end{bmatrix}, \quad \mathbf{D} = \begin{bmatrix} d_0 & d_{N-1} & d_{N-2} & \cdots & d_1 \\ d_1 & d_0 & d_{N-1} & \cdots & d_2 \\ d_2 & d_1 & d_0 & \cdots & d_3 \\ \vdots & \vdots & \vdots & & \vdots \\ d_{N-1} & d_{N-2} & d_{N-3} & \cdots & d_0 \end{bmatrix}.$$

(a) Show how to relate the matrix vector product $\mathbf{C}\mathbf{x}^T$ to the convolution product $\mathbf{c} * \mathbf{x}$ on $\mathbb{P}_N$ when $\mathbf{x} := (x_0, x_1, \dots, x_{N-1})$.

(b) Using (a), show that $\mathbf{C}\mathbf{D}$ is the circulant matrix corresponding to $\mathbf{c} * \mathbf{d}$.

(c) Show that $\mathbf{C}$ and $\mathbf{D}$ commute, *i.e.*, $\mathbf{C}\mathbf{D} = \mathbf{D}\mathbf{C}$.

▶ **EXERCISE 2.28:**   In this exercise you will derive *Taylor's formula* (from elementary calculus) by using certain convolution products.

(a) Let h be the Heaviside function of Ex. 2.3. Use the fundamental theorem of calculus to show that if g is continuously differentiable for $x \geq 0$ and $g(0) = 0$, then

$$g(x) = [(hg') * h](x), \qquad x \geq 0.$$

(b) Show that if f is twice continuously differentiable for $x \geq 0$, then

$$\left[((hf'') * h) * h \right](x) = f(x) - f(0) - xf'(0), \qquad x \geq 0.$$

Hint. Use (a) with $g(x) = f'(x) - f'(0)$ and with $g(x) := f(x) - f(0) - xf'(0)$.

(c) More generally, show that if f is n times continuously differentiable for $x \geq 0$, and we define $h_1 := h$, $h_2 := h * h$, $h_3 := h * h * h$, $\ldots$, then

$$[(hf^{(n)}) * h_n](x) = f(x) - f(0) - \frac{xf'(0)}{1!} - \frac{x^2f''(0)}{2!} - \cdots - \frac{x^{n-1}f^{(n-1)}(0)}{(n-1)!}, \quad x \geq 0.$$

(d) Using (c) and the compact formula for h_n obtained in Ex. 2.3, show that if f is n times continuously differentiable for $x \geq 0$, then

$$f(x) = \sum_{k=0}^{n-1} \frac{x^k f^{(k)}(0)}{k!} + \frac{1}{(n-1)!} \int_{u=0}^{x} f^{(n)}(u)(x-u)^{n-1} du, \quad x \geq 0.$$

Note. This formula holds if we allow $f^{(n)}$ to have isolated points of discontinuity where finite jumps occur. We can also shift the point of expansion from $x = 0$ to $x = a$ and remove the restriction $x \geq a$.

▶ **EXERCISE 2.29:**   Explain why:

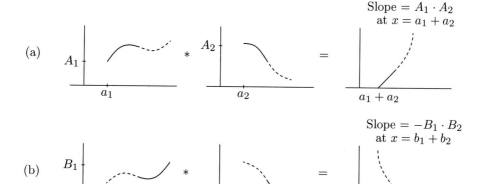

Note. You should assume that all of these functions are piecewise smooth. The 1st, 2nd factors in (a) vanish to the left of a_1, a_2 while the 1st, 2nd factors of (b) vanish to the right of b_1, b_2, respectively, *cf.* Ex. 2.23.

▶ **EXERCISE 2.30:**   Find the probability density functions

$$p_1[n] := p[n], \quad p_2[n] := (p * p)[n], \quad p_3[n] := (p * p * p)[n], \quad p_4[n] := (p * p * p * p)[n]$$

for throwing the integer n with 1, 2, 3, 4 fair dice when the six faces of each die are marked with the integers:

(a) 1,1,2,2,3,3;   (b) $-1,-1,-1,1,1,1$;   (c) 1,1,1,2,2,3.

Hint. Construct paper strips to facilitate a sliding strip computation of the convolution products ... or use a computer!

▶ **EXERCISE 2.31:**   The product $123 \times 456 = 56088$ can be obtained by computing $(3, 2, 1, 0, 0, 0) * (6, 5, 4, 0, 0, 0) = (18, 27, 28, 13, 4, 0)$ and then reducing the components of this convolution product on $\mathbb{P}_6$ to digits by using the base 10 carrying process

$$(18, 27, 28, 13, 4, 0) \nearrow\!\!\!\!\searrow (8, 28, 28, 13, 4, 0) \nearrow\!\!\!\!\searrow (8, 8, 30, 13, 4, 0)$$
$$\nearrow\!\!\!\!\searrow (8, 8, 0, 16, 4, 0) \nearrow\!\!\!\!\searrow (8, 8, 0, 6, 5, 0).$$

This exercise will help you understand this algorithm so that you can write a computer program to multiply large integers having thousands, millions, and even billions of digits.

We select a *base* $\beta = 2, 3, \ldots$, a *precision index* $M = 1, 2, \ldots$ and a corresponding vector length $N := 2M$, e.g., the above example uses $\beta = 10$, $M = 3$, and $N = 6$. An N-vector $\mathbf{a} = (a[0], a[1], \ldots, a[N-1])$ with integer components represents

$$\mathcal{I}(\mathbf{a}) := a[0] + a[1]\beta + a[2]\beta^2 + \cdots + a[N-1]\beta^{N-1}.$$

We say that $\mathbf{a}$ is *reduced* if $a[n] = 0, 1, \ldots, \beta - 1$ for each $n = 0, 1, \ldots, N - 1$, and we say that $\mathbf{a}$ is *half length* if $a[n] = 0$ for $n = M, M+1, \ldots, N-1$.

(a) Verify that $\mathcal{I}(\mathbf{a}) \leq \beta^M - 1$ when $\mathbf{a}$ is reduced and half length, with this bound being the best possible.

(b) Show that if $\mathbf{a}, \mathbf{b}$ are reduced and half length, then $\mathcal{I}(\mathbf{a}) \cdot \mathcal{I}(\mathbf{b}) = \mathcal{I}(\mathbf{a} * \mathbf{b})$, i.e., we can multiply large integers by convolving strings of "digits" on $\mathbb{P}_N$.

(c) Show that if $\mathbf{a}, \mathbf{b}$ are reduced and half length, then $\mathcal{I}(\mathbf{a} * \mathbf{b}) = \mathcal{I}(\mathbf{c})$ for some reduced N-vector $\mathbf{c}$, i.e., show that $\mathcal{I}(\mathbf{a}) \cdot \mathcal{I}(\mathbf{b})$ does not have more than N "digits". You can use the algorithm

$$\mathbf{c} := \mathbf{a} * \mathbf{b}$$

For $n = 0, 1, \ldots, N - 2$ do:

$$\left\lfloor \begin{array}{l} \text{carry} := \lfloor c[n]/\beta \rfloor \\[4pt] c[n] := c[n] - \text{carry} \cdot \beta \\[4pt] c[n+1] := c[n+1] + \text{carry} \end{array} \right.$$

to find $\mathbf{c}$. Here $\lfloor x \rfloor$ is the largest integer that does not exceed x.

(d) Show that if $\mathbf{a}, \mathbf{b}$ are reduced and half length, then $(\mathbf{a} * \mathbf{b})[n] \leq M(\beta - 1)^2$.

(e) Suppose that your computer can exactly represent every integer $1, 2, \ldots, P$ when doing routine floating point arithmetic. The above algorithm will allow you to multiply integers up to size $S := \beta^M - 1$ provided you choose β, M such that $M(\beta - 1)^2 \leq P$. What constraints does this impose on M and on S when:

  (i) $P = 2^{24} - 1$ (typical short precision) and $\beta = 10$? $\beta = 100$? $\beta = 2$?

  (ii) $P = 2^{56} - 1$ (typical long precision) and $\beta = 10$? $\beta = 100$? $\beta = 2$?

Note. In practice, small errors are introduced when the components of $\mathbf{a} * \mathbf{b}$ are computed by using (21) with the FFT (as described in Chapter 6). It can be shown that the modulus of each such error is bounded by a small multiple of $(\beta^2/P) \cdot M \cdot \log_2 M$. We can completely eliminate the error in a computed approximation to $(\mathbf{a} * \mathbf{b})[n]$ by rounding to the nearest integer if we choose β, M to ensure that the modulus of each error is less than $1/2$.

▶ **EXERCISE 2.32:** Let $\mathfrak{P}$ be a polygon in the complex plane with the vertices $v_0, v_1, \ldots, v_{N-1}$ (*i.e.*, $\mathfrak{P}$ consists of the N line segments joining v_0 to v_1, v_1 to $v_2, \ldots, v_{N-2}$ to v_{N-1}, and v_{N-1} to v_0). Let $\mathfrak{P}'$ be the polygon obtained from $\mathfrak{P}$ by using as vertices the N midpoints $w_0 := (v_{N-1} + v_0)/2$, $w_1 := (v_0 + v_1)/2$, $\ldots$, $w_{N-1} := (v_{N-2} + v_{N-1})/2$ of the sides of $\mathfrak{P}$ (*i.e.*, $\mathfrak{P}'$ consists of the N line segments joining w_0 to w_1, w_1 to $w_2, \ldots, w_{N-2}$ to w_{N-1}, and w_{N-1} to w_0). It is a simple matter to construct $\mathfrak{P}'$ when $\mathfrak{P}$ is given. The inverse, known as *Kasner's problem*, requires us to find $\mathfrak{P}$ when $\mathfrak{P}'$ is given. In this exercise you will solve Kasner's problem using the discrete convolution product.

(a) Sketch $\mathfrak{P}, \mathfrak{P}'$, and $\mathfrak{P}'' = (\mathfrak{P}')'$ in the cases where $\mathfrak{P}$ is a scalene triangle, a rectangle, a kite (*i.e.*, a quadrilateral having orthogonal diagonals that are of unequal length), and a pentagonal star (with $v_n = e^{4\pi i n/5}$).

(b) Let $\mathbf{v} = (v_0, v_1, \ldots, v_{N-1})$, $\mathbf{w} = (w_0, w_1, \ldots, w_{N-1})$, and $\mathbf{a} = (\frac{1}{2}, \frac{1}{2}, 0, 0, \ldots, 0)$ be regarded as functions on $\mathbb{P}_N$. Verify that the construction of $\mathfrak{P}'$ from $\mathfrak{P}$ requires us to compute $\mathbf{w} = \mathbf{a} * \mathbf{v}$ when $\mathbf{v}$ is given while the construction of $\mathfrak{P}$ from $\mathfrak{P}'$ requires us to solve the linear convolution equation $\mathbf{a} * \mathbf{v} = \mathbf{w}$ for $\mathbf{v}$ when $\mathbf{w}$ is given.

(c) When $N = 3, 5, 7, \ldots$ we regard $\mathbf{a} := (\frac{1}{2}, \frac{1}{2}, 0, 0, \ldots, 0)$, $\mathbf{b} := (1, -1, 1, -1, \ldots, 1, -1, 1)$, $\boldsymbol{\delta} := (1, 0, 0, \ldots, 0)$ as functions on $\mathbb{P}_N$. Verify that $\mathbf{a} * \mathbf{b} = \mathbf{b} * \mathbf{a} = \boldsymbol{\delta}$ (*i.e.*, that $\mathbf{b}$ is the *inverse* of $\mathbf{a}$ with respect to the convolution product). Use this fact to show the convolution equation $\mathbf{a} * \mathbf{v} = \mathbf{w}$ has a unique solution and then express the components of $\mathbf{v}$ in terms of the components of $\mathbf{w}$.

(d) When $N = 4, 6, 8, \ldots$ we set

$$\mathbf{b} := \frac{1}{4}(3, -1, -1, 3), \ \frac{1}{6}(5, -3, 1, 1, -3, 5), \ \frac{1}{8}(7, -5, 3, -1, -1, 3, -5, 7), \ \ldots$$

and regard $\mathbf{a} := (\frac{1}{2}, \frac{1}{2}, 0, 0, \ldots, 0)$, $\mathbf{c} := (1, -1, 1, -1, \ldots, 1, -1)$, $\boldsymbol{\delta} := (1, 0, \ldots, 0)$, as functions on $\mathbb{P}_N$. Verify that

$$\mathbf{a} * \mathbf{c} = \mathbf{c} * \mathbf{a} = 0, \quad \mathbf{a} * \mathbf{b} = \mathbf{b} * \mathbf{a} = \boldsymbol{\delta} - \frac{1}{N}\mathbf{c}, \quad \mathbf{c} * \mathbf{v} = \alpha\mathbf{c}$$

where $\alpha := v_0 - v_1 + v_2 - v_3 + \cdots + v_{N-2} - v_{N-1}$, and then use these relations to prove the following statements.

(i) If $\mathbf{c} * \mathbf{w} \neq 0$, then the convolution equation $\mathbf{a} * \mathbf{v} = \mathbf{w}$ has no solution.

(ii) If $\mathbf{c} * \mathbf{w} = 0$, then $\mathbf{v}$ is a solution of the convolution equation $\mathbf{a} * \mathbf{v} = \mathbf{w}$ if and only if $\mathbf{v}$ has the representation $\mathbf{v} = \mathbf{b} * \mathbf{w} + \beta\mathbf{c}$ for some choice of the scalar β.

(e) Find polygons $\mathfrak{P}'_0$, $\mathfrak{P}'_1$, $\mathfrak{P}'_\infty$ so that the inverse Kasner problem has no solution, exactly one solution, infinitely many solutions, respectively.

Note. A geometric technique for constructing $\mathfrak{P}$ from $\mathfrak{P}'$ can be found in Schoenberg, pp. 60–63.

▶ **EXERCISE 2.33:**   Let f, g be piecewise continuous functions on $\mathbb{R}$ with f being absolutely integrable and with g being bounded.

(a) Show that $f * g$ is well defined by (1).

   Hint. Write $|f(u)g(x - u)| \le B|f(u)|$, where B is a bound for $|g|$.

(b) Show that $f * g$ is continuous.

   Hint. Choose $a_1, x_1, \dots, a_K, x_K$ so that

$$g_0(x) := g(x) - \sum_{k=1}^{K} a_k h(x - x_k)$$

   is bounded and continuous on $\mathbb{R}$.

▶ **EXERCISE 2.34:**   In this exercise you will study differentiation of the convolution product. Let g be a continuous function on $\mathbb{R}$ with a piecewise continuous derivative g'. Let f be a piecewise continuous function on $\mathbb{R}$, and assume that $f(u)g(x-u)$, $f(u)g'(x-u)$ are absolutely integrable functions of u for each choice of x (so that $f * g$ and $f * g'$ are well defined). Show that $f * g$ is differentiable with $(f * g)'(x) = (f * g')(x)$ at each point x where $f * g'$ is continuous.

Hint. First show that

$$\frac{(f * g)(x + h) - (f * g)(x)}{h} - (f * g')(x) = \frac{1}{h} \int_{v=0}^{h} \{(f * g')(x + v) - (f * g')(x)\}dv.$$

▶ **EXERCISE 2.35:**   In this exercise you will study the associativity of the convolution product of functions on $\mathbb{Z}$.

(a) Let $\alpha \ne 0$ be a complex number and let f_1, f_2, f_3 be defined on $\mathbb{Z}$ by

$$f_1[n] := \alpha^n, \qquad f_2[n] := \begin{cases} 1 & \text{if } n = 0 \\ -\alpha & \text{if } n = 1 \\ 0 & \text{otherwise,} \end{cases} \qquad f_3[n] := \begin{cases} \alpha^n & n = 0, 1, 2, \dots \\ 0 & \text{otherwise.} \end{cases}$$

   Show that $f_1 * f_2$, $(f_1 * f_2) * f_3$, $f_2 * f_3$, $f_1 * (f_2 * f_3)$ are all well defined but $(f_1 * f_2) * f_3 \ne f_1 * (f_2 * f_3)$.

(b) Let f_1, f_2, f_3 be (arbitrary) functions on $\mathbb{Z}$. Show that

$$\{f_1 * (f_2 * f_3)\}[n] = \sum_{p=-\infty}^{\infty} \sum_{q=-\infty}^{\infty} f_1[p]f_2[n - p - q]f_3[q],$$

$$\{(f_1 * f_2) * f_3\}[n] = \sum_{q=-\infty}^{\infty} \sum_{p=-\infty}^{\infty} f_1[p]f_2[n - p - q]f_3[q].$$

   Although these double sums have exactly the same summands, the first summation is done by *columns* and the second is done by *rows*.

(c) A *sufficient* condition for the two double sums of (b) to exist and be equal is that

$$\sum_{p,q=-\infty}^{\infty} |f_1[p]f_2[n-p-q]f_3[q]| < \infty \quad \text{for all } n.$$

Use this result to verify that $(f_1 * f_2) * f_3 = f_1 * (f_2 * f_3)$ when:

  (i) f_1, f_2, f_3 all have finite pasts (*i.e.*, each is a translate of some function that vanishes when its argument is negative); or

  (ii) two of the three functions f_1, f_2, f_3 are support limited, *i.e.*, they vanish for all but finitely many values of their arguments; or

  (iii) two of the functions f_1, f_2, f_3 are absolutely summable and the third is bounded.

▶ **EXERCISE 2.36:** In this exercise you will study the associativity of the convolution product of piecewise continuous functions on $\mathbb{R}$.

(a) Let f_1, f_2, f_3 be defined on $\mathbb{R}$ by

$$f_1(x) := h(x) := \begin{cases} 1 & \text{if } x > 0 \\ 0 & \text{if } x < 0, \end{cases} \quad f_2(x) := (e^{-x^2/2})' = -x\,e^{-x^2/2}, \quad f_3(x) := h(-x).$$

Show that $f_1 * f_2$, $(f_1 * f_2) * f_3$, $f_2 * f_3$, and $f_1 * (f_2 * f_3)$ are all well defined but $(f_1 * f_2) * f_3 \neq f_1 * (f_2 * f_3)$.

(b) Let f_1, f_2, f_3 be (arbitrary) functions on $\mathbb{R}$ that are continuous except for finitely many points where finite jumps can occur. Show that

$$\{f_1 * (f_2 * f_3)\}(x) = \int_{u=-\infty}^{\infty} \int_{v=-\infty}^{\infty} f_1(u)f_2(x-u-v)f_3(v)dv\,du,$$

$$\{(f_1 * f_2) * f_3\}(x) = \int_{v=-\infty}^{\infty} \int_{u=-\infty}^{\infty} f_1(u)f_2(x-u-v)f_3(v)du\,dv.$$

Although the integrands are identical, the first integration uses vertical slices and the second uses horizontal slices of the plane, *cf.* Ex. 2.35(b).

(c) A *sufficient* condition for the two double integrals of (b) to exist and be equal is that

$$\iint_{\mathbb{R}^2} |f_1(u)f_2(x-u-v)f_3(v)|du\,dv < \infty \quad \text{for all } x.$$

Use this to verify that $(f_1 * f_2) * f_3 = f_1 * (f_2 * f_3)$ when:

  (i) f_1, f_2, f_3 all have finite pasts; or

  (ii) two of the three functions f_1, f_2, f_3 are support limited, *i.e.*, they vanish outside some finite interval; or

  (iii) two of the functions f_1, f_2, f_3 are absolutely integrable and the third is bounded.

▶ **EXERCISE 2.37:**   Let f be a complex-valued function defined on $\mathbb{R}$. We write $f \in S^0$ provided that f, f' are continuous except at a finite number of points (if any) where the one-sided limits $f(x+)$, $f(x-)$, $f'(x+)$, $f'(x-)$ exist and are finite. For $n = 1, 2, \ldots$ we write $f \in S^n$ provided that $f, f', \ldots, f^{(n-1)}$ are continuous and $f^{(n)} \in S^0$, e.g., $h(x)$, $xh(x)$, $x^2h(x)$, $\ldots$ lie in $S^0, S^1, S^2, \ldots$, respectively. In this exercise you will verify that convolution preserves and promotes smoothness by showing that if f, g have finite pasts and if $f \in S^m$, $g \in S^n$, then $f * g \in S^{m+n+1}$ [cf. Ex. 2.23(a)]. This increase of smoothness can be seen in the solutions of Exs. 2.3, 2.4, and 2.7.

(a) Let $x_1, \ldots, x_K$ be the points, if any, where $f^{(m)}$, $f^{(m+1)}$, $g^{(n)}$, $g^{(n+1)}$ have jump discontinuities, let

$$a_k := f^{(m)}(x_k+) - f^{(m)}(x_k-), \qquad b_k := f^{(m+1)}(x_k+) - f^{(m+1)}(x_k-),$$
$$c_k := g^{(n)}(x_k+) - g^{(n)}(x_k-), \qquad d_k := g^{(n+1)}(x_k+) - g^{(n+1)}(x_k-),$$

let $h_1 := h$, $h_2 := h * h$, $\ldots$, and let the functions f_0, g_0 be defined so that

$$f(x) = f_0(x) + \sum_{k=1}^{K} \{a_k h_{m+1}(x - x_k) + b_k h_{m+2}(x - x_k)\},$$

$$g(x) = g_0(x) + \sum_{k=1}^{K} \{c_k h_{n+1}(x - x_k) + d_k h_{n+2}(x - x_k)\}.$$

Show that f_0, g_0 have finite pasts and that $f_0 \in C^{m+1}$, $g_0 \in C^{n+1}$ (i.e., f_0 has $m+1$ continuous derivatives and g_0 has $n+1$ continuous derivatives).

(b) Show that $f_0 * g_0 \in C^{m+n+2}$.

(c) Show that $f_0 * h_p \in C^{m+p+1}$ and $g_0 * h_p \in C^{n+p+1}$, $p = 1, 2, \ldots$.

(d) Show that $h_p * h_q \in S^{p+q-1}$, $p = 1, 2, \ldots$.

(e) Finally, using (a)–(d) show that $f * g \in S^{m+n+1}$.

Note. An analogous argument can be used when f, g have finite futures.

Chapter 3

The Calculus for Finding Fourier Transforms of Functions on $\mathbb{R}$

3.1 Using the Definition to Find Fourier Transforms

Introduction

If you want to use Fourier analysis, you must develop basic skills for finding Fourier transforms of functions on $\mathbb{R}$. In principle, you can always use the defining integral from the analysis equation to obtain F when f is given. In practice, you will quickly discover that it is not so easy to find a transform such as

$$\int_{-\infty}^{\infty} \frac{\sin(\pi x)}{\pi x} e^{-2\pi i s x} dx = 2 \int_{0}^{\infty} \frac{\sin(\pi x)}{\pi x} \cos(2\pi s x) dx$$

by using the techniques of elementary integral calculus, *cf.* Ex. 1.1.

In this chapter we will present a *calculus* (*i.e.*, a computational process) for finding Fourier transforms of commonly used functions on $\mathbb{R}$. You will memorize a few Fourier transform pairs f, F and learn certain rules for modifying or combining known pairs to obtain new ones. It is analogous to memorizing that $(x^n)' = n x^{n-1}$, $(\sin x)' = \cos x$, $(e^x)' = e^x$, ... and then using the addition rule, product rule, quotient rule, chain rule, ... to find derivatives. You will need to spend a bit of time mastering the details, so don't despair when you see the multiplicity of drill exercises!

Once you learn to find Fourier transforms, you can immediately use Fourier's analysis and synthesis equations, Parseval's identity, and the Poisson relations to evaluate integrals and sums that cannot be found by more elementary methods. You will also need these skills when you study various applications of Fourier analysis in the second part of the course.

The box function

The *box function*

$$\Pi(x) := \begin{cases} 1 & \text{for } -\tfrac{1}{2} < x < \tfrac{1}{2} \\ 0 & \text{for } x < -\tfrac{1}{2} \text{ or } x > \tfrac{1}{2} \end{cases}$$

and the cardinal sine or *sinc function*

$$\operatorname{sinc}(s) := \frac{\sin(\pi s)}{\pi s}, \qquad s < 0 \text{ or } s > 0$$

are two of the most commonly used functions in Fourier analysis. We often simplify the definition of such functions by omitting the values at the singular points. When pressed, we use midpoint regularization and write

$$\Pi(x_0) = \frac{1}{2} \left\{ \lim_{x \to x_0+} \Pi(x) + \lim_{x \to x_0-} \Pi(x) \right\} = \frac{1}{2} \qquad \text{when } x_0 = \pm\tfrac{1}{2},$$

$$\operatorname{sinc}(0) = \frac{1}{2} \left\{ \lim_{s \to 0+} \operatorname{sinc}(s) + \lim_{s \to 0-} \operatorname{sinc}(s) \right\} = 1$$

to fill in the holes.

Since $f := \Pi$ is even, we can simplify the Fourier transform integral by writing

$$F(s) = \int_{-\infty}^{\infty} \Pi(x) e^{-2\pi i s x} dx = \int_{-\infty}^{\infty} \Pi(x) \cos(2\pi s x) dx.$$

We then use calculus to find

$$F(s) = \int_{-1/2}^{1/2} \cos(2\pi s x) dx = \left. \frac{\sin(2\pi s x)}{2\pi s} \right|_{x=-1/2}^{1/2} = \operatorname{sinc}(s),$$

cf. Fig. 3.1.

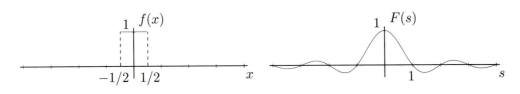

Figure 3.1. The box $f(x) = \Pi(x)$ and its Fourier transform $F(s) = \operatorname{sinc}(s)$.

The Heaviside step function

We define the *Heaviside step function*

$$h(x) := \begin{cases} 1 & \text{if } x > 0 \\ 0 & \text{if } x < 0, \end{cases}$$

cf. Ex. 2.3, and use h to create functions that vanish on a half line. For example,

$$\sin(2\pi x)e^{-x}h(x) = \begin{cases} \sin(2\pi x)e^{-x} & \text{if } x > 0 \\ 0 & \text{if } x < 0, \end{cases}$$

$$h(1 - x) = \begin{cases} 1 & \text{if } x < 1 \\ 0 & \text{if } x > 1, \end{cases}$$

$$h\left(\tfrac{1}{2} + x\right) h\left(\tfrac{1}{2} - x\right) = \Pi(x),$$

$$h(x) - h(-x) = \text{sgn}(x) := \begin{cases} 1 & \text{if } x > 0 \\ -1 & \text{if } x < 0, \end{cases}$$

$$\sum_{n=1}^{\infty} h(x - n) = \begin{cases} 0 & \text{if } x < 1 \\ 1 & \text{if } 1 < x < 2 \\ 2 & \text{if } 2 < x < 3 \\ \vdots \\ . \end{cases}$$

The truncated decaying exponential

We can find the Fourier transform of the truncated decaying exponential

$$f(x) := e^{-x}h(x)$$

by writing

$$F(s) = \int_{x=0}^{\infty} e^{-x}e^{-2\pi i s x}\,dx$$

$$= \lim_{L \to +\infty} \int_{0}^{L} \frac{d}{dx}\left\{\frac{-e^{-(1+2\pi i s)x}}{1 + 2\pi i s}\right\}dx$$

$$\overset{?}{=} \lim_{L \to +\infty} \frac{1 - e^{-(1+2\pi i s)L}}{1 + 2\pi i s}$$

$$= \frac{1}{1 + 2\pi i s}$$

$$= \frac{1}{1 + 4\pi^2 s^2} - \frac{2\pi i s}{1 + 4\pi^2 s^2}.$$

(We justify the step $\overset{?}{=}$ by applying the fundamental theorem of calculus to the real and imaginary parts of the integrand.) In this case we must display both $\operatorname{Re} F$ and $\operatorname{Im} F$ as shown in Fig. 3.2.

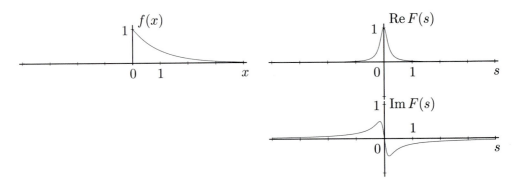

Figure 3.2. The truncated decaying exponential $f(x) := e^{-x}h(x)$ and its Fourier transform $F(s) = 1/(1 + 2\pi is)$.

The unit gaussian

We define the *unit gaussian*

$$f(x) := e^{-\pi x^2}$$

and use the fact that f is even to write

$$F(s) = \int_{-\infty}^{\infty} e^{-\pi x^2} e^{-2\pi isx} dx = \int_{-\infty}^{\infty} e^{-\pi x^2} \cos(2\pi sx) dx.$$

We will use an indirect argument to evaluate this integral.

Since the integrand and its derivative with respect to s rapidly approach 0 as $x \to \pm\infty$, we can write

$$F'(s) = \int_{-\infty}^{\infty} e^{-\pi x^2} \frac{\partial}{\partial s} \cos(2\pi sx) dx,$$

and in this way we see that

$$F'(s) + 2\pi s F(s) = \int_{-\infty}^{\infty} e^{-\pi x^2} \{(-2\pi x)\sin(2\pi sx) + (2\pi s)\cos(2\pi sx)\} dx$$

$$= \int_{-\infty}^{\infty} \frac{d}{dx} \{e^{-\pi x^2} \sin(2\pi sx)\} dx = 0.$$

It follows that

$$\frac{d}{ds}\{e^{\pi s^2}F(s)\} = e^{\pi s^2}\{F'(s) + 2\pi s F(s)\} = 0,$$

so that

$$e^{\pi s^2}F(s) = F(0), \qquad -\infty < s < \infty.$$

We use a familiar trick from multivariate calculus to find the positive constant

$$F(0) = \int_{-\infty}^{\infty} e^{-\pi x^2}\,dx.$$

We write

$$F(0)^2 = \left\{\int_{x=-\infty}^{\infty} e^{-\pi x^2}\,dx\right\}\left\{\int_{y=-\infty}^{\infty} e^{-\pi y^2}\,dy\right\}$$

$$= \int_{x=-\infty}^{\infty}\int_{y=-\infty}^{\infty} e^{-\pi(x^2+y^2)}\,dy\,dx$$

$$= \int_{r=0}^{\infty}\int_{\theta=0}^{2\pi} e^{-\pi r^2}\,r\,d\theta\,dr$$

$$= \int_{r=0}^{\infty} e^{-\pi r^2}\,2\pi r\,dr = 1,$$

and thereby see that $F(0) = 1$. In this way we obtain the Fourier transform

$$F(s) = e^{-\pi s^2}$$

that turns out to be the very same function as f, cf. Fig. 3.3.

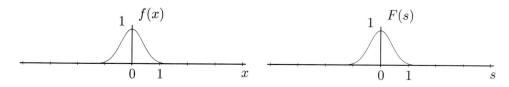

Figure 3.3. The unit gaussian $f(x) := e^{-\pi x^2}$ and its Fourier transform $F(s) := e^{-\pi s^2}$.

Summary

You will eventually memorize a number of the Fourier transform pairs f, F from Appendix 1. For now, make sure that you know that:

$$f(x) := \Pi(x) \qquad \text{has the FT} \quad F(s) = \text{sinc}(s)\,; \tag{1}$$

$$f(x) := e^{-x}h(x) \qquad \text{has the FT} \quad F(s) = 1/(1+2\pi i s)\,; \tag{2}$$

$$f(x) := e^{-\pi x^2} \qquad \text{has the FT} \quad F(s) = e^{-\pi s^2}\,. \tag{3}$$

3.2 Rules for Finding Fourier Transforms

Introduction

Throughout this section $f, f_1, f_2, \ldots$ and g will be suitably regular functions on $\mathbb{R}$ with the corresponding Fourier transforms $F, F_1, F_2, \ldots$ and G. If we obtain g from some modification of f, then there will be a corresponding modification of F that produces G, $cf.$ Fig. 3.4. Likewise, if we obtain g from some combination of f_1 and f_2, then there will be a corresponding combination of F_1 and F_2 that produces G. Such observations form the *rules* of our calculus for finding Fourier transforms. We will now state a number of such rules, give simple informal derivations, and illustrate how they are used.

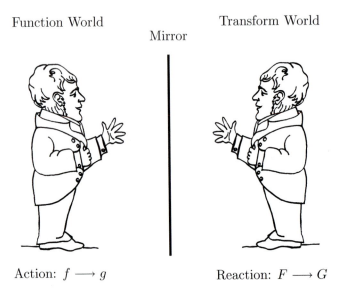

Function World   Mirror   Transform World

Action: $f \longrightarrow g$ Reaction: $F \longrightarrow G$

Figure 3.4. Action (raise left hand) and reaction (raise right hand) with mirror images is analogous to the mappings $f \to g$ and $F \to G$ of a Fourier transform rule.

Linearity

Let $c, c_1, c_2, \ldots$ be complex scalars. We verify the *scaling rule*

$$g(x) := c\,f(x) \quad \text{has the FT} \quad G(s) = c\,F(s) \tag{4}$$

by writing

$$G(s) := \int_{-\infty}^{\infty} c\,f(x)e^{-2\pi isx}\,dx = c \int_{-\infty}^{\infty} f(x)e^{-2\pi isx}\,dx = c\,F(s).$$

We verify the *addition rule*

$$g(x) := f_1(x) + f_2(x) \quad \text{has the FT} \quad G(s) = F_1(s) + F_2(s) \tag{5}$$

by writing

$$G(s) := \int_{-\infty}^{\infty} \{f_1(x) + f_2(x)\}e^{-2\pi isx}\,dx$$
$$= \int_{-\infty}^{\infty} f_1(x)e^{-2\pi isx}\,dx + \int_{-\infty}^{\infty} f_2(x)e^{-2\pi isx}\,dx = F_1(s) + F_2(s).$$

When taken together, (4) and (5) give the *linearity rule*

$$g(x) := c_1 f_1(x) + \cdots + c_m f_m(x) \quad \text{has the FT} \quad G(s) = c_1 F_1(s) + \cdots + c_m F_m(s). \tag{6}$$

You are familiar with this property from your work with derivatives and integrals.

Reflection and conjugation

We verify the *reflection rule*

$$g(x) := f(-x) \quad \text{has the FT} \quad G(s) = F(-s) \tag{7}$$

by writing

$$G(s) := \int_{-\infty}^{\infty} f(-x)e^{-2\pi isx}\,dx = \int_{-\infty}^{\infty} f(u)e^{-2\pi(-s)u}\,du = F(-s).$$

Example: Show that the *Laplace function*

$$g(x) := e^{-|x|} \quad \text{has the FT} \quad G(s) = 2/(1 + 4\pi^2 s^2). \tag{8}$$

Solution: We know that

$$f(x) := e^{-x}h(x) \quad \text{has the FT} \quad F(s) = 1/(1 + 2\pi is).$$

Since $g(x) = f(x) + f(-x)$, cf. Figs. 3.2 and 3.5, we can use the addition and reflection rules to see that

$$G(s) = \frac{1}{1 + 2\pi i s} + \frac{1}{1 - 2\pi i s} = \frac{2}{1 + 4\pi^2 s^2}.\qquad \blacksquare$$

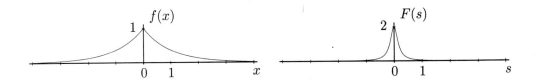

Figure 3.5. The Laplace function $f(x) = e^{-|x|}$ and its Fourier transform $F(s) = 2/(1 + 4\pi^2 s^2)$.

We verify the *conjugation rule*

$$g(x) := \overline{f(x)} \quad \text{has the FT} \quad G(s) = \overline{F(-s)} \qquad (9)$$

by writing

$$G(s) := \int_{-\infty}^{\infty} \overline{f(x)} e^{-2\pi i s x}\, dx = \overline{\int_{-\infty}^{\infty} f(x) e^{-2\pi i (-s) x}\, dx} = \overline{F(-s)}.$$

Example: Derive the *hermitian conjugation rule*

$$g(x) := \overline{f(-x)} \quad \text{has the FT} \quad G(s) = \overline{F(s)}. \qquad (10)$$

Solution: We use the reflection and conjugation rules in turn to see that

$$f(-x) \quad \text{has the FT} \quad F(-s),$$
$$\overline{f(-x)} \quad \text{has the FT} \quad \overline{F(--s)} = \overline{F(s)}. \qquad \blacksquare$$

You may find it instructive to use the rules (7), (9), and (10) to formulate answers for Ex. 1.2!

Translation and modulation

Let x_0 be a real parameter. We verify the *translation rule* (or *shift rule*)

$$g(x) := f(x - x_0) \quad \text{has the FT} \quad G(s) = e^{-2\pi i s x_0} \cdot F(s) \qquad (11)$$

by writing

$$G(s) := \int_{-\infty}^{\infty} f(x - x_0)e^{-2\pi i s x}\, dx = e^{-2\pi i s x_0} \int_{-\infty}^{\infty} f(u)e^{-2\pi i s u}\, du = e^{-2\pi i s x_0} F(s).$$

You will notice that the same algebraic sign is used with the two appearances of x_0 in (11). You may find it helpful to use the *same sign shift* mnemonic to remind yourself of this fact, *e.g.*,

$$\Pi(x+1) \quad \text{has the FT} \quad e^{+2\pi i s} \text{sinc } s,$$
$$\Pi(x-1) \quad \text{has the FT} \quad e^{-2\pi i s} \text{sinc } s.$$

Example: Find the Fourier transform of

$$g(x) := \begin{cases} 1 & \text{if } 0 < x < 1 \\ 2 & \text{if } 1 < x < 2 \\ 0 & \text{if } x < 0 \text{ or } x > 2. \end{cases}$$

Solution: We write

$$g(x) = \Pi\left(x - \frac{1}{2}\right) + 2\,\Pi\left(x - \frac{3}{2}\right)$$

and then use (6), (11), and (1) to obtain

$$G(s) = \{e^{-\pi i s} + 2e^{-3\pi i s}\}\text{sinc } s. \qquad \blacksquare$$

Let s_0 be a real parameter. We verify the *modulation rule* (or *transform shift rule*)

$$g(x) := e^{2\pi i s_0 x} \cdot f(x) \quad \text{has the FT} \quad G(s) = F(s - s_0) \qquad (12)$$

by writing

$$G(s) = \int_{-\infty}^{\infty} e^{2\pi i s_0 x} f(x)e^{-2\pi i s x}\, dx = \int_{-\infty}^{\infty} f(x)e^{-2\pi i (s - s_0)x}\, dx = F(s - s_0).$$

In this case *opposite* algebraic signs are associated with the two appearances of s_0.

Example: Find the Fourier transform of

$$g(x) := \begin{cases} \cos \pi x & \text{if } -\frac{1}{2} < x < \frac{1}{2} \\ 0 & \text{otherwise.} \end{cases}$$

Solution: We use Euler's formula for the cosine to write

$$g(x) = \frac{1}{2}\{e^{i\pi x} + e^{-i\pi x}\}\Pi(x)$$

$$= \frac{1}{2}e^{2\pi i(1/2)x}\Pi(x) + \frac{1}{2}e^{-2\pi i(1/2)x}\Pi(x),$$

and then use (6), (12), and (1) to obtain

$$G(s) = \frac{1}{2}\operatorname{sinc}\left(s - \frac{1}{2}\right) + \frac{1}{2}\operatorname{sinc}\left(s + \frac{1}{2}\right). \qquad \blacksquare$$

Dilation

Let $a \neq 0$ be a real parameter. We verify the *dilation rule* (or *similarity rule*)

$$g(x) := f(ax) \quad \text{has the FT} \quad G(s) = \frac{1}{|a|}F\left(\frac{s}{a}\right) \qquad (13)$$

by writing

$$G(s) := \int_{-\infty}^{\infty} f(ax)e^{-2\pi i s x}\,dx = \begin{cases} \dfrac{1}{a}\displaystyle\int_{-\infty}^{\infty} f(u)e^{-2\pi i(s/a)u}\,du & \text{if } a > 0 \\[2mm] \dfrac{1}{a}\displaystyle\int_{+\infty}^{-\infty} f(u)e^{-2\pi i(s/a)u}\,du & \text{if } a < 0 \end{cases}$$

$$= \frac{1}{|a|}F\left(\frac{s}{a}\right).$$

You will notice that the dilation factors a, $1/a$ that we use with f, F are reciprocals. If we compress one of these functions, then we stretch the other by the same amount.

Example: Find the Fourier transform of

$$g(x) := \begin{cases} 1 & \text{if } -1 < x < 1 \\ 0 & \text{if } x < -1 \text{ or } x > 1. \end{cases}$$

Solution: We observe that $g(x) = \Pi(x/2)$ and use the dilation rule to write

$$G(s) = 2\operatorname{sinc}(2s).$$

We obtain the same result (in a slightly different form) if we observe that $g(x) = \Pi(x + 1/2) + \Pi(x - 1/2)$ and use the translation rule to write

$$G(s) = (e^{\pi i s} + e^{-\pi i s})\operatorname{sinc} s = 2\cos \pi s \,\frac{\sin \pi s}{\pi s} = 2\,\frac{\sin(2\pi s)}{2\pi s}.$$

(The removable singularity at $x = 0$ has no effect on the integral that gives G.) $\blacksquare$

Example: Show that the *normal density*

$$g(x) := \frac{1}{\sqrt{2\pi}\sigma} e^{-(x-\mu)^2/2\sigma^2} \quad \text{has the FT} \quad G(s) = e^{-2\pi i s \mu} e^{-2\pi^2 \sigma^2 s^2}. \quad (14)$$

Here σ, μ are real parameters with $\sigma > 0$.

Solution: We use (3), (13), and (11) in turn to see that

$$e^{-\pi x^2} \qquad \text{has the FT} \qquad e^{-\pi s^2},$$

$$\frac{1}{\sqrt{2\pi}\sigma} e^{-x^2/2\sigma^2} = \frac{1}{\sqrt{2\pi}\sigma} e^{-\pi(x/\sqrt{2\pi}\sigma)^2} \qquad \text{has the FT} \quad e^{-\pi(\sqrt{2\pi}\sigma s)^2} = e^{-2\pi^2\sigma^2 s^2},$$

$$\frac{1}{\sqrt{2\pi}\sigma} e^{-(x-\mu)^2/2\sigma^2} \qquad \text{has the FT} \qquad e^{-2\pi i s\mu} e^{-2\pi^2\sigma^2 s^2}.$$

These functions and their transforms are shown in Fig. 3.6. ∎

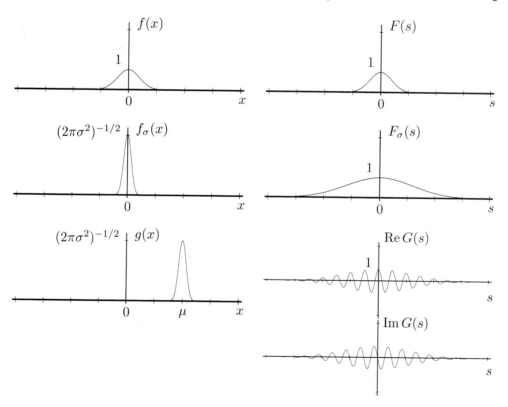

Figure 3.6. The unit gaussian f, the dilate f_σ, the translated dilate g, and the corresponding Fourier transforms F, F_σ, G.

Example: Find the Fourier transform of $g(x) := \cos(2\pi\beta x)e^{-\pi(x/\alpha)^2}$. Here α, β are parameters with $\alpha > 0$, $\beta > 0$.

Solution: Since

$$g(x) = \frac{1}{2}e^{2\pi i\beta x}e^{-\pi(x/\alpha)^2} + \frac{1}{2}e^{-2\pi i\beta x}e^{-\pi(x/\alpha)^2}$$

we can use the dilation and modulation rules to write

$$G(s) = \frac{\alpha}{2}e^{-\pi\alpha^2(s-\beta)^2} + \frac{\alpha}{2}e^{-\pi\alpha^2(s+\beta)^2},$$

cf. Fig. 3.7. ∎

Figure 3.7. The function $g(x) := \cos(2\pi\beta x)e^{-\pi x^2/\alpha^2}$ and its Fourier transform $G(s)$.

You must exercise some care when you use two of the rules in succession. For example, the functions

$$g_1(x) := f(a(x+b)), \qquad g_2(x) := f(ax+b) = f\left(a\left(x + \frac{b}{a}\right)\right)$$

are obtained when we a-dilate and b-translate as follows:

$$\begin{array}{ccc}
f(x) & \xrightarrow{a\text{-dilate}} & f(ax) & \xrightarrow{b\text{-translate}} & f(a(x+b)) \\
f(x) & \xrightarrow{b\text{-translate}} & f(x+b) & \xrightarrow{a\text{-dilate}} & f(ax+b).
\end{array}$$

Make sure that you understand why the correct Fourier transforms are given by

$$G_1(s) = \frac{e^{2\pi ibs}}{|a|}F\left(\frac{s}{a}\right), \qquad G_2(s) = \frac{e^{2\pi i(b/a)s}}{|a|}F\left(\frac{s}{a}\right),$$

cf. Ex. 3.1.

Inversion

In cases where Fourier's synthesis equation is valid we can formally verify the exceptionally powerful *inversion rule*

$$g(x) := F(x) \quad \text{has the FT} \quad G(s) = f(-s) \tag{15}$$

by simply writing

$$G(s) = \int_{-\infty}^{\infty} F(x)e^{-2\pi isx}dx = \int_{-\infty}^{\infty} F(x)e^{2\pi ix(-s)}dx = f(-s).$$

Of course, it takes the detailed analysis from Section 1.5 to show that this argument is valid when either f or F is a piecewise smooth function on $\mathbb{R}$ with small regular tails.

By applying the inversion rule to (1), (2), and (8) we find

$$g(x) = \operatorname{sinc} x \qquad\qquad \text{has the FT} \quad G(s) = \Pi(s), \tag{16}$$

$$g(x) = 1/(1 + 2\pi ix) \quad \text{has the FT} \quad G(s) = e^s h(-s), \tag{17}$$

$$g(x) = 2/(1 + 4\pi^2 x^2) \quad \text{has the FT} \quad G(s) = e^{-|s|}. \tag{18}$$

You will recall from your work on Ex. 1.1 that such Fourier transforms are not easily derived from the definition!

Derivative and power scaling

In cases where f is a suitably regular function on $\mathbb{R}$ we can verify the *derivative rule*

$$g(x) := f'(x) \quad \text{has the FT} \quad G(s) = 2\pi is \cdot F(s) \tag{19}$$

by using an integration by parts:

$$G(s) = \int_{-\infty}^{\infty} f'(x)e^{-2\pi isx}dx$$

$$\stackrel{?}{=} e^{-2\pi isx}f(x)\Big|_{x=-\infty}^{\infty} + 2\pi is\int_{-\infty}^{\infty} f(x)e^{-2\pi isx}dx$$

$$\stackrel{?}{=} 2\pi isF(s).$$

The argument is rigorous when f is a continuous function with a piecewise smooth derivative and both $f(x)$, $f'(x)$ [or both $F(s)$, $s \cdot F(s)$] satisfy the sufficient conditions for Fourier's representation as given in Section 1.5, *cf.* Ex. 3.41.

These hypotheses can also be used when we verify the *power scaling rule*

$$g(x) := x \cdot f(x) \quad \text{has the FT} \quad G(s) = (-2\pi i)^{-1} F'(s) \tag{20}$$

by writing

$$G(s) = \int_{-\infty}^{\infty} x \, f(x) e^{-2\pi i s x} \, dx$$

$$\overset{?}{=} \frac{1}{-2\pi i} \frac{d}{ds} \int_{-\infty}^{\infty} f(x) e^{-2\pi i s x} \, dx$$

$$= (-2\pi i)^{-1} F'(s).$$

Example: Show that

$$g(x) := 2x e^{-\pi x^2} \quad \text{has the FT} \quad G(s) = (-i) 2s e^{-\pi s^2}. \tag{21}$$

Solution: We can express g in terms of the unit gaussian by writing

$$g(x) = -\frac{1}{\pi} \frac{d}{dx} \{ e^{-\pi x^2} \} \quad \text{or} \quad g(x) = 2x \{ e^{-\pi x^2} \},$$

so we can use (3) with either the derivative rule or the power scaling rule to obtain

$$G(s) = -\frac{1}{\pi} 2\pi i s \, e^{-\pi s^2} \quad \text{or} \quad G(s) = 2(-2\pi i)^{-1} \frac{d}{ds} \{ e^{-\pi s^2} \}. \qquad \blacksquare$$

Example: Let $-\infty < \alpha < \infty$, $\beta > 0$, and $n = 0, 1, \ldots$. Show that

$$g_+(x) := \frac{1}{(x + \alpha + i\beta)^{n+1}} \quad \text{has the FT}$$

$$G_+(s) = \frac{-2\pi i}{n!} (-2\pi i s)^n h(s) e^{2\pi i \alpha s} \, e^{-2\pi \beta s},$$

$$g_-(x) := \frac{1}{(x + \alpha - i\beta)^{n+1}} \quad \text{has the FT} \tag{22}$$

$$G_-(s) = \frac{+2\pi i}{n!} (-2\pi i s)^n h(-s) e^{2\pi i \alpha s} \, e^{+2\pi \beta s}.$$

Solution: We use (17), (13), (4), (19), (4), (11), and (9) in turn to see that

$$\frac{1}{1 - 2\pi i x} \qquad \text{has the FT} \quad h(s)e^{-s},$$

$$\frac{1}{1 - 2\pi i (x/2\pi\beta)} \qquad \text{has the FT} \quad (2\pi\beta)h(s)e^{-2\pi\beta s},$$

$$\frac{1}{x + i\beta} \qquad \text{has the FT} \quad (-2\pi i)h(s)e^{-2\pi\beta s},$$

$$\frac{(-1)^n n!}{(x + i\beta)^{n+1}} \qquad \text{has the FT} \quad (-2\pi i)(2\pi i s)^n h(s)e^{-2\pi\beta s},$$

$$\frac{1}{(x + i\beta)^{n+1}} \qquad \text{has the FT} \quad (-2\pi i)\frac{(-2\pi i s)^n}{n!}h(s)e^{-2\pi\beta s},$$

$$\frac{1}{(x + \alpha + i\beta)^{n+1}} \qquad \text{has the FT} \quad (-2\pi i)\frac{(-2\pi i s)^n}{n!}h(s)e^{2\pi i \alpha s}e^{-2\pi\beta s},$$

$$\frac{1}{(x + \alpha - i\beta)^{n+1}} \qquad \text{has the FT} \quad (+2\pi i)\frac{(-2\pi i s)^n}{n!}h(-s)e^{2\pi i \alpha s}e^{2\pi\beta s}. \qquad \blacksquare$$

You can now use (22) in conjunction with a partial fraction decomposition to find the Fourier transform for any rational function $f(x) := p(x)/q(x)$, where p, q are polynomials with deg $q > $ deg p and q has no zeros on $\mathbb{R}$, cf. Ex. 3.12. It is surprising how much you can do with these simple rules!

Example: Find the Fourier transform of $f(x) := 1/(1 + x^4)$.

Solution: The function f has the partial fraction decomposition

$$f(x) = \frac{1}{4}\left\{\frac{\alpha + i\alpha}{x + \alpha + i\alpha} + \frac{-\alpha + i\alpha}{x - \alpha + i\alpha} + \frac{\alpha - i\alpha}{x + \alpha - i\alpha} + \frac{-\alpha - i\alpha}{x - \alpha - i\alpha}\right\}$$

where $\alpha := \sqrt{2}/2$, cf. Ex. 3.13. Using (22) we take Fourier transforms term by term to obtain

$$F(s) = \tfrac{1}{4}(\alpha + i\alpha)e^{2\pi i \alpha s}(-2\pi i)e^{-2\pi\alpha s}h(s) + \tfrac{1}{4}(-\alpha + i\alpha)e^{-2\pi i \alpha s}(-2\pi i)e^{-2\pi\alpha s}h(s)$$

$$+ \tfrac{1}{4}(\alpha - i\alpha)e^{2\pi i \alpha s}(2\pi i)e^{2\pi\alpha s}h(-s) + \tfrac{1}{4}(-\alpha - i\alpha)e^{-2\pi i \alpha s}(2\pi i)e^{2\pi\alpha s}h(-s). \qquad \blacksquare$$

Convolution and multiplication

In cases where f_1, f_2 are suitably regular functions on $\mathbb{R}$ we can formally verify the *convolution rule*

$$g(x) := (f_1 * f_2)(x) \quad \text{has the FT} \quad G(s) = F_1(s) \cdot F_2(s) \tag{23}$$

by writing

$$G(s) := \int_{x=-\infty}^{\infty} \int_{u=-\infty}^{\infty} f_1(u) f_2(x-u) e^{-2\pi i s x} du \, dx$$

$$\overset{?}{=} \int_{u=-\infty}^{\infty} \int_{x=-\infty}^{\infty} f_1(u) e^{-2\pi i s u} f_2(x-u) e^{-2\pi i s (x-u)} dx \, du$$

$$= \int_{u=-\infty}^{\infty} f_1(u) e^{-2\pi i s u} F_2(s) du$$

$$= F_1(s) F_2(s).$$

We must impose hypotheses to ensure that f_1, f_2, $f_1 * f_2$ have valid Fourier representations and to facilitate the above change in the limits of integration, *cf.* Ex. 3.44. Similar considerations apply when we formally verify the *multiplication rule*

$$g(x) := f_1(x) \cdot f_2(x) \quad \text{has the FT} \quad G(s) = (F_1 * F_2)(s) \tag{24}$$

by writing

$$G(s) := \int_{x=-\infty}^{\infty} f_1(x) f_2(x) e^{-2\pi i s x} dx$$

$$= \int_{x=-\infty}^{\infty} \int_{\sigma=-\infty}^{\infty} F_1(\sigma) e^{2\pi i \sigma x} f_2(x) e^{-2\pi i s x} d\sigma \, dx$$

$$\overset{?}{=} \int_{\sigma=-\infty}^{\infty} F_1(\sigma) \int_{x=-\infty}^{\infty} f_2(x) e^{-2\pi i (s-\sigma) x} dx \, d\sigma$$

$$= \int_{\sigma=-\infty}^{\infty} F_1(\sigma) F_2(s-\sigma) d\sigma.$$

Example: Show that

$$g(x) := \Lambda(x) \quad \text{has the FT} \quad G(s) = \text{sinc}^2 s, \tag{25}$$

$$g(x) = \text{sinc}^2 x \quad \text{has the FT} \quad G(s) = \Lambda(s) \tag{26}$$

where the *triangle function* is given by

$$\Lambda(x) := \begin{cases} 1 - |x| & \text{if } -1 < x < 1 \\ 0 & \text{otherwise.} \end{cases}$$

Solution: We recall that $\Lambda = \Pi * \Pi$ (as shown in Fig. 2.4), so we can use the convolution rule with (1) to obtain (25), *cf.* Fig. 3.8, and we can use the multiplication rule with (16) to obtain (26). ∎

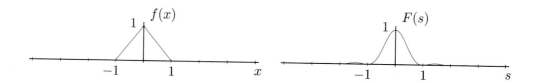

Figure 3.8. The triangle $f(x) = \Lambda(x) = (\Pi * \Pi)(x)$ and its Fourier transform $F(s) = \text{sinc}^2 s$.

Example: Find the Fourier transform of the piecewise linear function g shown in Fig. 3.9.

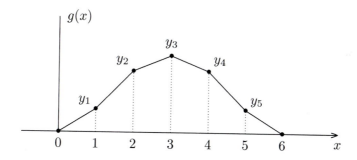

Figure 3.9. A continuous piecewise linear function g.

Solution: We can express g in the form

$$g(x) = y_1 \Lambda(x - 1) + y_2 \Lambda(x - 2) + y_3 \Lambda(x - 3) + y_4 \Lambda(x - 4) + y_5 \Lambda(x - 5).$$

(The left and right sides both vanish when $x \le 0$ or $x \ge 6$, both are linear on the intervals $[0, 1], [1, 2], \ldots, [5, 6]$, and both take the values $0, y_1, \ldots, y_5, 0$ when $n = 0, 1, \ldots, 5, 6$.) We use this with (6), (11), and (25) to write

$$G(s) = \{y_1 e^{-2\pi i s \cdot 1} + y_2 e^{-2\pi i s \cdot 2} + y_3 e^{-2\pi i s \cdot 3} + y_4 e^{-2\pi i s \cdot 4} + y_5 e^{-2\pi i s \cdot 5}\} \text{sinc}^2(s). \quad \blacksquare$$

Example: Find a continuous solution of the forced differential equation

$$y'(x) + y(x) = \Pi(x)$$

assuming that $y(x) = 0$ for all sufficiently large negative values of x, *cf.* Fig. 3.10.

Solution: Since $y'(x) = -y(x)$ when $x < -1/2$ or $x > 1/2$ we can conclude that both y and y' are piecewise smooth functions with small regular tails. We Fourier transform each term of the differential equation to obtain

$$(2\pi i s)Y(s) + Y(s) = \text{sinc } s$$

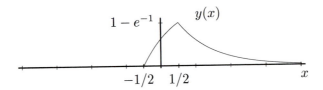

Figure 3.10. The solution of the differential equation $y'(x) + y(x) = \Pi(x)$ that vanishes when $x \leq -1/2$.

and thereby find

$$Y(s) = \text{sinc } s \cdot \frac{1}{1 + 2\pi i s}.$$

We then use (1), (2) and the convolution rule (23) to see that

$$y(x) = \int_{u=-\infty}^{\infty} \Pi(u) h(x - u) e^{-(x-u)} \, du$$

$$= \begin{cases} 0 & \text{for} \quad x < -1/2 \\ 1 - e^{-x-1/2} & \text{for} \ -1/2 \leq x \leq 1/2 \\ (e^{1/2} - e^{-1/2}) e^{-x} & \text{for } x > 1/2. \end{cases}$$ ∎

Summary

You have now seen the elements of the Fourier transform calculus for functions on $\mathbb{R}$. Exercises 3.2–3.12 provide the drill you will need to master this calculus, and Exs. 3.19–3.21 will help you visualize the meaning of these rules. You may refer to (1)–(26) (or to the tables in Appendices 1,2) as you are learning this material, but before you finish your study of this chapter you should memorize these basic identities.

As you find Fourier transforms you can freely use your knowledge of algebra, trigonometry, calculus, ... For example, you can easily verify that

$$f(x) := \frac{2 \sin^2(\pi x)}{\pi x}$$

has the equivalent representations

$$f(x) = 2\pi x \cdot \text{sinc}^2(x)$$
$$= i\{e^{-\pi i x} - e^{\pi i x}\} \text{sinc}(x)$$
$$= i\{2 \text{sinc}(2x) - 2e^{i\pi x} \text{sinc}(x)\}$$

and then use (26) and (20), (12) and (1), (12)–(13) and (1) to obtain the corresponding expressions

$$F(s) = i\left\{\frac{d}{ds}\Lambda(s)\right\}$$

$$= i\left\{\Pi\left(s+\frac{1}{2}\right) - \Pi\left(s-\frac{1}{2}\right)\right\}$$

$$= i\left\{\Pi\left(\frac{s}{2}\right) - 2\Pi\left(s-\frac{1}{2}\right)\right\}$$

for the Fourier transform. Of course, you can also verify that

$$f(x) = g(x) \cdot \mathrm{sinc}(x), \qquad g(x) := 2\sin(\pi x),$$

but it makes no sense to use (24) and write

$$F(s) = (G * \Pi)(s)$$

since you cannot find the Fourier transform G for g (at this point in the course). The rules apply only when the functions are suitably regular!

3.3 Selected Applications of the Fourier Transform Calculus

Evaluation of integrals and sums

The analysis and synthesis equations of Fourier, the Parseval and Plancherel identities, and the Poisson relations link suitably regular functions and their Fourier transforms. We will now present several examples to show how these links allow us to evaluate various integrals and sums that cannot be found with the usual techniques of elementary calculus.

Example: Find the value of the integral

$$I(\alpha) := \int_{-\infty}^{\infty} e^{-\alpha x^2}\,dx, \qquad \alpha > 0.$$

Solution: We use (3) with the dilation rule to see that

$$f(x) := e^{-\alpha x^2} = e^{-\pi(\sqrt{\alpha}\,x/\sqrt{\pi})^2} \quad \text{has the FT} \quad F(s) = \sqrt{\frac{\pi}{\alpha}}e^{-\pi^2 s^2/\alpha},$$

and then use the analysis equation to write

$$I(\alpha) = \int_{-\infty}^{\infty} f(x)e^{-2\pi i 0 x} dx = F(0) = \sqrt{\frac{\pi}{\alpha}}.$$

∎

Example: Find the value of the integral

$$I(\alpha) := \int_{-\infty}^{\infty} \frac{\sin(\pi x)\cos(2\pi\alpha x)}{\pi x} dx, \qquad -\infty < \alpha < \infty.$$

Solution: We know that the even function

$$f(x) = \mathrm{sinc}(\pi x) \quad \text{has the FT} \quad F(s) = \Pi(s),$$

so we can use Fourier's analysis equation to write

$$I(\alpha) = \int_{-\infty}^{\infty} f(x)e^{-2\pi i \alpha x} dx = F(\alpha) = \Pi(\alpha) = \begin{cases} 1 & \text{if } |\alpha| < \frac{1}{2} \\ \frac{1}{2} & \text{if } |\alpha| = \frac{1}{2} \\ 0 & \text{if } |\alpha| > \frac{1}{2}. \end{cases}$$

The midpoint regularization of Π is essential when $\alpha = \pm\frac{1}{2}$.

∎

Example: Find the value of the integral

$$I := \int_{-\infty}^{\infty} \frac{dx}{(1+x^2)^2}.$$

Solution: We use (18) and the dilation rule to see that

$$f(x) = \frac{1}{1+x^2} \quad \text{has the FT} \quad F(s) = \pi e^{-2\pi|s|}, \tag{27}$$

and then use Plancherel's identity (1.15) to write

$$I = \int_{-\infty}^{\infty} |f(x)|^2 dx = \int_{-\infty}^{\infty} |F(s)|^2 ds = 2\pi^2 \int_{0}^{\infty} e^{-4\pi s} ds = \frac{\pi}{2}.$$

∎

Example: Find the value of the integral

$$I := \int_{-\infty}^{\infty} \left(\frac{\sin x}{x}\right)^3 dx.$$

Solution: We use (16), (26), and the dilation rule to see that

$$f(x) := \frac{\sin x}{x} = \text{sinc}\left(\frac{x}{\pi}\right) \qquad \text{has the FT} \quad F(s) = \pi\Pi(\pi s),$$

$$g(x) := \left(\frac{\sin x}{x}\right)^2 = \text{sinc}^2\left(\frac{x}{\pi}\right) \quad \text{has the FT} \quad G(s) = \pi\Lambda(\pi s),$$

and then use Parseval's identity (1.11) to write

$$I = \int_{-\infty}^{\infty} f(x)\overline{g(x)}dx = \int_{-\infty}^{\infty} F(s)\overline{G(s)}ds = \pi^2 \int_{-1/2\pi}^{1/2\pi} \Lambda(\pi s)ds$$

$$= \pi^2 \frac{1}{2\pi}\left(1 + \frac{1}{2}\right) = \frac{3}{4}\pi.$$

[We use the formula for the area of a trapezoid to find the area under $\Lambda(\pi s)$ between $-1/2\pi$ and $+1/2\pi$.] ∎

You will observe from these examples that the Parseval identity or Plancherel identity allows us to exchange one integral for another that may (or may not!) be easier to evaluate.

Example: Find the value of the sum

$$S(p) := \sum_{m=-\infty}^{\infty} \frac{1}{1 + m^2 p^2}, \qquad p > 0.$$

Solution: We use (27) and the Poisson sum formula (1.45) to write

$$S(p) = \sum_{m=-\infty}^{\infty} f(mp) = \sum_{k=-\infty}^{\infty} \frac{1}{p} F\left(\frac{k}{p}\right) = \frac{\pi}{p} \sum_{k=-\infty}^{\infty} e^{-2\pi|k/p|}$$

$$= \frac{\pi}{p}\left\{2\sum_{k=0}^{\infty}(e^{-2\pi/p})^k - 1\right\} = \frac{\pi}{p}\coth\left(\frac{\pi}{p}\right).$$

The Poisson relation allows us to replace an intractable sum with a geometric progression that we can evaluate easily! ∎

Evaluation of convolution products

Now that you know how to find Fourier transforms, you can evaluate the convolution product $g := f_1 * f_2$ of suitably regular functions f_1, f_2 on $\mathbb{R}$ by

- finding the Fourier transforms F_1, F_2 of f_1, f_2,
- forming the product $G := F_1 \cdot F_2$, and
- finding the inverse Fourier transform g of G,

cf. (2.20). We will give three examples to illustrate this process.

Example: Evaluate the convolution integral

$$g(x) := \int_{u=-\infty}^{\infty} \frac{\sin \pi u}{\pi u} \frac{\sin \pi(x-u)}{\pi(x-u)} \, du.$$

Solution: The integral defines $g = f * f$ where

$$f(x) := \operatorname{sinc} x \quad \text{has the FT} \quad F(s) = \Pi(s).$$

Thus

$$G(s) = F(s) \cdot F(s) = \Pi(s) \cdot \Pi(s) = \Pi(s),$$

and by taking the inverse Fourier transform we find

$$g(x) = \operatorname{sinc} x.$$

(We can ignore the singular points at $s = \pm\frac{1}{2}$ when we write $\Pi \cdot \Pi = \Pi$ since both sides of this equation have the same midpoint regularization.) ∎

Example: Find the convolution product $g = f_1 * f_2$ of the normal densities

$$f_1(x) := \frac{1}{\sqrt{2\pi}\,\sigma_1} e^{-(x-\mu_1)^2/2\sigma_1^2}, \qquad f_2(x) := \frac{1}{\sqrt{2\pi}\,\sigma_2} e^{-(x-\mu_2)^2/2\sigma_2^2}.$$

Here $\sigma_1, \mu_1, \sigma_2, \mu_2$ are real parameters with $\sigma_1 > 0$, $\sigma_2 > 0$.

Solution: We use the transform pair (14) to find F_1, F_2 and write

$$\begin{aligned}
G(s) &= F_1(s) \cdot F_2(s) \\
&= e^{-2\pi i \mu_1 s} e^{-2\pi \sigma_1^2 s^2} \cdot e^{-2\pi i \mu_2 s} e^{-2\pi^2 \sigma_2^2 s^2} \\
&= e^{-2\pi i \mu s} e^{-2\pi^2 \sigma^2 s^2}
\end{aligned}$$

where $\mu := \mu_1 + \mu_2$, $\sigma^2 := \sigma_1^2 + \sigma_2^2$. We again use (14) to see that

$$g(x) = \frac{1}{\sqrt{2\pi}\,\sigma} e^{-(x-\mu)^2/2\sigma^2}.$$

You can find g by evaluating the convolution integral (*cf.* Ex. 2.6), but this indirect calculation is much easier! ∎

Example: Let $f_1 = f$, $f_2 = f * f$, $f_3 = f * f * f$, ... where f is the normal density

$$f(x) := \frac{e^{-(x-\mu)^2/2\sigma^2}}{\sqrt{2\pi}\,\sigma}$$

with $-\infty < \mu < \infty$ and $\sigma > 0$. Find a simple expression for $f_n(x)$.

Solution: We first use (14) to write

$$F_n(s) = F(s)^n = e^{-2\pi i s n\mu} e^{-2\pi^2 n\sigma^2 s^2},$$

and then use (14) to see that F_n is the Fourier transform of

$$f_n(x) = \frac{e^{-(x-n\mu)^2/2n\sigma^2}}{\sqrt{2\pi n}\,\sigma}.$$ ∎

The Hermite functions

The *Hermite polynomials* are defined by the Rodrigues formula

$$H_n(x) := (-1)^n e^{x^2} \{ \mathbf{D}^n e^{-x^2} \}, \qquad n = 0, 1, 2, \dots, \tag{28}$$

where $\mathbf{D} := d/dx$ is the derivative operator, *e.g.*,

$$H_0(x) = 1, \quad H_1(x) = -e^{x^2}\{e^{-x^2}\}' = 2x, \quad H_2(x) = e^{x^2}\{e^{-x^2}\}'' = 4x^2 - 2, \quad \dots .$$

These polynomials satisfy the two-term recursion

$$\begin{aligned}
H_n'(x) &= (-1)^n \mathbf{D}\{e^{x^2}\mathbf{D}^n e^{-x^2}\} \\
&= (-1)^n \{2xe^{x^2}\mathbf{D}^n e^{-x^2} + e^{x^2}\mathbf{D}^{n+1}e^{-x^2}\} \\
&= 2xH_n(x) - H_{n+1}(x)
\end{aligned} \tag{29}$$

and the three-term recursion

$$\begin{aligned}
H_{n+1}(x) :&= (-1)^{n+1}e^{x^2}\{\mathbf{D}^{n+1}e^{-x^2}\} \\
&= (-1)^n e^{x^2}\{\mathbf{D}^n(2xe^{-x^2})\} \\
&= (-1)^n e^{x^2}\{2x\mathbf{D}^n(e^{-x^2}) + n(2x)'\mathbf{D}^{n-1}(e^{-x^2})\} \\
&= 2xH_n(x) - 2nH_{n-1}(x).
\end{aligned} \tag{30}$$

[We use the Leibnitz differentiation rule (2.29) in the third step of (30).]
We will show that the *Hermite function*

$$f_n(x) := H_n(\sqrt{2\pi}\,x)e^{-\pi x^2} \quad \text{has the FT} \quad F_n(s) = (-i)^n H_n(\sqrt{2\pi}\,s)e^{-\pi s^2}, \tag{31}$$

i.e., that $f_0, f_1, f_2, f_3, \dots$ are eigenfunctions of the Fourier transform operator with the corresponding eigenvalues $1, -i, -1, i, \dots$, *cf.* (3), (21). Graphs of f_0, f_1, f_2, f_3

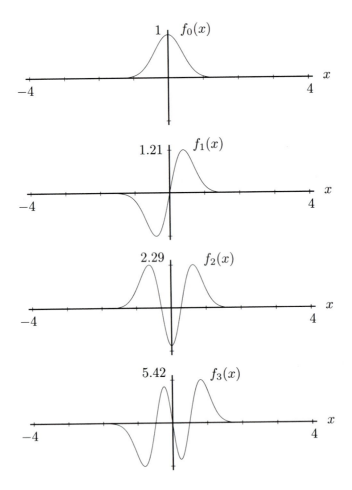

Figure 3.11. The Hermite functions $f_n(x) := H_n(\sqrt{2\pi x})e^{-\pi x^2}$, $n = 0, 1, 2, 3$.

are shown in Fig. 3.11. These functions appear in quantum mechanics, physical optics, statistics, ... We first use the two-term recursion (29) to write

$$
\begin{aligned}
f_{n+1}(x) :&= H_{n+1}(\sqrt{2\pi}\,x)e^{-\pi x^2} \\
&= \{2\sqrt{2\pi}\,x\,H_n(\sqrt{2\pi}\,x) - H_n'(\sqrt{2\pi}\,x)\}e^{-\pi x^2} \qquad (32) \\
&= \sqrt{2\pi}\,x f_n(x) - \frac{1}{\sqrt{2\pi}}f_n'(x),
\end{aligned}
$$

and then use the power scaling and derivative rules to obtain the corresponding relation

$$F_{n+1}(s) = \sqrt{2\pi}\,\frac{1}{-2\pi i}F'_n(s) - \frac{1}{\sqrt{2\pi}}2\pi i s\, F_n(s)$$

$$= (-i)\left\{\sqrt{2\pi}\,s\,F_n(s) - \frac{1}{\sqrt{2\pi}}F'_n(s)\right\}$$

for the Fourier transforms. Now if $F_n(s) = (-i)^n f_n(s)$ for some $n = 0, 1, 2, \ldots$, then we can use this relation with (32) to write

$$F_{n+1}(s) = (-i)\left\{\sqrt{2\pi}\,s(-i)^n f_n(s) - \frac{1}{\sqrt{2\pi}}(-i)^n f'_n(s)\right\} = (-i)^{n+1}f_{n+1}(s).$$

Since f_0 is the unit gaussian, $F_0(s) = (-i)^0 f_0(s)$, so the inductive proof of (31) is complete. Additional properties of the Hermite functions are developed in Ex. 3.37.

Smoothness and rates of decay

When $f(x), f'(x), \ldots, f^{(n)}(x)$ are suitably regular functions on $\mathbb{R}$ we can use the derivative rule to find the corresponding Fourier transforms $F(s), (2\pi i s)F(s)$, $\ldots, (2\pi i s)^n F(s)$ and thereby obtain the analysis equation

$$(2\pi i s)^n F(s) = \int_{x=-\infty}^{\infty} f^{(n)}(x)e^{-2\pi i s x}dx. \tag{33}$$

We majorize the integral by writing

$$|(2\pi i s)^n F(s)| \leq \int_{-\infty}^{\infty} |f^{(n)}(x)|dx,$$

and thereby see that $F(s)$ decays at least as fast as $1/s^n$ in the limit $s \to \pm\infty$.

Exercise 3.42 shows that the synthesis equation (33) holds when $f, f', \ldots, f^{(n-1)}$ are continuous, $f^{(n)}$ is piecewise continuous, and all of these functions are absolutely integrable. These hypotheses are a bit stronger than necessary for the $1/s^n$ decay rate, however, and we can verify that $s^n F(s) \to 0$ as $s \to \pm\infty$ by applying the Riemann-Lebesgue lemma (from Ex. 1.38) to the integral of (33).

In elementary applications of Fourier analysis we often work with polynomial splines that vanish outside some finite interval. In this case $f^{(n)}$ is piecewise constant for some $n = 0, 1, \ldots$, and we can use (33) to show that $F(s)$ decays to zero as fast as $1/s^{n+1}$ when $s \to \pm\infty$, cf. Ex. 3.43. For example, $f(x) := \Lambda(x)$ has a piecewise constant derivative and $F(s) = \text{sinc}^2(s)$ decays like $1/s^2$ as $s \to \pm\infty$.

When $f(x), (-2\pi i x)f(x), \ldots, (-2\pi i x)^n f(x)$ are suitably regular functions on $\mathbb{R}$ we can use the power scaling rule to find the corresponding Fourier transforms $F(s), F'(s), \ldots, F^{(n)}(s)$ and thereby obtain the analysis equation

$$F^{(n)}(s) := \frac{d^n}{ds^n} \int_{x=-\infty}^{\infty} f(x)e^{-2\pi i s x} dx = \int_{x=-\infty}^{\infty} (-2\pi i x)^n f(x)e^{-2\pi i s x} dx. \qquad (34)$$

In particular, if f is piecewise continuous and $f(x), xf(x), \ldots, x^n f(x)$ are all absolutely integrable (as in the case when f is a probability density with finite absolute moments of orders $0, 1, \ldots, n$), then $F, F', \ldots, F^{(n)}$ are continuous functions that vanish at $s = \pm\infty$, cf. Ex. 3.42.

We conclude this section with an observation of Fourier. Although we derived (33) with the understanding that $n = 0, 1, 2, \ldots$, the integral from the corresponding synthesis equation

$$f^{(n)}(x) = \int_{s=-\infty}^{\infty} (2\pi i s)^n F(s)e^{2\pi i s x} ds \qquad (35)$$

may be perfectly well defined for other values of n. For example, since

$$f(x) := e^{3\pi i s x} \text{ sinc } x \quad \text{has the FT} \quad F(s) = \Pi\left(s - \frac{3}{2}\right)$$

we can use (35) to synthesize the derivatives

$$f^{(n)}(x) = \int_{s=1}^{2} (2\pi i s)^n e^{2\pi i s x} ds, \quad n = 0, 1, \ldots,$$

the antiderivative

$$f^{(-1)}(x) = \int_{s=1}^{2} (2\pi i s)^{-1} e^{2\pi i s x} ds,$$

cf. Ex. 3.30, and the *fractional derivative*

$$f^{(1/2)}(x) = \int_{s=1}^{2} (2\pi i s)^{1/2} e^{2\pi i s x} ds.$$

References

R.N. Bracewell, *The Fourier Transform and Its Applications, 2nd ed.*, McGraw-Hill, New York, 1986.

Chapter 6 of this widely used text gives the rules for Fourier transforms of functions on $\mathbb{R}$ and Chapter 21 has a pictorial dictionary with 60 of the most frequently encountered Fourier transform pairs.

D.C. Champeney, *Fourier Transforms and Their Physical Applications*, Academic Press, New York, 1973.

Chapter 2 gives the rules for Fourier transforms of functions on $\mathbb{R}$ and a pictorial dictionary that has 53 transform pairs.

I.S. Gradshteyn and I.M. Ryzhiki, *Tables of Integrals, Series, and Products, 5th ed.* (edited by A. Jeffrey), Academic Press, New York, 1993. (A CD-ROM version is also available from the publisher.)

This highly evolved encyclopedia has a very large number of Fourier integrals that cannot be found by using the elementary Fourier transform calculus.

F. Oberhettinger, *Tables of Fourier Transforms and Fourier Transforms of Distributions*, Springer-Verlag, New York, 1990.

A few less common Fourier transforms can be found in this little reference.

A.V. Oppenheim, A.S. Willsky, and I.T. Young, *Signals and Systems*, Prentice Hall, Englewood Cliffs, NJ, 1983.

A practical introduction to the Fourier transform rules for functions on $\mathbb{R}$ is given in Chapter 4 of this electrical engineering text.

A. Pinkus and S. Zafrany, *Fourier Series and Integral Transforms*, Cambridge University Press, Cambridge, 1997.

Chapter 3 of this intermediate-level mathematics text has a nice exposition of the rules for Fourier transforms of functions on $\mathbb{R}$.

J.S. Walker, *Fourier Analysis*, Oxford University Press, New York, 1988.

Chapter 6 of this intermediate-level mathematics text develops the rules for Fourier transforms of functions on $\mathbb{R}$.

Exercise Set 3

▶ **EXERCISE 3.1:**   You will often have occasion to express the Fourier transform of $f(ax + b)$ in terms of the Fourier transform $F(s)$ of $f(x)$. (Here a, b are real parameters with $a \neq 0$.) This exercise will help you learn to do this correctly.

(a) Sketch the graphs of $\Pi(x)$, $\Pi(x - 3)$, and $\Pi(2x - 3) = \Pi(2(x - 3/2))$.

(b) Sketch the graphs of $\Pi(x)$, $\Pi(2x)$, and $\Pi(2(x - 3))$.

     Note. In (a), a right 3-translate is followed by a 2-dilate; in (b) a 2-dilate is followed by a right 3-translate. The *order* of these operations is important!

(c) Find the Fourier transforms of $b_1(x) := \Pi(2x - 3)$ and $b_2(x) := \Pi(2(x - 3))$ from (a) and (b).

(d) Set $b(x) := \Pi(2x)$ and check your answers to (c) by applying the translation rule to

$$b_1(x) = b\left(x - \frac{3}{2}\right), \quad b_2(x) = b(x - 3), \quad b_2(x) = b_1\left(x - \frac{3}{2}\right).$$

▶ **EXERCISE 3.2:**   Find the Fourier transform of each of the following functions.

(a) $f(x) = -\Pi(x + 1) + \Pi(x - 1)$

(b) $f(x) = \Pi(x + 1/2) + \Pi(x - 1/2)$

(c) $f(x) = \cos(4\pi x) \cdot \Pi(x)$

(d) $f(x) = \Pi(2x - 1)$

(e) $f(x) = x \cdot \Pi(x)$

(f) $f(x) = \text{sgn}(x) \cdot \Pi(x)$

(g) $f(x) = \Pi(2x) \cdot \Pi(3x)$

(h) $f(x) = \Pi(x + 1/4) \cdot \Pi(x - 1/4)$

(i) $f(x) = \Pi(x - 2) * \Pi(x + 3)$

(j) $f(x) = \Pi(x) * \Pi(x) * \Pi(x)$

(k) $f(x) = \Pi(x) * \Pi(2x) * \Pi(4x)$

(l) $f(x) = \Pi(x) \cdot e^{-x}$

Hint. Sketch the graph of f for (b), (f), (g), (h); use the modulation rule for (c); use the analysis equation for (l); .... .

▶ **EXERCISE 3.3:**   Find the Fourier transform of each of the following functions.

(a)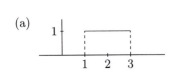

(b)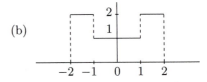

(c)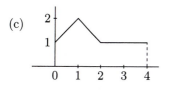

(d)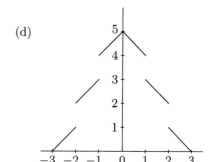

(e)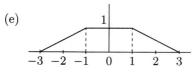

Hint. Synthesize these functions with scaled translates of dilates of $\Pi(x)$ and $\Lambda(x)$.

▶ **EXERCISE 3.4:**   Find the Fourier transform of each of the following functions.

(a) $f(x) = \cos(4\pi x) \cdot \mathrm{sinc}(x)$ (b) $f(x) = \sin(4\pi x) \cdot \mathrm{sinc}^2(x)$

(c) $f(x) = \mathrm{sinc}^3(x)$ (d) $f(x) = \mathrm{sinc}(2x) * \mathrm{sinc}(3x)$

(e) $f(x) = x \cdot \mathrm{sinc}^2(x)$ (f) $f(x) = [e^{-2\pi i x} \cdot \mathrm{sinc}(x)] * [e^{2\pi i x} \cdot \mathrm{sinc}(x)]$

▶ **EXERCISE 3.5:**   Find the Fourier transform of each of the following functions.

(a) $f(x) = e^{-x^2}$ (b) $f(x) = \cos(8\pi x) \cdot e^{-x^2}$

(c) $f(x) = e^{x-x^2}$ (d) $f(x) = 2x \cdot e^{-x^2}$

(e) $f(x) = (4x^2 - 2) \cdot e^{-x^2}$ (f) $f(x) = e^{-x^2} \cdot e^{-x^2}$

(g) $f(x) = e^{-x^2} * e^{-x^2}$ (h) $f(x) = \displaystyle\int_{-1/2}^{1/2} e^{-(x-u)^2}\, du$

Hint. Observe that (d), (e) are derivatives and (h) is a convolution product.

▶ **EXERCISE 3.6:**   Let $a > 0$, $b > 0$. Find the Fourier transform of each of the following functions.

(a) $f(x) = \cos(\pi a x)\, \Pi(x/2b)$ (b) $f(x) = \sin(\pi a x) \sin(\pi b x)/x$

(c) $f(x) = \cos(\pi a x) \sin(\pi b x)/x$ (d) $f(x) = \sin(\pi a x) \sin(\pi b x)/x^2$

▶ **EXERCISE 3.7:**   Let $a > 0$, $b > 0$. Find the Fourier transform of each of the following functions.

(a) $f(x) = 1/(x^2 + a^2)$ (b) $f(x) = \cos(\pi b x)/(x^2 + a^2)$

(c) $f(x) = x/(x^2 + a^2)$ (d) $f(x) = x/(x^2 + a^2)^2$

(e) $f(x) = \{1/(x^2 + a^2)\} * \{1/(x^2 + b^2)\}$ (f) $f(x) = \{1/(x^2 + a^2)\} \cdot \{1/(x^2 + b^2)\}$

Hint. When $a \neq b$ you can use partial functions for (f).

▶ **EXERCISE 3.8:**   Let a, b, c be real parameters with $a > 0$ and with $b^2 - 4ac < 0$. Find the Fourier transforms of each of the following functions.

(a) $f(x) = 1/(ax^2 + bx + c)$

(b) $f(x) = e^{-(ax^2 + bx + c)}$

Hint. Complete the square.

▶ **EXERCISE 3.9:**   Find the Fourier transform of each of the following functions.

(a) $f(x) = \cos(x) \, \Pi(x/\pi)$

(b) $f(x) = \Lambda(2x + 1) \cdot \Lambda(2x - 1)$

(c) $f(x) = \sin(10\pi x) \, e^{-5x} \, h(x)$

(d) $f(x) = \text{sinc}(4x) * \text{sinc}^2(2x)$

(e) $f(x) = 1/(x + i)$

(f) $f(x) = 1/(x + i)^3$

(g) $f(x) = e^{-|2\pi x|} \, \text{sinc}(x)$

(h) $f(x) = e^{-|2x + 5|}$

(i) $f(x) = e^{-3x^2}$

(j) $f(x) = 1/(x^2 + 2x + 2)$

(k) $f(x) = \displaystyle\int_{-\infty}^{\infty} e^{2\pi i s x - s^4} \, ds$

(l) $f(x) = \displaystyle\int_{-\infty}^{\infty} \frac{\cos(2\pi s x)}{1 + s^4} \, ds$

▶ **EXERCISE 3.10:**   Let $u(x) := e^{-\alpha x} \, h(x)$ where $\alpha > 0$ and $h(x)$ is the Heaviside step, and let $u_1 := u$, $u_2 := u * u$, $u_3 := u * u * u$, $\ldots$.

(a) Find the Fourier transform of u_1 and then use the convolution rule to deduce that u_{n+1} has the Fourier transform $U_{n+1}(s) = (\alpha + 2\pi i s)^{-n-1}$.

(b) Use the power scaling rule and the fact that

$$U_{n+1}(s) = \frac{(-2\pi i)^{-n}}{n!} \frac{d^n}{ds^n} (\alpha + 2\pi i s)^{-1}$$

to deduce that $u_{n+1}(x) = x^n \, e^{-\alpha x} \, h(x)/n!$, $n = 0, 1, \ldots$.

Note. Compare this calculation with the brute force analysis of Ex. 2.4.

▶ **EXERCISE 3.11:**   Let $g(x) := \text{sgn}(x) \, e^{-|x|}$.

(a) Write down the integral that defines the Fourier transform $G(s)$ and use the fact that g is odd to show that

$$G(s) = -2i \int_0^{\infty} e^{-x} \sin(2\pi s x) \, dx.$$

You can evaluate this integral with an integration by parts argument $\ldots$ or you can use the Fourier transform calculus.

(b) Observe that $g(x) = d(x) - d(-x)$ where $d(x) := e^{-x} \, h(x)$. Use this with the known Fourier transform of d to find G.

(c) Observe that $g = -f'$ where $f(x) := e^{-|x|}$. Use this with the known Fourier transform of f to find G.

▶ **EXERCISE 3.12:** From the partial fraction decomposition

$$f(x) := \frac{2}{1 + 4\pi^2 x^2} = \frac{1}{1 + 2\pi i x} + \frac{1}{1 - 2\pi i x}$$

we obtain the Fourier transform $F(s) = e^s\, h(-s) + e^{-s}\, h(s) = e^{-|s|}$. Use this procedure to find the Fourier transforms of

(a) $g_1(x) := \dfrac{2}{x^2 - 4x + 5}$; (b) $g_2(x) := \dfrac{2x - 4}{x^2 - 4x + 5}$.

Hint. Use the translation and dilation rules with the identity $g_1(x) = f\{(x-2)/2\pi\}$ to check your answer for (a).

▶ **EXERCISE 3.13:** In this exercise you will use an elementary (but tedious) argument to find the Fourier transform of $f(x) = 1/(1 + x^4)$. (An alternative computation of this Fourier transform is given in Section 7.5)

(a) Let $\alpha := \sqrt{2}/2$. Verify that f has the partial fraction decomposition

$$f(x) = \frac{1}{4}\left\{ \frac{\alpha + i\alpha}{x + \alpha + i\alpha} + \frac{-\alpha + i\alpha}{x - \alpha + i\alpha} + \frac{\alpha - i\alpha}{x + \alpha - i\alpha} + \frac{-\alpha - i\alpha}{x - \alpha - i\alpha} \right\}$$

Hint. You can obtain the roots of $x^4 + 1$ by deleting from the set of roots of $x^8 - 1 = (x^4 - 1)(x^4 + 1)$ the roots $\pm 1, \pm i$ of $x^4 - 1$.

(b) Using (a) and (22), show that

$$F(s) = \frac{1}{4}(\alpha + i\alpha)e^{2\pi i\alpha s}\,(-2\pi i)e^{-2\pi\alpha s}h(s) + \frac{1}{4}(-\alpha + i\alpha)e^{-2\pi i\alpha s}(-2\pi i)e^{-2\pi\alpha s}h(s)$$

$$+ \frac{1}{4}(\alpha - i\alpha)e^{2\pi i\alpha s}\,(+2\pi i)e^{2\pi\alpha s}h(-s) + \frac{1}{4}(-\alpha - i\alpha)e^{-2\pi i\alpha s}(+2\pi i)e^{2\pi\alpha s}h(-s).$$

(c) Using the fact that F (like f) must be even, show that

$$F(s) = \pi \sin\left\{ \sqrt{2}\,\pi|s| + \frac{\pi}{4} \right\} e^{-\sqrt{2}\pi|s|}.$$

(d) Use (c) (with the analysis equation (1.10)) to evaluate the following integrals.

(i) $\displaystyle\int_{-\infty}^{\infty} \frac{dx}{1 + x^4}$ (ii) $\displaystyle\int_{-\infty}^{\infty} \frac{\cos(\pi x)dx}{1 + x^4}$

▶ **EXERCISE 3.14:** Find $f_1 * f_2$ (as the inverse Fourier transform of $F_1 \cdot F_2$) when:

(a) $f_1(x) = 2\operatorname{sinc}(2x)$, $f_2(x) = 4\operatorname{sinc}(4x)$; (b) $f_1(x) = 2\operatorname{sinc}(2x)$, $f_2(x) = \operatorname{sinc}^2(x)$;

(c) $f_1(x) = f_2(x) = e^{-\pi x^2}$; (d) $f_1(x) = f_2(x) = 2/(1 + 4\pi^2 x^2)$;

(e) $f_1(x) = 1/(1 + 2\pi i x)$, $f_2(x) = 1/(1 - 2\pi i x)$.

▶ **EXERCISE 3.15:**   Let $a > 0$. Use the fact that $\text{sinc}(x)$, $\text{sinc}^2(x)$ have the Fourier transforms $\Pi(s)$, $\Lambda(s)$ together with a suitably chosen synthesis equation, Parseval's relation, ... to show that:

(a) $\displaystyle\int_{-\infty}^{\infty} \left(\frac{\sin ax}{x}\right) dx = \pi$;

(b) $\displaystyle\int_{-\infty}^{\infty} \left(\frac{\sin ax}{x}\right)^2 dx = a\pi$;

(c) $\displaystyle\int_{-\infty}^{\infty} \left(\frac{\sin ax}{x}\right)^3 dx = \frac{3a^2\pi}{4}$;

(d) $\displaystyle\int_{-\infty}^{\infty} \left(\frac{\sin ax}{x}\right)^4 dx = \frac{2a^3\pi}{3}$.

▶ **EXERCISE 3.16:**   Let a, b be real with $a > 0$. Use your knowledge of Fourier analysis to evaluate the following definite integrals.

(a) $\displaystyle\int_{0}^{\infty} \frac{\cos(bx)}{x^2 + a^2} dx$

(b) $\displaystyle\int_{0}^{\infty} \frac{\sin(ax)\cos(bx)}{x} dx$

(c) $\displaystyle\int_{0}^{\infty} e^{-ax}\cos(bx) dx$

(d) $\displaystyle\int_{0}^{\infty} \cos(bx)e^{-ax^2} dx$

▶ **EXERCISE 3.17:**   Let $a > 0$, $b > 0$. Use Parseval's identity to show that:

(a) $\displaystyle\int_{-\infty}^{\infty} \frac{dx}{(x^2 + a^2)(x^2 + b^2)} = \frac{\pi}{ab(a + b)}$;

(b) $\displaystyle\int_{-\infty}^{\infty} \frac{\sin(\pi ax)}{x(x^2 + b^2)} dx = \frac{\pi}{b^2}\{1 - e^{-\pi ab}\}$.

▶ **EXERCISE 3.18:**   Show that the n-translates of sinc are orthonormal *i.e.*,

$$\int_{-\infty}^{\infty} \text{sinc}(x - n) \cdot \text{sinc}(x - m) dx = \begin{cases} 1 & \text{if } n = m \\ 0 & \text{if } n \neq m, \end{cases} \qquad n, m = 0, \pm 1, \pm 2, \ldots .$$

▶ **EXERCISE 3.19:**   Let $V(s)$ be the Fourier transform of the Volkswagen function v.

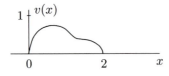

Express the Fourier transform of each of the following functions in terms of V.

(a)
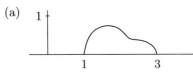

(b)
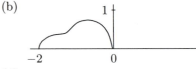

(c)
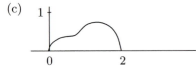

(d)
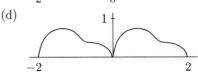

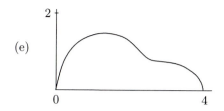

(e)

(f)

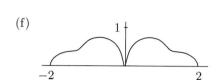

▶ **EXERCISE 3.20:**   Let f have as its Fourier transform the following function F.

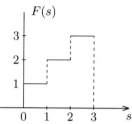

Sketch the graph of the Fourier transform of the function:

(a) $f(-x)$;

(b) $2f(2x)$;

(c) $e^{4\pi ix}f(x)$;

(d) $e^{6\pi ix}f(-x)$;

(e) $(1/2\pi i)f'(x)$;

(f) $(f*f)(x)$;

(g) $f(x) \cdot f(x)$;

(h) $f(x) \cdot \mathrm{sinc}(x)$;

(i) $\{f(x+\frac{1}{2}) - f(x-\frac{1}{2})\}/2i$.

▶ **EXERCISE 3.21:**   Let F be the Fourier transform of the following function f.

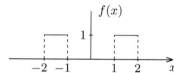

Sketch the graph of the function that has as its Fourier transform:

(a) $F(-s)$;

(b) $2\cos(2\pi s)F(s)$;

(c) $(-2\pi i)^{-1}F'(s)$;

(d) $(F*F)(s)$;

(e) $F(s)^2$;

(f) $(F*\mathrm{sinc})(s)$;

(g) $F(s) \cdot \mathrm{sinc}(s)$;

(h) $2^{-1}F(2^{-1}s)$;

(i) $\displaystyle\sum_{n=0}^{\infty} 2^{-n}F(2^{-n}s)$.

▶ **EXERCISE 3.22:**   Let F be the Fourier transform of a suitably regular function f on $\mathbb{R}$. Express each of the following in terms of f.

(a) $\displaystyle\int_{-\infty}^{\infty} F(-s)e^{2\pi isx}\,ds$;

(b) $\displaystyle\int_{-\infty}^{\infty} \overline{F(-s)}e^{2\pi isx}$;

(c) $\displaystyle\int_{-\infty}^{\infty} F(s-5)e^{2\pi isx}\,ds$;

(d) $\displaystyle\int_{-\infty}^{\infty} F(2s)e^{2\pi isx}\,ds$;

(e) $\displaystyle\int_{-\infty}^{\infty} s^2 F(s)e^{2\pi isx}\,ds$;

(f) $\displaystyle\int_{-\infty}^{\infty} \cos(2\pi s)F(s)e^{2\pi isx}\,ds$;

(g) $\displaystyle\int_{-\infty}^{\infty} F'''(s)e^{2\pi isx}\,ds$;

(h) $\displaystyle\int_{-\infty}^{\infty} sF(2s)e^{-2\pi isx}\,ds$;

(i) $\displaystyle\int_{-\infty}^{\infty} \tfrac{1}{2}[F(s)+\overline{F(s)}]e^{2\pi isx}\,ds$.

▶ **EXERCISE 3.23:**   Let f be a suitably regular function on $\mathbb{R}$ with the Fourier transform F. What can you deduce about F if you know that:

(a) $\displaystyle\int_{-\infty}^{\infty} f(x)dx = 1$?   (b) $\displaystyle\int_{-\infty}^{\infty} xf(x)dx = 1$?   (c) $\displaystyle\int_{-\infty}^{\infty} \cos(2\pi x)f(x)dx = 0$?

(d) $f'(0) = 0$?   (e) $\displaystyle\int_{-\infty}^{\infty} |f(x)|^2 dx = 1$?   (f) $\overline{f(x)} = f(-x)$?

▶ **EXERCISE 3.24:**   Let f be a suitably regular function on $\mathbb{R}$, let $a, a_0, a_1, \ldots$ be real, and let $h > 0$. Formulate the Fourier transform rule "$g(x) := \cdots$ has the FT $G(s) = \cdots$" when g is given by:

(a) $g(x) := \cos(2\pi ax)f(x)$;   (b) $g(x) := \sin(2\pi ax)f(x)$;

(c) $g(x) := \{f(x+h) - f(x-h)\}/2h$;   (d) $g(x) := \displaystyle\sum_{k=0}^{n} a_k f(x+kh)$;

(e) $g(x) := \displaystyle\sum_{k=0}^{n} a_k f^{(n)}(x)$;   (f) $g(x) := \displaystyle\int_{-h/2}^{h/2} f(x+u)du$.

Hint. You can write the integral from (f) as a convolution product.

▶ **EXERCISE 3.25:**   Let $f_1, f_2, \ldots, f_m$ be suitably regular functions on $\mathbb{R}$, and let $f := f_1 * f_2 * \cdots * f_m$. Let $a_1, a_2, \ldots, a_m$ be real, let $n_1, n_2, \ldots, n_m$ be nonnnegative integers, and let a be real with $a \neq 0$. Use rules from the Fourier transform calculus to derive the following identities.

(a) $f_1(x - a_1) * f_2(x - a_2) * \cdots * f_m(x - a_m) = f(x - a_1 - a_2 - \cdots - a_m)$

(b) $f_1^{(n_1)}(x) * f_2^{(n_2)}(x) * \cdots * f_m^{(n_m)}(x) = f^{(n_1+n_2+\cdots+n_m)}(x)$

(c) $f_1(ax) * f_2(ax) * \cdots * f_m(ax) = |a|^{1-m} f(ax)$

▶ **EXERCISE 3.26:**   The cross-correlation product $f_1 \star f_2$ of the suitably regular functions f_1, f_2 is defined by (2.5).

(a) Derive the cross-correlation rule:

$$g(x) := (f_1 \star f_2)(x) \quad \text{has the FT} \quad G(s) = \overline{F}_1(s) \cdot F_2(s).$$

(b) Specialize (a) to obtain the autocorrelation rule:

$$g(x) := (f \star f)(x) \quad \text{has the FT} \quad G(s) = |F(s)|^2.$$

(c) Use (b) to obtain Plancherel's identity

$$\int_{-\infty}^{\infty} |f(u)|^2 du = \int_{-\infty}^{\infty} |F(s)|^2 ds.$$

▶ **EXERCISE 3.27:** What can you infer about the Fourier transform F of the suitably regular function f on $\mathbb{R}$ if you know that f satisfies:

(a) the integral equation $\qquad\qquad f(x) = \displaystyle\int_{-\infty}^{\infty} f(u)\operatorname{sinc}(x-u)du?$

(b) the differential equation $\qquad\quad f'(x) + f(x) = e^{-\pi x^2}?$

(c) the difference equation $\qquad\quad f(x+h) - 2f(x) + f(x-h) = 0?$

(d) the dilation equation $\qquad\qquad f(x) = f(2x) + f(2x-1)?$

▶ **EXERCISE 3.28:** Use your knowledge of Fourier analysis to find a function f that satisfies the given integral equation.

(a) $\displaystyle\int_{u=0}^{\infty} f(u)\cos(2\pi ux)du = e^{-x}, \quad 0 \le x < \infty$

(b) $\displaystyle\int_{u=0}^{\infty} f(u)\sin(2\pi ux)du = \begin{cases} 1 & \text{if } 0 < x < 1 \\ 0 & \text{if } 1 < x < \infty \end{cases}$

(c) $\displaystyle\int_{u=-\infty}^{\infty} e^{-|x-u|}f(u)du = (1+|x|)e^{-|x|}, \quad -\infty < x < \infty$

Hint. Use the result of Ex. 2.5.

(d) $\displaystyle\int_{-\infty}^{\infty} f(u)f(x-u)du = e^{-\pi x^2}$

▶ **EXERCISE 3.29:** Let g be a piecewise smooth function on $\mathbb{R}$ with small regular tails, and suppose that we wish to find such a function f that satisfies the differential equation

$$-f''(x) + f(x) = g(x), \qquad -\infty < x < \infty.$$

(a) Fourier transform the differential equation and thereby show that any suitably regular solution can be written in the form

$$f(x) = \frac{1}{2}\int_{u=-\infty}^{\infty} e^{-|x-u|}g(u)du.$$

(b) Find the function f and sketch its graph when $g(x) := \Pi(x)$.

(c) Find the function f and sketch its graph when $g(x) := e^{-|x|}$.

Hint. Use the result of Ex. 2.5.

Note. When f is given by the integral of (a) and c_1, c_2 are constants, the function $f(x) + c_1 e^x + c_2 e^{-x}$ is also a solution of the differential equation. This general solution does not have small regular tails or a Fourier transform unless $c_1 = c_2 = 0$.

▶ **EXERCISE 3.30:** Let f be a suitably regular function on $\mathbb{R}$ with the Fourier transform F, let

$$g(x) := \int_{u=-\infty}^{x} f(u)du,$$

and assume that $g(x) \to 0$ as $x \to \pm\infty$ so rapidly that g has a Fourier transform G.

(a) State the antiderivative rule for expressing the Fourier transform G of g in terms of the Fourier transform F of f.

   Note. We must impose some condition on f to guarantee that $F(s)/s$ is well behaved in a neighborhood of $s = 0$.

(b) Using (a), show that the function

$$g(x) := \arctan(ax) - \arctan(bx), \quad a > b > 0$$

   has the Fourier transform

$$G(s) = \frac{i\{e^{-2\pi|s|/b} - e^{-2\pi|s|/a}\}}{2s}.$$

▶ **EXERCISE 3.31:** Use the Fourier transform calculus to prove the recursion relation for B-splines that is given in the note following Ex. 2.7.

▶ **EXERCISE 3.32:** Use your knowledge of Fourier analysis to construct continuously differentiable, absolutely integrable functions f, g on $\mathbb{R}$ such that $f(x) \cdot g(x) \neq 0$, and $(f * g)(x) = 0$ for every choice of x.

Hint. Consider $\operatorname{sinc}^2(x) + \operatorname{sinc}^2(\sqrt{2}\,x)$.

▶ **EXERCISE 3.33:** Let $\alpha > 0$, and let let $\nu = 1, 2, \ldots$. Verify that

$$f(x) := \frac{x^{\nu-1}e^{-x/\alpha}}{\alpha^{\nu}(\nu-1)!}h(x) \quad \text{has the FT} \quad F(s) = (1 + 2\pi i\alpha s)^{-\nu}$$

(a) by using (2) with the dilation and power scaling rules;

(b) by using the analysis equation to show that

$$F'(s) = -\frac{2\pi i\alpha\nu}{1 + 2\pi i\alpha s}F(s), \qquad F(0) = 1,$$

   and then solving this initial value problem.

Note. The argument from (b) can be used with any positive value of ν provided we replace $(\nu - 1)!$ with Euler's *gamma function*

$$\Gamma(\nu) := \int_{u=0}^{\infty} x^{\nu-1}e^{-x}dx$$

in the definition of f.

▶ **EXERCISE 3.34:** Many complex physical systems can be modeled by using a number of independent damped harmonic oscillators, with the response to an initial excitation having the form

$$y(t) := \sum_{m=1}^{M} e^{-\alpha_m t}\{A_m \cos(2\pi s_m t) + D_m \sin(2\pi s_m t)\}h(t).$$

Here A_m, D_m, $s_m > 0$, $\alpha_m > 0$ are parameters, $m = 1, 2, \ldots, M$. Such a function might represent the accoustical wave produced by ringing a large bell (*cf.* Fig. 11.10) or the free induction decay from an FT-NMR (*cf.* Fig. 1.16). In this exercise you will analyze the Fourier transform, *i.e.*, the *spectrum* of such a transient. Within this context, we use the independent variable t instead of x.

(a) Use the dilation rule and the modulation rule with (2) to show that

$$y_0(t) := e^{-\alpha_0 t + 2\pi i s_0 t}h(t) \quad \text{has the FT} \quad Y_0(s) = \frac{1 - i\,u(s)}{\alpha_0\{1 + u^2(s)\}}.$$

Here $u(s) := 2\pi(s - s_0)/\alpha_0$ and α_0, s_0 are real parameters with $\alpha_0 > 0$.

Note. The function y_0 satisfies the damped harmonic oscillator equation

$$y_0''(t) + 2\alpha_0 y_0'(t) + (4\pi^2 s_0^2 + \alpha_0^2)y_0(t) = 0.$$

(b) Plot Re Y_0, Im Y_0 as functions on $\mathbb{R}$, and plot $Y_0(s)$, $-\infty < s < \infty$, as an orbit in the complex plane $\mathbb{C}$, showing the effects of the parameters α, s_0.

Hint. Your graphs will resemble those of Fig. 1.15 and the orbit will be a circle.

(c) Plot the Fourier transforms of

$$y_c(t) := e^{-\alpha_0 t} \cos(2\pi s_0 t)h(t), \qquad y_s(t) := e^{-\alpha_0 t}\sin(2\pi s_0 t)h(t)$$

as functions on $\mathbb{R}$, showing the effects of the parameters α_0, s_0.

Hint. Use the conjugation rule and the analysis from (b).

(d) Describe the Fourier transform of y (as given above) in the case where $\alpha_1, \ldots, \alpha_M$ are small and positive, where $s_1, \ldots, s_M$ are positive and well separated, and where $A_1, \ldots, A_m, D_1, \ldots, D_m$ are nonnegative real numbers.

Note. To learn more about the use of such spectral methods in chemistry, *cf.* A.G. Marshall and F.R. Verdun, *Fourier Transforms in NMR, Optical, and Mass Spectroscopy*, Elsevier, New York, 1990.

▶ **EXERCISE 3.35:** In this exercise you will show that the Fresnel function

$$f(x) := e^{i\pi x^2} \quad \text{has the FT} \quad F(s) = \{(1 + i)/\sqrt{2}\}e^{-i\pi s^2}.$$

We will use this particular Fourier transform pair when we study diffraction in Section 9.4.

(a) Use the alternating series test from calculus to show that the integrals

$$\int_{-\infty}^{\infty} \sin(\pi x^2)dx, \qquad \int_{-\infty}^{\infty} \cos(\pi x^2)dx$$

are well defined and positive.

Hint. $\displaystyle \int_0^\infty \sin(\pi x^2)\,dx = \sum_{m=0}^\infty \int_{\sqrt{m}}^{\sqrt{m+1}} \sin(\pi x^2)\,dx.$

(b) Use polar coordinates to verify that

$$\int_{-\infty}^\infty e^{-\pi(\alpha-i)x^2}\,dx \cdot \int_{-\infty}^\infty e^{-\pi(\alpha-i)y^2}\,dy = \frac{i}{1+\alpha i}, \quad \alpha > 0,$$

and thereby show that

$$I := \int_{-\infty}^\infty e^{i\pi x^2}\,dx = \frac{1+i}{\sqrt{2}}.$$

Note. For alternative derivations, *cf.* Section 4.4 or R. Weinstock, *Amer. Math. Monthly* **97**(1990), 39–42.

(c) Use (b) and the identity $x^2 - 2xs = (x - s)^2 - s^2$ to show that

$$F(s) := \int_{-\infty}^\infty e^{i\pi x^2} e^{-2\pi i s x}\,dx = \frac{1+i}{\sqrt{2}} e^{-i\pi s^2}.$$

(d) Show that f has the Fourier representation (3).

Hint. Begin by using (c) to write $f(x) = \overline{F(x)/I}$.

Note. We can use the Fourier representation (3)–(4) for the Fresnel function even though neither f nor F have small tails:

$$|f(x)| = 1, \quad -\infty < x < \infty, \quad \text{and} \quad |F(s)| = 1, \quad -\infty < s < \infty.$$

▶ **EXERCISE 3.36:**   Use the Fourier transform pair of Ex. 3.35 to show that:
  (a) $f(x) := \cos(\pi x^2)$ has the FT $F(s) = \cos\{\pi(s^2 - 1/4)\}$;
  (b) $f(x) := \sin(\pi x^2)$ has the FT $F(s) = -\sin\{\pi(s^2 - 1/4)\}$;
  (c) $f(x) := |x|^{-1/2}$ has the FT $F(s) = |s|^{-1/2}$;
  (d) $f(x) := \operatorname{sgn}(x)|x|^{-1/2}$ has the FT $F(s) = -i\operatorname{sgn}(s)|s|^{-1/2}.$

Hint. Use the substitution $u^2 = 2|s|x$ in the analysis equation integrals for (c), (d).

▶ **EXERCISE 3.37:**   In this exercise you will use the generating function

$$g(x,t) := e^{-(x-t)^2}$$

to establish an orthogonality property of the Hermite functions (31).

(a) Find the Maclaurin series for $g(x,t)$ with respect to the argument t and thereby show that

$$e^{-(x-t)^2} = \sum_{n=0}^\infty H_n(x) e^{-x^2} \frac{t^n}{n!}.$$

(b) Use the identity

$$\int_{-\infty}^{\infty} H_m(x) e^{-(t-x)^2}\, dx = (-1)^m e^{-t^2} \int_{-\infty}^{\infty} e^{2tx} \{\mathbf{D}^m e^{-x^2}\}\, dx$$

and an integration by parts argument to show that

$$\int_{-\infty}^{\infty} H_m(x) e^{-(t-x)^2}\, dx = (2t)^m \sqrt{\pi}, \quad m = 0, 1, \ldots .$$

(c) Combine (a), (b) and thereby show that

$$\int_{-\infty}^{\infty} H_m(x) H_n(x) e^{-x^2}\, dx = \begin{cases} 2^n n! \sqrt{\pi} & \text{if } m = n \\ 0 & \text{if } m \neq n. \end{cases}$$

(d) Using (c), show that

$$\int_{-\infty}^{\infty} f_m(x) f_n(x)\, dx = \begin{cases} 2^n n! / \sqrt{2} & \text{if } m = n \\ 0 & \text{if } m \neq n. \end{cases}$$

Note. When f is a suitably regular function on $\mathbb{R}$ we can write

$$f(x) = \sum_{n=0}^{\infty} c_n f_n(x) \quad \text{with} \quad c_n := \frac{\sqrt{2}}{2^n n!} \int_{-\infty}^{\infty} f(x) f_n(x)\, dx, \quad n = 0, 1, \ldots .$$

Norbert Wiener used this series and the corresponding

$$F(s) = \sum_{n=0}^{\infty} (-i)^n c_n f_n(s)$$

to study the Fourier transform of square integrable functions on $\mathbb{R}$, *cf.* Norbert Wiener, *The Fourier Integral and Certain of Its Applications*, Dover, New York, 1958, pp. 46–71.

▶ **EXERCISE 3.38:**   This exercise will show you an alternative way to find the Fourier transform of the Hermite functions $f_n(x)$, $n = 0, 1, \ldots$ as given in (31).

(a) Suitably modify the expansion of Ex. 3.37(a) to show that

$$e^{-\pi x^2} \cdot e^{2\pi x t / \sqrt{\pi/2}} \cdot e^{-t^2} = \sum_{n=0}^{\infty} f_n(x) \frac{t^n}{n!} .$$

(b) Replace t by it and formally take Fourier transforms on a term-by-term basis to show that

$$e^{-\pi (s - t/\sqrt{\pi/2})^2} \cdot e^{t^2} = \sum_{n=0}^{\infty} (i)^n F_n(s) \frac{t^n}{n!}$$

where $F_n(s)$ is the Fourier transform of $f_n(x)$.

(c) Use (a)–(b) to show that
$$F_n(s) = (-i)^n f_n(s).$$

Note. We can make the informal argument of (b) rigorous by using the uniform bound $|f_n(x)|^2 \leq 2^{n+1} n!$ together with the theory of weak limits that will be developed in Section 7.6.

▶ **EXERCISE 3.39:**   Let f be a suitably regular function on $\mathbb{R}$, let s be a fixed real parameter, and let

$$z(x, s) := \int_0^x f(u) e^{-2\pi i s u} \, du, \qquad -\infty < x < \infty.$$

(a) Give a geometric interpretation of the differential relation
$$z(x + dx, s) = z(x, s) + e^{-2\pi i s x} \cdot \{f(x) dx\}.$$

(b) Give a geometric interpretation of the relation
$$F(s) = \lim_{x \to +\infty} z(x, s) - \lim_{x \to -\infty} z(x, s).$$

(c) Let $f(x) := \Pi(x)$. Show that

$$z(x, s) = \frac{1 - e^{-2\pi i s x_c}}{2\pi i s} \quad \text{where} \quad x_c := \begin{cases} -\frac{1}{2} & \text{if } x \leq -\frac{1}{2} \\ x & \text{if } -\frac{1}{2} \leq x \leq \frac{1}{2} \\ \frac{1}{2} & \text{if } x \geq \frac{1}{2}, \end{cases}$$

and plot the circular arcs $z(x, s)$, $-\infty < x < \infty$, that correspond to $s = 1/2, 1, 3/2$.

(d) Let $f(x) := e^{-x} h(x)$. Show that

$$z(x, s) = \begin{cases} 0 & \text{if } x \leq 0 \\ \dfrac{1 - e^{-x(1+2\pi i s)}}{1 + 2\pi i s} & \text{if } x > 0, \end{cases}$$

and plot the spiral $z(x, s)$, $-\infty < x < \infty$, that corresponds to $s = 1$.

▶ **EXERCISE 3.40:**   In this exercise you will show that no ordinary function δ can serve as a convolution product identity for functions on $\mathbb{R}$.

(a) Formally use the convolution rule to show that if $\delta * g = g$ when $g(x) := e^{-\pi x^2}$, then δ must have the Fourier transform $\Delta(s) := 1$, $-\infty < s < \infty$.

(b) What happens when you try to synthesize a function δ on $\mathbb{R}$ by using the Δ of (a)?

(c) Explain why $d_n(x) := n \operatorname{sinc}(nx)$ is a reasonable *approximate* convolution product identity when n is large and positive.

(d) Explain why $d_n(x) := n e^{-\pi n^2 x^2}$ is a reasonable *approximate* convolution product identity when n is large and positive.

▶ **EXERCISE 3.41:**   Fourier's representation is valid if either f or F is a piecewise smooth function with small regular tails. We must impose additional regularity conditions if we wish to use the derivative rule (19) or the power scaling rule (20). For example, we can use Fourier's representation with $g(x) := f'(x)$, $G(s) = 2\pi i s F(s)$ if either

- $f(x), f'(x)$ are both piecewise smooth functions with small regular tails and $f(x)$ is continuous, or
- $F(s), sF(s)$ are both piecewise smooth functions with small regular tails.

We can use Fourier's representation with $g(x) := (-2\pi i x)f(x)$, $G(s) = F'(s)$ if either

- $f(x), xf(x)$ are both piecewise smooth functions with small regular tails, or
- $F(s), F'(s)$ are both piecewise smooth functions with small regular tails and $F(s)$ is continuous.

In this exercise you are to prove these statements as follows.

(a) Let f be a continuous, absolutely integrable function on $\mathbb{R}$ with an absolutely integrable derivative f' that is defined and continuous except for finitely many points of $\mathbb{R}$. Show that

$$\int_{-\infty}^{\infty} f'(x)e^{-2\pi i s x}\,dx = 2\pi i s \int_{-\infty}^{\infty} f(x)e^{-2\pi i s x}\,dx.$$

Hint. Verify that $f(x) \to 0$ as $x \to \pm\infty$ [*cf.* (1.43)], and integrate by parts.

(b) Let f be defined and continuous except for finitely many points of $\mathbb{R}$, and assume that both $f(x)$ and $xf(x)$ are absolutely integrable on $\mathbb{R}$. Show that

$$F(s) := \int_{-\infty}^{\infty} f(x)e^{-2\pi i s x}\,dx$$

has a uniformly continuous derivative

$$G(s) := \int_{-\infty}^{\infty} (-2\pi i x)f(x)e^{-2\pi i s x}\,dx$$

that vanishes at $\pm\infty$.

Hint. Verify that

$$\left| \frac{F(s+h) - F(s)}{h} - G(s) \right| = 2\pi \left| \int_{-\infty}^{\infty} \left\{ \frac{e^{-2\pi i h x} - 1 + 2\pi i h x}{-2\pi i h x} \right\} x f(x)e^{-2\pi i s x}\,dx \right|$$

and then suitably modify the argument of Ex. 1.38(b) using the bounds

$$|e^{i\theta} - 1| \le |\theta|, \quad |e^{i\theta} - 1 - i\theta| \le \theta^2/2, \quad -\infty < \theta < \infty.$$

▶ **EXERCISE 3.42:**   In this exercise you will establish a link between the smoothness of an absolutely integrable function f and the *rate* at which the Fourier transform F goes to zero as $s \to \pm\infty$.

(a) Let $f, f', \ldots, f^{(m)}$ be absolutely integrable on $\mathbb{R}$, let $f, f', \ldots, f^{(m-1)}$ be continuous, and let $f^{(m)}$ be continuous except for finitely many points of $\mathbb{R}$. Show that $F(s)$, $sF(s), \ldots, s^m F(s)$ are uniformly continuous functions on $\mathbb{R}$ that vanish at $\pm\infty$.

Hint. Use the analysis of Exs. 3.41(a) and 1.38.

(b) Let $f(x)$, $xf(x), \ldots, x^m f(x)$ be absolutely integrable on $\mathbb{R}$ and let f be continuous except for finitely many points of $\mathbb{R}$. Show that $F(s)$, $F'(s), \ldots, F^{(m)}(s)$ are uniformly continuous functions on $\mathbb{R}$ that vanish at $\pm\infty$.

▶ **EXERCISE 3.43:**   Let $n = 0, 1, \ldots$, let $a_0 < a_1 < \cdots < a_M$, and let f be a piecewise polynomial function on $\mathbb{R}$ with

$$f(x) = 0 \quad \text{if} \quad x < a_0 \text{ or } x > a_M$$
$$f^{(n+1)}(x) = 0 \quad \text{if} \quad a_{m-1} < x < a_m, \quad m = 1, 2, \ldots, M.$$

(a) Assume that $n \geq 1$ and that $f, f', \ldots, f^{(n-1)}$ are all continuous. Show that $s^{n+1}F(s)$ is bounded and thereby prove that $F(s)$ approaches zero as fast as $1/s^{n+1}$ when $s \to \pm\infty$.

Hint. Begin with the analysis equation (33) for $f^{(n)}$.

(b) Let k be a positive integer and assume that $f, f', \ldots, f^{(k-1)}$ are all continuous. What can you infer about the rate that $F(s)$ approaches zero as $s \to \pm\infty$?

▶ **EXERCISE 3.44:**   Let f, g be piecewise smooth functions with small regular tails, and let F, G be the corresponding Fourier transforms. In this exercise you will show that the multiplication rule can be used with $f \cdot g$, $f \cdot G$, $F \cdot g$, $F \cdot G$, and the convolution rule can be used with $f * g$, $F * G$.

(a) Show that

$$\int_{x=-\infty}^{\infty} (f * g)(x)e^{-2\pi isx}\, dx = F(s) \cdot G(s).$$

Hint. Since $\displaystyle\iint |f(u)g(x-u)|\, du\, dx < \infty$, you can exchange the order of integration.

(b) Use Parseval's identity with $f(x)$ and $\gamma(x) := \overline{g(x_0 - x)}$ (*cf.* Ex. 1.40) to show that

$$\int_{s=-\infty}^{\infty} F(s) \cdot G(s)e^{+2\pi isx}\, ds = (f * g)(x).$$

Note. Together (a)–(b) establish the convolution rule for $f * g$ and the multiplication rule for $F \cdot G$.

(c) Use Parseval's identity to show that

$$\int_{x=-\infty}^{\infty} f(x) \cdot g(x) e^{-2\pi i s x} \, dx = (F * G)(s).$$

(d) Show that

$$\int_{s=-\infty}^{\infty} (F * G)(s) e^{+2\pi i s x} \, ds = f(x) \cdot g(x).$$

Hint. Use (c) with the representation theorem from Section 1.5.

Note. Together, (c)–(d) establish the convolution rule for $F*G$ and the multiplication rule for $f \cdot g$.

(e) Show that

$$\int_{x=-\infty}^{\infty} f(x) \cdot G(-x) e^{-2\pi i s x} \, dx = (F * g)(s).$$

Hint. Since $\displaystyle\iint |f(x)g(\sigma)| \, d\sigma \, dx < \infty$, you can exchange the order of integration.

Note. In this way we establish the product rule for $f(x) \cdot G(-x)$ [and likewise for $f(x) \cdot G(x)$, $F(x) \cdot g(x)$.] We need an additional hypothesis to obtain the convolution rule for $F*g$ (or for $f*G$). For example, if the function G is continuously differentiable except for finitely many points of $\mathbb{R}$, we can use (e) with the representation theorem of Ex. 1.39 to write

$$\int_{s=-\infty}^{\infty} (F * g)(s) e^{+2\pi i s x} \, ds = f(x) \cdot G(-x).$$

▶ **EXERCISE 3.45:** Construct an absolutely integrable function f on $\mathbb{R}$ such that:

$$\max_{|x| \geq L} f(x) = 1 \quad \text{for every } L > 0;$$

$$F, F', F'', \dots \quad \text{are uniformly continuous on } \mathbb{R};$$

$$F, F', F'', \dots \quad \text{all vanish at } \pm \infty.$$

Hint. You can use a sum of suitably translated and dilated triangles, *cf.* Ex. 3.42.

▶ **EXERCISE 3.46:** Let f be a piecewise smooth function on $\mathbb{R}$ that vanishes outside some finite interval. Show that

$$\sum_{m=-\infty}^{\infty} f(x - m) = 1$$

(*i.e.*, we can partition unity with the 1-translates of f) if and only if

$$F(k) = \begin{cases} 1 & \text{for } k = 0 \\ 0 & \text{for } k = \pm 1, \pm 2, \dots . \end{cases}$$

Note. The *B*-splines from Ex. 2.7 serve as examples.

▶ **EXERCISE 3.47:**   Use your knowledge of Fourier analysis to find a function f on $\mathbb{R}$ that has the specified properties.

(a) $\displaystyle\int_{-\infty}^{\infty} f(u)f(x-u)du = f(x), \quad -\infty < x < \infty,$

$\displaystyle\int_{-\infty}^{\infty} f(u)\,\text{sinc}(x-u)du = f(x), \quad -\infty < x < \infty, \quad \text{and}$

$\displaystyle\int_{-\infty}^{\infty} |f(x)|^2 dx = 1$

(b) $f(-x) = f(x), \quad -\infty < x < \infty,$

$\displaystyle\int_{-\infty}^{\infty} f(u)\,\text{sinc}(x-u)du = 0, \quad -\infty < x < \infty,$

$\displaystyle\int_{-\infty}^{\infty} |f(x)|dx < \infty, \quad \text{and,}$

$\displaystyle\int_{-\infty}^{\infty} |f(x)|^2 dx = 1$

Chapter 4

The Calculus for Finding Fourier Transforms of Functions on $\mathbb{T}_p$, $\mathbb{Z}$, and $\mathbb{P}_N$

4.1 Fourier Series

Introduction

Now that you know how to find Fourier transforms of functions on $\mathbb{R}$, you can quickly learn to find Fourier transforms of functions on $\mathbb{T}_p$, *i.e.*, to construct the *Fourier series*

$$f(x) = \sum_{k=-\infty}^{\infty} F[k]e^{2\pi ikx/p} \tag{1}$$

when f is given. In principle, you can always obtain F by evaluating the integrals from the analysis equation

$$F[k] = \frac{1}{p} \int_0^p f(x)e^{-2\pi ikx/p}dx, \qquad k = 0, \pm 1, \pm 2, \ldots, \tag{2}$$

but this is often quite tedious. We will present several other methods for finding these coefficients. You can then select the procedure that requires the least amount of work!

You will recall from your study of Chapter 1 that the synthesis equation (1) for f on $\mathbb{T}_p$ can be written as the analysis equation

$$\frac{f(-s)}{p} = \frac{1}{p} \sum_{n=-\infty}^{\infty} F[n]e^{-2\pi isn/p}$$

for F on $\mathbb{Z}$. In view of this duality, every Fourier series (1) simultaneously tells us that

$$f(x) \quad \text{has the FT} \quad F[k], \text{ and}$$

$$F[k] \quad \text{has the FT} \quad \frac{f(-s)}{p}. \tag{3}$$

Direct integration

You can evaluate the integrals (2) with the techniques from elementary calculus when the function f is a linear combination of segments of

$$x^n \cdot e^{\alpha x} \cdot e^{2\pi i \beta x}, \quad n = 0, 1, 2, \ldots, \quad -\infty < \alpha < \infty, \quad -\infty < \beta < \infty.$$

You will use the integration by parts formula

$$\int_{x=a}^{b} f(x)q(x)dx = f(x)q^{(-1)}(x)\Big|_{x=a}^{b} - \int_{x=a}^{b} f'(x)q^{(-1)}(x)dx$$

$$= \left\{ f(x)q^{(-1)}(x) - f'(x)q^{(-2)}(x) \right\}\Big|_{x=a}^{b} + \int_{x=a}^{b} f''(x)q^{(-2)}(x)dx \tag{4}$$

$$\vdots$$

for such calculations. Here $q^{(-1)}, q^{(-2)}, \ldots$ are successive antiderivatives of

$$q(x) = e^{2\pi i k x/p} \quad \text{or} \quad \cos(2\pi k x/p) \quad \text{or} \quad \sin(2\pi k x/p), \quad k = \pm 1, \pm 2, \ldots .$$

When f is a polynomial, the integrated term will eventually disappear from (4), and the resulting identity,

$$\int_{x=a}^{b} f(x)q(x)dx = f(x)q^{(-1)}(x) - f'(x)q^{(-2)}(x)$$

$$+ \cdots + (-1)^{n-1} f^{(n-1)}(x)q^{(-n)}(x)\Big|_{x=a}^{b}, \quad f^{(n)} \equiv 0, \tag{5}$$

is known as *Kronecker's rule*. The $k = 0$ integral is usually done separately.

Example: Find the Fourier series for the p-periodic sawtooth function f shown in Fig. 4.1.

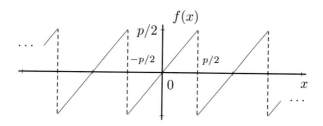

Figure 4.1. A p-periodic sawtooth function $f(x) := x$, $-p/2 < x < p/2$.

Solution: We use the odd symmetry of f with Kronecker's rule to write

$$F[k] := \frac{1}{p} \int_{x=-p/2}^{p/2} x\, e^{-2\pi i k x/p} dx$$

$$= \frac{-i}{p} \int_{x=-p/2}^{p/2} x \sin\left(\frac{2\pi k x}{p}\right) dx$$

$$= \frac{-2i}{p} \int_{0}^{p/2} x \sin\left(\frac{2\pi k x}{p}\right) dx$$

$$= \frac{-2i}{p} \left\{ x \frac{-p}{2\pi k} \cos\left(\frac{2\pi k x}{p}\right) + \left(\frac{p}{2\pi k}\right)^2 \sin\left(\frac{2\pi k x}{p}\right) \right\} \Big|_{0}^{p/2}$$

$$= (-1)^k \frac{ip}{2\pi k}, \qquad k = \pm 1, \pm 2, \dots .$$

Since f is odd, $F[0] = 0$ and thus

$$f(x) := x, \quad -\frac{p}{2} < x < \frac{p}{2} \quad \text{has the FS} \quad \sum_{k \neq 0} \frac{(-1)^{k+1} e^{2\pi i k x/p}}{2\pi i k/p}. \qquad (6)$$

We can also combine the $\pm k$ terms and use Euler's formula to write

$$f(x) = \frac{p}{\pi} \sum_{k=1}^{\infty} (-1)^{k+1} \frac{\sin(2\pi k x/p)}{k},$$

cf. Exs. 1.11 and 1.16(b). ∎

Example: Find the Fourier series for the p-periodic piecewise parabolic function f shown in Fig. 4.2.

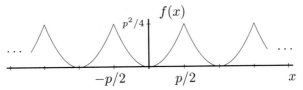

Figure 4.2. A p-periodic parabolic function $f(x) = x^2$, $-p/2 \le x \le p/2$.

Solution: We exploit the even symmetry of f as we compute

$$F[0] := \frac{1}{p} \int_0^p f(x)dx = \frac{2}{p} \int_0^{p/2} x^2 dx = \frac{p^2}{12},$$

and then use Kronecker's rule to find

$$F[k] := \frac{1}{p} \int_{-p/2}^{p/2} x^2 e^{-2\pi i k x/p} dx$$

$$= \frac{2}{p} \int_0^{p/2} x^2 \cos\left(\frac{2\pi k x}{p}\right) dx$$

$$= \frac{2}{p} \left\{ x^2 \left(\frac{p}{2\pi k}\right) \sin\left(\frac{2\pi k x}{p}\right) + 2x \left(\frac{p}{2\pi k}\right)^2 \cos\left(\frac{2\pi k x}{p}\right) \right.$$

$$\left. - \left(\frac{p}{2\pi k}\right)^3 \sin\left(\frac{2\pi k x}{p}\right) \right\} \Bigg|_{x=0}^{|p/2|}$$

$$= 2(-1)^{k+1} \left(\frac{p}{2\pi i k}\right)^2, \qquad k = \pm 1, \pm 2, \dots .$$

In this way we show that

$$f(x) = x^2, \quad -\frac{p}{2} \le x \le \frac{p}{2} \quad \text{has the FS} \quad \frac{p^2}{12} + \sum_{k \ne 0} \frac{(-1)^{k+1} 2 e^{2\pi i k x/p}}{(2\pi i k/p)^2}. \qquad (7)$$

∎

Elementary rules

There are rules for working with Fourier transforms of functions on $\mathbb{T}_p$ that are analogous to those for working with Fourier transforms of functions on $\mathbb{R}$. You will instantly recognize the linearity rule

$$g(x) := c_1 f_2(x) + \cdots + c_m f_m(x) \text{ has the FT } G[k] = c_1 F_1[k] + \cdots + c_m F_m[k], \quad (8)$$

the reflection and conjugation rules

$$g(x) := f(-x) \quad \text{has the FT} \quad G[k] = F[-k], \tag{9}$$
$$g(x) := \overline{f(x)} \quad \text{has the FT} \quad G[k] = \overline{F[-k]}, \tag{10}$$

the translation and modulation rules

$$g(x) := f(x - x_0) \qquad \text{has the FT} \quad G[k] = e^{-2\pi i k x_0/p} F[k], \quad -\infty < x_0 < x, \tag{11}$$
$$g(x) := e^{2\pi i k_0 x/p} f(x) \quad \text{has the FT} \quad G[k] = F[k - k_0], \quad k_0 = 0, \pm 1, \pm 2, \dots, \tag{12}$$

as well as the convolution and multiplication rules

$$g(x) := (f_1 * f_2)(x) \quad \text{has the FT} \quad G[k] = p\, F_1[k] \cdot F_2[k], \tag{13}$$
$$g(x) := f_1(x) \cdot f_2(x) \quad \text{has the FT} \quad G[k] = (F_1 * F_2)[k]. \tag{14}$$

You can always use these rules when the functions $f, f_1, f_2, \dots$ are piecewise smooth. The derivative rule

$$g(x) := f'(x) \quad \text{has the FT} \quad G[k] = \left(\frac{2\pi i k}{p}\right) \cdot F[k] \tag{15}$$

can be used when f is continuous and f' is piecewise smooth. The form of the complex exponential $e^{2\pi i k x/p}$ that we use for Fourier's representation of functions on $\mathbb{T}_p$ accounts for the form of the complex exponentials in (11)–(12) and for the multiplier in (15). The mnemonic *convolution gets the constant* will help you remember to include the factor p with (13) but not with (14).

Example: Derive the convolution rule (13).

Solution: When f_1, f_2 are piecewise smooth functions on $\mathbb{T}_p$ and $g := f_1 * f_2$, we can write

$$
\begin{aligned}
G[k] :&= \frac{1}{p} \int_{x=0}^{p} g(x) e^{-2\pi i k x/p} dx \\
&= \frac{1}{p} \int_{x=0}^{p} \int_{u=0}^{p} f_1(u) f_2(x-u) e^{-2\pi i k x/p} du\, dx \\
&= \frac{1}{p} \int_{u=0}^{p} f_1(u) e^{-2\pi i k u/p} \int_{x=0}^{p} f_2(x-u) e^{-2\pi i k(x-u)/p} dx\, du \\
&= \int_{u=0}^{p} f_1(u) e^{-2\pi i k u/p} F_2[k] du \\
&= p F_1[k] \cdot F_2[k].
\end{aligned}
$$

∎

Example: Show that

$$w_0(x) := \tfrac{1}{2} - x, \quad 0 < x < 1 \quad \text{has the FS} \quad \sum_{k \neq 0} \frac{e^{2\pi i k x}}{2\pi i k}. \tag{16}$$

Solution: From the graph of the sawtooth function f in Fig. 4.1 we see that

$$w_0(x) = -f\left(x - \tfrac{1}{2}\right)$$

when we set $p = 1$ in (6). We use the translation rule and the Fourier coefficients $F[k]$ from (6) to write

$$W_0[k] = -e^{-2\pi ik/2}\, F[k] = \begin{cases} 0 & \text{if } k = 0 \\ \dfrac{1}{2\pi ik} & \text{if } k = \pm 1, \pm 2, \dots, \end{cases}$$

and thereby obtain (16). ∎

Example: Show that

$$w_1(x) := \frac{-x^2}{2} + \frac{x}{2} - \frac{1}{12}, \quad 0 \le x \le 1, \quad \text{has the FS} \quad \sum_{k \ne 0} \frac{e^{2\pi ikx}}{(2\pi ik)^2}. \tag{17}$$

Solution: We can construct w_1 from the piecewise parabolic function of (7) by writing

$$w_1(x) = -\frac{1}{2}\left(x - \frac{1}{2}\right)^2 + \frac{1}{24} = -\frac{1}{2}f\left(x - \frac{1}{2}\right) + \frac{1}{24}$$

when $p = 1$. We then use the translation rule to obtain the Fourier coefficients

$$W_1[k] = \begin{cases} -\dfrac{1}{2}F[0] + \dfrac{1}{24} \\ -\dfrac{1}{2}e^{-2\pi ik/2}F[k] \end{cases} = \begin{cases} 0 & \text{if } k = 0 \\ \dfrac{1}{(2\pi ik)^2} & \text{if } k = \pm 1, \pm 2, \dots. \end{cases}$$

We can compute

$$W_1[0] := \int_0^1 w_1(x)\,dx = -\frac{1}{6} + \frac{1}{4} - \frac{1}{12} = 0,$$

observe that

$$w_1'(x) = w_0(x),$$

and use the derivative rule with (16) to obtain

$$W_1[k] = \frac{W_0[k]}{2\pi ik} = \frac{1}{(2\pi ik)^2}, \quad k = \pm 1, \pm 2, \dots.$$

We can verify that

$$(w_0 * w_0)(x) = \int_0^x \left(\frac{1}{2} - u\right)\left(\frac{1}{2} + u - x\right) du$$

$$+ \int_x^1 \left(\frac{1}{2} - u\right)\left(-\frac{1}{2} + u - x\right) dx = w_1(x), \quad 0 \le x \le 1,$$

and then use the convolution rule with (16) to write

$$W_1[k] = w_0[k] \cdot w_0[k] = \begin{cases} 0 & \text{if } k = 0 \\ \dfrac{1}{(2\pi i k)^2} & \text{if } k = \pm 1, \pm 2, \dots \end{cases}$$

There are many ways to use these rules! ∎

Poisson's relation

Let f be a piecewise smooth function on $\mathbb{R}$ that has small regular tails. We can use the *Poisson relation* (1.29) to see that

$$g(x) := \sum_{m=-\infty}^{\infty} f(x - mp) \quad (\text{with } f \text{ on } \mathbb{R} \text{ and } g \text{ on } \mathbb{T}_p)$$

has the FS $\qquad \displaystyle\sum_{k=-\infty}^{\infty} \frac{1}{p} F\left(\frac{k}{p}\right) e^{2\pi i k x / p}. \qquad (18)$

In particular, when we are given a piecewise smooth function g on $\mathbb{T}_p$ we can always choose a cutoff parameter $-\infty < a < \infty$, take

$$f(x) := \begin{cases} g(x) & \text{if } a \le x < a + p \\ 0 & \text{otherwise,} \end{cases}$$

find the Fourier transform F of f, and obtain the Fourier series for g from (18). The skills that you have acquired for finding Fourier transforms of functions on $\mathbb{R}$ can be used to find Fourier series for functions on $\mathbb{T}_p$! You may find it interesting to learn that this remarkable computational tool was discovered (but not published) by Gauss more than ten years before it appeared in a paper of Poisson, *cf.* C.F. Gauss, Schönes Theorem der Wahrscheinlichkeitsrechnung, *C.F. Gauss Werke, Band 8,* Königlichen Gesellschaften der Wissenschaften, Göttingen, 1900, pp. 88–89.

Example: Find the Fourier series for the p-periodic function g of Fig. 4.3.

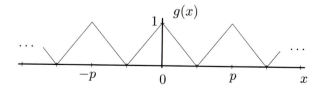

Figure 4.3. A p-periodic train of triangles.

Solution: We write

$$g(x) = \sum_{m=-\infty}^{\infty} \Lambda\left(\frac{x - mp}{p/2}\right),$$

use the Fourier transform calculus to see that

$$f(x) := \Lambda\left(\frac{2x}{p}\right) \quad \text{has the FT} \quad F(s) = \frac{p}{2}\operatorname{sinc}^2\left(\frac{ps}{2}\right),$$

and use Poisson's relation (18) to write

$$g(x) = \sum_{k=-\infty}^{\infty} \frac{1}{2}\operatorname{sinc}^2\left(\frac{k}{2}\right) e^{2\pi i kx/p}.$$

There is no need for the usual integration by parts!

We can tidy things up a bit by observing that

$$G[k] = \frac{1}{2}\operatorname{sinc}^2\left(\frac{k}{2}\right) = \begin{cases} \dfrac{1}{2} & \text{if } k = 0 \\ 0 & \text{if } k = \pm 2, \pm 4, \dots \\ \dfrac{2}{\pi^2 k^2} & \text{if } k = \pm 1, \pm 3, \dots \end{cases},$$

and writing

$$g(x) = \frac{1}{2} + \frac{4}{\pi^2}\left\{\frac{\cos(2\pi x/p)}{1^2} + \frac{\cos(6\pi x/p)}{3^2} + \frac{\cos(10\pi x/p)}{5^2} + \cdots\right\}. \quad \blacksquare$$

Example: Find the Fourier series for the p-periodic sawtooth function f shown in Fig. 4.1.

Solution: We use the power scaling rule to see that

$$q(x) := x \, \Pi\left(\frac{x}{p}\right)$$

has the Fourier transform

$$Q(s) = -\frac{1}{2\pi i}\frac{d}{ds}\left\{\frac{\sin(\pi p s)}{\pi s}\right\} = -\frac{1}{2\pi i}\left\{\frac{p\cos(\pi p s)}{s} - \frac{\sin(\pi p s)}{\pi s^2}\right\}, \quad s \neq 0$$

[with $Q(0) = 0$]. We use Poisson's relation to find

$$F[k] = \frac{1}{p}Q\left(\frac{k}{p}\right) = \begin{cases} 0 & \text{if } k = 0 \\ \dfrac{(-1)^{k+1}}{2\pi i k/p} & \text{if } k = \pm 1, \pm 2, \dots \end{cases},$$

and thereby obtain (6). $\blacksquare$

Example: Find the Fourier series for the p-periodic box train g shown in Fig. 4.4 when $0 < \alpha < p$.

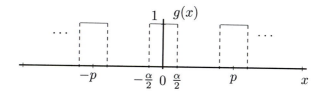

Figure 4.4. A p-periodic train of boxes.

Solution: You can find the Fourier transform of $f(x) := \Pi(x/\alpha)$ in your head and mentally use Poisson's relation to write

$$g(x) = \sum_{k=-\infty}^{\infty} \frac{\alpha}{p} \operatorname{sinc}\left(\frac{k\alpha}{p}\right) e^{2\pi i k x/p}.$$ ∎

Example: Use Poisson's relation to derive the translation rule (11) from the corresponding translation rule for functions on $\mathbb{R}$.

Solution: Let f be a piecewise smooth p-periodic function and let

$$d(x) := \begin{cases} f(x) & \text{if } 0 \le x < p \\ 0 & \text{otherwise,} \end{cases} \qquad g(x) := f(x - x_0).$$

By construction

$$f(x) = \sum_{m=-\infty}^{\infty} d(x - mp), \qquad g(x) = \sum_{m=-\infty}^{\infty} d(x - x_0 - mp),$$

and we can use Poisson's relation (twice) to write

$$G[k] = \frac{1}{p} e^{-2\pi i k x_0/p} D\left(\frac{k}{p}\right) = e^{-2\pi i k x_0/p} F[k].$$ ∎

You can derive all of the rules (8)–(14) in this way, *cf.* Ex. 4.8.

Example: Let $\alpha > 0$. Find the Fourier transform of the sampled gaussian

$$g_\alpha[n] := e^{-\pi(n\alpha/p)^2}, \quad n = 0, \pm 1, \pm 2, \dots .$$

Solution: We write $g_\alpha[n] = f(n/p)$ where the even function

$$f(x) := e^{-\pi(\alpha x)^2} \quad \text{has the FT} \quad F(s) = \frac{1}{\alpha} e^{-\pi(s/\alpha)^2}.$$

We now use the analysis equation (for functions on $\mathbb{Z}$), inversion rule (for functions on $\mathbb{R}$), and Poisson's relation to write

$$G_\alpha(s) := \frac{1}{p}\sum_{n=-\infty}^{\infty} f\left(\frac{n}{p}\right) e^{-2\pi isn/p} = \frac{1}{p}\sum_{k=-\infty}^{\infty} f\left(-\frac{k}{p}\right) e^{-2\pi iks/p}$$

$$= \sum_{m=-\infty}^{\infty} F(s - mp) = \frac{1}{\alpha}\sum_{m=-\infty}^{\infty} e^{-\pi(s-mp)^2/\alpha^2}.\qquad\blacksquare$$

Bernoulli functions and Eagle's method

We begin with

$$w_0(x) = \begin{cases} 0 & \text{if } x = 0, 1 \\ \dfrac{1}{2} - x & \text{if } 0 < x < 1 \end{cases} \qquad (19)$$

and take suitable antiderivatives to obtain polynomials

$$
\begin{aligned}
w_1(x) &= -\frac{x^2}{2} + \frac{x}{2} - \frac{1}{12}, & 0 \le x \le 1 \\
w_2(x) &= -\frac{x^3}{6} + \frac{x^2}{4} - \frac{x}{12}, & 0 \le x \le 1 \\
w_3(x) &= -\frac{x^4}{24} + \frac{x^3}{12} - \frac{x^2}{24} + \frac{1}{720}, & 0 \le x \le 1
\end{aligned}
\qquad (20)
$$

$$\vdots$$

with

$$w_n'(x) = w_{n-1}(x), \quad n = 1, 2, \ldots \text{ (and } x \ne 0, 1 \text{ when } n = 1) \qquad (21)$$

$$\int_0^1 w_n(x)\,dx = 0, \quad n = 0, 1, \ldots, \qquad (22)$$

cf. Ex. 4.22. We have found the Fourier representations (16), (17) for w_0 and w_1, and studied the convergence of these series in Section 1.5. We will now use (16) with (21)–(22) and the derivative rule to obtain the 1-periodic *Bernoulli functions*

$$w_n(x) := \sum_{\substack{k=-\infty \\ k \ne 0}}^{\infty} \frac{e^{2\pi ikx}}{(2\pi ik)^{n+1}}, \quad n = 0, 1, \ldots, \quad -\infty < x < \infty, \qquad (23)$$

cf. Fig. 4.5. You should try to remember the Fourier series (23) and the polynomial forms (19)–(20) for w_0, w_1, w_2. The most important thing about these functions is that they have been constructed so that

$$w_n^{(m)}(0+) - w_n^{(m)}(0-) = \begin{cases} 1 & \text{if } m = n \\ 0 & \text{if } m = 0, 1, \ldots, n-1, n+1, n+2, \ldots . \end{cases} \quad (24)$$

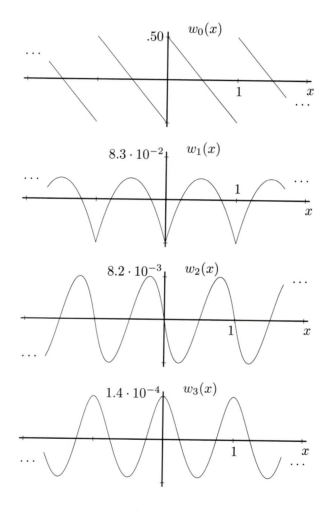

Figure 4.5. The first four Bernoulli functions.

You may recall that in Section 1.5 we used scaled translates of $w_0(x/p)$ and $w_1(x/p)$ to remove jump discontinuities from a p-periodic function f and its derivative. We will now use a generalization of this idea to construct Fourier series for piecewise polynomial functions on $\mathbb{T}_p$. The overall scheme is an elementary variation of *Eagle's method* that will be developed within a more general context in Section 7.5 and Ex. 7.75.

Example: Find the Fourier series for the p-periodic box train of Fig. 4.4.

Solution: The function g has jumps $+1, -1$ at $x = -\alpha/2,\ \alpha/2$. We remove these jumps [*cf.* (1.38)] by forming

$$c(x) := g(x) - w_0\left(\frac{x + \alpha/2}{p}\right) + w_0\left(\frac{x - \alpha/2}{p}\right).$$

Each term is a p-periodic broken line, so c is also such a function. In view of (24), neither c nor c' have jump discontinuities, so this p-periodic broken line must be a constant. Since w_0 has the average value 0, we can find this constant by evaluating the integral

$$c(x) = \frac{1}{p}\int_{-p/2}^{p/2} g(u)du = \frac{1}{p}\int_{-\alpha/2}^{\alpha/2} 1 \cdot du = \frac{\alpha}{p}.$$

In this way we see that

$$g(x) = \frac{\alpha}{p} + w_0\left(\frac{x + \alpha/2}{p}\right) - w_0\left(\frac{x - \alpha/2}{p}\right).$$

We now use this representation with (23) to obtain the Fourier series

$$g(x) = \frac{\alpha}{p} + \sum_{k \neq 0}\left\{\frac{e^{2\pi i k(x+\alpha/2)/p}}{2\pi i k} - \frac{e^{2\pi i k(x-\alpha/2)/p}}{2\pi i k}\right\}$$

$$= \sum_{k=-\infty}^{\infty} \frac{\alpha}{p}\,\mathrm{sinc}\left(\frac{\alpha k}{p}\right)e^{2\pi i k x/p}. \qquad\blacksquare$$

Example: Find the Fourier series for the p-periodic ramp train of Fig. 4.6.

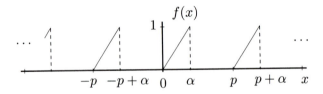

Figure 4.6. A p-periodic train of ramps.

Solution: The function f has a jump -1 at $x = \alpha$ and the derivative f' has jumps $1/\alpha, -1/\alpha$ at $x = 0, \alpha$. We remove these jumps by forming

$$c(x) := f(x) + w_0\left(\frac{x - \alpha}{p}\right) - \frac{p}{\alpha}w_1\left(\frac{x}{p}\right) + \frac{p}{\alpha}w_1\left(\frac{x - \alpha}{p}\right),$$

and with reasoning analogous to that used for the previous example, we conclude that the p-periodic piecewise polynomial function c is the constant

$$c(x) = \frac{1}{p}\int_0^p f(u)du = \frac{1}{p}\int_0^\alpha \frac{udu}{\alpha} = \frac{\alpha}{2p}.$$

In this way we deduce that

$$f(x) = \frac{\alpha}{2p} + \frac{p}{\alpha}w_1\left(\frac{x}{p}\right) - w_0\left(\frac{x - \alpha}{p}\right) - \frac{p}{\alpha}w_1\left(\frac{x - \alpha}{p}\right)$$

and use (23) to obtain the Fourier series

$$f(x) = \frac{\alpha}{2p} + \sum_{k \neq 0} \frac{1}{(2\pi ik)^2}\left\{\frac{p}{\alpha} - e^{-2\pi ik\alpha/p}\left(2\pi ik + \frac{p}{\alpha}\right)\right\}e^{2\pi ikx/p}.$$

You can look at Fig. 4.6 and immediately write down this series! ∎

Laurent series

In your study of calculus you learned to work with *power series* such as

$$e^z = 1 + z + \frac{z^2}{2!} + \frac{z^3}{3!} + \cdots,$$

$$\frac{1}{1 - z} = 1 + z + z^2 + z^3 + \cdots, \quad |z| < 1,$$

and you may remember developing the Laurent series

$$\frac{1 - r^2}{(1 - rz)(1 - r/z)} = \frac{1}{1 - rz} + \frac{r/z}{1 - r/z} = \sum_{k=-\infty}^{\infty} r^{|k|}z^k, \quad 0 < r < |z| < \frac{1}{r}$$

to solve Ex. 1.9. A *Laurent series*

$$C(z) = \sum_{k=-\infty}^{\infty} c_k z^k, \quad z \in \mathbb{C}, \quad a < |z| < b \tag{25}$$

is a complex power series that may contain terms with $z^{-1}, z^{-2}, \ldots$ as well as terms with $1, z, z^2, \ldots$. If the Laurent series (25) converges within some nondegenerate closed annulus that contains the unit circle (i.e., for $a \leq |z| \leq b$ where $a \leq 1$, $a < b$, $b \geq 1$), then you can produce a corresponding Fourier series by setting $z = e^{2\pi i x/p}$, i.e.,

$$f(x) := \mathcal{C}(e^{2\pi i x/p}) \quad \text{has the FS} \quad \sum_{k=-\infty}^{\infty} c_k e^{2\pi i k x/p}, \tag{26}$$

cf. Tolstov, pp. 105–112.

Example: Find the Fourier series for the function $f(x) := \cos\{e^{2\pi i x/p}\}$.

Solution: We set $z = e^{2\pi i x/p}$ in the Maclaurin series

$$\mathcal{C}(z) := \cos z = 1 - \frac{z^2}{2!} + \frac{z^4}{4!} - \frac{z^6}{6!} + \cdots$$

for the cosine function and thereby obtain the Fourier series

$$f(x) = 1 - \frac{e^{4\pi i x/p}}{2!} + \frac{e^{8\pi i x/p}}{4!} - \frac{e^{12\pi i x/p}}{6!} + \cdots . \qquad \blacksquare$$

Example: Find the Fourier series for $f(x) := \sin^2(2\pi x/p)$.

Solution: We define

$$\mathcal{C}(z) := \left(\frac{z - z^{-1}}{2i}\right)^2 = -\frac{1}{4}z^{-2} + \frac{1}{2} - \frac{1}{4}z^2$$

and write

$$f(x) = \mathcal{C}(e^{2\pi i x/p}) = -\frac{1}{4}e^{-4\pi i x/p} + \frac{1}{2} - \frac{1}{4}e^{4\pi i x/p}. \qquad \blacksquare$$

Example: Show that

$$f(x) := \frac{3}{5 - 4\cos(2\pi x/p)} \quad \text{has the FS} \quad \sum_{k=-\infty}^{\infty} \left(\frac{1}{2}\right)^{|k|} e^{2\pi i k x/p}. \tag{27}$$

Solution: We define

$$\mathcal{C}(z) := \frac{3}{5 - 4(z + z^{-1})/2} = \frac{3z}{(2z - 1)(2 - z)},$$

use the partial fraction decomposition

$$\mathcal{C}(z) = \frac{2}{2 - z} + \frac{1}{2z - 1} = \frac{1}{1 - (z/2)} + \frac{(1/2z)}{1 - (1/2z)}$$

and the formula for the sum of a geometric progression to obtain the Laurent series

$$\mathcal{C}(z) = \sum_{k=-\infty}^{\infty} \left(\frac{1}{2}\right)^{|k|} z^k, \quad \frac{1}{2} < |z| < 2,$$

and thereby find $F[k] = 2^{-|k|}$. ∎

Example: Show that the *Dirichlet kernel*

$$f(x) := \frac{\sin\{(2m+1)\pi x/p\}}{\sin(\pi x/p)} \quad \text{has the FS} \quad \sum_{k=-m}^{m} e^{2\pi ikx/p}, \quad m = 0, 1, \ldots . \quad (28)$$

Solution: We set $w := e^{i\pi x/p}$, $z := w^2 = e^{2\pi ix/p}$ and write

$$f(x) = \frac{(w^{2m+1} - w^{-(2m+1)})/2i}{(w - w^{-1})/2i} = \frac{z^{2m+1} - 1}{z^m(z-1)} = \sum_{k=-m}^{m} z^k = \sum_{k=-m}^{m} e^{2\pi ikx/p}. \quad ∎$$

In each of the previous three examples we were asked to find the Fourier series for a continuous rational function of $\cos(2\pi x/p)$ and $\sin(2\pi x/p)$. It is possible to find the Fourier series for any such function by using the Laurent series method together with manipulations that are familiar from elementary calculus. Additional details are best developed with concepts from complex analysis that are beyond the scope of this book.

Dilation and grouping rules

When f is a p-periodic function on $\mathbb{R}$ and $m = 1, 2, \ldots$, the dilate $f(mx)$ is p-periodic as well as p/m-periodic. We can verify the *dilation rule*

$$g(x) := f(mx) \quad \text{has the FT} \quad G[k] = \begin{cases} F[k/m] & \text{if } k = 0, \pm m, \pm 2m, \ldots \\ 0 & \text{otherwise} \end{cases} \quad (29)$$

by using the synthesis equation for f to write

$$g(x) = \sum_{\kappa=-\infty}^{\infty} F[\kappa] e^{2\pi i\kappa mx/p} = \sum_{m|k} F[k/m] e^{2\pi ikx/p}.$$

We obtain G by packing $m - 1$ zeros between successive components of F, cf. Fig. 4.7. The dilation rules for functions on $\mathbb{R}$ and for functions on $\mathbb{T}_p$ are quite different!

Let f be p-periodic and let $m = 1, 2, \ldots$. We sum the p/m translates of f to produce a p/m-periodic function

$$f_m(x) := f(x) + f\left(x - \frac{p}{m}\right) + f\left(x - \frac{2p}{m}\right) + \cdots + f\left(x - \frac{(m-1)p}{m}\right)$$

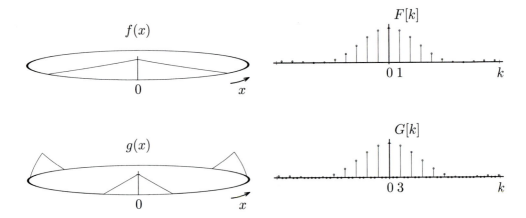

Figure 4.7. The functions $f(x)$, $g(x) := f(3x)$ on $\mathbb{T}_p$ and their Fourier transforms $F[k]$, $G[k]$ on $\mathbb{Z}$.

that has the p-periodic dilate $f_m(x/m)$, cf. Fig. 4.8. We verify the corresponding *grouping rule*

$$g(x) := \sum_{\ell=0}^{m-1} f\left(\frac{x}{m} - \ell\frac{p}{m}\right) \quad \text{has the FT} \quad G[k] = mF[mk] \tag{30}$$

by using the synthesis equation for f and the discrete orthogonality relation (1.19) to write

$$g(x) = \sum_{\ell=0}^{m-1} \sum_{\kappa=-\infty}^{\infty} F[\kappa] e^{2\pi i \kappa (x/m - \ell p/m)/p}$$

$$= \sum_{\kappa=-\infty}^{\infty} F[\kappa] e^{2\pi i \kappa x/mp} \cdot \sum_{\ell=0}^{m-1} e^{-2\pi i \kappa \ell/m}$$

$$= \sum_{\kappa=-\infty}^{\infty} F[\kappa] e^{2\pi i \kappa x/mp} \cdot \begin{cases} m & \text{if } \kappa = 0, \pm m, \pm 2m, \ldots \\ 0 & \text{otherwise} \end{cases}$$

$$= \sum_{k=-\infty}^{\infty} mF[mk] e^{2\pi i k x/p}.$$

You will observe that the Fourier transform G is a (scaled) dilate of F, so the dual of (30) gives the dilation rule for functions on $\mathbb{Z}$, cf. Appendix 2. The dilation rules for functions on $\mathbb{R}$ and for functions on $\mathbb{Z}$ are quite different!

We will not use the somewhat exotic rules (29)–(30) as much as the basic rules (8)–(15) [but (29)–(30) will be needed for our analysis of filter banks in Section 10.4].

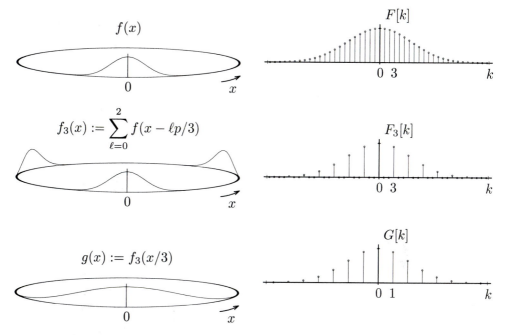

Figure 4.8. The functions $f(x)$, $f_3(x)$, $g(x) = f_3(x/3)$ and their Fourier transforms $F[k]$, $F_3[k]$, $G[k]$.

We have included (29)–(30) to help you understand some of the corresponding identities for working with DFTs that we will describe a bit later in this chapter.

Example: Let f be a suitably regular p-periodic function with the Fourier series

$$f(x) = \sum_{k=-\infty}^{\infty} F[k]e^{2\pi ikx/p},$$

and assume that f is p/m-periodic for some $m = 2, 3, \ldots$. Show that $F[k] = 0$ when $k \neq 0, \pm m, \pm 2m, \ldots$.

Solution: We can apply the translation rule (11) to

$$f\left(x - \frac{p}{m}\right) - f(x) = 0$$

and reach the conclusion from the resulting identity

$$(e^{-2\pi ik/m} - 1)F[k] = 0, \quad k = 0, \pm 1, \pm 2, \ldots .$$

We can also verify that

$$g(x) := \sum_{\ell=0}^{m-1} f\left(\frac{x}{m} - \ell\frac{p}{m}\right) = mf\left(\frac{x}{m}\right)$$

and reach the conclusion from the identity

$$\sum_{k=-\infty}^{\infty} F[k]e^{2\pi ikx/mp} = f\left(\frac{x}{m}\right) = \sum_{\kappa=-\infty}^{\infty} F[m\kappa]e^{2\pi im\kappa/mp}$$

that results from the synthesis equation for f and grouping rule (30).   ∎

4.2 Selected Applications of Fourier Series

Evaluation of sums and integrals

Now that you know how to find Fourier series, you can use the analysis and synthesis equations of Fourier or the identities of Parseval and Plancherel to evaluate many sums and integrals that cannot be found with the tools of calculus.

Example: Show that

$$1 - \frac{1}{3} + \frac{1}{5} - \frac{1}{7} + \cdots = \frac{\pi}{4}.$$

Solution: From the analysis in Section 1.5 we know that the piecewise linear function from Fig. 4.1 is represented by the symmetric form of its Fourier series (6). In particular, when $p = 1$ we have

$$x = \frac{1}{\pi}\sum_{k=1}^{\infty}(-1)^{k+1}\frac{\sin(2\pi kx)}{k}, \qquad -\frac{1}{2} < x < \frac{1}{2}.$$

We obtain the desired identity by setting $x = 1/4$.   ∎

Example: Find the value of the sum

$$S := 1 + \frac{1}{2^2} + \frac{1}{3^2} + \frac{1}{4^2} + \cdots .$$

Solution: We apply Plancherel's identity (1.16) to the Fourier series (6) (with $p = 1$) to obtain

$$S = \frac{4\pi^2}{2}\sum_{k\neq 0}\left|\frac{(-1)^{k+1}}{2\pi ik}\right|^2 = 2\pi^2\sum_{k=-\infty}^{\infty}|F[k]|^2$$

$$= 2\pi^2\int_{-1/2}^{1/2}|f(x)|^2\,dx = 4\pi^2\int_{0}^{1/2}x^2\,dx = \frac{\pi^2}{6}.$$ ∎

Example: Find the value of the integral

$$I_k := \int_0^1 \frac{3\cos(2\pi k x)dx}{5 - 4\cos(2\pi x)}, \quad k = 0, 1, \ldots .$$

Solution: Let $f(x) := 3/(5 - 4\cos 2\pi x)$. Using (27), we write

$$I_k = \int_0^1 \frac{3\,e^{-2\pi i k x}}{5 - 4\cos 2\pi x}dx = F[k] = \left(\frac{1}{2}\right)^{|k|}, \quad k = 0, \pm 1, \pm 2, \ldots . \qquad \blacksquare$$

The polygon function

Let $\mathfrak{P}_N$ be the regular N-gon with vertices $e^{2\pi i k/N}$, $k = 0, 1, \ldots, N-1$, in the complex plane, and let

$$p := 2N \sin\left(\frac{\pi}{N}\right)$$

be the corresponding perimeter. Let $f(x)$ be the p-periodic function that results when we specify the points of $\mathfrak{P}_N$ using the arclength x (measured counterclockwise from the vertex at 1) as a parameter. Thus Re f, Im f are piecewise linear functions that interpolate the points

$$\left(\frac{np}{N}, \cos\left(\frac{2\pi n}{N}\right)\right), \qquad \left(\frac{np}{N}, \sin\left(\frac{2\pi n}{N}\right)\right),$$

respectively, *cf.* Fig. 4.9. We will show how to use rules from the Fourier transform calculus to find the Fourier series for the *polygon function f*. (A Kronecker rule calculation is singularly unappealing!)

We define

$$g(x) := e^{-2\pi i x/p}\,f(x)$$

and use the modulation rule (12) to conclude that

$$G[k] = F[k+1], \quad k = 0, \pm 1, \pm 2, \ldots .$$

The function g is p/N-periodic (a clockwise rotation by $2\pi/N$ cancels the counterclockwise advance of x by p/N), so we can use the translation rule (11) to see that

$$G[k] = 0 \quad \text{when} \quad k \neq 0, \pm N, \pm 2N, \ldots .$$

The functions g, g' satisfy

$$g'(x) + \frac{2\pi i}{p}g(x) = e^{-2\pi i x/p}\,f'(x) = e^{-2\pi i x/p}\,C, \quad 0 < x < p/N$$

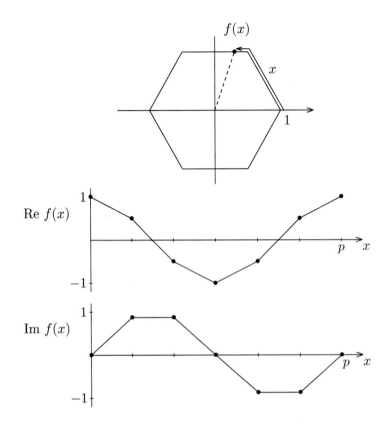

Figure 4.9. Construction of the polygon function f by parametrization of the regular N-gon $\mathfrak{P}_N$ and by piecewise linear interpolation of $\cos(2\pi x/p)$, $\sin(2\pi x/p)$ when $N = 6$.

where

$$C := \frac{f(p/N) - f(0)}{p/N} = \frac{e^{2\pi i/N} - 1}{p/N} = \frac{2\pi i}{p} e^{i\pi/N} \text{sinc}\left(\frac{1}{N}\right).$$

Since g, g' are p/N-periodic we can use the derivative rule (15) to write

$$\frac{2\pi im}{p/N} G[mN] + \frac{2\pi i}{p} G[mN]$$

$$= \frac{N}{p} \int_{x=0}^{p/N} C e^{-2\pi ix/p} e^{-2\pi imNx/p} dx$$

$$= -\frac{p}{2\pi i} \frac{|C|^2}{1 + mN}.$$

With a bit of algebra we find

$$G[mN] = \frac{\text{sinc}^2(1/N)}{(1+mN)^2}, \quad m = 0, \pm 1, \pm 2, \ldots ,$$

and thereby obtain the Fourier series

$$f(x) = \sum_{m=-\infty}^{\infty} \frac{\text{sinc}^2(1/N)}{(1+mN)^2} e^{2\pi i(1+mN)x/p}. \tag{31}$$

Rates of decay

Let $f, f', \ldots, f^{(n-1)}$ be continuous p-periodic functions and let $f^{(n)}$ be piecewise continuous and absolutely integrable on the interval $[0, p)$. You can use repeated integration by parts to write

$$\left(\frac{2\pi ik}{p}\right)^n F[k] = \frac{1}{p} \int_0^p f^{(n)}(x) e^{-2\pi ikx/p} dx \tag{32}$$

and thereby show that $F[k]$ decays to zero at least as fast as $1/k^n$ in the limit $k \to \pm\infty$, cf. Exs. 3.42 and 3.43. For variety, we will give an alternative argument based on Poisson's relation (18).

We define the B-splines

$$B_0 := \Pi, \quad B_1 = \Pi * \Pi, \quad B_2 = \Pi * \Pi * \Pi, \ldots ,$$

and recall that $B_{n+1}, B'_{n+1}, \ldots, B_{n+1}^{(n)}$ are continuous functions that vanish when $|x| > 1 + n/2$, cf. Ex. 2.7. We verify that the 1-translates of B_{n+1} form a *partition of unity* by using Poisson's relation (18) to write

$$\sum_{m=-\infty}^{\infty} B_{n+1}(x - m) = \sum_{k=-\infty}^{\infty} \text{sinc}^{n+2}(k) e^{2\pi ikx} = 1. \tag{33}$$

We define

$$g(x) := f(x) B_{n+1}\left(\frac{x}{p}\right)$$

on $\mathbb{R}$ and use (33) to write

$$f(x) = f(x) \sum_{m=-\infty}^{\infty} B_{n+1}\left(\frac{x}{p} - m\right)$$

$$= \sum_{m=-\infty}^{\infty} f(x - mp) B_{n+1}\left(\frac{x - mp}{p}\right)$$

$$= \sum_{m=-\infty}^{\infty} g(x - mp).$$

A second application of Poisson's relation now gives

$$F[k] = \frac{1}{p} G\left(\frac{k}{p}\right), \quad k = 0, \pm 1, \pm 2, \dots . \tag{34}$$

The function g on $\mathbb{R}$ has the same smoothness as the function f on $\mathbb{T}_p$, so we can use (34) to see that the Fourier coefficients $F[k]$ have the same rate of decay as the Fourier transform $G(s)$ of g, cf. Exs. 3.42 and 3.43.

Equidistribution of arithmetic sequences

Let x_0, γ be real numbers with $0 \le x_0 < 1$ and with γ being irrational. For $n = 1, 2, \dots$ let

$$x_n := x_0 + n\gamma + m_n$$

where the integer m_n is chosen to make $0 \le x_n < 1$. Thus $x_0, x_1, x_2, \dots$ is the arithmetic progression $x_0, x_0 + \gamma, x_0 + 2\gamma, \dots$ on the circle $\mathbb{T}_1$. We will show that these numbers are more or less evenly distributed around the circle. [You can use this idea to generate "random" numbers from $[0,1)$!] More specifically, we will show that

$$\lim_{N \to \infty} \frac{1}{N+1} \sum_{n=0}^{N} f(x_n) = \int_0^1 f(x)dx \tag{35}$$

when I is any interval from $\mathbb{T}_1$ and

$$f(x) := \begin{cases} 1 & \text{if } x \in I \\ 0 & \text{if } x \notin I \end{cases}$$

is the corresponding *indicator function*. This result is known as *Weyl's equidistribution theorem*.

When g is a suitably regular function on $\mathbb{T}_1$ we define

$$E_N\{g\} := \frac{1}{N+1} \sum_{n=0}^{N} g(x_n) - \int_0^1 g(x)dx,$$

and we set

$$e_k(x) := e^{2\pi i k x}, \quad k = 0, \pm 1, \pm 2, \dots .$$

We easily verify that

$$|E_N\{e_k\}| \le 2, \quad N = 1, 2, \dots, \quad k = 0, \pm 1, \pm 2, \dots, \tag{36}$$

$$E_N\{e_0\} = 0, \quad N = 1, 2, \dots, \text{ and} \tag{37}$$

$$E_N\{e_k\} = \frac{1}{N+1} \sum_{n=0}^{N} e^{2\pi i k(x_0 + n\gamma)} - \int_0^1 e^{2\pi i k x} dx$$

$$= \frac{e^{2\pi i k x_0}}{N+1} \frac{e^{2\pi i k(N+1)\gamma} - 1}{e^{2\pi i k \gamma} - 1} \qquad (38)$$

$$\to 0 \text{ as } N \to \infty \text{ when } k = \pm 1, \pm 2, \dots .$$

The identity (37) gives (35) when I has length 1. Henceforth we will assume that the length of this interval is less than 1.

Given any sufficiently small $\epsilon > 0$ we construct continuous piecewise linear functions $\ell_\epsilon, u_\epsilon$ as shown in Fig. 4.10. We observe that

$$E_N\{f\} \le \frac{1}{N+1} \sum_{n=0}^{N} u_\epsilon(x_n) - \int_0^1 f(x) dx \le E_N\{u_\epsilon\} + \epsilon.$$

There is an analogous lower bound, so

$$E_N\{\ell_\epsilon\} - \epsilon \le E_N\{f\} \le E_N\{u_\epsilon\} + \epsilon. \qquad (39)$$

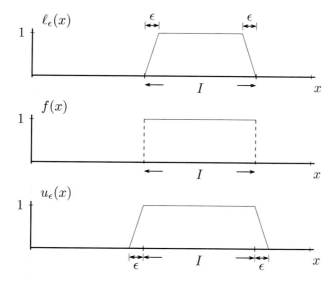

Figure 4.10. The functions ℓ_ϵ, f, u_ϵ used in the proof of Weyl's equidistribution theorem.

We use the Fourier representation for $u_\epsilon(x)$ and (36) to write

$$
|E_N\{u_\epsilon\}| = \left| E_N\left\{ \sum_{|k|\leq m} U_\epsilon[k]e_k + \sum_{|k|>m} |U_\epsilon[k]| e_k \right\} \right|
\tag{40}
$$

$$
\leq \sum_{|k|\leq m} |U_\epsilon[k]|\, |E_N\{e_k\}| + 2\sum_{|k|>m} |U_\epsilon[k]|.
$$

The Fourier coefficients of the continuous piecewise linear function $U_\epsilon[k]$ decay as fast as $1/k^2$, so the second term on the right of (40) will be less than ϵ when m is sufficiently large. We can then use (37)–(38) to see that the first term will be less than ϵ when N is sufficiently large. In conjunction with (39) and a similar analysis for ℓ_ϵ, this shows that

$$
|E_N\{f\}| \leq 3\epsilon
$$

when N is sufficiently large and thereby proves (35).

4.3 Discrete Fourier Transforms

Direct summation

You can find a number of commonly used discrete Fourier transforms by evaluating the finite sum from the analysis equation

$$
F[k] := \frac{1}{N}\sum_{n=0}^{N-1} f[n]e^{-2\pi ikn/N}.
$$

For example, you can verify that

$$
f[n] := \delta[n] \quad \text{has the FT} \quad F[k] = \frac{1}{N}\,,
\tag{41}
$$

$$
f[n] := 1 \quad\quad \text{has the FT} \quad F[k] = \delta[k]\,,
\tag{42}
$$

cf. Fig. 4.11, by evaluating such terms in your head (and using the discrete orthogonality relation (1.20) for (42).) Here

$$
\delta[n] := \begin{cases} 1 & \text{if } n = 0, \pm N, \pm 2N, \ldots \\ 0 & \text{otherwise} \end{cases}
$$

is the *discrete delta* [which serves as the identity (2.20) for the convolution product].

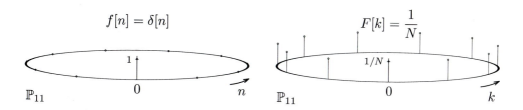

Figure 4.11. The discrete delta $f[n] = \delta[n]$ and its Fourier transform $F[k] = 1/N$ when $N = 11$.

When $m = 1, 2, \ldots$ evenly divides N, we define the *discrete comb*

$$c_m[n] := \begin{cases} 1 & \text{if } n = 0, \pm m, \pm 2m, \ldots \\ 0 & \text{otherwise} \end{cases} \tag{43}$$

on $\mathbb{P}_N$. The subscript m specifies the spacing between adjacent teeth and $m' := N/m$ gives the number of teeth. We verify that

$$f[n] := c_m[n] \quad \text{has the FT} \quad F[k] = \frac{1}{m} c_{N/m}[k] \tag{44}$$

by writing

$$C_m[k] := \frac{1}{N} \sum_{n=0}^{N-1} c_m[n] e^{-2\pi ikn/N} = \frac{1}{N} \sum_{n'=0}^{m'-1} e^{-2\pi ik \cdot mn'/mm'}$$

$$= \begin{cases} \dfrac{m'}{N} & \text{if } k = 0, \pm m', \pm 2m', \ldots \\ 0 & \text{otherwise} \end{cases} = \frac{1}{m} c_{m'}[k],$$

cf. Fig. 4.12. When we specialize (44) by setting $m = N$, $m = 1$, we obtain the DFTs (41), (42), respectively.

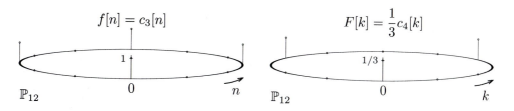

Figure 4.12. The discrete comb $f[n] = c_3[n]$ and its Fourier transform $F[k] = \frac{1}{3} c_4[k]$ when $N = 12$.

You can always use a brute force calculation to find a DFT when N is "small." (When N is "large" we do such calculations on a computer using one of the fast algorithms that we will describe in Chapter 6.)

Example: Find the DFT of $f := (1, 2, 3, 4)$.

Solution: Here $N = 4$, $e^{-2\pi i/N} = -i$, and $f[n] = n + 1$, $n = 0, 1, 2, 3$, so we can use the analysis equation to write

$$4F[0] = 1 \cdot (-i)^{0 \cdot 0} + 2 \cdot (-i)^{1 \cdot 0} + 3 \cdot (-i)^{2 \cdot 0} + 4 \cdot (-i)^{3 \cdot 0} = 10$$
$$4F[1] = 1 \cdot (-i)^{0 \cdot 1} + 2 \cdot (-i)^{1 \cdot 1} + 3 \cdot (-i)^{2 \cdot 1} + 4 \cdot (-i)^{3 \cdot 1} = -2 + 2i$$
$$4F[2] = 1 \cdot (-i)^{0 \cdot 2} + 2 \cdot (-i)^{1 \cdot 2} + 3 \cdot (-i)^{2 \cdot 2} + 4 \cdot (-i)^{3 \cdot 2} = -2$$
$$4F[3] = 1 \cdot (-i)^{0 \cdot 3} + 2 \cdot (-i)^{1 \cdot 3} + 3 \cdot (-i)^{2 \cdot 3} + 4 \cdot (-i)^{3 \cdot 3} = -2 - 2i.$$

In this way we see that

$$f := (1, 2, 3, 4) \quad \text{has the DFT} \quad F = \frac{1}{2}(5, -1 + i, -1, -1 - i). \qquad (45)$$

∎

Several discrete Fourier transforms can be found by suitably manipulating the formula

$$1 + z + z^2 + \cdots + z^{N-1} = \frac{z^N - 1}{z - 1}, \quad z \neq 1$$

for the sum of a geometric progression.

Example: Let r be a complex number. Show that

$$f[n] := r^n, \quad n = 0, 1, \ldots, N - 1 \quad \text{has the FT}$$

$$F[k] = \begin{cases} 1 & \text{if } e^{2\pi ik/N} = r \\ \dfrac{r^N - 1}{N(r\, e^{-2\pi ik/N} - 1)} & \text{otherwise.} \end{cases} \qquad (46)$$

Solution: We sum a geometric progression as we write

$$F[k] := \frac{1}{N} \sum_{n=0}^{N-1} (r\, e^{-2\pi ik/N})^n = \frac{1}{N} \begin{cases} N & \text{if } e^{2\pi ik/N} = r \\ \dfrac{(r\, e^{-2\pi ik/N})^N - 1}{r\, e^{-2\pi ik/N} - 1} & \text{otherwise} \end{cases}$$

and this gives (46). Figure 4.13 shows f, F when $r = .8$, $N = 100$. You might compare this illustration to Fig. 3.2. ∎

Other examples can be found in Exs. 4.26–4.28.

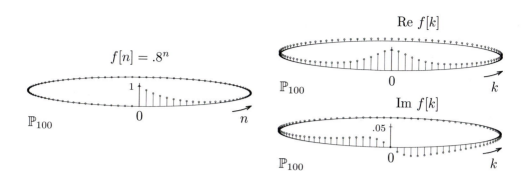

Figure 4.13. The function $f[n] = .8^n$, $n = 0, 1, \ldots, N - 1$ and its Fourier transform $F[k]$ when $N = 100$.

Basic rules

By now you will undoubtedly expect to see certain rules for working with Fourier transforms of functions on $\mathbb{P}_N$. You can always use the linearity rule

$$g[n] := c_1 f_1[n] + \cdots + c_m f_m[n] \text{ has the FT } G[k] = c_1 F_1[k] + \cdots + c_m F_m[k], \quad (47)$$

the reflection and conjugation rules

$$g[n] := f[-n] \quad \text{has the FT} \quad G[k] = F[-k], \tag{48}$$

$$g[n] := \overline{f[n]} \quad \text{has the FT} \quad G[k] = \overline{F[-k]}, \tag{49}$$

the translation and modulation rules

$$g[n] := f[n - n_0] \text{ has the FT } G[k] = e^{-2\pi i k n_0/N} F[k], \ n_0 = 0, \pm 1, \pm 2, \ldots, \quad (50)$$

$$g[n] := e^{2\pi i k_0 n/N} f[n] \text{ has the FT } G[k] = F[k - k_0], \ k_0 = 0, \pm 1, m2, \ldots, \quad (51)$$

the convolution and multiplication rules

$$g[n] := (f_1 * f_2)[n] \quad \text{has the FT} \quad G[k] = N F_1[k] \cdot F_2[k], \tag{52}$$

$$g[n] := f_1[n] \cdot f_2[n] \quad \text{has the FT} \quad G[k] = (F_1 * F_2)[k], \tag{53}$$

and the inversion rule

$$g[n] := F[k] \quad \text{has the FT} \quad G[k] = \frac{1}{N} f[-k]. \tag{54}$$

The form of the complex exponential $e^{2\pi i k n/N}$ that we use for Fourier's representation of functions on $\mathbb{P}_N$ accounts for the form of the complex exponentials in

(50)–(51). The *convolution* rule (52) gets the *constant, N*, and the *inversion* rule (54) gets the *inverse* constant, $1/N$, *cp.* (13) and (3).

Example: Let $k_0 = 0, \pm 1, \pm 2, \ldots$. Find the discrete Fourier transform of $f[n] := \cos(2\pi k_0 n/N)$.

Solution: We use Euler's formula to write

$$f[n] = \frac{1}{2} e^{2\pi i k_0 n/N} \cdot 1 + \frac{1}{2} e^{-2\pi i k_0 n/N} \cdot 1$$

and apply the modulation rule (51) to (42) to obtain

$$F[k] = \frac{1}{2}\delta[k - k_0] + \frac{1}{2}\delta[k + k_0],$$

cf. Fig. 4.14. Since f is both even and N-periodic, we can always replace k_0 by one of the integers $0, 1, \ldots, \lfloor N/2 \rfloor$. (The *floor function* $\lfloor x \rfloor$ is the largest integer that does not exceed the real argument x.) ∎

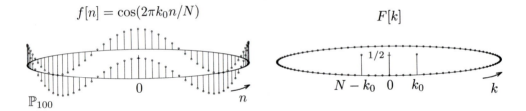

$f[n] = \cos(2\pi k_0 n/N)$ $F[k]$

Figure 4.14. The function $f[n] = \cos(2\pi k_0 n/N)$ and its Fourier transform $F[k]$ when $N = 100$ and $k_0 = 4$.

Example: Derive the inversion rule (54).

Solution: We use the analysis equation for $g := F$ and the synthesis equation for f to write

$$G[k] := \frac{1}{N} \sum_{n=0}^{N-1} F[n] e^{-2\pi i k n/N} = \frac{1}{N} \sum_{n=0}^{N-1} F[n] e^{2\pi i n(-k)/N} = \frac{1}{N} f[-k].$$ ∎

Example: Let $N = m \cdot m'$ where m, m' are positive integers and assume that f is an m-periodic function on $\mathbb{P}_N$. Show that $F[k] = 0$ when k is not a multiple of m'.

Solution: We can use the translation rule (50) and the periodicity

$$f[n + m] - f[n] = 0$$

to write

$$(e^{2\pi i k m/N} - 1)F[k] = (e^{2\pi i k/m'} - 1)F[k] = 0$$

and thereby conclude that $F[k] = 0$ when $m' \nmid k$.

We can reach the same conclusion by observing that

$$f = g * c_m$$

where

$$g[n] := \begin{cases} f[n] & \text{if } n = 0, 1, \ldots, m-1 \\ 0 & \text{if } n = m, m+1, \ldots, N-1, \end{cases}$$

and using the convolution rule (52) with (44) to write

$$F[k] = NG[k] \cdot \frac{1}{m} c_{m'}[k]. \qquad \blacksquare$$

Rules that link functions on $\mathbb{P}_N$ with functions on $\mathbb{P}_{N/m}$ and $\mathbb{P}_{m \cdot N}$

Let (a, b, c, d) represent a function on $\mathbb{P}_4$ and assume that

$$(a, b, c, d) \quad \text{has the DFT} \quad (A, B, C, D).$$

(Here a, b, c, d, A, B, C, D are complex scalars.) We will show that

$$(a, 0, b, 0, c, 0, d, 0) \quad \text{has the DFT} \quad \frac{1}{2}(A, B, C, D, A, B, C, D),$$

$$(a, 0, 0, b, 0, 0, c, 0, 0, d, 0, 0) \quad \text{has the DFT} \quad \frac{1}{3}(A, B, C, D, A, B, C, D, A, B, C, D),$$

$$\vdots$$

and that

$$(a, b, c, d, a, b, c, d) \quad \text{has the DFT} \quad (A, 0, B, 0, C, 0, D, 0),$$
$$(a, b, c, d, a, b, c, d, a, b, c, d) \quad \text{has the DFT} \quad (A, 0, 0, B, 0, 0, C, 0, 0, D, 0, 0),$$

$$\vdots$$

cf. Figs. 4.15 and 4.16. The first of these patterns illustrates the *zero packing rule* (or *upsampling rule*)

$$g[n] := \begin{cases} f[n/m], & n = 0, \pm m, \pm 2m, \ldots \\ 0 & \text{otherwise} \end{cases} \quad \text{(with } f \text{ on } \mathbb{P}_{N/m} \text{ and } g \text{ on } \mathbb{P}_N\text{)}$$

$$\text{has the FT} \quad G[k] = \frac{1}{m}F[k], \qquad (55)$$

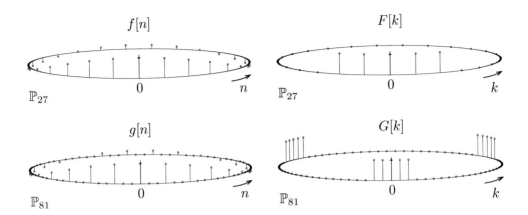

Figure 4.15. The effects of zero packing $(N = 81, m = 3)$ on a function f and its Fourier transform F.

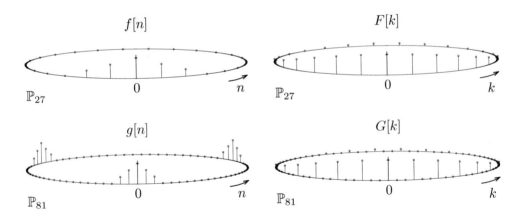

Figure 4.16. The effects of repetition $(N = 81, m = 3)$ on a function f and its Fourier transform F.

and the second illustrates the *repeat rule*

$$g[n] := f[n] \quad (\text{with } f \text{ on } \mathbb{P}_{N/m} \text{ and } g \text{ on } \mathbb{P}_N)$$

$$\text{has the FT} \quad G[k] = \begin{cases} F[k/m] & \text{if } k = 0, \pm m, \pm 2m, \ldots \\ 0 & \text{otherwise.} \end{cases} \tag{56}$$

We can easily prove (55) by using the analysis equation for g and the analysis equation for f to write

$$G[k] := \frac{1}{N} \sum_{n=0}^{N-1} g[n] e^{-2\pi i k n/N} = \frac{1}{N} \sum_{n'=0}^{N/m-1} f[n'] e^{-2\pi i k m n'/N} = \frac{1}{m} F[k].$$

We will use the zero packing rule to develop the FFT in Section 6.2.

For the proof of (56) we simply verify that we can synthesize g by writing

$$\sum_{k=0}^{N-1} G[k] e^{2\pi i k n/N} = \sum_{k'=0}^{N/m-1} F[k'] e^{2\pi i m k' n/N} = f[n] =: g[n].$$

You may wish to compare Fig. 4.7 with Figs. 4.15 and 4.16 as you visualize the meaning of these rules.

If you begin with the fact that

$$f = 1 \text{ on } \mathbb{P}_1 \quad \text{has the FT} \quad F = 1 \text{ on } \mathbb{P}_1$$

you can use the zero packing rule, repeat rule to obtain (41), (42). You can then derive (44) by applying the repeat rule to (41) or by applying the zero packing rule to (42), *cf.* Ex. 4.36.

The rules (55), (56) use processes that produce functions on $\mathbb{P}_N$ from functions on $\mathbb{P}_{N/m}$. We will now introduce two rules using processes that produce functions on $\mathbb{P}_N$ from functions on $\mathbb{P}_{m \cdot N}$, $m = 1, 2, \ldots$.

Let (a, b, c, d, e, f) represent a function on $\mathbb{P}_6$ and assume that

$$(a, b, c, d, e, f) \quad \text{has the DFT} \quad (A, B, C, D, E, F).$$

(As before, $a, \ldots, f, A, \ldots, F$ are complex scalars.) We will show that

$$(a + d, b + e, c + f) \quad \cdot \text{ has the DFT} \quad 2 \cdot (A, C, E),$$
$$(a + c + e, b + d + f) \quad \text{has the DFT} \quad 3 \cdot (A, D),$$

and that

$$(a, c, e) \quad \text{has the DFT} \quad (A + D, B + E, C + F),$$
$$(a, d) \quad \text{has the DFT} \quad (A + C + E, B + D + F).$$

The first of these patterns illustrates the *summation rule*

$$g[n] := \sum_{\ell=0}^{m-1} f[n - \ell N] \quad \text{(with } f \text{ on } \mathbb{P}_{m \cdot N} \text{ and } g \text{ on } \mathbb{P}_N\text{)} \tag{57}$$

$$\text{has the FT} \quad G[k] = m F[mk],$$

and the second illustrates the *decimation rule* (or *downsampling rule* or *sampling rule*)

$$g[n] := f[mn] \quad \text{(with } f \text{ on } \mathbb{P}_{m \cdot N} \text{ and } g \text{ on } \mathbb{P}_N\text{)}$$

$$\text{has the FT} \quad G[k] = \sum_{\ell=0}^{m-1} F[k - \ell N], \tag{58}$$

cf. Figs. 4.17 and 4.18.

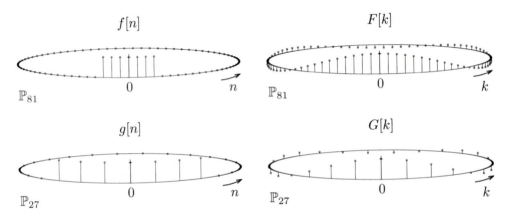

Figure 4.17. The effects of summation ($N = 27$, $m = 3$) on a function f and its Fourier transform F.

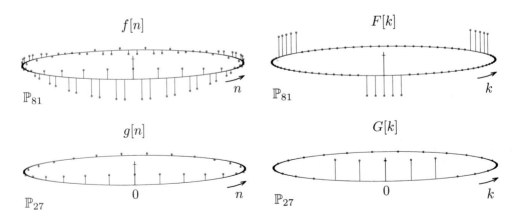

Figure 4.18. The effects of decimation ($N = 27$, $m = 3$) on a function f and its Fourier transform F.

We establish (57) by using the analysis equation for f to write

$$G[k] := \frac{1}{N} \sum_{n=0}^{N-1} \left\{ \sum_{\ell=0}^{m-1} f[n - \ell N] \right\} e^{-2\pi i k n / N}$$

$$= \frac{1}{N} \sum_{n=0}^{N-1} \sum_{\ell=0}^{m-1} f[n - \ell N] e^{-2\pi i k (n - \ell N) / N}$$

$$= \frac{1}{N} \sum_{n'=0}^{mN-1} f[n'] e^{-2\pi i m k n' / mN} = m \cdot F[mk].$$

We prove (58) by using the synthesis equation for f to write

$$\sum_{k=0}^{N-1} G[k] e^{2\pi i k n / N} = \sum_{k=0}^{N-1} \left\{ \sum_{\ell=0}^{m-1} F[k - \ell N] \right\} e^{2\pi i k n / N}$$

$$= \sum_{k=0}^{N-1} \sum_{\ell=0}^{m-1} F[k - \ell N] e^{2\pi i (k - \ell N) m n / mN}$$

$$= \sum_{k'=0}^{mN-1} F[k'] e^{2\pi i k' m n / mN} = f[mn] =: g[n].$$

You may wish to compare Fig. 4.8 with Figs. 4.17 and 4.18 as you visualize the meaning of these rules.

As stated in (57) and (58), the summation rule is a discrete form of N-periodization and the decimation rule is a discrete form of m-sampling as described in Section 1.4. With this in mind we form a totally discrete version of the Fourier-Poisson cube from Fig. 1.22, as shown in Fig. 4.19. This cube is a commuting diagram that uses decimation and summation to link Fourier transform pairs f, F on $\mathbb{P}_{NPQ}$; g, G on $\mathbb{P}_{NP}$; ϕ, Φ on $\mathbb{P}_{NQ}$; and γ, Γ on $\mathbb{P}_N$. You might find it instructive to express g, ϕ, γ in terms of f, to express γ in terms of g, ϕ, and to work out the relations that connect the corresponding Fourier transforms.

Dilation

Dilation is a straightforward process when we work with functions on $\mathbb{R}$, and the dilation rule (3.13) is easy to state, easy to use, and easy to prove. Things are considerably more complicated when we work with functions on $\mathbb{P}_N$, because in this context dilation can involve decimation as well as permutation. We will present two special versions of the dilation rule for functions on $\mathbb{P}_N$, and show how they can be combined to handle the general situation.

When we form the dilate $g[n] := f[mn]$ we will always assume that m is an integer. Since f is an N-periodic function on $\mathbb{Z}$, the indices mn can be taken modulo N, and we lose no generality if we assume that m takes one of the values $1, 2, \ldots, N$.

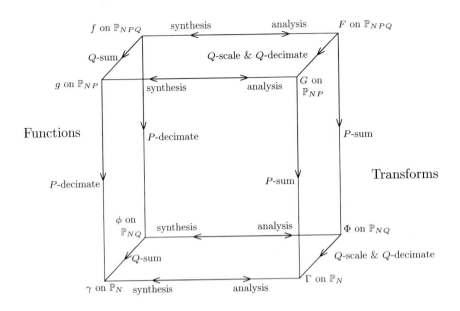

Figure 4.19. The discrete Fourier-Poisson cube.

We consider first the case where the dilation parameter $m = 1, 2, \dots, N - 1$ and the basic period $N = 2, 3, \dots$ are relatively prime, *i.e.*, the greatest common divisor of m and N is 1. When we reduce the (nonzero) dilated indices $m \cdot 1$, $m \cdot 2, \dots, m \cdot (N - 1)$ modulo N we get some rearrangement of the (nonzero) indices $1, 2, \dots, N - 1$, so there is some integer $m' = 1, 2, \dots, N - 1$ such that

$$mm' \equiv 1 \pmod{N},$$

cf. Ex. 4.51. For example, if $N = 7$, $m = 3$, and we reduce the indices

$$3 \cdot 1, \qquad 3 \cdot 2, \qquad 3 \cdot 3, \qquad 3 \cdot 4, \qquad 3 \cdot 5, \qquad 3 \cdot 6$$

modulo 7, we obtain the indices

$$3, \qquad 6, \qquad 2, \qquad 5, \qquad 1, \qquad 4$$

so $m' = 5$. In view of this discussion, we can verify the *P-dilation rule*

$$g[n] := f[mn] \quad \text{(with } \gcd(m, N) = 1\text{)}$$
$$\text{has the FT} \quad G[k] = F[m'k] \quad \text{where} \quad mm' \equiv 1 \pmod{N}, \tag{59}$$

by writing

$$G[k] := \frac{1}{N} \sum_{n=0}^{N-1} f[mn] e^{-2\pi i k n/N} = \frac{1}{N} \sum_{n=0}^{N-1} f[mn] e^{-2\pi i (km')(mn)/N}$$

$$= \frac{1}{N} \sum_{n'=0}^{N-1} f[n'] e^{-2\pi i (km')n'/N} = F[m'k].$$

The components of g, G are obtained by suitably permuting the components of f, F, respectively, *cf.* Fig. 4.20.

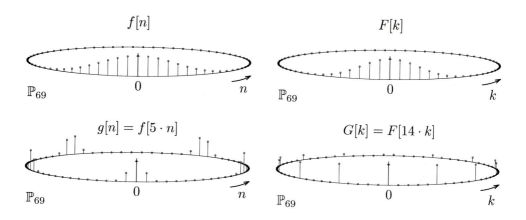

Figure 4.20. The effects of P-dilation ($N = 69, m = 5, m' = 14$) on a function f and its Fourier transform F.

We consider next the case where the dilation parameter $m = 1, 2, \ldots, N$ is a divisor of N so that $g[n] := f[mn]$ is obtained by repeating the decimated values $f[0], f[m], f[2m], \ldots, f[N-m]$, *cf.* Fig. 4.21. The resulting *D-dilation rule*

$$g[n] := f[mn] \quad \text{(with } m|N)$$

has the FT $\quad G[k] = \begin{cases} \displaystyle\sum_{\ell=0}^{m-1} F\left[\frac{k}{m} - \frac{\ell N}{m}\right] & \text{if } m|k \\ \\ 0 & \text{otherwise} \end{cases}$ (60)

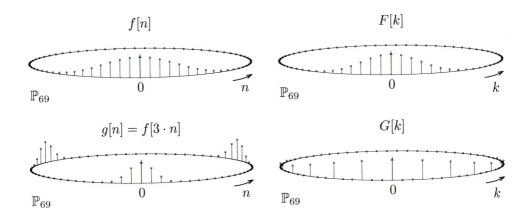

Figure 4.21. The effects of D-dilation ($N = 69$, $m = 3$) on a function f and its Fourier transform F.

can be verified by using (58) and (56) or by simply writing

$$
\sum_{k=0}^{N-1} G[k]e^{2\pi ikn/N} = \sum_{k'=0}^{N/m-1} \sum_{\ell=0}^{m-1} F\left[k' - \frac{\ell N}{m}\right]e^{2\pi ik'mn/N}
$$

$$
= \sum_{k'=0}^{N/m-1} \sum_{\ell=0}^{m-1} F\left[k' - \frac{\ell N}{m}\right]e^{2\pi i(k'-\ell N/m)mn/N}
$$

$$
= \sum_{k=0}^{N-1} F[k]e^{2\pi ikmn/N} = f[mn] =: g[n].
$$

We are now ready to consider the case where the dilation parameter m is an arbitrary integer $1, 2, \ldots, N$. We find a factorization

$$
m = d_1 d_2 \cdots d_r \mu
$$

where $d_1, d_2, \ldots, d_r$ are divisors of N and where μ, N are relatively prime. We can then write

$$
\begin{aligned}
f_1[n] &:= f[d_1 \cdot n] \\
f_2[n] &:= f_1[d_2 \cdot n] = f[d_1 d_2 \cdot n]
\end{aligned}
$$

$$
\vdots
$$

$$
\begin{aligned}
f_r[n] &:= f_{r-1}[d_r \cdot n] = f[d_1 d_2 \cdots d_r \cdot n] \\
g[n] &:= f_r[\mu \cdot n] = f[d_1 d_2 \cdots d_r \mu \cdot n].
\end{aligned}
$$

We use the D-dilation rule (60) to generate F_1 from F, F_2 from F_1, $\ldots$, and F_r from F_{r-1}. We then use the P-dilation rule (59) to generate G from F_r.

Poisson's relations

You know how to use the Poisson relation (1.29) [or (1.31)] to construct a Fourier series (18) from a function on $\mathbb{R}$ and its Fourier transform. We will now use the Poisson relations (1.29)–(1.32) to produce DFTs that are not easily found by other methods.

Suppose first that f is an absolutely summable function on $\mathbb{Z}$ with the Fourier transform F on $\mathbb{T}_p$. We can then use the *summation rule*

$$g[n] := \sum_{m=-\infty}^{\infty} f[n - mN] \quad \text{(with } f \text{ on } \mathbb{Z} \text{ and } g \text{ on } \mathbb{P}_N)$$

$$\text{has the FT} \quad G[k] = \frac{p}{N} f\left(k\frac{p}{N}\right)$$

(61)

that we establish by writing

$$G[k] := \frac{1}{N} \sum_{n=0}^{N-1} g[n] e^{-2\pi i kn/N} = \frac{1}{N} \sum_{n=0}^{N-1} \sum_{m=-\infty}^{\infty} f[n - mN] e^{-2\pi i kn/N}$$

$$= \frac{1}{N} \sum_{n=0}^{N-1} \sum_{m=-\infty}^{\infty} f[n - mN] e^{-2\pi i k(n-mN)/N} = \frac{1}{N} \sum_{n'=-\infty}^{\infty} f[n'] e^{-2\pi i kn'/N}$$

$$= \frac{p}{N} F\left(k\frac{p}{N}\right).$$

Example: Let m be a nonnegative integer with $2m + 1 \leq N$. Show that the centered box function

$$g_m[n] := \begin{cases} 1 & \text{if } n = 0, 1, \ldots, m \\ 0 & \text{if } n = m + 1, \ldots, N - m - 1 \\ 1 & \text{if } n = N - m, \ldots, N - 1 \end{cases}$$

(62)

$$\text{has the FT} \quad G[k] = \begin{cases} \dfrac{2m + 1}{N} & \text{if } k = 0 \\ \dfrac{\sin\{(2m + 1)k\pi/N\}}{N \sin\{k\pi/N\}} & \text{if } k = 1, 2, \ldots, N - 1. \end{cases}$$

Solution: We use (28) with (3) to see that

$$f[n] := \begin{cases} 1 & \text{if } n = 0, \pm 1, \ldots, \pm m \\ 0 & \text{otherwise} \end{cases} \quad \text{has the FT} \quad F(s) = \frac{1}{p} \frac{\sin\{(2m + 1)\pi s/p\}}{\sin(\pi s/p)},$$

and then use (61) (with m replaced by m') to obtain (62). ∎

We can use the inversion rule (54) to rewrite the sampling rule as follows. Let f be a function on $\mathbb{T}_p$ that has absolutely summable Fourier coefficients $F[k]$ (*e.g.*, as is the case when f is continuous and f' is piecewise smooth). We can then use the *sampling rule*

$$g[n] := f\left(\frac{np}{N}\right) \quad \text{(with } f \text{ on } \mathbb{T}_p \text{ and } g \text{ on } \mathbb{P}_N\text{)}$$

$$\text{has the FT} \quad G[k] = \sum_{m=-\infty}^{\infty} F[k - mN]. \tag{63}$$

Example: Find the discrete Fourier transform of

$$g[n] := \frac{3}{5 - 4\cos(2\pi n/N)}.$$

Solution: We use (27) and (63) to write

$$G[k] = \sum_{m=-\infty}^{\infty} \left(\tfrac{1}{2}\right)^{|k+mN|} = \sum_{m=0}^{\infty} \left(\tfrac{1}{2}\right)^{k+mN} + \sum_{m=1}^{\infty} \left(\tfrac{1}{2}\right)^{mN-k}$$

$$= \frac{\left(\tfrac{1}{2}\right)^{k} + \left(\tfrac{1}{2}\right)^{N-k}}{1 - \left(\tfrac{1}{2}\right)^{N}}, \quad k = 0, 1, \ldots, N-1. \qquad \blacksquare$$

The summation rule (61) and sampling rule (63) produce discrete Fourier transform pairs from functions on $\mathbb{Z}$ and $\mathbb{T}_p$. Suppose now that f is a smooth function on $\mathbb{R}$ with small regular tails. The function f and its Fourier transform are then linked by the Poisson sum formula (1.45), and we can use the *sample-sum rule*

$$g[n] := \sum_{m=-\infty}^{\infty} f\left(a \cdot \frac{n - mN}{\sqrt{N}}\right) \quad \text{(with } f \text{ on } \mathbb{R} \text{ and } g \text{ on } \mathbb{P}_N\text{)}$$

$$\text{has the FT} \quad G[k] = \frac{1}{|a|\sqrt{N}} \sum_{m=-\infty}^{\infty} F\left(\frac{1}{a} \cdot \frac{k - mN}{\sqrt{N}}\right). \tag{64}$$

Here a is a nonzero real dilation parameter.

We will establish (64) when $a = 1$. The extension to all real $a \neq 0$ is then done by using the dilation rule (3.13). We set $p := \sqrt{N}$ and use the Poisson sum formula (1.45) to verify the synthesis equation for g by writing

$$\sum_{k=0}^{N-1} G[k]e^{2\pi ikn/N} = \frac{1}{p} \sum_{k=0}^{N-1} \sum_{m=-\infty}^{\infty} F\left(\frac{k - mN}{p}\right) e^{2\pi i(k-mN)n/N}$$

$$= \sum_{k'=-\infty}^{\infty} \frac{1}{p} F\left(\frac{k'}{p}\right) e^{2\pi ik'(np/N)/p} = \sum_{m=-\infty}^{\infty} f\left(n\frac{p}{N} - mp\right)$$

$$= g[n].$$

Example: For $m = 0, 1, 2, \ldots$ we define the *discrete Hermite function*

$$g_m[n] := \sum_{\mu=-\infty}^{\infty} H_m \left\{ \sqrt{\frac{2\pi}{N}} (n - \mu N) \right\} e^{-\pi(n-\mu N)^2 / N} \tag{65}$$

where H_m is the Hermite polynomial (3.28), *cf.* Fig. 4.22. Show that

$$g_m[n] \quad \text{has the FT} \quad \frac{(-i)^m}{\sqrt{N}} g_m[k]. \tag{66}$$

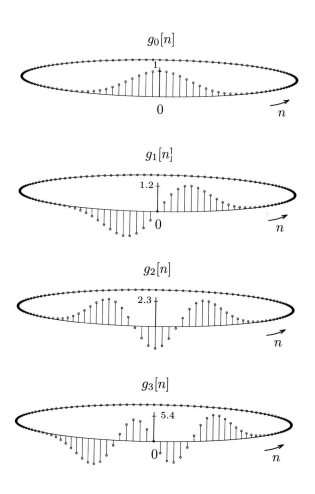

Figure 4.22. The discrete Hermite functions (64) when $m = 0, 1, 2, 3$ and $N = 128$.

Solution: We recall from (3.31) that

$$f_m(x) := H_m(\sqrt{2\pi}\,x)e^{-\pi x^2} \quad \text{has the FT} \quad (-i)^m f_m(s)$$

and use the sample-sum rule (64) (with $a = 1$) to see that

$$g_m[n] = \sum_{\mu=-\infty}^{\infty} f_m\left(\frac{n - \mu N}{\sqrt{N}}\right)$$

has the Fourier transform

$$G_m[k] = \frac{1}{\sqrt{N}} \sum_{\mu=-\infty}^{\infty} F_m\left(\frac{k - \mu N}{\sqrt{N}}\right) = \frac{(-i)^m}{\sqrt{N}} \sum_{\mu=-\infty}^{\infty} f_m\left(\frac{k - \mu N}{\sqrt{N}}\right) = \frac{(-i)^m}{\sqrt{N}} g_m[k].$$

In this way we construct eigenfunctions for the discrete Fourier transform operator, *cf.* Ex. 5.17.   ∎

4.4 Selected Applications of the DFT Calculus

The Euler-Maclaurin sum formula

In this section you will learn how to find a simple formula for any sum of the form

$$S := P(0) + P(1) + \cdots + P(N)$$

where P is a polynomial, *e.g.*, you will learn how to *derive* the formula

$$1^2 + 2^2 + \cdots + N^2 = N(N + 1)(2N + 1)/6 \tag{67}$$

from elementary calculus.

We begin by showing that the rth Bernoulli function (23) satisfies the identity

$$\frac{1}{2}w_r(0+) + \sum_{n=1}^{N-1} w_r\left(\frac{n}{N}\right) + \frac{1}{2}w_r(1-) = \frac{w_r(0)}{N^r}, \quad r = 0, 1, 2, \ldots. \tag{68}$$

When $r = 1, 2, \ldots$ the function w_r is continuous and piecewise smooth, so we can use the sampling rule (63) to find the Fourier transform of $w_r(n/N)$. In this way we see that

$$\sum_{n=0}^{N-1} w_r\left(\frac{n}{N}\right) = N \sum_{m=-\infty}^{\infty} W_r[0 - mN] = \frac{1}{N^r} \sum_{\substack{m=-\infty \\ m \neq 0}}^{\infty} \frac{1}{(2\pi i m)^{r+1}} = \frac{w_r(0)}{N^r}.$$

Since $w_0(x)$ is odd, the identity (68) also holds when $r = 0$ provided that we use midpoint regularization and define $w_0(0) := 0$.

We will now N-periodically extend the polynomial segment $P(x)$, $0 \le x < N$ to all of $\mathbb{R}$ and use Eagle's method to write

$$P(x) = C + J_0 \cdot w_0\left(\frac{x}{N}\right) + J_1 \cdot N w_1\left(\frac{x}{N}\right) + J_2 \cdot N^2 w_2\left(\frac{x}{N}\right) + \cdots, \quad 0 < x < N$$

where

$$C := \frac{1}{N} \int_0^N P(x)\,dx$$

is the average value of P on $0 \le x \le N$ and

$$J_r := P^{(r)}(0+) - P^{(r)}(N-)$$

is the jump in the rth derivative at the knot point $x = 0$, $r = 0, 1, \ldots$. We use this expression for $P(x)$ with (68) to see that

$$\frac{1}{2}P(0) + \sum_{n=1}^{N-1} P(n) + \frac{1}{2}P(N)$$
$$= N \cdot C + J_0 \cdot w_0(0) + J_1 \cdot w_1(0) + J_2 \cdot w_2(0) + \cdots .$$

Since $w_r(0) = 0$ when $r = 0, 2, 4, \ldots$ we can rearrange this result to obtain the *Euler-Maclaurin sum formula*

$$\sum_{n=0}^{N} P(n) = \frac{1}{2}\{P(0) + P(N)\} + \int_0^N P(x)\,dx$$
$$+ \{P'(0+) - P'(N-)\} \cdot w_1(0) + \{P'''(0+) - P'''(N-)\} \cdot w_3(0) + \cdots . \tag{69}$$

The universal constants

$$w_1(0) = \frac{-1}{12}, \quad w_3(0) = \frac{1}{720}, \quad w_5(0) = \frac{-1}{30240}, \quad w_7(0) = \frac{1}{1209600}, \ldots \tag{70}$$

can be obtained from the generating function of Ex. 4.22.

Example: Use the Euler-Maclaurin formula to derive (67).

Solution: We set $P(x) := x^2$ and use (69)–(70) to write

$$\sum_{n=0}^{N} n^2 = \frac{1}{2}\{0^2 + N^2\} + \int_0^N x^2\,dx - \frac{1}{12}\{0 - 2N\}$$
$$= \frac{N^2}{2} + \frac{N^3}{3} + \frac{N}{6} = \frac{N(N+1)(2N+1)}{6} . \quad \blacksquare$$

The discrete Fresnel function

We will show that the *discrete Fresnel function*

$$g[n] := e^{2\pi i n^2/N} \quad \text{(on } \mathbb{P}_N\text{)} \qquad \text{has the FT}$$

$$G[k] = \frac{(1+i)\{1 + (-1)^k (-i)^N\}}{2\sqrt{N}} \cdot e^{-\pi i k^2/2N}. \qquad (71)$$

This unusual Fourier transform pair, *cf.* Fig. 4.23, is used in number theory, optics, accoustics, communication theory, etc.

Re g and Im g Re G and Im G

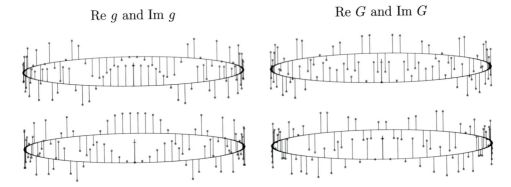

Figure 4.23. The discrete Fresnel function g and the Fourier transform G as given in (71) when $N = 91$.

We begin by defining the function

$$f(x) := e^{2\pi i x^2/N}, \qquad 0 \le x < N$$

on $\mathbb{T}_N$. The function f is continuous on $\mathbb{T}_N$ $[f(0+) = 1, f(0-) := f(N-) = 1]$ and the derivatives $f', f'', \ldots$ are defined and continuous except at the origin where

$$f'(0+) - f'(0-) := f'(0+) - f'(N-) = -4\pi i.$$

This being the case, we can use the sampling rule (63) to see that g has the Fourier transform

$$G[k] = \sum_{m=-\infty}^{\infty} F[k - mN]$$

where the Fourier coefficients are given by the analysis equation

$$F[k - mN] = \frac{1}{N} \int_0^N e^{2\pi i x^2/N} \cdot e^{-2\pi i (k-mN)x/N} dx.$$

We complete the square in the exponent of the integrand for the even m Fourier coefficients and thereby write

$$F[k - 2\mu N] = \frac{1}{N} \int_0^N e^{2\pi i x^2/N} \cdot e^{-2\pi i(k-2\mu N)x/N} dx$$

$$= e^{-2\pi i(k/2-\mu N)^2/N} \cdot \frac{1}{N} \int_0^N e^{2\pi i(x-k/2+\mu N)^2/N} dx$$

$$= e^{-\pi i k^2/2N} \cdot \frac{1}{N} \int_{\mu N-k/2}^{(\mu+1)N-k/2} e^{2\pi i u^2/N} du.$$

A similar computation for the odd m Fourier coefficients gives

$$F[k - (2\mu+1)N] = e^{-2\pi i(k/2-N/2-\mu N)^2/N} \cdot \frac{1}{N} \int_0^N e^{2\pi i(x-k/2+N/2+\mu N)^2/N} dx$$

$$= e^{-2\pi i(k^2/4+N^2/4-kN/2)/N} \cdot \frac{1}{N} \int_0^N e^{2\pi i(x-k/2+N/2+\mu N)^2/N} dx$$

$$= (-1)^k(-i)^N e^{-\pi i k^2/2N} \cdot \frac{1}{N} \int_{\mu N+N/2-k/2}^{(\mu+1)N+N/2-k/2} e^{2\pi i u^2/N} du.$$

We now use these expressions as we form the sum

$$G[k] = \sum_{\mu=-\infty}^{\infty} F[k - 2\mu N] + \sum_{\mu=-\infty}^{\infty} F[k - (2\mu+1)N]$$

$$= e^{-\pi i k^2/2N} \cdot \frac{1}{N} \left\{ \int_{-\infty}^{\infty} e^{2\pi i u^2/N} du + (-1)^k(-i)^N \int_{-\infty}^{\infty} e^{2\pi i u^2/N} du \right\}$$

$$= \{1 + (-1)^k(-i)^N\} e^{-\pi i k^2/2N} \cdot \sqrt{N} I$$

where

$$I := \int_{-\infty}^{\infty} e^{2\pi i t^2} dt.$$

This Fresnel integral is independent of N and k, so we can determine its value by setting $k = 0$, $N = 1$ in the formula for $G[k]$, and writing

$$1 = (1 - i) \cdot I.$$

In this way we find $I = (1 + i)/2$ and complete the verification of (71).

You may struggle a bit with the details of this argument, but you will have no trouble evaluating the *Gauss sum*

$$S_N := \sum_{n=0}^{N-1} e^{2\pi i n^2/N} = \sqrt{N} \cdot \begin{cases} 1 & \text{if } N = 1, 5, 9, \dots \\ 0 & \text{if } N = 2, 6, 10, \dots \\ i & \text{if } N = 3, 7, 11, \dots \\ 1+i & \text{if } N = 4, 8, 12, \dots \end{cases} \tag{72}$$

by using (71) to write

$$S_N = \sum_{n=0}^{N-1} g[n] = N \cdot G[0] = \frac{\sqrt{N}}{2} \cdot (1+i)\{1 + (-i)^N\}.$$

You may be interested to learn that it took Gauss several years to derive this result for his proof of the law of quadratic reciprocity in number theory. Exercises 4.51–4.59 will show you some of the other fascinating connections between number theory and discrete Fourier analysis.

References

W.L. Briggs and V.E. Henson, *The DFT*, SIAM, Philadelphia, 1995.

> Chapter 3 gives the rules for Fourier transforms of functions on $\mathbb{P}_N$, and the appendix has a pictorial dictionary of 15 DFT pairs. Chapter 9 includes a nice discussion of the Euler-Maclaurin sum formula.

J.W. Brown and R.V. Churchill, *Fourier Series and Boundary Value Problems, 5th ed.*, McGraw-Hill, New York, 1993.

> Elementary techniques for finding Fourier series are described in Chapter 2 of this highly evolved text for scientists and engineers.

V. Čížek, *Discrete Fourier Transforms and Their Applications*, Adam Hilger, Boston, 1986.

> Rules for Fourier transforms of functions on $\mathbb{P}_N$ are developed in Chapter 4.

R. Courant and F. John, *Introduction to Calculus and Analysis, Vol. I*, John Wiley & Sons, New York, 1965.

> Various Fourier series and a nice introduction to the Bernoulli functions can be found in Chapter 8 of this classic applied mathematics text.

P. Henrici, *Applied and Computational Complex Analysis, Vol. 3*, John Wiley & Sons, New York, 1986.

> Chapter 13 of this applied mathematics treatise has an exceptionally well written introduction to discrete Fourier analysis.

A.V. Oppenheim and R.W. Schafer, *Digital Signal Processing*, Prentice Hall, Englewood Cliffs, NJ, 1975.

> Chapter 3 introduces the Fourier transform calculus for functions on $\mathbb{P}_N$ within an engineering context.

M.R. Schroeder, *Number Theory in Science and Engineering*, 2nd ed., Springer-Verlag, New York, 1986.

Elementary number theory is used to evaluate Gauss sums and DFTs of Legendre sequences, discrete Fresnel functions, … in Chapter 15 of this fascinating text.

G.P. Tolstov, *Fourier Series* (English translation by R.A. Silverman), Prentice Hall, Englewood Cliffs, NJ, 1962; reprinted by Dover Publications, New York, 1976.

Integration is used to find a number of basic Fourier series in Chapter 1. Power series are used to obtain a few less common Fourier series in Chapter 4.

J.S. Walker, *Fourier Analysis*, Oxford University Press, New York, 1988.

Chapter 1 has an introduction to Fourier series. An elementary introduction to Poisson's relation is given in Chapter 12.

Exercise Set 4

▶ **EXERCISE 4.1:**   Let $0 < \alpha < p$ and let f be a p-periodic function with

$$f(x) := \begin{cases} 1 & \text{for } 0 < x < \alpha \\ 0 & \text{for } \alpha < x < p. \end{cases}$$

Find the Fourier series for f by using:

(a) Kronecker's rule (5) to evaluate

$$\frac{1}{p} \int_0^\alpha 1 \cdot e^{-2\pi i k x/p} dx;$$

(b) the known Fourier series (16) for the Bernoulli function w_0 and the identity

$$f(x) = w_0 \left(\frac{x}{p} \right) - w_0 \left(\frac{x - \alpha}{p} \right) + \frac{\alpha}{p}; \text{ and}$$

(c) Poisson's relation (18) and the identity

$$f(x) = \sum_{m=-\infty}^{\infty} \Pi \left(\frac{x - \alpha/2 - mp}{\alpha} \right).$$

▶ **EXERCISE 4.2:**   Use Poisson's relation to find the Fourier series for each of the following p-periodic functions on $\mathbb{R}$.

(a) $f(x) := |\sin(\pi x/p)|$

(b) $f(x) := \max\{\cos(2\pi x/p), 0\}$

(c) $f(x) := \displaystyle\sum_{m=-\infty}^{\infty} e^{-\pi(x-mp)^2}$

(d) $f(x) := \displaystyle\sum_{m=-\infty}^{\infty} e^{-a|x-mp|}, a > 0$

▶ **EXERCISE 4.3:**   Let $f(x) := x$ when $0 < x < 1$. Find the Fourier series for f if:

(a) f is 1-periodic;   (b) f is even and 2-periodic;   (c) f is odd and 2-periodic.

▶ **EXERCISE 4.4:**   Suitably manipulate the Fourier series (23) for the Bernoulli functions w_0, w_1, w_2 and thereby show that when $0 < x < 1$:

(a) $\displaystyle\sum_{k=1}^{\infty} \frac{\sin(2\pi kx)}{k} = \frac{\pi}{2}(1 - 2x)$;

(b) $\displaystyle\sum_{k=1}^{\infty} \frac{\cos(2\pi kx)}{k^2} = \frac{\pi^2}{6}(1 - 6x + 6x^2)$;

(c) $\displaystyle\sum_{k=1}^{\infty} \frac{\sin(2\pi kx)}{k^3} = \frac{\pi^3}{3}(x - 3x^2 + 2x^3)$;

(d) $\displaystyle\sum_{k=1}^{\infty} \frac{\cos(2\pi kx)}{k^4} = \frac{\pi^4}{90}(1 - 30x^2 + 60x^3 - 30x^4)$.

▶ **EXERCISE 4.5:**   Let $-\infty < s < \infty$ with $s \neq 0, \pm 1, \pm 2, \dots$. Derive each of the following Fourier series.

(a) $\displaystyle e^{2\pi isx} = \sum_{k=-\infty}^{\infty} \text{sinc}(k-s)e^{2\pi ikx}$, $-\frac{1}{2} < x < \frac{1}{2}$

(b) $\displaystyle \sin(2\pi sx) = \frac{2\sin(\pi s)}{\pi}\left\{ \frac{1 \cdot \sin(2\pi x)}{1^2 - s^2} - \frac{2 \cdot \sin(4\pi x)}{2^2 - s^2} + \cdots \right\}$, $-\frac{1}{2} < x < \frac{1}{2}$

(c) $\displaystyle \cos(2\pi sx) = \frac{2\sin(\pi s)}{\pi}\left\{ \frac{1}{2s} + \frac{s \cdot \cos(2\pi x)}{1^2 - s^2} - \frac{s \cdot \cos(4\pi x)}{2^2 - s^2} + \cdots \right\}$, $-\frac{1}{2} \leq x \leq \frac{1}{2}$

(d) $\displaystyle \csc(\pi s) = \frac{2}{\pi}\left\{ \frac{1}{2s} + \frac{s}{1^2 - s^2} - \frac{s}{2^2 - s^2} + \cdots \right\}$

(e) $\displaystyle \cot(\pi s) = \frac{2}{\pi}\left\{ \frac{1}{2s} - \frac{s}{1^2 - s^2} - \frac{s}{2^2 - s^2} - \cdots \right\}$

Note. The series of (d)–(e) are usually obtained by using tools from complex analysis!

▶ **EXERCISE 4.6:**   Let $-\infty < s < \infty$ and let f be a 2π-periodic function with

$$f(x) := \begin{cases} \cosh(2sx) & \text{for } |x| < \dfrac{\pi}{2} \\ -\cosh\{2s(x-\pi)\} & \text{for } |x - \pi| < \dfrac{\pi}{2} \end{cases}.$$

Find the Fourier series for f and thereby show that:

(a) $\displaystyle \cosh(2sx) = \frac{4\cosh(\pi s)}{\pi}\left\{ \frac{\cos x}{4s^2 + 1^2} - \frac{3 \cdot \cos 3x}{4s^2 + 3^2} + \frac{5 \cdot \cos 5x}{4s^2 + 5^2} - \cdots \right\}$, $-\frac{\pi}{2} < x < \frac{\pi}{2}$;

(b) $\displaystyle \text{sech}(\pi s) = \frac{4}{\pi}\left\{ \frac{1}{4s^2 + 1^2} - \frac{3}{4s^2 + 3^2} + \frac{5}{4s^2 + 5^2} - \cdots \right\}$, $-\infty < s < \infty$.

▶ **EXERCISE 4.7:**   Set $z = e^{2\pi ix/p}$ in a suitably chosen Laurent series (25) to find the Fourier series for the p-periodic function:

(a) $f(x) = \cos^3(2\pi x/p)$;

(b) $f(x) = 1/(1 - \alpha e^{2\pi ix/p})$, $|\alpha| < 1$;

(c) $f(x) = -\log\{1 - \alpha e^{2\pi ix/p}\}$, $|\alpha| < 1$;

(d) $f(x) = -\log\{1 - 2\alpha\cos(2\pi x/p) + \alpha^2\}$, $|\alpha| < 1$.

Note. When we set $\alpha = 1$ in (d) we obtain the Fourier series

$$-\log\left|2\sin\left(\frac{\pi x}{p}\right)\right| = \sum_{k=1}^{\infty}\frac{\cos(2\pi kx/p)}{k}, \qquad x \neq 0, \pm p, \pm 2p, \ldots .$$

▶ **EXERCISE 4.8:** Within the context of suitably regular p-periodic functions on $\mathbb{R}$, formally derive:

(a) the reflection rule;   (b) the conjugation rule;   (c) the translation rule;

(d) the modulation rule;   (e) the multiplication rule;   (f) the convolution rule;

(g) the derivative rule;   (h) the inversion rule.

Hint. You can use a direct argument ... or you can use Poisson's relation (18) to obtain the desired result from a corresponding rule from Chapter 3.

▶ **EXERCISE 4.9:** Let b, r be 2π-periodic functions on $\mathbb{R}$ with

$$b(x) := \begin{cases} 1 & \text{for } 0 < x < \pi \\ 0 & \text{for } \pi < x < 2\pi, \end{cases} \qquad r(x) := \begin{cases} x & \text{for } 0 < x < \pi \\ 0 & \text{for } \pi < x < 2\pi. \end{cases}$$

Verify that b, r have the Fourier series

$$b(x) = \frac{1}{2} + \sum_{k \neq 0}\frac{i[(-1)^k - 1]}{2\pi k}e^{ikx}, \qquad r(x) = \frac{\pi}{4} + \sum_{k \neq 0}\left\{\frac{i(-1)^k}{2k} + \frac{(-1)^k - 1}{2\pi k^2}\right\}e^{ikx},$$

and then use the Fourier transform calculus to find the Fourier series for

(a) $t(x) := (b * b)(x) = r(x) + r(-x)$;   (b) $f(x) := b(x) - b(-x) = 2b(x) - 1 = t'(x)$;

(c) $g(x) := r(x) + r(x - \pi)$;   (d) $j(x) := r(\pi - x) - r(\pi + x) = 2\pi w_0(x/2\pi)$;

(e) $d(x) := b(x - \pi/4) - b(x + \pi/4)$;   (f) $p_n(x) := b(x)\sin(nx), \quad n = 1, 2, \ldots .$

▶ **EXERCISE 4.10:** Let f be a suitably regular function on $\mathbb{T}_p$, let $-\infty < x_0 < \infty$, and let $m = 1, 2, 3, \ldots .$ What can you infer about the Fourier transform F if you know that:

(a) $\overline{f(x)} = f(x)$?   (b) $\overline{f(x)} = f(-x)$?   (c) $f(x + p/m) = f(x)$?

(d) $f(x + p/2) = -f(x)$?   (e) $f(x_0 + x) = f(x_0 - x)$?   (f) $f(x_0 + x) = -f(x_0 - x)$?

(g) $\displaystyle\int_0^p f(x)\,dx = 1$?   (h) $\displaystyle\int_0^p |f(x)|^2\,dx = 1$?

(i) $f(x) = \dfrac{1}{p}\displaystyle\int_0^p f(u)\dfrac{\sin\{(2m + 1)\pi(x - u)/p\}}{\sin\{\pi(x - u)/p\}}\,du$?

Hint. Do you recognize the Dirichlet kernel (28) in (i)?

▶ **EXERCISE 4.11:**   A 1-periodic function on $\mathbb{R}$ having the Fourier series

$$f(x) = \sum_{k=-\infty}^{\infty} c_k e^{2\pi i k x}$$

(with $F[k]$ replaced by c_k) takes the values

$$f(x) := \begin{cases} 1 & \text{if } 0 < x < 1/10 \\ 0 & \text{if } 1/10 < x < 1. \end{cases}$$

Sketch the graph of f on $[0, 1]$ and corresponding graphs of the functions represented by the following Fourier series.

(a) $\displaystyle\sum_{k=-\infty}^{\infty} c_{-k} e^{2\pi i k x}$ (b) $\displaystyle\sum_{k=-\infty}^{\infty} \{(-1)^k c_k\} e^{2\pi i k x}$ (c) $\displaystyle\sum_{k=-\infty}^{\infty} \left\{ \sum_{m=-\infty}^{\infty} c_m c_{k-m} \right\} e^{2\pi i k x}$

(d) $\displaystyle\sum_{k=-\infty}^{\infty} c_k^2 e^{2\pi i k x}$ (e) $\displaystyle\sum_{k=-\infty}^{\infty} \{i^k c_k\} e^{2\pi i k x}$ (f) $\displaystyle\sum_{k=-\infty}^{\infty} \{c_{k-5} - c_{k+5})/2i\} e^{2\pi i k x}$

(g) $\displaystyle\sum_{k=-\infty}^{\infty} c_{3k} e^{2\pi i k x}$ (h) $\displaystyle\sum_{k=-\infty}^{\infty} \{c_k + c_{-k}\} e^{2\pi i k x}$ (i) $\displaystyle\sum_{k=-\infty}^{\infty} \left\{ \sum_{m=-\infty}^{\infty} c_m c_{2k-m} \right\} e^{2\pi i k x}$

Hint. Write $(-1)^k = e^{2\pi i k (1/2)}$ when you analyze (b).

▶ **EXERCISE 4.12:**   Sketch the graph of the 6-periodic function on $\mathbb{R}$ that has the Fourier coefficients:

(a) $F[k] := (1/6) \operatorname{sinc}^2(k/6)$;

(b) $F_1[k] := (-1)^k F[k]$;

(c) $F_2[k] := \begin{cases} F[k/3] & \text{if } k = 0, \pm 3, \pm 6, \dots \\ 0 & \text{otherwise;} \end{cases}$

(d) $F_3[k] := 2F[2k]$;

(e) $F_4[k] := \displaystyle\sum_{\ell=-\infty}^{\infty} (1/6) \operatorname{sinc}(\ell/6) F[k - \ell]$;

(f) $F_5[k] := (2\pi i k/6) F[k]$;

Hint. Use Poisson's relation (18) to determine f from (a).

▶ **EXERCISE 4.13:**   Let 4-periodic functions f_1, f_2, f_3 on $\mathbb{R}$ be constructed from the 4-periodic function f on $\mathbb{R}$ as shown in the following graphs.

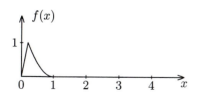

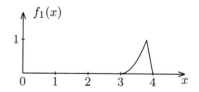

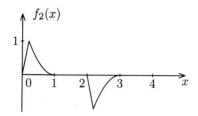

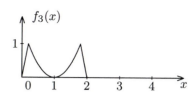

Find the Fourier series for f_1, f_2, f_3 using coefficients from $f(x) = \sum_{k=-\infty}^{\infty} c_k e^{2\pi i k x/4}$, *cp.* Ex. 4.11.

▶ **EXERCISE 4.14:**   Let $w_0, w_1, \ldots$ be the Bernoulli functions as defined by (23) (*cf.* Ex. 4.22).

(a) Show that $\displaystyle\sum_{k=1}^{\infty} 1/k^{2n} = \tfrac{1}{2}(2\pi)^{2n} \int_0^1 w_{n-1}^2(x)dx, \quad n = 1, 2, \ldots$.

(b) Show that $1/1^2 + 1/2^2 + 1/3^2 + \cdots = \pi^2/6$.

(c) Show that $1/1^4 + 1/2^4 + 1/3^4 + \cdots = \pi^4/90$.

▶ **EXERCISE 4.15:**   Let $N = 1, 2, \ldots$ and let $f(x) := (1 + e^{ix})^N$, $-\infty < x < \infty$.

(a) Find the Fourier series for f (using $p = 2\pi$).

(b) Use Plancherel's identity to show that

$$\binom{N}{0}^2 + \binom{N}{1}^2 + \cdots + \binom{N}{N}^2 = \binom{2N}{N}.$$

Hint. $[(1 + e^{ix})(1 + e^{-ix})]^N = [e^{ix/2} + e^{-ix/2}]^{2N}$

▶ **EXERCISE 4.16:**   An ellipse with eccentricity $0 < \epsilon < 1$ has the equation $r(\theta) = 1/(1 + \epsilon \cos \theta)$ in polar coordinates.

(a) Verify that the major and minor radii of this ellipse are given by

$$a := \tfrac{1}{2}\left\{[r(\theta)\cos\theta]_{\max} - [r(\theta)\cos\theta]_{\min}\right\} = (1-\epsilon^2)^{-1}, \quad b := [r(\theta)\sin\theta]_{\max} = (1-\epsilon^2)^{-1/2}.$$

(b) Show that $r(\theta) = \mathcal{C}(e^{i\theta})$ where $\mathcal{C}(z) := 2z/(\epsilon z^2 + 2z + \epsilon)$.

(c) Verify that $\mathcal{C}$ has the partial fraction decomposition

$$\mathcal{C}(z) = (1-\epsilon^2)^{-1/2}\left\{\frac{1}{1+\rho z} - \frac{(\rho/z)}{1+(\rho/z)}\right\} \quad \text{with} \quad \rho := \frac{\epsilon}{1+(1-\epsilon^2)^{1/2}}.$$

Hint. Observe that $-\rho$ and $-1/\rho$ are the roots of $\epsilon z^2 + 2z + \epsilon$.

(d) Show that $\mathcal{C}$ is given by the Laurent series

$$\mathcal{C}(z) = (1-\epsilon^2)^{-1/2}\sum_{k=-\infty}^{\infty}(-\rho)^{|k|}z^k, \quad \rho < |z| < 1/\rho.$$

(e) Use (b) and (d) to show that

$$r(\theta) = (1-\epsilon^2)^{-1/2}\sum_{k=-\infty}^{\infty}(-\rho)^{|k|}e^{ik\theta} = (1-\epsilon^2)^{-1/2}\left\{1 + 2\sum_{k=1}^{\infty}(-\rho)^k\cos k\theta\right\}.$$

(f) Use the Fourier series (e) to evaluate the integrals

$$\int_0^{\pi}\frac{\cos k\theta}{1+\epsilon\cos\theta}, \quad k = 0, 1, 2, \dots .$$

(g) Use The Fourier series (e) to show that

$$[r(\theta)]^2 = (1-\epsilon^2)^{-1}\sum_{k=-\infty}^{\infty}\left\{\sum_{m=-\infty}^{\infty}(-\rho)^{|m|}(-\rho)^{|k-m|}\right\}e^{ik\theta},$$

and thereby verify that the ellipse has the area

$$\frac{1}{2}\int_{\theta=0}^{2\pi}[r(\theta)]^2\,d\theta = \frac{\pi}{(1-\epsilon^2)^{3/2}}.$$

▶ **EXERCISE 4.17:**   Let $0 < \alpha < p$ and let f be a p-periodic function with

$$f(x) := \begin{cases} x^2 & \text{for } 0 < x < \alpha \\ 0 & \text{for } \alpha < x < p. \end{cases}$$

(a) Explain why

$$f(x) = 2p^2 w_2\left(\frac{x}{p}\right) - \alpha^2 w_0\left(\frac{x-\alpha}{p}\right) - 2\alpha p w_1\left(\frac{x-\alpha}{p}\right) - 2p^2 w_2\left(\frac{x-\alpha}{p}\right) + \frac{\alpha^3}{3p}$$

where $w_0, w_1, \ldots$ are the Bernoulli functions (23).

(b) Using (a) verify that

$$F[k] = \frac{2! \alpha^3}{p} \frac{R_2(2\pi i k \alpha/p)}{(2\pi i k \alpha/p)^3} e^{-2\pi i k \alpha/p}, \quad k = \pm 1, \pm 2, \ldots$$

where

$$R_n(z) := e^z - 1 - z - \frac{z^2}{2!} - \cdots - \frac{z^n}{n!}, \quad n = 1, 2, \ldots .$$

Note. This expression shows the pattern for $F[k]$ that results when x^2 is replaced by x^n, $n = 1, 2, \ldots$, in the definition of f, *cf.* Appendix 1.

▶ **EXERCISE 4.18:**   Let $a > 0$, b, c be real numbers with $a^2 + b^2 = c^2$, let $p > 0$, and let

$$f(x) := \sum_{k=-\infty}^{\infty} \frac{e^{2\pi i k x/p}}{k^2 + 2bk + c^2}, \quad -\infty < x < \infty.$$

(a) Show that f is the Fourier series of the p-periodic function f with

$$f(x) = \frac{\pi}{2a}\left(\frac{e^{\pi(a-ib)(2x/p-1)}}{\sinh\{\pi(a-ib)\}} + \frac{e^{-\pi(a+ib)(2x/p-1)}}{\sinh\{\pi(a+ib)\}}\right) \quad \text{when } 0 \le x \le p.$$

(b) Specialize (a) to the case where $x = 0$ and thereby show that

$$\sum_{k=-\infty}^{\infty} \frac{1}{k^2 + 2bk + c^2} = \frac{\pi}{2a}\left(\frac{e^{\pi(a+ib)}}{\sinh\{\pi(a+ib)\}} + \frac{e^{-\pi(a-ib)}}{\sinh\{\pi(a-ib)\}}\right).$$

(c) Specialize (c) to the case where $b = 0$ and thereby show that

$$\sum_{k=-\infty}^{\infty} \frac{1}{k^2 + c^2} = \frac{\pi}{c \tanh(\pi c)} .$$

(d) Using (c), show that

$$\sum_{k=1}^{\infty} \frac{1}{k^2} = \lim_{c \to 0+} \frac{1}{2}\left\{\sum_{k=-\infty}^{\infty} \frac{1}{k^2 + c^2} - \frac{1}{c^2}\right\} = \frac{\pi^2}{6} .$$

▶ **EXERCISE 4.19:**   The *Bessel function* J_k of the first kind with order $k = 0, \pm 1, \pm 2, \ldots$ and argument $-\infty < \alpha < \infty$ can be defined by writing

$$e^{i\alpha \sin x} =: \sum_{k=-\infty}^{\infty} J_k(\alpha)e^{ikx}, \quad -\infty < x < \infty.$$

i.e., $J_k(\alpha)$ is the kth Fourier coefficient of the 2π-periodic function $f(x, \alpha) := e^{i\alpha \sin x}$. Use this generating function and your knowledge of Fourier analysis to prove the following identities.

(a) $J_k(\alpha) = (1/\pi) \int_0^{\pi} \cos\{\alpha \sin x - kx\}dx$

   Hint. Analysis equation

(b) $1 = J_0^2(\alpha) + 2 \sum_{k=1}^{\infty} J_k(\alpha)^2$

   Hint. Plancherel identity

(c) $\cos\{\alpha \sin x\} = J_0(\alpha) + 2 \sum_{k=1}^{\infty} J_{2k}(\alpha)\cos(2kx)$
   $\sin\{\alpha \sin x\} = 2 \sum_{k=1}^{\infty} J_{2k+1}(\alpha)\sin\{(2k+1)x\}$

   Hint. Synthesis equation

(d) $k^m J_k(\alpha) \to 0$ as $k \to \infty$,
   $m = 0, \pm 1, \pm 2, \ldots$

   Hint. cf. (15)

(e) $2k J_k(\alpha) = \alpha J_{k-1}(\alpha) + \alpha J_{k+1}(\alpha)$
   Hint. $f_x(x, \alpha) = i\alpha \cos x \cdot f(x, \alpha)$

(f) $\overline{J_k(\alpha)} = J_k(\alpha)$
   Hint. $\overline{f(-x, \alpha)} = f(x, \alpha)$

(g) $J_{-k}(\alpha) = (-1)^k J_k(\alpha)$
   Hint. $f(x + \pi, \alpha) = f(-x, \alpha)$

(h) $J_k(-\alpha) = (-1)^k J_k(\alpha)$

(i) $\alpha^2 J''(\alpha) + \alpha J_k'(\alpha) + \alpha^2 J_k(\alpha) = k^2 J_k(\alpha)$
   Hint. $\alpha^2 f_{\alpha\alpha} + \alpha f_\alpha + \alpha^2 f = -f_{xx}$

(j) $2J_k'(\alpha) = J_{k-1}(\alpha) - J_{k+1}(\alpha)$

▶ **EXERCISE 4.20:**   Let f_1, f_2, and $f := f_1 * f_2$ be functions on $\mathbb{T}_5$ with

$$f_1(x) := \sum_{m=-\infty}^{\infty} \Pi(x - 5m), \qquad f_2(x) := \sum_{m=-\infty}^{\infty} \Pi\left(\frac{x - 5m}{2}\right).$$

(a) Sketch the graphs of f_1, f_2, and f.

(b) On the interval $[-5/2, 5/2]$, f can be expressed as the difference of two triangle functions. Use this fact and Poisson's relation (18) to find an expression for the Fourier coefficients of f.

(c) Use Poisson's relation (18) to find the Fourier coefficients of f_1, f_2 and then use the convolution rule to find another expression for the Fourier coefficients of f.

▶ **EXERCISE 4.21:**   Let f be real and 2π-periodic on $\mathbb{R}$. By using the transformation

$$x = r \cos\theta, \qquad y = r \sin\theta$$

we see that the graph $\mathfrak{G}$ of $r = f(\theta)$ (in polar coordinates) corresponds to the *orbit*

$$z(\theta) := r \cos\theta + ir \sin\theta = f(\theta)e^{i\theta}$$

in the complex plane, *e.g.*, the *cardioid* $r = 2(1 - \cos\theta)$ and the *3-leaf rose* $r = 2\sin 3\theta$ have the orbits $z = -1 + 2e^{i\theta} - e^{2i\theta}$ and $z = i\{e^{-2i\theta} - e^{4i\theta}\}$, respectively.

(a) Show that if z is hermitian, *i.e.*, $z(-\theta) = \overline{z(\theta)}$, then $\mathfrak{G}$ is symmetric about the x-axis. What is the corresponding property of the Fourier coefficients $Z[k]$?

(b) Show that if z is antihermitian, *i.e.*, $z(-\theta) = -\overline{z(\theta)}$, then $\mathfrak{G}$ is symmetric about the y-axis. What is the corresponding property of the Fourier coefficients $Z[k]$?

(c) Show that if $z(\theta + 2\pi/N) = z(\theta)e^{2\pi i\theta/N}$ for some (minimal) $N = 2, 3, \ldots$, then $\mathfrak{G}$ has an N-fold axis of symmetry. What is the corresponding property of the Fourier coefficients $Z[k]$?

Hint. Refer to the analysis of the polygon function of Fig. 4.9.

(d) More generally, show that if $z(\theta + 2\pi/N) = z(\theta)e^{2\pi im\theta/N}$ for some (minimal) $N = 2, 3, \ldots$ and some $m = 1, 2, \ldots, N - 1$ that is relatively prime to N, then $\mathfrak{G}$ has an N-fold axis of symmetry. What is the corresponding property of the Fourier coefficients $Z[k]$?

(e) Describe the symmetry of the *epicycloid* $z_n(\theta) := (n + 1)e^{i\theta} - e^{i(n+1)\theta}$ generated by a point on a circle of radius 1 that rolls *outside* a circle of radius $n = 1, 2, 3, \ldots$.

(f) Describe the symmetry of the *hypocycloid* $z_n(\theta) := (n - 1)e^{i\theta} + e^{-(n-1)i\theta}$ generated by a point on a circle of radius 1 that rolls *inside* a circle of radius $n = 1, 2, 3, \ldots$.

Note. You may wish to experiment with computer-generated plots of such decimated Fourier series, *e.g.*, try viewing $z(\theta) := e^{im\theta}\{1 + e^{6i\theta}/2 + e^{12i\theta}/3\}$, $m = -6, -5, \ldots, 6$.

▶ **EXERCISE 4.22:**   In this exercise you will develop a generating function for the Bernoulli functions (23), *cp.* Exs. 3.35 and 4.19. Let $\beta_0(x), \beta_1(x), \beta_2(x), \ldots$ be defined to be the coefficients of $1, u, u^2, \ldots$ in the Maclaurin series expansion

$$\frac{-ue^{xu}}{e^u - 1} =: \sum_{n=0}^{\infty} \beta_n(x)u^n.$$

(a) Find the first two terms of the u-power series for the left side and thereby show that $\beta_0(x) = -1$, $\beta_1(x) = \frac{1}{2} - x$.

(b) Differentiate both sides of the above relation with respect to x and (after some manipulation) show that $\beta_n'(x) = \beta_{n-1}(x)$, $n = 1, 2, \ldots$.

(c) Integrate both sides of the above relation from $x = 0$ to $x = 1$ and thereby show that

$$\int_0^1 \beta_n(x)dx = 0, \qquad n = 1, 2, \ldots .$$

Note. From (b) and (c) you can infer that $w_{n-1}(x) = \beta_n(x)$ when $0 < x < 1$, $n = 1, 2, \ldots$.

(d) Suitably manipulate the relation

$$-\frac{d^m}{dx^m}\left\{\frac{ue^{xu}}{e^u-1}\right\} = \sum_{n=0}^{\infty} \beta_n^{(m)}(x)u^n$$

and thereby show that when $n = 1, 2, \ldots$

$$\beta_n^{(m)}(0+) - \beta_n^{(m)}(1-) = \begin{cases} 1 & \text{if } m = n-1 \\ 0 & \text{otherwise.} \end{cases}$$

(e) Use the generating relation to deduce that $\beta_n(1-x) = (-1)^n \beta_n(x)$, $n = 1, 2, \ldots$, i.e., β_n is even or odd with respect to the point $x = \frac{1}{2}$.

Note. The *Bernoulli polynomials* $B_n(x) := -n!\,\beta_n(x)$, $n = 1, 2, \ldots$ are commonly used in numerical analysis, number theory, $\ldots$. For additional details, *cf.* K.S. Williams, Bernoulli's identity without calculus, *Math. Mag.* **70**(1997), 47–50.

▶ **EXERCISE 4.23:**   Let f be a continuous function on $[0, L]$ with $f(0) = f(L) = 0$, and assume that f' is piecewise smooth. In this exercise you will use Fourier analysis to derive *Wirtinger's inequality*.

(a) Odd extend f to $[-L, L]$ using $f(-x) := -f(x)$, $0 \leq x \leq L$, and then verify that the Fourier series

$$f(x) = \sum_{k \neq 0} c_k\, e^{2\pi ikx/2L}, \qquad f'(x) = \sum_{k \neq 0} d_k\, e^{2\pi ikx/2L}$$

are related in such a manner that $d_k = (\pi ik/L)c_k$, $k = \pm 1, \pm 2, \ldots$.

(b) Use Plancherel's identity (for f, f') together with the identity of (a) to show that

$$\int_0^L |f'(x)|^2\,dx \geq \frac{\pi^2}{L^2} \int_0^L |f(x)|^2\,dx$$

with equality if and only if f is a scalar multiple of $\sin(\pi x/L)$.

▶ **EXERCISE 4.24:**   A simple, piecewise smooth, positively oriented, closed curve $z(t)$, $0 \leq t \leq L$ in the complex plane is parametrized using arc length so that $|z'(t)| = 1$ at points where z' is continuous. In this exercise you will prove that the enclosed area, A, is maximized when the curve is a circle.

(a) Use Green's formula from calculus and Parseval's identity to show that

$$A = \frac{1}{2}\,\mathrm{Im}\int_0^L z'(t)\overline{z}(t)dt = \pi \sum_{k=-\infty}^{\infty} k\left|Z[k]\right|^2.$$

Here $Z[k]$, $k = 0, \pm 1, \pm 2, \ldots$ are the Fourier coefficients of z.

(b) Show that

$$L = \int_0^L |z'(t)|^2\,dt = \left(\frac{4\pi^2}{L}\right) \sum_{k=-\infty}^{\infty} k^2\left|Z[k]\right|^2.$$

(c) Show that

$$\frac{L^2}{4\pi} - A = \pi \sum_{k=-\infty}^{\infty} (k^2 - k) \left| Z[k] \right|^2,$$

and thereby prove that the enclosed area takes the maximum value $A = L^2/4\pi$ if and only if the curve is a circle $z(t) = Z[0] + Z[1]e^{2\pi it/L}$ with radius $\left| Z[1] \right| = L/2\pi$.

► **EXERCISE 4.25:** Let f be a function on $\mathbb{Z}$ that is both p_0-periodic and p_1-periodic where $p_0 > p_1$ are positive integers. In this exercise you will show that f is periodic with period $p = \gcd(p_0, p_1)$.

(a) Let m_0, m_1 be integers. Show that $f[m_0 p_0 + m_1 p_1 + n] = f[n]$, $n = 0, \pm 1, \pm 2, \ldots$.

(b) Let the positive integers $d_1, d_2, \ldots, d_k$ and $p_0 > p_1 > p_2 > \cdots > p_k$ be constructed by using the *Euclidean algorithm* so that

$$p_0 := d_1 \cdot p_1 + p_2, \quad p_1 := d_2 \cdot p_2 + p_3, \quad \cdots, \quad p_{k-2} := d_{k-1} \cdot p_{k-1} + p_k, \quad p_{k-1} := d_k \cdot p_k.$$

Show that p_k is the greatest common divisor of p_0, p_1.

Hint. Observe that $p_k \big| p_{k-1}$, $p_k \big| p_{k-2}$, $\ldots$, $p_k \big| p_0$.

(c) Show that f is p-periodic.

► **EXERCISE 4.26:** In this exercise you will show that the *ramp function* $f[n] := n$, $n = 0, 1, \ldots, N - 1$ on $\mathbb{P}_N$ has the Fourier transform

$$F[k] = \frac{1}{2} \begin{cases} N - 1 & \text{if } k = 0 \\ i \cot \left(\dfrac{k\pi}{N} \right) - 1 & \text{if } k = 1, 2, \ldots, N - 1. \end{cases}$$

(a) Verify that

$$\sum_{n=0}^{N-1} n z^n = z \frac{d}{dz} \sum_{n=0}^{N-1} z^n = \frac{(N-1)z^{N+1} - Nz^N + z}{(z-1)^2}, \quad z \neq 1,$$

and then use this identity with $z_k := e^{-2\pi ik/N}$ to derive the above expression for F by evaluating the discrete sum from the analysis equation.

(b) Verify that

$$f[n] - f[n-1] = 1 - N\delta[n]$$

and then use this identity with the translation rule to derive the above expression for F.

(c) Apply Plancherel's identity to f, F and thereby show that

$$\sum_{k=1}^{N-1} \cot^2 \left(\frac{k\pi}{N} \right) = \frac{(N-1)(N-2)}{3}, \quad N = 2, 3, \ldots .$$

▶ **EXERCISE 4.27:**   In this exercise you will find the Fourier transform of the signum function f on $\mathbb{P}_N$.

(a) Let $N := 2M$, $M = 1, 2, \ldots$, and let

$$f[n] := \begin{cases} 1 & \text{if } n = 1, 2, \ldots, M-1 \\ 0 & \text{if } n = 0, M \\ -1 & \text{if } n = -1, -2, \ldots, -(M-1). \end{cases}$$

Show that

$$F[k] = \begin{cases} -\dfrac{2i}{N} \cot\left(\dfrac{k\pi}{N}\right) & \text{if } k = \pm 1, \pm 3, \pm 5, \ldots \\ 0 & \text{if } k = 0, \pm 2, \pm 4, \ldots \ . \end{cases}$$

(b) Let $N := 2M + 1$, $M = 0, 1, \ldots$, and let

$$f[n] := \begin{cases} 1 & \text{if } n = 1, 2, \ldots, M \\ 0 & \text{if } n = 0 \\ -1 & \text{if } n = -1, -2, \ldots, -M. \end{cases}$$

Show that

$$F[k] = \begin{cases} -\dfrac{i}{N} \cot\left(\dfrac{k\pi}{2N}\right) & \text{if } k = \pm 1, \pm 3, \pm 5, \ldots, \pm(2M-1) \\ \dfrac{i}{N} \tan\left(\dfrac{k\pi}{2N}\right) & \text{if } k = 0, \pm 2, \pm 4, \ldots, \pm 2M. \end{cases}$$

Hint. The tricks from Ex. 4.26 may be useful.

▶ **EXERCISE 4.28:**   Let N be a "large" positive integer with a "small" divisor $m = 2, 3, \ldots$, let $0 < \alpha < 1$ with $\alpha^N \ll 1$, let

$$f[n] := \alpha^n, \quad n = 0, 1, \ldots, N-1,$$

be a function on $\mathbb{P}_N$, and let

$$f_m[n] := f\left[n - \frac{N}{m}\right], \quad n = 0, 1, \ldots, N-1,$$

be the cyclic translate. Show that we obtain m "circles" when we plot $1/F_m[k]$, $k = 0, 1, \ldots, N-1$, as points in the complex plane.

▶ **EXERCISE 4.29:**   Let $-\infty < \alpha < \infty$, let f be the function on $\mathbb{P}_N$ with

$$f[n, \alpha] := e^{2\pi i \alpha n / N}, \quad n = 0, 1, \ldots, N-1,$$

and let $F[k, \alpha]$ be the corresponding discrete Fourier transform.

(a) Explain why the samples $f[n, \alpha]$ always seem to come from sinusoids having a frequency parameter in the interval $-N/2 \le \alpha < N/2$. In (b)–(d) we will assume that this is the case.

Hint. Consider $f[n, \alpha + mN]$, $m = \pm 1, \pm 2, \ldots$, and *cf.* Ex. 1.17(d).

(b) Show that

$$F[k, \alpha] = e^{-\pi i (N-1)(k-\alpha)/N} \frac{\operatorname{sinc}(k - \alpha)}{\operatorname{sinc}\{(k - \alpha)/N\}} .$$

(c) Let $g[n, \alpha] := w[n] \cdot f[n, \alpha]$ so $G[k, \alpha] = (W * F)[k, \alpha]$. Here the nonnegative function w on $\mathbb{P}_N$ is a *window* that is used to smoothly turn on and turn off the samples of the complex exponential. What happens in the two extreme cases where $w[n] := 1$, $n = 0, 1, \dots, N - 1$, and where $w[n] := \delta[n - \lfloor N/2 \rfloor]$, $n = 0, 1, \dots, N - 1$?

(d) Let w be the *hanning window* $w[n] := \frac{1}{2}\{1 - \cos[\pi(2n + 1)/N]\}$. Show that

$$W[k] = -\tfrac{1}{4} e^{i\pi/N} \delta[k - 1] + \tfrac{1}{2} \delta[k] - \tfrac{1}{4} e^{-i\pi/N} \delta[k + 1]$$
$$G[k, \alpha] = -\tfrac{1}{4} e^{i\pi/N} F[k - 1, \alpha] + \tfrac{1}{2} F[k, \alpha] - \tfrac{1}{4} e^{-i\pi/N} F[k + 1, \alpha].$$

Note. A computer-generated plot of $|F[k, \alpha]|$ vs k will have a peak at $k = \lfloor \alpha \rfloor$ or $k = \lceil \alpha \rceil$, but as k moves away from α the graph will decay like $|k - \alpha|^{-1}$ when $\alpha \neq 0, \pm 1, \pm 2, \dots$. (Here $\lfloor \ \rfloor$ and $\lceil \ \rceil$ are the *floor* and *ceiling* functions, e.g., $\lfloor \pi \rfloor = 3$, $\lceil \pi \rceil = 4$.) In contrast, the graph of $|G[k, \alpha]|$ decays like $|k - \alpha|^{-3}$. Thus the hanning window helps us to resolve closely spaced peaks in the spectrum corresponding to a sum $c_1 f_1[n, \alpha_1] + c_2 f[n, \alpha_2] + \cdots + c_M f[n, \alpha_M]$, cf. Ex. 3.34.

▶ **EXERCISE 4.30:** Let f, g be defined on $\mathbb{Z}$, $\mathbb{P}_N$ by writing

$$f[n] := \binom{N - 1}{n}, \qquad g[n] := \sum_{m=-\infty}^{\infty} f[n - mN].$$

Here $N = 2, 3, \dots$ and the binomial coefficient vanishes when $n < 0$ or $n \geq N$.

(a) Show that f, g have the Fourier transforms

$$F(s) = p^{-1}(1 + e^{-2\pi i s/p})^{N-1}, \qquad G[k] = N^{-1}(1 + e^{-2\pi i k/N})^{N-1}.$$

(b) Apply Plancherel's identity to f and thereby show that

$$\sum_{n=0}^{N-1} \binom{N - 1}{n}^2 = 4^{N-1} \int_0^1 \{\cos \pi s\}^{2N-2} ds = \binom{2N - 2}{N - 1}.$$

Hint. cf. Ex. 4.15(b).

(c) Apply Plancherel's identity to g and thereby show that

$$\sum_{n=0}^{N-1} \binom{N - 1}{n}^2 = \frac{4^{N-1}}{N} \sum_{k=0}^{N-1} \left\{ \cos\left(\frac{k\pi}{N}\right) \right\}^{2N-2} .$$

▶ **EXERCISE 4.31:**   Find the Fourier transform of the function

$$g[n] := \sum_{m=-\infty}^{\infty} f[n - mN]$$

on $\mathbb{P}_N$ that is constructed from the function f on $\mathbb{Z}$ when:

(a) $f[n] := \begin{cases} 1 & \text{if } n = 0, \pm 1, \pm 2, \dots, \pm L \quad (\text{where } N \geq 2L + 1) \\ 0 & \text{otherwise}; \end{cases}$

(b) $f[n] := \begin{cases} 1 & \text{if } n = \pm L \quad (\text{where } N \geq 2L + 1) \\ 0 & \text{otherwise}; \end{cases}$

(c) $f[n] := \begin{cases} \alpha^n & \text{if } n = 0, 1, 2, \dots \quad (\text{where } 0 < \alpha < 1) \\ 0 & \text{if } n = -1, -2, \dots. \end{cases}$

▶ **EXERCISE 4.32:**   Let f, g be real functions on $\mathbb{P}_N$ and let $y := f + ig$. Show that we can obtain the Fourier transforms of f, g from the Fourier transform of y by writing

$$F[k] = \frac{1}{2}\{Y[k] + \overline{Y[-k]}\}, \quad G[k] = \frac{1}{2i}\{Y[k] - \overline{Y}[-k]\}, \quad k = 0, 1, \dots, N - 1.$$

▶ **EXERCISE 4.33:**   The 1st, 2nd, ... *backward differences* of a function f on $\mathbb{P}_N$ are defined by

$$\nabla f[n] := f[n] - f[n - 1], \quad \nabla^2 f[n] := \nabla f[n] - \nabla f[n - 1], \quad \dots.$$

(a) Show that $\nabla^m f \equiv 0$ for some $m = 1, 2, \dots$ if and only if f is a constant function.

 Hint. Express the Fourier transform of $\nabla^m f$ in terms of F.

 Note. If f is a function on $\mathbb{Z}$, then $\nabla^m f \equiv 0$ if and only if f is a polynomial of degree $m - 1$ or less.

(b) Let q be a function on $\mathbb{P}_N$. When is it possible to find a function f on $\mathbb{P}_N$ such that

$$\nabla^m f[n] = q[n]?$$

▶ **EXERCISE 4.34:**   The *discrete Bernoulli functions* $\mathfrak{w}_0, \mathfrak{w}_1, \mathfrak{w}_2, \dots$ on $\mathbb{P}_N$ have the properties

$$\sum_{n=0}^{N-1} \mathfrak{w}_m[n] = 0, \quad \nabla^{m+1}\mathfrak{w}_m[n] = \delta[n] - \frac{1}{N}$$

when ∇ is the backward difference operator of Ex. 4.33.

(a) Show that $\mathfrak{w}_m$ is uniquely determined by these two properties.

(b) Find a simple expression for the Fourier transform $\mathfrak{W}_m$ of $\mathfrak{w}_m$.

(c) Let $m = 0, 1, \ldots$ be chosen with $2m + 1 \leq N$. Express the box function

$$b_m[n] := \begin{cases} 1 & \text{if } n = 0, \pm 1, \ldots, \pm m \\ 0 & \text{if } m < |n| \leq N/2 \end{cases}$$

as a linear combination of translates of $\mathbb{W}_1$ (plus a constant).

Note. You can construct piecewise-polynomial functions on $\mathbb{P}_N$ from these discrete Bernoulli functions, but the theory is more complicated (and less useful) than that for the Bernoulli functions (23) on $\mathbb{T}_p$.

▶ **EXERCISE 4.35:** Given a function f on $\mathbb{P}_N$ (with $f[n] \neq 0$ for some n) we define

$$R\{f\} := \frac{\displaystyle\sum_{n=0}^{N-1} |f[n+1] - f[n]|^2}{4 \displaystyle\sum_{n=0}^{N-1} |f[n]|^2} \, .$$

(a) Use Plancherel's identity to express $R\{f\}$ in terms of the discrete Fourier transform of f.

(b) Using (a), show that $0 \leq R\{f\} \leq 1$.

(c) When is $R\{f\} = 0$? When is $R\{f\} = 1$?

(d) Prove the following *discrete Wirtinger inequality*, cf. Ex. 4.23: For any complex numbers $0 = f[0], f[1], \ldots, f[M-1], f[M] = 0$,

$$\sum_{n=0}^{M-1} |f[n+1] - f[n]|^2 \geq 4 \sin^2\left(\frac{\pi}{2M}\right) \sum_{n=0}^{M-1} |f[n]|^2$$

with equality if and only if $f[n] = C \sin(n\pi/M)$ for some constant C.

Hint. Odd extend f to create a function on $\mathbb{P}_{2M}$ and use (a).

▶ **EXERCISE 4.36:** The discrete Fourier transform of the 1-vector (1) is the 1-vector (1). Use this fact with the repeat rule and the zero packing rule to find the DFT of:

(a) $\delta[n] := \begin{cases} 1 & \text{if } n = 0 \\ 0 & \text{if } n = 1, 2, \ldots, N-1; \end{cases}$

(b) $u[n] := 1, \ n = 0, 1, \ldots, N-1;$

(c) $c_m[n] := \begin{cases} 1 & \text{if } m|n \\ 0 & \text{if } m \nmid n, \end{cases} \quad n = 0, 1, \ldots, N-1 \text{ when } m|N.$

Note. If you forget the constants $1/m$, 1 associated with the zero packing and repeat rules (55)–(56), you can use the analysis of (a)–(b) to determine them!

▶ **EXERCISE 4.37:**   Let (A, B, C, D) be the discrete Fourier transform of the vector (a, b, c, d). Find the discrete Fourier transform of

(a) (a, d, c, b);

(b) (c, d, a, b);

(c) $(a, -b, c, -d)$;

(d) (A, B, C, D);

(e) (a, b, c, d, a, b, c, d);

(f) $(a, 0, 0, b, 0, 0, c, 0, 0, d, 0, 0)$;

(g) (a, c);

(h) (b, d);

(i) $(d + b, a + c)$.

▶ **EXERCISE 4.38:**   Let (A, B, C, D), (E, F, G, H) be the discrete Fourier transforms of (a, b, c, d), (e, f, g, h), respectively. Use the zero packing and translation rules to find the discrete Fourier transform of

(a) (a,0,b,0,c,0,d,0);

(b) (0,e,0,f,0,g,0,h);

(c) (a,e,b,f,c,g,d,h).

Note. You can find the DFT of an N-vector by suitably combining the DFT's of two vectors having $N/2$ components. The recursive use of this idea leads to a fast algorithm for computing the DFT of a vector with $N = 2^m$ components, $m = 1, 2, \ldots$, cf. Section 6.2.

▶ **EXERCISE 4.39:**   Let f, g be 3-periodic, 4-periodic functions on $\mathbb{P}_{12}$, respectively.

(a) Show that $f * g = C$ where C is a constant.

   Hint. Determine the zero, nonzero structure of F, G.

(b) Express C in terms of $f[0]$, $f[1]$, $f[2]$ and $g[0], g[1], g[2], g[3]$.

▶ **EXERCISE 4.40:**   Given a vector f and $\mu = 1, 2, \ldots$ we form $f^{(\mu)}$ by concatenating μ copies of f and we form $f^{[\mu]}$ by inserting $\mu - 1$ zeros after each component of f, e.g.,

$$(a, b)^{(3)} = (a, b, a, b, a, b), \quad (a, b)^{[3]} = (a, 0, 0, b, 0, 0).$$

Let the vectors $f_1, \ldots, f_m$ with $N_1, \ldots, N_m$ components have the discrete Fourier transforms $F_1, \ldots, F_m$ and let N be the least common multiple of $N_1, \ldots, N_m$.

(a) Show that

$$g := f_1^{(N/N_1)} + \cdots + f_m^{(N/N_m)} \quad \text{has the FT} \quad G = F_1^{[N/N_1]} + \cdots + F_m^{[N/N_m]}.$$

(b) Show that

$$g := f^{[N/N_1]} + \cdots + f_m^{[N/N_m]} \quad \text{has the FT} \quad G = \frac{N_1}{N} F^{(N/N_1)} + \cdots + \frac{N_m}{N} F_m^{(N/N_m)}.$$

▶ **EXERCISE 4.41:**   In this exercise you will find all functions f on $\mathbb{P}_4$ that satisfy the convolution equation $f * f = (1, 0, 0, 0)$.

(a) Show that $f[n]$ is a solution if and only if $F[k] = \pm\frac{1}{4}$ for each $k = 0, 1, 2, 3$.

(b) Show that if $f[n]$ is a solution, then so is $-f[n]$, $f[-n]$, $f[n - 2]$ and $\overline{f[n]}$.

(c) Find all 16 solutions.

   Hint. $(1, 0, 0, 0)$, $\frac{1}{2}(-1, 1, 1, 1)$, $\frac{1}{2}(1, i, 1, -i)$, $\frac{1}{2}(0, 1 + i, 0, 1 - i)$.

► **EXERCISE 4.42:** Prove that δ is the only possible *multiplicative identity* (2.25) for the convolution product of functions on $\mathbb{P}_N$.

► **EXERCISE 4.43:** Let $a = (a_0, a_1, \dots, a_{N-1})$, $a' = (a'_0, a'_1, \dots, a'_{N-1})$ be complex vectors. We say that a' is a *multiplicative inverse* of a (with respect to the convolution product) if $a * a' = a' * a = \delta := (1, 0, 0, \dots, 0)$.

(a) How are the components of the Fourier transforms $A = (A_0, A_1, \dots, A_{N-1})$, $A' = (A'_0, A'_1, \dots, A'_{N-1})$ of a, a' related when $a * a' = \delta$?

(b) Formulate a necessary and sufficient condition for a to have a multiplicative inverse.

(c) Formulate a procedure for finding the multiplicative inverse of a (if such exists).

Note. You can test your procedure by finding the multiplicative inverse $(-5, 7, 1, 1)$ of $(1/24) \cdot (0, 1, 2, 3)$.

► **EXERCISE 4.44:** Let a, b, c be functions on $\mathbb{P}_N$. Find necessary and sufficient conditions for:

(a) $a * x = 0$ to have a solution $x \neq 0$;

(b) $a * x = b$ to have a unique solution x;

(c) $a * x * x + b * x + c = 0$ to have 2^N distinct solutions x;

Hint. Formulate your answers in terms of the discrete Fourier transforms A, B, C.

► **EXERCISE 4.45:** Let $a_{00}, a_{01}, a_{10}, a_{11}, b_0, b_1$ be complex N-vectors. Show that there are uniquely determined complex N-vectors x_0, x_1 such that

$$a_{00} * x_0 + a_{01} * x_1 = b_0$$
$$a_{10} * x_0 + a_{11} * x_1 = b_1$$

if and only if the N-vector determinant

$$d := \det \begin{vmatrix} a_{00} & a_{01} \\ a_{10} & a_{11} \end{vmatrix} := a_{00} * a_{11} - a_{10} * a_{01}$$

has a multiplicative inverse as defined in Ex. 4.43.

► **EXERCISE 4.46:** Let a, b be functions on $\mathbb{Z}$ with

$$a[0] = 5, \quad a[1] = 9, \quad a[2] = 2, \quad a[3] = 6,$$

$$b[0] = 15, \quad b[1] = 32, \quad b[2] = 35, \quad b[3] = 61, \quad b[4] = 23, \quad b[5] = 26, \quad b[6] = 6,$$

and with all other components being 0. Solve the convolution equation $a * x = b$ to find the unknown function x on $\mathbb{Z}$.

Hint. A, B are polynomials in $z := e^{-2\pi i s / P}$.

▶ **EXERCISE 4.47:**   Let $f = (f[0], f[1], \ldots, f[N-1])$ and let $m = 2, 3, \ldots$. Show that each of the following procedures can be used to generate a vector $g = (g[0], g[1], \ldots, g[mN-1])$ with $g[mn] = f[n]$, $n = 0, 1, \ldots, N-1$. Such interpolation schemes can be used when we want to draw a smooth curve through a modest number of data points. (Try each one on the asteroid orbit data from Ex. 1.19!)

(a) Insert $m-1$ zeros after each component of f, convolve the result with the mN component vector $(m, m-1, \ldots, 1, 0, \ldots, 0, 1, 2, \ldots, m-1)$, and scale the convolution product by $1/m$.

(b) Insert $(m-1)N$ zeros after the last component of $F = (F[0], F[1], \ldots, F[N-1])$, and take the inverse Fourier transform of the resulting mN component vector.

(c) Insert $(m-1)N$ zeros between the "middle" components $F\big[\lfloor N/2 \rfloor\big]$ and $F\big[\lfloor N/2 \rfloor + 1\big]$ of $F = \big(F[0], F[1], \ldots, F[N-1]\big)$ and take the inverse Fourier transform of the resulting mN component vector.

Note. The scheme (c) produces a smooth interpolant that can be calculated quickly by using the FFT.

▶ **EXERCISE 4.48:**   Let f be a suitably regular p-periodic function on $\mathbb{R}$, let $N = 1, 2, \ldots$, let $h := p/N$, and let

$$A[k] := \frac{1}{p} \sum_{n=0}^{N-1} h f(nh) e^{-2\pi i k n h/p}$$

be the trapezoid rule approximation for the corresponding Fourier coefficient $F[k]$, $k = 0, \pm 1, \pm 2, \ldots$.

(a) Show that $A[k] - F[k] = \sum_{\mu \neq 0} F[k - \mu N]$.

(b) Assume that $f, f', \ldots, f^{(m-1)}$ are continuous and that $f^{(m)}$ is piecewise smooth. Explain why

$$p \sum_{k=-\infty}^{\infty} \big|(2\pi i k/p)^m F[k]\big|^2 = \int_0^p \big|f^{(m)}(x)\big|^2 dx,$$

and use this identity to deduce that

$$\big|A[k] - F[k]\big| \leq C h^m \text{ when } |k| \leq N/2.$$

Here C is a constant (that depends on f and m but not on h.)

(c) Compare the error estimate of (b) with the error bound for trapezoid rule integration from your elementary calculus textbook.

▶ **EXERCISE 4.49:** Let $(x[n], y[n])$, $n = 0, 1, \ldots, N-1$ be the Cartesian coordinates of the vertices of a simple, positively oriented N-gon $\mathfrak{P}$, and let $z[n] := x[n] + iy[n]$, $n = 0, 1, \ldots, N - 1$.

(a) Show that the triangle with vertices 0, $ae^{i\alpha}$, $be^{i\beta}$ (where $a > 0$, $b > 0$, and $0 \leq \alpha < \beta < \pi$) has the area $\frac{1}{2} \operatorname{Im} \{(be^{i\beta})(\overline{ae^{i\alpha}})\}$.

(b) Show that $\mathfrak{P}$ has the area

$$A = \frac{1}{2} \operatorname{Im} \sum_{n=0}^{N-1} z[n+1]\overline{z[n]} = \frac{N}{2} \sum_{k=0}^{N-1} \sin\left(\frac{2\pi k}{N}\right) |Z[k]|^2.$$

(c) Show that

$$A = \frac{1}{2} \sum_{n=0}^{N-1} \{y[n+1] - y[n-1]\}x[n] = -iN \sum_{k=0}^{N-1} \sin\left(\frac{2\pi k}{N}\right) \overline{Y}[k]X[k].$$

▶ **EXERCISE 4.50:** Let $\mathfrak{P}$ be a simple N-gon with vertices $z[0], z[1], \ldots, z[N-1]$ (in the complex plane). The area, A, and and the perimeter,

$$L := \sum_{n=0}^{N-1} |z[n+1] - z[n]|,$$

are related in such a manner that

$$A \leq \frac{L^2}{4N \tan(\pi/N)}$$

with equality if and only if $\mathfrak{P}$ is a regular N-gon, *cp.* Ex. 4.24. Show that this is so by proving each of the following statements.

(a) If two adjacent sides of a simple polygon are not equal, the area can be increased without changing the perimeter.

Hint. Consider the set of all triangles having a fixed base and perimeter.

(b) If every side of the N-gon has length L/N, then

$$L^2 = N \sum_{n=0}^{N-1} |z[n+1] - z[n]|^2 = 4N^2 \sum_{k=0}^{N-1} \sin^2\left(\frac{k\pi}{N}\right) |Z[k]|^2.$$

(c) If every side of the N-gon has length L/N, then

$$L^2 - 4N \tan\left(\frac{\pi}{N}\right) A = 4N^2 \sum_{k=2}^{N-1} \left\{ \frac{\sin(k\pi/N)\sin([k-1]\pi/N)}{\cos(\pi/N)} \right\} |Z[k]|^2.$$

Hint. Use the area formula from Ex. 4.49(b).

(d) The area A is maximized when $z[n] = Z[0] + Z[1]e^{2\pi in/N}$, $n = 0, 1, \ldots, N - 1$.

▶ **EXERCISE 4.51:**   Let $m = 1, 2, \ldots$ and $N = 2, 3, \ldots$ be *relatively prime, i.e.,* there is no $d = 2, 3, \ldots$ such that $d|m$ and $d|N$. In this exercise you will show that there is some $m' = 1, 2, \ldots, N - 1$ such that $m \cdot m' \equiv 1 \pmod{N}$, a fact that is needed for the proof of the *P*-dilation rule (59). You can do this by verifying the following assertions.

(a) None of the integers $m, 2m, \ldots, (N - 1)m$ is divisible by N.

(b) No difference $km - \ell m$ with $0 < \ell < k < N$ is divisible by N.

(c) The remainders that result when we divide $m, 2m, \ldots, (N - 1)m$ by N include each of the integers $1, 2, \ldots, N - 1$.

▶ **EXERCISE 4.52:**   Let $m = 1, 2, \ldots$, let P be a prime, and assume that $P \nmid m$. Prove *Fermat's theorem*: $m^{P-1} \equiv 1 \pmod{P}$.

   Hint. Begin by using the result of Ex. 4.51 to show that

$$m \cdot 2m \cdot 3m \cdot \, \cdots \, \cdot (P - 1)m \equiv 1 \cdot 2 \cdot 3 \cdot \, \cdots \, \cdot (P - 1) \pmod{P}.$$

▶ **EXERCISE 4.53:**   Let f be a function on $\mathbb{P}_N$, let $m = 2, 3, \ldots, N - 1$, and let $g[n] := f[mn], \; n = 0, \pm 1, \pm 2, \ldots$. Prove the following assertions.

(a) If m and N are relatively prime, then $G[mk] = F[k], \; k = 0, \pm 1, \pm 2, \ldots$.

(b) If $m|N$, then $G[mk] = F_m[k], \; k = 0, \pm 1, \pm 2, \ldots$, where F_m is the discrete Fourier transform of the scaled and sampled function

$$f_m[n] := \begin{cases} mf[n] & \text{if } m|n \\ 0 & \text{otherwise} \end{cases}$$

   on $\mathbb{P}_N$.

▶ **EXERCISE 4.54:**   Let f be a function on $\mathbb{P}_N$, let $m = 2, 3, \ldots, N - 1$, and assume that m and N are relatively prime. Let

$$f_0[n] := f[n], \quad f_1[n] := f_0[mn], \quad f_2[n] := f_1[mn], \quad \cdots$$

be the successive *m*-dilates of f.

(a) Show that f_1 can be obtained by suitably permuting the components of f_0. (This is not the case when m, N have a common divisor $d = 2, 3, \ldots$.)

(b) Show that the above sequence of m-dilates is p-periodic, *i.e.*, $f_{k+p} = f_k$, for some $p = 2, 3, \ldots, N - 1$.

(c) Find the period $p = p(m)$ of (b) for each $m = 2, 3, \ldots, 12$ when $N = 13$.

   Hint. $p = 12$ for $m = 2, 6, 7, 11$.

▶ **EXERCISE 4.55:** Let P be a prime, let $n = 1, 2, \ldots, P - 1$, let $a_0, a_1, \ldots, a_n$ be integers, and assume that $a_0 \not\equiv 0 \pmod P$. Show that the polynomial congruence

$$a_0 x^n + a_1 x^{n-1} + \cdots + a_{n-1} x + a_n \equiv 0 \pmod P$$

has at most n distinct solutions $x = 0, 1, \ldots, P - 1$.

Hint. If x and x_0 are such solutions, then

$$a_0 (x^n - x_0^n) + a_1 (x^{n-1} - x_0^{n-1}) + \cdots + a_{n-1} (x - x_0)$$
$$= (x - x_0)\{a_0 x^{n-1} + \cdots\} \equiv 0 \pmod P.$$

Use this (with Ex. 4.51) to prove the result by mathematical induction.

▶ **EXERCISE 4.56:** Let $P = 3, 5, 7, 11, \ldots$ be an odd prime.

(a) Show that the polynomial congruence $x^{(P-1)/2} \equiv 1 \pmod P$ has precisely $(P-1)/2$ distinct solutions

$$x \equiv 1^2, 2^2, \ldots, \left[\frac{P-1}{2}\right]^2 \pmod P$$

from $0, 1, \ldots, P - 1$.

Hint. Use the results of Exs. 4.51, 4.52, and 4.55.

(b) Show that the polynomial congruence $x^{(P-1)/2} \equiv -1 \pmod P$ has precisely $(P-1)/2$ distinct solutions from $0, 1, \ldots, P - 1$.

Hint. Use Ex. 4.55 with (a) and the identity

$$x^{P-1} - 1 = \{x^{(P-1)/2} - 1\} \cdot \{x^{(P-1)/2} + 1\}.$$

Note. When $x \equiv k^2 \pmod N$ for some $k = 1, 2, \ldots, N - 1$ we say that x is a *quadratic residue* modulo N, and if $x \not\equiv 0, 1^2, \ldots, (N-1)^2$ we say that x is a *quadratic nonresidue* modulo N. This exercise shows that there are precisely $(P - 1)/2$ quadratic residues and $(P - 1)/2$ quadratic nonresidues among $x = 1, 2, \ldots, P - 1$ when P is an odd prime.

▶ **EXERCISE 4.57:** Let $P = 3, 5, 7, 11, \ldots$ be an odd prime, and let the P-periodic *Legendre function* be defined on $\mathbb{Z}$ by writing

$$\ell[n] := \begin{cases} 0 & \text{if } n \equiv 0 \pmod P \\ 1 & \text{if } n \equiv 1^2, 2^2, \ldots, (P-1)^2 \pmod P \\ -1 & \text{otherwise.} \end{cases}$$

In this exercise you will find the Fourier transform L of ℓ.

(a) Use Ex. 4.56 to show that $\ell[n] \equiv n^{(P-1)/2} \pmod P$, $n = 0, \pm 1, \pm 2, \ldots$.

(b) Verify that ℓ has the multiplicative property $\ell[k \cdot n] = \ell[k] \cdot \ell[n]$, $k, n = 0, \pm 1, \pm 2, \ldots$.

(c) Show that if $k \equiv 1^2, 2^2, \ldots, (P-1)^2 \pmod P$, then $\ell[k \cdot n] = \ell[n]$, $n = 0, \pm 1, \pm 2, \ldots$, i.e., ℓ is not changed by k-dilation.

(d) Show that $L[k] = L[1] \cdot \ell[k]$, $k = 0, \pm1, \pm2, \ldots$, and thereby show that ℓ is an eigenfunction of the discrete Fourier transform, *cp.* (66).

Hint. Use (b) to write $\ell[n] = \ell[kn] \cdot \ell[k]$ when $k \equiv 1, 2, \ldots, P - 1 \pmod{P}$.

(e) Show that

$$
L[1] = \begin{cases} \dfrac{1}{\sqrt{P}} & \text{if } P \equiv 1 \pmod 4 \\[2mm] \dfrac{-i}{\sqrt{P}} & \text{if } P \equiv 3 \pmod 4. \end{cases}
$$

Hint. Use (72) with the identity

$$
PL[1] = \sum_{n=0}^{P-1} \{\ell[n] + 1\} e^{-2\pi i n/P} = \sum_{n=0}^{P-1} e^{-2\pi i n^2/P}.
$$

▶ **EXERCISE 4.58:**   Let m, M be positive integers, let $N := m \cdot M$, and let $f_m[n] := e^{2\pi i m n^2/N}$. Show that $|F_m|$ is a comb with

$$
\sqrt{M}\,|F_m[k]| = \begin{cases} \sqrt{2} & \text{if } k = 0, \pm2m, \pm4m, \ldots \text{ and } M = 4, 8, 12, \ldots \\ \sqrt{2} & \text{if } k = \pm m, \pm3m, \pm5m, \ldots \text{ and } M = 2, 6, 10, \ldots \\ 1 & \text{if } k = 0, \pm m, \pm2m, \ldots \text{ and } M = 1, 3, 5, \ldots \\ 0 & \text{otherwise.} \end{cases}
$$

Hint. Use the repeat rule (56) and the formula (71) for F_1.

▶ **EXERCISE 4.59:**   Let $P = 3, 5, 7, 11, \ldots$ be an odd prime and let r be a *primitive root* modulo P, i.e., $r^0, r^1, \ldots, r^{P-2}$ are congruent modulo P to some rearrangement of the integers $1, 2, \ldots, P - 1$. (For example, $r = 3$ is a primitive root modulo 7 since $3^0, 3^1, \ldots, 3^5$ have the remainders $1, 3, 2, 6, 4, 5$ when divided by 7.) From Ex. 4.52 we know that $r^{P-1} \equiv 1 \pmod{P}$ so the discrete function

$$
f[n] := e^{2\pi i r^n/P}, \qquad n = 0, \pm1, \pm2, \ldots
$$

is N-periodic with period $N := P - 1$.

(a) Show that f has the autocorrelation product $f \star f = P \cdot \delta - 1$.

(b) Show that the discrete Fourier transform of f has the modulus

$$
|F[k]| = \begin{cases} \dfrac{1}{P-1} & \text{if } k \equiv 0 \pmod N \\[2mm] \dfrac{\sqrt{P}}{P-1} & \text{otherwise.} \end{cases}
$$

Hint. Use (a) and the correlation product rule.

Note. This result has been used as the basis for a design of concert halls having desirable accoustical properties, *cf.* M.R. Schroeder, pp. 163–167.

Chapter 5

Operator Identities Associated with Fourier Analysis

5.1 The Concept of an Operator Identity

Introduction

In the preceding two chapters we developed a calculus for finding Fourier transforms of functions on $\mathbb{R}$, $\mathbb{T}_p$, $\mathbb{Z}$, and $\mathbb{P}_N$, and the corresponding rules are succinctly stated in Appendix 2. Each of these rules involves a pair of function-to-function mappings, *i.e.*, a pair of *operators*. In this chapter we will study these operators and the elementary relations that link them to one another. The change in emphasis will enable us to characterize the symmetry properties associated with Fourier analysis, to deepen and unify our understanding of the transformation rules that we use so often in practice, and to facilitate our study of the related sine, cosine, Hartley, and Hilbert transforms. Later on, we will use operators to describe fast algorithms for computing the DFT, to describe fast algorithms for computing with wavelets, to analyze thin lens systems in optics, etc.

Operators applied to functions on $\mathbb{P}_N$

It is easy to illustrate these ideas when we work with functions defined on $\mathbb{P}_N$, *i.e.*, with functions that can be identified with complex N-vectors. From linear algebra we know that any *linear mapping* $\mathcal{A} : \mathbb{C}^N \to \mathbb{C}^N$ can be represented by an $N \times N$ matrix of complex coefficients. In particular, the discrete Fourier transform operator $\mathcal{F}$ defined by the analysis equation

$$
\mathcal{F} \begin{bmatrix} f[0] \\ f[1] \\ \vdots \\ f[N-1] \end{bmatrix} := \begin{bmatrix} N^{-1} \sum\limits_{n=0}^{N-1} f[n] e^{-2\pi i \cdot 0 \cdot n/N} \\ N^{-1} \sum\limits_{n=0}^{N-1} f[n] e^{-2\pi i \cdot 1 \cdot n/N} \\ \vdots \\ N^{-1} \sum\limits_{n=0}^{N-1} f[n] e^{-2\pi i \cdot (N-1) \cdot n/N} \end{bmatrix},
$$

is a linear operator that is represented by the $N \times N$ complex matrix

$$
\mathcal{F} := \frac{1}{N} \begin{bmatrix} 1 & 1 & 1 & \cdots & 1 \\ 1 & \omega & \omega^2 & \cdots & \omega^{N-1} \\ 1 & \omega^2 & \omega^4 & \cdots & \omega^{2N-2} \\ \vdots & \vdots & \vdots & & \vdots \\ 1 & \omega^{N-1} & \omega^{2N-2} & \cdots & \omega^{(N-1)(N-1)} \end{bmatrix}, \quad \omega := e^{-2\pi i/N}. \tag{1}
$$

The reflection operator $\mathbf{R}$, defined by writing

$$
\mathbf{R} f[n] := f[-n], \quad n = 0, \pm 1, \pm 2, \ldots
$$

(when we think of f as being an N-periodic function on $\mathbb{Z}$), or by writing

$$
\mathbf{R} \begin{bmatrix} f[0] \\ f[1] \\ f[2] \\ \vdots \\ f[N-1] \end{bmatrix} := \begin{bmatrix} f[0] \\ f[N-1] \\ f[N-2] \\ \vdots \\ f[1] \end{bmatrix}
$$

(when we think of f as being a column N-vector) is also a linear operator that is represented by the $N \times N$ real matrix

$$
\mathbf{R} := \begin{bmatrix} 1 & 0 & 0 & \cdots & 0 & 0 \\ 0 & 0 & 0 & \cdots & 0 & 1 \\ 0 & 0 & 0 & \cdots & 1 & 0 \\ \vdots & \vdots & \vdots & & \vdots & \vdots \\ 0 & 0 & 1 & \cdots & 0 & 0 \\ 0 & 1 & 0 & \cdots & 0 & 0 \end{bmatrix}. \tag{2}
$$

Using such notation we can formulate the rules of discrete Fourier analysis in terms of certain operator identities. For example, the inversion rule

$$
g[n] := F[n] \quad \text{has the FT} \quad G[k] = N^{-1} f[-k]
$$

of (4.52) can be written in the *argument-free* form

$$g := \mathcal{F}f \quad \text{has the FT} \quad \mathcal{F}g = N^{-1}\mathbf{R}f,$$

which is more succinctly expressed by writing

$$\mathcal{F}(\mathcal{F}f) = N^{-1}\mathbf{R}f$$

when f is any function on $\mathbb{P}_N$. In this way we see that the inversion rule of discrete Fourier analysis is equivalent to the *function-free* operator identity

$$\mathcal{F}^2 = N^{-1}\mathbf{R},$$

which links the matrices (1) and (2). Operator identities that are equivalent to the other rules for taking discrete Fourier transforms can be found in an analogous fashion.

Blanket hypotheses

In Chapter 1 we proved that Fourier's representation can be used for every function on $\mathbb{P}_N$ and for large classes of functions defined on $\mathbb{Z}$, $\mathbb{T}_p$, and $\mathbb{R}$. We will presently introduce a number of operators $\mathcal{A}, \mathcal{A}_1, \mathcal{A}_2, \mathcal{A}_3, \ldots$ that are useful in Fourier analysis. Each of these operators will be defined on some complex linear space of such suitably regular functions, and each will have a range that is contained in some complex linear space of such functions.

When two operators $\mathcal{A}_1, \mathcal{A}_2$ share a common domain and

$$\mathcal{A}_1 f = \mathcal{A}_2 f$$

for all functions f in that domain, we will write

$$\mathcal{A}_1 = \mathcal{A}_2,$$

thereby defining operator equality. Given an operator $\mathcal{A}$ and a complex scalar α, we define the scalar products $\alpha\mathcal{A}, \mathcal{A}\alpha$ by setting

$$(\alpha\mathcal{A})f := \alpha(\mathcal{A}f)$$
$$(\mathcal{A}\alpha)f := \mathcal{A}(\alpha f)$$

for all functions f in the domain of $\mathcal{A}$. Analogously, given operators $\mathcal{A}_1, \mathcal{A}_2$ having a common domain, we define the operator sum $\mathcal{A}_1 + \mathcal{A}_2$ by setting

$$(\mathcal{A}_1 + \mathcal{A}_2)f := (\mathcal{A}_1 f) + (\mathcal{A}_2 f)$$

for all functions f in the common domain. Finally, when the range of $\mathcal{A}_2$ is included in the domain of $\mathcal{A}_1$, we define the operator product $\mathcal{A}_1\mathcal{A}_2$ by setting

$$(\mathcal{A}_1\mathcal{A}_2)f := \mathcal{A}_1(\mathcal{A}_2 f)$$

for all functions f in the domain of $\mathcal{A}_2$. We will let $\mathbf{I}$ denote the identity operator and define

$$\mathcal{A}^0 := \mathbf{I}, \quad \mathcal{A}^1 := \mathcal{A}, \quad \mathcal{A}^2 := \mathcal{A}\mathcal{A}, \quad \mathcal{A}^3 := \mathcal{A}(\mathcal{A}^2), \quad \cdots$$

in cases where the range of $\mathcal{A}$ is contained in its domain.

We will always work in a context where operator addition and operator multiplication are associative, $i.e.$,

$$\mathcal{A}_1 + (\mathcal{A}_2 + \mathcal{A}_3) = (\mathcal{A}_1 + \mathcal{A}_2) + \mathcal{A}_3$$
$$\mathcal{A}_1(\mathcal{A}_2\mathcal{A}_3) = (\mathcal{A}_1\mathcal{A}_2)\mathcal{A}_3,$$

and where scalar multiplication and operator multiplication distribute over operator addition, $i.e.$,

$$\alpha(\mathcal{A}_1 + \mathcal{A}_2) = (\alpha\mathcal{A}_1) + (\alpha\mathcal{A}_2)$$
$$(\mathcal{A}_1 + \mathcal{A}_2)\alpha = \mathcal{A}_1\alpha + \mathcal{A}_2\alpha$$
$$\mathcal{A}_1(\mathcal{A}_2 + \mathcal{A}_3) = (\mathcal{A}_1\mathcal{A}_2) + (\mathcal{A}_1\mathcal{A}_3)$$
$$(\mathcal{A}_1 + \mathcal{A}_2)\mathcal{A}_3 = (\mathcal{A}_1\mathcal{A}_3) + (\mathcal{A}_2\mathcal{A}_3).$$

Operator addition is always commutative, $i.e.$,

$$\mathcal{A}_1 + \mathcal{A}_2 = \mathcal{A}_2 + \mathcal{A}_1,$$

but operator multiplication is not, $i.e.$, we will frequently encounter situations where

$$\mathcal{A}_1\mathcal{A}_2 \neq \mathcal{A}_2\mathcal{A}_1.$$

All of our operators will be real homogeneous, $i.e.$,

$$\mathcal{A}\alpha = \alpha\mathcal{A}$$

when the scalar α is real, but in a few cases we will work with nonlinear operators (such as complex conjugation) for which

$$\mathcal{A}\alpha = \overline{\alpha}\mathcal{A}.$$

Most of our operators will be linear, however, and satisfy the familiar rules from the algebra of complex $N \times N$ matrices.

With this preparation, we are now ready to introduce a number of operators that are useful in the study of Fourier analysis. Our presentation will focus on the various relations that link these operators with little or no regard for the precise definition of regularity that is used to specify the underlying function spaces. While such technical considerations are of unquestioned value, they lie beyond the scope of this introductory text.

5.2 Operators Generated by Powers of $\mathcal{F}$

Powers of $\mathcal{F}$

We define the Fourier transform operator $\mathcal{F}$ by writing

$$
\begin{aligned}
(\mathcal{F}f)(s) &:= \int_{x=-\infty}^{\infty} f(x)e^{-2\pi i s x}dx, && -\infty < s < \infty, \\
(\mathcal{F}f)[k] &:= p^{-1}\int_{x=0}^{p} f(x)e^{-2\pi i k x/p}dx, && k = 0, \pm 1, \pm 2, \ldots, \\
(\mathcal{F}f)(s) &:= p^{-1}\sum_{n=-\infty}^{\infty} f[n]e^{-2\pi i s n/p}, && 0 \le s < p, \\
(\mathcal{F}f)[k] &:= N^{-1}\sum_{n=0}^{N-1} f[n]e^{-2\pi i k n/N}, && k = 0, 1, \ldots, N-1,
\end{aligned}
\tag{3}
$$

and we define the reflection operator $\mathbf{R}$ by writing

$$
\begin{aligned}
(\mathbf{R}f)(x) &:= f(-x) && \text{for } -\infty < x < \infty, \\
(\mathbf{R}f)(x) &:= f(-x) = \begin{cases} f(0) & \text{for } x = 0 \\ f(p-x) & \text{for } 0 < x < p, \end{cases} \\
(\mathbf{R}f)[n] &:= f[-n] && \text{for } n = 0, \pm 1, \pm 2, \ldots, \\
(\mathbf{R}f)[n] &:= f[-n] = \begin{cases} f[0] & \text{for } n = 0 \\ f[N-n] & \text{for } n = 1, 2, \ldots, N-1 \end{cases}
\end{aligned}
\tag{4}
$$

when f is a suitably regular function on $\mathbb{R}$, $\mathbb{T}_p$, $\mathbb{Z}$, $\mathbb{P}_N$, respectively.

We could introduce subscripts $\mathbb{R}$, $\mathbb{T}_p$, $\mathbb{Z}$, $\mathbb{P}_N$ to specify one of the operators from (3) or from (4). For example, the inversion rules for functions on $\mathbb{R}$, $\mathbb{T}_p$, $\mathbb{Z}$, $\mathbb{P}_N$ correspond to the operator identities

$$
\mathcal{F}_{\mathbb{R}}\mathcal{F}_{\mathbb{R}} = \mathbf{R}_{\mathbb{R}}, \quad \mathcal{F}_{\mathbb{Z}}\mathcal{F}_{\mathbb{T}_p} = p^{-1}\mathbf{R}_{\mathbb{T}_p}, \quad \mathcal{F}_{\mathbb{T}_p}\mathcal{F}_{\mathbb{Z}} = p^{-1}\mathbf{R}_{\mathbb{Z}}, \quad \mathcal{F}_{\mathbb{P}_N}\mathcal{F}_{\mathbb{P}_N} = N^{-1}\mathbf{R}_{\mathbb{P}_N}.
$$

Likewise, the reflection operators of (4) are involutions (*i.e.*, if we reflect twice we return to the original function), so we have

$$\mathbf{R}_{\mathbb{R}}\mathbf{R}_{\mathbb{R}} = \mathbf{I}_{\mathbb{R}}, \quad \mathbf{R}_{\mathbb{T}_p}\mathbf{R}_{\mathbb{T}_p} = \mathbf{I}_{\mathbb{T}_p}, \quad \mathbf{R}_{\mathbb{Z}}\mathbf{R}_{\mathbb{Z}} = \mathbf{I}_{\mathbb{Z}}, \quad \mathbf{R}_{\mathbb{P}_N}\mathbf{R}_{\mathbb{P}_N} = \mathbf{I}_{\mathbb{P}_N}.$$

We will drop such cumbersome subscripts, however, and simply write

$$\boldsymbol{\mathcal{F}}^2 = \beta^{-1}\mathbf{R}, \tag{5}$$

$$\mathbf{R}^2 = \mathbf{I} \tag{6}$$

for these two sets of equations with the understanding that the *universal constant* β takes the values

$$\beta := 1, p, p, N \tag{7}$$

when the operators on either side of (5),(6) are (initially) applied to suitably regular functions defined on $\mathbb{R}$, $\mathbb{T}_p$, $\mathbb{Z}$, $\mathbb{P}_N$, respectively. You will quickly learn to use the context to interpret the uncluttered expressions from (5), (6), and similar relations.

We will freely combine operator identities such as (5)–(6) using the algebraic properties discussed previously. For example, we have

$$\boldsymbol{\mathcal{F}}\mathbf{R} = \boldsymbol{\mathcal{F}}(\beta\boldsymbol{\mathcal{F}}^2) = \beta\boldsymbol{\mathcal{F}}^3 = (\beta\boldsymbol{\mathcal{F}}^2)\boldsymbol{\mathcal{F}} = \mathbf{R}\boldsymbol{\mathcal{F}} \tag{8}$$

and

$$\boldsymbol{\mathcal{F}}^4 = (\boldsymbol{\mathcal{F}}^2)(\boldsymbol{\mathcal{F}}^2) = (\beta^{-1}\mathbf{R})(\beta^{-1}\mathbf{R}) = \beta^{-2}\mathbf{R}^2 = \beta^{-2}\mathbf{I}. \tag{9}$$

By using (9) together with (5) and (6) we obtain the expressions

$$\boldsymbol{\mathcal{F}}^{-1} = \beta^2\boldsymbol{\mathcal{F}}^3 = \beta\mathbf{R}\boldsymbol{\mathcal{F}} = \beta\boldsymbol{\mathcal{F}}\mathbf{R} \tag{10}$$

for the inverse of the Fourier transform operator.

From (9) and (10) it follows that

$$\boldsymbol{\mathcal{F}}^{4m} = \beta^{-2m}\mathbf{I}, \quad \boldsymbol{\mathcal{F}}^{4m+1} = \beta^{-2m}\boldsymbol{\mathcal{F}}, \quad \boldsymbol{\mathcal{F}}^{4m+2} = \beta^{-2m}\boldsymbol{\mathcal{F}}^2, \quad \boldsymbol{\mathcal{F}}^{4m+3} = \beta^{-2m}\boldsymbol{\mathcal{F}}^3,$$
$$m = 0, \pm 1, \pm 2, \ldots \, .$$

This being the case, any polynomial

$$\mathcal{C}(\boldsymbol{\mathcal{F}}) := c_0\mathbf{I} + c_1\boldsymbol{\mathcal{F}} + c_2\boldsymbol{\mathcal{F}}^2 + \cdots + c_n\boldsymbol{\mathcal{F}}^n$$

of degree $n = 1, 2, \ldots$ in $\boldsymbol{\mathcal{F}}$ can be written as a polynomial

$$\mathcal{A}(\boldsymbol{\mathcal{F}}) = a_0\mathbf{I} + a_1\boldsymbol{\mathcal{F}} + a_2\boldsymbol{\mathcal{F}}^2 + a_3\boldsymbol{\mathcal{F}}^3$$

of degree 3 or less with

$$a_0 := c_0 + \beta^{-2}c_4 + \beta^{-4}c_8 + \cdots, \qquad a_1 := c_1 + \beta^{-2}c_5 + \beta^{-4}c_9 + \cdots,$$
$$a_2 := c_2 + \beta^{-2}c_6 + \beta^{-4}c_{10} + \cdots, \qquad a_3 := c_3 + \beta^{-2}c_7 + \beta^{-4}c_{11} + \cdots.$$

We will now introduce a number of operators of this form. Of course, any two such operators must commute with one another, *cf.* Ex. 5.1.

The even and odd projection operators

The even, odd parts of a function f are obtained by writing

$$\tfrac{1}{2}\{f(x) + f(-x)\}, \qquad \tfrac{1}{2}\{f(x) - f(-x)\}, \qquad -\infty < x < \infty,$$

$$\tfrac{1}{2}\{f(x) + f(p - x)\}, \qquad \tfrac{1}{2}\{f(x) - f(p - x)\}, \qquad 0 \le x < p,$$

$$\tfrac{1}{2}\{f[n] + f[-n]\}, \qquad \tfrac{1}{2}\{f[n] - f[-n]\}, \qquad n = 0, \pm 1, \pm 2, \dots,$$

$$\tfrac{1}{2}\{f[n] + f[N - n]\}, \qquad \tfrac{1}{2}\{f[n] - f[N - n]\}, \qquad n = 0, 1, \dots, N - 1$$

when f is defined on $\mathbb{R}$, $\mathbb{T}_p$, $\mathbb{Z}$, $\mathbb{P}_N$, respectively. You will recognize these as the functions

$$\mathbf{P}_e f, \qquad \mathbf{P}_o f$$

where the even and odd projection operators are succinctly defined by writing

$$\mathbf{P}_e := \tfrac{1}{2}(\mathbf{I} + \mathbf{R}), \qquad \mathbf{P}_o := \tfrac{1}{2}(\mathbf{I} - \mathbf{R}). \tag{11}$$

From the definitions (11) and the involutory property (6) of $\mathbf{R}$, we immediately obtain the projection relations

$$\mathbf{P}_e + \mathbf{P}_o = \mathbf{I}, \quad \mathbf{P}_e^2 = \mathbf{P}_e, \quad \mathbf{P}_o^2 = \mathbf{P}_o, \quad \mathbf{P}_e \mathbf{P}_o = \mathbf{P}_o \mathbf{P}_e = \mathbf{0}, \tag{12}$$

and the identities

$$\mathbf{P}_e - \mathbf{P}_o = \mathbf{R}, \quad \mathbf{P}_e \mathbf{R} = \mathbf{R} \mathbf{P}_e = \mathbf{P}_e, \quad \mathbf{P}_o \mathbf{R} = \mathbf{R} \mathbf{P}_o = -\mathbf{P}_o \tag{13}$$

that link $\mathbf{P}_e, \mathbf{P}_o$, and $\mathbf{R}$.

Example: Show that $\mathbf{P}_e^2 = \mathbf{P}_e$.

Solution: We use (11), (6), (11) in turn to write

$$\mathbf{P}_e^2 := \tfrac{1}{2}(\mathbf{I} + \mathbf{R})\tfrac{1}{2}(\mathbf{I} + \mathbf{R}) = \tfrac{1}{4}\{\mathbf{I}^2 + 2\mathbf{R} + \mathbf{R}^2\}$$

$$= \tfrac{1}{4}\{\mathbf{I} + 2\mathbf{R} + \mathbf{I}\} = \tfrac{1}{2}\{\mathbf{I} + \mathbf{R}\} =: \mathbf{P}_e. \qquad \blacksquare$$

The normalized exponential transform operators

The presence of the scale factor β in (5) and (9) [as well as in the Plancherel identities (1.15)–(1.18)] is an indication that the operator $\mathcal{F}$ and its iterates change the overall size of a function f defined on $\mathbb{T}_p$, $\mathbb{Z}$, or $\mathbb{P}_N$ (except when $p = 1$, $N = 1$). For this reason we are motivated to introduce the *normalized* exponential transform operators

$$\mathbf{E}_- := \beta^{1/2} \mathcal{F}, \qquad \mathbf{E}_+ := \beta^{-1/2} \mathcal{F}^{-1} \tag{14}$$

observing that we then have

$$(\mathbf{E}_\pm f)(s) = \int_{x=-\infty}^{\infty} f(x)e^{\pm 2\pi i s x}dx, \qquad -\infty < s < \infty,$$

$$(\mathbf{E}_\pm f)[k] = p^{-1/2}\int_{x=0}^{p} f(x)e^{\pm 2\pi i k x/p}dx, \qquad k = \pm 1, \pm 2, \ldots,$$

$$(\mathbf{E}_\pm f)(s) = p^{-1/2}\sum_{n=-\infty}^{\infty} f[n]e^{\pm 2\pi i s n/p}, \qquad 0 \le s < p,$$

$$(\mathbf{E}_\pm f)[k] = N^{-1/2}\sum_{n=0}^{N-1} f[n]e^{\pm 2\pi i k n/N}, \qquad k = 0, 1, \ldots, N-1,$$

when f is a suitably regular function on $\mathbb{R}$, $\mathbb{T}_p$, $\mathbb{Z}$, $\mathbb{P}_N$, respectively. The symbols $\mathbf{E}_-$, $\mathbf{E}_+$ remind us of the $-i$, $+i$ exponential kernels we use to construct these operators. Although it is a bit of a nuisance to include the scale factors $\beta^{\pm 1/2}$, it is quite easy to recall and use the scalar-free operator identities

$$\mathbf{E}_+\mathbf{E}_- = \mathbf{I}, \qquad \mathbf{E}_-\mathbf{E}_+ = \mathbf{I}, \tag{15}$$

$$\mathbf{E}_-^2 = \mathbf{R}, \qquad \mathbf{E}_+^2 = \mathbf{R}, \tag{16}$$

$$\mathbf{R}\mathbf{E}_- = \mathbf{E}_+, \qquad \mathbf{E}_-\mathbf{R} = \mathbf{E}_+,$$
$$\tag{17}$$
$$\mathbf{R}\mathbf{E}_+ = \mathbf{E}_-, \qquad \mathbf{E}_+\mathbf{R} = \mathbf{E}_-,$$

which can be derived from (14) by using (5), (6), and (8).

The factorization $\mathbf{I} = \mathbf{E}_+\mathbf{E}_-$ is a succinct way to summarize the analysis and synthesis equations of Fourier. Indeed, we can recover a suitably regular function f from the Fourier transform

$$F := (\beta^{-1/2}\mathbf{E}_-)f$$

by applying the operator $\beta^{1/2}\mathbf{E}_+$, i.e.,

$$(\beta^{+1/2}\mathbf{E}_+)F = (\beta^{+1/2}\mathbf{E}_+)(\beta^{-1/2}\mathbf{E}_-)f = f.$$

The normalized cosine transform and sine transform operators

After viewing the Euler identities

$$\cos\theta = \frac{1}{2}\{e^{i\theta} + e^{-i\theta}\}, \qquad \sin\theta = \frac{1}{2i}\{e^{i\theta} - e^{-i\theta}\}$$

we are motivated to define the normalized cosine transform and sine transform operators

$$\mathbf{C} := \frac{1}{2}\{\mathbf{E}_+ + \mathbf{E}_-\}, \qquad \mathbf{S} := \frac{1}{2i}\{\mathbf{E}_+ - \mathbf{E}_-\}. \tag{18}$$

We immediately deduce the operator identities

$$\mathbf{E}_+ = \mathbf{C} + i\mathbf{S}, \qquad \mathbf{E}_- = \mathbf{C} - i\mathbf{S} \tag{19}$$

(which remind us of the Euler identities $e^{i\theta} = \cos\theta + i\sin\theta$, $e^{-i\theta} = \cos\theta - i\sin\theta$),

$$\mathbf{C}^2 = \mathbf{P}_e, \qquad \mathbf{S}^2 = \mathbf{P}_o, \tag{20}$$
$$\mathbf{C}^2 + \mathbf{S}^2 = \mathbf{I}, \qquad \mathbf{C}^2 - \mathbf{S}^2 = \mathbf{R}, \tag{21}$$

and the commutation relations

$$\begin{aligned} \mathbf{SC} = \mathbf{CS} = 0, \\ \mathbf{CR} = \mathbf{RC} = \mathbf{C}, \\ \mathbf{SR} = \mathbf{RS} = -\mathbf{S}, \end{aligned} \tag{22}$$

cf. Ex. 5.3.

Example: Show that $\mathbf{C}^2 + \mathbf{S}^2 = \mathbf{I}$.

Solution: We use (18) with (15) and (16) to write

$$\mathbf{C}^2 = \frac{1}{2}(\mathbf{E}_+ + \mathbf{E}_-)\frac{1}{2}(\mathbf{E}_+ + \mathbf{E}_-) = \frac{1}{4}\{\mathbf{E}_+^2 + \mathbf{E}_+\mathbf{E}_- + \mathbf{E}_-\mathbf{E}_+ + \mathbf{E}_-^2\} = \frac{1}{2}(\mathbf{I} + \mathbf{R}),$$

$$\mathbf{S}^2 = \frac{1}{2i}(\mathbf{E}_+ - \mathbf{E}_-)\frac{1}{2i}(\mathbf{E}_+ - \mathbf{E}_-) = -\frac{1}{4}\{\mathbf{E}_+^2 - \mathbf{E}_+\mathbf{E}_- - \mathbf{E}_-\mathbf{E}_+ + \mathbf{E}_-^2\} = \frac{1}{2}(\mathbf{I} - \mathbf{R}),$$

and thereby obtain both of the relations (21). ∎

The factorization $\mathbf{P}_e = \mathbf{CC}$ succinctly summarizes the analysis and synthesis equations that are associated with the cosine transform. Indeed, we can recover any suitably regular *even* function f from

$$F_c := \mathbf{C}f$$

by applying the operator $\mathbf{C}$, *i.e.*,

$$\mathbf{C}F_c = \mathbf{CC}f = \mathbf{P}_e f = f.$$

Analogously, we can recover any suitably regular *odd* function f from the sine transform

$$F_s := \mathbf{S}f$$

by applying the operator $\mathbf{S}$, *i.e.*,

$$\mathbf{S}F_s = \mathbf{SS}f = \mathbf{P}_o f = f,$$

cf. Exs. 1.3 and 1.12.

The normalized Hartley transform operators

The normalized Hartley transform operators

$$\mathbf{H}_+ := \mathbf{C} + \mathbf{S}, \qquad \mathbf{H}_- := \mathbf{C} - \mathbf{S} \tag{23}$$

are real analogues of the operators $\mathbf{E}_+, \mathbf{E}_-$ from (19). These operators are involutory, *i.e.*,

$$\mathbf{H}_+^2 = \mathbf{I}, \qquad \mathbf{H}_-^2 = \mathbf{I}, \tag{24}$$

and they satisfy the commutation relations

$$\mathbf{H}_+ \mathbf{H}_- = \mathbf{H}_- \mathbf{H}_+ = \mathbf{R}, \tag{25}$$

$$\mathbf{H}_+ \mathbf{R} = \mathbf{R}\mathbf{H}_+ = \mathbf{H}_-, \tag{26}$$

$$\mathbf{H}_- \mathbf{R} = \mathbf{R}\mathbf{H}_- = \mathbf{H}_+,$$

cf. (15) and (16).

Example: Show that $\mathbf{H}_+^2 = \mathbf{I}$.

Solution: We begin with (23) and then use (22), (21) in turn to write

$$\mathbf{H}_+^2 = (\mathbf{C} + \mathbf{S})(\mathbf{C} + \mathbf{S}) = \mathbf{C}^2 + \mathbf{C}\mathbf{S} + \mathbf{S}\mathbf{C} + \mathbf{S}^2 = \mathbf{C}^2 + \mathbf{S}^2 = \mathbf{I}. \qquad \blacksquare$$

The factorization $\mathbf{I} = \mathbf{H}_+ \mathbf{H}_+$ leads to the *symmetric* synthesis-analysis equations

$$f = \mathbf{H}_+ F_h, \qquad F_h := \mathbf{H}_+ f$$

for Hartley analysis. The real-valued function

$$\operatorname{cas}\theta := \cos\theta + \sin\theta$$

(with cas being an abbreviation for <u>c</u>os <u>a</u>nd <u>s</u>in) that is used in the construction of a Hartley transform plays a role analogous to the complex-valued function

$$\operatorname{cis}\theta := \cos\theta + i\sin\theta$$

(with cis being an abbreviation for cos and i sin) that is used to construct a Fourier transform. These synthesis-analysis equations take the form

$$f(x) = \int_{s=-\infty}^{\infty} F_h(s)\mathrm{cas}(2\pi sx)ds, \qquad F_h(s) = \int_{x=-\infty}^{\infty} f(x)\mathrm{cas}(2\pi sx)dx,$$

$$f(x) = p^{-1/2} \sum_{k=-\infty}^{\infty} F_h[k]\mathrm{cas}(2\pi kx/p), \quad F_h[k] = p^{-1/2}\int_{x=0}^{p} f(x)\mathrm{cas}(2\pi kx/p),$$
$$\tag{27}$$

$$f[n] = p^{-1/2}\int_{s=0}^{p} F_h(s)\mathrm{cas}(2\pi sn/p), \qquad F_h(s) = p^{-1/2}\sum_{n=-\infty}^{\infty} f[n]\mathrm{cas}(2\pi sn/p),$$

$$f[n] = N^{-1/2}\sum_{k=0}^{N-1} F_h[k]\mathrm{cas}(2\pi kn/N), \quad F_h[k] = N^{-1/2}\sum_{n=0}^{N-1} f[n]\mathrm{cas}(2\pi kn/N),$$

when f is a suitably regular function on $\mathbb{R}$, $\mathbb{T}_p$, $\mathbb{Z}$, $\mathbb{P}_N$, respectively. The Hartley transform F_h is always real valued when f is real valued. This is a very nice property that the Fourier transform does not have.

Connections

From (11) and (22) we see that the operators $\mathbf{C}, \mathbf{S}$ (like the functions cos, sin) are even, odd in the sense that

$$\begin{aligned} \mathbf{P}_e\mathbf{C} = \mathbf{C}, \qquad \mathbf{P}_o\mathbf{C} = \mathbf{0}, \\ \mathbf{P}_e\mathbf{S} = \mathbf{0}, \qquad \mathbf{P}_o\mathbf{S} = \mathbf{S}, \end{aligned} \tag{28}$$

so when we apply $\mathbf{P}_e, \mathbf{P}_o$ to the transform operators of (19) and (23) we find

$$\mathbf{P}_e\mathbf{E}_\pm = \mathbf{C}, \qquad \mathbf{P}_o\mathbf{E}_\pm = \pm i\,\mathbf{S}, \tag{29}$$
$$\mathbf{P}_e\mathbf{H}_\pm = \mathbf{C}, \qquad \mathbf{P}_o\mathbf{H}_\pm = \pm\mathbf{S}. \tag{30}$$

With the aid of these identities we easily obtain the relations

$$\mathbf{E}_- = \beta^{1/2}\mathcal{F} = \mathbf{C} - i\,\mathbf{S} = (\mathbf{P}_e - i\,\mathbf{P}_o)\mathbf{H}_+ \tag{31}$$
$$\mathbf{C} = \mathbf{P}_e\mathbf{H}_+ = \mathbf{P}_e\mathbf{E}_- = \beta^{1/2}\mathbf{P}_e\mathcal{F} \tag{32}$$
$$\mathbf{S} = \mathbf{P}_o\mathbf{H}_+ = i\,\mathbf{P}_o\mathbf{E}_- = \beta^{1/2}i\,\mathbf{P}_o\mathcal{F} \tag{33}$$
$$\mathbf{H}_+ = \mathbf{C} + \mathbf{S} = (\mathbf{P}_e + i\,\mathbf{P}_o)\mathbf{E}_- = \beta^{1/2}(\mathbf{P}_e + i\,\mathbf{P}_o)\mathcal{F} \tag{34}$$

that link the transforms associated with $\mathcal{F}$, $\mathbf{E}_-$, $\mathbf{C}$, $\mathbf{S}$, and $\mathbf{H}_+$.

Example: Find the Hartley transform (27) of $f(x) := e^{2\pi i x}\Pi(x)$.

Solution: We use the calculus from Chapter 3 to find the Fourier transform

$$(\mathcal{F}f)(s) = \text{sinc}(s - 1),$$

and we then use $\mathbf{P}_e, \mathbf{P}_o$ to obtain

$$(\mathbf{C}f)(s) = (\mathbf{P}_e\mathcal{F}f)(s) = \frac{1}{2}\{\text{sinc}(s - 1) + \text{sinc}(s + 1)\},$$

$$(\mathbf{S}f)(s) = (i\,\mathbf{P}_o\mathcal{F}f)(s) = \frac{i}{2}\{\text{sinc}(s - 1) - \text{sinc}(s + 1)\}.$$

We add these to produce

$$(\mathbf{H}_+f)(s) = (\mathbf{C}f)(s) + (\mathbf{S}f)(s)$$
$$= \frac{1}{2}\{\text{sinc}(s - 1) + \text{sinc}(s + 1)\} + \frac{i}{2}\{\text{sinc}(s - 1) - \text{sinc}(s + 1)\}. \qquad \blacksquare$$

Example: Find the Hartley series (27) for the 4-periodic function f with

$$f(x) := \begin{cases} 1 & \text{if } 0 < x < 1 \\ 0 & \text{if } 1 < x < 4. \end{cases}$$

Solution: The function f is generated by summing the 4-translates of $\Pi(x - 1/2)$, so we can use Poisson's relation to find the Fourier coefficients

$$(\mathcal{F}f)[k] = \frac{1}{4}e^{-2\pi i(1/2)(k/4)}\text{sinc}\left(\frac{k}{4}\right)$$
$$= \frac{1}{4}\left\{\cos\left(\frac{k\pi}{4}\right) - i\sin\left(\frac{k\pi}{4}\right)\right\}\text{sinc}\left(\frac{k}{4}\right).$$

We use $\mathbf{P}_e, \mathbf{P}_o$ (with the $\beta^{1/2}$ factor) to obtain

$$(\mathbf{C}f)[k] = 4^{1/2}(\mathbf{P}_e\mathcal{F}f)[k] = \frac{1}{2}\cos\left(\frac{k\pi}{4}\right)\text{sinc}\left(\frac{k}{4}\right),$$

$$(\mathbf{S}f)[k] = 4^{1/2}(i\,\mathbf{P}_o\mathcal{F}f)[k] = \frac{1}{2}\sin\left(\frac{k\pi}{4}\right)\text{sinc}\left(\frac{k}{4}\right),$$

and form

$$(\mathbf{H}_+f)[k] = (\mathbf{C}f)[k] + (\mathbf{S}f)[k] = \frac{1}{2}\text{cas}\left(\frac{k\pi}{4}\right)\text{sinc}\left(\frac{k}{4}\right).$$

In this way we find the Hartley series

$$f(x) = \frac{1}{4}\sum_{k=-\infty}^{\infty}\text{cas}\left(\frac{k\pi}{4}\right)\text{sinc}\left(\frac{k}{4}\right)\text{cas}\left(\frac{2\pi k x}{4}\right)$$

[which requires an additional factor of $\beta^{-1/2} = 1/2$, cf. (27)]. $\qquad \blacksquare$

Tag notation

Newton used the symbol $\dot{f}$ for the derivative of the function f. The *tag*, `, indicates that the function f has been processed by applying the derivative operator, **D**. The superscript prime, $'$, is used for the same purpose in elementary calculus. We will use the caret, $^\wedge$, the klicka, $^\vee$, and, the tilde, $^\sim$, as superscripts to show that $\mathcal{F}, \mathbf{R}, \mathbf{H}_+$ have been applied to a function, *i.e.*, we write

$$f' := \mathbf{D}f, \quad f^\wedge := \mathcal{F}f, \quad f^\vee := \mathbf{R}f, \quad f^\sim := \mathbf{H}_+ f. \tag{35}$$

Operator identities such as

$$f^{\wedge\wedge} = \beta^{-1} f^\vee, \quad f^{\wedge\wedge\wedge\wedge} = \beta^{-2} f, \quad f^{\vee\vee} = f, \quad f^{\vee\wedge} = f^{\wedge\vee}, \quad f^{\sim\sim} = f$$

can be expressed compactly with this notation. A string of tags is always applied from left to right, *e.g.*,

$$f^{\vee'\wedge} := \left((f^\vee)' \right)^\wedge.$$

In contrast, a string of operators is always applied from right to left, *e.g.*,

$$\mathcal{F}\mathbf{D}\mathbf{R}f := \mathcal{F}(\mathbf{D}(\mathbf{R}f)).$$

5.3 Operators Related to Complex Conjugation

The bar and dagger operators

We define the bar and dagger operators

$$\mathcal{B}f := \overline{f}, \qquad \mathcal{D}f := \overline{\mathbf{R}f} \tag{36}$$

using an overbar to denote the complex conjugate. Like $\mathbf{R}$, the operators $\mathcal{B}, \mathcal{D}$ are involutory, *i.e.*,

$$\mathcal{B}^2 = \mathbf{I}, \qquad \mathcal{D}^2 = \mathbf{I}, \tag{37}$$

and the operators $\mathbf{R}, \mathcal{B}, \mathcal{D}$ commute, *i.e.*,

$$\mathbf{R}\mathcal{B} = \mathcal{B}\mathbf{R} = \mathcal{D}, \qquad \mathcal{B}\mathcal{D} = \mathcal{D}\mathcal{B} = \mathbf{R}, \qquad \mathcal{D}\mathbf{R} = \mathbf{R}\mathcal{D} = \mathcal{B}. \tag{38}$$

Example: Show that $\mathcal{B}\mathcal{D} = \mathbf{R}$.

Solution: When f is a function on $\mathbb{R}$ we have

$$(\mathcal{B}\mathcal{D}f)(x) := \mathcal{B}(\overline{f(-x)}) := \overline{\overline{f(-x)}} = f(-x) =: (\mathbf{R}f)(x).$$

The same argument works when f is a function on $\mathbb{T}_p$, $\mathbb{Z}$, or $\mathbb{P}_N$. ∎

We will use the bar, $^-$, and dagger, †, as superscripts to show that $\mathcal{B}, \mathcal{D}$ have been applied, *i.e.*,

$$f^- := \mathcal{B}f, \qquad f^\dagger := \mathcal{D}f. \tag{39}$$

We can use this tag notation to rewrite (37)–(38) in the form

$$f^{--} = f, \qquad f^{\dagger\dagger} = f,$$
$$f^{-\vee} = f^{\vee-} = f^\dagger, \qquad f^{\dagger-} = f^{-\dagger} = f^\vee, \qquad f^{\vee\dagger} = f^{\dagger\vee} = f^-.$$

The operators $\mathcal{B}, \mathcal{D}$ are additive, *i.e.*,

$$\mathcal{B}(f_1 + f_2) = (\mathcal{B}f_1) + (\mathcal{B}f_2), \qquad \mathcal{D}(f_1 + f_2) = (\mathcal{D}f_1) + (\mathcal{D}f_2),$$

but not homogeneous since

$$\mathcal{B}(\alpha f) = \bar\alpha f^- = \bar\alpha(\mathcal{B}f), \qquad \mathcal{D}(\alpha f) = \bar\alpha f^\dagger = \bar\alpha(\mathcal{D}f)$$

agree with $\alpha(\mathcal{B}f), \alpha(\mathcal{D}f)$ only in those cases where f is the zero function or the scalar α is real. We express this lack of homogeneity by writing

$$\mathcal{B}\alpha = \bar\alpha\mathcal{B}, \qquad \mathcal{D}\alpha = \bar\alpha\mathcal{D} \tag{40}$$

when α is any complex scalar. Thus neither $\mathcal{B}$ nor $\mathcal{D}$ is linear. In particular, neither of these operators can be represented by $N \times N$ matrices when we work with functions defined on $\mathbb{P}_N$.

Since the kernel functions cos, sin, cas that are used to construct the operators **C**, **S**, $\mathbf{H}_\pm$ are all real valued it is easy to see that $\mathcal{B}$, $\mathcal{D}$ commute with these operators, *i.e.*,

$$\begin{aligned}
\mathcal{B}\,\mathbf{C} &= \mathbf{C}\mathcal{B}, & \mathcal{D}\,\mathbf{C} &= \mathbf{C}\mathcal{D}, \\
\mathcal{B}\,\mathbf{S} &= \mathbf{S}\mathcal{B}, & \mathcal{D}\,\mathbf{S} &= \mathbf{S}\mathcal{D}, \\
\mathcal{B}\mathbf{H}_\pm &= \mathbf{H}_\pm\mathcal{B}, & \mathcal{D}\mathbf{H}_\pm &= \mathbf{H}_\pm\mathcal{D}.
\end{aligned} \tag{41}$$

In contrast, by using (19), (41), and (40), we find

$$\mathbf{E}_\pm\mathcal{B} = (\mathbf{C} \pm i\,\mathbf{S})\mathcal{B} = \mathcal{B}\mathbf{C} \pm i\,\mathcal{B}\mathbf{S} = \mathcal{B}\mathbf{C} \mp \mathcal{B}i\,\mathbf{S} = \mathcal{B}(\mathbf{C} \mp i\,\mathbf{S}) = \mathcal{B}\mathbf{E}_\mp. \tag{42}$$

We use this relation with (17) and (38) to obtain the commutation identities

$$\mathbf{E}_\pm\mathcal{B} = \mathcal{D}\mathbf{E}_\pm, \qquad \mathbf{E}_\pm\mathcal{D} = \mathcal{B}\mathbf{E}_\pm. \tag{43}$$

Since $\mathcal{F}$ is a real scalar multiple of $\mathbf{E}_-$ we also have

$$\mathcal{F}\mathcal{B} = \mathcal{D}\mathcal{F}, \qquad \mathcal{F}\mathcal{D} = \mathcal{B}\mathcal{F}. \tag{44}$$

The real, imaginary, hermitian, and antihermitian projection operators

We define the real and imaginary projection operators

$$\mathbf{P}_r := \tfrac{1}{2}(\mathbf{I} + \mathcal{B}), \qquad \mathbf{P}_i := \tfrac{1}{2}(\mathbf{I} - \mathcal{B}) \tag{45}$$

and note that

$$\mathbf{P}_r f = \tfrac{1}{2}(f + f^-)$$

is the real part of f and

$$\mathbf{P}_i f = \frac{1}{2}(f - f^-) = i\left\{\frac{1}{2i}(f - f^-)\right\}$$

is the pure imaginary part of f *including* the factor i, cf. Ex. 5.10. Likewise, we define the hermitian and antihermitian projection operators

$$\mathbf{P}_h := \tfrac{1}{2}(\mathbf{I} + \mathcal{D}), \qquad \mathbf{P}_a := \tfrac{1}{2}(\mathbf{I} - \mathcal{D}) \tag{46}$$

so that

$$\mathbf{P}_h f = \tfrac{1}{2}(f + f^\dagger), \qquad \mathbf{P}_a f = \tfrac{1}{2}(f - f^\dagger)$$

are the hermitian, antihermitian parts of f, respectively. Since $\mathcal{B}$, $\mathcal{D}$ are both involutory, the projection relations

$$\mathbf{P}_r + \mathbf{P}_i = \mathbf{I}, \quad \mathbf{P}_r^2 = \mathbf{P}_r, \quad \mathbf{P}_i^2 = \mathbf{P}_i, \quad \mathbf{P}_r\mathbf{P}_i = \mathbf{P}_i\mathbf{P}_r = 0 \tag{47}$$

$$\mathbf{P}_h + \mathbf{P}_a = \mathbf{I}, \quad \mathbf{P}_h^2 = \mathbf{P}_h, \quad \mathbf{P}_a^2 = \mathbf{P}_a, \quad \mathbf{P}_h\mathbf{P}_a = \mathbf{P}_a\mathbf{P}_h = 0 \tag{48}$$

analogous to (12) and the identities

$$\mathbf{P}_r - \mathbf{P}_i = \mathcal{B}, \quad \mathbf{P}_r\mathcal{B} = \mathcal{B}\mathbf{P}_r = \mathbf{P}_r, \quad \mathbf{P}_i\mathcal{B} = \mathcal{B}\mathbf{P}_i = -\mathbf{P}_i \tag{49}$$

$$\mathbf{P}_h - \mathbf{P}_a = \mathcal{D}, \quad \mathbf{P}_h\mathcal{D} = \mathcal{D}\mathbf{P}_h = \mathbf{P}_h, \quad \mathbf{P}_a\mathcal{D} = \mathcal{D}\mathbf{P}_a = -\mathbf{P}_a \tag{50}$$

analogous to (13) are easily verified. You should have no difficulty remembering (or with minimal effort deriving) such relations when they are needed.

Symmetric functions

The commuting operators $\mathbf{R}$, $\mathcal{B}$, $\mathcal{D}$ have been used to define the six commuting projection operators $\mathbf{P}_e$, $\mathbf{P}_o$, $\mathbf{P}_r$, $\mathbf{P}_i$, $\mathbf{P}_h$, $\mathbf{P}_a$ of (11), (45), and (46). Common symmetry properties used in Fourier analysis can be formulated in terms of these operators. Indeed, a function f is said to be even, odd, real, pure imaginary, hermitian, antihermitian according as

$$\mathbf{P}_e f = f, \quad \mathbf{P}_o f = f, \quad \mathbf{P}_r f = f, \quad \mathbf{P}_i f = f, \quad \mathbf{P}_h f = f, \quad \mathbf{P}_a f = f,$$

or equivalently as

$$\mathbf{R}f = f, \quad \mathbf{R}f = -f, \quad \mathcal{B}f = f, \quad \mathcal{B}f = -f, \quad \mathcal{D}f = f, \quad \mathcal{D}f = -f,$$

respectively.

Four other symmetries can be formulated using products of $\mathbf{P}_e$, $\mathbf{P}_o$, $\mathbf{P}_r$, $\mathbf{P}_i$. A function f is said to be real-even, real-odd, pure imaginary-even, pure imaginary-odd, according as

$$\mathbf{P}_r\mathbf{P}_e f = f, \quad \mathbf{P}_r\mathbf{P}_o f = f, \quad \mathbf{P}_i\mathbf{P}_e f = f, \quad \mathbf{P}_i\mathbf{P}_o f = f,$$

respectively. A routine calculation [*cf.* Ex. 5.13(a)] shows that every product of powers of $\mathbf{P}_e$, $\mathbf{P}_o$, $\mathbf{P}_r$, $\mathbf{P}_i$, $\mathbf{P}_h$, $\mathbf{P}_a$ reduces to $\mathbf{0}$, $\mathbf{I}$ or one of the ten projections $\mathbf{P}_e$, $\mathbf{P}_o$, $\mathbf{P}_r$, $\mathbf{P}_i$, $\mathbf{P}_h$, $\mathbf{P}_a$, $\mathbf{P}_r\mathbf{P}_e$, $\mathbf{P}_r\mathbf{P}_o$, $\mathbf{P}_i\mathbf{P}_e$, $\mathbf{P}_i\mathbf{P}_o$, given above. In this way we verify that the symmetry list of Ex. 1.2 is complete.

Less common symmetries can be generated from these operators in other ways. For example, a symmetry is associated with each of the projections

$$\mathbf{I} - \mathbf{P}_r\mathbf{P}_e, \quad \mathbf{I} - \mathbf{P}_r\mathbf{P}_o, \quad \mathbf{I} - \mathbf{P}_i\mathbf{P}_e, \quad \mathbf{I} - \mathbf{P}_i\mathbf{P}_o.$$

These are complementary to $\mathbf{P}_r\mathbf{P}_e$, $\mathbf{P}_r\mathbf{P}_o$, $\mathbf{P}_i\mathbf{P}_e$, $\mathbf{P}_i\mathbf{P}_o$ in the same way that $\mathbf{P}_o$, $\mathbf{P}_i$, $\mathbf{P}_a$ are complementary to $\mathbf{P}_e$, $\mathbf{P}_r$, $\mathbf{P}_h$, respectively, *cf.* Ex. 5.13(b).

Symmetric operators

Operators, like functions, can possess certain symmetry properties. We say that an operator $\mathcal{A}$ preserves the symmetry of being even, odd when

$$\mathcal{A}\mathbf{P}_e = \mathbf{P}_e\mathcal{A}, \qquad \mathcal{A}\mathbf{P}_o = \mathbf{P}_o\mathcal{A},$$

respectively. Using (11), we see that $\mathcal{A}$ preserves these symmetries if and only if

$$\mathcal{A}\mathbf{R} = \mathbf{R}\mathcal{A}.$$

Each of the transform operators $\mathcal{F}$, $\mathbf{C}$, $\mathbf{S}$, $\mathbf{H}_\pm$ commutes with $\mathbf{R}$ so each of these operators preserves the symmetries of being even and odd.

Analogously, we say that $\mathcal{A}$ preserves the symmetry of being real, pure imaginary when $\mathcal{A}$ commutes with the projections $\mathbf{P}_r$, $\mathbf{P}_i$, respectively, in which case

$$\mathcal{A}\mathcal{B} = \mathcal{B}\mathcal{A}.$$

Likewise, we say that $\mathcal{A}$ preserves the symmetry of being hermitian, antihermitian when $\mathcal{A}$ commutes with the projections $\mathbf{P}_h$, $\mathbf{P}_a$, respectively, in which case

$$\mathcal{A}\mathcal{D} = \mathcal{D}\mathcal{A}.$$

The transform operators $\mathbf{C}$, $\mathbf{S}$, $\mathbf{H}_{\pm}$ commute with $\mathcal{B}$, $\mathcal{D}$ so they preserve all four of these symmetries. On the other hand, we see from (44) that the Fourier transform operator $\mathcal{F}$ does not commute with either $\mathcal{B}$ or $\mathcal{D}$, so $\mathcal{F}$ cannot preserve any of these symmetries.

Using the commutation relations (38) we find

$$\mathbf{P}_r\mathbf{P}_e = \tfrac{1}{4}(\mathbf{I} + \mathbf{R} + \mathcal{B} + \mathcal{D}), \qquad \mathbf{P}_r\mathbf{P}_o = \tfrac{1}{4}(\mathbf{I} - \mathbf{R} + \mathcal{B} - \mathcal{D}),$$
$$\mathbf{P}_i\mathbf{P}_e = \tfrac{1}{4}(\mathbf{I} + \mathbf{R} - \mathcal{B} - \mathcal{D}), \qquad \mathbf{P}_i\mathbf{P}_o = \tfrac{1}{4}(\mathbf{I} - \mathbf{R} - \mathcal{B} + \mathcal{D}), \tag{51}$$

and thereby conclude that $\mathcal{A}$ preserves the symmetries of being real-even, real-odd, pure imaginary-even, pure imaginary-odd if and only if $\mathcal{A}$ commutes with the operators

$$\mathbf{R} + \mathcal{B} + \mathcal{D}, \qquad -\mathbf{R} + \mathcal{B} - \mathcal{D}, \qquad \mathbf{R} - \mathcal{B} - \mathcal{D}, \qquad -\mathbf{R} - \mathcal{B} + \mathcal{D},$$

respectively. Since the transform operators $\mathbf{C}$, $\mathbf{S}$, $\mathbf{H}_{\pm}$ commute with $\mathbf{R}$, $\mathcal{B}$, $\mathcal{D}$, they preserve all four of these product symmetries. On the other hand, by using (44) we verify that $\mathcal{F}$ commutes with both $\mathbf{R} + \mathcal{B} + \mathcal{D}$ and $\mathbf{R} - \mathcal{B} - \mathcal{D}$ but not with either $-\mathbf{R} + \mathcal{B} - \mathcal{D}$ or $-\mathbf{R} - \mathcal{B} + \mathcal{D}$. Thus $\mathcal{F}$ preserves the symmetries of being real-even and pure imaginary-even but preserves neither the symmetry of being real-odd nor the symmetry of being pure imaginary-odd.

In summary, we have shown that the Fourier transform operator $\mathcal{F}$ preserves only four of the ten common symmetries of Fourier analysis while the real operators $\mathbf{C}$, $\mathbf{S}$, $\mathbf{H}_{\pm}$ preserve all ten of them.

5.4 Fourier Transforms of Operators

The basic definition

In the previous two chapters we formulated a number of rules for taking Fourier transforms. Each of these rules has the form

$$g := \mathcal{A}f \quad \text{has the FT} \quad g^{\wedge} = \mathcal{A}^{\wedge}f^{\wedge} \tag{52}$$

where $\mathcal{A}$, $\mathcal{A}^{\wedge}$ are certain operators. For example, the reflection rule

$$g := \mathbf{R}f \quad \text{has the FT} \quad g^{\wedge} = \mathbf{R}f^{\wedge}$$

has this form with $\mathcal{A} = \mathcal{A}^{\wedge} = \mathbf{R}$, and the conjugation rule

$$g := \mathcal{B}f \quad \text{has the FT} \quad g^{\wedge} = \mathcal{D}f^{\wedge}$$

has this form with $\mathcal{A} = \mathcal{B}$ and $\mathcal{A}^\wedge = \mathcal{D}$. The rule (52) is valid provided that

$$\mathcal{F}(\mathcal{A}f) = \mathcal{A}^\wedge(\mathcal{F}f)$$

whenever f is a function in the domain of $\mathcal{A}$, and in this way we see that (52) is equivalent to the operator identity

$$\mathcal{F}\mathcal{A} = \mathcal{A}^\wedge\mathcal{F}.$$

When $\mathcal{A}$, $\mathcal{A}^\wedge$ are related in this manner, *i.e.*, when

$$\mathcal{A}^\wedge := \mathcal{F}\mathcal{A}\mathcal{F}^{-1} \tag{53}$$

we will say that the operator $\mathcal{A}^\wedge$ is the *Fourier transform* of the operator $\mathcal{A}$. The notation has been chosen so that we can write the transformation rule (52) in the easily remembered form

$$(\mathcal{A}f)^\wedge = \mathcal{A}^\wedge f^\wedge. \tag{54}$$

We will use the *commuting diagram* of Fig. 5.1 to visualize the two equivalent ways we can form the function (54).

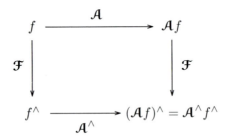

Figure 5.1. The commuting diagram for the rule (52).

Example: Show that

$$\mathbf{R}^\wedge = \mathbf{R}, \qquad \mathcal{B}^\wedge = \mathcal{D}, \qquad \mathcal{D}^\wedge = \mathcal{B}. \tag{55}$$

Solution: We use (8) and (44) with the definition (53) to write

$$\mathbf{R}^\wedge = \mathcal{F}\mathbf{R}\mathcal{F}^{-1} = \mathbf{R}\,\mathcal{F}\mathcal{F}^{-1} = \mathbf{R},$$
$$\mathcal{B}^\wedge = \mathcal{F}\mathcal{B}\mathcal{F}^{-1} = \mathcal{D}\,\mathcal{F}\mathcal{F}^{-1} = \mathcal{D},$$
$$\mathcal{D}^\wedge = \mathcal{F}\mathcal{D}\mathcal{F}^{-1} = \mathcal{B}\,\mathcal{F}\mathcal{F}^{-1} = \mathcal{B}.$$

These operator transforms correspond to the reflection rule, the conjugation rule, and the hermitian conjugation rule. ∎

Example: Let a be a nonzero real parameter. Find the Fourier transform of the dilation operator

$$(\mathbf{S}_a f)(x) := f(ax) \tag{56}$$

that is applied to suitably regular functions on $\mathbb{R}$.

Solution: We know that

$$f(ax) \quad \text{has the FT} \quad \frac{1}{|a|} f^\wedge \left(\frac{s}{a}\right),$$

i.e.,

$$\mathbf{S}_a f \quad \text{has the FT} \quad \frac{1}{|a|} \mathbf{S}_{1/a} f^\wedge,$$

so

$$\mathbf{S}_a^\wedge = |a|^{-1} \mathbf{S}_{1/a}. \tag{57}$$

∎

Example: The multiplication and convolution operators

$$\mathbf{M}_g f := g \cdot f, \qquad \mathbf{C}_g f := g * f \tag{58}$$

are defined when both g and f are suitably regular functions on $\mathbb{R}$, $\mathbb{T}_p$, $\mathbb{Z}$, or $\mathbb{P}_N$. Show that

$$\mathbf{C}_g = \beta \, \mathbf{\mathcal{F}}^{-1} \mathbf{M}_{g^\wedge} \mathbf{\mathcal{F}}. \tag{59}$$

This factorization corresponds to the indirect scheme for finding convolution products as given in (2.20) and (2.21).

Solution: The convolution rule

$$(g * f)^\wedge = \beta \, g^\wedge \cdot f^\wedge$$

from (2.16)–(2.19) corresponds to the operator identity

$$\mathbf{\mathcal{F}} \mathbf{C}_g = \beta \mathbf{M}_{g^\wedge} \mathbf{\mathcal{F}},$$

which we can rearrange to produce (59). ∎

Example: Let the repeat and zero packing operators

$$\mathbf{\mathcal{R}}_2 := \begin{bmatrix} 1 & 0 & 0 \\ 0 & 1 & 0 \\ 0 & 0 & 1 \\ 1 & 0 & 0 \\ 0 & 1 & 0 \\ 0 & 0 & 1 \end{bmatrix}, \qquad \mathbf{\mathcal{Z}}_2 := \begin{bmatrix} 1 & 0 & 0 \\ 0 & 0 & 0 \\ 0 & 1 & 0 \\ 0 & 0 & 0 \\ 0 & 0 & 1 \\ 0 & 0 & 0 \end{bmatrix}$$

be applied to 3-component column vectors (*i.e.*, functions on $\mathbb{P}_3$), and let the decimation and summation operators

$$\Xi_2 := \begin{bmatrix} 1 & 0 & 0 & 0 & 0 & 0 \\ 0 & 0 & 1 & 0 & 0 & 0 \\ 0 & 0 & 0 & 0 & 1 & 0 \end{bmatrix}, \qquad \Sigma_3 := \begin{bmatrix} 1 & 0 & 0 & 1 & 0 & 0 \\ 0 & 1 & 0 & 0 & 1 & 0 \\ 0 & 0 & 1 & 0 & 0 & 1 \end{bmatrix}$$

be applied to 6-component column vectors (*i.e.*, functions on $\mathbb{P}_6$). Show that $\mathcal{R}_2$, $\mathcal{Z}_2$ have the Fourier transforms

$$\mathcal{R}_2^{\wedge} = \mathcal{Z}_2, \qquad \mathcal{Z}_2^{\wedge} = \tfrac{1}{2}\mathcal{R}_2$$

that correspond to the rules (4.56), (4.55) when $N = 6$, and that Ξ_2, Σ_3 have the Fourier transforms

$$\Xi_2^{\wedge} = \Sigma_3, \qquad \Sigma_3^{\wedge} = 2\,\Xi_2,$$

which correspond to the rules (4.58), (4.57) when $N = 3$.

Solution: Let $\omega := e^{-2\pi i/6}$. We use the fact that $\omega^3 = -1$, $\omega^6 = 1$ to verify that

$$\mathcal{F}\mathcal{R}_2 = \frac{1}{6}\begin{bmatrix} 1 & 1 & 1 & 1 & 1 & 1 \\ 1 & \omega & \omega^2 & \omega^3 & \omega^4 & \omega^5 \\ 1 & \omega^2 & \omega^4 & \omega^6 & \omega^8 & \omega^{10} \\ 1 & \omega^3 & \omega^6 & \omega^9 & \omega^{12} & \omega^{15} \\ 1 & \omega^4 & \omega^8 & \omega^{12} & \omega^{16} & \omega^{20} \\ 1 & \omega^5 & \omega^{10} & \omega^{15} & \omega^{20} & \omega^{25} \end{bmatrix}\begin{bmatrix} 1 & 0 & 0 \\ 0 & 1 & 0 \\ 0 & 0 & 1 \\ 1 & 0 & 0 \\ 0 & 1 & 0 \\ 0 & 0 & 1 \end{bmatrix}$$

$$= \frac{1}{3}\begin{bmatrix} 1 & 1 & 1 \\ 0 & 0 & 0 \\ 1 & \omega^2 & \omega^4 \\ 0 & 0 & 0 \\ 1 & \omega^4 & \omega^8 \\ 0 & 0 & 0 \end{bmatrix} = \frac{1}{3}\begin{bmatrix} 1 & 0 & 0 \\ 0 & 0 & 0 \\ 0 & 1 & 0 \\ 0 & 0 & 0 \\ 0 & 0 & 1 \\ 0 & 0 & 0 \end{bmatrix}\begin{bmatrix} 1 & 1 & 1 \\ 1 & \omega^2 & \omega^4 \\ 1 & \omega^4 & \omega^8 \end{bmatrix} = \mathcal{Z}_2\mathcal{F},$$

and thereby show that

$$\mathcal{R}_2^{\wedge} := \mathcal{F}\mathcal{R}_2\mathcal{F}^{-1} = \mathcal{Z}_2.$$

We use the transpose of this matrix identity and the symmetry of $\mathcal{F}$ to write

$$\Xi_2^{\wedge} := \mathcal{F}\Xi_2\mathcal{F}^{-1} = \mathcal{F}\mathcal{Z}_2^T\mathcal{F}^{-1} = (\mathcal{F}^{-1}\mathcal{Z}_2\mathcal{F})^T = \mathcal{R}_2^{\wedge T} = \Sigma_3.$$

Finally, we use the complex conjugates of these two identities to write

$$\mathcal{R}_2 = \overline{\mathcal{R}}_2 = \overline{\mathcal{F}}^{-1}\mathcal{Z}_2\overline{\mathcal{F}} = (6\mathcal{F})\mathcal{Z}_2(3\mathcal{F})^{-1} = 2\mathcal{Z}_2^{\wedge}$$

$$\Xi_2 = \overline{\Xi}_2 = \overline{\mathcal{F}}^{-1}\Sigma_3\overline{\mathcal{F}} = (3\mathcal{F})\Sigma_3(6\mathcal{F})^{-1} = \tfrac{1}{2}\Sigma_3^{\wedge}. \qquad \blacksquare$$

Algebraic properties

By using (53) (and the linearity of $\mathcal{F}$) we see that the process of taking the Fourier transform of an operator is *linear*, *i.e.*,

$$
\begin{aligned}
(\alpha_1 \mathcal{A}_1 + \alpha_2 \mathcal{A}_2)^\wedge &= \mathcal{F}(\alpha_1 \mathcal{A}_1 + \alpha_2 \mathcal{A}_2)\mathcal{F}^{-1} \\
&= \alpha_1 (\mathcal{F}\mathcal{A}_1 \mathcal{F}^{-1}) + \alpha_2 (\mathcal{F}\mathcal{A}_2 \mathcal{F}^{-1}) \\
&= \alpha_1 \mathcal{A}_1^\wedge + \alpha_2 \mathcal{A}_2^\wedge,
\end{aligned}
$$

and *multiplicative*, *i.e.*,

$$
\begin{aligned}
(\mathcal{A}_1 \mathcal{A}_2)^\wedge &= \mathcal{F}(\mathcal{A}_1 \mathcal{A}_2)\mathcal{F}^{-1} \\
&= (\mathcal{F}\mathcal{A}_1 \mathcal{F}^{-1})(\mathcal{F}\mathcal{A}_2 \mathcal{F}^{-1}) \\
&= \mathcal{A}_1^\wedge \mathcal{A}_2^\wedge.
\end{aligned}
$$

More generally, if $\alpha_1, \alpha_2, \ldots, \alpha_m$ are scalars and $\mathcal{A}_1, \mathcal{A}_2, \ldots, \mathcal{A}_m$ are operators, we have

$$
(\alpha_1 \mathcal{A}_1 + \alpha_2 \mathcal{A}_2 + \ldots + \alpha_m \mathcal{A}_m)^\wedge = \alpha_1 \mathcal{A}_1^\wedge + \alpha_2 \mathcal{A}_2^\wedge + \ldots + \alpha_m \mathcal{A}_m^\wedge, \tag{60}
$$

$$
(\mathcal{A}_1 \mathcal{A}_2 \; \cdots \; \mathcal{A}_m)^\wedge = \mathcal{A}_1^\wedge \mathcal{A}_2^\wedge \; \cdots \; \mathcal{A}_m^\wedge, \tag{61}
$$

provided that these sums and products are all well defined.

Example: Show that

$$
\mathbf{P}_r^\wedge = \mathbf{P}_h, \qquad \mathbf{P}_i^\wedge = \mathbf{P}_a, \qquad \mathbf{P}_h^\wedge = \mathbf{P}_r, \qquad \mathbf{P}_a^\wedge = \mathbf{P}_i. \tag{62}
$$

Solution: We use (45), (46), and (55) to write

$$
\mathbf{P}_r^\wedge := \tfrac{1}{2}(\mathbf{I} + \mathcal{B})^\wedge = \tfrac{1}{2}(\mathbf{I}^\wedge + \mathcal{B}^\wedge) = \tfrac{1}{2}(\mathbf{I} + \mathcal{D}) =: \mathbf{P}_h,
$$

$$
\mathbf{P}_h^\wedge := \tfrac{1}{2}(\mathbf{I} + \mathcal{D})^\wedge = \tfrac{1}{2}(\mathbf{I}^\wedge + \mathcal{D}^\wedge) = \tfrac{1}{2}(\mathbf{I} + \mathcal{B}) =: \mathbf{P}_r.
$$

A similar argument is used for the other two identities. ∎

We can use (60)–(61) to move from the *rule*-based calculus of Chapters 3, 4 to an *operator*-based calculus. We again construct a table of functions $f_1, f_2, \ldots$ with known Fourier transforms $f_1^\wedge, f_2^\wedge, \ldots$ (Appendix 1), but instead of listing various rules (Appendix 2) we construct an equivalent table of elementary operators $\mathcal{A}_1, \mathcal{A}_2, \ldots$ with known Fourier transforms $\mathcal{A}_1^\wedge, \mathcal{A}_2^\wedge, \ldots$ (Appendix 3). It is then possible to find the Fourier transform of any function g that can be generated by applying some operator in the algebra generated by $\mathcal{A}_1, \mathcal{A}_2, \ldots$ to some function in the linear space spanned by $f_1, f_2, \ldots$. For example, it is possible to find the Fourier transform of

$$
g := \mathcal{A}_1(2\mathcal{A}_2 - 3\mathcal{A}_3)(f_1 + 3f_2)
$$

by writing

$$g^\wedge = 2\mathcal{A}_1^\wedge \mathcal{A}_2^\wedge f_1^\wedge - 3\mathcal{A}_1^\wedge \mathcal{A}_3^\wedge f_1^\wedge + 6\mathcal{A}_1^\wedge \mathcal{A}_2^\wedge f_2^\wedge - 9\mathcal{A}_1^\wedge \mathcal{A}_3^\wedge f_2^\wedge.$$

The rules force us to focus on detailed point-by-point manipulations; the corresponding operator notation helps us develop a global view of the corresponding mappings.

Duality

There is often a close connection between the Fourier transform rule

$$\mathcal{A}f \quad \text{has the FT} \quad \mathcal{A}^\wedge f^\wedge$$

and the *dual* rule

$$\mathcal{A}^\wedge f \quad \text{has the FT} \quad \mathcal{A}^{\wedge\wedge} f^\wedge.$$

(Notice that the same operator $\mathcal{A}^\wedge$ appears on the right side of the rule and on the left side of the dual rule.)

The simplest situation occurs when the operator $\mathcal{A}$ commutes with $\mathcal{F}$ (as is the case when $\mathcal{A} = \mathbf{R}, \mathbf{P}_e, \mathbf{P}_o, \mathcal{F}, \ldots$.) We can then use (53) to see that $\mathcal{A}^\wedge = \mathcal{A}$ and thereby conclude that the rule and its dual are identical. In particular, the reflection rule and the inversion rule are each self dual.

There is a fundamental relation that links the operators $\mathcal{A}^{\wedge\wedge}$ and $\mathcal{A}$. We use (53) with (5) and (6) to write

$$\begin{aligned}
\mathcal{A}^{\wedge\wedge} &= (\mathcal{F}^2)\mathcal{A}(\mathcal{F}^2)^{-1} \\
&= (\beta_r^{-1}\mathbf{R})\mathcal{A}(\beta_d^{-1}\mathbf{R})^{-1} \\
&= (\beta_d/\beta_r)\mathbf{R}\mathcal{A}\mathbf{R}
\end{aligned} \qquad (63)$$

where β_d, β_r are the $1, p, p, N$ factors we associate with functions that lie in the domain, range of $\mathcal{A}$, respectively. This is one situation where our notation must account for differences in the domain and range of $\mathcal{A}$. Figure 5.2 shows an alternative derivation of this factorization that is based on the identities (10) and (6).

Example: For real values of the parameter a and for suitably regular functions on $\mathbb{R}$, we define the translation and modulation operators

$$(\mathcal{T}_a f)(x) := f(x + a), \qquad (\mathcal{E}_a f)(x) := e^{2\pi i a x} \cdot f(x). \qquad (64)$$

Use (63) to show that

$$(\mathcal{T}_a f)^\wedge = \mathcal{E}_a f^\wedge \quad \text{has the dual} \quad (\mathcal{E}_a f)^\wedge = \mathcal{T}_{-a} f^\wedge.$$

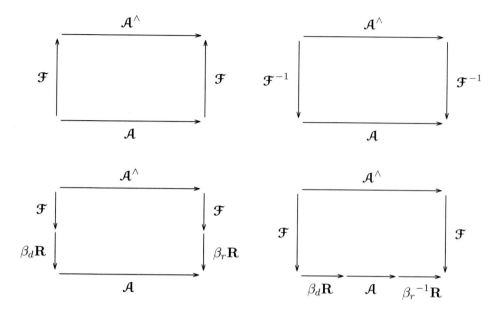

Figure 5.2. Pictorial derivation of the factorization (63) from the commuting diagram of Fig. 5.1.

Solution: From the translation rule (3.11) and from the modulation rule (3.12) we already know that

$$\mathcal{T}_a^{\wedge} = \mathcal{E}_a, \qquad \mathcal{E}_a^{\wedge} = \mathcal{T}_{-a} \tag{65}$$

so that

$$\mathcal{T}_a^{\wedge\wedge} = \mathcal{T}_{-a}.$$

We use (63) (with $\beta_d = \beta_r = 1$) to derive this result by writing

$$(\mathcal{T}_a^{\wedge\wedge} f)(x) = \mathbf{R}\mathcal{T}_a \mathbf{R}\{f(x)\} = \mathbf{R}\mathcal{T}_a\{f(-x)\} = \mathbf{R}\{f(-(x+a))\}$$
$$= f(x - a) = (\mathcal{T}_{-a} f)(x)$$

whenever f is a suitably regular function on $\mathbb{R}$. We must use *opposite* signs with the two parameters in the modulation rule (3.12) because $\mathbf{R}\mathcal{T}_a\mathbf{R} = \mathcal{T}_{-a}$!   ∎

Example: For $m = 1, 2, \ldots$ and for functions f on $\mathbb{P}_M$ we use the repeat and zero packing operators

$$(\mathcal{R}_m f)[n] := f[n], \qquad (\mathcal{Z}_m f)[n] := \begin{cases} f\left[\dfrac{n}{m}\right] & \text{if } m|n \\ 0 & \text{otherwise} \end{cases} \tag{66}$$

to construct functions on $\mathbb{P}_{M \cdot m}$. Use (63) to show that

$$(\mathcal{R}_m f)^{\wedge} = \mathcal{Z}_m f^{\wedge} \quad \text{has the dual} \quad (\mathcal{Z}_m f)^{\wedge} = m^{-1}\mathcal{R}_m f^{\wedge}. \tag{67}$$

Solution: From the repeat rule (4.54) and the zero packing rule (4.53) we already know that

$$\mathcal{R}_m^\wedge = \mathcal{Z}_m, \qquad\qquad \mathcal{Z}_m^\wedge = m^{-1}\mathcal{R}_m$$

so that

$$\mathcal{R}_m^{\wedge\wedge} = m^{-1}\mathcal{R}_m.$$

It is easy to derive this result from (63). We have $\beta_d = M$, $\beta_r = mM$ and

$$\mathbf{R}\,\mathcal{R}_m = \mathcal{R}_m\mathbf{R}$$

(with the left $\mathbf{R}$ reflecting mM-vectors and the right $\mathbf{R}$ reflecting M-vectors). Thus

$$\mathcal{R}_m^{\wedge\wedge} = \left(\frac{M}{mM}\right)\mathbf{R}\,\mathcal{R}_m\mathbf{R} = m^{-1}\mathcal{R}_m\mathbf{R}^2 = m^{-1}\mathcal{R}_m.$$

The nuisance factor m^{-1} is just the β_d/β_r ratio!   ∎

Example: For $p > 0$ and for suitably regular function f on $\mathbb{R}$ we use the summation and sampling operators

$$(\boldsymbol{\Sigma}_p f)(x) := \sum_{m=-\infty}^{\infty} f(x + mp), \qquad (\boldsymbol{\Xi}_p f)[n] := f(np)$$

to construct functions on $\mathbb{T}_p$ and $\mathbb{Z}$. Derive the dual (1.31) of Poisson's relation

$$(\boldsymbol{\Sigma}_p f)^\wedge = p^{-1}\boldsymbol{\Xi}_{1/p}f^\wedge$$

from (1.29).

Solution: We will use (63) with $\beta_d = 1$, $\beta_r = p$, and the commutation relation

$$\mathbf{R}\boldsymbol{\Sigma}_p = \boldsymbol{\Sigma}_p\mathbf{R}.$$

(The left $\mathbf{R}$ is applied to functions on $\mathbb{T}_p$ and the right $\mathbf{R}$ is applied to functions on $\mathbb{R}$.) We write

$$\boldsymbol{\Sigma}_p^{\wedge\wedge} = \left(\frac{1}{p}\right)\mathbf{R}\boldsymbol{\Sigma}_p\mathbf{R} = p^{-1}\boldsymbol{\Sigma}_p\mathbf{R}^2 = p^{-1}\boldsymbol{\Sigma}_p,$$

and in this way we obtain the desired dual rule

$$(\boldsymbol{\Xi}_{1/p}f)^\wedge = \boldsymbol{\Sigma}_p f^\wedge.$$ ∎

After you have assimilated the concept of an operator transform (and taken a fresh look at Fig. 3.4) you may find it helpful to spend a few minutes studying the $\mathcal{A}, \mathcal{A}^\wedge$ pairs that are tabulated in Appendix 3. There is no better way to develop an overview of the various rules from the Fourier transform calculus!

5.5 Rules for Hartley Transforms

The Hartley transform of a function on $\mathbb{R}$, $\mathbb{T}_p$, $\mathbb{Z}$, and $\mathbb{P}_N$ is defined by (27). We have already shown that you can convert a Fourier transform to the corresponding Hartley transform by using (34), *i.e.*,

$$
\begin{aligned}
f^{\sim} &= \beta^{1/2}\{\mathbf{P}_e + i\,\mathbf{P}_o\}f^{\wedge} \\
&= \beta^{1/2}\left\{\frac{1}{2}(f^{\wedge} + f^{\wedge\vee}) + \frac{i}{2}(f^{\wedge} - f^{\wedge\vee})\right\} \\
&= \beta^{1/2}\{\gamma f^{\wedge} + \gamma^{-}f^{\wedge\vee}\}
\end{aligned}
\tag{68}
$$

where

$$
\gamma := \frac{1+i}{2}, \qquad \gamma^{-} = \frac{1-i}{2}\ .
$$

In this section we will show how you can use operator notation to derive a set of rules for working with Hartley transforms.

We define the Hartley transform of an operator $\mathcal{A}$ by writing

$$
\mathcal{A}^{\sim} := \mathbf{H}_{+}\mathcal{A}\mathbf{H}_{+}^{-1} = \mathbf{H}_{+}\mathcal{A}\mathbf{H}_{+},
\tag{69}
$$

cf. (53). We can then use the corresponding Hartley transform rule

$$
g := \mathcal{A}f \quad \text{has the Hartley transform} \quad g^{\sim} = \mathcal{A}^{\sim}f^{\sim}
$$

i.e.,

$$
(\mathcal{A}f)^{\sim} = \mathcal{A}^{\sim}f^{\sim},
\tag{70}
$$

cf. (55). The dual rule takes the form

$$
g := \mathcal{A}^{\sim}f \quad \text{has the Hartley transform} \quad g^{\sim} = \mathcal{A}f^{\sim}
$$

since we always have

$$
\mathcal{A}^{\sim\sim} = \mathbf{H}_{+}^{2}\mathcal{A}\mathbf{H}_{+}^{2} = \mathcal{A}\ .
$$

Using the defining relation (69), it is easy to see that

$$
\mathcal{A}^{\sim} = \mathcal{A} \quad \text{when} \quad \mathbf{H}_{+}\mathcal{A} = \mathcal{A}\mathbf{H}_{+}.
\tag{71}
$$

We noted earlier that $\mathcal{A}$ commutes with $\mathbf{H}_{+}$ when

$$
\mathcal{A} = \mathcal{B}, \mathcal{D}, \mathbf{P}_e, \mathbf{P}_o, \mathbf{P}_r, \mathbf{P}_i, \mathbf{P}_h, \mathbf{P}_a, \mathbf{H}_{+}, \ldots
\tag{72}
$$

so the corresponding self-dual Hartley transform rules are particularly simple.

In all other cases, we will use the operator identities

$$\mathbf{H}_+ = \qquad \beta^{1/2}(\mathbf{P}_e + i\,\mathbf{P}_o)\boldsymbol{\mathcal{F}} = \beta^{1/2}(\gamma\mathbf{I} + \gamma^-\mathbf{R})\boldsymbol{\mathcal{F}}$$

$$\mathbf{H}_+ = \beta^{-1/2}\boldsymbol{\mathcal{F}}^{-1}(\mathbf{P}_e - i\,\mathbf{P}_o) = \beta^{-1/2}\boldsymbol{\mathcal{F}}^{-1}(\gamma^-\mathbf{I} + \gamma\mathbf{R})$$

for the left, right $\mathbf{H}_+$ in (69) to obtain the fundamental relation

$$\boldsymbol{\mathcal{A}}^\sim = (\beta_r/\beta_d)^{1/2}(\gamma\mathbf{I} + \gamma^-\mathbf{R})\boldsymbol{\mathcal{A}}^\wedge(\gamma^-\mathbf{I} + \gamma\mathbf{R}). \tag{73}$$

We must again use notation that allows us to distinguish between the β associated with the domain and the β associated with the range of the operator $\boldsymbol{\mathcal{A}}$. Since the list (72) includes all of the inhomogeneous operators that we commonly encounter, we will further simplify (73) by using the identities $\gamma\gamma = i/2$, $\gamma\gamma^- = \gamma^-\gamma = 1/2$, $\gamma^-\gamma^- = -i/2$ to see that

$$\boldsymbol{\mathcal{A}}^\sim = \tfrac{1}{2}(\beta_r/\beta_d)^{1/2}\{\boldsymbol{\mathcal{A}}^\wedge + \mathbf{R}\boldsymbol{\mathcal{A}}^\wedge\mathbf{R} + i\,\boldsymbol{\mathcal{A}}^\wedge\mathbf{R} - i\,\mathbf{R}\boldsymbol{\mathcal{A}}^\wedge\} \quad \text{when} \quad \boldsymbol{\mathcal{A}}\,i = i\,\boldsymbol{\mathcal{A}}. \tag{74}$$

The identity (74) shows how to produce $\boldsymbol{\mathcal{A}}^\sim$ from $\boldsymbol{\mathcal{A}}^\wedge$, *i.e.*, how to find a Hartley transform rule from the corresponding Fourier transform rule.

When the operator $\boldsymbol{\mathcal{A}}$ preserves even and odd symmetry the four-term sum from (74) collapses, and we find

$$\boldsymbol{\mathcal{A}}^\sim = (\beta_r/\beta_d)^{1/2}\boldsymbol{\mathcal{A}}^\wedge \quad \text{when} \quad \boldsymbol{\mathcal{A}}\mathbf{R} = \mathbf{R}\boldsymbol{\mathcal{A}} \quad \text{and} \quad \boldsymbol{\mathcal{A}}\,i = i\,\boldsymbol{\mathcal{A}}. \tag{75}$$

In this way we see that the various dilation, grouping, decimation, repeat, summation, and sampling rules for finding Hartley transforms are (apart from a scale factor) identical to the corresponding rules for finding Fourier transforms.

When $\boldsymbol{\mathcal{A}}$ exchanges even and odd symmetry the sum (74) again collapses, but this time we find

$$\boldsymbol{\mathcal{A}}^\sim = -i(\beta_r/\beta_d)^{1/2}\mathbf{R}\boldsymbol{\mathcal{A}}^\wedge \quad \text{when} \quad \boldsymbol{\mathcal{A}}\mathbf{R} = -\mathbf{R}\boldsymbol{\mathcal{A}} \quad \text{and} \quad \boldsymbol{\mathcal{A}}\,i = i\,\boldsymbol{\mathcal{A}}. \tag{76}$$

We can use this identity to obtain the derivative and power scaling rules for Hartley transforms.

Example: Let f be a suitably regular function on $\mathbb{R}$. Use (74) to derive the derivative and power scaling rules

$$f'(x) \qquad \text{has the Hartley transform} \quad -2\pi s \cdot f^\sim(-s), \tag{77}$$

$$x \cdot f(x) \quad \text{has the Hartley transform} \quad (2\pi)^{-1} \cdot f^{\sim\prime}(-s). \tag{78}$$

Solution: From the derivative rule (3.19) we know that the derivative and power scaling operators

$$(\mathbf{D}f)(x) := f'(x), \qquad (\boldsymbol{\mathcal{P}}f)(x) := 2\pi i x \cdot f(x) \tag{79}$$

are related in such a manner that

$$\mathbf{D}^\wedge = \mathcal{P}.$$

Now

$$\mathbf{DR}\{f(x)\} = \mathbf{D}\{f(-x)\} = -f'(-x) = -\mathbf{R}\{f'(x)\} = -\mathbf{RD}\{f(x)\},$$

so we can use (76) to write

$$\mathbf{D}^\sim = -i\,\mathbf{RD}^\wedge = -i\,\mathbf{R}\mathcal{P}.$$

In this way we see that $f'(x)$ has the Hartley transform

$$-i\,\mathbf{R}\mathcal{P}\{f^\sim(s)\} = -i\,\mathbf{R}\{2\pi i s \cdot f^\sim(s)\} = -2\pi s \cdot f^\sim(-s).$$

We use the reflection rule with the dual

$$-2\pi x \cdot f(-x) \quad \text{has the Hartley transform} \quad f^{\sim\prime}(s)$$

to obtain the power scaling rule. For iterates of $\mathbf{D}$, $\mathcal{P}$, cf. Ex. 5.37. ∎

The full four-term sum from (74) must be used to derive the translation, modulation, convolution, and multiplication rules for Hartley transforms, so these rules are a bit more complicated than the corresponding rules for Fourier transforms.

Example: Find the translation rule for Hartley transforms of functions on $\mathbb{R}$.

Solution: We use (74) with (64)–(65) to write

$$\begin{aligned}
\mathcal{T}_a^\sim &= \frac{1}{2}\{\mathcal{T}_a^\wedge + \mathbf{R}\mathcal{T}_a^\wedge\mathbf{R} + i\,\mathcal{T}_a^\wedge\mathbf{R} - i\,\mathbf{R}\mathcal{T}_a^\wedge\} \\
&= \frac{1}{2}\{\mathcal{E}_a + \mathbf{R}\mathcal{E}_a\mathbf{R} + i\,\mathcal{E}_a\mathbf{R} - i\,\mathbf{R}\mathcal{E}_a\} \\
&= \frac{1}{2}\{\mathcal{E}_a + \mathcal{E}_{-a}\} - \frac{1}{2i}\{\mathcal{E}_a - \mathcal{E}_{-a}\}\mathbf{R}
\end{aligned}$$

and thereby show that

$$g(x) = f(x+a) \quad \text{has the Hartley transform}$$
$$g^\sim(s) = \cos(2\pi as)f^\sim(s) - \sin(2\pi as)f^\sim(-s). \tag{80}$$

∎

Example: Derive the convolution rule

$$(f*g)^\sim[k] = \frac{N^{1/2}}{2}\{f^\sim[k]g^\sim[k] + f^\sim[k]g^\sim[-k] + f^\sim[-k]g^\sim[k] - f^\sim[-k]g^\sim[-k]\} \tag{81}$$

for Hartley transforms of functions on $\mathbb{P}_N$.

Solution: We ue (74) with (58)–(59), the condensed notation

$$\langle g \rangle := \mathbf{M}_g,$$

the identity,

$$\mathbf{R}\langle g \rangle = \langle g^\vee \rangle \mathbf{R},$$

and (32)–(33) to write

$$\mathcal{C}_{\tilde{g}} = \frac{1}{2}\{\mathcal{C}_g^\wedge + \mathbf{R}\mathcal{C}_g^\wedge\mathbf{R} + i\,\mathcal{C}_g^\wedge\mathbf{R} - i\,\mathbf{R}\mathcal{C}_g^\wedge\}$$

$$= \frac{N}{2}\{\langle g^\wedge \rangle + \mathbf{R}\langle g^\wedge \rangle\mathbf{R} + i\langle g^\wedge \rangle\mathbf{R} - i\,\mathbf{R}\langle g^\wedge \rangle\}$$

$$= \frac{N}{2}\{\langle g^\wedge \rangle + \langle g^{\wedge\vee} \rangle + i\langle g^\wedge \rangle\mathbf{R} - i\langle g^{\wedge\vee} \rangle\mathbf{R}\}$$

$$= N\{\langle \mathbf{P}_e g^\wedge \rangle + \langle i\,\mathbf{P}_o g^\wedge \rangle\mathbf{R}\}$$

$$= N^{1/2}\{\langle \mathbf{P}_e g^\sim \rangle + \langle \mathbf{P}_o g^\sim \rangle\mathbf{R}\}.$$

This identity is equivalent to (81).   ∎

You will find additional Hartley transform rules in Exs. 5.37–5.39. Similar arguments can be used to develop rules for working with the transforms associated with **S** and **C**, *cf.* Ex. 5.40.

5.6 Hilbert Transforms

Defining relations

We define the odd *signum function* on $\mathbb{R}$, $\mathbb{T}_p$, $\mathbb{Z}$, $\mathbb{P}_N$ by writing

$$\text{sgn}(x) := \begin{cases} 1 & \text{if } x > 0 \\ 0 & \text{if } x = 0 \\ -1 & \text{if } x < 0, \end{cases}$$

(82a)

$$\text{sgn}(x) := \begin{cases} 1 & \text{if } 0 < x < p/2 \\ 0 & \text{if } x = 0,\ p/2 \\ -1 & \text{if } p/2 < x < p, \end{cases}$$

$$\text{sgn}[n] := \begin{cases} 1 & \text{if } n = 1, 2, \dots \\ 0 & \text{if } n = 0 \\ -1 & \text{if } n = -1, -2, \dots, \end{cases}$$

$$\text{sgn}[n] := \begin{cases} 1 & \text{if } n = 1, 2, \dots, \lfloor (N-1)/2 \rfloor \\ 0 & \text{if } n = 0 \text{ or } N/2 \text{ (when } N \text{ is even)} \\ -1 & \text{if } n = N-1, N-2, \dots, N - \lfloor (N-1)/2 \rfloor, \end{cases} \tag{82b}$$

respectively. We then form the Hilbert transform operator

$$\mathcal{H} := -i\,\mathcal{F}^{-1}\mathbf{M}_{\text{sgn}}\mathcal{F}, \tag{83}$$

which acts by applying the factor $-i \cdot \text{sgn}$ in the Fourier transform domain. We use the tag, $^{\#}$, as a superscript to show that $\mathcal{H}$ has been applied, *i.e.*,

$$f^{\#} := \mathcal{H}f.$$

Thus when f is a suitably regular function on $\mathbb{R}$, $\mathbb{T}_p$, $\mathbb{Z}$, $\mathbb{P}_N$ we have

$$f^{\#}(x) = -i \int_{s=-\infty}^{\infty} \text{sgn}(s) f^{\wedge}(s) e^{2\pi i s x} ds,$$

$$f^{\#}(x) = -i \sum_{k=-\infty}^{\infty} \text{sgn}[k] f^{\wedge}[k] e^{2\pi i k x/p},$$

$$f^{\#}[n] = -i \int_{s=0}^{p} \text{sgn}(s) f^{\wedge}(s) e^{2\pi i s n/p} ds,$$

$$f^{\#}[n] = -i \sum_{k=0}^{N-1} \text{sgn}[k] f^{\wedge}[k] e^{2\pi i k n/N},$$

respectively.

Operator identities

We will now develop operator identities that give the basic rules for working with Hilbert transforms.

As a first step we use (83) to write

$$\mathcal{H}^2 = -\mathcal{F}^{-1}\mathbf{M}_{\text{sgn}^2}\mathcal{F},$$

and thereby see that

$$\mathcal{H}^2 f = -f \quad \text{when} \quad f^{\wedge} = \text{sgn}^2 \cdot f^{\wedge}. \tag{84}$$

In particular, $-\mathcal{H}$ serves as an inverse for $\mathcal{H}$ when we work with suitably regular functions on $\mathbb{R}$, with absolutely summable functions on $\mathbb{Z}$, with piecewise smooth functions on $\mathbb{T}_p$ that have

$$\int_0^p f(x)dx = 0,$$

or with functions on $\mathbb{P}_N$ that have

$$\sum_{n=0}^{N-1} f[n] = 0 \quad \text{and} \quad \sum_{n=0}^{N-1}(-1)^n f[n] = 0 \quad \text{(when N is even)}.$$

The symmetry-preserving properties of $\mathcal{H}$ follow from the identities

$$\mathcal{H}\mathbf{R} = -\mathbf{R}\,\mathcal{H}, \tag{85}$$

$$\mathcal{H}\mathcal{B} = \mathcal{B}\,\mathcal{H}, \tag{86}$$

$$\mathcal{H}\mathbf{D} = -\mathbf{D}\,\mathcal{H}, \tag{87}$$

e.g., (85) shows that even functions have odd Hilbert transforms and vice versa, (86) shows that real functions have real Hilbert transforms, and (87) shows that hermitian functions have antihermitian Hilbert transforms and vice versa.

Example: Show that $\mathcal{H}$ commutes with $\mathcal{B}$.

Solution: We use (83), (44), and the odd parity of sgn to write

$$\begin{aligned}
\mathcal{H}\mathcal{B} &= -i\,\mathcal{F}^{-1}\mathbf{M}_{\mathrm{sgn}}\mathcal{F}\mathcal{B} \\
&= -i\,\mathcal{F}^{-1}\mathbf{M}_{\mathrm{sgn}}\mathbf{D}\,\mathcal{F} \\
&= i\,\mathcal{F}^{-1}\mathbf{D}\,\mathbf{M}_{\mathrm{sgn}}\mathcal{F} \\
&= i\,\mathcal{B}\mathcal{F}^{-1}\mathbf{M}_{\mathrm{sgn}}\mathcal{F} \\
&= \mathcal{B}\,\mathcal{H}.
\end{aligned}$$

∎

We use the factorization (59) with the factorization (83) to verify that

$$\mathcal{H}\,\mathcal{C}_g = \mathcal{C}_g\mathcal{H}. \tag{88}$$

This operator identity then leads to the convolution rule

$$(f * g)^\# = f^\# * g = f * g^\# \tag{89}$$

for Hilbert transforms.

When we work with suitably regular functions on $\mathbb{R}$, the translation, derivative, and dilation rules correspond to factorizations

$$\mathcal{T}_a = \mathcal{F}^{-1}\mathcal{E}_a\mathcal{F},$$

$$\mathbf{D} = \mathcal{F}^{-1}\mathbf{M}_q\mathcal{F}, \quad q(s) := 2\pi i s,$$

$$\mathcal{S}_a = \mathcal{F}^{-1}|a|^{-1}\mathcal{S}_{1/a}\mathcal{F}, \quad a < 0 \text{ or } a > 0$$

that are analogous to those of (59) and (83). We use these in conjunction with (83) to obtain commutation relations

$$\mathcal{H}\mathcal{T}_a = \mathcal{T}_a\mathcal{H}, \quad -\infty < a < \infty \tag{90}$$

$$\mathcal{H}\mathbf{D} = \mathbf{D}\mathcal{H} \tag{91}$$

$$\mathcal{H}\mathcal{S}_a = \begin{cases} \mathcal{S}_a\mathcal{H}, & a > 0 \\ -\mathcal{S}_a\mathcal{H}, & a < 0, \end{cases} \tag{92}$$

which give translation, derivative, and dilation rules for Hilbert transforms of functions on $\mathbb{R}$.

The Kramers-Kronig relations

Let F be a suitably regular function on $\mathbb{R}$ that vanishes on the half line $s \le 0$, and let

$$f(x) := \int_0^\infty F(s)e^{2\pi isx}ds.$$

Since $\text{sgn} \cdot F = F$, we can write

$$\mathcal{H}f = -i\,\mathcal{F}^{-1}\mathbf{M}_{\text{sgn}}\mathcal{F}f = -i\,\mathcal{F}^{-1}(\text{sgn} \cdot F) = -i\,\mathcal{F}^{-1}F = -i\,f$$

or equivalently,

$$\mathcal{H}\{f_R + i\,f_I\} = -i\{f_R + i\,f_I\}$$

when f_R, f_I are the real and imaginary parts of f. Since $\mathcal{H}$ commutes with $\mathcal{B}$, $\mathcal{H}f_R, \mathcal{H}f_I$ are real, so we can equate the real and imaginary parts of this identity to obtain the *Kramers-Kronig* relations

$$\mathcal{H}f_R = f_I, \quad \mathcal{H}f_I = -f_R \quad \text{when} \quad F(s) = 0 \text{ for } s \le 0. \tag{93}$$

With analogous arguments, we also find

$$\mathcal{H}F_R = -F_I, \quad \mathcal{H}F_I = F_R \quad \text{when} \quad f(x) = 0 \text{ for } x \le 0. \tag{94}$$

You can use (93) or (94) and your knowledge of Fourier analysis to generate a number of Hilbert transform pairs. You can then obtain additional Hilbert transform pairs by using the rules that correspond to (84)–(88) and (90)–(92).

Example: Derive the following Hilbert transforms:

$$g(x): \quad \frac{1}{1+x^2} \qquad \frac{x}{1+x^2} \qquad \frac{-2x}{(1+x^2)^2} \qquad \frac{1}{1+(x-1)^2}$$

$$g^{\#}(x): \quad \frac{x}{1+x^2} \qquad \frac{-1}{1+x^2} \qquad \frac{1-x^2}{(1+x^2)^2} \qquad \frac{x-1}{1+(x-1)^2}$$

Solution: We know from (3.22) that

$$f(x) := \frac{1}{1 - 2\pi i x} = \frac{1}{1 + (2\pi x)^2} + \frac{2\pi i x}{1 + (2\pi x)^2} \quad \text{has the FT} \quad F(s) = h(s)e^{-s},$$

so we can use (93) to see that

$$f_R(x) := \frac{1}{1 + (2\pi x)^2} \quad \text{has the Hilbert transform} \quad f_I(x) := \frac{2\pi x}{1 + (2\pi x)^2}.$$

We now use (92), (84), (91), (90) to fill in the above table with

$$g_1 := \mathbf{S}_{1/2\pi} f_R \quad g_2 := \mathcal{H} g_1 \quad g_3 := \mathbf{D} g_1 \quad g_4 := \mathcal{T}_{-1} g_1$$
$$g_1^{\#} = \mathbf{S}_{1/2\pi} f_I \quad g_2^{\#} = \mathcal{H} g_1^{\#} \quad g_3^{\#} = \mathbf{D} g_1^{\#} \quad g_4^{\#} = \mathcal{T}_{-1} g_1^{\#}. \qquad \blacksquare$$

The analytic function

Let f be a suitably regular function on $\mathbb{T}_p$ with the Fourier representation

$$f(x) = \sum_{k=-\infty}^{\infty} a_k e^{2\pi i k x / p}. \tag{95}$$

We combine f with the Hilbert transform (or *conjugate function*)

$$f^{\#}(x) = \sum_{k=-\infty}^{\infty} -i \operatorname{sgn}[k] a_k e^{2\pi i k x / p} \tag{96}$$

to form

$$f(x) + i f^{\#}(x) = \sum_{k=-\infty}^{\infty} a_k \{1 + \operatorname{sgn}[k]\} e^{2\pi i k x / p}$$

$$= a_0 + 2 \sum_{k=1}^{\infty} a_k e^{2\pi i k x / p}$$

$$= A(e^{2\pi i x / p})$$

where
$$A(z) := a_0 + 2a_1 z + 2a_2 z^2 + 2a_3 z^3 + \cdots \tag{97}$$

is said to be the *analytic function* for the f of (95).

Since $a_{-1}, a_{-2}, \ldots$ do not appear in (97), we cannot recover f from $\mathcal{A}$. On the other hand, we *can* generate a Hilbert transform pair

$$g(x) := \frac{1}{2}\{\mathcal{A}(e^{2\pi i x/p}) + \mathcal{A}(e^{-2\pi i x/p})\},$$

$$g^{\#}(x) := \frac{1}{2i}\{\mathcal{A}(e^{2\pi i x/p}) - \mathcal{A}(e^{-2\pi i x/p})\}$$

from any power series $\mathcal{A}$ that converges on the unit circle.

Example: Find the Hilbert transforms and the analytic functions that correspond to

$$f_k(x) := \cos(2\pi k x), \quad g_k(x) := \sin(2\pi k x), \quad k = 1, 2, \ldots .$$

Solution: The function f_k on $\mathbb{T}_1$ has the Fourier representation

$$f_k(x) = \tfrac{1}{2}e^{-2\pi i k x} + \tfrac{1}{2}e^{2\pi i k x},$$

so we use (96) and (84) to write

$$f_k^{\#}(x) = (-i)\left\{-\tfrac{1}{2}e^{-2\pi i k x} + \tfrac{1}{2}e^{2\pi i k x}\right\} = \sin(2\pi k x),$$
$$g_k^{\#}(x) = f_k^{\#\#}(x) = -\cos(2\pi k x).$$

Knowing $f, f^{\#}, g, g^{\#}$ we use Euler's relations to see that

$$f_k(x) + i\, f_k^{\#}(x) = e^{2\pi i k x}, \qquad g_k(x) + i\, g_k^{\#}(x) = -i\, e^{2\pi i k x}$$

and thereby obtain the corresponding analytic functions

$$\mathcal{A}_f(z) = z^k, \qquad \mathcal{A}_g(z) = -i\, z^k. \qquad \blacksquare$$

References

R.N. Bracewell, *The Fourier Transform and Its Applications, 2nd ed.*, McGraw-Hill, New York, 1986.

Operator notation is used from time to time in this well-known text.

R.N. Bracewell, *The Hartley Transform*, Oxford University Press, New York, 1986.

Chapter 3 develops the rules for Hartley transforms.

F.B. Hildebrand, *Introduction to Numerical Analysis*, McGraw-Hill, New York, 1956.

> Chapter 5 of this classic applied mathematics text contains an informal elementary introduction to the use of operators in numerical analysis.

W. Kaplan, *Operational Methods for Linear Systems*, Addison-Wesley, Reading, MA, 1962.

> Chapter 2 of this text introduces scientists and engineers to the operators of systems analysis.

L. Mirsky, *An Introduction to Linear Algebra*, Clarendon Press, Oxford, 1955; reprinted by Dover Publications, New York, 1982.

> Most linear algebra books include some discussion of linear operators and the algebra of matrices. Mirsky gives a detailed elementary exposition of these topics in Chapters 3, 4.

Exercise Set 5

▶ **EXERCISE 5.1:**   Let $\mathcal{A}_1 := \mathcal{Q}_1(\mathcal{F})$, $\mathcal{A}_2 := \mathcal{Q}_2(\mathcal{F})$ where $\mathcal{Q}_1, \mathcal{Q}_2$ are polynomials. Carefully explain why $\mathcal{A}_1\mathcal{A}_2 = \mathcal{A}_2\mathcal{A}_1$.

▶ **EXERCISE 5.2:**   Let $\alpha > 0$ and let $f(x) := e^{-\alpha x} h(x)$.

(a) Sketch $\mathbf{P}_e f$ and $\mathbf{P}_o f$.

(b) Explain why $\mathcal{F}$ commutes with $\mathbf{P}_e$ and $\mathbf{P}_o$.

(c) Use the identities $\mathcal{F}(\mathbf{P}_e f) = \mathbf{P}_e(\mathcal{F}f)$ and $\mathcal{F}(\mathbf{P}_o f) = \mathbf{P}_o(\mathcal{F}f)$ to find the Fourier transforms of $\mathbf{P}_e f$, $\mathbf{P}_o f$.

▶ **EXERCISE 5.3:**   Show that the reflection operator $\mathbf{R}$ and the normalized cosine transform and sine transform operators $\mathbf{C}, \mathbf{S}$ satisfy the commutation relations (22).

▶ **EXERCISE 5.4:**   Let $\mathcal{F}$ be the Fourier transform operator that is applied to suitably regular functions on $\mathbb{R}$.

(a) Use the Maclaurin series for $\cos x$, $\sin x$ to show that

$$\cos(\mathcal{F}) = a_0\mathbf{I} - a_2\mathbf{R}, \qquad \sin(\mathcal{F}) = (a_1\mathbf{I} - a_3\mathbf{R})\mathcal{F}$$

where a_0, a_1, a_2, a_3 are certain constants.

(b) Express the cosine transform operator $\mathbf{C}$ and the sine transform operator $\mathbf{S}$ in terms of $\mathbf{I}$, $\mathbf{R}$, $\mathcal{F}$, $\mathbf{R}\mathcal{F}$.

▶ **EXERCISE 5.5:** Let α, γ be scalars. Find a simple expression for the inverse of each of the following operators.

(a) $\alpha \mathbf{I} + \gamma \mathbf{R}, \qquad \alpha^2 - \gamma^2 \neq 0$

(b) $\alpha \mathbf{P}_e + \gamma \mathbf{P}_o, \qquad \alpha \gamma \neq 0$

(c) $\alpha \mathbf{E}_+ + \gamma \mathbf{E}_-, \qquad \alpha^2 - \gamma^2 \neq 0$

(d) $\alpha \mathbf{C} + \gamma \mathbf{S}, \qquad \alpha \gamma \neq 0$

(e) $\alpha \mathbf{H}_+ + \gamma \mathbf{H}_-, \qquad \alpha^2 - \gamma^2 \neq 0$

▶ **EXERCISE 5.6:** A suitably regular p-periodic function f can be expressed in the alternative forms

$$f(x) = \frac{a_0}{2} + \sum_{k=1}^{\infty} \left\{ a_k \cos\left(\frac{2\pi k x}{p}\right) + b_k \sin\left(\frac{2\pi k x}{p}\right) \right\}$$

$$= \sum_{k=-\infty}^{\infty} c_k \operatorname{cis}\left(\frac{2\pi k x}{p}\right) = \sum_{k=-\infty}^{\infty} d_k \operatorname{cas}\left(\frac{2\pi k x}{p}\right)$$

where $\operatorname{cis}\theta := \cos\theta + i\sin\theta$, $\operatorname{cas}\theta := \cos\theta + \sin\theta$. We use the coefficients to define functions a, b, c, d on $\mathbb{Z}$:

$$a[k] := \begin{cases} a_k & \text{if } k \geq 0 \\ a_{-k} & \text{if } k < 0, \end{cases} \quad b[k] := \begin{cases} b_k & \text{if } k > 0 \\ 0 & \text{if } k = 0 \\ -b_{-k} & \text{if } k < 0, \end{cases} \quad c[k] := c_k, \quad d[k] := d_k.$$

(a) Show how to obtain a, b, c, d from $\mathbf{C}f$ and $\mathbf{S}f$.

   Hint. Begin with $c = \mathcal{F}f = (1/\sqrt{p})\{\mathbf{C}f - i\,\mathbf{S}f\}$.

(b) Express a_k, b_k, c_k, d_k in terms of suitable definite integrals involving f.

(c) Express a, b in terms of c and in terms of d.

   Hint. Use (a) and the operators $\mathbf{P}_e, \mathbf{P}_o, \mathbf{R}$, e.g., write $a = 2\mathbf{P}_e c = c + \mathbf{R}c$.

(d) Express c in terms of d and in terms of a, b.

(e) Express d in terms of a, b and in terms of c.

(f) What can you infer about a, b, c, d if you know that f is real valued? pure imaginary valued?

(g) What can you infer about a, b, c, d if you know that f is even? odd?

(h) What can you infer about a, b, c, d if you know that f is hermitian? antihermitian?

Note. You may wish to compare the present operator-based analysis to the component-based analysis you used for Ex. 1.11.

▶ **EXERCISE 5.7:**   Let $f(x) := \sum\limits_{m=-\infty}^{\infty} g(x - mp)$ where $p > 0$ and where g is a

suitably regular function on $\mathbb{R}$ with the Hartley transform

$$g^\sim(s) := \int_{-\infty}^{\infty} g(x)\text{cas}(2\pi sx)dx.$$

(a) Find a Poisson formula that enables you to express the coefficients of the Hartley
series

$$f(x) = \sum_{k=-\infty}^{\infty} d_k \, \text{cas}\left(\frac{2\pi kx}{p}\right)$$

in terms of $g^\sim$.

Hint. Use a direct argument ... or manipulate the identity $d = \sqrt{p}(\mathbf{P}_e + i\,\mathbf{P}_o)\mathcal{F}f$.

(b) Using (a), find the Hartley series for

$$f(x) := \begin{cases} 1 & \text{if} \quad 0 < x < p/4 \\ 0 & \text{if} \quad p/4 < x < p. \end{cases}$$

▶ **EXERCISE 5.8:**   Let f be the function on $\mathbb{P}_N$ with components $f[n] := n$,
$n = 0, 1, \dots, N - 1$.

(a) Find the components $(\mathbf{C}f)[k]$, $(\mathbf{S}f)[k]$, $k = 0, 1, \dots, N - 1$, of the normalized cosine
transform, sine transform of f.

Hint. Use (32), (33) with the Fourier transform of f that is given in Ex. 4.26.

(b) Show that the normalized Hartley transform (27) of f has the components

$$f^\sim[k] = \frac{N^{1/2}}{2} \begin{cases} N - 1 & \text{if} \quad k = 0 \\ -1 - \cot\left(\dfrac{\pi k}{N}\right) & \text{if} \quad k = 1, 2, \dots, N - 1. \end{cases}$$

▶ **EXERCISE 5.9:**   Let $N = 2, 3, \dots$, let f be the N-periodic discrete function with

$$f[n] := \begin{cases} 1 & \text{if } n = 0, 1 \\ 0 & \text{if } n = 2, 3, \dots, N - 1, \end{cases}$$

and let $f_1 := f$, $f_2 := f * f$, $f_3 := f * f * f$, $\dots$.

(a) Show that $f_m^\wedge[k] = N^{-1}(1 + e^{-2\pi ik/N})^m$.

(b) Show that $f_m^\sim[k] = 2^m N^{-1/2} \cos^m(\pi k/N)\text{cas}(\pi mk/N)$.

Hint. Use (a) with (34).

▶ **EXERCISE 5.10:**   Let the complex-valued function f have the representations

$$f = f_r + f_i = f_R + if_I$$

where $f_r := f_R$ and f_I are real valued, and where $f_i := if_I$ is pure imaginary. (Thus f_i is the imaginary part of f *with* the i and f_I is the imaginary part of f *without* the i.) Let the operators $\mathbf{P}_r$, $\mathbf{P}_i$, $\mathbf{P}_I$ be defined by writing

$$\mathbf{P}_r f := f_r, \qquad \mathbf{P}_i f := f_i, \qquad \mathbf{P}_I f := f_I.$$

(a) Express $\mathbf{P}_r$, $\mathbf{P}_i$, and $\mathbf{P}_I$ in terms of the complex conjugation operator $\mathcal{B}$.

(b) Find simple expressions for the products $\mathbf{P}_r\mathbf{P}_r$, $\mathbf{P}_I\mathbf{P}_I$, $\mathbf{P}_r\mathbf{P}_I$, $\mathbf{P}_I\mathbf{P}_r$.

Note. If you compare your answers from (b) with (12), (47), (48) you will understand why we have chosen to work with $\mathbf{P}_r$, $\mathbf{P}_i$ in the book.

▶ **EXERCISE 5.11:**   A group of Fourier analysis students are learning to use a computer subroutine, *DFT*. After initializing the integer variable $N = 1, 2, \ldots$ and the N-component real arrays f_R, f_I, you can use the command

$$DFT(N, f_R, f_I, F_R, F_I)$$

to tell the computer to load the N-component real arrays F_R, F_I with the real and imaginary parts of the discrete Fourier transform

$$F[k] = F_R[k] + iF_I[k], \quad k = 0, 1, \ldots, N-1$$

of the complex vector

$$f[n] := f_R[n] + if_I[n], \quad n = 0, 1, \ldots, N-1.$$

(a) "I think we can use *DFT* to compute an *inverse* Fourier transform," says one student, who suggests the pseudocode

$$DFT(N, f_R, f_I, F_R, F_I)$$
$$f_R := F_R$$
$$f_I := F_I$$
$$DFT(N, f_R, f_I, F_R, F_I)$$
$$f_R := F_R$$
$$f_I := F_I$$
$$DFT(N, f_R, f_I, F_R, F_I)$$
$$F_R := N^2 \cdot F_R$$
$$F_I := N^2 \cdot F_I.$$

"That's terribly inefficient," says a second student. "The pseudocode

$$DFT(N, f_R, f_I, F_R, F_I)$$

$$\text{For } k = 1, 2, \ldots, \lfloor N/2 \rfloor$$

$$\quad\quad \text{Swap } F_R[k] \text{ and } F_R[N-k]$$

$$\quad\quad \text{Swap } F_I[k] \text{ and } F_I[N-k]$$

$$F_R := N \cdot F_R$$

$$F_I := N \cdot F_I$$

will do the job with only *one DFT* computation!"

"I don't like componentwise operations," says a third student, who recommends the pseudocode

$$f_I := -f_I$$

$$DFT(N, f_R, f_I, F_R, F_I)$$

$$F_I := -N \cdot F_I$$

$$F_R := N \cdot F_R.$$

Write down the operator identities that underlie these three approaches to computing an inverse Fourier transform. Which one do you think is the best? Why?

(b) "I need to compute the Fourier transforms of two *real N*-vectors, f, g," said one student. "I guess I'll use the pseudocode

$$DFT(N, f, 0, F_R, F_I)$$

$$DFT(N, g, 0, G_R, G_I)$$

to do the job."

"I think you could get by with just one call to DFT," says a second student. "You can write

$$DFT(N, f, g, T_R, T_I)$$

and then suitably process T_R, T_I to find F_R, F_I, G_R, G_I."

"That will never work," says a third student, "because the Fourier transform scrambles the real and imaginary parts; you cannot get both F and G without using DFT twice."

Give operator identities which show that the second student is right! Supply corresponding pseudocode [analogous to that in (a)] for computing F_R, F_I, G_R, G_I from T_R, T_I.

▶ **EXERCISE 5.12:**   Verify the following identities.

(a) $f^{\wedge} = f^{\vee\wedge\vee} = f^{-\wedge\dagger} = f^{\dagger\wedge-}$ (b) $f^{\wedge\vee} = f^{\vee\wedge} = f^{-\wedge-} = f^{\dagger\wedge\dagger}$

(c) $f^{\wedge-} = f^{\vee\wedge\dagger} = f^{-\wedge\vee} = f^{\dagger\wedge}$ (d) $f^{\wedge\dagger} = f^{\vee\wedge-} = f^{-\wedge} = f^{\dagger\wedge\vee}$

▶ **EXERCISE 5.13:** The identity, reflection, bar, and dagger operators are used to construct the projection operators $\mathbf{P}_e$, $\mathbf{P}_o$, $\mathbf{P}_r$, $\mathbf{P}_i$, $\mathbf{P}_h$, $\mathbf{P}_a$, cf. (11), (45), and (46).

(a) Show that there are exactly 10 distinct nonzero products that can be formed from a string of one or more of these projection operators.

Hint. In addition to the above six you will find $\mathbf{P}_{re} := \mathbf{P}_r\mathbf{P}_e$, $\mathbf{P}_{ro} := \mathbf{P}_r\mathbf{P}_o$, $\mathbf{P}_{ie} := \mathbf{P}_i\mathbf{P}_e$, $\mathbf{P}_{io} := \mathbf{P}_i\mathbf{P}_o$.

(b) Let $\mathbf{P}$ be one of the operators $\mathbf{P}_{re}$, $\mathbf{P}_{ro}$, $\mathbf{P}_{ie}$, $\mathbf{P}_{io}$, and let $\mathbf{Q} := \mathbf{I} - \mathbf{P}$. Verify that $\mathbf{P}^2 = \mathbf{P}$, $\mathbf{Q}^2 = \mathbf{Q}$, $\mathbf{PQ} = \mathbf{QP} = \mathbf{0}$, cp. (12), (47), (48).

(c) We associate *hermitian* symmetry with $\mathbf{P}_h$ since f is hermitian if and only if $\mathbf{P}_h f = f$. Likewise we associate *real* symmetry with $\mathbf{P}_h^{\wedge} = \mathbf{P}_r$. Make a table showing $\mathbf{P}$, the symmetry associated with $\mathbf{P}$, and the symmetry associated with $\mathbf{P}^{\wedge}$ for the 10 projection operators $\mathbf{P}$ of (a).

▶ **EXERCISE 5.14:** Let $\mathbf{A}$ be a linear operator applied to the functions on $\mathbb{P}_N$. Such an operator can be represented by an $N \times N$ matrix $\{a_{mn}\}_{m,n=0}^{N-1}$, i.e.,

$$(\mathbf{A}f)[m] := \sum_{n=0}^{N-1} a_{mn} f[n], \qquad m = 0, 1, \dots, N-1.$$

(a) Show that $\mathbf{A}$ commutes with $\mathbf{R}$, $\mathcal{B}$, $\mathcal{D}$ if and only if $a_{mn} = a_{-m,-n}$, $a_{mn} = \overline{a_{mn}}$, $a_{mn} = \overline{a_{-m,-n}}$, respectively. Here matrix indices are taken modulo N, i.e., $a_{-m,-n} := a_{N-m,N-n}$ when $m, n = 1, 2, \dots, N$.

(b) What must be true of a_{mn} if $\mathbf{A}$ preserves the symmetry of being even? real? hermitian?

▶ **EXERCISE 5.15:** In this exercise you will develop properties of the four *fundamental projections*

$$\mathbf{Q}_k := \frac{1}{4} \sum_{\ell=0}^{3} (-i)^{k\ell} \mathbf{E}_+^{\ell} = \frac{1}{4} \sum_{\ell=0}^{3} (+i)^{k\ell} \mathbf{E}_-^{\ell}$$

when these operators are applied to a suitably regular function f defined on $\mathbb{R}$ or $\mathbb{P}_N$. (In these two cases the domain of $\mathbf{E}_+, \mathbf{E}_-$ is the same as the range!)

(a) Use the definition to show that

$$\mathbf{Q}_0 + \mathbf{Q}_1 + \mathbf{Q}_2 + \mathbf{Q}_3 = \mathbf{I}, \qquad \mathbf{Q}_k \mathbf{Q}_\ell = \begin{cases} \mathbf{Q}_k & \text{when } k = \ell \\ \mathbf{0} & \text{when } k \neq \ell. \end{cases}$$

(b) Verify that $\mathbf{Q}_0 = \frac{1}{2}(\mathbf{P}_e + \mathbf{C})$, $\mathbf{Q}_1 = \frac{1}{2}(\mathbf{P}_o + \mathbf{S})$, $\mathbf{Q}_2 = \frac{1}{2}(\mathbf{P}_e - \mathbf{C})$, $\mathbf{Q}_3 = \frac{1}{2}(\mathbf{P}_o - \mathbf{S})$, and thereby show that each of the projections $\mathbf{Q}_0$, $\mathbf{Q}_1$, $\mathbf{Q}_2$, $\mathbf{Q}_3$ is real with $\mathbf{Q}_0$, $\mathbf{Q}_2$ preserving even symmetry and with $\mathbf{Q}_1$, $\mathbf{Q}_3$ preserving odd symmetry.

(c) Show that $\mathbf{E}_+\mathbf{Q}_k = (+i)^k\mathbf{Q}_k$, $\mathbf{E}_-\mathbf{Q}_k = (-i)^k\mathbf{Q}_k$, $k = 0, 1, 2, 3$.

(d) Use (a) and (c) to show that f has the decomposition $f = f_0 + f_1 + f_2 + f_3$ where

$$\mathbf{E}_+ f_k = (+i)^k f_k, \quad \mathbf{E}_- f_k = (-i)^k f_k, \quad \mathcal{F} f_k = (-i)^k \beta^{-1/2} f_k, \quad k = 0, 1, 2, 3,$$

i.e., f_0, f_1, f_2, f_3 are eigenfunctions of the operators $\mathbf{E}_+$, $\mathbf{E}_-$, $\mathcal{F}$ (or they are zero functions). Here $\beta = 1$, N when f is defined on $\mathbb{R}$, $\mathbb{P}_N$, respectively.

Hint. $f_k := \mathbf{Q}_k f$.

(e) Show that if we know f and $f^\wedge$, we can find the projections of (d) using

$$f_k(x) = \tfrac{1}{4}\{f(x) + i^k f^\wedge(x) + (-1)^k f(-x) + (-i)^k f^\wedge(-x)\}$$

when f, $f^\wedge$ are defined on $\mathbb{R}$ and

$$f_k[n] = \tfrac{1}{4}\{f[n] + \sqrt{N}i^k f^\wedge[n] + (-1)^k f[-n] + \sqrt{N}(-i)^k f^\wedge[-n]\}$$

when f, $f^\wedge$ are defined on $\mathbb{P}_N$.

▶ **EXERCISE 5.16:**   In this exercise you will show how the fundamental projections $\mathbf{Q}_0$, $\mathbf{Q}_1$, $\mathbf{Q}_2$, $\mathbf{Q}_3$ (from Ex. 5.15) can be used to represent an arbitrary polynomial in the Fourier transform operator $\mathcal{F}$ in cases where $\mathcal{F}$ is applied to functions defined on $\mathbb{R}$ or on $\mathbb{P}_N$. Since $\mathcal{F}^4$ is a scalar multiple of the identity operator, $\mathbf{I}$, we can assume that the polynomial has the form $\mathcal{P}(x) := c_0 + c_1 x + c_2 x^2 + c_3 x^3$ where c_0, c_1, c_2, c_3 are complex numbers.

(a) Show that $\mathcal{P}(\mathbf{E}_+)\mathbf{Q}_k = \mathcal{P}((+i)^k)\mathbf{Q}_k$, $\mathcal{P}(\mathbf{E}_-)\mathbf{Q}_k = \mathcal{P}((-i)^k)\mathbf{Q}_k$, $k = 0, 1, 2, 3$.

Hint. Use Ex. 5.15(c).

(b) Show that

$$\mathcal{P}(\mathbf{E}_+) = \mathcal{P}(1)\mathbf{Q}_0 + \mathcal{P}(i)\mathbf{Q}_1 + \mathcal{P}(-1)\mathbf{Q}_2 + \mathcal{P}(-i)\mathbf{Q}_3$$
$$\mathcal{P}(\mathbf{E}_-) = \mathcal{P}(1)\mathbf{Q}_0 + \mathcal{P}(-i)\mathbf{Q}_1 + \mathcal{P}(-1)\mathbf{Q}_2 + \mathcal{P}(i)\mathbf{Q}_3.$$

Hint. Apply $\mathcal{P}(\mathbf{E}_\pm)$ to the identity $\mathbf{I} = \mathbf{Q}_0 + \mathbf{Q}_1 + \mathbf{Q}_2 + \mathbf{Q}_3$ from Ex. 5.15(a).

(c) Express $\mathbf{I}$, $\mathbf{P}_e$, $\mathbf{P}_o$, $\mathbf{R}$, $\mathbf{C}$, $\mathbf{S}$, $\mathbf{H}_+$, $\mathbf{E}_+$, $\mathbf{E}_-$ in terms of $\mathbf{Q}_0$, $\mathbf{Q}_1$, $\mathbf{Q}_2$, $\mathbf{Q}_3$.

Hint. You can show that $\mathbf{R} = \mathbf{Q}_0 - \mathbf{Q}_1 + \mathbf{Q}_2 - \mathbf{Q}_3$ by setting $\mathcal{P}(x) := x^2$ in (b) or by suitably manipulating the identities of Ex. 5.15(b).

(d) A polynomial in $\mathcal{F}$ can be written as a linear combination of $\mathbf{I}$, $\mathbf{E}_+$, $\mathbf{E}_+^2 = \mathbf{R}$, and $\mathbf{E}_+^3 = \mathbf{E}_-$. Show that

$$A_0\mathbf{I} + A_1\mathbf{E}_+ + A_2\mathbf{E}_+^2 + A_3\mathbf{E}_+^3 = a_0\mathbf{Q}_0 + a_1\mathbf{Q}_1 + a_2\mathbf{Q}_2 + a_3\mathbf{Q}_3$$

if and only if the coefficients A_k, a_n are related in such a manner that

$$(A_0, A_1, A_2, A_3) = (a_0, a_1, a_2, a_3)^\wedge.$$

(e) Let

$$\mathbf{A} := A_0\mathbf{I} + A_1\mathbf{E}_+ + A_2\mathbf{E}_+^2 + A_3\mathbf{E}_+^3 = a_0\mathbf{Q}_0 + a_1\mathbf{Q}_1 + a_2\mathbf{Q}_2 + a_3\mathbf{Q}_3,$$

$$\mathbf{B} := B_0\mathbf{I} + B_1\mathbf{E}_+ + B_2\mathbf{E}_+^2 + B_3\mathbf{E}_+^3 = b_0\mathbf{Q}_0 + b_1\mathbf{Q}_1 + b_2\mathbf{Q}_2 + b_3\mathbf{Q}_3.$$

Show that

$$\mathbf{AB} = \mathbf{BA} = a_0b_0\mathbf{Q}_0 + a_1b_1\mathbf{Q}_1 + a_2b_2\mathbf{Q}_2 + a_3b_3\mathbf{Q}_3 = C_0\mathbf{I} + C_1\mathbf{E}_+ + C_2\mathbf{E}_+^2 + C_3\mathbf{E}_+^3$$

where

$$(C_0, C_1, C_2, C_3) := (A_0, A_1, A_2, A_3) * (B_0, B_1, B_2, B_3).$$

(f) Show that $\mathbf{A} := a_0\mathbf{Q}_0 + a_1\mathbf{Q}_1 + a_2\mathbf{Q}_2 + a_3\mathbf{Q}_3$ is an involution (*i.e.*, $\mathbf{A}^2 = \mathbf{I}$) if and only if $\mathbf{A}$ is one of the 16 operators

$$\pm(\mathbf{P}_e \pm \mathbf{P}_o), \ \pm(\mathbf{P}_e \pm \mathbf{S}), \ \pm(\mathbf{C} \pm \mathbf{P}_o), \ \pm(\mathbf{C} \pm \mathbf{S}).$$

Note. The Hartley transform operators, $\pm\mathbf{H}_\pm$, are the only such involutions that actually *transform* both the even and the odd parts of functions to which they are applied!

(g) Let A_0, A_1, A_2, A_3 be complex numbers and let $\mathbf{A} := A_0\mathbf{I} + A_1\mathbf{E}_+ + A_2\mathbf{E}_+^2 + A_3\mathbf{E}_+^3$. Devise a procedure for finding the corresponding coefficients B_0, B_1, B_2, B_3 for the inverse operator $\mathbf{B} = B_0\mathbf{I} + B_1\mathbf{E}_+ + B_2\mathbf{E}_+^2 + B_3\mathbf{E}_+^3$ when such an inverse exists.

▶ **EXERCISE 5.17:** In this exercise you will determine the multiplicities n_0, n_1, n_2, n_3 of the eigenvalues $\lambda = 1/\sqrt{N}, -i/\sqrt{N}, -1/\sqrt{N}, i/\sqrt{N}$ of the $N \times N$ discrete Fourier transform matrix $\mathcal{F}$, or equivalently, of the eigenvalues $\lambda = 1, -i, -1, i$ of the normalized operator $\mathbf{E}_-$.

(a) Show that $\mathbf{E}_- f = (-i)^k f$ if and only if $\mathbf{Q}_k f = f$, $k = 0, 1, 2, 3$, *cf.* Ex. 5.15.

(b) Show that n_k is the multiplicity of the eigenvalue $\lambda = 1$ of $\mathbf{Q}_k$, $k = 0, 1, 2, 3$.

(c) Use (b) and Ex. 5.15(a) to show that $n_0 + n_1 + n_2 + n_3 = N$.

(d) Use (b) and Ex. 5.15(b) to show that $n_0 + n_2 = \lfloor N/2 \rfloor + 1$.

Hint. Since $\mathbf{Q}_0 + \mathbf{Q}_2 = \mathbf{P}_e$, $n_0 + n_2$ is the dimension of the space of even functions on $\mathbb{P}_N$.

(e) Show that $n_0 - in_1 - n_2 + in_3 = 1, 0, -i, 1 - i$ when $N \equiv 1, 2, 3, 4 \pmod 4$.

Hint. Recall from linear algebra that the *trace* $\sum a_{nn}$ of a matrix $A = \{a_{mn}\}_{m,n=0}^{N-1}$ gives the sum of its eigenvalues, and use the expression (4.72) for the Gauss sum.

(f) Using (c)–(d), show that n_0, n_1, n_2, n_3 are given by the following table.

N	n_0	n_1	n_2	n_3
$4m$	$m+1$	m	m	$m-1$
$4m+1$	$m+1$	m	m	m
$4m+2$	$m+1$	m	$m+1$	m
$4m+3$	$m+1$	$m+1$	$m+1$	m

Note. $\mathbf{Q}_2 = \mathbf{0}$ if $N = 1$, $\mathbf{Q}_1 = \mathbf{0}$ if $N = 1, 2$, and $\mathbf{Q}_3 = \mathbf{0}$ if $N = 1, 2, 3, 4$.

▶ **EXERCISE 5.18:**   The following operators are applied to functions on $\mathbb{P}_N$. Find each eigenvalue and determine its multiplicity.

(a) **C**   (b) **S**   (c) **R**   (d) **H**$_+$

Hint. When $\mathbf{E}_- f = \lambda f$, you can write $\mathbf{E}_+ f = \mathbf{E}_-^3 f = \lambda^3 f$ and $\mathbf{E}_+ f = \overline{\lambda} f$. Use this with the results of Ex. 5.17.

▶ **EXERCISE 5.19:**   Let the translation, exponential modulation operators $\mathcal{T}_a, \mathcal{E}_a$ from (64) be applied to functions on $\mathbb{R}$, and let **R**, $\mathcal{B}$, $\mathcal{D}$ be the reflection, bar, and dagger operators. (Here $-\infty < a < \infty$.) Show that these operators satisfy the following relations.

(a) $\mathbf{R}\mathcal{T}_a = \mathcal{T}_{-a}\mathbf{R}$ (b) $\mathcal{B}\mathcal{T}_a = \mathcal{T}_a\mathcal{B}$ (c) $\mathcal{D}\mathcal{T}_a = \mathcal{T}_{-a}\mathcal{D}$

(d) $\mathbf{R}\mathcal{E}_a = \mathcal{E}_{-a}\mathbf{R}$ (e) $\mathcal{B}\mathcal{E}_a = \mathcal{E}_{-a}\mathcal{B}$ (f) $\mathcal{D}\mathcal{E}_a = \mathcal{E}_a\mathcal{D}$

Note. Appendix 3 shows how to define translation and exponential modulation operators for functions on $\mathbb{T}_p$, $\mathbb{Z}$, $\mathbb{P}_N$, and you can use the above identities within such settings.

▶ **EXERCISE 5.20:**   Let the dilation, translation, exponential modulation, derivative, and power scaling operators $\mathcal{S}_b$, $\mathcal{T}_a$, $\mathcal{E}_a$, **D**, $\mathcal{P}$ from (56), (64), and (79) be applied to suitably regular functions on $\mathbb{R}$. (Here $-\infty < a < \infty$ and $-\infty < b < \infty$ with $b \neq 0$.) Show that these operators satisfy the following relations.

(a) $\mathcal{T}_a\mathcal{S}_b = \mathcal{S}_b\mathcal{T}_{a \cdot b}$ (b) $\mathcal{E}_a\mathcal{S}_b = \mathcal{S}_b\mathcal{E}_{a/b}$ (c) $\mathbf{D}\mathcal{S}_b = b\mathcal{S}_b\mathbf{D}$ (d) $\mathcal{P}\mathcal{S}_b = b^{-1}\mathcal{S}_b\mathcal{P}$

▶ **EXERCISE 5.21:**   Find the 4×4 matrix that corresponds to each of the following operators (when these operators are applied to functions on $\mathbb{P}_4$).

(a) **E**$_+$, **E**$_-$, *cf.* (14)   (b) **R**, **P**$_e$, **P**$_o$ *cf.* (3) and (11)

(c) **C**, **S**, *cf.* (18)   (d) **H**$_+$, **H**$_-$, *cf.* (23)

(e) **Q**$_0$, **Q**$_1$, **Q**$_2$, **Q**$_3$, *cf.* Ex. 5.15(b)   (f) $\mathcal{M}_g$, $\mathbf{C}_g$, *cf.* (58)

(g) $\mathcal{T}_0$, $\mathcal{T}_1$, $\mathcal{T}_2$, $\mathcal{T}_3$ (h) $\mathcal{E}_0$, $\mathcal{E}_1$, $\mathcal{E}_2$, $\mathcal{E}_3$

Note. If you wish, you can verify that these matrices satisfy identities such as $\mathbf{C}^2 + \mathbf{S}^2 = \mathbf{I}$, $\mathbf{R}\mathbf{E}_+ = \mathbf{E}_-$, $\mathbf{H}_+^2 = \mathbf{I}$, $\mathcal{T}_1^2 = \mathcal{T}_2$, $\ldots$.

▶ **EXERCISE 5.22:**   Let $\mathcal{F}$ be the discrete Fourier transform matrix (1). The inversion rule corresponds to the relation $\mathcal{F}^2 = N^{-1}\mathbf{R}$ where **R** is the matrix (2). What property of $\mathcal{F}$ corresponds to:

(a) the reflection rule?   (b) the translation rule?   (c) Parseval's identity?

Hint. Use the identity $\overline{\mathcal{F}} = \mathcal{B}\mathcal{F}\mathcal{B} = \mathbf{R}\mathcal{F}$ for (c).

▶ **EXERCISE 5.23:**   A Fourier transform rule and its dual are often quite dissimilar. Explain why we do not get substantially new rules from the dual of the dual and from the dual of the dual of the dual.

Hint. Observe that $(\mathbf{R}\mathcal{A}\mathbf{R})^{\wedge} = \mathbf{R}\mathcal{A}^{\wedge}\mathbf{R}$.

▶ **EXERCISE 5.24:** Let $\mathbf{X}$ be the operator that exchanges the real and imaginary parts of a complex-valued function, i.e.,

$$\mathbf{X}f := f_I + if_R \qquad \text{when} \qquad f = f_R + if_I$$

(with f_R, f_I being real-valued).

(a) Show that $\mathbf{X} = i\mathcal{B}$ [where $\mathcal{B}$ is the bar operator (36)].

(b) Show that $\mathbf{X}^2 = \mathbf{I}$ by using the definition of $\mathbf{X}$.

(c) Show that $\mathbf{X}^2 = \mathbf{I}$ by using the identity of (a).

(d) Find a simple expression for the operator $\mathbf{X}^\wedge$.

(e) Let $f = f_R + if_I$, $f^\wedge = F_R + iF_I$ where f_R, f_I, F_R, F_I are real valued. Express the Fourier transform rule associated with $\mathbf{X}$ in terms of f_R, f_I, F_R, F_I.

▶ **EXERCISE 5.25:** Find the Fourier transform of each of the following operators that are applied to suitably regular functions on $\mathbb{R}$.

(a) The averaging operator

$$(\mathbf{A}_h f)(x) := \frac{1}{2h} \int_{x-h}^{x+h} f(u)du, \quad h > 0.$$

Hint. Use (53) and (59).

(b) The low-pass filter operator

$$(\mathbf{L}_\sigma f)(x) := \int_{s=-\sigma}^{\sigma} f^\wedge(s)e^{2\pi isx}ds = \int_{s=-\sigma}^{\sigma} \left\{ \int_{u=-\infty}^{\infty} f(u)e^{-2\pi isu}du \right\} e^{2\pi isx}ds,$$

which removes from the Fourier representation of f all sinusoids having frequencies outside the band $-\sigma < s < \sigma$.

(c) The high-pass filter operator

$$(\mathbf{H}_\sigma f)(x) := \int_{|s|>\sigma} f^\wedge(s)e^{2\pi isx}ds = \int_{|s|>\sigma} \left\{ \int_{u=-\infty}^{\infty} f(u)e^{-2\pi isu} \right\} e^{2\pi isx}ds,$$

which removes from the Fourier representation of f all sinusoids having frequencies within the band $-\sigma < s < \sigma$.

▶ **EXERCISE 5.26:** Let $h > 0$, $\alpha > 0$, and let the operators $\mathbf{D}$, $\mathcal{T}_h$, $\mathcal{S}_\alpha$ be defined by (35), (64), and (56). Let $c_0, c_1, \ldots, c_n$ be complex scalars and let

$$P(x) := c_0 + c_1 x + \cdots + c_n x^n.$$

Find the Fourier transform of:

(a) The differential operator $\{P(\mathbf{D})f\}(x) := c_0 f(x) + c_1 f'(x) + \cdots + c_n f^{(n)}(x)$;

(b) The difference operator $\{P(\mathcal{T}_h)f\}(x) := c_0 f(x) + c_1 f(x+h) + \cdots + c_n f(x+nh)$;

(c) The dilation operator $\{\mathcal{S}_\alpha P(\mathcal{T}_h)f\}(x) := c_0 f(\alpha x) + c_1 f(\alpha x+h) + \cdots + c_n f(\alpha x+nh)$.

▶ **EXERCISE 5.27:**   Let $p > 0$, $q > 0$, and let the operator

$$(\mathbf{J}f)(x) := f(px/q)$$

map the function f on $\mathbb{T}_p$ to the function $\mathbf{J}f$ on $\mathbb{T}_q$.

(a) Find the Fourier transform $\mathbf{J}^\wedge$ of the operator $\mathbf{J}$ and draw a commuting diagram to illustrate the corresponding mappings.

(b) Find $\mathbf{J}^{\wedge\wedge}$ by taking the Fourier transform of $\mathbf{J}^\wedge F$ where $F := f^\wedge$ is a function on $\mathbb{Z}$.

   Hint. Assume that the Fourier transforms of F, $\mathbf{J}^\wedge F$ are functions on $\mathbb{T}_p$, $\mathbb{T}_q$, respectively.

(c) Find $\mathbf{J}^{\wedge\wedge}$ by simplifying the operator identity $\mathbf{J}^{\wedge\wedge} = (\beta_d/\beta_r)\mathbf{R}\mathbf{J}\mathbf{R}$.

▶ **EXERCISE 5.28:**   A linear operator $\mathbf{A}$ is said to be translation invariant (LTI), modulation invariant (LMI) provided that $\mathbf{A}$ commutes with every translation operator $\mathcal{T}_h$, every exponential modulation operator $\mathcal{E}_h$, respectively.

(a) Show that $\mathbf{A}$ is LTI if and only if $\mathbf{A}^\wedge$ is LMI.

(b) Show that the multiplication operator $\mathcal{M}_g$ is LMI.

(c) Using (a)–(b), show that the convolution operator $\mathcal{C}_g$ is LTI.

(d) Let $\mathbf{A}$ be an invertible LMI operator. Show that $\mathbf{A}^{-1}$ is also LMI.

(e) Let $\mathbf{A}$ be an invertible LTI operator. Show that $\mathbf{A}^{-1}$ is also LTI.

▶ **EXERCISE 5.29:**   Let $\mathbf{A}$ be an operator that is applied to the functions on $\mathbb{P}_N$.

(a) Show that $\mathbf{A}$ is LMI (*cf.* Ex. 5.28) if and only if $\mathbf{A} = \mathcal{M}_{\mathbf{A}u}$, i.e., $\mathbf{A}f := (\mathbf{A}u) \cdot f$, where u is the function on $\mathbb{P}_N$ with $u[n] = 1$, $n = 0, 1, \ldots, N-1$.

   Hint. Use matrix representations for $\mathbf{A}$ and $\mathcal{E}_m$.

(b) Show that $\mathbf{A}$ is LTI (*cf.* Ex. 5.28) if and only if $\mathbf{A} = \mathcal{C}_{\mathbf{A}\delta}$, i.e., $\mathbf{A}f := (\mathbf{A}\delta) * f$, where δ is the function on $\mathbb{P}_N$ with $\delta[0] = 1$ and $\delta[n] = 0$ for $n = 0, 1, \ldots, N-1$.

   Hint. Use Ex. 5.28(a) with the representation of (a).

(c) Let $e_k[n] := e^{2\pi i k n/N}$. Show that $\mathbf{A}$ is LTI if and only if $e_0, e_1, \ldots, e_{N-1}$ are eigenfunctions of $\mathbf{A}$.

(d) Let $\mathbf{A}$ be an LTI operator with *impulse response* r and *frequency response* λ, i.e., $r[n] := (\mathbf{A}\delta)[n]$, $n = 0, 1, \ldots, N-1$ as in (b) and $\mathbf{A}e_k = \lambda[k] \cdot e_k$, $k = 0, 1, \ldots, N-1$ as in (c). How are the functions r and λ related?

▶ **EXERCISE 5.30:**   The following operators are applied to functions on $\mathbb{P}_N$. Find their eigenvalues and corresponding eigenfunctions.

(a) The translation operator $\mathcal{T}_m$.

(b) The negative discrete Laplacian $\mathbf{A}$ with $(\mathbf{A}f)[n] := -f[n-1] + 2f[n] - f[n+1]$.

   Hint. $\lambda_k = 4\sin^2(k\pi/N)$.

▶ **EXERCISE 5.31:** Let f be a function on $\mathbb{P}_N$ with $f^\wedge[k] \neq 0$ for $k = 0, 1, \ldots, N-1$ and let $\mathcal{T} := \mathcal{T}_1$ be the unit translation operator. Show how to find coefficients $c_0, c_1, \ldots, c_N$ so that

$$\{c_0 \mathbf{I} + c_1 \mathcal{T} + c_2 \mathcal{T}^2 + \cdots + c_{N-1} \mathcal{T}^{N-1}\} f = g$$

when the function g on $\mathbb{P}_N$ is specified. (Thus $f, \mathcal{T}f, \mathcal{T}^2 f, \ldots, \mathcal{T}^{N-1} f$ form a basis for the N-dimensional linear space of complex functions defined on $\mathbb{P}_N$.)

Hint. $\{g^\wedge / f^\wedge\}^\wedge$

Note. N. Weiner proved the following generalization of this result: Let f be an absolutely integrable function on $\mathbb{R}$ with $f^\wedge(s) \neq 0$, $-\infty < s < \infty$. Let g be an absolutely integrable function on $\mathbb{R}$, and let $\epsilon > 0$ be given. Then there exist real numbers $a_1 < a_2 < \cdots < a_N$ and complex scalars $c_1, c_2, \ldots, c_N$ such that

$$\int_{-\infty}^{\infty} \left| g(x) - \sum_{n=1}^{N} c_n (\mathcal{T}_{a_n} f)(x) \right| dx < \epsilon.$$

▶ **EXERCISE 5.32:** When we work with functions on $\mathbb{R}$ and use the dilation operator (56) with $a > 1$:

    (i) $(\mathcal{S}_a f)(x) := f(ax)$ is a "compressed" version of f;

    (ii) $(\mathcal{S}_{1/a} f)(x) := f(x/a)$ is a "stretched" version of f;

    (iii) $\mathcal{S}_a \mathcal{S}_{1/a} = \mathbf{I}$; and

    (iv) $\mathcal{S}_a^\wedge \mathcal{S}_{1/a}^\wedge = \mathbf{I}^\wedge$ simplifies to $\mathcal{S}_{1/a} \mathcal{S}_a = \mathbf{I}$.

Write down the operator definitions that correspond to (i)–(ii) and the operator identities that correspond to (iii)–(iv) when we work with functions on:

    (a) $\mathbb{T}_p$;   (b) $\mathbb{Z}$;   (c) $\mathbb{P}_N$.

Hint. Use $m = 2, 3, \ldots$ in place of $a > 1$. Give two answers for (c), one using the dilation operator $\mathcal{S}_m$ [with $\gcd(m, N) = 1$] and one using the decimation operator Ξ_m (with $m|N$). All of the required operators are defined in Appendix 3.

▶ **EXERCISE 5.33:** Sketch graphs for the functions at the four corners of the following commuting diagrams. All of the required operators are defined in Appendix 3.

(a)

$$
\begin{array}{ccc}
f & \overset{\mathcal{S}_3}{\underset{\mathcal{G}_3}{\rightleftarrows}} & \mathcal{S}_3 f \\
\mathcal{F} \downarrow & & \downarrow \mathcal{F} \\
f^\wedge & \overset{\mathcal{Z}_3}{\underset{\mathcal{S}_3}{\rightleftarrows}} & \mathcal{Z}_3 f^\wedge
\end{array}
$$

$$f(x) := \sum_{\ell=-\infty}^{\infty} \Pi\left(\frac{x - 12\ell}{3}\right) \quad \text{(on } \mathbb{T}_{12})$$

$$f \; \underset{\Sigma_3}{\overset{\mathcal{R}_3}{\rightleftarrows}} \; \mathcal{R}_3 f$$

(b) $\quad \mathcal{F} \Big\downarrow \qquad\qquad \Big\downarrow \mathcal{F} \qquad f[n] := \begin{cases} 1 & \text{if } n = 0, 1, 11 \\ 0 & \text{if } n = 2, 3, \dots, 10 \end{cases} \quad (\text{on } \mathbb{P}_{12})$

$$f^\wedge \; \underset{\Xi_3}{\overset{\mathcal{Z}_3}{\rightleftarrows}} \; \mathcal{Z}_3 f^\wedge$$

▶ **EXERCISE 5.34:**   When $a > 0$ the sample-sum operator

$$(\mathcal{X}_a f)[n] := \sum_{m=-\infty}^{\infty} f\left(\frac{a}{\sqrt{N}}[n - mN]\right)$$

[from (4.64)] maps a suitably regular function f on $\mathbb{R}$ to a function $\mathcal{X}_a f$ on $\mathbb{P}_N$.

(a) Show that

$$\mathcal{X}_a = \Sigma_N \Xi_{a/\sqrt{N}}$$

where Ξ_p is the p-sampling operator for functions on $\mathbb{R}$ and Σ_N is the N-summation operator for functions on $\mathbb{Z}$, as defined in Appendix 3.

(b) Use (a) and the multiplicative property (61) of the operator Fourier transform to verify that

$$\mathcal{X}_a^\wedge = \frac{1}{a\sqrt{N}} \Xi_{1/a\sqrt{N}} \Sigma_{\sqrt{N}/a}$$

where now Σ_q is the q-summation operator for functions on $\mathbb{R}$ and $\Xi_{q/N}$ is the q/N-sampling operator for functions on $\mathbb{T}_q$, $q := \sqrt{N}/a$, as defined in Appendix 3.

(c) Using (b), show that

$$\mathcal{X}_a^\wedge = \frac{1}{a\sqrt{N}} \mathcal{X}_{1/a}.$$

(d) Verify that $\mathcal{X}_a^{\wedge\wedge} = N^{-1}\mathcal{X}_a$ by using the identity from (c).

(e) Verify that $\mathcal{X}_a^{\wedge\wedge} = N^{-1}\mathcal{X}_a$ by using the identity (63).

▶ **EXERCISE 5.35:**   Let $m = 1, 2, \dots$. The *end* padding operator

$$(\mathcal{P}_m f)[n] := \begin{cases} f[n] & \text{if } n = 0, 1, \dots, N - 1 \\ 0 & \text{if } n = N, N + 1, \dots, mN - 1 \end{cases}$$

maps a function f on $\mathbb{P}_N$ to a function f on $\mathbb{P}_{mN}$.

(a) Show that $\mathcal{F} = m\Xi_m \mathcal{F} \mathcal{P}_m$ when Ξ_m is the m-decimation operator for functions on $\mathbb{P}_{mN}$.

   Hint. You must show that $(m\mathcal{F}\mathcal{P}_m f)[mn] = \mathcal{F}f[n]$, *cf.* Ex. 4.47(b).

(b) Write the operator identity of (a) in terms of matrices in the case where $m = 2$ and $N = 3$.

(c) Show that $\Xi_m \mathcal{F}^{-1} \mathcal{P}_m \mathcal{F} = \mathbf{I}$.

    Hint. Use (a) and the identity $\Xi_m \mathbf{R} = \mathbf{R} \Xi_m$.

(d) Using (c) explain why $g := \mathcal{F}^{-1} \mathcal{P}_m \mathcal{F} f$ is said to interpolate f.

    Note. Form strings of zeros with N or $2N$ or $3N \ldots$ zeros in each string and with a total of $(m-1)N$ zeros in all of the strings. Insert these zero strings before, between, or after the components $f[0], f[1], \ldots, f[N-1]$ of an f on $\mathbb{P}_N$ to produce a function $\mathcal{P}'_m f$ on $\mathbb{P}_{mN}$. The operator identities of (a), (c) will hold with $\mathcal{P}'_m$ in place of $\mathcal{P}_m$, cf. Ex. 4.47(c).

▶ **EXERCISE 5.36:**   Show that $\mathbf{A}$ distributes over the convolution product, *i.e.*,

$$\mathbf{A}(f * g) = (\mathbf{A}f) * (\mathbf{A}g) \text{ for all } f, g$$

if and only if

$$\mathbf{A}^{\wedge}(f^{\wedge} \cdot g^{\wedge}) = (\beta_r / \beta_d)(\mathbf{A}^{\wedge} f^{\wedge}) \cdot (\mathbf{A}^{\wedge} g^{\wedge}) \text{ for all } f^{\wedge}, g^{\wedge}.$$

Use this result to verify that the following operators distribute over the convolution product. All of the required operators are defined in Appendix 3.

(a) $\mathbf{R}$, $\mathcal{B}$, $\mathbf{D}$ for functions on $\mathbb{R}$, $\mathbb{T}_p$, $\mathbb{Z}$, $\mathbb{P}_N$ (b) $\mathcal{E}_a$, $|a|\mathcal{S}_a$, Σ_p for functions on $\mathbb{R}$

(c) $\mathcal{E}_m$, $\mathcal{S}_m$, $\mathcal{G}_m$ for functions on $\mathbb{T}_p$ (d) $\mathcal{E}_m$, $\mathcal{Z}_m$, Σ_N for functions on $\mathbb{Z}$

(e) $\mathcal{E}_m$, $\mathcal{S}_m$ (when m, N are relatively prime), $\Sigma_{N/m}$ (when $m|N$), $\mathcal{Z}_m$, $m^{-1}\mathcal{R}_m$ for functions on $\mathbb{P}_N$

▶ **EXERCISE 5.37:**   Let $\mathbf{R}$, $\mathbf{D}$, $\mathcal{P}$, be the reflection, derivative, and power scaling operators for functions on $\mathbb{R}$, as given in (4) and (79).

(a) Use the identities $\mathbf{D}^{\sim} = i \mathcal{P} \mathbf{R}$, $\mathcal{P} \mathbf{R} = -\mathbf{R} \mathcal{P}$ to show that

$$(\mathbf{D}^n)^{\sim} = (-1)^{n(n-1)/2} (i \, \mathcal{P})^n \mathbf{R}^n, \qquad n = 0, 1, 2, \ldots .$$

(b) Use (a) to show that $g(x) := f^{(n)}(x)$ has the Hartley transform

$$g^{\sim}(s) = (-1)^{n(n+1)/2}(2\pi s)^n f^{\sim}((-1)^n s) = (2\pi s)^n \begin{cases} f^{\sim}(s) & \text{if } n = 0, 4, 8, \ldots \\ -f^{\sim}(-s) & \text{if } n = 1, 5, 9, \ldots \\ -f^{\sim}(s) & \text{if } n = 2, 6, 10, \ldots \\ f^{\sim}(-s) & \text{if } n = 3, 7, 11, \ldots \end{cases}$$

    when f is a suitably regular function on $\mathbb{R}$.

(c) Rearrange the identity of (a) and thereby show that $g(x) := x^n f(x)$ has the Hartley transform

$$g^{\sim}(s) = (-1)^{n(n-1)/2}(2\pi)^{-n}(f^{\sim})^{(n)}((-1)^n s) = (2\pi)^{-n} \begin{cases} (f^{\sim})^{(n)}(s) & \text{if } n = 0, 4, 8, \ldots \\ (f^{\sim})^{(n)}(-s) & \text{if } n = 1, 5, 9, \ldots \\ -(f^{\sim})^{(n)}(s) & \text{if } n = 2, 6, 10, \ldots \\ -(f^{\sim})^{(n)}(-s) & \text{if } n = 3, 7, 11, \ldots . \end{cases}$$

    Hint. The identity $\mathbf{D}^{\sim\sim} = \mathbf{D}$ may be useful.

▶ **EXERCISE 5.38:**   In this exercise you will derive expressions for the Hartley transforms of various operators using a compact tag notation. For this purpose we define

$$\mathcal{A}^\wedge := \mathcal{F}\mathcal{A}\mathcal{F}^{-1} = \mathbf{E}_-\mathcal{A}\mathbf{E}_+$$

$$\mathcal{A}^\vee := \mathbf{R}\mathcal{A}\mathbf{R}$$

$$\mathcal{A}^\sim := \mathbf{H}_+\mathcal{A}\mathbf{H}_+$$

$$\mathcal{A}^\square := (\mathbf{P}_e + i\,\mathbf{P}_o)\mathcal{A}(\mathbf{P}_e - i\,\mathbf{P}_o) = \tfrac{1}{4}\big\{(1+i)\mathbf{I} + (1-i)\mathbf{R}\big\}\mathcal{A}\big\{(1-i)\mathbf{I} + (1+i)\mathbf{R}\big\}$$

$$\mathcal{A}^\diamond := \tfrac{1}{2}\{\mathcal{A} + \mathbf{R}\mathcal{A}\mathbf{R} + \mathcal{A}\mathbf{R} - \mathbf{R}\mathcal{A}\}.$$

(a) Explain why the transformations associated with the tags $^\wedge$, $^\vee$, $^\sim$, $^\square$, $^\diamond$, all commute, i.e., $\mathcal{A}^{\wedge\vee} = \mathcal{A}^{\vee\wedge}, \mathcal{A}^{\wedge\sim} = \mathcal{A}^{\sim\wedge}, \ldots, \mathcal{A}^{\square\diamond} = \mathcal{A}^{\diamond\square}$.

(b) You know that $\mathcal{A}^{\wedge\wedge} = \mathcal{A}^\vee$, $\mathcal{A}^{\vee\vee} = \mathcal{A}$, $\mathcal{A}^{\sim\sim} = \mathcal{A}$. Show that $\mathcal{A}^{\square\square} = \mathcal{A}^\vee$, $\mathcal{A}^{\diamond\diamond} = \mathcal{A}$.

(c) Derive the identities $\mathcal{A}^\sim = \mathcal{A}^{\wedge\square}$, $\mathcal{A}^\wedge = \mathcal{A}^{\sim\vee\square}$ that show how the $^\square$ transformation can be used to change the Fourier transform of an operator to the Hartley transform, and vice versa.

(d) Show that if $\mathcal{A}$ is a linear operator and $\mathcal{A}$ commutes with $\mathbf{R}$, then $\mathcal{A}^\square = \mathcal{A}$, and $\mathcal{A}^\sim = \mathcal{A}^\wedge$. Give examples of operators $\mathcal{A}$ for which this is the case.

(e) Use (c) and the identities $\mathcal{T}_a^\wedge = \mathcal{E}_a$, $\mathcal{E}_a^\wedge = \mathcal{T}_{-a}$ from Appendix 3 to show that $\mathcal{T}_a^\sim = \mathcal{E}_a^\square$, $\mathcal{E}_a^\sim = \mathcal{T}_{-a}^\square$.

(f) Show that $\mathbf{M}_{g^\wedge}^\square = \beta^{-1/2}\mathbf{M}_{g^\sim}^\diamond$.

Hint: It is sufficient to show that $\mathbf{M}_{g^\wedge}^\square f = \beta^{-1/2}\mathbf{M}_{g^\sim}^\diamond f$ when f is even and when f is odd.

(g) Use (c),(f), the identity $\mathcal{C}_g^\wedge = \beta\mathbf{M}_{g^\wedge}$, and (b) in turn to show that

$$\mathcal{C}_g^\sim = \beta^{1/2}\mathbf{M}_{g^\sim}^\diamond, \qquad \mathbf{M}_g^\sim = \beta^{-1/2}\mathcal{C}_{g^\sim}^\diamond.$$

▶ **EXERCISE 5.39:**   In this exercise you will develop properties of the products

$$f \circledast g := \tfrac{1}{2}\{f*g + f*g^\vee + f^\vee*g - f^\vee*g^\vee\}, \quad f \odot g := \tfrac{1}{2}\{f\cdot g + f\cdot g^\vee + f^\vee\cdot g - f^\vee\cdot g^\vee\}$$

(These products are commutative and associative, and both of them distribute over addition.)

(a) Show that

$$f*g = \tfrac{1}{2}\{f \circledast g + f \circledast g^\vee + f^\vee \circledast g - f^\vee \circledast g^\vee\}, \quad f\cdot g = \tfrac{1}{2}\{f \odot g + f \odot g^\vee + f^\vee \odot g - f^\vee \odot g^\vee\}.$$

(b) Verify the following eight product rules for working with Fourier and Hartley transforms.

$$(f * g)^\wedge = \beta f^\wedge \cdot g^\wedge, \qquad\qquad (f * g)^\sim = \beta^{1/2} f^\sim \odot g^\sim$$
$$(f \cdot g)^\wedge = f^\wedge * g^\wedge, \qquad\qquad (f \cdot g)^\sim = \beta^{-1/2} f^\sim \circledast g^\sim$$
$$(f \circledast g)^\wedge = \beta f^\wedge \odot g^\wedge, \qquad\qquad (f \circledast g)^\sim = \beta^{1/2} f^\sim \cdot g^\sim$$
$$(f \odot g)^\wedge = f^\wedge \circledast g^\wedge, \qquad\qquad (f \odot g)^\sim = \beta^{-1/2} f^\sim * g^\sim$$

Note. The operators $\mathcal{K}_g f := g \circledast f$, $\mathbf{N}_g f := g \odot f$ have the transforms $\mathcal{K}_g^\wedge = \beta \mathbf{N}_{g^\wedge}$, $\mathbf{N}_g^\wedge = \mathcal{K}_{g^\wedge}$, $\mathcal{K}_g^\sim = \beta^{1/2} \mathbf{M}_{g^\sim}$, $\mathbf{N}_g^\sim = \beta^{-1/2} \mathbf{C}_{g^\sim}$.

▶ **EXERCISE 5.40:** In this exercise you will use operator methods to deduce rules for working with (normalized) cosine transforms and sine transforms. These transforms will be denoted by using the superscripts ^{c}, ^{s} as tags, *i.e.*,

$$f^c := \mathbf{C} f, \qquad f^s := \mathbf{S} f.$$

(a) Show that when $f^\wedge$ is known, we can write

$$f^c = \beta^{1/2} \mathbf{P}_e f^\wedge, \qquad f^s = i\, \beta^{1/2} \mathbf{P}_o f^\wedge.$$

(b) Show that we can find the cosine and sine transforms of $\mathcal{A} f$ when we know the cosine and sine transforms of f by using the relations

$$(\mathcal{A} f)^c = \mathcal{A}_{cc} f^c + \mathcal{A}_{cs} f^s, \qquad (\mathcal{A} f)^s = \mathcal{A}_{sc} f^c + \mathcal{A}_{ss} f^s$$

where

$$\mathcal{A}_{cc} := \mathbf{C}\,\mathcal{A}\,\mathbf{C}, \quad \mathcal{A}_{cs} := \mathbf{C}\,\mathcal{A}\,\mathbf{S}, \quad \mathcal{A}_{sc} := \mathbf{S}\,\mathcal{A}\,\mathbf{C}, \quad \mathcal{A}_{ss} := \mathbf{S}\,\mathcal{A}\,\mathbf{S}.$$

(c) Show that if $\mathcal{A}$ commutes with $\mathbf{S}$, $\mathbf{C}$ (as is the case when $\mathcal{A} = \mathbf{R}$, $\mathcal{B}$, $\mathcal{D}$, $\mathbf{P}_e$, $\mathbf{P}_o$, $\mathbf{P}_r$, $\mathbf{P}_i$, $\mathbf{P}_h$, $\mathbf{P}_a$, ...), then

$$(\mathcal{A} f)^c = \mathcal{A}(f^c), \qquad (\mathcal{A} f)^s = \mathcal{A}(f^s).$$

(d) Show that when $\mathcal{A}$ is linear we can write

$$\mathcal{A}_{cc} = \mathbf{P}_e \mathcal{A}^\wedge \mathbf{P}_e, \quad \mathcal{A}_{cs} = -i\, \mathbf{P}_e \mathcal{A}^\wedge \mathbf{P}_o, \quad \mathcal{A}_{sc} = i\, \mathbf{P}_o \mathcal{A}^\wedge \mathbf{P}_e, \quad \mathcal{A}_{ss} = \mathbf{P}_o \mathcal{A}^\wedge \mathbf{P}_o.$$

(e) Use (b) and (d) to show that if $\mathcal{A}$ is a linear operator that commutes with $\mathbf{R}$, then

$$(\mathcal{A} f)^c = \mathcal{A}^\wedge f^c, \qquad (\mathcal{A} f)^s = \mathcal{A}^\wedge f^s.$$

In particular, the various dilation, grouping, decimation, zero packing, repeat, summation, and sampling rules are identical to those for taking Fourier transforms.

(f) Use (b) and (d) to obtain the translation rules

$$(\mathcal{T}_a f)^c = \frac{1}{2}(\mathcal{E}_a + \mathcal{E}_{-a})f^c + \frac{1}{2i}(\mathcal{E}_a - \mathcal{E}_{-a})f^s, \quad (\mathcal{T}_a f)^s = -\frac{1}{2i}(\mathcal{E}_a - \mathcal{E}_{-a})f^c + \frac{1}{2}(\mathcal{E}_a + \mathcal{E}_{-a})f^s.$$

(g) Use (b) and (d) to obtain the modulation rules

$$(\mathcal{E}_a f)^c = (\mathbf{P}_e \mathcal{T}_{-a})f^c - i(\mathbf{P}_e \mathcal{T}_{-a})f^s, \qquad (\mathcal{E}_a f)^s = i(\mathbf{P}_o \mathcal{T}_{-a})f^c + (\mathbf{P}_o \mathcal{T}_{-a})f^s.$$

(h) Use (b) and (d) to obtain the convolution rules

$$(f * g)^c = \beta^{1/2}\{g^c \cdot f^c - g^s \cdot f^s\}, \qquad (f * g)^s = \beta^{1/2}\{g^s \cdot f^c + g^c \cdot f^s\}.$$

(i) Use (b) and (d) to obtain the multiplication rules

$$(f \cdot g)^c = \beta^{-1/2}\{g^c * f^c - g^s * f^s\}, \qquad (f \cdot g)^s = \beta^{-1/2}\{g^s * f^c + g^c * f^s\}.$$

▶ **EXERCISE 5.41:**   In this exercise you will derive rules for taking Hilbert transforms of suitably regular functions on $\mathbb{R}$. For example, we can use (84) to deduce the operator identity $\mathcal{H}^2 = -\mathbf{I}$ that corresponds to the inversion rule

$$g(x) := f^\#(x) \quad \text{has the Hilbert transform} \quad g^\#(x) = -f(x).$$

Use operator identities from the text to formulate an analogously stated version of the Hilbert transform

(a) reflection rule,   (b) conjugation rule,   (c) translation rule,

(d) derivative rule,   (e) dilation rule,   (f) convolution rule.

▶ **EXERCISE 5.42:**   Let $f_0(x) := e^{\pi i x}\operatorname{sinc} x$ be a function on $\mathbb{R}$. Make a table with the columns

Formula for f Graph of $f^\wedge$ Graph of $f^{\#\wedge}$ Formula for $f^\#$

and then supply the entries for each of the following.

(a) $f = f_0$ (b) $f = \mathcal{B}f_0$ (c) $f = \mathbf{P}_e f_0$

(d) $f = \mathbf{P}_i f_0$ (e) $f = \mathcal{S}_{1/2}\mathbf{P}_e f_0$ (f) $f = f_0 * (\mathbf{P}_e f_0)$

Hint. Use the Hilbert transform rules from Ex. 5.41.

▶ **EXERCISE 5.43:**   Let f, g be suitably regular functions on $\mathbb{R}$. Use the Parseval and Plancherel relations to derive the corresponding identities

$$\int_{-\infty}^{\infty} \overline{f(x)}g(x)\,dx = \int_{-\infty}^{\infty} \overline{(\mathcal{H}f)}(x)(\mathcal{H}g)(x)\,dx, \qquad \int_{-\infty}^{\infty} |f(x)|^2\,dx = \int_{-\infty}^{\infty} |(\mathcal{H}f)(x)|^2\,dx$$

that link f, g to their Hilbert transforms.

▶ **EXERCISE 5.44:** Let f, g be suitably regular functions on $\mathbb{R}$. Verify that

$$f * g = -f^{\#} * g^{\#}$$

(a) by using the convolution rule for Fourier transforms [with (83)];

(b) by using the convolution rule (89) for Hilbert transforms [with (84)].

▶ **EXERCISE 5.45:** For a suitably regular function f on $\mathbb{R}$ we define

$$(\mathbf{P}_{+}f)(x) := \int_{0}^{\infty} f^{\wedge}(s)e^{2\pi isx}\,dx, \qquad (\mathbf{P}_{-}f)(x) := \int_{-\infty}^{0} f^{\wedge}(s)e^{2\pi isx}\,ds.$$

(a) Use the Kramers-Kronig relations (93) to find operator identities linking $\mathbf{P}_{+}$, $\mathbf{P}_{r}$, $\mathbf{P}_{i}$, $\mathcal{H}$.

(b) Find corresponding operator identities linking $\mathbf{P}_{-}$, $\mathbf{P}_{r}$, $\mathbf{P}_{i}$, $\mathcal{H}$.

▶ **EXERCISE 5.46:** Let F be a piecewise smooth function on $[-\sigma, \sigma]$, let

$$f(x) := \int_{s=-\sigma}^{\sigma} F(s)e^{2\pi isx}\,ds,$$

let $\beta \geq \sigma$, and let

$$g(x) := \cos(2\pi\beta x)f(x).$$

(a) Find a simple expression for the Hilbert transform, $g^{\#}$, of g.

(b) Sketch the curve $g + ig^{\#}$, [i.e., $(x, g(x), g^{\#}(x))$, $-\infty < x < \infty$], in the particular case where $f(x) := \Lambda(x)$ and $\beta = 5$.

▶ **EXERCISE 5.47:** Let the Hilbert transform operator $\mathcal{H}$ be applied to a suitably regular function f on $\mathbb{T}_{p}$ with

$$f(x) = a_0 + \sum_{k=1}^{\infty}\left\{a_k \cos\left(\frac{2\pi kx}{p}\right) + b_k \sin\left(\frac{2\pi kx}{p}\right)\right\}.$$

(a) Express $\mathcal{H}f$ in terms of the coefficients a_k, b_k.

(b) Set certain coefficients to zero in the series for f and $\mathcal{H}f$ and thereby obtain series for $\mathcal{H}\mathbf{P}_{e}f$ and $\mathcal{H}\mathbf{P}_{o}f$.

(c) Find the Hilbert transform rules

$$g := \mathbf{P}_{e}f \quad \text{has the Hilbert transform} \quad g^{\#} = \cdots ,$$
$$g := \mathbf{P}_{o}f \quad \text{has the Hilbert transform} \quad g^{\#} = \cdots .$$

(d) Use the rules (c) to find $\mathcal{H}\mathbf{P}_{e}f$ and $\mathcal{H}\mathbf{P}_{o}f$.

▶ **EXERCISE 5.48:**   Let $\mathcal{H}$ be the Hilbert transform operator (83) for suitably regular functions on $\mathbb{Z}$, and let the function sgn(x) on $\mathbb{T}_p$ be defined by (82).

(a) Show that $i\,\text{sgn}^{\wedge}[k] = \begin{cases} \dfrac{2}{\pi k} & \text{if } k = \pm 1, \pm 3, \pm 5, \dots \\ 0 & \text{otherwise.} \end{cases}$

(b) Show that $\mathcal{H} g = i\,\text{sgn}^{\wedge} * g$ when g is a suitably regular function on $\mathbb{Z}$.

   Note. cp. Ex. 7.55.

(c) Use (a) and (b) to show that

$$(\mathcal{H} g)[k] = \frac{2}{\pi} \sum_{\ell=-\infty}^{\infty} \frac{g[k - 2\ell - 1]}{2\ell + 1} \,.$$

(d) Let f be a real piecewise smooth function on $\mathbb{T}_p$ that vanishes on the half circle
$-p/2 < x < 0$, and let

$$a[k] := \frac{2}{p} \int_0^{p/2} f(x) \cos\left(\frac{2\pi kx}{p}\right) dx, \qquad b[k] := \frac{2}{p} \int_0^{p/2} f(x) \sin\left(\frac{2\pi kx}{p}\right) dx$$

   be the coefficients for the cosine series and for the sin series for f on the interval
   $0 < x < p/2$. Show that
$$\mathcal{H} a = b, \qquad \mathcal{H} b = -a,$$

   and express these Kramers-Kronig relations by using the representation of $\mathcal{H}$ that is
   given in (c).

▶ **EXERCISE 5.49:**   Let $\mathcal{H}$ be the Hilbert transform operator (83) for functions on
$\mathbb{P}_N$ and let the function sgn$[n]$ on $\mathbb{P}_N$ be given by (82).

(a) Show that $\mathcal{H} \delta$ and $i\,\text{sgn}^{\wedge}$ have the same Fourier transforms and thereby prove that

$$\mathcal{H} \delta = i\,\text{sgn}^{\wedge}.$$

   Note. The components of $\text{sgn}^{\wedge}$ are given in Ex. 4.27.

(b) Show that $\mathcal{H} f = i\,\text{sgn}^{\wedge} * f$.

   Note. cp. Exs. 5.4 and 7.55.

Chapter 6

The Fast Fourier Transform

6.1 Pre-FFT Computation of the DFT

Introduction

In this chapter we will study the problem of computing the components

$$F[k] := \frac{1}{N} \sum_{n=0}^{N-1} e^{-2\pi i k n/N} f[n], \quad k = 0, 1, \ldots, N-1$$

of the discrete Fourier transform of given complex numbers $f[0], f[1], \ldots, f[N-1]$. We write these relations in the compact form

$$\mathbf{F} = \boldsymbol{\mathcal{F}} \mathbf{f}$$

where

$$\mathbf{f} := \begin{bmatrix} f[0] \\ f[1] \\ \vdots \\ f[N-1] \end{bmatrix}, \quad \mathbf{F} := \begin{bmatrix} F[0] \\ F[1] \\ \vdots \\ F[N-1] \end{bmatrix}$$

are complex N-component column vectors and where the $N \times N$ DFT matrix

$$\boldsymbol{\mathcal{F}} := \frac{1}{N} \begin{bmatrix} 1 & 1 & 1 & \cdots & 1 \\ 1 & \omega & \omega^2 & \cdots & \omega^{N-1} \\ 1 & \omega^2 & \omega^4 & \cdots & \omega^{2N-2} \\ \vdots & \vdots & \vdots & & \vdots \\ 1 & \omega^{N-1} & \omega^{2N-2} & \cdots & \omega^{(N-1)(N-1)} \end{bmatrix}$$

is expressed in terms of powers of

$$\omega := e^{-2\pi i/N} = \cos(2\pi/N) - i\,\sin(2\pi/N).$$

We will use indices $0, 1, \ldots, N-1$ (rather than $1, 2, \ldots, N$) for the rows of vectors and for the rows and columns of matrices. When it is necessary, we will use a subscript to specify the size of a matrix, e.g., $\mathbf{I}_8$, $\mathcal{F}_{16}$ will denote the 8×8 identity matrix and the 16×16 DFT matrix, respectively.

Given an $N \times N$ matrix

$$\mathbf{A} := \{a_{kn}\}_{k,n=0}^{N-1}$$

and an N-vector

$$\mathbf{b} := \{b_n\}_{n=0}^{N-1},$$

we can evaluate the components of

$$\mathbf{c} := \mathbf{Ab} := \left\{ \sum_{n=0}^{N-1} a_{kn} b_n \right\}_{k=0}^{N-1}$$

by using the algorithm

> For $k = 0, 1, \ldots, N-1$ do:
> $\quad S := 0$
> $\quad$ For $n = 0, 1, \ldots, N-1$ do:
> $\quad\quad S := S + a_{kn} \cdot b_n$
> $\quad c_k := S.$

The cost of this computation is approximately N^2 operations when we define an *operation* to be the work we do as we execute the statement

$$S := S + a_{kn} \cdot b_n$$

from the inner loop. [More specifically, we fetch a_{kn}, b_n, and the "old" value of S from storage; we form the product $a_{kn} \cdot b_n$ and the sum $S + (a_{kn} \cdot b_n)$; and we store this result as the "new" value of S.] Of course, complex arithmetic requires more effort than real arithmetic, and by using the real-imaginary decomposition

$$\begin{aligned} S_R + i\,S_I &= S_R + i\,S_I + (a_R + i\,a_I)(b_R + i\,b_I) \\ &= \{(S_R + a_R \cdot b_R) - a_I \cdot b_I\} + i\{(S_I + a_R \cdot b_I) + a_I \cdot b_R\} \end{aligned}$$

we verify that

$$\text{1 complex operation} = \text{4 real operations.}$$

In this way we see that the naive matrix-vector computation of an N-point DFT requires approximately N^2 complex operations or $4N^2$ real operations.

Horner's algorithm for computing the DFT

In practice we always exploit the structure of $\mathcal{F}$ when we compute a DFT. (It makes no sense to generate and store the N^2 elements of $\mathcal{F}$ since each element is one of the N complex numbers $N^{-1}, N^{-1}\omega, \ldots, N^{-1}\omega^{N-1}$.) For example,

$$N\,F[k] = f[0] + f[1]\omega^k + f[2](\omega^k)^2 + \cdots + f[N-1](\omega^k)^{N-1}$$

is a polynomial with coefficients $f[0], f[1], \ldots, f[N-1]$ and argument $z := \omega^k$. This being the case, we can use Horner's algorithm to evaluate $N\,F[k]$, e.g., when $N = 4$ we write

$$4F[k] = f[0] + (\omega^k)\{f[1] + (\omega^k)\{f[2] + (\omega^k)f[3]\}\}$$

and evaluate $F[k]$ by computing in turn

$$z := e^{-2\pi i k/4}$$
$$S_1 := f[3]$$
$$S_2 := f[2] + z \cdot S_1$$
$$S_3 := f[1] + z \cdot S_2$$
$$S_4 := f[0] + z \cdot S_3$$
$$F[k] := S_4/4.$$

Algorithm 6.1 is a natural generalization. This *Horner algorithm* is easy to use, easy to code, and numerically stable. It requires approximately N^2 complex operations to produce an N-point DFT (just like the above matrix·vector algorithm).

$$z := 1$$
$$\omega := e^{-2\pi i/N}$$
For $k = 0, 1, \ldots, N-1$ do:
$\quad S := f[N-1]$
$\quad$ For $\ell = 2, 3, \ldots, N$ do:
$\quad\quad S := f[N-\ell] + z \cdot S$
$\quad F[k] := S/N$
$\quad z = z \cdot \omega$

Algorithm 6.1. Computation of the DFT with Horner's scheme.

Other preFFT methods for computing the DFT

The numerical task of evaluating a real trigonometric polynomial lies at the heart of applied Fourier analysis, *cf.* Ex. 6.2, and many schemes for carrying out such computations were devised in the century and a half that links Fourier to the development of digital computers. Kelvin, Michelson and Stratton, and others invented special-purpose mechanical analog computers for this purpose near the close of the 19th century, *cf.* Ex. 1.45. The most influential numerical analysis text from the first half of the 20th century included well-designed flow charts for 12-point and for 24-point harmonic analysis, *cf.* Ex. 6.4 and Appendix 4. Physicists confronted with the task of analyzing X-ray diffraction data devised a clever paper strip procedure for finding Fourier coefficients by adding parallel columns of integers, *cf.* Ex. 6.5. Such methods proved to be satisfactory in cases where there was a small fixed N and where only 2- to 3-digit accuracy was required. For more exacting calculations, a real version of the Horner algorithm was developed, *cf.* Ex. 6.1, and this was the accepted standard prior to the development of the Cooley-Tukey algorithm.

How big is $4N^2$?

You cannot appreciate the scientific revolution that was initiated by the FFT until you understand what it means to *pay* $4N^2$ real operations to *purchase* an N-point DFT at *Horner's market*. You should have no difficulty doing the arithmetic for one 3-digit real operation $S := S + a \cdot b$ in 10^2 sec. (This allows plenty of time for you to check your work.) If you could sustain this rate of computation, you could generate a 12-point DFT in about

$$4 \cdot 12^2 \text{ operations} \cdot \frac{10^2 \text{ sec}}{\text{operation}} \cdot \frac{1 \text{ min}}{60 \text{ sec}} \cdot \frac{1 \text{ hr}}{60 \text{ min}} = 16 \text{ hr.}$$

What would motivate *you* to carry out such a task? (You have seen the problem that motivated Gauss in Ex. 1.19!)

Digital computers changed the unit cost, but the curse of $4N^2$ remained. In the 1950s a digital computer that was capable of performing 10^3 operations/sec could do a 100-point DFT in about

$$4 \cdot 100^2 \text{ operations} \cdot \frac{1 \text{ sec}}{10^3 \text{ operations}} = 40 \text{ sec,}$$

but a 1000-point DFT took 4000 sec = 1.1 hr. Today a PC that is capable of 10^7 operations/sec can do a 1000-point DFT in about

$$4 \cdot 1000^2 \text{ operations} \cdot \frac{1 \text{ sec}}{10^7 \text{ operations}} = .4 \text{ sec.}$$

This seems fast, but if we use Horner's algorithm to compute one such DFT for each frame of a 1000-frame movie (*cf.* Section 9.5) we must wait 400 sec = 6.7 min to see the result.

The announcement of a fast algorithm for the DFT

In 1965 James W. Cooley and John W. Tukey published a new and substantially faster algorithm for computing the DFT of an N-vector on a digital computer. They showed that when N is a composite integer with the factorization

$$N = P_1 P_2 \cdots P_m$$

(where $P_1, P_2, \ldots, P_m$ are chosen from the integers $2, 3, 4, \ldots$), then it is possible to reduce the cost for computing the DFT of an N-vector from

$$N^2 = N \cdot \{P_1\ P_2 \cdots P_m\} \quad \text{to} \quad N \cdot \{(P_1 - 1) + (P_2 - 1) + \cdots + (P_m - 1)\}$$

complex operations. The cost reduction is most dramatic in cases where $P_1 = P_2 = \cdots = P_m = 2$) when we pass from

$$N^2 = 2^{2m} \quad \text{to} \quad 2^m \cdot m = N \log_2 N$$

complex operations. For example, when $N = 1024 = 2^{10}$ we reduce the cost from

$$N^2 = 1,048,576 \quad \text{to} \quad N \log_2 N = 10,240$$

complex operations. The new algorithm cut the price of a 1024-point DFT calculation by a factor of 100! Such dramatic reductions in computational cost made it practical to do Fourier analysis on a digital computer, and this helps to explain why

J.W. Cooley and J.W. Tukey, An algorithm for the machine computation of complex Fourier series, *Math. Comp.* **19**(1965), 297–301

is the most frequently cited mathematics paper that has ever been written.

In the next section we will give an elementary derivation of the FFT using the rules from the DFT calculus. Later on we will show how to derive the FFT by factoring the matrix $\boldsymbol{\mathcal{F}}$. Alternative derivations are developed in Exs. 6.24, and 6.25. Perhaps the most natural is that of Gauss. It is now clear that the most important algorithm of the 20th century was created by Gauss in 1805, published in his collected works in 1866, and completely forgotten until a decade after the appearance of the Cooley-Tukey algorithm, when Herman Goldstine, who was writing a history of numerical analysis, discovered Gauss's terse (neo-Latin!) description of a fast way to compute a DFT, *cf.* M.T. Heideman, D.H. Johnson, and C.S. Burrus, Gauss and history of the fast Fourier transform, *Arch. Hist. Exact Sci.* **34**(1985), 265–277.

6.2 Derivation of the FFT via DFT Rules

Decimation-in-time

From your study of the DFT calculus you will recall that when

$$(a, b, c, d) \quad \text{has the DFT} \quad (A, B, C, D)$$

we can use the translation rule (4.50) to see that

$$(d, a, b, c) \quad \text{has the DFT} \quad (A, \omega B, \omega^2 C, \omega^3 D), \quad \omega := e^{-2\pi i/4},$$

and we can use the zero packing rule (4.55) to see that

$$(a, 0, b, 0, c, 0, d, 0) \quad \text{has the DFT} \quad \tfrac{1}{2}(A, B, C, D, A, B, C, D).$$

The same relations are expressed by the commuting diagrams of Fig. 6.1. We use a 4-component initial vector to illustrate the action of the translation, exponential modulation, zero packing, and repeat operators $\mathfrak{T}_{-1}$, $\mathfrak{E}_{-1}$, $\mathfrak{Z}_2$, $\mathfrak{R}_2$. The modulation parameter must be changed to $\omega := e^{-2\pi i/N}$ when $\mathfrak{E}_{-1}$ is applied to an N-component vector.

$$\begin{array}{ccc}
(a, b, c, d) & \xrightarrow{\ \mathfrak{F}\ } & (A, B, C, D) \\
\Big\downarrow \mathfrak{T}_{-1} & & \Big\downarrow \mathfrak{E}_{-1} \\
(d, a, b, c) & \xrightarrow{\ \mathfrak{F}\ } & (A, \omega B, \omega^2 C, \omega^3 D), \ \omega := e^{-2\pi i/4}
\end{array}$$

$$\begin{array}{ccc}
(a, b, c, d) & \xrightarrow{\ \mathfrak{F}\ } & (A, B, C, D) \\
\Big\downarrow \mathfrak{Z}_2 & & \Big\downarrow \tfrac{1}{2}\mathfrak{R}_2 \\
(a, 0, b, 0, c, 0, d, 0) & \xrightarrow{\ \mathfrak{F}\ } & \tfrac{1}{2}(A, B, C, D, A, B, C, D)
\end{array}$$

Figure 6.1. Commuting diagrams for the translation and zero packing rules.

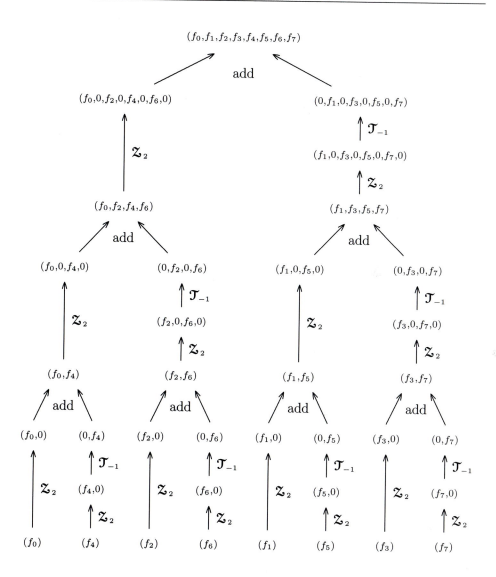

Figure 6.2. Assembly of $(f_0, f_1, \ldots, f_7)$ from $(f_0), \ldots, (f_7)$.

Figure 6.2 shows how we can assemble an 8-component vector $(f_0, f_1, \ldots, f_7)$ from the eight 1-component vectors

$$(f_0), \ (f_4), \ (f_2), \ (f_6), \ (f_1), \ (f_5), \ (f_3), \ (f_7)$$

by using the mappings $\mathcal{Z}_2$, $\mathcal{T}_{-1}$ together with vector addition. We will now replace every vector and every operator in Fig. 6.2 with its Fourier transform, *cf.* Fig. 6.3.

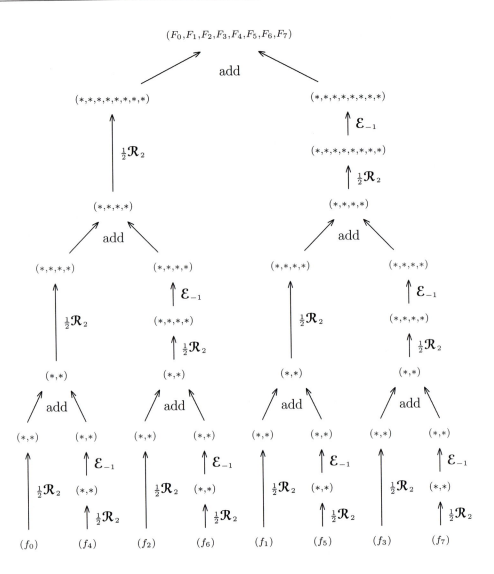

Figure 6.3. Generation of $(F_0, F_1, \ldots, F_7)$ from $(f_0), \ldots, (f_7)$.

The resulting diagram shows how to generate the DFT

$$(F_0, F_1, \ldots, F_7) := (f_0, f_1, \ldots, f_7)^\wedge$$

from

$$(f_0)^\wedge = (f_0), \ (f_4)^\wedge = (f_4), \ \ldots, \ (f_3)^\wedge = (f_3), \ (f_7)^\wedge = (f_7)$$

by using the operators $\mathcal{R}_2$ and $\mathcal{E}_{-1}$ together with vector addition. (We use an asterisk, $*$, to denote vector components from the intermediate stages of this process.)

It is easy to analyze the work required to compute a discrete Fourier transform in this manner. We must perform $N - 1$ complex multiplications when we apply $\mathcal{E}_{-1}$ to an N-vector, so the 7 appearances of $\mathcal{E}_{-1}$ in Fig. 6.3 correspond to

$$(8 - 1) + 2 \cdot (4 - 1) + 4 \cdot (2 - 1) = 8 \cdot 3 - 7$$

complex multiplications. We can perform a single real scaling by $1/8$ at the beginning or end of the process instead of repeatedly applying the factors of $1/2$, and this takes 16 real multiplications. The 7 vector additions correspond to

$$8 + 2 \cdot 4 + 4 \cdot 2 = 8 \cdot 3$$

complex additions. Since the effort required to apply $\mathcal{R}_2$ is negligible, the DFT can be computed by expending approximately $8 \cdot 3$ complex operations.

The natural generalization of this analysis shows that it is possible to compute the DFT of an N-vector with no more than $N \log_2 N$ complex operations when $N = 2^m$, $m = 1, 2, \ldots$. The new scheme is $N^2/(N \log_2 N)$ times faster than the naive algorithms from the preceding section, and this explains why we say that Fig. 6.3 gives a *fast Fourier transform* or FFT.

Decimation-in-frequency

In practice the components of the data vector are often samples of some time-varying signal, so the index n used with f measures *time* and the index k used with $f^\wedge$ measures *frequency*, cf. Ex. 1.17. For this reason the FFT of Fig. 6.3 is said to be based on a *decimation-in-time*, as shown in Fig. 6.2. We will now develop an alternative scheme that is based on a *decimation-in-frequency*.

You will recall that when

$$(a, b, c, d, e, f, g, h) \quad \text{has the DFT} \quad (A, B, C, D, E, F, G, H)$$

we can use the modulation rule (4.51) to see that

$$(a, \omega b, \omega^2 c, \omega^3 d, \omega^4 e, \omega^5 f, \omega^6 g, \omega^7 h) \quad \text{has the DFT} \quad (B, C, D, E, F, G, H, A),$$
$$\omega := e^{-2\pi i/8},$$

and we can use the summation rule (4.57) to see that

$$\tfrac{1}{2}(a + e, b + f, c + g, d + h) \quad \text{has the DFT} \quad (A, C, E, G),$$

cf. Fig. 6.4. We use an 8-component initial vector to illustrate the action of the modulation, translation, summation, and decimation operators $\mathcal{E}_{-1}$, $\mathcal{J}_1$, $\mathbf{\Sigma}_4$, and $\mathbf{\Xi}_2$.

$$(a,b,c,d,e,f,g,h) \quad \xrightarrow{\quad \mathcal{F} \quad} \quad (A,B,C,D,E,F,G,H)$$

$$\downarrow \mathcal{E}_{-1} \qquad\qquad\qquad\qquad \downarrow \mathcal{T}_1$$

$$(a, \omega b, \omega^2 c, \omega^3 d, \omega^4 e, \omega^5 f, \omega^6 g, \omega^7 h), \quad \xrightarrow{\quad \mathcal{F} \quad} \quad (B,C,D,E,F,G,H,A)$$

$$\omega := e^{-2\pi i/8}$$

$$(a,b,c,d,e,f,g,h) \quad \xrightarrow{\quad \mathcal{F} \quad} \quad (A,B,C,D,E,F,G,H)$$

$$\downarrow \tfrac{1}{2}\Sigma_4 \qquad\qquad\qquad\qquad \downarrow \Xi_2$$

$$\tfrac{1}{2}(a+e, b+f, c+g, d+h) \quad \xrightarrow{\quad \mathcal{F} \quad} \quad (A,C,E,G)$$

Figure 6.4. Commuting diagrams for the modulation and summation rules.

Figure 6.5 shows how we can disassemble an 8-component vector $(F_0, F_1, \ldots, F_8) := f^\wedge$ into the eight 1-component vectors

$$(F_0), (F_4), (F_6), (F_2), (F_1), (F_5), (F_3), (F_7)$$

by using the mappings Ξ_2 and $\mathcal{T}_1$. We will now replace every vector and every operator in Fig. 6.5 with its inverse Fourier transform. The resulting diagram of Fig. 6.6 shows how to generate

$$(F_0) = (F_0)^\wedge, (F_4) = (F_4)^\wedge, \ldots, (F_3) = (F_3)^\wedge, (F_7) = (F_7)^\wedge$$

from the data vector

$$(f_0, f_1, \ldots, f_7)$$

by using the operators Σ_4, Σ_2, Σ_1, $\mathcal{E}_{-1}$. (We again use an asterisk, $*$, to denote vector components from the intermediate stages of this process.)

You should have no difficulty showing that we expend no more than $8 \cdot 3$ complex operations as we compute an 8-point DFT by using the mappings of Fig. 6.6 (provided that we perform a single scaling by $1/8$ to account for all of the $1/2$'s). The natural generalization of this analysis allows us to compute the DFT of an N-vector with no more than $N \log_2 N$ complex operations when $N = 2^m$, $m = 1, 2, \ldots$. We now have *two* fast ways to produce a DFT!

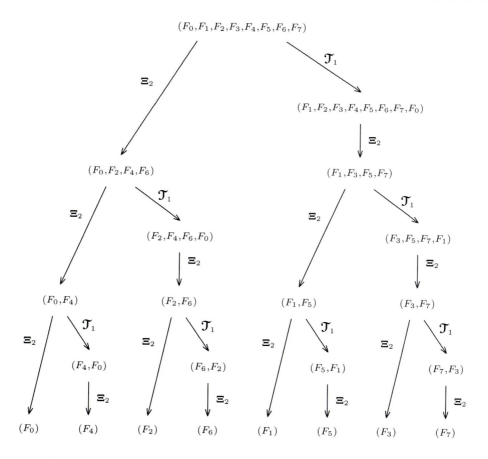

Figure 6.5. Disassembly of $(F_0, F_1, \ldots, F_7)$ into $(F_1), \ldots, (F_7)$.

Recursive algorithms

The computational trees of Figs. 6.3 and 6.6 are formed by suitably connecting a number of identical mapping structures as shown in Fig. 6.7. This being the case, you can write code for *one* of these building blocks and then use this code *recursively* to calculate a DFT. Exercises 6.6 and 6.7 will help you sort out the details.

Such schemes lead to stable algorithms. If we work on a computer having unit roundoff ϵ, our computed approximation F_k^a to F_k satisfies a bound of the form

$$|F_k^a - F_k| \leq (2 \, \log_2 N + 3) \cdot \max_{0 \leq n < N} |f_n| \cdot \epsilon + O(\epsilon^2), \quad k = 0, 1, \ldots, N - 1.$$

In contrast, when we use the Horner algorithm we find

$$|F_k^a - F_k| \leq \{(3/2)N + 2\} \cdot \max_{0 \leq n < N} |f_n| \cdot \epsilon + O(\epsilon^2), \quad k = 0, 1, \ldots, N - 1$$

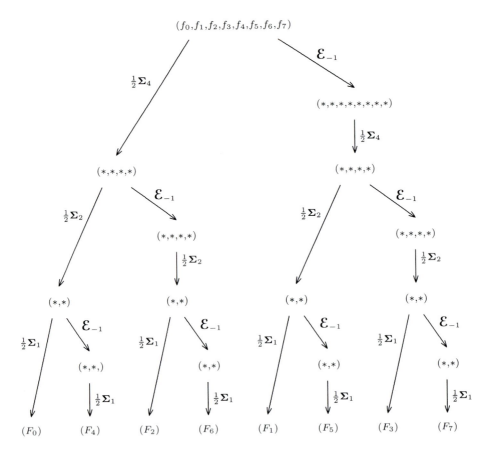

Figure 6.6. Generation of $(F_0), \ldots, (F_7)$ from $(f_0, f_1, \ldots, f_7)$.

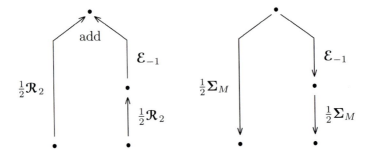

Figure 6.7. The mapping structures that give the trees of Fig. 6.3 and Fig. 6.6.

(*cf.* P. Henrici, *Applied and Computational Complex Analysis, Vol. 3*, John Wiley & Sons, New York, 1986, pp. 9–14). This analysis suggests (and experience confirms) that the FFT gives more accurate results than the slow algorithms described earlier.

If you work in a computing environment that does not permit recursion or if you want to write a somewhat more efficient nonrecursive code, you must create data structures to keep track of the vectors that appear in the large trees that generalize those of Figs. 6.3 and 6.6. We will presently introduce matrix factorizations to simplify this process.

6.3 The Bit Reversal Permutation

Introduction

We encounter the permutation

$$(f_0, f_1, f_2, f_3, f_4, f_5, f_6, f_7) \rightarrow (f_0, f_4, f_2, f_6, f_1, f_5, f_3, f_7)$$

when we use the assembly of Fig. 6.2 or the disassembly of Fig. 6.5. If we express the subscripts in base 2 notation, we see why this mapping is known as the *bit reversal permutation*.

$$
\begin{aligned}
0 &= (000)_2 \rightarrow (000)_2 = 0 \\
1 &= (001)_2 \rightarrow (100)_2 = 4 \\
2 &= (010)_2 \rightarrow (010)_2 = 2 \\
3 &= (011)_2 \rightarrow (110)_2 = 6 \\
4 &= (100)_2 \rightarrow (001)_2 = 1 \\
5 &= (101)_2 \rightarrow (101)_2 = 5 \\
6 &= (110)_2 \rightarrow (011)_2 = 3 \\
7 &= (111)_2 \rightarrow (111)_2 = 7
\end{aligned}
$$

It is easy to see how this mapping originates. At the first stage of the process we map

$$(f_0, f_1, f_2, f_3, f_4, f_5, f_6, f_7) \quad \text{to} \quad (f_0, f_2, f_4, f_6, f_1, f_3, f_5, f_7).$$

In so doing we right cyclically permute the three base 2 index bits, placing the low-order bit (0 for an even index, 1 for an odd index) in the high-order position (0 for the first half, 1 for the second half), *i.e.*, $f[(b_3 b_2 b_1)_2]$ is placed in position $(b_1 b_3 b_2)_2$. At the second stage we map

$$(f_0, f_2, f_4, f_6, f_1, f_3, f_5, f_7) \quad \text{to} \quad (f_0, f_4, f_2, f_6, f_1, f_5, f_3, f_7).$$

The net effect this time is to right cyclically permute (*i.e.*, interchange) the two low-order index bits. Thus we move the component $f[(b_3b_2b_1)_2]$ that we previously placed in position $(b_1b_3b_2)_2$ to position $(b_1b_2b_3)_2$.

Analogously, when $N = 16$, we successively move the component $f[(b_4b_3b_2b_1)_2]$ from its original position $(b_4b_3b_2b_1)_2$ to positions $(b_1b_4b_3b_2)_2$, $(b_1b_2b_4b_3)_2$, and $(b_1b_2b_3b_4)_2$. The process generalizes, and when $N = 2^m$, $m = 1, 2, \ldots$, the overall effect is to move $f[(b_mb_{m-1}\cdots b_2b_1)_2]$ to position $(b_1b_2\cdots b_{m-1}b_m)_2$ for every choice of $b_1, b_2, \ldots, b_m = 0, 1$. Of course, when $r = (b_1b_2\cdots b_m)_2$ is the bit-reversed form of $n = (b_mb_{m-1}\cdots b_1)_2$, then n is also the bit-reversed form of r, so the permutation can be carried out simply by swapping the values of $f[r]$ and $f[n]$. The positions of $f[0] = f[(0\ 0\ \cdots\ 0)_2]$ and $f[N-1] = f[(1\ 1\ \cdots\ 1)_2]$ are never changed by this process.

We will now develop several algorithms for performing the bit-reversal permutation. This discussion will help you understand the corresponding section of an FFT code (*e.g.*, the lines between statements 8 and 20 in the FFT code given in Appendix 5). If you have no interest in such details, you can skip the remainder of this section.

A naive algorithm

Let $N = 2^m$ for some $m = 1, 2, \ldots$. Elementary algorithms for applying the bit reversal permutation to an N-vector $\mathbf{f}$ often have the following structure.

> For $n = 1, 2, \ldots, N - 2$ do:
> > Find the integer $r = (b_1b_2\cdots b_{m-1}b_m)_2$ that
> > corresponds to $n = (b_mb_{m-1}\cdots b_2b_1)_2$
> > If $r > n$, then swap $f[r]$ and $f[n]$

When we are given an index

$$n = (b_mb_{m-1}\ldots b_2b_1)_2 = b_1 + b_2 \cdot 2 + b_3 \cdot 2^2 + \cdots + b_m \cdot 2^{m-1}$$

we can generate in turn the bits $b_1, b_2, b_3, \ldots$ and compute the corresponding Horner sequence $b_1, 2b_1 + b_2, 2(2b_1 + b_2) + b_3, \ldots$ for

$$r = (b_1b_2\ldots b_{m-1}b_m)_2 = b_m + b_{m-1} \cdot 2 + b_{m-2} \cdot 2^2 + \cdots + b_1 \cdot 2^{m-1}.$$

For example, when $m = 5$ and $n = 11 = (01011)_2$, this leads to the following calculation of $r = (11010)_2 = 26$.

$$11 \div 2 = 5 \text{ with remainder } b_1 := 1, \quad r_1 := 1$$
$$5 \div 2 = 2 \text{ with remainder } b_2 := 1, \quad r_2 := 2r_1 + b_2 = 3$$
$$2 \div 2 = 1 \text{ with remainder } b_3 := 0, \quad r_3 := 2r_2 + b_3 = 6$$
$$1 \div 2 = 0 \text{ with remainder } b_4 := 1, \quad r_4 := 2r_3 + b_4 = 13$$
$$0 \div 2 = 0 \text{ with remainder } b_5 := 0, \quad r_5 := 2r_4 + b_5 = 26$$

In this way we see that Algorithm 6.2 performs the bit reversal permutation. The cost is approximately $3N \log_2 N$ integer operations.

For $n = 1, 2, \ldots, N - 2$ do:
 $\quad r := 0$
 $\quad d := n$
 $\quad$ For $k = 1, 2, \ldots, m$ do:
 $\qquad q := \lfloor d/2 \rfloor$
 $\qquad b := d - 2q$
 $\qquad r := 2r + b$
 $\qquad d := q$
 $\quad$ If $r > n$, then swap $f[r]$ and $f[n]$

Algorithm 6.2. A naive scheme for applying the bit reversal permutation to $f[0], f[1], \ldots, f[N-1]$ when $N = 2^m$.

The reverse carry algorithm

When we execute Algorithm 6.2, each computation of the bit-reversed index $r = r[n]$ from the index n is done *ab initio*. Since we compute $r[1], r[2], \ldots, r[N-2]$ in turn, there is some advantage to be gained from a recursive scheme that uses a known value of $r[n]$ to facilitate the computation of $r[n+1]$. If n is even, this can be done with a simple addition as illustrated in the following calculation of $r[23] = 29$ from $r[22] = 13$ when $N = 2^5$.

$$
\begin{array}{ccc}
 & \text{Mirror} & \\
n = (10110)_2 & \vdots & (01101)_2 = r[n] \\
+1 = (00001)_2 & \vdots & +(10000)_2 = N/2 \\
\hline
n+1 = (10111)_2 & \vdots & (11101)_2 = r[n+1]
\end{array}
$$

Indeed, when $r[n] < N/2$ we always have

$$r[n+1] = r[n] + N/2.$$

If n is odd, (*i.e.* $r[n] \geq N/2$), it is still possible to generate $r[n+1]$ from $r[n]$, but we must now mirror the *carrying process* associated with addition in base 2 arithmetic as illustrated in the following calculation of $r[12] = 6$ from $r[11] = 26$ when $N = 2^5$.

Mirror

$$n = (01011)_2 \qquad (11010)_2 = r[n]$$

$$+1 = (00001)_2 \qquad + (10000)_2 = N/2$$

carry $\quad \uparrow \qquad\qquad \uparrow \qquad$ reverse carry

Mirror

$$n - 1 = (01010)_2 \qquad (01010)_2 = r[n] - N/2$$

$$+2 = (00010)_2 \qquad + (01000)_2 = N/4$$

carry $\quad \uparrow \qquad\qquad \uparrow \qquad$ reverse carry

Mirror

$$n - 1 - 2 = (01000)_2 \qquad (00010)_2 = r[n] - N/2 - N/4$$

$$+4 = (00100)_2 \qquad + (00100)_2 = N/8$$

$$n + 1 = (01100)_2 \qquad (00110)_2 = r[n + 1]$$

Algorithm 6.3 uses this *reverse carry* process to perform the bit reversal permutation. It costs approximately $4N$ integer additions (including those used for comparisons), N divisions by 2, and slightly less than $N/2$ swaps to execute this algorithm.

$$r := 0$$

For $n = 1, 2, \ldots, N - 2$ do:

$\qquad k := N/2$

$\qquad$ While $r \geq k$ do:

$\qquad\qquad r := r - k$

$\qquad\qquad k := k/2$

$\qquad r := r + k$

$\qquad$ If $r > n$, then swap $f[r]$ and $f[n]$

Algorithm 6.3. The reverse carry scheme for applying the bit reversal permutation to $f[0], f[1], \ldots, f[N - 1]$ when $N = 2^m$.

The Bracewell-Buneman algorithm

The reverse carry algorithm for generating a complete set of n, r pairs for a given $N = 2^m$ is based on the idea of generating $r[n+1]$ from $r[n]$. We will now show that it is possible to produce a more efficient algorithm by using storage to gain speed.

Let $m = 1, 2, \ldots$. For each $n = 0, 1, 2, \ldots, 2^m - 1$ we form $r_m[n]$ by reversing the bits of n so that

$$r_m[(b_m b_{m-1} \ldots b_2 b_1)_2] := (b_1 b_2 \ldots b_{m-1} b_m)_2,$$

e.g., $r_2[3] = (11)_2 = 3$, $r_3[3] = (110)_2 = 6$, $r_4[3] = (1100)_2 = 12, \ldots$. Now since

$$\begin{aligned}
r_{m+1}[(b_{m+1} b_m \ldots b_2 b_1)_2] &= (b_1 b_2 \ldots b_m b_{m+1})_2 \\
&= 2 \cdot (b_1 b_2 \ldots b_m)_2 + b_{m+1} \\
&= 2 r_m[(b_m b_{m-1} \ldots b_2 b_1)_m] + b_{m+1},
\end{aligned}$$

we can write

$$r_{m+1}[n] = \begin{cases} 2 r_m[n] & \text{if } n = 0, 1, \ldots, 2^m - 1 \\ 2 r_m[n - 2^m] + 1 & \text{if } n = 2^m, 2^m + 1, \ldots, 2^{m+1} - 1. \end{cases}$$

This identity shows that we can get the *left half* of row $m + 1$ from the table

n:	0	1	2	3	4	5	6	7	8	9	10	11	12	13	14	15
$r_1[n]$:	0	1														
$r_2[n]$:	0	2	1	3												
$r_3[n]$:	0	4	2	6	1	5	3	7								
$r_4[n]$:	0	8	4	12	2	10	6	14	1	9	5	13	3	11	7	15

by doubling row m, and we can then get the *right half* of row $m + 1$ by adding 1 to each component of the left half. This idea underlies the following algorithm for generating the bit-reversed indices $r_m[0], r_m[1], \ldots, r_m[N - 1]$ when $N = 2^m$.

$$r[0] := 0$$
$$M := 1$$

While $M < N$ do:

> For $k = 0, 1, \ldots, M - 1$ do:
>
> > $T := 2r[k]$
> >
> > $r[k] := T$
> >
> > $r[k + M] := T + 1$
>
> $M := 2M$

After initializing the array r, we can perform the bit reversal permutation using

> For $n = 1, 2, \ldots, N - 2$ do:
> └ If $r[n] > n$, then swap $f[n]$ and $f[r[n]]$.

There is really no reason why we should fetch the bit-reversed index for each $n = 1, 2, \ldots, N - 2$ since we only perform a swap when $r_m[n] > n$. This observation led R. Bracewell and O. Buneman to devise a more efficient scheme that generates only those bit-reversed pairs n, r for which $r > n$. The algorithm gains additional speed by using a small integer storage array with at most $\sqrt{2N}$ components.

The Bracewell-Buneman algorithm uses a certain left-right decomposition of the base 2 representations of n, r. We illustrate the idea with $N = 2^6$. A 6-bit index n has the representation

$$n := (b_6 b_5 b_4 b_3 b_2 b_1)_2 = 8p + q$$

where

$$p := (b_6 b_5 b_4)_2, \quad q := (b_3 b_2 b_1)_2.$$

It follows that

$$r_6[n] = (b_1 b_2 b_3 b_4 b_5 b_6)_2 = 8r_3[q] + r_3[p].$$

We want to find the 6-bit reversed pairs n, r with $r > n$. In view of this decomposition, all such pairs can be obtained from 3-bit integers p, q with

$$8r_3[q] + r_3[p] > 8p + q$$

or equivalently, with

$$r_3[q] > p.$$

(In the case where $r_3[q] = p$, we also have $r_3[p] = q$ so that $r_6[n] = n$.) It follows that every bit-reversed pair n, r with $r > n$ occurs precisely once in the list

$$n = 8p + q, \quad r = 8r_3[q] + r_3[p] : q = 1, 2, \ldots, 7 \text{ and } p = 0, 1, \ldots, r_3[q] - 1.$$

More generally, when $N = 2^m$ and $m = 2\mu$ is even, we obtain every bit-reversed pair n, r with $r > n$ precisely once in the list

$$n = 2^\mu p + q, \quad r = 2^\mu r_\mu[q] + r_\mu[p] : q = 1, 2, \ldots, 2^\mu - 1 \text{ and } p = 0, 1, \ldots, r_\mu[q] - 1.$$

This left-right representation of n, r must be modified slightly when there is an odd number of index bits. For example, when $N = 2^7$ we write

$$n := (b_7 b_6 b_5 b_4 b_3 b_2 b_1)_2 = 8p + q$$

where

$$p := (b_7 b_6 b_5 b_4)_2, \quad q := (b_3 b_2 b_1)_2, = (0 b_3 b_2 b_1)_2,$$

and observe that

$$r_7[n] = (b_1 b_2 b_3 b_4 b_5 b_6 b_7)_2 = 16 r_3[q] + r_4[p] = 8 r_4[q] + r_4[p].$$

Thus every bit-reversed pair n, r with $r > n$ occurs precisely once in the list

$$n = 8p + q, \quad r = 8 r_4[q] + r_4[p] : q = 1, 2, \ldots, 7, \; p = 0, 1, \ldots, r_4[q] - 1.$$

More generally, when $N = 2^m$ and $m = 2\mu + 1$ is odd, we obtain every bit-reversed pair n, r with $r > n$ precisely once in the list

$$n = 2^\mu p + q, \quad r = 2^\mu r_{\mu+1}[q] + r_{\mu+1}[p] : q = 1, 2, \ldots, 2^\mu - 1, \; p = 0, 1, \ldots, r_{\mu+1}[q] - 1.$$

These even m and odd m representations for n, r are used to derive the following algorithm for the bit reversal permutation when $N = 2^m$ with $m = 2\mu + \lambda$ for some $\mu = 1, 2, \ldots$ and some $\lambda = 0, 1$.

> For $q = 1, 2, \ldots, 2^\mu - 1$ do:
>> For $p = 0, 1, \ldots, r_{\mu+\lambda}[q] - 1$ do:
>>> $n' := 2^\mu p + q$
>>>
>>> $r' := 2^\mu r_{\mu+\lambda}[q] + r_{\mu+\lambda}[p]$
>>>
>>> Swap $f[n']$ and $f[r']$

To avoid the repeated computation of $r_{\mu+\lambda}[0], r_{\mu+\lambda}[1], \ldots, r_{\mu+\lambda}[2^\mu - 1]$, we can generate these indices with the efficient double-add one algorithm and store them in an auxiliary integer array (with $2^{\mu+\lambda} \leq \sqrt{2N}$ components) as part of the initialization process. We can also improve efficiency by doing part of the computation of n', r' outside the inner loop. In this way we obtain the exceptionally fast Bracewell-Buneman Algorithm 6.4 for the bit reversal permutation. After sorting out the details of this algorithm you should have no difficulty reading the corresponding lines of the FFT code from Appendix 5.

$$\mu_+ := \lfloor (m+1)/2 \rfloor \qquad (\text{so } \mu_+ := \mu + \lambda)$$

$$M := 1$$

$$r[0] := 0$$

For $\nu = 1, 2, \ldots, \mu_+$ do:

> For $k = 0, 1, \ldots, M-1$ do:
>
> > $T := 2r[k]$
> >
> > $r[k] := T$
> >
> > $r[k+M] := T+1$
>
> $M := M + M$

If m is odd, then $M := M/2$ \qquad $(\text{so } M := 2^\mu)$

For $q = 1, 2, \ldots, M-1$ do:

> $n' := q - M$
>
> $r'' := r[q] \cdot M$
>
> For $p = 0, 1, \ldots, r[q] - 1$ do:
>
> > $n' := n' + M$ \qquad $(\text{so } n' := Mp + q)$
> >
> > $r' = r'' + r[p]$ \qquad $(\text{so } r' := Mr[q] + r[p])$
> >
> > Swap $f[n']$ and $f[r']$

Algorithm 6.4. The exceptionally efficient Bracewell-Buneman scheme for applying the bit reversal permutation to $f[0], f[1], \ldots, f[N-1]$ when $N = 2^m$.

6.4 Sparse Matrix Factorization of $\mathcal{F}$ When $N = 2^m$

Introduction

In this section we will produce a factorization

$$\mathcal{F} = \mathcal{M}_m \, \mathcal{M}_{m-1} \cdots \mathcal{M}_1 \mathbf{B}$$

of the matrix $\mathcal{F}$ when $N = 2^m$. Each of the matrices $\mathbf{B}, \mathcal{M}_1, \ldots, \mathcal{M}_m$ will have a very small number of nonzero elements in each row. (Such matrices are said to be *sparse.*) With minimal effort we can then compute

$$\mathbf{f}_0 := \mathbf{B}\mathbf{f}, \ \ \mathbf{f}_1 := \mathcal{M}_1\mathbf{f}_0, \ \ \mathbf{f}_2 := \mathcal{M}_2\mathbf{f}_1, \ldots, \mathbf{f}_m := \mathcal{M}_m\mathbf{f}_{m-1},$$

and thereby obtain

$$\mathbf{f}_m = \mathcal{M}_m\mathcal{M}_{m-1}\cdots\mathcal{M}_1\mathbf{B}\mathbf{f} = \mathcal{F}\mathbf{f}.$$

There is no reason to save the string of vectors that we generate during this process, so we carry out the calculation with an *in-place* algorithm

$$\mathbf{f} := \mathbf{B}\mathbf{f}$$
$$\text{For } \mu = 1, 2, \ldots, m \text{ do:}$$
$$\quad \mathrel{\rotatebox[origin=c]{90}{$\lceil$}} \mathbf{f} := \mathcal{M}_\mu \mathbf{f}$$

that successively overwrites the original $\mathbf{f}$ with $\mathbf{f}_0, \mathbf{f}_1, \ldots, \mathbf{f}_m$. The matrices $\mathbf{B}, \mathcal{M}_1, \ldots, \mathcal{M}_m$ have simple structures that facilitate the writing of readable code.

The zipper identity

For clarity we will take $N = 8$ and set $\omega := e^{-2\pi i/8}$ as we derive the critical relation that gives the complete sparse factorization. Since the kth column of a matrix product $\mathbf{A}_1\mathbf{A}_2$ can be found by applying $\mathbf{A}_1$ to the kth column of $\mathbf{A}_2$, we easily verify that

$$\mathcal{F}_8 = \frac{1}{8}\left[\begin{array}{cccc|cccc} 1 & 1 & 1 & 1 & 1 & 1 & 1 & 1 \\ 1 & w^2 & w^4 & w^6 & w & w^3 & w^5 & w^7 \\ 1 & w^4 & w^8 & w^{12} & w^2 & w^6 & w^{10} & w^{14} \\ 1 & w^6 & w^{12} & w^{18} & w^3 & w^9 & w^{15} & w^{21} \\ \hline 1 & w^8 & w^{16} & w^{24} & w^4 & w^{12} & w^{20} & w^{28} \\ 1 & w^{10} & w^{20} & w^{30} & w^5 & w^{15} & w^{25} & w^{35} \\ 1 & w^{12} & w^{24} & w^{36} & w^6 & w^{18} & w^{30} & w^{42} \\ 1 & w^{14} & w^{28} & w^{42} & w^7 & w^{21} & w^{35} & w^{49} \end{array}\right]\left[\begin{array}{cccccccc} 1 & 0 & 0 & 0 & 0 & 0 & 0 & 0 \\ 0 & 0 & 1 & 0 & 0 & 0 & 0 & 0 \\ 0 & 0 & 0 & 0 & 1 & 0 & 0 & 0 \\ 0 & 0 & 0 & 0 & 0 & 0 & 1 & 0 \\ \hline 0 & 1 & 0 & 0 & 0 & 0 & 0 & 0 \\ 0 & 0 & 0 & 1 & 0 & 0 & 0 & 0 \\ 0 & 0 & 0 & 0 & 0 & 1 & 0 & 0 \\ 0 & 0 & 0 & 0 & 0 & 0 & 0 & 1 \end{array}\right].$$

Next, we use the identities

$$\omega^2 = e^{-2\pi i/4}, \quad \omega^4 = -1, \quad \omega^8 = 1$$

to express each of the 4×4 blocks from the left matrix factor in terms of $\mathcal{F}_4$, *i.e.*,

$$\frac{1}{4}\begin{bmatrix} 1 & 1 & 1 & 1 \\ 1 & \omega^2 & \omega^4 & \omega^6 \\ 1 & \omega^4 & \omega^8 & \omega^{12} \\ 1 & \omega^6 & \omega^{12} & \omega^{18} \end{bmatrix} = \mathcal{F}_4, \quad \frac{1}{4}\begin{bmatrix} 1 & 1 & 1 & 1 \\ \omega & \omega^3 & \omega^5 & \omega^7 \\ \omega^2 & \omega^6 & \omega^{10} & \omega^{14} \\ \omega^3 & \omega^9 & \omega^{15} & \omega^{21} \end{bmatrix} = \begin{bmatrix} 1 & & & \\ & \omega & & \\ & & \omega^2 & \\ & & & \omega^3 \end{bmatrix} \mathcal{F}_4,$$

$$\frac{1}{4}\begin{bmatrix} 1 & \omega^8 & \omega^{16} & \omega^{24} \\ 1 & \omega^{10} & \omega^{20} & \omega^{30} \\ 1 & \omega^{12} & \omega^{24} & \omega^{36} \\ 1 & \omega^{14} & \omega^{28} & \omega^{42} \end{bmatrix} = \mathcal{F}_4, \quad \frac{1}{4}\begin{bmatrix} \omega^4 & \omega^{12} & \omega^{20} & \omega^{28} \\ \omega^5 & \omega^{15} & \omega^{25} & \omega^{35} \\ \omega^6 & \omega^{18} & \omega^{30} & \omega^{42} \\ \omega^7 & \omega^{21} & \omega^{35} & \omega^{49} \end{bmatrix} = \begin{bmatrix} -1 & & & \\ & -\omega & & \\ & & -\omega^2 & \\ & & & -\omega^3 \end{bmatrix} \mathcal{F}_4,$$

with the missing matrix elements all being 0. In this way we produce the factorization

$$\mathcal{F}_8 = \frac{1}{2}\left[\begin{array}{cccc|cccc} 1 & & & & 1 & & & \\ & 1 & & & & \omega & & \\ & & 1 & & & & \omega^2 & \\ & & & 1 & & & & \omega^3 \\ \hline 1 & & & & -1 & & & \\ & 1 & & & & -\omega & & \\ & & 1 & & & & -\omega^2 & \\ & & & 1 & & & & -\omega^3 \end{array}\right] \left[\begin{array}{c|c} \mathcal{F}_4 & \mathbf{0}_4 \\ \hline \mathbf{0}_4 & \mathcal{F}_4 \end{array}\right] \begin{bmatrix} 1 & 0 & 0 & 0 & 0 & 0 & 0 & 0 \\ 0 & 0 & 1 & 0 & 0 & 0 & 0 & 0 \\ 0 & 0 & 0 & 0 & 1 & 0 & 0 & 0 \\ 0 & 0 & 0 & 0 & 0 & 0 & 1 & 0 \\ 0 & 1 & 0 & 0 & 0 & 0 & 0 & 0 \\ 0 & 0 & 0 & 1 & 0 & 0 & 0 & 0 \\ 0 & 0 & 0 & 0 & 0 & 1 & 0 & 0 \\ 0 & 0 & 0 & 0 & 0 & 0 & 0 & 1 \end{bmatrix}.$$

The same argument gives the *zipper identity*

$$\mathcal{F}_{2M} = \frac{1}{2}\mathbf{Q}_{2M}\left[\begin{array}{c|c} \mathcal{F}_M & \mathbf{0}_M \\ \hline \mathbf{0}_M & \mathcal{F}_M \end{array}\right] \mathbf{S}_{2M}, \qquad M = 1, 2, \dots \tag{1}$$

where

$$\mathbf{Q}_{2M} := \left[\begin{array}{ccccc|ccccc} 1 & & & & & 1 & & & & \\ & 1 & & & & & \omega & & & \\ & & 1 & & & & & \omega^2 & & \\ & & & \ddots & & & & & \ddots & \\ & & & & 1 & & & & & \omega^{M-1} \\ \hline 1 & & & & & -1 & & & & \\ & 1 & & & & & -\omega & & & \\ & & 1 & & & & & -\omega^2 & & \\ & & & \ddots & & & & & \ddots & \\ & & & & 1 & & & & & -\omega^{M-1} \end{array}\right] \quad \text{with } \omega := e^{-2\pi i/(2M)}, \tag{2}$$

and where we obtain the *shuffle permutation*

$$\mathbf{S}_{2M} := [\boldsymbol{\delta}_0, \boldsymbol{\delta}_M, \boldsymbol{\delta}_1, \boldsymbol{\delta}_{M+1}, \boldsymbol{\delta}_2, \boldsymbol{\delta}_{M+2}, \dots, \boldsymbol{\delta}_{M-1}, \boldsymbol{\delta}_{2M-1}] \tag{3}$$

by performing a perfect 2-shuffle of the $2M$ columns $\delta_0, \delta_1, \ldots \delta_{2M-1}$ of the identity

$$\mathbf{I}_{2M} := [\delta_0, \delta_1, \delta_2, \ldots, \delta_{M-1} \mid \delta_M, \delta_{M+1}, \ldots, \delta_{2M-1}]. \tag{4}$$

Exponent notation

Given an $M \times M$ matrix $\mathbf{A}$ and the $M \times M$ zero matrix $\mathbf{0}$ we define

$$\mathbf{A}^{(1)} := \mathbf{A}, \quad \mathbf{A}^{(2)} := \begin{bmatrix} \mathbf{A} & \mathbf{0} \\ \mathbf{0} & \mathbf{A} \end{bmatrix}, \quad \mathbf{A}^{(3)} := \begin{bmatrix} \mathbf{A} & \mathbf{0} & \mathbf{0} \\ \mathbf{0} & \mathbf{A} & \mathbf{0} \\ \mathbf{0} & \mathbf{0} & \mathbf{A} \end{bmatrix}, \ldots \tag{5}$$

thereby producing $M \times M$, $2M \times 2M$, $3M \times 3M$, ... block diagonal matrices with $1, 2, 3, \ldots$ copies of $\mathbf{A}$ along the diagonal. We easily verify the power rule

$$[\mathbf{A}^{(p)}]^{(q)} = \mathbf{A}^{(pq)}, \quad p, q = 1, 2, \ldots \tag{6}$$

and the product rule

$$[\mathbf{AB}]^{(p)} = \mathbf{A}^{(p)}\mathbf{B}^{(p)}, \quad p = 1, 2, \ldots \tag{7}$$

for this new *exponent*. In addition, we have

$$[\alpha\mathbf{A}]^{(p)} = \alpha\mathbf{A}^{(p)}, \quad p = 1, 2, \ldots \tag{8}$$

when α is any complex scalar,

$$[\mathbf{A}^T]^{(p)} = [\mathbf{A}^{(p)}]^T, \quad p = 1, 2, \ldots \tag{9}$$

(where the tag, ^{T}, denotes the matrix transpose) and

$$[\mathbf{A}^{-1}]^{(p)} = [\mathbf{A}^{(p)}]^{-1}, \quad p = 1, 2, \ldots \tag{10}$$

when $\mathbf{A}$ is nonsingular.

Sparse matrix factorization of $\mathcal{F}$

Using the exponent (5) we write the zipper identity (1) in the compact form

$$\mathcal{F}_{2M} = \frac{1}{2}\mathbf{Q}_{2M}\mathcal{F}_M^{(2)}\mathbf{S}_{2M}, \quad M = 1, 2, \ldots . \tag{11}$$

This identity makes it possible for us to factor $\mathcal{F}_N$ when $N = 2^m$. For example, we *unzip* $\mathcal{F}_{16}$ by using (11) and (6)–(8) to write

$$\mathcal{F}_{16} = \frac{1}{2}\mathbf{Q}_{16}\mathcal{F}_8^{(2)}\mathbf{S}_{16}$$

$$= \frac{1}{2}\mathbf{Q}_{16}\left[\frac{1}{2}\mathbf{Q}_8\mathcal{F}_4^{(2)}\mathbf{S}_8\right]^{(2)}\mathbf{S}_{16}$$

$$= \frac{1}{4}\mathbf{Q}_{16}\mathbf{Q}_8^{(2)}\mathcal{F}_4^{(4)}\mathbf{S}_8^{(2)}\mathbf{S}_{16}$$

$$= \frac{1}{4}\mathbf{Q}_{16}\mathbf{Q}_8^{(2)}\left[\frac{1}{2}\mathbf{Q}_4\mathcal{F}_2^{(2)}\mathbf{S}_4\right]^{(4)}\mathbf{S}_8^{(2)}\mathbf{S}_{16}$$

$$= \frac{1}{8}\mathbf{Q}_{16}\mathbf{Q}_8^{(2)}\mathbf{Q}_4^{(4)}\mathcal{F}_2^{(8)}\mathbf{S}_4^{(4)}\mathbf{S}_8^{(2)}\mathbf{S}_{16}$$

$$= \frac{1}{8}\mathbf{Q}_{16}\mathbf{Q}_8^{(2)}\mathbf{Q}_4^{(4)}\left[\frac{1}{2}\mathbf{Q}_2\mathcal{F}_1^{(2)}\mathbf{S}_2\right]^{(8)}\mathbf{S}_4^{(4)}\mathbf{S}_8^{(2)}\mathbf{S}_{16}$$

$$= \frac{1}{16}\mathbf{Q}_{16}\mathbf{Q}_8^{(2)}\mathbf{Q}_4^{(4)}\mathbf{Q}_2^{(8)}\mathcal{F}_1^{(16)}\mathbf{S}_2^{(8)}\mathbf{S}_4^{(4)}\mathbf{S}_8^{(2)}\mathbf{S}_{16}$$

$$= \frac{1}{16}\mathbf{Q}_{16}\mathbf{Q}_8^{(2)}\mathbf{Q}_4^{(4)}\mathbf{Q}_2^{(8)}\mathbf{B}_{16}$$

where

$$\mathbf{B}_{16} := \mathbf{S}_2^{(8)}\mathbf{S}_4^{(4)}\mathbf{S}_8^{(2)}\mathbf{S}_{16}$$

and we use the fact that

$$\mathcal{F}_1^{(16)} = [1]^{(16)} = \mathbf{I}_{16}.$$

Analogously,

$$\mathcal{F}_{2^m} = \frac{1}{2^m}\mathbf{Q}_{2^m}\mathbf{Q}_{2^{m-1}}^{(2)}\mathbf{Q}_{2^{m-2}}^{(4)}\cdots\mathbf{Q}_2^{(2^{m-1})}\mathbf{B}_{2^m}, \quad m = 1, 2, \ldots \tag{12}$$

where

$$\mathbf{B}_{2^m} := \mathbf{S}_2^{(2^{m-1})}\mathbf{S}_4^{(2^{m-2})}\cdots\mathbf{S}_{2^{m-1}}^{(2)}\mathbf{S}_{2^m}. \tag{13}$$

The action of $\mathbf{B}_{2^m}$

The shuffle permutation $\mathbf{S}_8$ maps $\mathbf{f} := (f_0, \ldots, f_7)^T$ to

$$\mathbf{S}_8\mathbf{f} = [\delta_0, \delta_4, \delta_1, \delta_5, \delta_2, \delta_6, \delta_3, \delta_7]\mathbf{f}$$

$$= f_0\,\delta_0 + f_1\,\delta_4 + f_2\,\delta_1 + f_3\,\delta_5 + f_4\,\delta_2 + f_5\,\delta_6 + f_6\,\delta_3 + f_7\,\delta_7$$

$$= (f_0, f_2, f_4, f_6, f_1, f_3, f_5, f_7)^T,$$

and the shuffle permutation $\mathbf{S}_4$ maps the half vectors $(f_0, f_2, f_4, f_6)^T$, $(f_1, f_3, f_5, f_7)^T$ from $\mathbf{S}_8\mathbf{f}$ to $(f_0, f_4, f_2, f_6)^T$, $(f_1, f_5, f_3, f_7)^T$, respectively. Since $\mathbf{S}_2 = \mathbf{I}_2$, we have

$$\mathbf{B}_8\mathbf{f} := \mathbf{S}_2^{(4)}\mathbf{S}_4^{(2)}\mathbf{S}_8\mathbf{f} = (f_0, f_4, f_2, f_6, f_1, f_5, f_3, f_7)^T,$$

i.e., $\mathbf{B}_8$ is the bit reversal permutation for 8-component vectors.

We can also follow the permutation process by keeping track of the index of the cell that contains the component of $\mathbf{f}$ that is originally found in cell n. For example, when $N = 16$ we find

$$f[(b_4b_3b_2b_1)_2] \overset{\mathbf{S}_{16}}{\nearrow} \text{cell } (b_1b_4b_3b_2)_2 \overset{\mathbf{S}_8^{(2)}}{\nearrow} \text{cell } (b_1b_2b_4b_3)_2 \overset{\mathbf{S}_4^{(4)}}{\nearrow} \text{cell } (b_1b_2b_3b_4)_2,$$

so

$$\mathbf{B}_{16} := \mathbf{S}_2^{(8)}\mathbf{S}_4^{(4)}\mathbf{S}_8^{(2)}\mathbf{S}_{16}$$

is the bit reversal permutation for the 16-component vectors. Analogously, we see that (13) is the bit reversal permutation for vectors with 2^m components, $m = 1, 2, \ldots$.

An FFT algorithm

The factorization (12) corresponds to a fast algorithm for computing the DFT of any vector of length $N = 2^m$. The in-place computation of

$$\begin{bmatrix} f_0 \\ f_1 \\ \vdots \\ f_{2M-1} \end{bmatrix} := \mathbf{Q}_{2M} \begin{bmatrix} f_0 \\ f_1 \\ \vdots \\ f_{2M-1} \end{bmatrix} := \left[\begin{array}{ccccc|ccccc} 1 & & & & & 1 & & & & \\ & 1 & & & & & \omega & & & \\ & & 1 & & & & & \omega^2 & & \\ & & & \ddots & & & & & \ddots & \\ & & & & 1 & & & & & \omega^{M-1} \\ \hline 1 & & & & & -1 & & & & \\ & 1 & & & & & -\omega & & & \\ & & 1 & & & & & -\omega^2 & & \\ & & & \ddots & & & & & \ddots & \\ & & & & 1 & & & & & -\omega^{M-1} \end{array}\right] \begin{bmatrix} f_0 \\ f_1 \\ f_2 \\ \vdots \\ f_{M-1} \\ \hline f_M \\ f_{M+1} \\ f_{M+2} \\ \vdots \\ f_{2M-1} \end{bmatrix}$$

(with $\omega := e^{-2\pi i/(2M)}$) can be done with the one-loop algorithm

For $\lambda = 0, 1, \ldots, M - 1$ do:

$$\left\lfloor\; \begin{bmatrix} f_\lambda \\ f_{\lambda+M} \end{bmatrix} := \begin{bmatrix} 1 & \omega^\lambda \\ 1 & -\omega^\lambda \end{bmatrix} \begin{bmatrix} f_\lambda \\ f_{\lambda+M} \end{bmatrix}. \right. \tag{14}$$

More generally, the in-place computation of

$$\mathbf{f} := \mathbf{Q}_{2M}^{(K)} \mathbf{f}$$

when $N = K \cdot 2M$, $i.e.$, the computation of

$$\begin{bmatrix} f_{\kappa \cdot 2M} \\ f_{\kappa \cdot 2M+1} \\ \vdots \\ f_{\kappa \cdot 2M+2M-1} \end{bmatrix} := \mathbf{Q}_{2M} \begin{bmatrix} f_{\kappa \cdot 2M} \\ f_{\kappa \cdot 2M+1} \\ \vdots \\ f_{\kappa \cdot 2M+2M-1} \end{bmatrix} \quad \text{for each } \kappa = 0, 1, \dots, K-1$$

can be done with the two-loop algorithm

> For $\lambda = 0, 1, \dots, M-1$ do:
>> For $\kappa = 0, 1, \dots, K-1$ do:
>>> $$\begin{bmatrix} f_{\kappa \cdot 2M+\lambda} \\ f_{\kappa \cdot 2M+\lambda+M} \end{bmatrix} := \begin{bmatrix} 1 & \omega^\lambda \\ 1 & -\omega^\lambda \end{bmatrix} \begin{bmatrix} f_{\kappa \cdot 2M+\lambda} \\ f_{\kappa \cdot 2M+\lambda+M} \end{bmatrix}.$$

We fetch or generate ω^λ only once when we order the loops in this way. We must do such a computation for each $\mathbf{Q}$-factor of $\mathcal{F}_{2^m}$. In addition, we must carry out the initial bit reversal permutation and the final scaling by $1/2^m$. In this way we see that the matrix factorization (12) gives us the in-place FFT of Algorithm 6.5. This algorithm corresponds to the decimation-in-time FFT of Fig. 6.3, but it takes a bit of effort to verify that this is in fact the case, $cf.$ Ex. 6.16(a).

Perform the bit reversal permutation on $f[0], f[1], \dots, f[2^m - 1]$.
For $\mu = 1, 2, \dots, m$ do:
> $\omega := e^{-2\pi i/2^\mu}$
> $U := 1$
> For $\lambda = 0, 1, \dots, 2^{\mu-1} - 1$ do:
>> For $\kappa = 0, 1, \dots, 2^{m-\mu} - 1$ do:
>>> $$\begin{bmatrix} f[\kappa \cdot 2^\mu + \lambda] \\ f[\kappa \cdot 2^\mu + \lambda + 2^{\mu-1}] \end{bmatrix} := \begin{bmatrix} 1 & U \\ 1 & -U \end{bmatrix} \begin{bmatrix} f[\kappa \cdot 2^\mu + \lambda] \\ f[\kappa \cdot 2^\mu + \lambda + 2^{\mu-1}] \end{bmatrix}$$
>> $U := \omega U$

For $k = 0, 1, \dots, 2^m - 1$ do:
> $f[k] := f[k]/2^m$

Algorithm 6.5. Naive decimation-in-time FFT based on (12) when $N = 2^m$.

An alternative FFT algorithm

Since the DFT matrix $\mathcal{F}_{2^m}$ is symmetric we can use (12) and (9) to write

$$
\begin{aligned}
\mathcal{F}_{2^m} &= \mathcal{F}_{2^m}^T \\
&= \left[\frac{1}{2^m}\, \mathbf{Q}_{2^m}\, \mathbf{Q}_{2^{m-1}}^{(2)}\, \cdots\, \mathbf{Q}_{2}^{(2^{m-1})}\, \mathbf{B}_{2^m} \right]^T \\
&= \frac{1}{2^m}\, \mathbf{B}_{2^m}^T \left[\mathbf{Q}_{2}^{(2^{m-1})} \right]^T \cdots \left[\mathbf{Q}_{2^{m-1}}^{(2)} \right]^T \left[\mathbf{Q}_{2^m} \right]^T \\
&= \frac{1}{2^m}\, \mathbf{B}_{2^m} \left[\mathbf{Q}_{2}^T \right]^{(2^{m-1})} \cdots \left[\mathbf{Q}_{2^{m-1}}^T \right]^{(2)} \left[\mathbf{Q}_{2^m}^T \right].
\end{aligned}
\tag{15}
$$

In the last step we use the fact that

$$
\mathbf{B}_{2^m}^T = \mathbf{B}_{2^m}, \quad m = 1, 2, \ldots .
\tag{16}
$$

[Two applications of $\mathbf{B}_{2^m}$ return a vector to its original state, so

$$
\mathbf{B}_{2^m}\, \mathbf{B}_{2^m} = \mathbf{I}_{2^m} ,
$$

and the columns of $\mathbf{B}_{2^m}$, *i.e.*, the permuted columns of $\mathbf{I}_{2^m}$, are orthonormal, so

$$
\mathbf{B}_{2^m}^T\, \mathbf{B}_{2^m} = \mathbf{I}_{2^m}.
$$

The symmetry (16) follows from these two relations.]

We can easily move from the matrix factorization (15) to an in-place FFT. At the heart of the computation is the vector replacement

$$
\begin{bmatrix} f_0 \\ f_1 \\ \vdots \\ f_{2M-1} \end{bmatrix} := \mathbf{Q}_{2M}^T \begin{bmatrix} f_0 \\ f_1 \\ \vdots \\ f_{2M-1} \end{bmatrix} :=
\left[\begin{array}{cccccc|cccccc}
1 & & & & & & 1 & & & & & \\
& 1 & & & & & & 1 & & & & \\
& & 1 & & & & & & 1 & & & \\
& & & \ddots & & & & & & \ddots & & \\
& & & & 1 & & & & & & 1 & \\
\hline
1 & & & & & & -1 & & & & & \\
& \omega & & & & & & -\omega & & & & \\
& & \omega^2 & & & & & & -\omega^2 & & & \\
& & & \ddots & & & & & & \ddots & & \\
& & & & \omega^{M-1} & & & & & & -\omega^{M-1} &
\end{array} \right]
\begin{bmatrix} f_0 \\ f_1 \\ f_2 \\ \vdots \\ f_{M-1} \\ \hline f_M \\ f_{M+1} \\ f_{M+2} \\ \vdots \\ f_{2M-1} \end{bmatrix}
$$

(with $\omega := e^{-2\pi i/(2M)}$), which can be done with the one-loop algorithm

$$\text{For } \lambda = 0, 1, \ldots, M-1 \text{ do:}$$
$$\left\lfloor \begin{bmatrix} f_\lambda \\ f_{\lambda+M} \end{bmatrix} := \begin{bmatrix} 1 & 1 \\ \omega^\lambda & -\omega^\lambda \end{bmatrix} \begin{bmatrix} f_\lambda \\ f_{\lambda+M} \end{bmatrix}, \right.$$

cf. (14). By combining this *inner loop* with suitable *outer loops*, we obtain the FFT of Algorithm 6.6. This algorithm corresponds to the decimation-in-frequency FFT of Fig. 6.6, *cf.* Ex. 6.16(b).

$$\text{For } \mu = m, m-1, \ldots, 1 \text{ do:}$$
$$w := e^{-2\pi i/2^\mu}$$
$$U := 1$$
$$\text{For } \lambda = 0, 1, \ldots, 2^{\mu-1} - 1 \text{ do:}$$
$$\text{For } \kappa = 0, 1, \ldots, 2^{m-\mu} - 1 \text{ do:}$$
$$\left\lfloor \begin{bmatrix} f[\kappa \cdot 2^\mu + \lambda] \\ f[\kappa \cdot 2^\mu + \lambda + 2^{\mu-1}] \end{bmatrix} := \begin{bmatrix} 1 & 1 \\ U & -U \end{bmatrix} \begin{bmatrix} f[\kappa \cdot 2^\mu + \lambda] \\ f[\kappa \cdot 2^\mu + \lambda + 2^{\mu-1}] \end{bmatrix} \right.$$
$$U := wU$$

Perform the bit reversal permutation on $f[0], f[1], \ldots, f[2^m - 1]$.

$$\text{For } k = 0, 1, \ldots, 2^m - 1 \text{ do:}$$
$$\left\lfloor f[k] := f[k]/2^m \right.$$

Algorithm 6.6. Naive decimation-in-frequency FFT based on (15) when $N = 2^m$.

Precomputation of $s_k := \sin(2\pi k/N)$

Algorithms 6.5 and 6.6 make use of the complex numbers

$$e^{-2\pi i k/N} = \cos\left(\frac{2\pi k}{N}\right) - i \sin\left(\frac{2\pi k}{N}\right), \quad k = 0, 1, \ldots, N-1,$$

in the innermost loop. We can use storage to gain speed by precomputing the real numbers

$$c_k := \cos\left(\frac{2\pi k}{N}\right), \quad s_k := \sin\left(\frac{2\pi k}{N}\right), \quad k = 0, 1, \ldots, N-1,$$

and retrieving them as needed. In practice, we find it preferable to generate and store only $s_0, s_1, \ldots, s_{N/4}$ since we can use the symmetries

$$s_k = s_{N/2-k} = -s_{N/2+k} = -s_{N-k}, \qquad\qquad k = 0, 1, \ldots, N/4$$
$$c_k = -c_{N/2-k} = -c_{N/2+k} = c_{N-k} = s_{N/4-k}, \qquad k = 0, 1, \ldots, N/4$$

to obtain all of the other values of s_k, c_k when $N = 2^m$. This enables us to reduce the required storage by a factor of 8.

The usual numerical procedures for evaluating $\sin\theta$, $0 \leq \theta \leq \pi/2$, require approximately 5, 8 real operations for 8, 16 decimal places of accuracy. We will describe a clever scheme of Buneman that makes it possible to compute $s_0, s_1, \ldots, s_{N/4}$ by using only slightly more than $N/4$ real operations. (Some alternative procedures are described in Ex. 6.19.) We begin with

$$s_0 := \sin(0) = 0, \qquad s_{N/8} := \sin(\pi/4) = 1/\sqrt{2}, \qquad s_{N/4} := \sin(\pi/2) = 1,$$

and then use the trigonometric identity

$$\sin(\alpha) = \frac{1}{2}\sec(\beta)\{\sin(\alpha - \beta) + \sin(\alpha + \beta)\}$$

to pass from this coarse grid to finer ones. Indeed, since

$$\sin\left(\frac{\pi}{8}\right) = \frac{1}{2}\sec\left(\frac{\pi}{8}\right)\left\{\sin\left(\frac{\pi}{8} - \frac{\pi}{8}\right) + \sin\left(\frac{\pi}{8} + \frac{\pi}{8}\right)\right\}$$
$$\sin\left(\frac{3\pi}{8}\right) = \frac{1}{2}\sec\left(\frac{\pi}{8}\right)\left\{\sin\left(\frac{3\pi}{8} - \frac{\pi}{8}\right) + \sin\left(\frac{3\pi}{8} + \frac{\pi}{8}\right)\right\}$$

we can use the known values of $s_{kN/8}$, $k = 0, 1, 2$, to generate

$$s_{N/16} = \frac{1}{2}\sec\left(\frac{\pi}{8}\right)(s_0 + s_{N/8}),$$
$$s_{3N/16} = \frac{1}{2}\sec\left(\frac{\pi}{8}\right)(s_{N/8} + s_{2N/8}).$$

We can then use the known values of $s_{kN/16}$, $k = 0, 1, 2, 3, 4$, to generate

$$s_{N/32} = \frac{1}{2}\sec\left(\frac{\pi}{16}\right)(s_0 + s_{N/16}),$$
$$s_{3N/32} = \frac{1}{2}\sec\left(\frac{\pi}{16}\right)(s_{N/16} + s_{2N/16}),$$
$$s_{5N/32} = \frac{1}{2}\sec\left(\frac{\pi}{16}\right)(s_{2N/16} + s_{3N/16}),$$
$$s_{7N/32} = \frac{1}{2}\sec\left(\frac{\pi}{16}\right)(s_{3N/16} + s_{4N/16}),$$

etc. Buneman even devised a clever recursion

$$h_3 = 1/\sqrt{2}, \quad h_{\mu+1} = \{2 + 1/h_\mu\}^{-1/2}, \quad \mu = 3, 4, \ldots, \tag{17}$$

for computing the half secants

$$h_\mu := \frac{1}{2} \sec\left(\frac{2\pi}{2^\mu}\right), \quad \mu = 3, 4, \ldots, \tag{18}$$

that are used in this process, *cf.* Ex. 6.18. These ideas are used in Algorithm 6.7 and in the FFT code from Appendix 5.

$$s[0] := 0$$
$$s[N/8] := 1/\sqrt{2}$$
$$s[N/4] := 1$$
$$h := 1/\sqrt{2}$$
$$k := N/8$$

While $k > 1$ do:

$\quad\Big|\quad h := \{2 + 1/h\}^{-1/2}$

$\quad\Big|\quad \ell := k/2$

$\quad\Big|\quad$ For $j = \ell, \ell + k, \ell + 2k, \ldots, N/4 - \ell$ do:

$\quad\Big|\quad\quad\mathsf{L}\ s[j] = h \cdot (s[j - \ell] + s[j + \ell])$

$\quad\Big|\quad k := k/2$

Algorithm 6.7. Buneman's clever scheme for generating $s[k] := \sin(2\pi k/N)$, $k = 0, 1, \ldots, N/4$, when $N = 2^m$.

Application of Q_{4M}

We will now use (12) to develop an algorithm that minimizes the time spent in retrieving the precomputed s_k's. We observe that the application of the matrix

$$\mathbf{Q}_2 := \begin{bmatrix} 1 & 1 \\ 1 & -1 \end{bmatrix} \tag{19}$$

can be accomplished by simply adding and subtracting vector components, and to exploit this possibility we will apply $\mathbf{Q}_2^{(N/2)}$ *outside* the inner loop of our FFT. The remaining factors $\mathbf{Q}_{4M}^{(N/4M)}$, $4M = 4, 8, 16, \ldots$ can be applied in a unified manner. The pattern is well illustrated by using $4M = 16$ with

$$
\mathbf{Q}_{16} := \left[
\begin{array}{cccc|cccc|cccc|cccc}
1 & & & & & & & & 1 & & & & & & & \\
& 1 & & & & & & & & c_1-is_1 & & & & & & \\
& & 1 & & & & & & & & c_2-is_2 & & & & & \\
& & & 1 & & & & & & & & c_3-is_3 & & & & \\
\hline
& & & & 1 & & & & & & & & -i & & & \\
& & & & & 1 & & & & & & & & -c_3-is_3 & & \\
& & & & & & 1 & & & & & & & & -c_2-is_2 & \\
& & & & & & & 1 & & & & & & & & -c_1-is_1 \\
\hline
1 & & & & & & & & -1 & & & & & & & \\
& 1 & & & & & & & & -c_1+is_1 & & & & & & \\
& & 1 & & & & & & & & -c_2+is_2 & & & & & \\
& & & 1 & & & & & & & & -c_3+is_3 & & & & \\
\hline
& & & & 1 & & & & & & & & i & & & \\
& & & & & 1 & & & & & & & & c_3+is_3 & & \\
& & & & & & 1 & & & & & & & & c_2+is_2 & \\
& & & & & & & 1 & & & & & & & & c_1+is_1
\end{array}
\right] \tag{20}
$$

and

$$
c_\lambda := \cos\left(\frac{2\pi\lambda}{16}\right), \quad s_\lambda := \sin\left(\frac{2\pi\lambda}{16}\right), \quad \lambda = 1, 2, 3.
$$

Each c_λ, s_λ pair appears exactly four times in the matrix (20), and after examining the structure it is easy to see that we can compute

$$
\mathbf{f} := \mathbf{Q}_{4M}\mathbf{f}, \quad M = 1, 2, \ldots
$$

by using the algorithm

$$
\begin{bmatrix} f_0 \\ f_M \\ f_{2M} \\ f_{3M} \end{bmatrix} := \begin{bmatrix} 1 & 0 & 1 & 0 \\ 0 & 1 & 0 & -i \\ 1 & 0 & -1 & 0 \\ 0 & 1 & 0 & i \end{bmatrix} \begin{bmatrix} f_0 \\ f_M \\ f_{2M} \\ f_{3M} \end{bmatrix}
$$

For $\lambda = 1, 2, \ldots, M - 1$ do:

$$
\left|
\begin{array}{l}
s := \sin(2\pi\lambda/4M) \\
c := \cos(2\pi\lambda/4M) \\
\begin{bmatrix} f_\lambda \\ f_{2M-\lambda} \\ f_{2M+\lambda} \\ f_{4M-\lambda} \end{bmatrix} := \begin{bmatrix} 1 & 0 & c-is & 0 \\ 0 & 1 & 0 & -c-is \\ 1 & 0 & -c+is & 0 \\ 0 & 1 & 0 & c+is \end{bmatrix} \begin{bmatrix} f_\lambda \\ f_{2M-\lambda} \\ f_{2M+\lambda} \\ f_{4M-\lambda} \end{bmatrix}.
\end{array}
\right. \tag{21}
$$

Now that you understand the structure of the innermost loop, you should be able to see how the factors of $\mathcal{F}_{2^m}$ from (12) are applied as we execute Algorithm 6.8.

Perform the bit reversal permutation on $f[0], f[1], \ldots, f[N-1]$

  by using Algorithm 6.4   (apply B_N)

Precompute $s[\kappa] := \sin(2\pi\kappa/N)$, $\kappa = 0, 1, \ldots, N/4$ using Algorithm 6.7

For $\kappa = 0, 1, \ldots, 2^{m-1} - 1$ do: (apply $Q_2^{(N/2)}$)

$$\begin{bmatrix} f[2\kappa] \\ f[2\kappa+1] \end{bmatrix} := \begin{bmatrix} 1 & 1 \\ 1 & -1 \end{bmatrix} \begin{bmatrix} f[2\kappa] \\ f[2\kappa+1] \end{bmatrix}$$

For $\mu = 2, 3, \ldots, m$ do: (apply $Q_{2^\mu}^{(N/2^\mu)}$)

  $M := 2^{\mu-2}$

  For $\kappa = 0, 1, \ldots, 2^{m-\mu} - 1$ do:

$$\begin{bmatrix} f[\kappa\cdot 4M] \\ f[\kappa\cdot 4M+M] \\ f[\kappa\cdot 4M+2M] \\ f[\kappa\cdot 4M+3M] \end{bmatrix} := \begin{bmatrix} 1 & 0 & 1 & 0 \\ 0 & 1 & 0 & -i \\ 1 & 0 & -1 & 0 \\ 0 & 1 & 0 & i \end{bmatrix} \cdot \begin{bmatrix} f[\kappa\cdot 4M] \\ f[\kappa\cdot 4M+M] \\ f[\kappa\cdot 4M+2M] \\ f[\kappa\cdot 4M+3M] \end{bmatrix}$$

  For $\lambda = 1, 2, \ldots, M - 1$ do:

    Fetch $s := \sin(2\pi\lambda/4M)$ from $s[\lambda \cdot 2^{m-\mu}]$

    Fetch $c := \sin(2\pi(M - \lambda)/4M)$ from $s[(M - \lambda) \cdot 2^{m-\mu}]$

    For $\kappa = 0, 1, \ldots, 2^{m-\mu} - 1$ do:

$$\begin{bmatrix} f[\kappa\cdot 4M+\lambda] \\ f[\kappa\cdot 4M+2M-\lambda] \\ f[\kappa\cdot 4M+2M+\lambda] \\ f[\kappa\cdot 4M+4M-\lambda] \end{bmatrix} := \begin{bmatrix} 1 & 0 & c-is & 0 \\ 0 & 1 & 0 & -c-is \\ 1 & 0 & -c+is & 0 \\ 0 & 1 & 0 & c+is \end{bmatrix} \cdot \begin{bmatrix} f[\kappa\cdot 4M+\lambda] \\ f[\kappa\cdot 4M+2M-\lambda] \\ f[\kappa\cdot 4M+2M+\lambda] \\ f[\kappa\cdot 4M+4M-\lambda] \end{bmatrix}$$

For $k = 0, 1, \ldots, 2^m - 1$ do: (apply $1/N$)

  $f[k] := f[k]/2^m$

Algorithm 6.8. A decimation-in-time FFT that is based on (12) and (21) when $N = 2^m$.

A corresponding FORTRAN code is given in Appendix 5. The code uses separate real arrays to hold the real and imaginary parts of $f_0, f_1, \ldots, f_{N-1}$ with the three vector replacement statements being unpacked and written in terms of real arithmetic in the natural manner. The two essential complex products

$$(c - is)f[\kappa \cdot 4M + 2M + \lambda], \qquad (c + is)f[\kappa \cdot 4M + 4M - \lambda]$$

and the four essential complex sums from the inner loop require 8 real multiplications and 12 real additions. Since this loop is executed a total of

$$(m - 1) \cdot (M - 1) \cdot \frac{N}{4M} \approx \frac{1}{4}N \log_2 N$$

times, the cost of executing the algorithm is approximately $2N \log_2 N$ real multiplications and $3N \log_2 N$ real additions. (This is a bit more precise than an operation count.) Improvements in the running time can be made by using a different radix, as described in Ex. 6.22.

At this point you have a basic understanding of the FFT when $N = 2^m$. In the following sections we will introduce important generalizations that make a good algorithm even better!

6.5 Sparse Matrix Factorization of H When $N = 2^m$

The zipper identity and factorization

We will use ideas from the preceding section to develop Bracewell's fast algorithm for applying the discrete Hartley transform matrix

$$\mathbf{H}_N := N^{-1/2} \left\{ \mathrm{cas}\left(\frac{2\pi kn}{N} \right) \right\}_{k,n=0}^{N-1}$$

when $N = 2^m$ for some $m = 1, 2, \ldots$, cf. (5.27). In this case the zipper identity takes the form

$$\mathbf{H}_{2M} = \frac{1}{\sqrt{2}} \mathbf{T}_{2M} \mathbf{H}_M^{(2)} \mathbf{S}_{2M}, \quad M = 1, 2, \ldots, \tag{22}$$

where $\mathbf{T}_{2M}$ is a sparse *real* matrix having the 2×2 block structure

$$\mathbf{T}_{2M} := \begin{bmatrix} \mathbf{I}_M & \mathbf{X}_M \\ \mathbf{I}_M & -\mathbf{X}_M \end{bmatrix}, \qquad M = 1, 2, \ldots \tag{23}$$

with

$$\mathbf{X}_M := \begin{bmatrix} 1 & 0 & 0 & \ldots & 0 \\ \hline 0 & c_1 & & & \\ 0 & & c_2 & & \\ \vdots & & & \ddots & \\ 0 & & & & c_{M-1} \end{bmatrix} + \begin{bmatrix} 0 & 0 & & \ldots & 0 & 0 \\ \hline 0 & & & & & s_1 \\ 0 & & & & s_2 & \\ \vdots & & & \iddots & & \\ 0 & s_{M-1} & & & & \end{bmatrix} \tag{24}$$

and

$$c_k := \cos\left(\frac{2k\pi}{2M}\right), \quad s_k := \sin\left(\frac{2k\pi}{2M}\right), \qquad k = 1, \ldots, M-1,$$

cf. Ex. 6.17. By using (22) and (13), we immediately obtain the sparse real factorization

$$\mathbf{H}_{2^m} = 2^{-m/2}\mathbf{T}_{2^m}\mathbf{T}_{2^{m-1}}^{(2)}\mathbf{T}_{2^{m-2}}^{(4)} \cdots \mathbf{T}_2^{(2^{m-1})}\mathbf{B}_{2^m} \tag{25}$$

that leads to a *fast Hartley transform* or FHT.

Application of T_{4M} using precomputed s_k's

We will use (25) to develop an FHT that is analogous to the FFT of Algorithm 6.8. The matrix

$$\mathbf{T}_2 := \begin{bmatrix} 1 & 1 \\ 1 & -1 \end{bmatrix}$$

is identical to the matrix (19) for the FFT. The structure of $\mathbf{T}_{4M}$, $M = 1, 2, \ldots$ is well illustrated by the matrix

$$\mathbf{T}_{16} := \begin{bmatrix}
1 & & & & 1 & & & & & & & & & & & \\
& 1 & & & & c_1 & & & & & & & s_1 & & & \\
& & 1 & & & & c_2 & & & & & & & s_2 & & \\
& & & 1 & & & & c_3 & & s_3 & & & & & & \\
& & & & 1 & & & & 1 & & & & & & & \\
& & & & & 1 & & & & s_3 & -c_3 & & & & & \\
& & & & & & 1 & & & s_2 & & -c_2 & & & & \\
& & & & & & & 1 & s_1 & & & -c_1 & & & & \\
1 & & & & -1 & & & & & & & & & & & \\
& 1 & & & & -c_1 & & & & & & & -s_1 & & & \\
& & 1 & & & & -c_2 & & & & & & & -s_2 & & \\
& & & 1 & & & & -c_3 & & -s_3 & & & & & & \\
& & & & 1 & & & & -1 & & & & & & & \\
& & & & & 1 & & & & -s_3 & c_3 & & & & & \\
& & & & & & 1 & & & -s_2 & & c_2 & & & & \\
& & & & & & & 1 & -s_1 & & & c_1 & & & &
\end{bmatrix} \tag{26}$$

with

$$c_\lambda := \cos\left(\frac{2\pi\lambda}{16}\right), \quad s_\lambda := \sin\left(\frac{2\pi\lambda}{16}\right), \quad \lambda = 1, 2, 3.$$

Each c_λ, s_λ pair appears exactly four times in the matrix (26), and after examining the structure it is easy to see that we can compute

$$\mathbf{f} := \mathbf{T}_{4M}\mathbf{f}, \qquad M = 1, 2, \ldots$$

by using the algorithm

$$
\begin{bmatrix} f_0 \\ f_M \\ f_{2M} \\ f_{3M} \end{bmatrix} := \begin{bmatrix} 1 & 0 & 1 & 0 \\ 0 & 1 & 0 & 1 \\ 1 & 0 & -1 & 0 \\ 0 & 1 & 0 & -1 \end{bmatrix} \begin{bmatrix} f_0 \\ f_M \\ f_{2M} \\ f_{3M} \end{bmatrix}
$$

For $\lambda = 1, 2, \ldots, M - 1$ do:

$$
\begin{aligned}
& s := \sin(2\pi\lambda/4M) \qquad\qquad\qquad\qquad\qquad (27) \\
& c := \cos(2\pi\lambda/4M)
\end{aligned}
$$

$$
\begin{bmatrix} f_\lambda \\ f_{2M-\lambda} \\ f_{2M+\lambda} \\ f_{4M-\lambda} \end{bmatrix} := \begin{bmatrix} 1 & 0 & c & s \\ 0 & 1 & s & -c \\ 1 & 0 & -c & -s \\ 0 & 1 & -s & c \end{bmatrix} \begin{bmatrix} f_\lambda \\ f_{2M-\lambda} \\ f_{2M+\lambda} \\ f_{4M-\lambda} \end{bmatrix}.
$$

After examining the parallel structures of (12) and (25), it is easy to see that we can convert the FFT of Algorithm 6.8 into the FHT of Algorithm 6.9 by using (27) instead of (21) for the inner loop and replacing 2^m by $2^{m/2}$ for the final scaling.

Although the FFT and FHT algorithms are quite similar, the FHT has two important advantages in cases where we work with real data. Since the 4×4 matrices that appear in the FHT algorithm have real elements, it is possible to use a single *real* array of length N to hold the components of $\mathbf{f}$. Each of the 4-vector replacements in the inner loop of the FHT requires computation of the 4 real products

$$
c \cdot f[\kappa \cdot 4M + 2M + \lambda], \ \ s \cdot f[\kappa \cdot 4M + 2M + \lambda], \ \ c \cdot f[\kappa \cdot 4M + 4M - \lambda], \ \ s \cdot f[\kappa \cdot 4M + 4M - \lambda],
$$

and 6 real additions. Since this loop is executed approximately $(N/4) \log_2 N$ times, the total cost of the computation is approximately $N \log_2 N$ real multiplications and $1.5 N \log_2 N$ real additions. When we use the FFT we must provide storage for both the real and imaginary parts of $\mathbf{f}$ and the total cost is approximately $2N \log_2 N$ real multiplications and $3N \log_2 N$ real additions. In this way we see that Bracewell's FHT uses half as much storage and half as many operations as the FFT when we work with real data. [Perhaps this will help you understand why the minor differences between Algorithm 6.8 and Algorithm 6.9 are protected by the first U.S. patent ever issued for a mathematical algorithm, *cf.* E.N. Zalta, Are algorithms patentable? *Notices AMS* **35**(1988), 796–799.]

The FHT can be used to cut the cost of other kinds of computations that are often done with the FFT. For example, if we wish to generate the DFT $\mathbf{f}^\wedge$ of a given real N-vector $\mathbf{f}$, we can use the FHT to compute $\mathbf{f}^\sim$ and then use the identity

$$
\mathbf{f}^\wedge = \frac{1}{\sqrt{N}} (\mathbf{P}_e - i\,\mathbf{P}_0)\mathbf{f}^\sim,
$$

Perform the bit reversal permutation on $f[0], f[1], \ldots, f[N-1]$
  by using Algorithm 6.4   (apply B_N)

Precompute $s[\kappa] := \sin(2\pi\kappa/N)$, $\kappa = 0, 1, \ldots, N/4$ using Algorithm 6.7

For $\kappa = 0, 1, \ldots, 2^{m-1} - 1$ do: (apply $T_2^{(N/2)}$)

$$\begin{bmatrix} f[2\kappa] \\ f[2\kappa+1] \end{bmatrix} := \begin{bmatrix} 1 & 1 \\ 1 & -1 \end{bmatrix} \begin{bmatrix} f[2\kappa] \\ f[2\kappa+1] \end{bmatrix}$$

For $\mu = 2, 3, \ldots, m$ do: (apply $T_{2^\mu}^{(N/2^\mu)}$)

  $M := 2^{\mu-2}$

  For $\kappa = 0, 1, \ldots, 2^{m-\mu} - 1$ do:

$$\begin{bmatrix} f[\kappa \cdot 4M] \\ f[\kappa \cdot 4M + M] \\ f[\kappa \cdot 4M + 2M] \\ f[\kappa \cdot 4M + 3M] \end{bmatrix} := \begin{bmatrix} 1 & 0 & 1 & 0 \\ 0 & 1 & 0 & 1 \\ 1 & 0 & -1 & 0 \\ 0 & 1 & 0 & -1 \end{bmatrix} \cdot \begin{bmatrix} f[\kappa \cdot 4M] \\ f[\kappa \cdot 4M + M] \\ f[\kappa \cdot 4M + 2M] \\ f[\kappa \cdot 4M + 3M] \end{bmatrix}$$

  For $\lambda = 1, 2, \ldots, M - 1$ do:

    Fetch $s := \sin(2\pi\lambda/4M)$ from $s[\lambda \cdot 2^{m-\mu}]$

    Fetch $c := \sin(2\pi(M - \lambda)/4M)$ from $s[(M - \lambda) \cdot 2^{m-\mu}]$

    For $\kappa = 0, 1, \ldots, 2^{m-\mu} - 1$ do:

$$\begin{bmatrix} f[\kappa \cdot 4M + \lambda] \\ f[\kappa \cdot 4M + 2M - \lambda] \\ f[\kappa \cdot 4M + 2M + \lambda] \\ f[\kappa \cdot 4M + 4M - \lambda] \end{bmatrix} := \begin{bmatrix} 1 & 0 & c & s \\ 0 & 1 & s & -c \\ 1 & 0 & -c & -s \\ 0 & 1 & -s & c \end{bmatrix} \cdot \begin{bmatrix} f[\kappa \cdot 4M + \lambda] \\ f[\kappa \cdot 4M + 2M - \lambda] \\ f[\kappa \cdot 4M + 2M + \lambda] \\ f[\kappa \cdot 4M + 4M - \lambda] \end{bmatrix}$$

For $k = 0, 1, \ldots, 2^m - 1$ do: (apply $1/\sqrt{N}$)
  $f[k] := f[k]/2^{m/2}$

Algorithm 6.9. A decimation-in-time FHT that is based on (25)
and (27) when $N = 2^m$.

cf. (5.31). The cost of computing $\mathbf{f}^\wedge$ in this way is about half of that required by the direct application of the FFT. (An equally efficient alternative is given in Ex. 6.8.) Analogously, the computation of the discrete convolution product $\mathbf{f} * \mathbf{g}$ of given real N-vectors $\mathbf{f}, \mathbf{g}$ can be done by using the FHT in conjunction with the identity from (5.81).

6.6 Sparse Matrix Factorization of $\mathcal{F}$ When $N = P_1 P_2 \cdots P_m$

Introduction

Let $N = P_1 P_2 \cdots P_m$ where $m = 2, 3, \ldots$ and $P_\mu = 2, 3, \ldots$ for each $\mu = 1, 2, \ldots, m$. In this section we will use a generalization of the zipper identity (11) to factor $\mathcal{F}_N$. The factorization will facilitate the development of an algorithm that allows us to compute an N-point DFT with approximately

$$N\{(P_1 - 1) + (P_2 - 1) + \cdots + (P_m - 1)\}$$

complex operations. In cases where N is highly composite, this gives us a fast Fourier transform.

The zipper identity for $\mathcal{F}_{MP}$

The structure of the generalized zipper identity is well illustrated using $N = 12$ with the factors $M = 4$, $P = 3$. We form the 12×12 permutation matrix

$$\mathbf{S}_{4,3} := [\boldsymbol{\delta}_0, \boldsymbol{\delta}_4, \boldsymbol{\delta}_8, \boldsymbol{\delta}_1, \boldsymbol{\delta}_5, \boldsymbol{\delta}_9, \boldsymbol{\delta}_2, \boldsymbol{\delta}_6, \boldsymbol{\delta}_{10}, \boldsymbol{\delta}_3, \boldsymbol{\delta}_7, \boldsymbol{\delta}_{11}]$$

by performing a perfect 3-shuffle of the columns $\boldsymbol{\delta}_0, \boldsymbol{\delta}_1, \ldots, \boldsymbol{\delta}_{11}$ of the 12×12 identity matrix, $\mathbf{I}_{12}$, and observe that

$$\mathcal{F}_{12} = \frac{1}{12} \left[\begin{array}{cccc|cccc|cccc}
1 & 1 & 1 & 1 & 1 & 1 & 1 & 1 & 1 & 1 & 1 & 1 \\
1 & \omega^3 & \omega^6 & \omega^9 & \omega & \omega^4 & \omega^7 & \omega^{10} & \omega^2 & \omega^5 & \omega^8 & \omega^{11} \\
1 & \omega^6 & \omega^{12} & \omega^{18} & \omega^2 & \omega^8 & \omega^{14} & \omega^{20} & \omega^4 & \omega^{10} & \omega^{16} & \omega^{22} \\
1 & \omega^9 & \omega^{18} & \omega^{27} & \omega^3 & \omega^{12} & \omega^{21} & \omega^{30} & \omega^6 & \omega^{15} & \omega^{24} & \omega^{33} \\
\hline
1 & \omega^{12} & \omega^{24} & \omega^{36} & \omega^4 & \omega^{16} & \omega^{28} & \omega^{40} & \omega^8 & \omega^{20} & \omega^{32} & \omega^{44} \\
1 & \omega^{15} & \omega^{30} & \omega^{45} & \omega^5 & \omega^{20} & \omega^{35} & \omega^{50} & \omega^{10} & \omega^{25} & \omega^{40} & \omega^{55} \\
1 & \omega^{18} & \omega^{36} & \omega^{54} & \omega^6 & \omega^{24} & \omega^{42} & \omega^{60} & \omega^{12} & \omega^{30} & \omega^{48} & \omega^{66} \\
1 & \omega^{21} & \omega^{42} & \omega^{63} & \omega^7 & \omega^{28} & \omega^{49} & \omega^{70} & \omega^{14} & \omega^{35} & \omega^{56} & \omega^{77} \\
\hline
1 & \omega^{24} & \omega^{48} & \omega^{72} & \omega^8 & \omega^{32} & \omega^{56} & \omega^{80} & \omega^{16} & \omega^{40} & \omega^{64} & \omega^{88} \\
1 & \omega^{27} & \omega^{54} & \omega^{81} & \omega^9 & \omega^{36} & \omega^{63} & \omega^{90} & \omega^{18} & \omega^{45} & \omega^{72} & \omega^{99} \\
1 & \omega^{30} & \omega^{60} & \omega^{90} & \omega^{10} & \omega^{40} & \omega^{70} & \omega^{100} & \omega^{20} & \omega^{50} & \omega^{80} & \omega^{110} \\
1 & \omega^{33} & \omega^{66} & \omega^{99} & \omega^{11} & \omega^{44} & \omega^{77} & \omega^{110} & \omega^{22} & \omega^{55} & \omega^{88} & \omega^{121}
\end{array}\right] \mathbf{S}_{4,3}$$

$$
= \frac{1}{3}
\left[
\begin{array}{cccc|cccc|cccc}
1 & & & & 1 & & & & 1 & & & \\
& 1 & & & & \omega & & & & \omega^2 & & \\
& & 1 & & & & \omega^2 & & & & \omega^4 & \\
& & & 1 & & & & \omega^3 & & & & \omega^6 \\
\hline
1 & & & & \omega^4 & & & & \omega^8 & & & \\
& 1 & & & & \omega^5 & & & & \omega^{10} & & \\
& & 1 & & & & \omega^6 & & & & \omega^{12} & \\
& & & 1 & & & & \omega^7 & & & & \omega^{14} \\
\hline
1 & & & & \omega^8 & & & & \omega^{16} & & & \\
& 1 & & & & \omega^9 & & & & \omega^{18} & & \\
& & 1 & & & & \omega^{10} & & & & \omega^{20} & \\
& & & 1 & & & & \omega^{11} & & & & \omega^{22}
\end{array}
\right]
\boldsymbol{\mathcal{F}}_4^{(3)} \mathbf{S}_{4,3}
$$

with $\omega := e^{-2\pi i/12}$. This factorization process generalizes in a natural manner, and in this way we obtain the zipper identity

$$
\boldsymbol{\mathcal{F}}_{MP} = \frac{1}{P} \mathbf{Q}_{M,P} \boldsymbol{\mathcal{F}}_M^{(P)} \mathbf{S}_{M,P}, \quad M, P = 1, 2, \ldots. \tag{28}
$$

The $MP \times MP$ matrix

$$
\mathbf{Q}_{M,P} :=
\begin{bmatrix}
\mathbf{W}_{0,0} & \mathbf{W}_{0,1} & \cdots & \mathbf{W}_{0,P-1} \\
\mathbf{W}_{1,0} & \mathbf{W}_{1,1} & \cdots & \mathbf{W}_{1,P-1} \\
\vdots & & & \\
\mathbf{W}_{P-1,0} & \mathbf{W}_{P-1,1} & \cdots & \mathbf{W}_{P-1,P-1}
\end{bmatrix}
\tag{29}
$$

is formed from the $M \times M$ diagonal blocks

$$
\mathbf{W}_{k,\ell} = \omega^{k\ell M}
\begin{bmatrix}
1 & & & & \\
& \omega^\ell & & & \\
& & \omega^{2\ell} & & \\
& & & \ddots & \\
& & & & \omega^{(M-1)\ell}
\end{bmatrix},
\tag{30}
$$

using powers of $\omega := e^{-2\pi i/MP}$, and the permutation matrix

$$
\mathbf{S}_{M,P} := [\boldsymbol{\delta}_0, \boldsymbol{\delta}_M, \boldsymbol{\delta}_{2M}, \ldots, \boldsymbol{\delta}_{(P-1)M}, \boldsymbol{\delta}_1, \boldsymbol{\delta}_{1+M}, \boldsymbol{\delta}_{1+2M}, \ldots, \boldsymbol{\delta}_{1+(P-1)M},
$$
$$
\ldots, \boldsymbol{\delta}_{M-1}, \boldsymbol{\delta}_{2M-1}, \boldsymbol{\delta}_{3M-1}, \ldots, \boldsymbol{\delta}_{PM-1}]
\tag{31}
$$

is obtained by performing a perfect P-shuffle of the columns $\boldsymbol{\delta}_0, \boldsymbol{\delta}_1, \ldots, \boldsymbol{\delta}_{MP-1}$ of the identity matrix $\mathbf{I}_{MP}$.

You will observe that $\mathbf{Q}_{M,2}$ and $\mathbf{S}_{M,2}$ are identical to the matrices $\mathbf{Q}_{2M}$ and $\mathbf{S}_{2M}$ from (2)–(3). A single subscript (specifying the size of the matrix) was all that we needed for the derivation of (12). Within the present context we must specify both of the factors M, P, and we do this by using them as subscripts (with the *order* being important!) The product of these subscripts now gives the size of the matrix.

Factorization of $\mathcal{F}_{P_1 P_2 \cdots P_m}$

We use the zipper identity (28) with $P = P_m$, $P = P_{m-1}, \ldots, P = P_1$ in turn [together with (6)–(8)] to write

$$
\begin{aligned}
\mathcal{F}_{P_1 P_2 \cdots P_m} &= \frac{1}{P_m} \mathbf{Q}_{P_1 \cdots P_{m-1}, P_m} \cdot \mathcal{F}^{(P_m)}_{P_1 \cdots P_{m-1}} \cdot \mathbf{S}_{P_1 \cdots P_{m-1}, P_m} \\
&= \frac{1}{P_{m-1} P_m} \mathbf{Q}_{P_1 \cdots P_{m-1}, P_m} \cdot \mathbf{Q}^{(P_m)}_{P_1 \cdots P_{m-2}, P_{m-1}} \cdot \mathcal{F}^{(P_{m-1} P_m)}_{P_1 \cdots P_{m-2}} \\
&\qquad\qquad\qquad\qquad \cdot \mathbf{S}^{(P_m)}_{P_1 \cdots P_{m-2}, P_{m-1}} \cdot \mathbf{S}_{P_1 \cdots P_{m-1}, P_m} \qquad (32) \\
&\quad\vdots \\
&= \frac{1}{P_1 P_2 \cdots P_m} \mathbf{Q}_{P_1 \cdots P_{m-1}, P_m} \cdot \mathbf{Q}^{(P_m)}_{P_1 \cdots P_{m-2}, P_{m-1}} \cdot \\
&\qquad \cdots \cdot \mathbf{Q}^{(P_3 \cdots P_m)}_{P_1, P_2} \cdot \mathbf{Q}^{(P_2 \cdots P_m)}_{1, P_1} \cdot \mathbf{S}_{P_1, P_2, \ldots, P_m}
\end{aligned}
$$

where the final factor

$$
\mathbf{S}_{P_1, P_2, \ldots, P_m} := \mathbf{S}^{(P_3 \cdots P_m)}_{P_1, P_2} \mathbf{S}^{(P_4 \cdots P_m)}_{P_1 P_2, P_3} \cdots \mathbf{S}^{(P_m)}_{P_1 \cdots P_{m-2}, P_{m-1}} \mathbf{S}_{P_1 \cdots P_{m-1}, P_m} \qquad (33)
$$

is a permutation matrix. This factorization depends on our choice of $P_1, P_2, \ldots, P_m$ (e.g., when $N = 12$ we can use $P_1 = 2$, $P_2 = 2$, $P_3 = 3$; $P_1 = 2$, $P_2 = 3$, $P_3 = 2$; $P_1 = 3$, $P_2 = 2$, $P_3 = 2$; $P_1 = 2$, $P_2 = 6$; $P_1 = 6$, $P_2 = 2$; $P_1 = 3$, $P_2 = 4$; or $P_1 = 4$, $P_2 = 3$).

From (29)–(30) we see that each row of $\mathbf{Q}_{M,P}$ has exactly P nonzero entries with the first of these being a 1. This being the case, it will take no more than $P - 1$ complex operations per component to apply $\mathbf{Q}_{M,P}$ to a compatible vector, and the same is true of $\mathbf{Q}^{(K)}_{M,P}$, $K = 1, 2, \ldots$. Thus we expend approximately

$$
N(P_1 - 1) + N(P_2 - 1) + \cdots + N(P_m - 1)
$$

complex operations as we apply the factors of (32). Of course, this reduces to $N \log_2 N$ when $P_1 = P_2 = \cdots = P_m = 2$.

An FFT

We will use (32) to develop a fast algorithm for computing the DFT. At the heart of an in-place computation we must carry out the vector replacement

$$
\mathbf{f} := \mathbf{Q}_{M,P} \mathbf{f}.
$$

We will place the nonzero elements of $\mathbf{Q}_{M,P}$ that lie in rows $\lambda, \lambda + M, \lambda + 2M, \ldots,$ $\lambda + (P-1)M$ in the matrix

$$
\mathbf{\Omega}_{\lambda,M,P} := \begin{bmatrix}
1 & \omega^{\lambda} & \omega^{2\lambda} & \cdots & \omega^{(P-1)\lambda} \\
1 & \omega^{\lambda+M} & \omega^{2(\lambda+M)} & \cdots & \omega^{(P-1)(\lambda+M)} \\
1 & \omega^{\lambda+2M} & \omega^{2(\lambda+2M)} & \cdots & \omega^{(P-1)(\lambda+2M)} \\
\vdots & & & & \\
1 & \omega^{\lambda+(P-1)M} & \omega^{2[\lambda+(P-1)M]} & \cdots & \omega^{(P-1)[\lambda+(P-1)M]}
\end{bmatrix},
$$

$\omega := e^{-2\pi i/MP}$, and thereby see that the replacement can be done by writing

$$
\text{For } \lambda = 0, 1, \ldots, M-1 \text{ do:}
$$

$$
\begin{bmatrix}
f_{\lambda} \\
f_{\lambda+M} \\
f_{\lambda+2M} \\
\vdots \\
f_{\lambda+(P-1)M}
\end{bmatrix}
:= \mathbf{\Omega}_{\lambda,M,P}
\begin{bmatrix}
f_{\lambda} \\
f_{\lambda+M} \\
f_{\lambda+2M} \\
\vdots \\
f_{\lambda+(P-1)M}
\end{bmatrix}.
$$

You should now be able to see how the FFT of Algorithm 6.10 successively applies the factors

$$
\mathbf{Q}_{1,P_1}^{(N/P_1)}, \mathbf{Q}_{P_1,P_2}^{(N/P_1 P_2)}, \mathbf{Q}_{P_1 P_2, P_3}^{(N/P_1 P_2 P_3)}, \ldots, \mathbf{Q}_{P_1 \cdots P_{m-1}, P_m}^{(N/P_1 \cdots P_m)}
$$

from (32).

A direct application of the $P \times P$ matrix $\mathbf{\Omega}_{\lambda,M,P}$ in the inner loop would require $P(P-1)$ complex multiplications and the same number of complex additions. In practice, there are clever tricks that we can use to reduce this effort, and most of these exploit the fact that the matrix has the factorization

$$
\mathbf{\Omega}_{\lambda,M,P} = P\mathcal{F}_P
\begin{bmatrix}
1 & & & & \\
& \omega^{\lambda} & & & \\
& & \omega^{2\lambda} & & \\
& & & \ddots & \\
& & & & \omega^{(P-1)\lambda}
\end{bmatrix}, \quad \omega := e^{-2\pi i/MP}. \tag{34}
$$

For example, when $P = 2$ we can generate

$$
\begin{bmatrix} f_0' \\ f_1' \end{bmatrix} := \mathbf{\Omega}_{\lambda,M,2} \begin{bmatrix} f_0 \\ f_1 \end{bmatrix} = \begin{bmatrix} 1 & 1 \\ 1 & -1 \end{bmatrix} \begin{bmatrix} 1 & \\ & \omega^{\lambda} \end{bmatrix} \begin{bmatrix} f_0 \\ f_1 \end{bmatrix}
$$

by computing in turn

$$
t := \omega^{\lambda} \cdot f_1, \quad f_0' := f_0 + t, \quad f_1' := f_0 - t.
$$

Permute $f[0], f[1], \ldots, f[N-1]$ using $\mathbf{S}_{P_1, P_2, \ldots, P_m}$

$M := 1$

$P := 1$

$K := N$

For $\mu = 1, 2, \ldots, m$ do:

$\quad M := M \cdot P$

$\quad P := P_\mu$

$\quad K := K/P$

$\quad \omega := e^{-2\pi i / MP}$

$\quad$ For $\lambda = 0, 1, \ldots, M-1$ do:

$\quad\quad$ For $\kappa = 0, 1, \ldots, K-1$ do:

$$
\begin{bmatrix}
f[\lambda & +\kappa MP] \\
f[\lambda+ & M+\kappa MP] \\
& \vdots \\
f[\lambda+(P-1)M+\kappa MP]
\end{bmatrix}
:= \boldsymbol{\Omega}_{\lambda, M, P}
\begin{bmatrix}
f[\lambda & +\kappa MP] \\
f[\lambda+ & M+\kappa MP] \\
& \vdots \\
f[\lambda+(P-1)M+\kappa MP]
\end{bmatrix}
$$

For $k = 0, 1, \ldots, N-1$ do:

$\quad f[k] := f[k]/N$

Algorithm 6.10. Naive decimation-in-time FFT based on (32) when $N = P_1 P_2 \cdots P_m$.

This eliminates one complex multiplication and reduces the per component cost from 4 real multiplications and 4 real additions to 2 real multiplications and 3 real additions. Analogously, when $P = 4$ we can generate

$$
\begin{bmatrix}
f_0' \\
f_1' \\
f_2' \\
f_3'
\end{bmatrix}
:=
\begin{bmatrix}
1 & 1 & 1 & 1 \\
1 & -i & -1 & i \\
1 & -1 & 1 & -1 \\
1 & i & -1 & -i
\end{bmatrix}
\begin{bmatrix}
1 & & & \\
& \omega^\lambda & & \\
& & \omega^{2\lambda} & \\
& & & \omega^{3\lambda}
\end{bmatrix}
\begin{bmatrix}
f_0 \\
f_1 \\
f_2 \\
f_3
\end{bmatrix}
$$

by computing in turn

$$
\begin{aligned}
t_1 &:= \omega^\lambda \cdot f_1, & t_2 &:= \omega^{2\lambda} \cdot f_2, & t_3 &:= \omega^{3\lambda} \cdot f_3, \\
s_1 &:= f_0 + t_2, & s_2 &:= t_1 + t_3, & d_1 &:= f_0 - t_2, & d_2 &:= t_1 - t_3, \\
f_0' &:= s_1 + s_2, & f_1' &:= d_1 - i\,d_2, & f_2' &:= s_1 - s_2, & f_3' &:= d_1 + i\,d_2.
\end{aligned}
$$

If we neglect the multiplications by i [since we can compute

$$(a + ib) + i(c + id) = (a - d) + i(b + c)$$

by performing two real additions that are equivalent to one complex addition] we see that this eliminates 9 complex multiplications and 4 complex additions. The per component cost for applying $\boldsymbol{\Omega}_{\lambda,M,4}$ is thus reduced from 12 real multiplications and 12 real additions to 3 real multiplications and 11/2 real additions. When N is a power of 4, this leads to an FFT that uses approximately

$$N \log_2 N \cdot \{1.5 \text{ real multiplications } + 2.75 \text{ real additions}\}.$$

In contrast, Algorithm 6.8 uses

$$N \log_2 N \cdot \{2 \text{ real multiplications } + 3 \text{ real additions}\}.$$

Exercise 6.22 shows how to reduce the cost to

$$N \log_2 N \cdot \{1.33 \text{ real multiplications } + 2.75 \text{ real additions}\}$$

when N is a power of 8.

The permutation $S_{Q,P}$

If we lay down the components of $\mathbf{f} = (f_0, f_1, \ldots, f_{11})^T$ in 4 rows

$$
\begin{array}{ccc}
f_0 & f_1 & f_2 \\
f_3 & f_4 & f_5 \\
f_6 & f_7 & f_8 \\
f_9 & f_{10} & f_{11}
\end{array}
$$

and then pick them up by columns we produce the perfect 4-shuffle

$$(f_0, f_3, f_6, f_9, f_1, f_4, f_7, f_{10}, f_2, f_5, f_8, f_{11})^T$$

of $\mathbf{f}$. The analogous perfect P-shuffle of the columns $\boldsymbol{\delta}_0, \boldsymbol{\delta}_1, \ldots, \boldsymbol{\delta}_{PQ-1}$ of the $PQ \times PQ$ identity matrix gives the *shuffle permutation*

$$
\begin{aligned}
\mathbf{S}_{Q,P} := [&\boldsymbol{\delta}_0, \boldsymbol{\delta}_Q, \boldsymbol{\delta}_{2Q}, \ldots, \boldsymbol{\delta}_{(P-1)Q}, \boldsymbol{\delta}_1, \boldsymbol{\delta}_{1+Q}, \boldsymbol{\delta}_{1+2Q}, \ldots, \boldsymbol{\delta}_{1+(P-1)Q}, \\
& \ldots, \boldsymbol{\delta}_{Q-1}, \boldsymbol{\delta}_{2Q-1}, \boldsymbol{\delta}_{3Q-1}, \ldots, \boldsymbol{\delta}_{QP-1}].
\end{aligned}
\tag{35}
$$

Using this definition it is easy to verify that

$$\boldsymbol{\delta}_{q+pQ} \text{ appears in column } p + qP \text{ of } \mathbf{S}_{Q,P},$$

$$\mathbf{S}_{Q,P} \boldsymbol{\delta}_{p+qP} = \boldsymbol{\delta}_{q+pQ}, \tag{36}$$

$$\mathbf{S}_{Q,P} \text{ puts } f_{p+qP} \text{ into position } q + pQ, \tag{37}$$

(for all $p = 0, 1, \ldots, P-1$, $q = 0, 1, \ldots, Q-1$), and to see that

$$\mathbf{S}_{Q,P}\mathbf{f} = (f_0, f_P, f_{2P}, \ldots, f_{(Q-1)P}, f_1, f_{1+P}, f_{1+2P}, f_{1+(Q-1)P},$$
$$\ldots, f_{P-1}, f_{2P-1}, f_{3P-1}, \ldots, f_{QP-1})^T \tag{38}$$

is a perfect Q-shuffle of $\mathbf{f}$. You will note that the subscripts on the right-hand side of (35) mostly jump by Q (when $P > 2$) while those on the right-hand side of (38) mostly jump by P (when $Q > 2$).

Using (36) we see that

$$\mathbf{S}_{P,Q}\mathbf{S}_{Q,P} = \mathbf{I}_{PQ},$$

and thereby infer that

$$\mathbf{S}_{Q,P}^{-1} = \mathbf{S}_{P,Q}. \tag{39}$$

Since $\mathbf{S}_{P,Q}$ is a permutation matrix, this implies that

$$\mathbf{S}_{Q,P}^{T} = \mathbf{S}_{P,Q}, \tag{40}$$

cf. Ex. 6.13. In particular,

$$\mathbf{f}^T\mathbf{S}_{Q,P} = (\mathbf{S}_{Q,P}^T\mathbf{f})^T = (\mathbf{S}_{P,Q}\mathbf{f})^T$$
$$= (f_0, f_Q, f_{2Q}, \ldots, f_{(P-1)Q}, f_1, f_{1+Q}, f_{1+2Q}, \ldots, f_{1+(P-1)Q},$$
$$\ldots, f_{Q-1}, f_{2Q-1}, f_{3Q-1}, \ldots, f_{PQ-1}),$$

[and this is precisely the action that we required during the derivation of the zipper identity (28)!].

The permutation $S_{P_1,P_2,\ldots,P_m}$

Using (37) (with f_n replaced by $f[n]$) we see that

$\mathbf{S}_{Q,P}$ maps $f[p + qP]$ to position $q + pQ$ when
$$p = 0, 1, \ldots, P-1, \quad q = 0, 1, \ldots, Q-1 \text{ and } N = PQ. \tag{41}$$

More generally,

$\mathbf{S}_{Q,P}^{(R)}$ maps $f[p + qP + rPQ]$ to position $q + pQ + rPQ$ when
$$p = 0, 1, \ldots, P-1, \quad q = 0, 1, \ldots, Q-1, \quad r = 0, 1, \ldots, R-1 \tag{42}$$
$$\text{and } N = PQR.$$

We will use (41)–(42) to determine the action of the permutation (33) that is needed for Algorithm 6.10.

The pattern is well illustrated by the case where $m = 4$ and $N = P_1 P_2 P_3 P_4$. We choose an index $n = 0, 1, \ldots, N - 1$, write this index in the form

$$n = p_4 + p_3 P_4 + p_2 P_3 P_4 + p_1 P_2 P_3 P_4$$

for suitably chosen

$$p_1 = 0, 1, \ldots, P_1 - 1, \quad p_2 = 0, 1, \ldots, P_2 - 1,$$
$$p_3 = 0, 1, \ldots, P_3 - 1, \quad p_4 = 0, 1, \ldots, P_4 - 1,$$

and then follow $f[n]$ as we apply in turn the factors $\mathbf{S}_{P_1 P_2 P_3, P_4}$, $\mathbf{S}_{P_1 P_2, P_3}^{(P_4)}$, $\mathbf{S}_{P_1, P_2}^{(P_3 P_4)}$ of $\mathbf{S}_{P_1, P_2, P_3, P_4}$. Using (41) we see that the initial application of $\mathbf{S}_{P_1 P_2 P_3, P_4}$ moves $f[n]$ from position

$$n = p_4 + (p_3 + p_2 P_3 + p_1 P_2 P_3) P_4$$

to position

$$n' = (p_3 + p_2 P_3 + p_1 P_2 P_3) + p_4 (P_3 P_2 P_1).$$

Using (42) we see that the subsequent application of $\mathbf{S}_{P_1 P_2, P_3}^{(P_4)}$ moves $f[n]$ from position

$$n' = p_3 + (p_2 + p_1 P_2) P_3 + p_4 (P_3 P_2 P_1)$$

to position

$$n'' = (p_2 + p_1 P_2) + p_3 (P_2 P_1) + p_4 (P_3 P_2 P_1).$$

Again using (42) we see that the final application of $\mathbf{S}_{P_1, P_2}^{(P_3 P_4)}$ moves $f[n]$ from position

$$n'' = p_2 + p_1 P_2 + (p_3 + p_4 P_3)(P_2 P_1)$$

to position

$$r = p_1 + p_2 P_1 + p_3 P_2 P_1 + p_4 P_3 P_2 P_1.$$

This argument can be used for any $m = 2, 3, 4, \ldots$, so if

$$n = p_m + p_{m-1} P_m + p_{m-2} P_{m-1} P_m + \cdots + p_1 P_2 P_3 \cdots P_m \tag{43}$$

for some choice of

$$p_1 = 0, 1, \ldots, P_1 - 1, \quad p_2 = 0, 1, \ldots, P_2 - 1, \ldots, p_m = 0, 1, \ldots, P_m - 1,$$

then $\mathbf{S}_{P_1, P_2, \ldots, P_m}$ maps $f[n]$ to position

$$r = p_1 + p_2 P_1 + p_3 P_2 P_1 + \cdots + p_m P_{m-1} P_{m-2} \cdots P_1. \tag{44}$$

(The choice $P_1 = P_2 = \ldots = P_m = 2$ gives the bit reversal permutation $\mathbf{B}_{2^m}$.)

A suitable generalization of Algorithm 6.2 for bit reversal can be used to find the r that corresponds to n. For example, when $P_1 = 3$, $P_2 = 5$, $P_3 = 7$, $P_4 = 11$ and we are given the index

$$n = 1153 = 9 + 6 \cdot (11) + 4 \cdot (7 \cdot 11) + 2 \cdot (5 \cdot 7 \cdot 11)$$

we can produce the corresponding index

$$r = 2 + 4 \cdot (3) + 6 \cdot (5 \cdot 3) + 9 \cdot (7 \cdot 5 \cdot 3) = 1049$$

with the following calculation.

$$
\begin{aligned}
1153 \div 11 &= 104 \ \text{with remainder } p_4 := 9, & r_1 &:= 9 \\
104 \div 7 &= 14 \ \ \text{with remainder } p_3 := 6, & r_2 &:= 7r_1 + 6 = 69 \\
14 \div 5 &= 2 \ \ \ \text{with remainder } p_2 := 4, & r_3 &:= 5r_2 + 4 = 349 \\
2 \div 3 &= 0 \ \ \ \text{with remainder } p_1 := 2, & r_4 &:= 3r_3 + 2 = 1049
\end{aligned}
$$

Algorithm 6.11 uses these ideas to generate the index r when $P_1, P_2, \ldots, P_m$ and $n = 0, 1, \ldots, N - 1$ are given. By reversing the order of $P_1, P_2, \ldots, P_m$, Algorithm 6.12 allows us to generate the index n that corresponds to a given $r = 0, 1, \ldots, N-1$. The identity

$$\mathbf{S}_{P_1, P_2, \ldots, P_m}^{-1} = \mathbf{S}_{P_m, P_{m-1}, \ldots, P_1}, \tag{45}$$

which expresses the relationship between these two algorithms, generalizes (39).

$$
\begin{aligned}
&r := 0 \\
&d := n \\
&\text{For } \mu = m, m - 1, \ldots, 1 \text{ do:} \\
&\quad \left|
\begin{aligned}
&q := \lfloor d/P_\mu \rfloor \\
&p := d - P_\mu \cdot q \\
&r := p + P_\mu \cdot r \\
&d := q
\end{aligned}
\right.
\end{aligned}
$$

Algorithm 6.11. Computation of the r from (44) when the n from (43) is given.

$$n := 0$$
$$d := r$$

For $\mu = 1, 2, \ldots, m$ do:

$$q := \lfloor d/P_\mu \rfloor$$
$$p := d - P_\mu \cdot q$$
$$n := p + P_\mu \cdot n$$
$$d := q$$

Algorithm 6.12. Computation of the n from (43) when the r from (44) is given.

If we can choose the factors $P_1, P_2, \ldots, P_m$ so that

$$P_1 = P_m, \quad P_2 = P_{m-1}, \quad P_3 = P_{m-2}, \ldots, P_m = P_1, \tag{46}$$

then (45) shows that $\mathbf{S}_{P_1, P_2, \ldots, P_m}$ is self reciprocal, *i.e.*, the symmetry (46) makes it possible to carry out the permutation by simply swapping $f[n]$ and $f[r]$ whenever $r > n$. For example, when $N = 12$ and we take $P_1 = 2$, $P_2 = 3$, $P_3 = 2$ we find

n:	0	1	2	3	4	5	6	7	8	9	10	11
$r[n]$:	0	6	2	8	4	10	1	7	3	9	5	11

and we can apply $\mathbf{S}_{2,3,2}$ by swapping $f[1]$ and $f[6]$, $f[3]$ and $f[8]$, $f[5]$ and $f[10]$.

In the absence of the symmetry (46) this is never the case. For example, when $N = 12$ and we take $P_1 = 2$, $P_2 = 2$, $P_3 = 3$ we find

n:	0	1	2	3	4	5	6	7	8	9	10	11
$r[n]$:	0	4	8	2	6	10	1	5	9	3	7	11

so the application of $\mathbf{S}_{2,2,3}$ requires us to cyclically permute

$$f[1], \ f[4], \text{ and } f[6]; \ f[2], \ f[8], \ f[9], \text{ and } f[3]; \ f[5], \ f[10], \text{ and } f[7].$$

If we reverse the order of the factors and take $P_1 = 3$, $P_2 = 2$, $P_3 = 2$ we find

n:	0	1	2	3	4	5	6	7	8	9	10	11
$r[n]$:	0	6	3	9	1	7	4	10	2	8	5	11

so the application of $\mathbf{S}_{3,2,2}$ requires us to cyclically permute

$$f[1], \ f[6], \text{ and } f[4]; \ f[2], \ f[3], \ f[9], \text{ and } f[8]; \ f[5], \ f[7], \text{ and } f[10].$$

The cycles are the same as those for $\mathbf{S}_{2,2,3} = \mathbf{S}_{3,2,2}^{-1}$, but the direction is reversed!

Algorithm 6.13 provides a simple scheme for applying the permutation $\mathbf{S}_{P_1, P_2, \ldots, P_m}$. This algorithm uses an auxiliary array. If you wish to do the permutation without using two data arrays, you can precompute the indices for the various cycles. For example, if you wish to apply $\mathbf{S}_{2,2,3}$ you can precompute

$$(-1, 6, 4, -2, 3, 9, 8, -5, 7, 10)$$

(with negative indices marking the start of each cycle) and then perform the permutation by writing

$$
\begin{aligned}
T &:= f[1], & f[1] &:= f[6], & f[6] &:= f[4], & f[4] &:= T, \\
T &:= f[2], & f[2] &:= f[3], & f[3] &:= f[9], & f[9] &:= f[8], & f[8] &:= T, \\
T &:= f[5], & f[5] &:= f[7], & f[7] &:= f[10], & f[10] &:= T.
\end{aligned}
$$

In cases where $P_1, P_2, \ldots, P_m$ have the symmetry (46), you can develop a very efficient generalization of the Bracewell-Buneman algorithm that does the permutation in place by performing the necessary swaps. This is particularly effective when $P_1 = P_2 = \cdots = P_m$.

For $n = 0, 1, \ldots, N - 1$ do:
> $r := 0$
> $d := n$
> For $\mu = m, m - 1, \ldots, 1$ do:
> > $q := \lfloor d/P_\mu \rfloor$
> > $p := d - P_\mu \cdot q$
> > $r := p + P_\mu \cdot r$
> > $d := q$
> $g[r] := f[n]$

For $n = 0, 1, \ldots, N - 1$
> $f[n] := g[n]$

Algorithm 6.13. A naive scheme for applying the permutation $\mathbf{S}_{P_1, P_2, \ldots, P_m}$ when $f[0], f[1], \ldots, f[N-1]$ and $N = P_1 P_2 \cdots P_m$ are given.

Closely related factorizations of $\mathcal{F}, \mathbf{H}$

Since $\mathcal{F}$ is symmetric, we can transpose (32) to produce the factorization

$$
\begin{aligned}
\mathcal{F}_{P_1 P_2 \cdots P_m} = {} & \frac{1}{P_1 P_2 \cdots P_m} \mathbf{S}_{P_m, P_{m-1}, \ldots, P_1} \cdot [\mathbf{Q}_{1, P_1}^T]^{(P_2 \cdots P_m)} \\
& \cdot [\mathbf{Q}_{P_1, P_2}^T]^{(P_3 \cdots P_m)} \cdot \ldots \cdot [\mathbf{Q}_{P_1 \cdots P_{m-2}, P_{m-1}}^T]^{(P_m)} \cdot \mathbf{Q}_{P_1 \cdots P_{m-1}, P_m}^T,
\end{aligned} \tag{47}
$$

which gives the Gentleman-Sande version of the FFT, *cf.* (15) and Ex. 6.16(b).

There are factorizations of the Hartley transform matrix $\mathbf{H}_{P_1 P_2 \cdots P_m}$ that correspond to (32) and (47). These can be derived with the zipper identity of Ex. 6.23.

6.7 Kronecker Product Factorization of $\mathcal{F}$

Introduction

In the years following the publication of the Cooley-Tukey algorithm, dozens of other FFTs were discovered by Gentleman and Sande, Pease, Stockham, Single-ton, Burrus, de Boor, Winograd, Temperton, and many others. These variations facilitated the development of more elegant codes, more efficient access of data, somewhat smaller operation counts, etc. Most of these FFTs can be derived from the Cooley-Tukey factorization (32) by using a mathematical construct known as the Kronecker product (or direct product or tensor product). We will develop some of the basic properties of the Kronecker product and show how they are used to produce useful sparse factorizations of the matrix $\mathcal{F}$. You should consult Van Loan's book and the references cited therein for a comprehensive treatment of this topic.

The Kronecker product

Let $\mathbf{A}, \mathbf{B}$ be $K \times L$, $M \times N$ matrices with elements

$$
\begin{aligned}
A[k, \ell], \quad & k = 0, 1, \ldots, K - 1, \quad \ell = 0, 1, \ldots, L - 1, \\
B[m, n], \quad & m = 0, 1, \ldots, M - 1, \quad n = 0, 1, \ldots, N - 1.
\end{aligned}
$$

We define the *Kronecker product* $\mathbf{A} \otimes \mathbf{B}$ to be the $KM \times LN$ matrix

$$
\mathbf{A} \otimes \mathbf{B} := \begin{bmatrix}
A[0, 0]\mathbf{B} & A[0, 1]\mathbf{B} & \cdots & A[0, L-1]\mathbf{B} \\
A[1, 0]\mathbf{B} & A[1, 1]\mathbf{B} & \cdots & A[1, L-1]\mathbf{B} \\
\vdots & \vdots & & \vdots \\
A[K-1, 0]\mathbf{B} & A[K-1, 1]\mathbf{B} & \cdots & A[K-1, L-1]\mathbf{B}
\end{bmatrix} \tag{48}
$$

with the elements

$$(A \otimes B)[m + kM, n + \ell N] := A[k, \ell]B[m, n]. \tag{49}$$

Each block of $\mathbf{A} \otimes \mathbf{B}$ is a scalar multiple of $\mathbf{B}$, and the elements of $\mathbf{A}$ serve as the scale factors, *e.g.*,

$$\begin{bmatrix} 1 \\ 2 \end{bmatrix} \otimes [3, 5] = \begin{bmatrix} 1\,[3,5] \\ 2\,[3,5] \end{bmatrix} = \begin{bmatrix} 3 & 5 \\ 6 & 10 \end{bmatrix},$$

$$[3 \quad 5] \otimes \begin{bmatrix} 1 \\ 2 \end{bmatrix} = \begin{bmatrix} 3\begin{bmatrix} 1 \\ 2 \end{bmatrix} & 5\begin{bmatrix} 1 \\ 2 \end{bmatrix} \end{bmatrix} = \begin{bmatrix} 3 & 5 \\ 6 & 10 \end{bmatrix},$$

$$\begin{bmatrix} 1 & 2 & 3 \\ 4 & 5 & 6 \end{bmatrix} \otimes \begin{bmatrix} 1 & 0 \\ 0 & 1 \end{bmatrix} = \begin{bmatrix} 1 & 0 & 2 & 0 & 3 & 0 \\ 0 & 1 & 0 & 2 & 0 & 3 \\ 4 & 0 & 5 & 0 & 6 & 0 \\ 0 & 4 & 0 & 5 & 0 & 6 \end{bmatrix},$$

$$\begin{bmatrix} 1 & 0 \\ 0 & 1 \end{bmatrix} \otimes \begin{bmatrix} 1 & 2 & 3 \\ 4 & 5 & 6 \end{bmatrix} = \begin{bmatrix} 1 & 2 & 3 & 0 & 0 & 0 \\ 4 & 5 & 6 & 0 & 0 & 0 \\ 0 & 0 & 0 & 1 & 2 & 3 \\ 0 & 0 & 0 & 4 & 5 & 6 \end{bmatrix}.$$

You will immediately recognize the special Kronecker product

$$\mathbf{A}^{(P)} = \mathbf{I}_P \otimes \mathbf{A} \tag{50}$$

from (5) and recall how the corresponding algebraic identities

$$\mathbf{I}_Q \otimes (\mathbf{I}_P \otimes \mathbf{A}) = \mathbf{I}_{PQ} \otimes \mathbf{A} \tag{51}$$
$$\mathbf{I}_P \otimes (\mathbf{AB}) \quad = (\mathbf{I}_P \otimes \mathbf{A})(\mathbf{I}_P \otimes \mathbf{B}) \quad \text{when } \mathbf{AB} \text{ is defined} \tag{52}$$
$$\mathbf{I}_P \otimes (\alpha\mathbf{A}) \quad = \alpha(\mathbf{I}_P \otimes \mathbf{A}) \tag{53}$$
$$(\mathbf{I}_P \otimes \mathbf{A})^T \quad = \mathbf{I}_P \otimes \mathbf{A}^T \tag{54}$$

from (6)–(9) were used to derive the sparse factorizations (12), (15), (32), and (47). Indeed, the exponent notation (5) provides a gentle introduction to the use of identities such as (51)–(54).

It is a fairly simple matter to verify that (48) has the algebraic properties

$$\mathbf{A} \otimes (\mathbf{B} \otimes \mathbf{C}) = (\mathbf{A} \otimes \mathbf{B}) \otimes \mathbf{C}, \tag{55}$$
$$(\mathbf{A} + \mathbf{B}) \otimes \mathbf{C} = (\mathbf{A} \otimes \mathbf{C}) + (\mathbf{B} \otimes \mathbf{C}) \quad \text{when } \mathbf{A} + \mathbf{B} \text{ is defined}, \tag{56}$$
$$\mathbf{C} \otimes (\mathbf{A} + \mathbf{B}) = (\mathbf{C} \otimes \mathbf{A}) + (\mathbf{C} \otimes \mathbf{B}) \quad \text{when } \mathbf{A} + \mathbf{B} \text{ is defined}, \tag{57}$$

that we would demand from any binary operation that carries the name *product*, but there are situations (as illustrated above) where

$$\mathbf{A} \otimes \mathbf{B} \neq \mathbf{B} \otimes \mathbf{A}.$$

Example: Verify (55) when $\mathbf{A}, \mathbf{B}, \mathbf{C}$ are 2×2 matrices.

Solution: We use (48) four times as we write

$$
\begin{aligned}
\mathbf{A} \otimes (\mathbf{B} \otimes \mathbf{C}) &= \begin{bmatrix} a_{11}\mathbf{B} \otimes \mathbf{C} & a_{12}\mathbf{B} \otimes \mathbf{C} \\ a_{21}\mathbf{B} \otimes \mathbf{C} & a_{22}\mathbf{B} \otimes \mathbf{C} \end{bmatrix} \\[2mm]
&= \begin{bmatrix} a_{11}\begin{bmatrix} b_{11}\mathbf{C} & b_{12}\mathbf{C} \\ b_{21}\mathbf{C} & b_{22}\mathbf{C} \end{bmatrix} & a_{12}\begin{bmatrix} b_{11}\mathbf{C} & b_{12}\mathbf{C} \\ b_{21}\mathbf{C} & b_{22}\mathbf{C} \end{bmatrix} \\ a_{21}\begin{bmatrix} b_{11}\mathbf{C} & b_{12}\mathbf{C} \\ b_{21}\mathbf{C} & b_{22}\mathbf{C} \end{bmatrix} & a_{22}\begin{bmatrix} b_{11}\mathbf{C} & b_{12}\mathbf{C} \\ b_{21}\mathbf{C} & b_{22}\mathbf{C} \end{bmatrix} \end{bmatrix} \\[2mm]
&= \begin{bmatrix} a_{11}\mathbf{B} & a_{12}\mathbf{B} \\ a_{21}\mathbf{B} & a_{22}\mathbf{B} \end{bmatrix} \otimes \mathbf{C} \\[2mm]
&= (\mathbf{A} \otimes \mathbf{B}) \otimes \mathbf{C}.
\end{aligned}
$$

You can use a similar argument to prove (55) or you can use Ex. 6.26(a). ∎

The Kronecker product (48) also has the special properties

$$(\alpha\mathbf{A}) \otimes \mathbf{B} = \alpha(\mathbf{A} \otimes \mathbf{B}) = \mathbf{A} \otimes (\alpha\mathbf{B}), \tag{58}$$

$$(\mathbf{A} \otimes \mathbf{B})(\mathbf{C} \otimes \mathbf{D}) = (\mathbf{AC}) \otimes (\mathbf{BD}) \text{ when } \mathbf{AC} \text{ and } \mathbf{BD} \text{ are defined}, \tag{59}$$

$$(\mathbf{A} \otimes \mathbf{B})^T = \mathbf{A}^T \otimes \mathbf{B}^T, \tag{60}$$

$$(\mathbf{A} \otimes \mathbf{B})^{-1} = \mathbf{A}^{-1} \otimes \mathbf{B}^{-1} \text{ when } \mathbf{A}^{-1} \text{ and } \mathbf{B}^{-1} \text{ are defined}, \tag{61}$$

$$\mathbf{I}_P \otimes \mathbf{I}_Q = \mathbf{I}_{PQ} \text{ when } P = 1, 2, \ldots, \; Q = 1, 2, \ldots. \tag{62}$$

Example: Verify the multiplication rule (59) when $\mathbf{A}, \mathbf{B}, \mathbf{C}, \mathbf{D}$ are 2×2 matrices.

Solution: Using (48) and the definition of the matrix product we write

$$
\begin{aligned}
(\mathbf{A} \otimes \mathbf{B})(\mathbf{C} \otimes \mathbf{D}) &= \begin{bmatrix} a_{11}\mathbf{B} & a_{12}\mathbf{B} \\ a_{21}\mathbf{B} & a_{22}\mathbf{B} \end{bmatrix} \begin{bmatrix} c_{11}\mathbf{D} & c_{12}\mathbf{D} \\ c_{21}\mathbf{D} & c_{22}\mathbf{D} \end{bmatrix} \\[2mm]
&= \begin{bmatrix} (a_{11}c_{11} + a_{12}c_{21})\mathbf{BD} & (a_{11}c_{12} + a_{12}c_{22})\mathbf{BD} \\ (a_{21}c_{11} + a_{22}c_{21})\mathbf{BD} & (a_{21}c_{12} + a_{22}c_{22})\mathbf{BD} \end{bmatrix} \\[2mm]
&= (\mathbf{AC}) \otimes (\mathbf{BD}).
\end{aligned}
$$

You can use a similar argument to prove (59) or you can use Ex. 6.26(b). ∎

Rearrangement of Kronecker products

Let $\mathbf{x}, \mathbf{y}$ be column vectors with 2, 3 components. When we use (38) to perform a perfect 2-shuffle of the components of $\mathbf{x} \otimes \mathbf{y}$ we find

$$
\mathbf{S}_{2,3}(\mathbf{x} \otimes \mathbf{y}) = \mathbf{S}_{2,3} \begin{bmatrix} x_1\mathbf{y} \\ x_2\mathbf{y} \end{bmatrix} = \mathbf{S}_{2,3} \begin{bmatrix} x_1 y_1 \\ x_1 y_2 \\ x_1 y_3 \\ \hline x_2 y_1 \\ x_2 y_2 \\ x_2 y_3 \end{bmatrix} = \begin{bmatrix} x_1 y_1 \\ x_2 y_1 \\ \hline x_1 y_2 \\ x_2 y_2 \\ \hline x_1 y_3 \\ x_2 y_3 \end{bmatrix} = \begin{bmatrix} y_1\mathbf{x} \\ y_2\mathbf{x} \\ y_3\mathbf{x} \end{bmatrix} = \mathbf{y} \otimes \mathbf{x}.
$$

More generally, when $\mathbf{x}, \mathbf{y}$ have P, Q components for some $P, Q = 1, 2, \ldots$ an analogous argument shows that

$$
\mathbf{S}_{P,Q}(\mathbf{x} \otimes \mathbf{y}) = \mathbf{y} \otimes \mathbf{x}. \tag{63}
$$

Using (63) we can easily derive the commutation relation

$$
\mathbf{S}_{P,Q}(\mathbf{A} \otimes \mathbf{B}) = (\mathbf{B} \otimes \mathbf{A})\mathbf{S}_{P,Q}, \tag{64}
$$

which holds when $\mathbf{A}$ is $P \times P$ and $\mathbf{B}$ is $Q \times Q$. Indeed, if we are given any P, Q component column vectors $\mathbf{x}, \mathbf{y}$ we can use (59) and (63) to write

$$
\begin{aligned}
\{\mathbf{S}_{P,Q}(\mathbf{A} \otimes \mathbf{B})\}(\mathbf{x} \otimes \mathbf{y}) &= \mathbf{S}_{P,Q}\{(\mathbf{Ax} \otimes \mathbf{By})\} \\
&= (\mathbf{By}) \otimes (\mathbf{Ax}) \\
&= (\mathbf{B} \otimes \mathbf{A})(\mathbf{y} \otimes \mathbf{x}) \\
&= \{(\mathbf{B} \otimes \mathbf{A})\mathbf{S}_{P,Q}\}(\mathbf{x} \otimes \mathbf{y}).
\end{aligned}
$$

Since the set of all such vectors $\mathbf{x} \otimes \mathbf{y}$ spans the linear space of PQ component column vectors, *cf.* Ex. 6.29, this establishes (64).

Routine manipulations of Kronecker product factorizations make use of (55), (59) and (63) or (64). We will give two examples to illustrate how this is done.

Example: Verify the shuffle permutation identity

$$\mathbf{S}_{P,QR} = \mathbf{S}_{PR,Q}\mathbf{S}_{PQ,R} \tag{65}$$

Solution: Let $\mathbf{p}, \mathbf{q}, \mathbf{r}$ be any P, Q, R component column vectors. Using (63) and the associativity (55) of the Kronecker product, we see that

$$\begin{aligned}
\mathbf{S}_{P,QR}(\mathbf{p} \otimes (\mathbf{q} \otimes \mathbf{r})) &= (\mathbf{q} \otimes \mathbf{r}) \otimes \mathbf{p} \\
&= \mathbf{q} \otimes (\mathbf{r} \otimes \mathbf{p}) \\
&= \mathbf{S}_{PR,Q}((\mathbf{r} \otimes \mathbf{p}) \otimes \mathbf{q}) \\
&= \mathbf{S}_{PR,Q}(\mathbf{r} \otimes (\mathbf{p} \otimes \mathbf{q})) \\
&= \mathbf{S}_{PR,Q}\mathbf{S}_{PQ,R}((\mathbf{p} \otimes \mathbf{q}) \otimes \mathbf{r}) \\
&= \mathbf{S}_{PR,Q}\mathbf{S}_{PQ,R}(\mathbf{p} \otimes (\mathbf{q} \otimes \mathbf{r})).
\end{aligned}$$

Since the set of all such vectors $\mathbf{p} \otimes (\mathbf{q} \otimes \mathbf{r})$ spans the linear space of PQR component column vectors, this establishes (65). ∎

Example: Show that $\mathbf{B}_{16} = (\mathbf{S}_{2,2} \otimes \mathbf{I}_4)(\mathbf{S}_{2,4} \otimes \mathbf{I}_2)\mathbf{S}_{2,8}$.

Solution: Let $\mathbf{a}$, $\mathbf{b}$, $\mathbf{c}$, $\mathbf{d}$ be any 2-component column vectors. Using (13) and (50) together with (63) (and the associativity of the Kronecker product) we write

$$\begin{aligned}
\mathbf{B}_{16}&(\mathbf{a} \otimes \mathbf{b} \otimes \mathbf{c} \otimes \mathbf{d}) \\
&= (\mathbf{I}_4 \otimes \mathbf{S}_{2,2})(\mathbf{I}_2 \otimes \mathbf{S}_{4,2})(\mathbf{S}_{8,2})(\mathbf{a} \otimes \mathbf{b} \otimes \mathbf{c} \otimes \mathbf{d}) \\
&= (\mathbf{I}_4 \otimes \mathbf{S}_{2,2})(\mathbf{I}_2 \otimes \mathbf{S}_{4,2})(\mathbf{d} \otimes \mathbf{a} \otimes \mathbf{b} \otimes \mathbf{c}) \\
&= (\mathbf{I}_4 \otimes \mathbf{S}_{2,2})(\mathbf{d} \otimes \mathbf{c} \otimes \mathbf{a} \otimes \mathbf{b}) \\
&= \mathbf{d} \otimes \mathbf{c} \otimes \mathbf{b} \otimes \mathbf{a}.
\end{aligned}$$

Analogously,

$$\begin{aligned}
(\mathbf{S}_{2,2} \otimes \mathbf{I}_4)&(\mathbf{S}_{2,4} \otimes \mathbf{I}_2)(\mathbf{S}_{2,8})(\mathbf{a} \otimes \mathbf{b} \otimes \mathbf{c} \otimes \mathbf{d}) \\
&= (\mathbf{S}_{2,2} \otimes \mathbf{I}_4)(\mathbf{S}_{2,4} \otimes \mathbf{I}_2)(\mathbf{b} \otimes \mathbf{c} \otimes \mathbf{d} \otimes \mathbf{a}) \\
&= (\mathbf{S}_{2,2} \otimes \mathbf{I}_4)(\mathbf{c} \otimes \mathbf{d} \otimes \mathbf{b} \otimes \mathbf{a}) \\
&= \mathbf{d} \otimes \mathbf{c} \otimes \mathbf{b} \otimes \mathbf{a},
\end{aligned}$$

and this establishes the above identity for $\mathbf{B}_{16}$. The same argument show that

$$\mathbf{B}_{2^m} = (\mathbf{S}_{2,2} \otimes \mathbf{I}_{2^{m-2}})(\mathbf{S}_{2,4} \otimes \mathbf{I}_{2^{m-3}}) \cdots (\mathbf{S}_{2,2^{m-2}} \otimes \mathbf{I}_2)\mathbf{S}_{2,2^{m-1}}. \tag{66}$$

∎

Parallel and vector operations

Let $\mathbf{A}$, $\mathbf{B}$ be $P \times P$, $Q \times Q$ matrices and let the PQ component column vector $\mathbf{x}$ be partitioned into Q component column blocks $\mathbf{x}_0, \mathbf{x}_1, \ldots, \mathbf{x}_{P-1}$. When we evaluate

$$(\mathbf{I}_P \otimes \mathbf{B})\mathbf{x} = \begin{bmatrix} \mathbf{B} & & & \\ & \mathbf{B} & & \\ & & \ddots & \\ & & & \mathbf{B} \end{bmatrix} \begin{bmatrix} \mathbf{x}_0 \\ \mathbf{x}_1 \\ \vdots \\ \mathbf{x}_{P-1} \end{bmatrix}$$

we can process $\mathbf{x}_0, \mathbf{x}_1, \ldots, \mathbf{x}_{P-1}$ separately to produce $\mathbf{B}\mathbf{x}_0, \mathbf{B}\mathbf{x}_1, \ldots, \mathbf{B}\mathbf{x}_{P-1}$. On the other hand, when we evaluate

$$(\mathbf{A} \otimes \mathbf{I}_Q)\mathbf{x} = \begin{bmatrix} A[0,0]\mathbf{I}_Q & A[0,1]\mathbf{I}_Q & \cdots & A[0,P-1]\mathbf{I}_Q \\ A[1,0]\mathbf{I}_Q & A[1,1]\mathbf{I}_Q & \cdots & A[1,P-1]\mathbf{I}_Q \\ \vdots & \vdots & & \vdots \\ A[P-1,0]\mathbf{I}_Q & A[P-1,1]\mathbf{I}_Q & \cdots & A[P-1,P-1]\mathbf{I}_Q \end{bmatrix} \begin{bmatrix} \mathbf{x}_0 \\ \mathbf{x}_1 \\ \vdots \\ \mathbf{x}_{P-1} \end{bmatrix}$$

we must scale and sum the vectors $\mathbf{x}_0, \mathbf{x}_1, \ldots, \mathbf{x}_{P-1}$ as we compute

$$A[k,0]\mathbf{x}_0 + A[k,1]\mathbf{x}_1 + \cdots + A[k,P-1]\mathbf{x}_{P-1}, \quad k = 0, 1, \ldots, P-1.$$

For this reason we refer to the computation of $(\mathbf{I}_P \otimes \mathbf{B})\mathbf{x}$ as a *parallel operation* and to the computation of $(\mathbf{A} \otimes \mathbf{I}_Q)\mathbf{x}$ as a *vector operation*. As you might surmise, computers have been designed to do such computations with great efficiency.

If we want to evaluate $(\mathbf{A} \otimes \mathbf{B})\mathbf{x}$ on a computer that excels at parallel operations, we use (59), (64), and (39) to produce a factorization

$$\mathbf{A} \otimes \mathbf{B} = (\mathbf{A} \otimes \mathbf{I}_Q)(\mathbf{I}_P \otimes \mathbf{B}) = \mathbf{S}_{Q,P}(\mathbf{I}_Q \otimes \mathbf{A})\mathbf{S}_{P,Q}(\mathbf{I}_P \otimes \mathbf{B}),$$

which allows us to apply $\mathbf{A} \otimes \mathbf{B}$ to $\mathbf{x}$ with parallel operations and sorts. If we use a computer that excels at vector operations we might prefer the factorization

$$\mathbf{A} \otimes \mathbf{B} = (\mathbf{A} \otimes \mathbf{I}_Q)\mathbf{S}_{Q,P}(\mathbf{B} \otimes \mathbf{I}_P)\mathbf{S}_{P,Q},$$

which uses vector operations and sorts. In this way we can modify the elementary factorizations (12) and (32) to produce algorithms that run efficiently on a particular computer.

Stockham's autosort FFT

We use the commutation relation (64) to write (11) in the alternative form

$$\mathcal{F}_{2M} = \frac{1}{2}\mathbf{Q}_{2M}\mathbf{S}_{M,2}(\mathcal{F}_M \otimes \mathbf{I}_2).$$

We then use this new zipper identity with (59) and (62) to factor $\mathcal{F}_{16}$ by writing

$$\begin{aligned}
\mathcal{F}_{16} &= \frac{1}{2}\mathbf{Q}_{16}\mathbf{S}_{8,2}(\mathcal{F}_8 \otimes \mathbf{I}_2) \\
&= \frac{1}{4}\mathbf{Q}_{16}\mathbf{S}_{8,2}((\mathbf{Q}_8\mathbf{S}_{4,2}(\mathcal{F}_4 \otimes \mathbf{I}_2)) \otimes \mathbf{I}_2) \\
&= \frac{1}{4}\mathbf{Q}_{16}\mathbf{S}_{8,2}(\mathbf{Q}_8 \otimes \mathbf{I}_2)(\mathbf{S}_{4,2} \otimes \mathbf{I}_2)(\mathcal{F}_4 \otimes \mathbf{I}_4) \qquad (67) \\
&\;\;\vdots \\
&= \frac{1}{16}(\mathbf{Q}_{16} \otimes \mathbf{I}_1)(\mathbf{S}_{8,2} \otimes \mathbf{I}_1)(\mathbf{Q}_8 \otimes \mathbf{I}_2)(\mathbf{S}_{4,2} \otimes \mathbf{I}_2) \\
&\qquad\qquad (\mathbf{Q}_4 \otimes \mathbf{I}_4)(\mathbf{S}_{2,2} \otimes \mathbf{I}_4)(\mathbf{Q}_2 \otimes \mathbf{I}_8)(\mathbf{S}_{1,2} \otimes \mathbf{I}_8).
\end{aligned}$$

This shows the pattern for Stockham's autosort factorization of $\mathcal{F}_N$ with vector operations when $N = 2^m$. We must provide an additional array of storage to carry out the rearrangement of the data that takes place just before we apply $\mathbf{Q}_2 \otimes \mathbf{I}_{N/2}, \mathbf{Q}_4 \otimes \mathbf{I}_{N/4}, \dots$, but this eliminates the need for the bit reversal permutation.

An alternative Stockham factorization uses only parallel operations. We begin with the Cooley-Tukey factorization

$$\mathcal{F}_{16} = \frac{1}{16}\mathbf{Q}_{16}(\mathbf{I}_2 \otimes \mathbf{Q}_8)(\mathbf{I}_4 \otimes \mathbf{Q}_4)(\mathbf{I}_8 \otimes \mathbf{Q}_2)(\mathbf{S}_{2,2} \otimes \mathbf{I}_4)(\mathbf{S}_{2,4} \otimes \mathbf{I}_2)\mathbf{S}_{2,8}$$

with the expression (66) for $\mathbf{B}_{16}$. When $\mathbf{A}$, $\mathbf{B}$ are $P \times P$, $Q \times Q$ we can use (59) to obtain the commutation rule

$$(\mathbf{I}_P \otimes \mathbf{B})(\mathbf{A} \otimes \mathbf{I}_Q) = \mathbf{A} \otimes \mathbf{B} = (\mathbf{A} \otimes \mathbf{I}_Q)(\mathbf{I}_P \otimes \mathbf{B}), \qquad (68)$$

which allows us to rearrange the above factors of $\mathcal{F}_{16}$ and write

$$\begin{aligned}
\mathcal{F}_{16} = \frac{1}{16}&(\mathbf{I}_1 \otimes \mathbf{Q}_{16})(\mathbf{S}_{2,1} \otimes \mathbf{I}_8)(\mathbf{I}_2 \otimes \mathbf{Q}_8)(\mathbf{S}_{2,2} \otimes \mathbf{I}_4) \\
&\cdot (\mathbf{I}_4 \otimes \mathbf{Q}_4)(\mathbf{S}_{2,4} \otimes \mathbf{I}_2)(\mathbf{I}_8 \otimes \mathbf{Q}_2)(\mathbf{S}_{2,8} \otimes \mathbf{I}_1).
\end{aligned} \qquad (69)$$

This shows the pattern for the factorization of $\mathcal{F}_N$ when $N = 2^m$.

There are analogous Stockham factorizations of $\mathcal{F}_N$ when $N = P_1 P_2 \cdots P_m$, and we can use these (or the corresponding transposed factorizations) as frameworks for developing FFT codes, *cf.* Ex. 6.34. You now have all of the concepts that you will need to sort out the details!

References

R. Bracewell, *The Hartley Transform*, Oxford University Press, New York, 1986.

> The FHT as seen by its creator.

E.O. Brigham, *The Fast Fourier Transform and Its Applications*, Prentice Hall, Englewood Cliffs, NJ, 1988.

> A pictorial, intuitive approach to FFT algorithms with selected applications in science and engineering.

J.W. Cooley and J.W. Tukey, An algorithm for the machine computation of complex Fourier series, *Math. Comp.* **19**(1965), 297–301.

> This initial presentation of the FFT is the most frequently cited mathematics paper ever written ... and you can actually read it!

R. Tolimieri, M. An, and C. Lu, *Algorithms for Discrete Fourier Transform and Convolution*, Springer-Verlag, New York, 1989.

> A mature (but exceptionally readable) exposition of fast algorithms for Fourier analysis.

C. Van Loan, *Computational Frameworks for the Fast Fourier Transform*, SIAM, Philadelphia, 1992.

> The definitive mathematical exposition of the FFT (using MATLAB).

J.S. Walker, *Fast Fourier Transforms, 2nd ed.*, CRC Press, Boca Raton, FL, 1996.

> An elementary introduction to the FFT with selected applications to PDEs, optics, etc., illustrated with PC software created by the author.

Exercise Set 6

▶ **EXERCISE 6.1:** Many computational tasks of Fourier analysis require us to find the value of some polynomial

$$P(c_0, c_1, \ldots, c_{N-1}; z) := c_0 + c_1 z + c_2 z^2 + \cdots + c_{N-1} z^{N-1}$$

when the complex coefficients $c_0, c_1, \ldots, c_{N-1}$ and a point $z = e^{i\theta}$ on the unit circle of the complex plane are given. The naive Horner algorithm

$$P := c_{N-1}$$

For $n = N - 2, N - 3, \ldots, 0$ do:

$$\llcorner P := c_n + z \cdot P$$

$$P(c_0, c_1, \ldots, c_{N-1}; z) := P$$

uses complex arithmetic. In this exercise you will study alternative algorithms that use real arithmetic. For this purpose we define

$$C(\alpha_0, \alpha_1, \ldots, \alpha_m; \theta) := \alpha_0 + \alpha_1 \cos\theta + \alpha_2 \cos 2\theta + \cdots + \alpha_m \cos m\theta$$
$$S(\alpha_0, \alpha_1, \ldots, \alpha_m; \theta) := \qquad \alpha_1 \sin\theta + \alpha_2 \sin 2\theta + \cdots + \alpha_m \sin m\theta$$

where $m = 0, 1, 2, \ldots$ and where the coefficients $\alpha_0, \alpha_1, \ldots, \alpha_m$ are real.

(a) Let $c_n = a_n + ib_n$ where a_n, b_n are real, $n = 0, 1, \ldots, N-1$. Show that

$$P(c_0, c_1, \ldots, c_{N-1}; e^{i\theta}) = \{C(a_0, a_1, \ldots, a_{N-1}; \theta) - S(b_0, b_1, \ldots, b_{N-1}; \theta)\}$$
$$+ i\{C(b_0, b_1, \ldots, b_{N-1}; \theta) + S(a_0, a_1, \ldots, a_{N-1}; \theta)\}.$$

(b) Set $P = R + iX$ with R, X real in the above Horner algorithm and thereby show that $C(\alpha_0, \alpha_1, \ldots, \alpha_{N-1}; \theta)$, $S(\alpha_0, \alpha_1, \ldots, \alpha_{N-1}; \theta)$ can be computed as follows.

$$c := \cos\theta$$
$$s := \sin\theta$$
$$R := \alpha_{N-1}$$
$$X := 0$$

For $n = N-2, N-3, \ldots, 1, 0$ do:

$$\left| \begin{array}{l} T := \alpha_n + c \cdot R - s \cdot X \\ X := \qquad s \cdot R + c \cdot X \\ R := T \end{array} \right.$$

$$C(\alpha_0, \alpha_1, \ldots, \alpha_{N-1}; \theta) := R$$
$$S(\alpha_0, \alpha_1, \ldots, \alpha_{N-1}; \theta) := X.$$

(c) How many real additions and multiplications must be done when we compute a C, S pair by using the algorithm of (b)? How many must be done when we evaluate the real and imaginary parts of $P(c_0, c_1, \ldots, c_{N-1}; e^{i\theta})$ by using the decomposition of (a) together with the algorithm of (b)?

(d) Show that

$$C(\alpha_0, \alpha_1, \ldots, \alpha_{m-1}, A, B; \theta) = C(\alpha_0, \alpha_1, \ldots, \alpha_{m-2}, \alpha_{m-1} - B, A + 2B\cos\theta; \theta),$$
$$S(\alpha_0, \alpha_1, \ldots, \alpha_{m-1}, A, B; \theta) = S(\alpha_0, \alpha_1, \ldots, \alpha_{m-2}, \alpha_{m-1} - B, A + 2B\cos\theta; \theta).$$

Hint. Use the trigonometric identities

$$\cos\{(m+1)\theta\} = 2\cos\theta\,\cos(m\theta) - \cos\{(m-1)\theta\},$$
$$\sin\{(m+1)\theta\} = 2\cos\theta\,\sin(m\theta) - \sin\{(m-1)\theta\}.$$

(e) Use (d) to verify that $C(\alpha_0, \alpha_1, \ldots, \alpha_{N-1}; \theta)$ and $S(\alpha_0, \alpha_1, \ldots, \alpha_{N-1}; \theta)$ can be found by using the following real algorithm of Goertzel.

$$A := 0$$
$$B := 0$$
$$u := 2 \cos \theta$$

For $n = N - 1, N - 2, \ldots, 0$ do:

$$
\begin{vmatrix}
T := A \\
A := \alpha_n - B \\
B := T + u \cdot B
\end{vmatrix}
$$

$$C(\alpha_0, \alpha_1, \ldots, \alpha_{N-1}; \theta) := A + B \cdot \cos \theta$$
$$S(\alpha_0, \alpha_1, \ldots, \alpha_{N-1}; \theta) := \qquad B \cdot \sin \theta$$

(f) How does the cost of Goertzel's algorithm compare with that of (b)?

▶ **EXERCISE 6.2:** Suppose that you have created efficient macros *cpoly*, *spoly* for computing

$$cpoly(\alpha_0, \alpha_1, \ldots, \alpha_m; \theta) := \sum_{k=0}^{m} \alpha_k \cos(k\theta), \quad spoly(\alpha_0, \alpha_1, \ldots, \alpha_m; \theta) := \sum_{k=1}^{m} \alpha_k \sin(k\theta)$$

when the real numbers $\alpha_0, \alpha_1, \ldots, \alpha_m, \theta$ are given, *e.g.*, as described in Ex. 6.1(e). Show how to use *cpoly*, *spoly* to do the following computational tasks.

(a) Let the complex N-vector $\mathbf{f}$ have the discrete Fourier transform $\mathbf{F}$, and let

$$f[n] = f_R[n] + if_I[n], \qquad F[k] = F_R[k] + iF_I[k]$$

where f_R, f_I, F_R, F_I are real valued. Compute $F_R[k]$, $F_I[k]$ when $f_R[n]$, $f_I[n]$, $n = 0, 1, \ldots, N - 1$ are given.

Hint. $F_R[k] = N^{-1} \cdot cpoly(f_R[0], \ldots, f_R[N-1]; 2\pi k/N) + \cdots$.

(b) Let y be the trigonometric polynomial

$$
y(x) :=
\begin{cases}
\dfrac{a_0}{2} + \displaystyle\sum_{k=1}^{M} \left\{ a_k \cos\left(\dfrac{2\pi kx}{p}\right) + b_k \sin\left(\dfrac{2\pi kx}{p}\right) \right\} & \text{if } N = 2M + 1 \\[4ex]
\dfrac{a_0}{2} + \displaystyle\sum_{k=1}^{M-1} \left\{ a_k \cos\left(\dfrac{2\pi kx}{p}\right) + b_k \sin\left(\dfrac{2\pi kx}{p}\right) \right\} + \dfrac{a_M}{2} \cos\left(\dfrac{2\pi Mx}{p}\right) \\[2ex]
\qquad\qquad\qquad\qquad\qquad\qquad\qquad\qquad \text{if } N = 2M.
\end{cases}
$$

Compute values for a_k, b_k that will make $y(np/N) = y_n$, $n = 0, 1, \ldots, N - 1$ when N and the real numbers $y_0, y_1, \ldots, y_{N-1}$ are given.

▶ **EXERCISE 6.3:**   Suppose that you have created efficient macros *flip*, *bar*, *chopm*, *ft* for computing

$$flip(x_0, x_1, x_2, \ldots, x_{N-1}) := (x_0, x_{N-1}, x_{N-2}, \ldots, x_1),$$
$$bar(x_0, x_1, x_2, \ldots, x_{N-1}) := (\bar{x}_0, \bar{x}_1, \bar{x}_2, \ldots, \bar{x}_{N-1}),$$
$$chop_m(x_0, x_1, x_2, \ldots, x_{N-1}) := (x_0, x_1, x_2, \ldots, x_{m-1}), \quad m = 1, 2, \ldots, N, \text{ and}$$
$$ft(x_0, x_1, x_2, \ldots, x_{N-1}) := (X_0, X_1, X_2, \ldots, X_{N-1}).$$

Here $\bar{x}_n$ is the complex conjugate of x_n and

$$X_k := \frac{1}{N} \sum_{n=0}^{N-1} x_n e^{-2\pi i k n / N}, \qquad k = 0, 1, \ldots, N - 1.$$

Show how to use *flip*, *bar*, *chopm*, *ft* (and the binary vector operations of componentwise addition, subtraction, multiplication) to carry out the following computational tasks.

(a) Compute the vector $\mathbf{f}$ when the complex N-vector $\mathbf{F} := \mathbf{f}^\wedge$ is given.

(b) Compute the convolution product $\mathbf{f} * \mathbf{g}$ when the complex N-vectors $\mathbf{f}, \mathbf{g}$ are given.

(c) Compute the autocorrelation $\mathbf{f} \star \mathbf{f}$ when the complex N-vector $\mathbf{f}$ is given.

(d) Let $N = 2M$ and let $a_0, b_0, a_1, b_1, \ldots, a_M, b_M$ be given with $b_0 = b_M = 0$. Compute the vector $\mathbf{y} = (y_0, y_1, \ldots, y_{N-1})$ of samples

$$y_n := \frac{a_0}{2} + \sum_{k=1}^{M-1} \left\{ a_k \cos\left(\frac{2\pi k n}{N}\right) + b_k \sin\left(\frac{2\pi k n}{N}\right) \right\} + \frac{a_M}{2} \cos\left(\frac{2\pi M n}{N}\right),$$

$$n = 0, 1, \ldots, N - 1.$$

(e) Let $\mathbf{y} = (y_0, y_1, \ldots, y_{N-1})$ be given with $N = 2M$. Compute $\mathbf{a} = (a_0, a_1, \ldots, a_M)$, $\mathbf{b} = (b_0, b_1, \ldots, b_M)$ so that a_k, b_k and y_n are related as in (d).

   Hint. Use the result of Ex. 1.16.

▶ **EXERCISE 6.4:**   Let $\mathbf{f} := (f_0, f_1, \ldots, f_{11})$ be a complex vector with the discrete Fourier transform $\mathbf{F} = (F_0, F_1, \ldots, F_{11})$, and let $\alpha := \frac{1}{2}(1 + \sqrt{3}\, i)$, $\beta := \frac{1}{2}(\sqrt{3} + i)$. Use the summation rule from (4.57) to derive the following expressions for $F_0, F_1, \ldots, F_{11}$.

(a) $12F_0 = (f_0 + f_2 + f_4 + f_6 + f_8 + f_{10}) + (f_1 + f_3 + f_5 + f_7 + f_9 + f_{11})$

   $12F_6 = (f_0 + f_2 + f_4 + f_6 + f_8 + f_{10}) - (f_1 + f_3 + f_5 + f_7 + f_9 + f_{11})$

(b) $12F_4 = (f_0 + f_3 + f_6 + f_9) - \alpha (f_1 + f_4 + f_7 + f_{10}) - \bar{\alpha} (f_2 + f_5 + f_8 + f_{11})$

   $12F_8 = (f_0 + f_3 + f_6 + f_9) - \bar{\alpha} (f_1 + f_4 + f_7 + f_{10}) - \alpha (f_2 + f_5 + f_8 + f_{11})$

(c) $12F_3 = (f_0 + f_4 + f_8) - i (f_1 + f_5 + f_9) - (f_2 + f_6 + f_{10}) + i (f_3 + f_7 + f_{11})$

   $12F_9 = (f_0 + f_4 + f_8) + i (f_1 + f_5 + f_9) - (f_2 + f_6 + f_{10}) - i (f_3 + f_7 + f_{11})$

(d) $\quad 12F_2 = \{(f_0 + f_6) - (f_3 + f_9)\} + \overline{\alpha}\{(f_1 + f_7) - (f_4 + f_{10})\}$
$$- \alpha\{(f_2 + f_8) - (f_5 + f_{11})\}$$
$12F_{10} = \{(f_0 + f_6) - (f_3 + f_9)\} + \alpha\{(f_1 + f_7) - (f_4 + f_{10})\}$
$$- \overline{\alpha}\{(f_2 + f_8) - (f_5 + f_{11})\}$$

(e) $\quad 12F_1 = (f_0 - f_6) + \overline{\beta}(f_1 - f_7) + \overline{\alpha}(f_2 - f_8) - i(f_3 - f_9)$
$$- \alpha(f_4 - f_{10}) - \beta(f_5 - f_{11})$$
$12F_{11} = (f_0 - f_6) + \beta(f_1 - f_7) + \alpha(f_2 - f_8) + i(f_3 - f_9)$
$$- \overline{\alpha}(f_4 - f_{10}) - \overline{\beta}(f_5 - f_{11})$$
$12F_5 = (f_0 - f_6) - \beta(f_1 - f_7) + \alpha(f_2 - f_8) - i(f_3 - f_9)$
$$- \overline{\alpha}(f_4 - f_{10}) + \overline{\beta}(f_5 - f_{11})$$
$12F_7 = (f_0 - f_6) - \overline{\beta}(f_1 - f_7) + \overline{\alpha}(f_2 - f_8) + i(f_3 - f_9)$
$$- \alpha(f_4 - f_{10}) + \beta(f_5 - f_{11})$$

Note. A real version of this analysis was used to produce the Whittaker-Robinson flow chart for 12-point harmonic analysis, *cf.* Appendix 4.

▶ **EXERCISE 6.5:** In this (historical) exercise you will learn about the *paper strip* method for doing Fourier analysis. Suppose that we wish to compute 3-digit approximations for

$$y_n := a_0 + \sum_{k=1}^{M}\left\{a_k \cdot \cos\left(\frac{2\pi kn}{N}\right) + b_k \cdot \sin\left(\frac{2\pi kn}{N}\right)\right\}$$

when the real coefficients $a_0, a_1, b_1, \ldots, a_M, b_M$ are given, *cf.* Ex. 6.1 and Ex. 6.2. We create paper strips labeled

$$100 \cdot C_k, \quad 200 \cdot C_k, \quad \ldots, \quad 900 \cdot C_k, \qquad 100 \cdot S_k, \quad 200 \cdot S_k, \quad \ldots, \quad 900 \cdot S_k,$$
$$10 \cdot C_k, \quad 20 \cdot C_k, \quad \ldots, \quad 90 \cdot C_k, \qquad 10 \cdot S_k, \quad 20 \cdot S_k, \quad \ldots, \quad 90 \cdot S_k,$$
$$1 \cdot C_k, \quad 2 \cdot C_k, \quad \ldots, \quad 9 \cdot C_k, \qquad 1 \cdot S_k, \quad 2 \cdot S_k, \quad \ldots, \quad 9 \cdot S_k,$$

with $k = 0, 1, \ldots, M$ for the C's and $k = 1, 2, \ldots, M$ for the S's. Each strip has N identically sized cells indexed with $n = 0, 1, \ldots, N - 1$. In the nth cell of the strip labeled $\alpha \cdot C_k, \alpha \cdot S_k$ is the integer closest to $\alpha \cdot \cos(2\pi kn/N), \alpha \cdot \sin(2\pi kn/N)$, respectively. For example, when $N = 12$ we produce the strip

$100 \cdot C_1$	100	87	50	0	-50	-87	-100	-87	-50	0	50	87

and print the negative

$-100 \cdot C_1$	-100	-87	-50	0	50	87	100	87	50	0	-50	-87

on the reverse side. To evaluate

$$y_n = .352 - .826 \cdot \cos\left(\frac{2\pi n}{N}\right) + .074 \cdot \cos\left(\frac{4\pi n}{N}\right) + .169 \cdot \sin\left(\frac{2\pi n}{N}\right)$$

we align the strips labeled

$$
\begin{array}{cccc}
300 \cdot C_0, & -800 \cdot C_1, & & 100 \cdot S_1, \\
50 \cdot C_0, & -20 \cdot C_1, & 70 \cdot C_2, & 60 \cdot S_1, \\
2 \cdot C_0, & -6 \cdot C_1, & 4 \cdot C_2, & 9 \cdot S_1,
\end{array}
$$

add the (signed) integers in the N columns, apply the scale factor 10^{-3} to these sums, and thereby obtain $y_0, y_1, \ldots, y_{N-1}$.

(a) Let $\langle \alpha \rangle$ denote the rounded value of the real number α, i.e.,

$$\langle \alpha \rangle := \begin{cases} \lceil \alpha \rceil & \text{if } \lceil \alpha \rceil - \alpha \leq \frac{1}{2} \\ \lfloor \alpha \rfloor & \text{if } \lceil \alpha \rceil - \alpha > \frac{1}{2}. \end{cases}$$

How big is the error

$$10^{-3} \cdot \left\{ 826 \cdot \cos\left(\frac{2\pi}{12}\right) - \left\langle 800 \cdot \cos\left(\frac{2\pi}{12}\right)\right\rangle - \left\langle 20 \cdot \cos\left(\frac{2\pi}{12}\right)\right\rangle - \left\langle 6 \cdot \cos\left(\frac{2\pi}{12}\right)\right\rangle \right\}$$

associated with the C_1 terms of the above sum when $n = 1$?

Hint. Use a calculator!

(b) How many C strips and how many S strips must we create if we wish to compute 3-digit discrete Fourier transforms for arbitrary $N = 12$ component vectors?

(c) What is the maximum number of strips that we must fetch when we evaluate a particular set of y_n's (to 3 digits) when $N = 12$ and $M = 6$?

(d) Use the paper strip method to compute a 3-digit approximation to the discrete Fourier transform of $\mathbf{f} = (0, 6, 12, 18, 24, 30)$.

Hint. Use a calculator to create the C strips and the S strips that you need for this particular task. You can use Ex. 4.26 to find the exact DFT

$$\mathbf{F} = (15, -3, -3, -3, -3, -3) + i\sqrt{3} \cdot (0, 3, 1, 0, -1, -3).$$

(e) Show that the cost for computing a d-digit approximation to a complex N-point discrete Fourier transform is about $2N^2 d$ integer additions (if we neglect the cost of fetching and filing the paper strips!).

▶ **EXERCISE 6.6:**   In this exercise you will analyze a recursive algorithm for computing the discrete Fourier transform of a complex vector $\mathbf{f} := (f_0, f_1, \ldots, f_{N-1})$ when $N = 2^m$ for some $m = 0, 1, \ldots$. We define

$$
\begin{aligned}
even(f_0, f_1, \ldots, f_{N-1}) &:= (f_0, f_2, f_4, \ldots, f_{N-2}) \\
odd(f_0, f_1, \ldots, f_{N-1}) &:= (f_1, f_3, f_5, \ldots, f_{N-1}) \\
two(f_0, f_1, \ldots, f_{N-1}) &:= \tfrac{1}{2}(f_0, f_1, \ldots, f_{N-1}, f_0, f_1, \ldots, f_{N-1}) \\
mod(f_0, f_1, \ldots, f_{N-1}) &:= (f_0, \omega f_1, \ldots, \omega^{N-1} f_{N-1}), \quad \omega := e^{-2\pi i/N}.
\end{aligned}
$$

We use these mappings to define *rfft* recursively by writing:

> If $N = 1$, then
>> $rfft(\mathbf{f}) := \mathbf{f}$;
>
> else
>> $rfft(\mathbf{f}) := mod(two(rfft(odd(\mathbf{f})))) + two(rfft(even(\mathbf{f})))$.

(a) Write down the operator identity that underlies this computation of $\mathbf{f}^\wedge = rfft(\mathbf{f})$.

Hint. Examine Figs. 6.1, 6.2, 6.7, and observe that the macros *two*, *mod* are based on the repeat and exponential modulation operators.

(b) How many times will *rfft* be "called" in the process of computing the DFT of a vector with $N = 2^m$ components?

(c) How many simultaneously active copies of *rfft* must be used during the computation of (b), *i.e.*, how deep is the recursion?

▶ **EXERCISE 6.7:** Let $\mathbf{f}$, *even*, *odd*, *mod* be as in Ex. 6.6, let

$$sum(f_0, f_1, \ldots, f_{N-1}) := \tfrac{1}{2}(f_0 + f_{N/2}, f_1 + f_{N/2+1}, \ldots, f_{N/2-1} + f_{N-1}),$$

let *bit* be the bit reversal permutation, and let & denote the associative operation of vector concatenation, *e.g.*, $(1,2) \,\&\, (3,4) := (1,2,3,4)$.

(a) Let the operator b be defined recursively by writing:

> If $N = 1$, then
>> $b(\mathbf{f}) := \mathbf{f}$;
>
> else
>> $b(\mathbf{f}) := b(even(\mathbf{f})) \,\&\, b(odd(\mathbf{f}))$.

Show that $b(\mathbf{f}) = bit(\mathbf{f})$.

(b) Let the operator *pfft* be defined recursively by writing:

> If $N = 1$, then
>> $pfft(\mathbf{f}) := \mathbf{f}$;
>
> else
>> $pfft(\mathbf{f}) := pfft(sum(\mathbf{f})) \,\&\, pfft(sum(mod(\mathbf{f})))$.

Show that $\mathbf{f}^\wedge = bit(pfft(\mathbf{f}))$.

Hint. Examine Figs. 6.4, 6.5, 6.7, and observe that the macros *sum*, *mod* are based on the discrete summation and exponential modulation operators.

▶ **EXERCISE 6.8:**   Let $\mathbf{f} := (f_0, f_1, \ldots, f_{2M-1})$ be a real vector with $N := 2M$ components, $M = 1, 2, \ldots$, and let $\mathbf{g} := \mathbf{g}_0 + i\mathbf{g}_1$ where

$$\mathbf{g}_0 := (f_0, f_2, f_4, \ldots, f_{2M-2}), \qquad \mathbf{g}_1 := (f_1, f_3, f_5, \ldots, f_{2M-1}).$$

(a) Using the projection operators $\mathbf{P}_h$, $\mathbf{P}_a$ express $\mathbf{g}_0^\wedge$, $\mathbf{g}_1^\wedge$ in terms of $\mathbf{g}^\wedge$.

Hint. cf. Ex. 5.11(b).

(b) Using the operators $\mathcal{R}_2$, $\mathcal{E}_{-1}$ (as in Fig. 6.7) express $\mathbf{f}^\wedge$ in terms of $\mathbf{g}_0^\wedge$, $\mathbf{g}_1^\wedge$.

(c) Combine the expressions of (a), (b) and thereby express $\mathbf{f}^\wedge$ in terms of $\mathbf{g}^\wedge$.

(d) Using (c), show that it is possible to compute the DFT of a real vector having $2M$ components with only slightly more effort than that required to compute the DFT of a complex vector having M components.

Note. This procedure allows us to compute the DFT of a real N-vector in half the time required to compute the DFT of a complex N-vector when $N = 2^m$, $m = 1, 2, \ldots$ (and when we use an FFT to generate the DFTs.)

▶ **EXERCISE 6.9:**   In this exercise you will learn one way to compute the discrete Fourier transform of a large vector f that must be split into segments for processing. We again use & to denote the associative operation of vector concatenation, cf. Ex. 6.7, and we use various operators from Appendix 3.

(a) Let $N = 2, 4, 6, \ldots$, let $\mathbf{f}_0$, $\mathbf{f}_1$ be N-vectors, let $\mathbf{f} := \mathbf{f}_0 \,\&\, \mathbf{f}_1$ (so that $\mathbf{f}$ has $2N$ components), and let $\mathbf{e}_1$ be the N-vector with $e_1[n] := e^{-2\pi in/2N}$, $n = 0, 1, \ldots, N-1$. Show that we can find the $2N$ component Fourier transform $\mathbf{F} := \mathbf{f}^\wedge$ by computing in turn the N-vectors

$$\mathbf{g}_0 := \mathcal{F}\{(\Xi_2 \mathbf{f}_0) \,\&\, (\Xi_2 \mathbf{f}_1)\}, \qquad \mathbf{g}_1 := \mathcal{M}_{e_1}\mathcal{F}\{(\Xi_2 \mathcal{T}_1 \mathbf{f}_0) \,\&\, (\Xi_2 \mathcal{T}_1 \mathbf{f}_1)\},$$

$$\mathbf{F}_0 := \tfrac{1}{2}\{\mathbf{g}_0 + \mathbf{g}_1\}, \qquad \mathbf{F}_1 := \tfrac{1}{2}\{\mathbf{g}_0 - \mathbf{g}_1\},$$

and taking $\mathbf{F} = \mathbf{F}_0 \,\&\, \mathbf{F}_1$.

Hint. Suitably use $\mathcal{Z}_2$, $\mathcal{T}_{-1}$ to obtain $\mathbf{f}$ from

$$(\Xi_2 \mathbf{f}_0) \,\&\, (\Xi_2 \mathbf{f}_1) = (f[0], f[2], f[4], \ldots, f[2N-2]),$$
$$(\Xi_2 \mathcal{T}_1 \mathbf{f}_0) \,\&\, (\Xi_2 \mathcal{T}_1 \mathbf{f}_1) = (f[1], f[3], f[5], \ldots, f[2N-1]),$$

apply the Fourier transform operator, and simplify the result.

(b) Let $N = 4, 8, 12, \ldots$, let $\mathbf{f}_0$, $\mathbf{f}_1$, $\mathbf{f}_2$, $\mathbf{f}_3$ be N-vectors, let $\mathbf{f} := \mathbf{f}_0 \,\&\, \mathbf{f}_1 \,\&\, \mathbf{f}_2 \,\&\, \mathbf{f}_3$ (so that $\mathbf{f}$ has $4N$-components), and let $\mathbf{e}_0$, $\mathbf{e}_1$, $\mathbf{e}_2$, $\mathbf{e}_3$ be the N-vectors with $e_k[n] := e^{-2\pi ikn/4N}$, $n = 0, 1, \ldots, N-1$. Show that we can find the $4N$-component Fourier transform $\mathbf{F} := \mathbf{f}^\wedge$ by computing in turn the N-vectors

$$\mathbf{g}_0 := \quad\ \ \mathcal{F}\{\Xi_4 \mathbf{f}_0) \quad\ \&\, (\Xi_4 \mathbf{f}_1) \quad\ \&\, (\Xi_4 \mathbf{f}_2) \quad\ \&\, (\Xi_4 \mathbf{f}_3)\},$$

$$\mathbf{g}_1 := \mathcal{M}_{e_1}\mathcal{F}\{\Xi_4 \mathcal{T}_1 \mathbf{f}_0) \,\&\, (\Xi_4 \mathcal{T}_1 \mathbf{f}_1) \,\&\, (\Xi_4 \mathcal{T}_1 \mathbf{f}_2) \,\&\, (\Xi_4 \mathcal{T}_1 \mathbf{f}_3)\},$$

$$\mathbf{g}_2 := \mathcal{M}_{e_2}\mathcal{F}\{\Xi_4 \mathcal{T}_2 \mathbf{f}_0) \,\&\, (\Xi_4 \mathcal{T}_2 \mathbf{f}_1) \,\&\, (\Xi_4 \mathcal{T}_2 \mathbf{f}_2) \,\&\, (\Xi_4 \mathcal{T}_2 \mathbf{f}_3)\},$$

$$\mathbf{g}_3 := \mathcal{M}_{e_3}\mathcal{F}\{\Xi_4 \mathcal{T}_3 \mathbf{f}_0) \,\&\, (\Xi_4 \mathcal{T}_3 \mathbf{f}_1) \,\&\, (\Xi_4 \mathcal{T}_3 \mathbf{f}_2) \,\&\, (\Xi_4 \mathcal{T}_3 \mathbf{f}_3)\},$$

$$\mathbf{F}_0 := \tfrac{1}{4}\{\mathbf{g}_0 + \mathbf{g}_1 + \mathbf{g}_2 + \mathbf{g}_3\}, \qquad \mathbf{F}_1 := \tfrac{1}{4}\{\mathbf{g}_0 - i\mathbf{g}_1 - \mathbf{g}_2 + i\mathbf{g}_3\},$$
$$\mathbf{F}_2 := \tfrac{1}{4}\{\mathbf{g}_0 - \mathbf{g}_1 + \mathbf{g}_2 - \mathbf{g}_3\}, \qquad \mathbf{F}_3 := \tfrac{1}{4}\{\mathbf{g}_0 + i\mathbf{g}_1 - \mathbf{g}_2 - i\mathbf{g}_3\},$$

and taking $\mathbf{F} = \mathbf{F}_0 \,\&\, \mathbf{F}_1 \,\&\, \mathbf{F}_2 \,\&\, \mathbf{F}_3$.

Note. After writing the last 4 equations in the compact form

$$\mathbf{F}_k = \frac{1}{4} \sum_{n=0}^{3} e^{-2\pi i k n/4}\, \mathbf{g}_n,$$

you should be able to figure out a corresponding algorithm for computing the (mN)-component Fourier transform $\mathbf{F} := \mathbf{f}^\wedge$ of $\mathbf{f} := \mathbf{f}_0 \,\&\, \mathbf{f}_1 \,\&\, \cdots \,\&\, \mathbf{f}_{m-1}$, when $\mathbf{f}_0, \mathbf{f}_1, \ldots, \mathbf{f}_{m-1}$ are N-vectors and $m = 2, 3, \ldots$ is a divisor of N.

▶ **EXERCISE 6.10:** In this exercise you will use the DFT calculus to show that there is a fast algorithm for computing the discrete Fourier transform of an N-vector $\mathbf{f}$ when N is highly composite but not necessarily a power of 2.

(a) Let $N = MP$ where $M, P = 2, 3, \ldots$, let $\mathbf{f} := (f_0, f_1, \ldots, f_{N-1})$, and let

$$\mathbf{g}_0 := \Xi_P \mathbf{f}, \quad \mathbf{g}_1 := \Xi_P \mathcal{T}_1 \mathbf{f}, \ldots, \quad \mathbf{g}_{P-1} := \Xi_P \mathcal{T}_{P-1} \mathbf{f}$$

where Ξ_P is the decimation operator and $\mathcal{T}_m$ is the translation operator. Explain why

$$\mathbf{f} = \mathcal{Z}_P \mathbf{g}_0 + \mathcal{T}_{-1} \mathcal{Z}_P \mathbf{g}_1 + \cdots + \mathcal{T}_{-P+1} \mathcal{Z}_P \mathbf{g}_{P-1},$$

and then use this identity to show that

$$P\mathbf{f}^\wedge = \mathcal{R}_P \mathbf{g}_0^\wedge + \mathcal{E}_{-1} \mathcal{R}_P \mathbf{g}_1^\wedge + \cdots + \mathcal{E}_{-P+1} \mathcal{R}_P \mathbf{g}_{P-1}^\wedge.$$

Here $\mathcal{Z}_P$, $\mathcal{R}_P$, $\mathcal{E}_m$ are the zero packing, repeat, and exponential modulation operators.

(b) Within the context of (a), show that we can compute $P\mathbf{f}^\wedge$ by computing $\mathbf{g}_0^\wedge, \mathbf{g}_1^\wedge, \ldots, \mathbf{g}_{P-1}^\wedge$ and then expending $(P-1)N$ complex multiplications and a like number of complex additions to combine these vectors.

(c) Let $N = P_1 P_2 \cdots P_m$ where $P_\mu = 2, 3, \ldots$ for each $\mu = 1, 2, \ldots, m$ and $m = 2, 3, \ldots$. Using (b), show that we can compute the discrete Fourier transform of an N-vector $\mathbf{f}$ by expending approximately $(P_1 + P_2 + \cdots + P_m)N$ complex multiplications and a like number of complex additions.

(d) Let $N = P_1 P_2 \cdots P_m$ as in (c). Show that

$$P_1 + P_2 + \cdots + P_m \geq mN^{1/m} \geq (1.8841\ldots) \cdot \log_2 N.$$

In this way you show that the algorithm of (b)–(c) requires at least $1.88N \log_2 N$ complex operations.

Hint. Minimize $x_1 + x_2 + \cdots + x_m$ subject to the constraints $x_1 > 0$, $x_2 > 0$, $\ldots$, $x_m > 0$, $x_1 x_2 \cdots x_m = N$, and then minimize $\mu N^{1/\mu} / \log_2 N$, $\mu > 0$.

▶ **EXERCISE 6.11:**   Let r_m be the bit reversal map for the integers $0, 1, \ldots, N-1$ when $N := 2^m$, i.e., $r_m[(b_m b_{m-1} \cdots b_1)_2] := (b_1 b_2 \cdots b_m)_2$ where $b_\mu = 0, 1$ for each $\mu = 1, 2, \ldots, m$.

(a) For how many indices $n = 0, 1, \ldots, 2^m - 1$ is $r_m[n] = n$?

(b) How many swaps $f[n] \leftrightarrow f[r[n]]$ must be performed when we apply the bit reversal permutation $\mathbf{B}_N$ to an N-vector $\mathbf{f}$?

▶ **EXERCISE 6.12:**   Suppose that you wish to apply the bit reversal permutation to the complex N-vector $\mathbf{f}$ when $N := 2^m$, i.e., overwrite $f[0], f[1], \ldots, f[N-1]$ with $f[r_m[0]], f[r_m[1]], \ldots, f[r_m[N-1]]$. Here r_m is the bit reversal map from Ex. 6.11.

(a) Show that the naive bit reversal permutation Algorithm 6.2 requires approximately:

$3N \log_2 N$ integer multiplies, $2N \log_2 N$ integer additions, and $2N$ array read/writes.

(b) Show that the Bracewell-Buneman Algorithm 6.4 requires approximately:

N integer multiplies, N integer additions, and $2.5N$ array read/writes

after the bit reversed indices have been precomputed and stored.

▶ **EXERCISE 6.13:**   Let $p_0, p_1, \ldots, p_{N-1}$ be some rearrangement of the indices $0, 1, \ldots, N-1$, and let the $N \times N$ permutation matrix $\mathbf{P}$ be defined by

$$\mathbf{P}(f[0], f[1], \ldots, f[N-1])^T := (f[p_0], f[p_1], \ldots, f[p_{N-1}])^T.$$

(a) What are the elements of $\mathbf{P}$?

(b) Show that $\mathbf{P}^T \mathbf{P} = \mathbf{P}\mathbf{P}^T = \mathbf{I}$.

(c) What must be true of $p_0, p_1, \ldots, p_{N-1}$ if $\mathbf{P}^2 = \mathbf{I}$?

▶ **EXERCISE 6.14:**   Let $\mathcal{F}_{16}$ have the sparse matrix factorization (12) with $N = 2^4$.

(a) Write down each nonzero element (either a 1 or some power of $\omega := e^{-2\pi i/16}$) of the five matrix factors.

(b) Write down each nonzero element of the five matrix factors that result when we simplify

$$\mathcal{F}_{16} = \mathcal{F}_{16}^T = 16^{-1} \{ \mathbf{Q}_{16} \mathbf{Q}_8^{(2)} \mathbf{Q}_4^{(4)} \mathbf{Q}_2^{(8)} \mathbf{B}_{16} \}^T.$$

(c) Show that when we simplify

$$\mathcal{F}_{16} = 16^{-1} \overline{\mathcal{F}_{16}}^{-1} = \{ \overline{\mathbf{Q}_{16} \mathbf{Q}_8^{(2)} \mathbf{Q}_4^{(4)} \mathbf{Q}_2^{(8)} \mathbf{B}_{16}} \}^{-1}$$

we obtain the factorization of (b).

Hint. Observe that $\mathbf{Q}_{2M}^{-1} = \frac{1}{2} \overline{\mathbf{Q}}_{2M}^T$.

► **EXERCISE 6.15:** In this exercise you will learn how *Mason flow diagrams* are used to represent FFT algorithms. We use the flow diagram

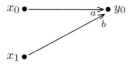

for a scalar computation of the form $y_0 := ax_0 + bx_1$. The values of x_0, x_1 are taken from the input nodes at the feet of the arrows, scaled by the complex numbers a, b written near the heads of the arrows, and added to produce the output y_0. To avoid unnecessary clutter we sometimes delete a factor of 1, *e.g.*, we use the diagram

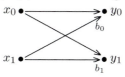

for the *butterfly relations* $y_0 := x_0 + b_0 x_1$, $y_1 := x_0 + b_1 x_1$.

(a) Convince yourself that the FFT corresponding to the factorization

$$\mathcal{F}_4 = \frac{1}{4} \begin{bmatrix} 1 & 0 & 1 & 0 \\ 0 & 1 & 0 & \omega \\ 1 & 0 & -1 & 0 \\ 0 & 1 & 0 & -\omega \end{bmatrix} \begin{bmatrix} 1 & 1 & 0 & 0 \\ 1 & -1 & 0 & 0 \\ 0 & 0 & 1 & 1 \\ 0 & 0 & 1 & -1 \end{bmatrix} \begin{bmatrix} 1 & 0 & 0 & 0 \\ 0 & 0 & 1 & 0 \\ 0 & 1 & 0 & 0 \\ 0 & 0 & 0 & 1 \end{bmatrix}, \quad \omega = e^{-2\pi i/4}$$

can be represented by the following Mason flow diagram.

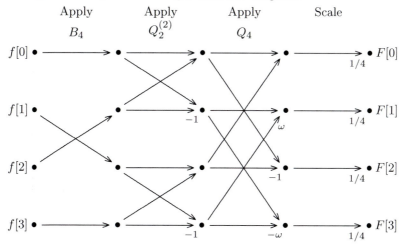

(b) Construct the Mason flow diagram associated with the factorization

$$\mathcal{F}_8 = 8^{-1} \mathbf{Q}_8 \mathbf{Q}_4^{(2)} \mathbf{Q}_2^{(4)} \mathbf{B}_8.$$

(c) Construct the Mason flow diagram associated with the factorization

$$\mathcal{F}_8 = 8^{-1} \mathbf{B}_8 (\mathbf{Q}_2^T)^{(4)} (\mathbf{Q}_4^T)^{(2)} \mathbf{Q}_8^T.$$

▶ **EXERCISE 6.16:**   Let $N = 2^m$ for some $m = 1, 2, \ldots$. A tree with 1 vector of length N, 2 of length $N/2$, 4 of length $N/4$, ... is produced when we use the Fourier transform calculus to create a fast algorithm for computing the DFT of an N-component vector, *cf.* Figs. 6.3 and 6.6. In this exercise you will show that these trees correspond to factorizations of $\mathcal{F}_N$ that provide structures for storing the intermediate results of the computations.

(a) Let $N = 2M$. Use the decimation-in-time identity

$$\mathbf{f} = \mathcal{Z}_2 \Xi_2 \mathbf{f} + \mathcal{T}_{-1} \mathcal{Z}_2 \Xi_2 \mathcal{T}_1 \mathbf{f}$$

i.e., $(f_0, f_1, \ldots, f_{N-1}) = (f_0, 0, f_2, 0, \ldots, f_{N-2}, 0) + (0, f_1, 0, f_3, \ldots, 0, f_{N-1})$ to derive the zipper identity (11) which leads to the factorization (12).

(b) Let $N = 2M$. Use the decimation-in-frequency identity

$$\mathbf{f}^\wedge = \mathcal{Z}_2 \Xi_2 \mathbf{f}^\wedge + \mathcal{T}_{-1} \mathcal{Z}_2 \Xi_2 \mathcal{T}_1 \mathbf{f}^\wedge = \mathcal{Z}_2 \left\{ \tfrac{1}{2} \Sigma_M \mathbf{f} \right\}^\wedge + \mathcal{T}_{-1} \mathcal{Z}_2 \left\{ \tfrac{1}{2} \Sigma_M \mathcal{E}_{-1} \mathbf{f} \right\}^\wedge$$

to derive the zipper identity

$$\mathcal{F}_{2M} = \tfrac{1}{2} \mathbf{S}_{2M}^T \mathcal{F}_M^{(2)} \mathbf{Q}_{2M}^T,$$

which leads to the factorization (15).

▶ **EXERCISE 6.17:**   In this exercise you will establish the zipper identity (22) that gives the factorization (25) for Bracewell's FHT.

(a) Let $\mathbf{X}_M$ be given by (24) and let

$$\mathbf{R}_M := \left[\operatorname{cas} \left\{ \frac{2\pi k(2\ell + 1)}{2M} \right\} \right]_{k,\ell=0}^{M-1}.$$

Show that (22) follows from the identity

$$\sqrt{M}\, \mathbf{X}_M \mathbf{H}_M = \mathbf{R}_M.$$

(b) Prove the matrix identity of (a).

Hint. Use $\operatorname{cas}(\alpha + \beta) = \cos(\alpha)\operatorname{cas}(\beta) + \sin(\alpha)\operatorname{cas}(-\beta)$ to establish the equality

$$\cos\left\{ \frac{\pi k}{M} \right\} \operatorname{cas}\left\{ \frac{2\pi k\ell}{M} \right\} + \sin\left\{ \frac{\pi k}{M} \right\} \operatorname{cas}\left\{ \frac{2\pi (M-k)\ell}{M} \right\} = \operatorname{cas}\left\{ \frac{2\pi k(2\ell + 1)}{2M} \right\}$$

of the elements in row $k = 0, 1, \ldots, M - 1$ and column $\ell = 0, 1, \ldots, M - 1$.

▶ **EXERCISE 6.18:**   Buneman's precomputation of $\sin(2\pi k/2^m)$, $k = 0, 1, \ldots, 2^{m-2}$ makes use of the half secants $h_3, h_4, \ldots$ from (18). Verify that h_μ can be generated by means of the recursion (17).

► **EXERCISE 6.19:** Suppose that you wish to compute and store

$$\omega_k := \cos\left(\frac{2\pi k}{N}\right) - i\sin\left(\frac{2\pi k}{N}\right), \qquad k = 0, 1, \dots, N-1$$

for use in an FFT code. Approximately how many real additions and multiplications are required for each of the following?

(a) Use direct calls of sin, cos for each ω_k. Assume that sin and cos can be computed to 8-place, 16-place accuracy by using 5, 8 real additions and a like number of real multiplications.

(b) Use Buneman's Algorithm 6.7 to generate the real and imaginary parts of the ω_k's.

(c) Use direct calls of sin, cos to generate

$$\omega_{2^\mu} = \cos\left(\frac{2\pi 2^\mu}{N}\right) - i\sin\left(\frac{2\pi 2^\mu}{N}\right), \qquad \mu = 0, 1, \dots, m-1$$

and then compute

$$\begin{bmatrix} \omega_2 \\ \omega_3 \end{bmatrix} = \omega_2 \cdot \begin{bmatrix} 1 \\ \omega_1 \end{bmatrix}, \qquad \begin{bmatrix} \omega_4 \\ \omega_5 \\ \omega_6 \\ \omega_7 \end{bmatrix} = \omega_4 \cdot \begin{bmatrix} 1 \\ \omega_1 \\ \omega_2 \\ \omega_3 \end{bmatrix}, \qquad \dots \; .$$

(d) Use direct calls of sin, cos to generate

$$\omega_1 = \cos\left(\frac{2\pi}{N}\right) - i\sin\left(\frac{2\pi}{N}\right)$$

and then compute

$$\omega_2 = \omega_1 \cdot \omega_1, \quad \omega_3 = \omega_1 \cdot \omega_2, \quad \omega_4 = \omega_1 \cdot \omega_3, \quad \dots \; .$$

Note. If we work on a computer with unit roundoff ϵ, the size of the errors associated with (a)–(d) will be of $O(\epsilon)$, $O(m\epsilon)$, $O(m\epsilon)$, $O(N\epsilon)$.

► **EXERCISE 6.20:** Let $\mathbf{f}, \mathbf{g}$ be N component vectors with $N = 2^m$ for some $m = 0, 1, 2, \dots$.

(a) Show that the convolution product of $\mathbf{f}, \mathbf{g}$ is given by

$$\mathbf{f} * \mathbf{g} = N^2 \mathbf{R}_N \mathcal{F}_N\{(\mathcal{F}_N \mathbf{f}) \circ (\mathcal{F}_N \mathbf{g})\}$$

where $\mathbf{R}_N, \mathcal{F}_N$ is the reflection operator, $\mathcal{F}_N$ is the discrete Fourier transform operator, and the vector product $\circ$ is taken componentwise.

(b) Explain why $(\mathbf{Pf}) \circ (\mathbf{Pg}) = \mathbf{P}(\mathbf{f} \circ \mathbf{g})$ when $\mathbf{P}$ is any permutation operator.

(c) Use (a) and (b) with (12) and (15) to show that

$$\mathbf{f} * \mathbf{g} = N^{-1} \mathbf{R}_N \mathfrak{Q}_N\{(\mathfrak{Q}_N^T \mathbf{f}) \circ (\mathfrak{Q}_N^T \mathbf{g})\}$$

where

$$\mathfrak{Q}_N := \mathbf{Q}_N \mathbf{Q}_{N/2}^{(2)} \mathbf{Q}_{N/4}^{(4)} \cdots \mathbf{Q}_2^{(N/2)}.$$

This gives us a fast algorithm for computing $\mathbf{f} * \mathbf{g}$ that does not make use of the bit reversal permutation.

▶ **EXERCISE 6.21:**   In this exercise you will study Bluestein's clever scheme for finding the DFT of a vector $\mathbf{f}$ with N components, $N = 2, 3, 4, \ldots$, by using an FFT that works with vectors having 2^m components, $m = 1, 2, \ldots$.

(a) Show that the components of the DFT are given by Bluestein's identity

$$F[k] = N^{-1} e^{-\pi i k^2/N} \sum_{n=0}^{N-1} \{f[n] e^{-\pi i n^2/N}\} e^{\pi i (k-n)^2/N}.$$

(b) Let $M = 2^m$, $m = 1, 2, \ldots$ be chosen so that $2N - 1 \leq M < 4N - 2$, and let

$$u[n] := \begin{cases} f[n] e^{-\pi i n^2/N} & \text{if } n = 0, 1, \ldots, N-1 \\ 0 & \text{if } n = N, N+1, \ldots, M-1 \end{cases}$$

$$v[n] := \begin{cases} e^{\pi i n^2/N} & \text{if } n = 0, 1, \ldots, N-1 \\ 0 & \text{if } n = N, N+1, \ldots, M-N-1 \\ e^{\pi i (M-n)^2/N} & \text{if } n = M-N+1, M-N+2, \ldots, M-1 \end{cases}$$

be defined on $\mathbb{P}_M$. Verify that the sums from (a) can be expressed in terms of the convolution product $\mathbf{u} * \mathbf{v}$ by writing

$$\sum_{n=0}^{N-1} \{f[n] e^{-\pi i n^2/N}\} e^{\pi i (k-n)^2/N} = (u * v)[k], \quad k = 0, 1, \ldots, N-1.$$

Hint. The choice $M \geq 2N - 1$ was made to avoid the wraparound effect illustrated in Fig. 2.8. The indices $k = N, N+1, \ldots, M-1$ are not used with this identity.

(c) Using (a), (b), and the identity of Ex. 6.20(a), show that we can compute the DFT of $\mathbf{f}$ by expending approximately $6N \log_2 N$ complex operations when N is large. Bluestein's scheme gives us an FFT in cases where N is not a power of 2!

Hint. Use the operation count $(M/2) \log_2 M$ for applying $\boldsymbol{\mathcal{F}}_M$ and neglect N in comparison to $N \log_2 N$.

▶ **EXERCISE 6.22:**   We must compute the two complex products

$$(c - is) \cdot f[\kappa \cdot 4M + 2M + \lambda], \quad (c + is) \cdot f(\kappa \cdot 4M + 4M - \lambda]$$

and 4 complex additions in the innermost loop of the radix 2 FFT given in Algorithm 6.8, so the whole algorithm uses approximately

$$N \log_2 N \cdot \{2 \text{ real multiplications } + 3 \text{ real additions}\}.$$

In this exercise you will show that this operation count can be reduced by 18% when $N = 2^m$ is large.

(a) Let $\mathbf{x}$, $\mathbf{y}$ be real 8-vectors. Show how to evaluate the real and imaginary parts of the components of the complex 8-vector $8\boldsymbol{\mathcal{F}}_8(\mathbf{x} + i\mathbf{y})$ by using 4 real multiplications and 52 real additions.

Hint. Define the real 8-vectors **r**, **s**, **t**, **u**, **v**, **w** by

$$\mathbf{r} + i\mathbf{s} := \mathbf{Q}_2^{(4)}\mathbf{B}_8(\mathbf{x} + i\mathbf{y}), \qquad \mathbf{t} + i\mathbf{u} := \mathbf{Q}_4^{(2)}(\mathbf{r} + i\mathbf{s}), \qquad \mathbf{v} + i\mathbf{w} := \mathbf{Q}_8(\mathbf{t} + i\mathbf{u}),$$

and note that we do not need to multiply when we compute

$$i(a + ib) = -b + ia, \qquad (1 + i)(a + ib) = (a - b) + i(a + b).$$

(b) Show that the application of the diagonal matrix of *twiddle factors* from (34) can be done by using 28 real multiplications and 14 real additions (when $P = 8$).

(c) Show that the use of (a) and (b) within the inner loop of Algorithm 6.10 makes it possible for us to generate the DFT of a complex vector having $N = 8^n$ components by expending approximately

$$N \log_2 N \cdot \{1.33 \text{ real multiplications} + 2.75 \text{ real additions}\}.$$

Note. The cost factors in { } increase only slightly when $P_1 = \cdots = P_{m-1} = 8$, $P_m = 2$ or 4, and m is "large."

▶ **EXERCISE 6.23:**   Let P, M be positive integers, let $N = MP$, and let $\mathbf{H}_N$ be the $N \times N$ Hartley transform matrix.

(a) Derive the zipper identity $\mathbf{H}_{MP} = P^{-1/2}\mathbf{T}_{M,P}\mathbf{H}_M^{(P)}\mathbf{S}_{M,P}$ where

$$\mathbf{T}_{M,P} := \begin{bmatrix} \mathbf{X}_{0,0} & \mathbf{X}_{0,1} & \cdots & \mathbf{X}_{0,P-1} \\ \mathbf{X}_{1,0} & \mathbf{X}_{1,1} & \cdots & \mathbf{X}_{1,P-1} \\ \vdots & & & \\ \mathbf{X}_{P-1,0} & \mathbf{X}_{P-1,1} & \cdots & \mathbf{X}_{P-1,P-1} \end{bmatrix}$$

is an $MP \times MP$ matrix formed from the $M \times M$ blocks

$$\mathbf{X}_{k,\ell} := \begin{bmatrix} c_{\ell k M} & & & & \\ & c_{\ell k M + \ell} & & & \\ & & c_{\ell k M + 2\ell} & & \\ & & & \ddots & \\ & & & & c_{\ell k M + (M-1)\ell} \end{bmatrix} - \begin{bmatrix} s_{\ell k M} & & & & \\ & & & & s_{\ell k M + \ell} \\ & & & s_{\ell k M + 2\ell} & \\ & & \cdot^{\cdot^{\cdot}} & & \\ & s_{\ell k M + (M-1)\ell} & & & \end{bmatrix}$$

using

$$c_\nu := \cos\left(\frac{2\pi\nu}{MP}\right), \qquad s_\nu := \sin\left(\frac{2\pi\nu}{MP}\right), \qquad \nu = 0, 1, \ldots, MP,$$

and $\mathbf{S}_{M,P}$ is the shuffle permutation (31).

(b) Use (a) with $P = P_m$, $P = P_{m-1}, \ldots, P = P_1$ in turn to find a sparse factorization of $\mathbf{H}_{P_1 P_2 \cdots P_m}$.

Hint. Examine (32) before you begin.

▶ **EXERCISE 6.24:**   In this (historical) exercise you will learn how Gauss derived the FFT using *interpolation* by trigonometric polynomials.

(a) Let the complex numbers $y_0, y_1, \ldots, y_{N-1}$ and the corresponding uniformly spaced real abscissas $t_n := t_0 + n \cdot T/N$, $n = 0, 1, \ldots, N-1$, be given. Here $-\infty < t_0 < \infty$, $T > 0$, and $N = 1, 2, \ldots$. Show that the T-periodic trigonometric polynomial

$$y(t) := \sum_{k=0}^{N-1} c_k \, e^{2\pi i k t/T}$$

interpolates the data, *i.e.*, $y(t_n) = y_n$, $n = 0, 1, \ldots, N-1$, when

$$c_k := e^{-2\pi i k t_0/T} \cdot Y_k \text{ with } Y_k := \frac{1}{N} \sum_{n=0}^{N-1} e^{-2\pi i k n/N} y_n, \quad k = 0, 1, \ldots, N-1.$$

Thus the interpolation problem can be solved by computing the DFT of $y_0, y_1, \ldots, y_{N-1}$ and then using a suitable exponential modulation to account for the choice of the initial abscissa t_0 and the period T. Approximately N^2 complex operations are required to solve the problem in this way.

(b) Suppose now that N is composite with the factorization $N = MP$ for some $M = 2, 3, \ldots$ and $P = 2, 3, \ldots$. Gauss observed that the coefficients $c_0, c_1, \ldots, c_{N-1}$ of (a) can be found by solving $P + M$ *smaller* interpolation problems as follows.

(i) For each $r = 0, 1, \ldots, P-1$ find the M coefficients of the T-periodic trigonometric polynomial

$$y_r(t) := \sum_{\mu=0}^{M-1} c_\mu^{(r)} e^{2\pi i \mu t/T},$$

which passes through the M points

$$(t_r, y_r), \ (t_{r+P}, y_{r+P}), \ (t_{r+2P}, y_{r+2P}), \ \ldots, \ (t_{r+(M-1)P}, y_{r+(M-1)P}).$$

(The motivation for solving such interpolation problems with decimated data sets is developed in Ex. 1.19.)

(ii) For each $\mu = 0, 1, \ldots, M-1$ find the p coefficients of the T/M-periodic trigonometric polynomial

$$C_\mu(t) := \sum_{\lambda=0}^{P-1} C_{\mu\lambda} \, e^{2\pi i \lambda M t/T},$$

which takes the values $c_\mu^{(0)}, c_\mu^{(1)}, \ldots, c_\mu^{(P-1)}$ at the corresponding points $t_0, t_1, \ldots, t_{P-1}$ (and, in view of the periodicity, at the corresponding points $t_{mP}, t_{mP+1}, \ldots, t_{mP+P-1}$ as well, $m = 1, 2, \ldots, M-1$.)

(iii) For each $\mu = 0, 1, \ldots, M - 1$ and each $\lambda = 0, 1, \ldots, P - 1$ set

$$c_{\mu + \lambda M} := C_{\mu \lambda},$$

and with $c_0, c_1, \ldots, c_{N-1}$ thus defined, take

$$y(t) = \sum_{k=0}^{N-1} c_k \, e^{2\pi i k t / T}$$

as the solution to the interpolation problem of (a). Show that this procedure produces the c_k's of (a). The analysis of Ex. 1.19 is something that most of us would do in the process of solving the particular interpolation problem associated with the orbit of Pallas; the above generalization is the work of a genius!

Hint. Use (i), (ii), (iii) in turn to show that

$$y_{r+mP} = \sum_{\mu=0}^{M-1} c_\mu^{(r)} \, e^{2\pi i \mu t_{r+mP}/T} = \sum_{\mu=0}^{M-1} C_\mu(t_{r+mP}) \, e^{2\pi i \mu t_{r+mP}/T}$$

$$= \sum_{\mu=0}^{M-1} \sum_{\lambda=0}^{P-1} C_{\mu\lambda} \, e^{2\pi i (\mu + \lambda M) t_{r+mP}/T} = \sum_{k=0}^{N-1} c_k \, e^{2\pi i k t_{r+mP}/T}$$

for each $r = 0, 1, \ldots, P - 1$ and for each $m = 0, 1, \ldots, M - 1$.

(c) Let $N = P_1 P_2 \cdots P_m$ where $P_\mu = 2, 3, \ldots$ for each $\mu = 1, 2, \ldots, m$. Show that the recursive use of the Gauss procedure allows us to solve the N-point interpolation problem (and thus compute an N-point DFT) using approximately $N \cdot (P_1 + P_2 + \cdots + P_m)$ complex operations.

Note. With tongue in cheek, Gauss observed that *"Experience will teach the user that this method will greatly lessen the tedium of mechanical calculations."*

Hint: When $N = P_1 M_1$ the Gauss procedure reduces the cost from N^2 to $P_1 M_1^2 + M_1 P_1^2$ operations. When $M_1 = P_2 M_2$ each of the length M_1 interpolation problems can be solved by using the Gauss procedure, thereby reducing the cost to $M_1 P_1^2 + P_1 [M_2 P_2^2 + P_2 M_2^2]$ operations.

(d) Specialize (a)–(b) to the case where $t_0 = 0$ and $N = 2M$, and thereby show that Gauss's procedure produces the matrix factorization

$$\mathcal{F}_{2M} = \frac{1}{2} \begin{bmatrix} \mathbf{I}_M & \mathbf{I}_M \\ \mathbf{I}_M & -\mathbf{I}_M \end{bmatrix} \begin{bmatrix} \mathbf{I}_M & \\ & \Lambda_M \end{bmatrix} \begin{bmatrix} \mathcal{F}_M & \\ & \mathcal{F}_M \end{bmatrix} \mathbf{S}_{2M}$$

where $\mathbf{S}_{2M}$ is the shuffle permutation (3) and

$$\Lambda_M := \begin{bmatrix} 1 & & & & \\ & \omega & & & \\ & & \omega^2 & & \\ & & & \ddots & \\ & & & & \omega^{M-1} \end{bmatrix}, \qquad \omega := e^{-2\pi i / 2M}$$

is the diagonal matrix of twiddle factors. [The familiar zipper identity (11) is obtained by multiplying the first pair of matrices from the Gauss factorization.]

▶ **EXERCISE 6.25:**   In this (historical) exercise you will learn how Cooley and Tukey derived the FFT by using *nested summation*.

(a) Let $N = 2^3$, let $\omega := e^{-2\pi i/8}$, and let each index $n, k = 0, 1, \ldots, 7$ be expressed in base 2 form, *i.e.*,

$$n = (\nu_3\nu_2\nu_1)_2 := \nu_3 \cdot 2^2 + \nu_2 \cdot 2 + \nu_1, \qquad k = (\kappa_3\kappa_2\kappa_1)_2 := \kappa_3 \cdot 2^2 + \kappa_2 \cdot 2 + \kappa_1$$

with each bit ν_j, κ_j taking the values $0, 1$. Corresponding multiindices are used to specify the components of an N-vector $\mathbf{f}$ and its DFT $\mathbf{F}$, *i.e.*, we write

$$f[\nu_3, \nu_2, \nu_1] := f[n], \qquad F[\kappa_3, \kappa_2, \kappa_1] := F[k]$$

when n, k have the above base 2 representations. Show that

$$F[\kappa_3, \kappa_2, \kappa_1] = \frac{1}{8} \sum_{\nu_1=0}^{1} \sum_{\nu_2=0}^{1} \sum_{\nu_3=0}^{1} f[\nu_3, \nu_2, \nu_1]\omega^{(4\nu_3+2\nu_2+\nu_1)(4\kappa_3+2\kappa_2+\kappa_1)}$$

$$= \frac{1}{8} \sum_{\nu_1=0}^{1} \left\{ \sum_{\nu_2=0}^{1} \left\{ \sum_{\nu_3=0}^{1} f[\nu_3, \nu_2, \nu_1]\omega^{4\nu_3\kappa_1} \right\} \omega^{2\nu_2(2\kappa_2+\kappa_1)} \right\} \omega^{\nu_1(4\kappa_3+2\kappa_2+\kappa_1)}.$$

Hint: $\displaystyle\sum_{n=0}^{7} = \sum_{\nu_1=0}^{1} \sum_{\nu_2=0}^{1} \sum_{\nu_3=0}^{1}$

(b) Using (a), verify that the DFT of an 8-vector can be obtained by performing the following computations.

$$f_1[\kappa_1, \nu_2, \nu_1] := \sum_{\nu_3=0}^{1} f[\nu_3, \nu_2, \nu_1]\omega^{4\nu_3\kappa_1}, \qquad\qquad \nu_1, \nu_2, \kappa_1 = 0, 1$$

$$f_2[\kappa_1, \kappa_2, \nu_1] := \sum_{\nu_2=0}^{1} f_1[\kappa_1, \nu_2, \nu_1]\omega^{2\nu_2(2\kappa_2+\kappa_1)}, \qquad \nu_1, \kappa_1, \kappa_2 = 0, 1$$

$$f_3[\kappa_1, \kappa_2, \kappa_3] := \sum_{\nu_1=0}^{1} f_2[\kappa_1, \kappa_2, \nu_1]\omega^{\nu_1(4\kappa_3+2\kappa_2+\kappa_1)}, \quad \kappa_1, \kappa_2, \kappa_3 = 0, 1$$

$$f_4[\kappa_3, \kappa_2, \kappa_1] := f_3[\kappa_1, \kappa_2, \kappa_3], \qquad\qquad\qquad\qquad \kappa_1, \kappa_2, \kappa_3 = 0, 1$$

$$F[\kappa_3, \kappa_2, \kappa_1] := 8^{-1}f_4[\kappa_3, \kappa_2, \kappa_1], \qquad\qquad\qquad \kappa_1, \kappa_2, \kappa_3 = 0, 1.$$

(c) Show that approximately 8 complex operations must be expended to generate each of the arrays $\mathbf{f}_1, \mathbf{f}_2, \mathbf{f}_3$ in (b).

Note. The Cooley-Tukey derivation of (a)–(c) extends at once to the case where $N = 2^m$, $m = 1, 2, \ldots$. When $N = P_1 P_2 \ldots P_m$ with $P_\mu = 2, 3, 4, \ldots$ for each $\mu = 1, 2, \ldots, m$ we write

$$n = \nu_1 + P_1\nu_2 + P_1P_2\nu_3 + \cdots + (P_1P_2 \cdots P_{m-1})\nu_n$$

$$k = \kappa_m + P_m\kappa_{m-1} + P_mP_{m-1}\kappa_{m-2} + \cdots + (P_mP_{m-1} \cdots P_2)\kappa_1$$

where ν_j, $\kappa_j = 0, 1, \ldots, P_j - 1$, $j = 1, 2, \ldots, m$ and use the multiindex notation

$$f[\nu_1, \nu_2, \ldots, \nu_m] := f[n], \qquad F[\kappa_1, \kappa_2, \ldots, \kappa_m] := F[k]$$

to derive the mixed radix Cooley-Tukey algorithm.

▶ **EXERCISE 6.26:**   In this exercise you will use the definition (48) to deduce three properties of the Kronecker product.

(a) Let $\mathbf{A}, \mathbf{B}, \mathbf{C}$ be $K \times L$, $M \times N$, $P \times Q$ matrices. Use (48) to show that

$$\{\mathbf{A} \otimes (\mathbf{B} \otimes \mathbf{C})\}\big[(p + mP) + k(MP), \ (q + nQ) + \ell(NQ)\big]$$
$$= \mathbf{A}[k, \ell] \cdot \mathbf{B}[m, n] \cdot \mathbf{C}[p, q]$$
$$= \{(\mathbf{A} \otimes \mathbf{B}) \otimes \mathbf{C}\}\big[p + (m + kM)P, \ q + (n + \ell N)Q\big]$$

when $k = 0, 1, \ldots, K-1$, $\ell = 0, 1, \ldots, L-1$, $m = 0, 1, \ldots, M-1$, $n = 0, 1, \ldots, N-1$, $p = 0, 1, \ldots, P-1$, $q = 0, 1, \ldots, Q-1$ and thereby obtain the associative rule (55).

(b) Let $\mathbf{A}, \mathbf{B}, \mathbf{C}, \mathbf{D}$ be $K \times L$, $M \times N$, $L \times P$, $N \times Q$ matrices (so that $\mathbf{AC}$ and $\mathbf{BD}$ are defined). Use (48) to verify that

$$(\mathbf{A} \otimes \mathbf{B})[m + kM, n + \ell N] = \mathbf{A}[k, \ell] \cdot \mathbf{B}[m, n]$$
$$(\mathbf{C} \otimes \mathbf{D})[n + \ell N, q + pQ] = \mathbf{C}[\ell, p] \cdot \mathbf{D}[n, q]$$

and thereby obtain the product rule (59).

(c) Establish the transpose rule (60).

Note. Such subscript-based arguments are convincing, but not memorable!

▶ **EXERCISE 6.27:**   Let $\mathbf{A}, \mathbf{B}$ be square matrices with inverses $\mathbf{A}^{-1}, \mathbf{B}^{-1}$. Show that the Kronecker product $\mathbf{A} \otimes \mathbf{B}$ has the inverse (61).

Hint. Make use of the product rule (59) proved in Ex. 6.26.

▶ **EXERCISE 6.28:**   Let $\mathbf{A}, \mathbf{B}$ be matrices with the eigenvectors $\mathbf{a}, \mathbf{b}$ and the corresponding eigenvalues α, β. Show that $\mathbf{a} \otimes \mathbf{b}$ is an eigenvector of $\mathbf{A} \otimes \mathbf{B}$ with the eigenvalue $\alpha\beta$.

▶ **EXERCISE 6.29:**   Let $\mathbf{e}_0 := (1, 0)^{\mathrm{T}}$, $\mathbf{e}_1 := (0, 1)^{\mathrm{T}}$.

(a) Form the 8 Kronecker products $\mathbf{e}_{\beta_3} \otimes \mathbf{e}_{\beta_2} \otimes \mathbf{e}_{\beta_1}$ with $\beta_\mu = 0, 1$ for $\mu = 1, 2, 3$.

(b) Give a simple description of $\mathbf{e}_{\beta_m} \otimes \mathbf{e}_{\beta_{m-1}} \otimes \cdots \otimes \mathbf{e}_{\beta_1}$ when $\beta_\mu = 0, 1$ for $\mu = 1, 2, \ldots, m$, and thereby prove that these Kronecker products form a basis for $\mathbb{C}^N$ when $N = 2^m$.

Hint. Let $n := (\beta_m \beta_{m-1} \cdots \beta_1)_2$.

(c) For $P = 2, 3, \ldots$ we create the P-vectors

$$\mathbf{e}_{P,0} := (1, 0, 0, \ldots, 0)^{\mathrm{T}}, \ \mathbf{e}_{P,1} := (0, 1, 0, \ldots, 0)^{\mathrm{T}}, \ \ldots, \ \mathbf{e}_{P,P-1} := (0, 0, \ldots, 0, 1)^{\mathrm{T}}.$$

Let $N = P_1 P_2 \cdots P_m$ with $P_\mu = 2, 3, \ldots$ for each $\mu = 1, 2, \ldots, m$. Show that the Kronecker products

$$\mathbf{e}_{P_m, p_m} \otimes \mathbf{e}_{P_{m-1}, p_{m-1}} \otimes \cdots \otimes \mathbf{e}_{P_1, p_1}, \quad p_\mu = 0, 1, \ldots, P_\mu - 1, \quad \mu = 1, 2, \ldots, m,$$

form a basis for $\mathbb{C}^N$.

▶ **EXERCISE 6.30:** An $N \times N$ matrix $\mathbf{P}$ is said to be a permutation matrix if every row and every column of $\mathbf{P}$ has precisely one 1 with the other $N - 1$ elements being 0.

(a) Show that the Kronecker product $\mathbf{P} := \mathbf{P}_1 \otimes \mathbf{P}_2 \otimes \cdots \otimes \mathbf{P}_m$ of permutation matrices $\mathbf{P}_1, \mathbf{P}_2, \ldots, \mathbf{P}_m$ is also a permutation matrix.

(b) Find a simple expression for the inverse of the matrix $\mathbf{P}$ from (a).

   Hint. Use Ex. 6.13(b) and (61).

▶ **EXERCISE 6.31:** Let P, Q, R be positive integers. In this exercise you will show that

$$\mathbf{S}_{QR,P} \mathbf{S}_{RP,Q} \mathbf{S}_{PQ,R} = \mathbf{I}_{PQR}$$

and then use this identity to obtain a more general relation of the same form.

(a) Use (63) [and (55)] to verify that

$$\mathbf{S}_{QR,P} \mathbf{S}_{RP,Q} \mathbf{S}_{PQ,R} (\mathbf{p} \otimes \mathbf{q} \otimes \mathbf{r}) = \mathbf{p} \otimes \mathbf{q} \otimes \mathbf{r}$$

   when $\mathbf{p}, \mathbf{q}, \mathbf{r}$ are column vectors with P, Q, R components, and thereby prove the identity.

(b) Use (41) to show that when we apply $\mathbf{S}_{QR,P} \mathbf{S}_{RP,Q} \mathbf{S}_{PQ,R}$ to a PQR component column vector $\mathbf{f}$, the position of the component $f[r + qR + pRQ]$ is unchanged. This gives a second proof of the identity.

(c) Rearrange the identity and thereby show that

$$\mathbf{S}_{N/P_1, P_1} \mathbf{S}_{N/P_2, P_2} = \mathbf{S}_{N/P_1 P_2, P_1 P_2}.$$

   Here $N := P_1 P_2 \cdots P_m$ where $P_1, P_2, \ldots, P_m$ are positive integers and $m \geq 2$.

(d) Use (c) to show that

$$\mathbf{S}_{N/P_1, P_1} \mathbf{S}_{N/P_2, P_2} \cdots \mathbf{S}_{N/P_m, P_m} = \mathbf{I}_{P_1 P_2 \cdots P_m},$$

   and thereby generalize the identity of (a)–(b).

▶ **EXERCISE 6.32:** Verify that each of the following matrices have the same effect on the vector $\mathbf{a} \otimes \mathbf{b} \otimes \mathbf{c} \otimes \mathbf{d}$ (when $\mathbf{a}$, $\mathbf{b}$, $\mathbf{c}$, $\mathbf{d}$ are 2-component column vectors), and thereby obtain 4 factorizations of the bit reversal permutation $\mathbf{B}_{16}$.

(a) $(\mathbf{I}_8 \otimes \mathbf{S}_{1,2})(\mathbf{I}_4 \otimes \mathbf{S}_{2,2})(\mathbf{I}_2 \otimes \mathbf{S}_{4,2})(\mathbf{I}_1 \otimes \mathbf{S}_{8,2})$

(b) $(\mathbf{I}_1 \otimes \mathbf{S}_{2,8})(\mathbf{I}_2 \otimes \mathbf{S}_{2,4})(\mathbf{I}_4 \otimes \mathbf{S}_{2,2})(\mathbf{I}_8 \otimes \mathbf{S}_{2,1})$

(c) $(\mathbf{S}_{8,2} \otimes \mathbf{I}_1)(\mathbf{S}_{4,2} \otimes \mathbf{I}_2)(\mathbf{S}_{2,2} \otimes \mathbf{I}_4)(\mathbf{S}_{1,2} \otimes \mathbf{I}_8)$

(d) $(\mathbf{S}_{2,1} \otimes \mathbf{I}_8)(\mathbf{S}_{2,2} \otimes \mathbf{I}_4)(\mathbf{S}_{2,4} \otimes \mathbf{I}_2)(\mathbf{S}_{2,8} \otimes \mathbf{I}_1)$

Hint. Use (59) and (63), freely associating the factors of the Kronecker product to facilitate the computation.

▶ **EXERCISE 6.33:** This exercise will introduce you to the remarkable two-loop FFTs of Pease and Glassman.

(a) Show how to rearrange the Cooley-Tukey factorization

$$16\,\mathcal{F}_{16} = \mathbf{Q}_{16}(\mathbf{I}_2 \otimes \mathbf{Q}_8)(\mathbf{I}_4 \otimes \mathbf{Q}_4)(\mathbf{I}_8 \otimes \mathbf{Q}_2)\mathbf{B}_{16}$$

to obtain the factorization

$$16\,\mathcal{F}_{16} = \{(\mathbf{Q}_2 \otimes \mathbf{I}_8)(\mathbf{\Delta}_{16} \otimes \mathbf{I}_1)\mathbf{S}_{8,2}\}\{(\mathbf{Q}_2 \otimes \mathbf{I}_8)(\mathbf{\Delta}_8 \otimes \mathbf{I}_2)\mathbf{S}_{8,2}\}$$
$$\{(\mathbf{Q}_2 \otimes \mathbf{I}_8)(\mathbf{\Delta}_4 \otimes \mathbf{I}_4)\mathbf{S}_{8,2}\}\{(\mathbf{Q}_2 \otimes \mathbf{I}_8)(\mathbf{\Delta}_2 \otimes \mathbf{I}_8)\mathbf{S}_{8,2}\}\mathbf{B}_{16}.$$

The $2M \times 2M$ matrix

$$\mathbf{\Delta}_{2M} := \mathrm{diag}\{1, 1, \ldots, 1;\ 1, \omega, \omega^2, \ldots, \omega^{M-1}\}, \quad \omega := e^{-2\pi i/2M}$$

has been chosen so that

$$\mathbf{Q}_{2M} := (\mathbf{Q}_2 \otimes \mathbf{I}_m)\mathbf{\Delta}_{2M}.$$

This factorization can be generalized to the case where $N = 2^m$, $m = 1, 2, \ldots$.

(b) Let $M = N/2$ and let $K = 1, 2, \ldots, 2^m$. When we analyze

$$\mathbf{g} := (\mathbf{Q}_2 \otimes \mathbf{I}_M)(\mathbf{\Delta}_K \otimes \mathbf{I}_{N/K})\mathbf{S}_{M,2}\mathbf{f}$$

we find that $\mathbf{g}_m$ and $\mathbf{g}_{m+M}$ depend on two components of $\mathbf{f}$. Sort out the details!

(c) Use (a) and (b) to develop a fast two-loop algorithm for evaluating the DFT of a vector $\mathbf{f}$ with $N = 2^m$ components.

Note. The same matrices $\mathbf{Q}_2 \otimes \mathbf{I}_M$, $\mathbf{S}_{M,2}$ appear at each stage of the computation; only the scale factors in the second half of the diagonal matrix $\mathbf{\Delta}_K \otimes \mathbf{I}_{N/K}$ change from step to step. You could design a special-purpose computer that uses hardware instead of software to do these mappings!

▶ **EXERCISE 6.34:**   In this exercise you will develop the Stockham sparse factorization of $\mathcal{F}_{P_1 P_2 \cdots P_m}$ that uses parallel operations. An elegant FORTRAN code for the resulting algorithm can be found in C. de Boor, The FFT as nested multiplication, with a twist, *SIAM J. Sci. Stat. Comp.* **1**(1980), 173–177.

(a) Show that

$$\mathbf{S}_{P_1, P_2, P_3, P_4} := (\mathbf{I}_{P_3 P_4} \otimes \mathbf{S}_{P_1, P_2})(\mathbf{I}_{P_4} \otimes \mathbf{S}_{P_1 P_2, P_3})\mathbf{S}_{P_1 P_2 P_3, P_4}$$

has the alternative representation

$$\mathbf{S}_{P_1, P_2, P_3, P_4} = (\mathbf{S}_{P_3, P_4} \otimes \mathbf{I}_{P_1 P_2})(\mathbf{S}_{P_2, P_3 P_4} \otimes \mathbf{I}_{P_1})\mathbf{S}_{P_1, P_2 P_3 P_4} \, .$$

Hint. Apply both matrices to an arbitrary $\mathbf{p}_1 \otimes \mathbf{p}_2 \otimes \mathbf{p}_3 \otimes \mathbf{p}_4$ where $\mathbf{p}_1$, $\mathbf{p}_2$, $\mathbf{p}_3$, $\mathbf{p}_4$ are column vectors with P_1, P_2, P_3, P_4 components.

(b) Derive the Stockham factorization

$$\begin{aligned}
P_1 P_2 P_3 P_4 \cdot \mathcal{F}_{P_1 P_2 P_3 P_4} =& (\mathbf{I}_1 \otimes \mathbf{Q}_{P_1 P_2 P_3, P_4})(\mathbf{S}_{P_4, 1} \otimes \mathbf{I}_{P_1 P_2 P_3}) \\
& (\mathbf{I}_{P_4} \otimes \mathbf{Q}_{P_1 P_2, P_3})(\mathbf{S}_{P_3, P_4} \otimes \mathbf{I}_{P_1 P_2}) \\
& (\mathbf{I}_{P_3 P_4} \otimes \mathbf{Q}_{P_1, P_2})(\mathbf{S}_{P_2, P_3 P_4} \otimes \mathbf{I}_{P_1}) \\
& (\mathbf{I}_{P_2 P_3 P_4} \otimes \mathbf{Q}_{1, P_1})(\mathbf{S}_{P_1, P_2 P_3 P_4} \otimes \mathbf{I}_1).
\end{aligned}$$

This factorization can be generalized to the case where N has $m = 2, 3, \ldots$ factors.

Hint. Begin with the Cooley-Tukey factorization (32), replace $\mathbf{S}_{P_1, P_2, P_3, P_4}$ with the second representation from (a), and use the commutation identity (68).

(c) When $\mathbf{x}_0, \mathbf{x}_1, \mathbf{x}_2, \mathbf{x}_3$ are 3-component column vectors, it is easy to verify that

$$(\mathbf{S}_{2,2} \otimes \mathbf{I}_3) \begin{bmatrix} \mathbf{x}_0 \\ \mathbf{x}_1 \\ \mathbf{x}_2 \\ \mathbf{x}_3 \end{bmatrix} = \begin{bmatrix} \mathbf{x}_0 \\ \mathbf{x}_2 \\ \mathbf{x}_1 \\ \mathbf{x}_3 \end{bmatrix},$$

i.e., $\mathbf{S}_{2,2}$ shuffles the blocks $\mathbf{x}_0, \mathbf{x}_1, \mathbf{x}_2, \mathbf{x}_3$. Describe the action of the permutation $\mathbf{S}_{P,Q} \otimes \mathbf{I}_R$ that appears in the factorization of (b).

Chapter 7

Generalized Functions on $\mathbb{R}$

7.1 The Concept of a Generalized Function

Introduction

Let $y(t)$ be the displacement at time t of a mass m that is attached to a spring having the force constant k as shown in Fig. 7.1. We assume that the mass is at rest in its equilibrium position [*i.e.*, $y(t) = 0$] for all $t \leq 0$. At time $t = 0$ we begin to subject the mass to an impulsive driving force

$$f_\epsilon(t) := \begin{cases} p/\epsilon & \text{if } 0 < t < \epsilon \\ 0 & \text{otherwise.} \end{cases} \tag{1}$$

When the duration $\epsilon > 0$ is "small," this force simulates the tap of a hammer that transfers the momentum

$$\int_0^\epsilon f_\epsilon(t)dt = p$$

to the mass and "rapidly" changes its velocity from $y'(0) = 0$ to $y'(\epsilon) \approx p/m$.

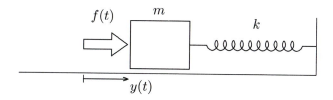

Figure 7.1. An undamped mass-spring system with the displacement function $y(t)$ and the forcing function $f(t)$.

You should have no trouble verifying that

$$
y_\epsilon(t) := \frac{p}{m\omega^2\epsilon}
\begin{cases}
0 & \text{if } t \leq 0 \\
1 - \cos\omega t & \text{if } 0 \leq t \leq \epsilon \\
\sin(\omega\epsilon)\sin(\omega t) - [1 - \cos(\omega\epsilon)]\cos(\omega t) & \text{if } t \geq \epsilon
\end{cases}
\tag{2}
$$

is a twice continuously differentiable function that satisfies the forced differential equation

$$
m\, y''(t) + k\, y(t) = f_\epsilon(t) \tag{3}
$$

for the motion (except at the points $t = 0$, $t = \epsilon$ where y_ϵ'' is not defined), *cf.* Ex. 7.1. Here

$$
\omega := \sqrt{k/m}
$$

so that $\sin(\omega t)$, $\cos(\omega t)$ are solutions of the unforced differential equation

$$
m\, y''(t) + k\, y(t) = 0.
$$

Now as $\epsilon \to 0+$, the response function (2) has the pointwise limit

$$
y_0(t) := \lim_{\epsilon \to 0+} y_\epsilon(t) = \frac{p}{m\omega}
\begin{cases}
0 & \text{if } t \leq 0 \\
\sin(\omega t) & \text{if } t \geq 0,
\end{cases}
\tag{4}
$$

and it is natural to think of y_0 as the response of the system to an impulse

$$
f_0(t) = \lim_{\epsilon \to 0+} f_\epsilon(t) \tag{5}
$$

of strength

$$
\int_{-\infty}^{\infty} f_0(t)\,dt = \lim_{\epsilon \to 0+} \int_{-\infty}^{\infty} f_\epsilon(t)\,dt = p \tag{6}
$$

that acts only at time $t = 0$, *cf.* Fig. 7.2. The physical intuition is certainly valid, and such arguments have been used by physicists and engineers (*e.g.*, Euler, Fourier, Maxwell, Heaviside, Dirac) for more than two centuries. And for most of this time such arguments have been suspect! After all, the *limit* (5) gives us a *function* f_0 that vanishes everywhere, so how can the *integral* (6) have a value $p \neq 0$?

Such anomalies cannot be resolved within a context where *function, integral,* and *limit* have the usual definitions from elementary calculus, but in what is now regarded as one of the most stunning achievements of 20th-century mathematics, Laurent Schwartz developed a perfectly rigorous theory of *generalized functions* (or *distributions*) for analyzing such phenomena. We will present an elementary introduction to the generalized functions that have Fourier transforms, *cf.* Fig. 1.33. (Such generalized functions correspond to the *tempered distributions* of Schwartz.) As you master these ideas you will acquire certain computational skills that are absolutely essential in modern science and engineering.

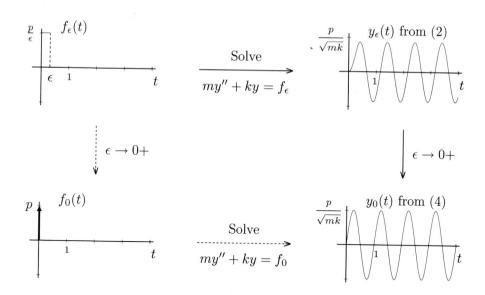

Figure 7.2. Imagination of an impulsive driving function $f_0 = \lim\limits_{\epsilon \to 0+} f_\epsilon$ that produces the response $y_0 = \lim\limits_{\epsilon \to 0+} y_\epsilon$ to the mass-spring system of Fig. 7.1.

Functions and functionals

In Fourier analysis we often work with ordinary functions that map $\mathbb{R}$ to $\mathbb{C}$. Such a *function* f is properly defined when we have some rule for producing the complex number $f(x)$ that corresponds to an arbitrary x from the domain $\mathbb{R}$. Common examples include

$$f(x) := e^{-\pi x^2}, \quad f(x) := \operatorname{sinc} x, \quad f(x) := e^{2\pi i x}.$$

We will now introduce certain functionals that map $\mathbb{S}$ to $\mathbb{C}$. Here $\mathbb{S}$ is a linear space of exceptionally well behaved ordinary complex-valued functions on $\mathbb{R}$. (A precise definition will be given in the next section.) Such a *functional* f is properly defined when we have some rule for producing the complex number $f\{\phi\}$ that corresponds to an arbitrary function ϕ from the domain $\mathbb{S}$, *cf.* Fig. 7.3. Common examples include

$$f\{\phi\} := \int_{-\infty}^{\infty} e^{-\pi x^2} \phi(x)dx, \quad f\{\phi\} := \phi(0), \quad f\{\phi\} := \sum_{n=-\infty}^{\infty} \phi(n).$$

We use braces { } to remind ourselves that the argument ϕ is a function instead of a number.

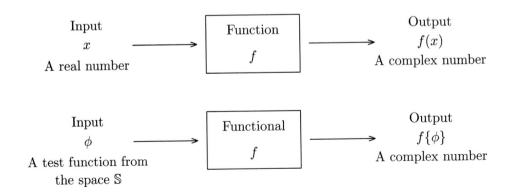

Figure 7.3. The mappings that correspond to a function and to a functional.

We will often find it convenient to use the same name for a function and for a closely related functional. For example, when we are given a suitably regular function f on $\mathbb{R}$ we will define the corresponding *fundamental functional*

$$f\{\phi\} := \int_{-\infty}^{\infty} f(x)\phi(x)dx \tag{7}$$

on $\mathbb{S}$. The formula (7) shows us how to find the value of the functional f at the argument ϕ by suitably processing the function f. The same formula allows us to evaluate the function f at any argument x_0 (where the function f is continuous) by suitably processing the functional f. Indeed, for each $n = 1, 2, \ldots$ we construct a nonnegative function ϕ_n from $\mathbb{S}$ such that

$$\phi_n(x) = 0 \quad \text{when} \quad |x - x_0| > 1/n,$$
$$\int_{x_0-1/n}^{x_0+1/n} \phi_n(x)dx = 1,$$

and we then use the integral mean value theorem from calculus to see that

$$f(x_0) = \lim_{n\to\infty} \int_{-\infty}^{\infty} f(x)\phi_n(x)dx = \lim_{n\to\infty} f\{\phi_n\}, \tag{8}$$

cf. Fig. 7.4 and Ex. 7.2. (A similar argument was used to prove the Weierstrass theorem in Section 1.3.)

We routinely identify function with functional when we make measurements with devices that report local averages. For example, suppose that we use a thermometer to measure the temperature, $f(x_0)$, at the coordinate x_0 within a certain

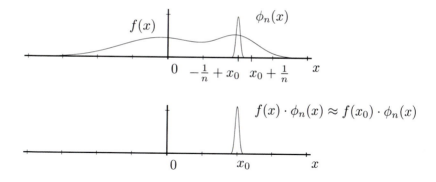

Figure 7.4. Visualization of the analysis that leads to (8).

gas column. Since the bulb of the measuring device occupies some nonzero volume of space, the instrument cannot possibly determine the temperature at the dimensionless "point" x_0. At best, the thermometer gives some local average

$$\int_{-\infty}^{\infty} f(x)\phi(x)dx$$

with the function ϕ being determined by the position, size, shape, composition, ... of the bulb. Similar considerations apply when we use a photometer, magnetometer, pressure gauge, ...

The class of fundamental functionals (7) is a large one, but there are many functionals that do not have this form. The Dirac *delta functional*

$$\delta\{\phi\} := \phi(0) \tag{9}$$

is the most important. It is easy to show that there is no ordinary piecewise continuous function δ for which

$$\delta\{\phi\} = \int_{-\infty}^{\infty} \delta(x)\phi(x)dx \tag{10}$$

whenever ϕ is a function from $\mathbb{S}$, *cf.* Ex. 7.3, but we will invent a *generalized function* δ that corresponds to the functional (9). We then use (9) to assign meaning to the "integral" on the right side of (10), *i.e.*,

$$\int_{-\infty}^{\infty} \delta(x)\phi(x)dx := \delta\{\phi\} := \phi(0).$$

You will often encounter a generalized function f and a deceptively familiar expression

$$\int_{-\infty}^{\infty} f(x)\phi(x)dx$$

as you study the remainder of this text. In each case there is a corresponding well-defined functional f, and you can assign meaning to the "integral" by using

$$\int_{-\infty}^{\infty} f(x)\phi(x)dx := f\{\phi\}$$

instead of the definition of integral that you learned in a calculus or analysis class. At first it may seem a bit confusing to associate a new meaning with an old expression, but you will quickly discover that the use of this integral notation will minimize the time that it takes to learn how to do analysis with generalized functions. You will remember an analogous situation from calculus. The Leibnitz notation dy/dx for a derivative is not an ordinary "quotient," but it can be manipulated like one when you use the chain rule or when you use substitution to evaluate integrals.

Schwartz functions

A complex-valued function ϕ on $\mathbb{R}$ is said to be a *Schwartz function* provided that

$$\phi, \phi', \phi'', \dots \quad \text{are all defined and continuous on } \mathbb{R}, \text{ and}$$

$$\lim_{x \to \pm\infty} x^n \phi^{(m)}(x) = 0 \quad \text{for each } m = 0, 1, 2, \dots \text{ and } n = 0, 1, 2, \dots . \quad (11)$$

(Here $\phi^{(0)} := \phi$, $\phi^{(1)} := \phi'$, $\phi^{(2)} := \phi''$, $\dots$.) The first condition forces ϕ and its derivatives to be exceptionally smooth, and the second forces ϕ and its derivatives to have exceptionally small tails at $\pm\infty$. Indeed, $\phi^{(m)}(x)$ goes to 0 faster than $1/x^n$ goes to 0 as $x \to \pm\infty$ in the sense that

$$\lim_{x \to \pm\infty} \frac{\phi^{(m)}(x)}{1/x^n} = 0, \quad n = 1, 2, \dots .$$

We will let $\mathbb{S}$ denote the linear space of all such Schwartz functions.

Example: Show that $\phi(x) := e^{-x^2}$ is a Schwartz function.

Solution: The function ϕ and its derivatives

$$\phi'(x) = -2xe^{-x^2}, \quad \phi''(x) = (4x^2 - 2)e^{-x^2}, \quad \dots$$

are continuous, and by using (3.28) we write

$$\lim_{x \to \pm\infty} x^n \phi^{(m)}(x) = \lim_{x \to \pm\infty} (-1)^m x^n H_m(x)e^{-x^2} = 0$$

for each $m = 0, 1, 2, \dots$ and $n = 0, 1, 2, \dots .$ ∎

Example: Let $a < b$ be given. Construct a Schwartz function ϕ such that

$$\phi(x) > 0 \text{ if } a < x < b \text{ and } \phi(x) = 0 \text{ for } x \le a \text{ or } x \ge b.$$

Solution: We define

$$v(x) := \begin{cases} 0 & \text{if } x \le 0 \\ e^{-1/x} & \text{if } x > 0, \end{cases} \tag{12}$$

and after observing that

$$\lim_{x \to 0+} v(x) = \lim_{x \to 0+} e^{-1/x} = \lim_{u \to +\infty} e^{-u} = 0,$$

$$\lim_{x \to 0+} v'(x) = \lim_{x \to 0+} \frac{1}{x^2} e^{-1/x} = \lim_{u \to +\infty} u^2 e^{-u} = 0,$$

$$\lim_{x \to 0+} v''(x) = \lim_{x \to 0+} \left(\frac{1}{x^4} - \frac{2}{x^3} \right) e^{-1/x} = \lim_{u \to +\infty} (u^4 - 2u^3) e^{-u} = 0,$$

$$\vdots$$

we conclude that $v, v', v'', \dots$ are well defined and continuous on $\mathbb{R}$. This being the case,

$$\phi(x) := v(x - a)v(b - x) = \begin{cases} e^{-(b-a)/\{(x-a)(b-x)\}} & \text{if } a < x < b \\ 0 & \text{otherwise} \end{cases} \tag{13}$$

and its derivatives are defined and continuous on $\mathbb{R}$. By construction, $\phi(x) = 0$ when $x \le a$ or $x \ge b$, so (11) also holds. ∎

Example: Let $a < b < c < d$ be given. Construct a Schwartz function ϕ such that

$$\phi(x) = 0 \text{ for } x \le a, \ \phi(x) = 1 \text{ for } b \le x \le c, \ \phi(x) = 0 \text{ for } x \ge d,$$
$$\phi'(x) > 0 \text{ for } a < x < b, \ \phi'(x) < 0 \text{ for } c < x < d. \tag{14}$$

Such a *mesa function* is shown in Fig. 7.5.

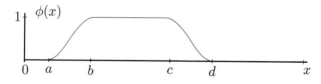

Figure 7.5. The mesa function (15) with the properties (14).

Solution: We construct a scaled antiderivative

$$g(x; a, b) := \frac{\int_{-\infty}^{x} v(u - a)v(b - u)du}{\int_{a}^{b} v(u - a)v(b - u)du}$$

for the Schwartz function from the preceding example with

$$g(x; a, b) = 0 \;\; \text{for} \;\; x \leq a, \;\; g(x; a, b) = 1 \;\; \text{for} \;\; x \geq b,$$
$$g'(x; a, b) > 0 \;\; \text{for} \;\; a < x < b.$$

The product

$$\phi(x) := g(x; a, b)g(-x; -d, -c) \tag{15}$$

[analogous to the product $h(x - a)h(d - x)$ of Heaviside functions] is a Schwartz function that has all of the properties (14). ∎

When $\phi \in \mathbb{S}$ and $m = 0, 1, 2, \ldots$ we can use (11) to see that

$$B_m := \max_{-\infty < x < \infty} |\phi^{(m)}(x)|, \quad C_m := \max_{-\infty < x < \infty} |x^2 \phi^{(m)}(x)|$$

are well defined and finite. We bound $\phi^{(m)}$ by writing

$$|\phi^{(m)}(x)| \leq \begin{cases} B_m & \text{if } x^2 B_m \leq C_m \\ C_m/x^2 & \text{otherwise,} \end{cases}$$

and thereby show that

$$\int_{-\infty}^{\infty} |\phi^{(m)}(x)|dx < \infty.$$

Since the smooth function ϕ has small regular tails, we can use the analysis from Section 1.5 to see that the Fourier transform

$$\phi^{\wedge}(s) := \int_{-\infty}^{\infty} \phi(x)e^{-2\pi i s x}dx$$

is well defined, continuous, and absolutely integrable on $\mathbb{R}$, and that Fourier's representation

$$\phi(x) = \int_{-\infty}^{\infty} \phi^{\wedge}(s)e^{2\pi i s x}ds$$

is valid at every point. Moreover, a Schwartz function satisfies all of the "extra" hypotheses we introduced in Chapter 3 when we described the derivative rule, the power scaling rule, etc. We can freely use *all* of the rules from the Fourier transform calculus when we work with Schwartz functions!

The class $\mathbb{S}$ has a number of *closure properties* that we will use from time to time. It is fairly easy to show that if $\phi(x)$ is a Schwartz function, then so is

$$
\begin{aligned}
& c\,\phi(x), && c \in \mathbb{C}; \\
& \phi(x - x_0), && -\infty < x_0 < \infty; \\
& \phi(ax), && a < 0 \text{ or } a > 0; \\
& e^{2\pi i s_0 x}\phi(x), && -\infty < s_0 < \infty; \\
& x^n\phi(x), && n = 0, 1, 2, \ldots ; \\
& \phi^{(m)}(x), && m = 0, 1, 2, \ldots ; \text{ and} \\
& \overline{\phi}(x).
\end{aligned}
$$

Example: Let $\phi \in \mathbb{S}$, let $n = 1, 2, \ldots$, and let $\psi(x) := x^n\phi(x)$. Show that $\psi \in \mathbb{S}$.

Solution: We use the Leibnitz rule (2.29) to see that

$$
\psi^{(m)}(x) = \sum_{k=0}^{\min(n,m)} \binom{m}{k}(x^n)^{(k)}\phi^{(m-k)}(x), \quad m = 0, 1, 2, \ldots,
$$

is continuous, and then use (11) to infer that

$$
|x^p\psi^{(m)}(x)| \le \sum_{k=0}^{\min(n,m)} \binom{m}{k}n!|x^{n-k+p}\phi^{(m-k)}(x)|
$$

$$
\to 0 \quad \text{as} \quad x \to \pm\infty
$$

when $p = 0, 1, 2, \ldots$. ∎

It takes a bit more effort to show that if ϕ is a Schwartz function, then so is the Fourier transform

$$
\phi^{\wedge}(s) := \int_{-\infty}^{\infty} \phi(x)e^{-2\pi i s x}dx,
$$

cf. Ex. 7.5, the quotient

$$
\psi(x) := \begin{cases} \phi(x)/(x - x_0) & \text{if } x \ne x_0 \\ \phi'(x_0) & \text{if } x = x_0 \end{cases}
$$

$$
= \int_{t=0}^{1} \phi'(tx + (1 - t)x_0)dt, \quad -\infty < x_0 < \infty,
\tag{16}
$$

in those cases where $\phi(x_0) = 0$, and the antiderivative

$$
\psi(x) := \int_{-\infty}^{x} \phi(u)du
\tag{17}
$$

in those cases where

$$\int_{-\infty}^{\infty} \phi(u)du = 0. \tag{18}$$

Example: Show that (17) is a Schwartz function when ϕ is a Schwartz function satisfying (18).

Solution: Since $\psi' = \phi$, $\psi'' = \phi', \ldots$ it is enough to show that $x^n \psi(x) \to 0$ as $x \to \pm\infty$ for each $n = 0, 1, 2, \ldots$. We define

$$D_n := \max_{-\infty < x < \infty} |x^{n+2}\phi(x)|$$

and obtain the desired limits from the bounds

$$|x^n \psi(x)| = \left| x^n \int_{-\infty}^{x} \phi(u)du \right| \leq \left| x^n \int_{-\infty}^{x} D_n u^{-n-2} du \right| = \frac{D_n}{(n+1)|x|} \ , \ x < 0,$$

$$|x^n \psi(x)| = \left| x^n \int_{x}^{\infty} \phi(u)du \right| \leq \left| x^n \int_{x}^{\infty} D_n u^{-n-2} du \right| = \frac{D_n}{(n+1)x} \ , \ x > 0. \quad \blacksquare$$

The class $\mathbb{S}$ is also closed under certain binary operations. Indeed, if $\phi_1, \phi_2 \in \mathbb{S}$, then

$$\phi_1 + \phi_2, \quad \phi_1 \cdot \phi_2, \ \text{and} \ \phi_1 * \phi_2$$

are all Schwartz functions, *cf.* Ex. 7.6.

Functionals for generalized functions

A complex-valued function g on $\mathbb{R}$ is said to be *slowly growing* if

$$\lim_{x \to \pm\infty} \frac{g(x)}{x^n} = 0$$

for some choice of $n = 0, 1, 2, \ldots$, *e.g.*, the functions

$$x^3 + x, \quad e^{-\pi x^2}, \quad \sin x, \quad x \ln |x|$$

are slowly growing while

$$e^x, \quad e^{-x}, \quad e^{x^2}$$

are not. In this section we will use functions that are both continuous and slowly growing (CSG) to construct functionals for generalized functions on $\mathbb{R}$.

We will frequently form the product $g \cdot \phi$ of a CSG function g and a Schwartz function ϕ. Such a product is always a bounded, continuous, absolutely integrable function that vanishes at $\pm\infty$. The continuity of the product follows from that of g

and ϕ. Since g is CSG there is some $m = 0, 1, 2, \ldots$ and some $M_1 > 0$ (depending on g) such that

$$\left| \frac{g(x)}{(1+x^2)^m} \right| \leq M_1, \qquad -\infty < x < \infty,$$

and since ϕ is a Schwartz function there is some $M_2 > 0$ (depending on m and ϕ) such that

$$|(1+x^2)^{m+1}\phi(x)| \leq M_2, \qquad -\infty < x < \infty.$$

The resulting bound

$$|g(x)\phi(x)| = \left| \frac{g(x)}{(1+x^2)^m} \right| \cdot |(1+x^2)^{m+1}\phi(x)| \cdot \frac{1}{1+x^2} \leq \frac{M_1 M_2}{1+x^2}$$

shows that $g \cdot \phi$ is bounded and absolutely integrable with $\lim\limits_{x \to \pm\infty} g(x)\phi(x) = 0$. In view of this discussion we will always represent a CSG function g with the fundamental functional

$$g\{\phi\} := \int_{-\infty}^{\infty} g(x)\phi(x)dx, \qquad \phi \in \mathbb{S}.$$

Now if g happens to have an ordinary CSG derivative g', then a careful integration by parts allows us to write

$$\int_{-\infty}^{\infty} g'(x)\phi(x)dx = \lim_{\substack{L \to -\infty \\ U \to +\infty}} \int_{L}^{U} g'(x)\phi(x)dx$$

$$= \lim_{\substack{L \to -\infty \\ U \to +\infty}} \left\{ g(x)\phi(x) \Big|_{L}^{U} - \int_{L}^{U} g(x)\phi'(x)dx \right\}$$

$$= -\int_{-\infty}^{\infty} g(x)\phi'(x)dx, \qquad \phi \in \mathbb{S}.$$

In this way we see that g' can be represented by the functional

$$g'\{\phi\} := -\int_{-\infty}^{\infty} g(x)\phi'(x)dx, \qquad \phi \in \mathbb{S}. \tag{19}$$

You will notice that the integral from (19) is well defined (since $\phi' \in \mathbb{S}$ when $\phi \in \mathbb{S}$), even in cases where the CSG function g is not differentiable or in cases where g is differentiable but the integrand $g'(x)\phi(x)$ from the fundamental functional for g' is not (Riemann or Lebesgue) integrable. We will use the functional (19) to represent the *generalized derivative* of the CSG function g.

Analogously, when $g, g', \ldots, g^{(n)}$ are all CSG for some $n = 1, 2, \ldots$ we find

$$\int_{-\infty}^{\infty} g^{(n)}(x)\phi(x)dx = -\int_{-\infty}^{\infty} g^{(n-1)}(x)\phi'(x)dx = (-1)^2 \int_{-\infty}^{\infty} g^{(n-2)}(x)\phi''(x)dx$$

$$= \cdots = (-1)^n \int_{-\infty}^{\infty} g(x)\phi^{(n)}(x)dx, \quad \phi \in \mathbb{S},$$

so we can use the functional

$$g^{(n)}\{\phi\} := (-1)^n \int_{-\infty}^{\infty} g(x)\phi^{(n)}(x)dx, \quad \phi \in \mathbb{S}, \tag{20}$$

to represent $g^{(n)}$. The integral from (20) is well defined (since $\phi^{(n)} \in \mathbb{S}$ when $\phi \in \mathbb{S}$), even in cases where the CSG function g does not have ordinary derivatives $g', g'', \ldots, g^{(n)}$ or in cases where $g^{(n)}$ exists but $g^{(n)}(x)\phi(x)$ is not integrable. We will use (20) to represent the *generalized nth derivative* of the CSG function g.

The time has come for a very important definition. We will say that f is a *generalized function* if $f = g^{(n)}$ for some choice of the CSG function g and for some nonnegative integer n. In keeping with our previous discussion, we will use the *integral notation*

$$\int_{-\infty}^{\infty} f(x)\phi(x)dx := f\{\phi\} := g^{(n)}\{\phi\} := (-1)^n \int_{-\infty}^{\infty} g(x)\phi^{(n)}(x)dx, \quad \phi \in \mathbb{S}, \tag{21}$$

when we work with f.

We will also routinely use function notation within this context, and this necessitates a new understanding of equality. Given generalized functions f_1, f_2 and $a < b$ we will say that

$$f_1(x) = f_2(x) \text{ for } a < x < b \text{ provided that}$$
$$f_1\{\phi\} = f_2\{\phi\} \text{ for each } \phi \in \mathbb{S} \text{ with } \phi(x) = 0 \text{ when } x < a \text{ or } x > b. \tag{22}$$

In the case where $a = -\infty$ and $b = +\infty$ we simply write

$$f_1 = f_2 \text{ provided that } f_1\{\phi\} = f_2\{\phi\} \text{ for each } \phi \in \mathbb{S}. \tag{23}$$

We will also write

$$f = f_0 \text{ or } f(x) = f_0(x) \text{ for } -\infty < x < \infty$$

when the generalized function f is the fundamental functional

$$f\{\phi\} = \int_{-\infty}^{\infty} f_0(x)\phi(x)dx, \quad \phi \in \mathbb{S}$$

of a CSG function f_0 or the ordinary derivative $f_0 = g'$ of a CSG function g that is continuously differentiable except for certain isolated points of $\mathbb{R}$, *e.g.*, $f_0(x) = \operatorname{sgn}(x)$, $\lfloor x \rfloor$, $\log |x|$.

7.2 *Common Generalized Functions*

Introduction

In this section you will learn to recognize a few generalized functions that will be needed for the study of sampling, PDEs, wavelets, probability, diffraction, etc. Later on you will master the rules for manipulating these "functions" and find that it is really very easy to do Fourier analysis within this new context.

Functions from calculus

The ordinary *power function*

$$p_n(x) := x^n, \qquad n = 0, 1, 2, \ldots, \tag{24}$$

is CSG, so we can use the fundamental functional

$$p_n\{\phi\} := \int_{-\infty}^{\infty} x^n \phi(x) dx, \qquad \phi \in \mathbb{S} \tag{25}$$

to obtain a corresponding generalized function. Using the definition (19), we construct the functional

$$p_n'\{\phi\} := -\int_{-\infty}^{\infty} x^n \phi'(x) dx, \qquad \phi \in \mathbb{S}$$

for the generalized derivative, and use a careful integration by parts to show that

$$p_n'\{\phi\} = n p_{n-1}\{\phi\}, \qquad \phi \in \mathbb{S}.$$

In this way we see that p_n has the generalized derivative

$$p_n' = n\, p_{n-1}$$

that corresponds to the differentiation rule $(x^n)' = n\, x^{n-1}$ from calculus.

Analogously, the generalized functions

$$s\{\phi\} := \int_{-\infty}^{\infty} \sin(x)\phi(x) dx, \quad c\{\phi\} := \int_{-\infty}^{\infty} \cos(x)\phi(x) dx, \qquad \phi \in \mathbb{S}$$

have the generalized derivatives

$$s' = c, \qquad c' = -s$$

corresponding to the rules $\sin'(x) = \cos(x)$, $\cos'(x) = -\sin(x)$ from calculus.

The ordinary exponential function is *not* slowly growing, and the functional

$$e\{\phi\} := \int_{-\infty}^{\infty} e^x \phi(x) dx$$

is *not* defined for every $\phi \in \mathbb{S}$. Indeed,

$$\psi(x) := \begin{cases} e^{-1/x} e^{-\sqrt{x}} & \text{for } x > 0 \\ 0, & \text{for } x \le 0 \end{cases}$$

is a Schwartz function with

$$e\{\psi\} = \int_0^{\infty} (e^x e^{-1/x} e^{-\sqrt{x}}) dx = +\infty.$$

There is no generalized function of the form (20) that corresponds to e^x (or to $e^{(\alpha+i\beta)x}$, $-\infty < \alpha < \infty$, $-\infty < \beta < \infty$ when $\alpha \neq 0$).

Dirac's delta function

The ramp function

$$r(x) := \begin{cases} x & \text{for } x > 0 \\ 0 & \text{for } x \le 0 \end{cases} \tag{26}$$

is CSG, so we represent r by the fundamental functional

$$r\{\phi\} := \int_{-\infty}^{\infty} r(x)\phi(x) dx = \int_0^{\infty} x\,\phi(x) dx, \qquad \phi \in \mathbb{S}.$$

The generalized derivative r' is represented by

$$r'\{\phi\} := -\int_{-\infty}^{\infty} r(x)\phi'(x) dx = -\int_0^{\infty} x\,\phi'(x) dx$$

$$= -x\,\phi(x)\Big|_0^{\infty} + \int_0^{\infty} \phi(x) dx = \int_0^{\infty} \phi(x) dx, \quad \phi \in \mathbb{S},$$

so we can write

$$r' = h \tag{27}$$

where h is the Heaviside step function, *cf.* Fig. 7.6.

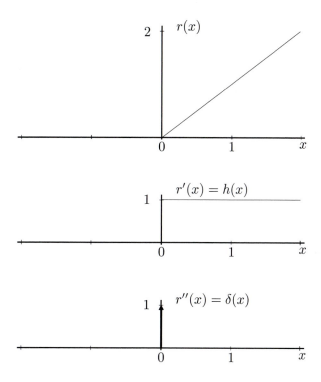

Figure 7.6. The CSG ramp r from (26) and its generalized derivatives $r' = h$, $r'' = \delta$.

We define Dirac's *delta function*

$$\delta := r'', \tag{28}$$

and determine the corresponding functional by writing

$$\delta\{\phi\} := r''\{\phi\} := (-1)^2 \int_{-\infty}^{\infty} r(x)\phi''(x)dx$$
$$= \int_{0}^{\infty} x\,\phi''(x)dx = \int_{0}^{\infty} \frac{d}{dx}[x\,\phi'(x) - \phi(x)]dx$$
$$= \phi(0), \qquad \phi \in \mathbb{S},$$

as given by (9). We graphically represent δ by using an arrow of unit length as shown in Fig. 7.6. The generalized nth derivative of δ,

$$\delta^{(n)} := r^{(n+2)}, \qquad n = 0, 1, 2, \dots,$$

is obtained from the functional

$$\delta^{(n)}\{\phi\} := r^{(n+2)}\{\phi\} := (-1)^{n+2}\int_{-\infty}^{\infty} r(x)\phi^{(n+2)}(x)dx$$

$$= (-1)^n \int_0^{\infty} x\,\phi^{(n+2)}(x)dx = (-1)^n \int_0^{\infty} \frac{d}{dx}\left[x\,\phi^{(n+1)}(x) - \phi^{(n)}(x)\right]dx \quad (29)$$

$$= (-1)^n \phi^{(n)}(0), \quad n = 0, 1, 2, \ldots, \quad \phi \in \mathbb{S}.$$

From (22) and (29) we see that

$$\begin{aligned} \delta^{(n)}(x) &= 0 \quad \text{for} \quad -\infty < x < 0, \ n = 0, 1, 2, \ldots, \\ \delta^{(n)}(x) &= 0 \quad \text{for} \quad 0 < x < \infty, \ n = 0, 1, 2, \ldots, \end{aligned} \tag{30}$$

but these "values" do not determine the "integral"

$$\int_{-\infty}^{\infty} \delta^{(n)}(x)\phi(x)dx := \delta^{(n)}\{\phi\} = (-1)^n \phi^{(n)}(0), \quad \phi \in \mathbb{S}. \tag{31}$$

Example: Evaluate

$$\int_{-\infty}^{\infty} \delta'(x)x\,e^{-x^2}dx.$$

Solution: Using (31) we find

$$\int_{-\infty}^{\infty} \delta'(x)x\,e^{-x^2}dx := -\int_{-\infty}^{\infty} \delta(x)\left(x\,e^{-x^2}\right)'dx = -e^{-x^2} + 2x^2 e^{-x^2}\bigg|_{x=0} = -1$$

(even though the "integrand" seems to vanish at each "point" x). ∎

Example: We define the normalized truncated power function

$$\sigma_n(x) := \begin{cases} x^n/n! & \text{if } x > 0 \\ 0, & \text{if } x \le 0, \end{cases} \quad n = 0, 1, 2, \ldots . \tag{32}$$

Find the generalized derivatives.

Solution: The first n derivatives

$$\sigma_n' = \sigma_{n-1}, \quad \sigma_n'' = \sigma_{n-2}, \quad \ldots, \quad \sigma_n^{(n-1)} = \sigma_1 = r, \quad \sigma_n^{(n)} = \sigma_0 = h \tag{33}$$

are *ordinary* functions, but we must use the delta function as we write

$$\sigma_n^{(n+1)} = \delta, \quad \sigma_n^{(n+2)} = \delta', \quad \sigma^{(n+3)} = \delta'', \ldots . \tag{34}$$

∎

The comb III

The piecewise linear antiderivative

$$q(x) := \int_0^x \tau(u)du \tag{35}$$

of the slowly growing floor function

$$\tau(x) := \lfloor x \rfloor := m \quad \text{when} \quad m \le x < m+1 \quad \text{and} \quad m = 0, \pm 1, \pm 2, \ldots \tag{36}$$

is continuous, *cf.* Fig. 7.7. This antiderivative is also slowly growing [as we see by integrating the inequality $u - 1 \le \tau(u) \le u$ from $u = 0$ to $u = x$], so we can use the fundamental functional

$$q\{\phi\} := \int_{-\infty}^{\infty} q(x)\phi(x)dx, \qquad \phi \in \mathbb{S}. \tag{37}$$

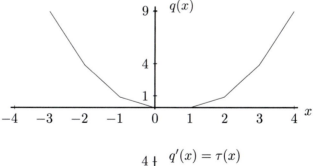

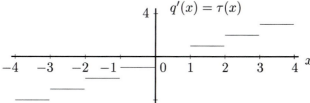

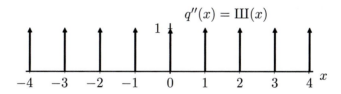

Figure 7.7. The CSG "parabola" q from (35) and its generalized derivatives $q' = \tau$, $q'' = \text{III}$.

The ordinary derivative of $q(x)$ has jump discontinuities at $x = 0, \pm 1, \pm 2, \ldots$, but we can use

$$q'(x) = m \quad \text{for} \quad m < x < m+1 \quad \text{and} \quad m = 0, \pm 1, \pm 2, \ldots,$$

$$\lim_{m \to \pm \infty} q(m)\phi(m) = 0 \quad \text{when} \quad \phi \in \mathbb{S},$$

$$\sum_{m=-\infty}^{\infty} [q(m)\phi(m) - q(m+1)\phi(m+1)] = 0 \quad \text{when} \quad \phi \in \mathbb{S}$$

with a careful integration by parts to write

$$q'\{\phi\} : = -\int_{-\infty}^{\infty} q(x)\phi'(x)dx = \sum_{m=-\infty}^{\infty} \int_{m}^{m+1} -q(x)\phi'(x)dx$$

$$= \sum_{m=-\infty}^{\infty} \left\{ q(m)\phi(m) - q(m+1)\phi(m+1) + \int_{m}^{m+1} m\,\phi(x)dx \right\} \quad (38)$$

$$= \int_{-\infty}^{\infty} \tau(x)\phi(x)dx, \qquad \phi \in \mathbb{S}.$$

This allows us to identify the generalized derivative q' with the function τ of (36).

The second generalized derivative

$$\text{Ш} := q'' \tag{39}$$

is known as the *comb function* (from the appearance of the graph in Fig. 7.7), the *shah function* (from the name of the Cyrillic letter Ш), or the *sampling function* (from the identity

$$\text{Ш}\{\phi\} = \sum_{n=-\infty}^{\infty} \phi(n), \quad \phi \in \mathbb{S} \tag{40}$$

that specifies its action). We derive (40) by using (38) (with ϕ replaced by ϕ') and the fundamental theorem of calculus to write

$$\text{Ш}\{\phi\} : = (-1)^2 \int_{-\infty}^{\infty} q(x)\phi''(x)dx = \sum_{m=-\infty}^{\infty} -m \int_{m}^{m+1} \phi'(x)dx$$

$$= \sum_{m=-\infty}^{\infty} [-m\,\phi(m+1) + m\,\phi(m)]$$

$$= \sum_{m=-\infty}^{\infty} \{\phi(m) + [(m-1)\phi(m) - m\,\phi(m+1)]\}$$

$$= \sum_{m=-\infty}^{\infty} \phi(m), \qquad \phi \in \mathbb{S}.$$

[Since $m^2\phi(m)$ is bounded, the last sum converges absolutely.] Analogously,

$$\text{III}^{(n)}\{\phi\} = (-1)^n \sum_{m=-\infty}^{\infty} \phi^{(n)}(m), \quad \phi \in \mathbb{S}, \quad n = 0, 1, 2, \ldots . \tag{41}$$

From (41) we see that

$$\text{III}^{(n)}(x) = 0 \text{ for } m < x < m+1, \quad m = 0, \pm 1, \pm 2, \ldots , \tag{42}$$

but these "values" do not determine the "integral"

$$\int_{-\infty}^{\infty} \text{III}^{(n)}(x)\phi(x)dx := \text{III}^{(n)}\{\phi\} = (-1)^n \sum_{m=-\infty}^{\infty} \phi^{(n)}(m), \quad \phi \in \mathbb{S}. \tag{43}$$

The functions $\mathbf{x^{-1}, x^{-2}, \ldots}$

The ordinary function $\log|x|$ has an integrable singularity at the origin, but the antiderivative

$$\ell(x) := \int_0^x \log|u|du = x\log|x| - x \tag{44}$$

is CSG, so we write

$$\ell\{\phi\} := \int_{-\infty}^{\infty} (x\log|x| - x)\phi(x)dx, \quad \phi \in \mathbb{S}.$$

We use a careful integration by parts to show that

$$\ell'\{\phi\} = \int_{-\infty}^{\infty} \log|x|\phi(x)dx, \quad \phi \in \mathbb{S}, \tag{45}$$

and we identify $\ell'(x)$ with the ordinary function $\log|x|$, cf. Fig. 7.8.

The ordinary functions $x^{-1}, x^{-2}, x^{-3}, \ldots$ can be obtained from the ordinary derivatives of $\ell(x)$ by writing

$$x^{-1} = \ell''(x), \quad x^{-2} = -\ell'''(x), \quad x^{-3} = \ell''''(x)/2!, \quad \ldots .$$

We obtain corresponding generalized functions by using the generalized derivatives of ℓ to *define* the inverse power functions

$$p_{-1} := \ell'', \quad p_{-2} := -\ell''', \quad p_{-3} := \ell''''/2!, \quad \ldots . \tag{46}$$

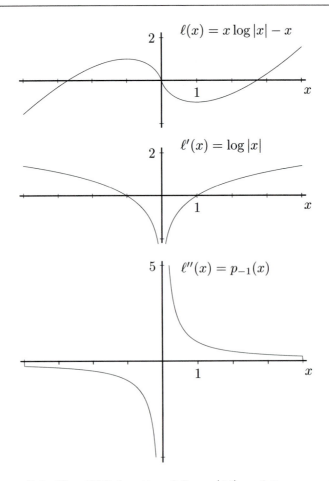

Figure 7.8. The CSG function ℓ from (44) and its generalized derivatives ℓ', $\ell'' = p_{-1}$.

We must use some care, however, because the improper integral for the fundamental functional

$$\int_{-\infty}^{\infty} \frac{\phi(x)}{x^n}\,dx \tag{47}$$

is *not* well defined unless the Schwartz function ϕ has a zero of multiplicity n at the origin. [We did not have this problem with the functionals (25) for the power functions (24).]

Given $\phi \in \mathbb{S}$ we define

$$\phi_1(x) := \phi(x) - \phi(0)$$
$$\phi_2(x) := \phi(x) - \phi(0) - x\,\phi'(0)$$
$$\phi_3(x) := \phi(x) - \phi(0) - x\,\phi'(0) - x^2\phi''(0)/2! \tag{48}$$

$$\vdots$$

and observe that

$$\phi_n^{(n)}(x) = \phi^{(n)}(x), \quad n = 1, 2, \dots . \tag{49}$$

We use Taylor's formula (*cf.* Ex. 2.28) with the fact that

$$\phi_n(0) = \phi_n'(0) = \cdots = \phi_n^{(n-1)}(0) = 0$$

to infer that there is a continuous function ψ_n such that $\phi_n(x) = x^n \psi_n(x)$, $-\infty < x < \infty$, and thereby verify that

$$\lim_{x \to 0} \log |x| \phi_1(x) = 0, \tag{50}$$

$$\lim_{x \to 0} x^{1-n} \phi_n(x) = 0, \ n = 2, 3, \dots . \tag{51}$$

We will use (48)–(51) to obtain simple expressions that describe the action of the inverse power functions $p_{-1}, p_{-2}, p_{-3}, \dots$.

We begin by showing that

$$p_{-1}\{\phi\} = \lim_{L \to \infty} \int_{-L}^{L} \frac{\phi(x) - \phi(0)}{x} dx, \quad \phi \in \mathbb{S}. \tag{52}$$

Indeed, using (46), (44), (45) (with ϕ replaced by ϕ'), (49), and (50) in turn we write

$$p_{-1}\{\phi\} := \ell''\{\phi\} := (-1)^2 \int_{-\infty}^{\infty} (x \log |x| - x) \phi''(x) dx$$

$$= - \int_{-\infty}^{\infty} \log |x| \phi_1'(x) dx$$

$$= \lim_{L \to +\infty} \lim_{\epsilon \to 0+} \left\{ - \int_{-L}^{-\epsilon} \log |x| \phi_1'(x) dx - \int_{\epsilon}^{L} \log |x| \phi_1'(x) dx \right\}$$

$$= \lim_{L \to +\infty} \lim_{\epsilon \to 0+} \left\{ - \log |x| \phi_1(x) \Big|_{-L}^{-\epsilon} + \int_{-L}^{-\epsilon} \frac{\phi_1(x)}{x} dx \right.$$

$$\left. - \log |x| \phi_1(x) \Big|_{\epsilon}^{L} + \int_{\epsilon}^{L} \frac{\phi_1(x)}{x} dx \right\}$$

$$= \lim_{L \to +\infty} \left\{ - \log |x| \phi_1(x) \Big|_{-L}^{L} + \int_{-L}^{L} \frac{\phi_1(x)}{x} dx \right\}$$

$$= \lim_{L \to +\infty} \int_{-L}^{L} \frac{\phi_1(x)}{x} dx,$$

and in conjunction with the defining relation (48) for ϕ_1 this gives (52).

An alternative representation,

$$p_{-1}\{\phi\} = \int_0^\infty \frac{\phi(x) - \phi(-x)}{x}\,dx, \quad \phi \in \mathbb{S}, \tag{53}$$

is obtained by using

$$\int_{-L}^0 \frac{\phi(x) - \phi(0)}{x}\,dx = -\int_0^L \frac{\phi(-x) - \phi(0)}{x}\,dx$$

for the left half of the integral (52).

An analogous argument gives corresponding expressions for the functionals $p_{-2}, p_{-3}, \ldots$. We simply replace ϕ by the $\phi_2, \phi_3, \ldots$ from (48), i.e.,

$$p_{-2}\{\phi\} = \lim_{L \to +\infty} \int_{-L}^L \frac{\phi(x) - \phi(0) - x\,\phi'(0)}{x^2}\,dx, \quad \phi \in \mathbb{S},$$

$$p_{-3}\{\phi\} = \lim_{L \to +\infty} \int_{-L}^L \frac{\phi(x) - \phi(0) - x\,\phi'(0) - (x^2/2)\phi''(0)}{x^3}\,dx, \quad \phi \in \mathbb{S}, \tag{54}$$

$$\vdots$$

By using (52) and (54) we see that

$$\begin{aligned} p_{-n}(x) &= x^{-n} \quad \text{for} \quad -\infty < x < 0, \quad n = 1, 2, \ldots, \\ p_{-n}(x) &= x^{-n} \quad \text{for} \quad 0 < x < \infty, \quad n = 1, 2, \ldots, \end{aligned} \tag{55}$$

but these "values" do not determine the "integral"

$$\int_{-\infty}^\infty p_{-n}(x)\phi(x)\,dx := p_{-n}\{\phi\}$$

$$= \lim_{L \to +\infty} \int_{-L}^L x^{-n}\left\{\phi(x) - \sum_{k=0}^{n-1} x^k \phi^{(k)}(0)/k!\right\}dx, \quad \phi \in \mathbb{S}. \tag{56}$$

Summary

At this point you should understand how a CSG function g is used with the functional (20) and an elementary notion of integration to produce a generalized function $f = g^{(n)}$, $n = 0, 1, 2, \ldots$. Ordinary CSG functions such as

$$x^2 - x, \quad e^{2\pi i x}, \quad x^2 \sin(x), \quad e^{i\pi x^2}, \quad x \log|x|$$

can be represented with $n = 0$, but we must use $n \geq 1$ when we construct the functionals (9), (40), (52) for δ, III, p_{-1}.

In cases where f is a slowly growing ordinary function that is *locally integrable,* i.e.,

$$\int_a^b f(x)dx$$

is well defined for every choice of $-\infty < a < b < \infty$, we can construct a CSG antiderivative

$$g(x) := \int_0^x f(u)du$$

and use the fundamental functional for f to represent the generalized derivative g'. We used such antiderivatives (26), (35), (44) for

$$r'(x) = h(x), \quad q'(x) = \lfloor x \rfloor, \quad \ell'(x) = \log|x|$$

to define δ, III, p_{-1}. A thorough study of this construction of generalized functions from slowly growing, locally integrable functions is best done within a context that includes Lebesgue's theory of integration. After you master the use of δ, III, p_{-1} and a few less exotic generalized functions you may wish to explore such ideas in more detail.

7.3 *Manipulation of Generalized Functions*

Introduction

It is one thing to define generalized functions and quite another to use them for some worthwhile purpose! In this section you will learn to form a linear combination

$$c_1 f_1(x) + c_2 f_2(x), \qquad c_1, c_2 \in \mathbb{C}$$

of generalized functions f_1, f_2 and to construct the

translate	$f(x - x_0)$,	$-\infty < x_0 < \infty$,
dilate	$f(ax)$,	$a < 0$ or $a > 0$,
derivative	$f'(x)$,	
Fourier transform	$f^\wedge(x)$,	
product	$\alpha(x) \cdot f(x)$,	(when α is suitably regular), and
convolution product	$(\beta * f)(x)$	(when β is suitably regular)

for a generalized function f. In each case there is a corresponding functional of the form (20). The new understanding of sum, translate, dilate, ... will reduce to the familiar classical concept when f is a suitably regular ordinary function on $\mathbb{R}$, but the extensions will facilitate a far more powerful (and useful!) form of Fourier analysis than that developed in Chapters 1–5.

The linear space $\mathbb{G}$

Given generalized functions f_1, f_2 and complex scalars c_1, c_2 we define

$$f := c_1 f_1 + c_2 f_2$$

by using the functional

$$f\{\phi\} := c_1 f_1\{\phi\} + c_2 f_2\{\phi\}, \qquad \phi \in \mathbb{S}. \tag{57}$$

The same relation can be expressed by using the integral notation

$$\int_{-\infty}^{\infty} [c_1 f_1(x) + c_2 f_2(x)]\phi(x)dx = c_1 \int_{-\infty}^{\infty} f_1(x)\phi(x)dx + c_2 \int_{-\infty}^{\infty} f_2(x)\phi(x)dx, \tag{58}$$
$$\phi \in \mathbb{S}.$$

We will verify that (57) defines a generalized function.

A CSG function g always has the CSG antiderivatives

$$g^{(-1)}(x) := \int_0^x g(u)du, \quad g^{(-2)}(x) := \int_0^x g^{(-1)}(u)du,$$
$$g^{(-3)}(x) := \int_0^x g^{(-2)}(u)du, \ldots . \tag{59}$$

Now if

$$f_1 = g_1^{(n_1)}, \qquad f_2 = g_2^{(n_2)}$$

where g_1, g_2 are CSG and n_1, n_2 are nonnegative integers, we will define $n := \max(n_1, n_2)$ and use antiderivatives to construct the CSG function

$$g := c_1 g_1^{(n_1 - n)} + c_2 g_2^{(n_2 - n)}.$$

A routine calculation then shows that

$$g^{(n)}\{\phi\} = c_1 g_1^{(n_1)}\{\phi\} + c_2 g_2^{(n_2)}\{\phi\} = f\{\phi\}, \qquad \phi \in \mathbb{S}.$$

In this way we see that the set of generalized functions is a *linear space* that we will call $\mathbb{G}$.

Translate, dilate, derivative, and Fourier transform

When f is a suitably regular ordinary function on $\mathbb{R}$ and $\phi \in \mathbb{S}$ we can use a change of variable, an integration by parts, or Parseval's identity [*cf.* Ex. 1.24(a)] to write

$$\int_{-\infty}^{\infty} f(x+x_0)\phi(x)dx = \int_{-\infty}^{\infty} f(x)\phi(x-x_0)dx, \quad -\infty < x_0 < \infty, \tag{60}$$

$$\int_{-\infty}^{\infty} f(ax)\phi(x)dx = \int_{-\infty}^{\infty} f(x)|a|^{-1}\phi\left(\frac{x}{a}\right)dx, \quad a < 0 \text{ or } a > 0, \tag{61}$$

$$\int_{-\infty}^{\infty} f'(x)\phi(x)dx = \int_{-\infty}^{\infty} f(x)[-\phi'(x)]dx, \tag{62}$$

$$\int_{-\infty}^{\infty} f^{\wedge}(x)\phi(x)dx = \int_{-\infty}^{\infty} f(x)\phi^{\wedge}(x)dx. \tag{63}$$

You can use these identities when f is a generalized function and $\phi \in \mathbb{S}$ provided you regard each of these "integrals" as notation for a corresponding functional! Such manipulations are valid because we *define* the translate, dilate, derivative, and Fourier transform

$$f_1(x) := f(x + x_0), \ \ f_2(x) := f(ax), \ \ f_3(x) := f'(x), \ \ f_4(x) := f^{\wedge}(x)$$

of a generalized function f by using the functionals

$$f_1\{\phi\} := f\{\phi_1\}, \ \ f_2\{\phi\} := f\{\phi_2\}, \ \ f_3\{\phi\} := f\{\phi_3\}, \ \ f_4\{\phi\} := f\{\phi_4\}$$

where

$$\phi_1(x) := \phi(x - x_0), \ \ \phi_2(x) := |a|^{-1}\phi\left(\frac{x}{a}\right), \ \ \phi_3(x) := -\phi'(x), \ \ \phi_4(x) = \phi^{\wedge}(x)$$

are Schwartz functions when $\phi \in \mathbb{S}$. [You will find these Schwartz functions on the right hand side of (60)–(63).] Of course, (62) is perfectly consistent with (20). We can express these definitions more succinctly by writing

$$(\mathcal{T}_{x_0}f)\{\phi\} := f\{\mathcal{T}_{-x_0}\phi\}, \quad (\mathcal{S}_a f)\{\phi\} := f\{|a|^{-1}\mathcal{S}_{1/a}\phi\},$$

$$(\mathbf{D}f)\{\phi\} := f\{-\mathbf{D}\phi\}, \quad (\mathcal{F}f)\{\phi\} := f\{\mathcal{F}\phi)\}, \quad \phi \in \mathbb{S},$$

but you will usually find that it is much easier to remember the equivalent integral notation of (60)–(63). The following examples will show you how to manipulate generalized functions and introduce you to the art of *definition chasing* that is always used within this context. (To help you sort out the details, we will provide terse justifications for the links in the first few deductive chains.)

Example: Find the linear functional that represents $\delta(x - x_0)$ when $-\infty < x_0 < \infty$.

Solution: The functional for $f(x) := \delta(x - x_0)$ is

$$f\{\phi\} = \int_{-\infty}^{\infty} \delta(x - x_0)\phi(x)dx \qquad \text{Integral notation}$$

$$:= \int_{-\infty}^{\infty} \delta(x)\phi(x + x_0)dx \qquad \text{Change of variable using (60)}$$

$$:= \phi(x_0), \quad \phi \in \mathbb{S}. \qquad \text{Action of } \delta \text{ from (9)}$$

The corresponding "integral"

$$\int_{-\infty}^{\infty} \delta(x - x_0)\phi(x)dx = \phi(x_0), \qquad \phi \in \mathbb{S} \qquad (64)$$

is known as the *sifting relation* for δ. No ordinary function has this property! ∎

Example: Find the linear functional that represents $\delta(ax)$ when $a < 0$ or $a > 0$.

Solution: The functional for $f(x) := \delta(ax)$ is

$$f\{\phi\} = \int_{-\infty}^{\infty} \delta(ax)\phi(x)dx \qquad \text{Integral notation}$$

$$:= \frac{1}{|a|} \int_{-\infty}^{\infty} \delta(x)\phi\left(\frac{x}{a}\right)dx \qquad \text{Change of variable using (61)}$$

$$= |a|^{-1}\phi(0), \quad \phi \in \mathbb{S}. \qquad \text{Action of } \delta \text{ from (9)}$$

The *functionals* for $\delta(ax)$ and $|a|^{-1}\delta(x)$ are identical, so we write

$$\delta(ax) = |a|^{-1}\delta(x) \quad \text{when} \quad a < 0 \text{ or } a > 0. \qquad (65)$$

∎

Example: Find the linear functional that represents $\delta''(x)$.

Solution: The functional for $f(x) := \delta''(x)$ is

$$f\{\phi\} = \int_{-\infty}^{\infty} \delta''(x)\phi(x)dx \qquad \text{Integral notation}$$

$$:= -\int_{-\infty}^{\infty} \delta'(x)\phi'(x)dx \qquad \text{Integrate by parts using (62)}$$

$$:= (-1)^2 \int_{-\infty}^{\infty} \delta(x)\phi''(x)dx \qquad \text{Integrate by parts using (62)}$$

$$= (-1)^2\phi''(0), \quad \phi \in \mathbb{S}. \qquad \text{Action of } \delta \text{ from (9)}$$

Of course, this agrees with (31). ∎

Example: Find the Fourier transform of δ.

Solution: The functional for $\delta^\wedge(s)$ is

$$\delta^\wedge\{\phi\} = \int_{-\infty}^{\infty} \delta^\wedge(s)\phi(s)ds \qquad\qquad \text{Integral notation}$$

$$: = \int_{-\infty}^{\infty} \delta(x)\phi^\wedge(x)dx \qquad\qquad \text{Parseval's identity (63)}$$

$$= \phi^\wedge(0) \qquad\qquad \text{Action of } \delta$$

$$=: \int_{-\infty}^{\infty} 1 \cdot \phi(s)ds, \quad \phi \in \mathbb{S}. \qquad\qquad \text{Analysis equation for } \phi$$

In view of this representation we write

$$\delta^\wedge(s) = 1 \tag{66}$$

and regard $\delta^\wedge$ as an ordinary CSG function on $\mathbb{R}$, *cf.* Fig. 7.9. ∎

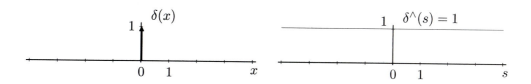

Figure 7.9. The Dirac delta function and its Fourier transform.

Example: Find the Fourier transform of III.

Solution: We use a special version

$$\sum_{n=-\infty}^{\infty} \phi(n) = \sum_{k=-\infty}^{\infty} \phi^\wedge(k), \quad \phi \in \mathbb{S} \tag{67}$$

of the Poisson sum formula (1.45) as we write

$$\amalg^{\wedge}\{\phi\} = \int_{-\infty}^{\infty} \amalg^{\wedge}(s)\phi(s)ds \qquad\qquad \text{Integral notation}$$

$$:= \int_{-\infty}^{\infty} \amalg(x)\phi^{\wedge}(x)dx \qquad\qquad \text{Parseval's identity (63)}$$

$$:= \sum_{n=-\infty}^{\infty} \phi^{\wedge}(n) \qquad\qquad\quad \text{Action of } \amalg \text{ from (40)}$$

$$= \sum_{n=-\infty}^{\infty} \phi(n) \qquad\qquad\quad \text{Poisson sum formula (67)}$$

$$=: \int_{-\infty}^{\infty} \amalg(s)\phi(s)ds, \quad \phi \in \mathbb{S}. \qquad \text{Action of } \amalg \text{ from (40)}$$

The functionals for $\amalg$ and $\amalg^{\wedge}$ are identical, so we write

$$\amalg^{\wedge}(s) = \amalg(s), \qquad\qquad\qquad\qquad (68)$$

cf. Fig. 7.10. ∎

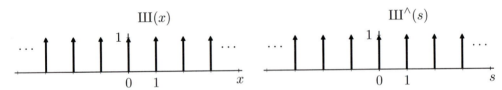

Figure 7.10. The comb function and its Fourier transform.

Example: Find the Fourier transform of the power function $p_1(x) := x$.

Solution: The functional for $p_1^{\wedge}(s)$ is

$$p_1^{\wedge}\{\phi\} = \int_{-\infty}^{\infty} p_1^{\wedge}(s)\phi(s)ds \qquad\qquad \text{Integral notation}$$

$$:= \int_{-\infty}^{\infty} p_1(x)\phi^{\wedge}(x)dx \qquad\qquad \text{Parseval's identity (63)}$$

$$:= \int_{-\infty}^{\infty} x\,\phi^{\wedge}(x)dx \qquad\qquad\quad \text{Action of } p_1 \text{ from (25)}$$

$$= \frac{1}{2\pi i}\phi'(0) \qquad\qquad\qquad \text{Synthesis equation for } \phi'$$

$$=: \frac{-1}{2\pi i}\delta'\{\phi\}, \quad \phi \in \mathbb{S}, \qquad \text{Action of } \delta' \text{ from (29)}$$

so we write

$$p_1^\wedge(s) = \frac{-1}{2\pi i}\delta'(s).$$

The function $p_1(x) := x$ does not have an *ordinary* Fourier transform, but it does have a *generalized* Fourier transform! ■

Example: Find the Fourier transform of the inverse power function p_{-1}.

Solution: The functional for $p_{-1}^\wedge(s)$ is

$$p_{-1}^\wedge\{\phi\} = \int_{-\infty}^{\infty} p_{-1}^\wedge(s)\phi(s)\,ds \qquad\qquad \text{Integral notation}$$

$$:= \int_{-\infty}^{\infty} p_{-1}(x)\phi^\wedge(x)\,dx \qquad\qquad \text{Parseval's identity (63)}$$

$$= \int_{0}^{\infty} \frac{\phi^\wedge(x) - \phi^\wedge(-x)}{x}\,dx \qquad\qquad \text{Action of } p_{-1} \text{ from (53)}$$

$$= \int_{x=0}^{\infty}\int_{s=-\infty}^{\infty} \phi(s)\left(\frac{e^{-2\pi isx} - e^{2\pi isx}}{x}\right)ds\,dx \qquad \text{Analysis equation for } \phi$$

$$= -\pi i \int_{x=-\infty}^{\infty}\int_{s=-\infty}^{\infty} \phi(s)\frac{\sin(2\pi sx)}{\pi x}\,ds\,dx \qquad \text{Euler's identity}$$

$$\stackrel{?}{=} -\pi i \int_{s=-\infty}^{\infty} \phi(s) \int_{x=-\infty}^{\infty} \frac{\sin(2\pi sx)}{\pi x}\,dx\,ds$$

$$= -\pi i \int_{s=-\infty}^{\infty} \phi(s)\,\mathrm{sgn}(s) \int_{u=-\infty}^{\infty} \mathrm{sinc}(u)\,du\,ds \qquad \text{Change of variable}$$

$$= \int_{s=-\infty}^{\infty} [-\pi i\,\mathrm{sgn}(s)]\phi(s)\,ds, \quad \phi \in \mathbb{S}. \qquad \text{Synthesis equation for } \Pi$$

The exchange of the order of integration at $\stackrel{?}{=}$ is valid when $\phi \in \mathbb{S}$, *cf.* Ex. 7.17. The functional for $p_{-1}^\wedge$ is identical to the fundamental functional for the slowly growing function $-\pi i\,\mathrm{sgn}(s)$, so we write

$$p_{-1}^\wedge(s) = -\pi i\,\mathrm{sgn}(s), \qquad\qquad (69)$$

cf. Fig. 7.11. ■

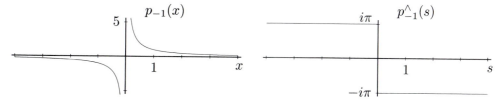

Figure 7.11. The inverse power function p_{-1} and its Fourier transform.

The identities (64)–(69) are of fundamental importance, and we will use them repeatedly in the remainder of the text. You should learn them at this time!

Each of the functionals (60)–(63) can be expressed in the form (20), *i.e.*, the translate, dilate, derivative, and Fourier transform of a generalized function (as defined above) are also generalized functions. We will show that this is true for the Fourier transform and leave the other three cases for Ex. 7.21.

Let $f = \gamma^{(\mu)}$ where γ is CSG and μ is a nonnegative integer. Replace γ by the antiderivative $\gamma^{(-2)}$ from (59), if necessary (and augment μ accordingly) to ensure that γ is twice continuously differentiable. Choose $p = 0, 1, 2, \ldots$ to ensure that

$$\gamma_p(x) := \frac{\gamma(x)}{(1 + 4\pi^2 x^2)^p}$$

has small regular tails, and let Γ_p be the corresponding bounded, continuous Fourier transform on $\mathbb{R}$. Given $\phi \in \mathbb{S}$ we use the derivative rule, the power scaling rule, and Parseval's identity to write

$$\int_{-\infty}^{\infty} f(s)\phi^\wedge(s)ds := \int_{-\infty}^{\infty} \gamma^{(\mu)}(s)\phi^\wedge(s)ds$$

$$:= (-1)^\mu \int_{-\infty}^{\infty} \gamma(s)\mathbf{D}^\mu \phi^\wedge(s)ds$$

$$:= \int_{-\infty}^{\infty} \gamma_p(s)(1 + 4\pi^2 s^2)^p(-\mathbf{D})^\mu \phi^\wedge(s)ds$$

$$= \int_{-\infty}^{\infty} \Gamma_p(x)(1 - \mathbf{D}^2)^p[(2\pi i x)^\mu \phi(x)]dx$$

where $\mathbf{D}$ is the derivative operator. In this way we see that the functional $f^\wedge$ from (63) is a linear combination of the functionals

$$\int_{-\infty}^{\infty} x^\ell \Gamma_p(x)\phi^{(k)}(x), \quad \ell = 0, 1, \ldots, \mu, \quad k = 0, 1, \ldots, 2p,$$

which have the form (20). Since $\mathbb{G}$ is a linear space, this shows that $f^\wedge \in \mathbb{G}$, *i.e.*, that $f^\wedge = g^{(n)}$ for some CSG function g and some nonnegative integer n.

Reflection and conjugation

When f is a suitably regular ordinary function on $\mathbb{R}$ and $\phi \in \mathbb{S}$ we can write

$$\int_{-\infty}^{\infty} f^\vee(x)\phi(x)dx = \int_{-\infty}^{\infty} f(x)\phi^\vee(x)dx, \tag{70}$$

$$\int_{-\infty}^{\infty} f^-(x)\phi(x)dx = \overline{\int_{-\infty}^{\infty} f(x)\phi^-(x)dx}, \tag{71}$$

$$\int_{-\infty}^{\infty} f^{\dagger}(x)\phi(x)dx = \overline{\int_{-\infty}^{\infty} f(x)\phi^{\dagger}(x)dx}, \tag{72}$$

where

$$f^{\vee}(x) := f(-x), \quad f^{-}(x) := \overline{f(x)}, \quad f^{\dagger}(x) := \overline{f(-x)}. \tag{73}$$

You can use each of these identities when f is a generalized function and $\phi \in \mathbb{S}$ since we *define* $f^{\vee}$, f^{-}, $f^{\dagger}$ by using the functionals

$$f^{\vee}\{\phi\} := f\{\phi^{\vee}\}, \quad f^{-}\{\phi\} := \overline{f\{\phi^{-}\}}, \quad f^{\dagger}\{\phi\} := \overline{f\{\phi^{\dagger}\}}.$$

[Of course, (70) can be obtained by setting $a = -1$ in (61).]

A generalized function f is said to be

<blockquote>even,   odd,   real,   pure imaginary,   hermitian,   antihermitian</blockquote>

when

$$f^{\vee} = f, \quad f^{\vee} = -f, \quad f^{-} = f, \quad\quad f^{-} = -f, \quad\quad f^{\dagger} = f, \quad\quad f^{\dagger} = -f,$$

respectively. Thus, δ is even and real since

$$\delta^{\vee}\{\phi\} := \delta\{\phi^{\vee}\} = \phi^{\vee}(0) = \phi(0) = \delta\{\phi\}, \quad \phi \in \mathbb{S}, \text{ and}$$

$$\delta^{-}\{\phi\} := \overline{\delta\{\phi^{-}\}} = \overline{\phi^{-}(0)} = \phi(0) = \delta\{\phi\}, \quad \phi \in \mathbb{S}.$$

Example: Let f be a generalized function. Show that f' is odd, even when f is even, odd, respectively.

Solution: The desired conclusion follows from the identity

$$f'^{\vee}\{\phi\} = -f^{\vee'}\{\phi\}, \quad \phi \in \mathbb{S},$$

that we establish by writing

$$\int_{-\infty}^{\infty} f'^{\vee}(x)\phi(x)dx := \int_{-\infty}^{\infty} f'(x)\phi^{\vee}(x)dx := -\int_{-\infty}^{\infty} f(x)\phi^{\vee'}(x)dx$$

$$= \int_{-\infty}^{\infty} f(x)\phi'^{\vee}(x)dx =: \int_{-\infty}^{\infty} f^{\vee}(x)\phi'(x)dx$$

$$=: -\int_{-\infty}^{\infty} f^{\vee'}(x)\phi(x)dx, \quad \phi \in \mathbb{S}. \quad\blacksquare$$

Example: Let f be a generalized function. Show that

$$f^{\vee\wedge} = f^{\wedge\vee}, \quad f^{-\wedge} = f^{\wedge\dagger}, \quad f^{\dagger\wedge} = f^{\wedge-} \tag{74}$$

and thereby extend the observations of Ex. 1.2 to the present context.

Solution: Since $\phi^{\wedge\vee} = \phi^{\vee\wedge}$ when $\phi \in \mathbb{S}$ we can use (63) and (70) to write

$$f^{\vee\wedge}\{\phi\} = f\{\phi^{\wedge\vee}\} = f\{\phi^{\vee\wedge}\} = f^{\wedge\vee}\{\phi\}, \quad \phi \in \mathbb{S}.$$

The other two identities are proved in a similar fashion. ∎

Multiplication and convolution

When f, α, β are suitably regular ordinary functions on $\mathbb{R}$ and $\phi \in \mathbb{S}$, we can write

$$\int_{-\infty}^{\infty} [\alpha(x) \cdot f(x)]\phi(x)dx = \int_{-\infty}^{\infty} f(x)[\phi(x) \cdot \alpha(x)]dx \tag{75}$$

and then use

$$\int_{v=-\infty}^{\infty} \int_{u=-\infty}^{\infty} f(u)\beta(v-u)\phi(v)du\,dv \overset{?}{=} \int_{u=-\infty}^{\infty} \int_{v=-\infty}^{\infty} f(u)\beta(v-u)\phi(v)dv\,du$$

to write

$$\int_{-\infty}^{\infty} [\beta * f](x)\phi(x)dx = \int_{-\infty}^{\infty} f(x)[\phi * \beta^{\vee}](x)dx. \tag{76}$$

You can use (75) when f is a generalized function and α is an ordinary function with the property that

$$\phi \cdot \alpha \in \mathbb{S} \quad \text{when} \quad \phi \in \mathbb{S} \tag{77}$$

because we *define* the product $\alpha \cdot f$ by using the functional

$$[\alpha \cdot f]\{\phi\} := f\{\phi \cdot \alpha\}, \quad \phi \in \mathbb{S}$$

which corresponds to (75). Likewise, you can use (76) when f is a generalized function and β is a generalized function with the property that

$$\phi * \beta \in \mathbb{S} \quad \text{when} \quad \phi \in \mathbb{S} \tag{78}$$

because we *define* the convolution product $\beta * f$ by using the functional

$$[\beta * f]\{\phi\} := f\{\phi * \beta^{\vee}\}, \quad \phi \in \mathbb{S}$$

that corresponds to (76).

We use the Leibnitz formula

$$(\phi \cdot \alpha)^{(m)} = \sum_{k=0}^{m} \binom{m}{k} \alpha^{(k)} \cdot \phi^{(m-k)}$$

to see that (77) holds when the functions

$$\alpha, \alpha', \alpha'', \ldots \text{ are all CSG,} \tag{79}$$

e.g., as is the case when

$$\alpha(x) = \phi(x) \qquad \text{where } \phi \in \mathbb{S};$$
$$\alpha(x) = x^n \qquad \text{where } n = 0, 1, 2, \ldots ;$$
$$\alpha(x) = e^{2\pi i s_0 x} \qquad \text{where } -\infty < s_0 < \infty;$$
$$\alpha(x) = e^{i\pi \rho x^2} \qquad \text{where } -\infty < \rho < \infty; \text{ and}$$
$$\alpha(x) = \mathcal{P}(x)/\mathcal{Q}(x) \qquad \text{where } \mathcal{P}, \mathcal{Q} \text{ are algebraic or trigonometric}$$
$$\text{polynomials and } \mathcal{Q} \text{ has no real zeros.}$$

Since you might reasonably expect to find

$$\beta * \phi \stackrel{?}{=} [\beta^\wedge \cdot \phi^\wedge]^{\wedge\vee},$$

you will not be surprised to learn that (78) holds when the functions

$$\beta^\wedge, \beta^{\wedge\prime}, \beta^{\wedge\prime\prime}, \ldots \quad \text{are all CSG,} \tag{80}$$

[*cf.* (92)]. The sufficient conditions (79), (80) will serve for all of the subsequent applications and guarantee that the functionals (75), (76) have the form (20), *cf.* Ex. 7.21.

We will show how (75)–(76) allow us to simplify certain expressions involving

$$\delta_{x_0}^{(k)}(x) := \delta^{(k)}(x - x_0), \quad -\infty < x_0 < \infty, \quad k = 0, 1, 2, \ldots . \tag{81}$$

Example: Let $\alpha, \alpha', \alpha'', \ldots$ be CSG. Show that

$$\alpha(x) \cdot \delta(x - x_0) = \alpha(x_0)\delta(x - x_0).$$

Solution: We use (75) and the sifting relation (64) as we write

$$\int_{-\infty}^{\infty} [\alpha(x) \cdot \delta(x - x_0)]\phi(x)dx := \int_{-\infty}^{\infty} \delta(x - x_0)[\phi(x)\alpha(x)]dx$$
$$= \phi(x_0)\alpha(x_0)$$
$$= \int_{-\infty}^{\infty} [\alpha(x_0)\delta(x - x_0)]\phi(x), \quad \phi \in \mathbb{S}. \qquad \blacksquare$$

You can use an analogous argument to show that

$$\alpha(x) \cdot \delta^{(k)}(x - x_0) = \sum_{\ell=0}^{k} \binom{k}{\ell}(-1)^{k-\ell}\alpha^{(k-\ell)}(x_0)\delta^{(\ell)}(x - x_0), \tag{82}$$

cf. Ex. 7.13.

Example: Let f be a generalized function. Show that

$$\delta * f = f \tag{83}$$

(*i.e.*, show that δ serves as an identity for the convolution product, *cf.* Ex. 3.40).

Solution: We first use (76) to show that

$$\psi * \delta = \psi \quad \text{when} \quad \psi \in \mathbb{S}$$

by writing

$$\int_{-\infty}^{\infty} [\psi * \delta](x)\phi(x)dx : = \int_{-\infty}^{\infty} \delta(x)[\phi * \psi^\vee](x)dx$$

$$= [\phi * \psi^\vee](0)$$

$$= \int_{-\infty}^{\infty} \psi(x)\phi(x)dx, \quad \phi \in \mathbb{S}.$$

We then use (76) a second time to write

$$\int_{-\infty}^{\infty} [\delta * f](x)\phi(x)dx : = \int_{-\infty}^{\infty} f(x)[\phi * \delta^\vee](x)dx$$

$$= \int_{-\infty}^{\infty} f(x)[\phi * \delta](x)dx$$

$$= \int_{-\infty}^{\infty} f(x)\phi(x)dx, \quad \phi \in \mathbb{S}. \qquad \blacksquare$$

You can use an analogous argument to show that

$$[\psi * \delta_{x_0}^{(k)}](x) = \psi^{(k)}(x - x_0), \quad -\infty < x_0 < \infty, \quad k = 0, 1, \ldots \tag{84}$$

when ψ is a Schwartz function and

$$[\delta_{x_0}^{(k)} * f](x) = f^{(k)}(x - x_0), \quad -\infty < x_0 < \infty, \quad k = 0, 1, \ldots \tag{85}$$

when f is a generalized function, *cf.* Ex. 7.13.

Example: Let g be CSG and let $\psi \in \mathbb{S}$. Show that

$$\psi * g^{(n)} = [\psi * g]^{(n)} = \psi^{(n)} * g, \quad n = 0, 1, 2, \ldots . \tag{86}$$

Solution: We chase definitions as we write

$$\int_{-\infty}^{\infty} [\psi * g^{(n)}](x)\phi(x)dx := \int_{-\infty}^{\infty} g^{(n)}(x)[\phi * \psi^{\vee}](x)dx$$

$$:= (-1)^n \int_{-\infty}^{\infty} g(x)[\phi * \psi^{\vee}]^{(n)}(x)dx$$

$$= (-1)^n \int_{-\infty}^{\infty} g(x)[\phi^{(n)} * \psi^{\vee}](x)dx$$

$$=: (-1)^n \int_{-\infty}^{\infty} [\psi * g](x)\phi^{(n)}(x)dx$$

$$=: \int_{-\infty}^{\infty} [\psi * g]^{(n)}(x)\phi(x)dx, \quad \phi \in \mathbb{S},$$

$$\int_{-\infty}^{\infty} [\psi * g^{(n)}](x)\phi(x)dx = (-1)^n \int_{-\infty}^{\infty} g(x)[\phi * \psi^{\vee}]^{(n)}(x)dx$$

$$= \int_{-\infty}^{\infty} g(x)[\phi * \psi^{(n)\vee}](x)dx$$

$$=: \int_{-\infty}^{\infty} (\psi^{(n)} * g)(x)\phi(x)dx, \quad \phi \in \mathbb{S}. \quad \blacksquare$$

Let f be a generalized function and let $\psi \in \mathbb{S}$. We set $f = g^{(n)}$ where g is CSG and n is a nonnegative integer, and use (86) to write

$$\psi * f = \psi^{(n)} * g, \ (\psi * f)' = \psi^{(n+1)} * g, \ (\psi * f)'' = \psi^{(n+2)} * g, \ \dots . \tag{87}$$

Since the convolution product of the CSG function g and a Schwartz function is always CSG (*cf.* Ex. 7.18), this shows that $\psi * f$ and all of its derivatives are CSG.

Example: Show that $x \cdot p_{-1}(x) = 1$ when p_{-1} is the inverse power function.

Solution: We use (53) as we write

$$\int_{-\infty}^{\infty} [x \cdot p_{-1}(x)]\phi(x)dx := \int_{-\infty}^{\infty} p_{-1}(x)[\phi(x) \cdot x]dx$$

$$= \int_{0}^{\infty} \frac{x\phi(x) - (-x)\phi(-x)}{x}dx$$

$$= \int_{-\infty}^{\infty} 1 \cdot \phi(x)dx, \quad \phi \in \mathbb{S}. \quad \blacksquare$$

You can extend this argument to establish the generalized *power rule*

$$x^m \cdot p_n(x) = p_{m+n}(x), \quad m = 0, 1, 2, \dots, \quad n = 0, \pm 1, \pm 2, \dots, \tag{88}$$

cf. Ex. 7.16.

We have chosen the decidedly asymmetric notation $\alpha \cdot f$, $\beta * f$ to help you remember that these products are only defined when α, β satisfy (77), (78). We can restore some symmetry by setting

$$f \cdot \alpha := \alpha \cdot f, \qquad f * \beta := \beta * f$$

(to make the products commute), but you will need to remember the restrictions. Many products, *e.g.*,

$$\delta \cdot \delta, \quad \delta \cdot p_{-1}, \quad 1 * 1, \quad x * x^2$$

are not defined, and the familiar associativity relations

$$[f_1 \cdot f_2] \cdot f_3 \stackrel{?}{=} f_1 \cdot [f_2 \cdot f_3],$$
$$[f_1 * f_2] * f_3 \stackrel{?}{=} f_1 * [f_2 * f_3]$$

do not always hold within this context. The insightful little examples of Schwartz,

$$[\delta(x) \cdot x] \cdot p_{-1}(x) = [0] \cdot p_{-1}(x) = 0$$
$$\delta(x) \cdot [x \cdot p_{-1}(x)] = \delta(x) \cdot [1] = \delta(x)$$

and

$$[1 * \delta'(x)] * \text{sgn}(x) = [0] * \text{sgn}(x) = 0$$
$$1 * [\delta'(x) * \text{sgn}(x)] = 1 * [2\delta(x)] = 2,$$

will help you see what can go wrong, *cf.* Ex. 2.36.

Division

Let g be a generalized function and let $\alpha, \alpha', \alpha'', \ldots$ be CSG. If $1/\alpha$, $(1/\alpha)', (1/\alpha)'', \ldots$ are also CSG, we can solve the linear equation

$$\alpha \cdot f = g$$

by setting
$$f := (1/\alpha) \cdot g.$$

We verify this by writing

$$\int_{-\infty}^{\infty} [\alpha(x) \cdot [(1/\alpha(x)) \cdot g(x)]] \phi(x) dx$$

$$:= \int_{-\infty}^{\infty} [(1/\alpha(x)) \cdot g(x)][\phi(x) \cdot \alpha(x)] dx$$

$$:= \int_{-\infty}^{\infty} g(x)[[\phi(x) \cdot \alpha(x)] \cdot (1/\alpha(x))] dx$$

$$= \int_{-\infty}^{\infty} g(x)\phi(x) dx, \quad \phi \in \mathbb{S}.$$

We will show that this solution is unique. Indeed, if $\alpha \cdot f_1 = g$ and $\alpha \cdot f_2 = g$, then $f_0 := f_1 - f_2$ is a solution of the homogeneous equation $\alpha \cdot f_0 = 0$ with

$$\int_{-\infty}^{\infty} f_0(x)\phi(x)dx = \int_{-\infty}^{\infty} f_0(x)\big[[\phi(x) \cdot (1/\alpha(x))] \cdot \alpha(x)\big]dx$$

$$=: \int_{-\infty}^{\infty} [\alpha(x) \cdot f_0(x)][\phi(x) \cdot (1/\alpha(x))]dx$$

$$= 0, \quad \phi \in \mathbb{S}.$$

Example: Find the generalized solution of $e^{i\pi x^2} \cdot f(x) = e^{-\pi x^2}$.

Solution: The functions

$$\alpha(x) := e^{i\pi x^2}, \qquad 1/\alpha(x) = e^{-i\pi x^2}$$

and all of their derivatives are CSG, so we can write

$$f(x) = e^{-i\pi x^2} \cdot e^{-\pi x^2}. \qquad \blacksquare$$

You must use some care when you form such quotients. The following example shows what can happen when $1/\alpha$ is continuous but not slowly growing.

Example: Show that there is no generalized solution of $e^{-\pi x^2} \cdot f(x) = 1$.

Solution: Let $\alpha(x) := e^{-\pi x^2}$ and assume that f is a generalized solution. Since $\alpha^{\wedge} = \alpha = \alpha^{\vee}$, we can write

$$\int_{-\infty}^{\infty} (\alpha * f^{\wedge})(x)\phi(x)dx := \int_{-\infty}^{\infty} f^{\wedge}(x)[(\phi * \alpha)(x)]dx$$

$$:= \int_{-\infty}^{\infty} f(x)[(\phi * \alpha)^{\wedge}(x)]dx$$

$$= \int_{-\infty}^{\infty} f(x)[\phi^{\wedge}(x) \cdot \alpha(x)]dx$$

$$=: \int_{-\infty}^{\infty} [\alpha(x) \cdot f(x)]\phi^{\wedge}(x)dx$$

$$= \int_{-\infty}^{\infty} 1 \cdot \phi^{\wedge}(x)dx$$

$$= \int_{-\infty}^{\infty} \delta(x)\phi(x)dx, \quad \phi \in \mathbb{S},$$

i.e., $\alpha * f^{\wedge} = \delta$. This is impossible since the convolution product of the Schwartz function α and the generalized function $f^{\wedge}$ is CSG but δ is not, *cf.* (87) and Ex. 7.3. $\blacksquare$

We will now consider the homogeneous equation

$$(x - x_0)^n \cdot f(x) = 0 \tag{89}$$

where $-\infty < x_0 < \infty$ and $n = 1, 2, \dots$. In this case $\alpha(x) := (x - x_0)^n$ and its derivatives are CSG but $1/\alpha(x)$ has a singularity at $x = x_0$. Using (82) we see that

$$(x - x_0)^n \cdot \delta^{(k)}(x - x_0) = 0 \quad \text{when} \quad k = 0, 1, \dots, n-1,$$

so the generalized function

$$f(x) = c_0 \delta(x - x_0) + c_1 \delta'(x - x_0) + \cdots + c_{n-1} \delta^{(n-1)}(x - x_0) \tag{90}$$

satisfies (89) for every choice of the constants $c_0, c_1, \dots, c_{n-1}$. It takes a bit more effort to show that every generalized solution has this form, cf. Ex. 7.23. A few examples will show you how to use this result.

Example: Find all generalized solutions of $x \cdot f(x) = 1$.

Solution: We have shown that $x \cdot p_{-1}(x) = 1$, cf. (88). We use (89)–(90) (with $x_0 = 0$, $n = 1$) to solve the homogeneous equation

$$x \cdot [f(x) - p_{-1}(x)] = 0$$

by writing

$$f(x) = p_{-1}(x) + c\,\delta(x)$$

where c is an arbitrary constant. ∎

Example: Find all generalized solutions of $x^2 \cdot f(x) = \sin^2(\pi x)$.

Solution: We solve the homogeneous equation

$$x^2 \cdot [f(x) - \pi^2 \text{sinc}^2(x)] = 0$$

by writing

$$f(x) = \pi^2 \text{sinc}^2(x) + c_0 \delta(x) + c_1 \delta'(x)$$

where c_0, c_1 are arbitrary constants. ∎

Example: Find all generalized solutions of $(x^2 - 1) \cdot f(x) = 1$.

Solution: We formally write

$$\frac{1}{x^2 - 1} = \frac{1}{2}\left[\frac{1}{x - 1} - \frac{1}{x + 1}\right] = \frac{1}{2}[p_{-1}(x - 1) - p_{-1}(x + 1)],$$

and use translates of (88) to see that

$$(x^2 - 1) \cdot \frac{1}{2}[p_{-1}(x - 1) - p_{-1}(x + 1)]$$

$$= \frac{1}{2}[(x + 1)(x - 1)p_{-1}(x - 1) - (x - 1)(x + 1)p_{-1}(x + 1)]$$

$$= \frac{1}{2}[(x + 1) \cdot 1 - (x - 1) \cdot 1] = 1.$$

We then solve the homogeneous equation

$$(x^2 - 1) \cdot \left[f(x) - \frac{1}{2}p_{-1}(x - 1) + \frac{1}{2}p_{-1}(x + 1) \right] = 0$$

to obtain

$$f(x) = \frac{1}{2}p_{-1}(x - 1) - \frac{1}{2}p_{-1}(x + 1) + c\,\delta(x - 1) + d\,\delta(x + 1)$$

where c, d are arbitrary constants, *cf.* Ex. 7.24.   ∎

7.4 Derivatives and Simple Differential Equations

Differentiation rules

The following *differentiation rules* can be used with generalized functions f, f_1, f_2.

$$
\begin{aligned}
[c_1 f_1(x) + c_2 f_2(x)]' &= c_1 f_1'(x) + c_2 f_2'(x), & c_1, c_2 &\in \mathbb{C} \\
[f(x - x_0)]' &= f'(x - x_0), & -\infty &< x_0 < \infty \\
[f(ax)]' &= a\,f'(ax), & a < 0 &\text{ or } a > 0 \\
[\alpha(x) \cdot f(x)]' &= \alpha(x) \cdot f'(x) + \alpha'(x) \cdot f(x), & \alpha, \alpha', \alpha'', &\ldots \text{ are CSG} \\
[(\beta * f)(x)]' &= (\beta' * f)(x) = (\beta * f')(x), & \beta^\wedge, \beta^{\wedge\prime}, \beta^{\wedge\prime\prime}, &\ldots \text{ are CSG}
\end{aligned}
$$

You can establish such rules by using the familiar differentiation rules from calculus and chasing definitions.

Example: Show that $(\alpha \cdot f)' = \alpha \cdot f' + \alpha' \cdot f$.

Solution: When f is a generalized function and $\alpha, \alpha', \alpha'', \ldots$ are CSG

$$\int_{-\infty}^{\infty} [\alpha(x) \cdot f(x)]' \phi(x) dx := - \int_{-\infty}^{\infty} f(x) [\phi'(x) \alpha(x)] dx$$

$$= - \int_{-\infty}^{\infty} f(x) \left([\phi(x) \alpha(x)]' - \phi(x) \alpha'(x) \right) dx$$

$$=: \int_{-\infty}^{\infty} \left(f'(x) [\phi(x) \alpha(x)] + f(x) [\phi(x) \alpha'(x)] \right) dx$$

$$=: \int_{-\infty}^{\infty} [\alpha(x) \cdot f'(x) + \alpha'(x) \cdot f(x)] \phi(x) dx, \quad \phi \in \mathbb{S}. \quad \blacksquare$$

Example: Use the product rule to find the second derivative of the ramp $r(x) := x \cdot h(x)$ where h is the Heaviside function.

Solution: We use the Leibnitz formula

$$r''(x) = (x)'' \cdot h(x) + 2(x)' \cdot h'(x) + x \cdot h''(x)$$

(which follows from the above product rule) together with the identities

$$h' = \delta, \quad h'' = \delta', \quad \ldots$$

from (34) to write
$$r''(x) = 2\delta(x) + x \cdot \delta'(x).$$

We use (82) (with $x_0 = 0$, $k = 1$) to reduce this to the $r'' = \delta$ of (28). $\quad \blacksquare$

Derivatives of piecewise smooth functions with jumps

We will often have occasion to form the generalized derivative of a slowly growing ordinary function f that has a continuous slowly growing ordinary derivative except for certain isolated points $x_1, x_2, \ldots, x_m$ where f and f' can have jump discontinuities. The simplest such function

$$h(x - x_1) = \begin{cases} 1 & \text{if } x > x_1 \\ 0 & \text{if } x < x_1 \end{cases}$$

has the generalized derivative

$$h'(x - x_1) = \delta(x - x_1)$$

even though $h'(x - x_1) = 0$ for $x < x_1$ and for $x > x_1$, *cf.* Fig. 7.6. Now if f has the jump

$$J_\mu := f(x_\mu +) - f(x_\mu -)$$

at the point $x = x_\mu$, $\mu = 1, 2, \ldots, m$, the piecewise smooth function

$$f_0(x) := f(x) - \sum_{\mu=1}^{m} J_\mu h(x - x_\mu)$$

will be continuous. We can then write

$$f'(x) = f_0'(x) + \sum_{\mu=1}^{m} J_\mu \delta(x - x_\mu),$$

with the generalized function f_0' being represented by the fundamental functional of the ordinary derivative, *cf.* Fig. 7.12. [The fact that $f_0'(x)$ is not defined at the points $x_1, x_2, \ldots, x_m$ is of no consequence.] We will give three examples to illustrate this process.

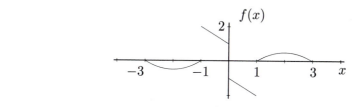

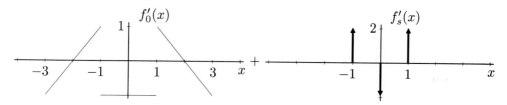

Figure 7.12. A piecewise smooth function f and its generalized derivative $f' = f_0' + f_s'$.

Example: Let $f(x) := \Pi(x)$. Find $f', f'', \ldots$.

Solution: We observe that

$$f(x) = h(x + \tfrac{1}{2}) - h(x - \tfrac{1}{2})$$

and write

$$f'(x) = \delta(x + \tfrac{1}{2}) - \delta(x - \tfrac{1}{2}).$$

The term $\delta(x + \tfrac{1}{2})$ comes from the jump $+1$ at the left edge of the box and the term $-\delta(x - \tfrac{1}{2})$ comes from the jump -1 at the right edge of the box. Knowing f', we find

$$f^{(n+1)}(x) = \delta^{(n)}(x + \tfrac{1}{2}) - \delta^{(n)}(x - \tfrac{1}{2}), \quad n = 0, 1, \ldots . \qquad \blacksquare$$

Example: Let $f(x) := e^{-x}h(x)$. Find $f', f'', \ldots$.

Solution: The function f has a jump $+1$ at $x = 0$, so we have

$$f'(x) = -e^{-x}h(x) + \delta(x) = -f(x) + \delta(x).$$

[We can verify this directly by writing

$$\int_{-\infty}^{\infty} f'(x)\phi(x)dx := -\int_{-\infty}^{\infty} f(x)\phi'(x)dx$$

$$= -\int_{0}^{\infty} e^{-x}\phi'(x)dx$$

$$= \phi(0) - \int_{0}^{\infty} e^{-x}\phi(x)dx$$

$$= \int_{-\infty}^{\infty} [\delta(x) - f(x)]\phi(x)dx, \quad \phi \in \mathbb{S}.]$$

Knowing $f' = -f + \delta$, we compute in turn

$$f''(x) = f(x) - \delta(x) + \delta'(x), \quad f'''(x) = -f(x) + \delta(x) - \delta'(x) + \delta''(x), \quad \ldots \quad \blacksquare$$

Example: Show that the function (4) is a solution of the differential equation

$$m\, y_0''(t) + k\, y_0(t) = p\, \delta(t)$$

when $m > 0$, $k > 0$, and $\omega := \sqrt{k/m}$, cf. Figs. 7.1 and 7.2.

Solution: We compute in turn

$$y_0'(t) = \begin{cases} 0 & \text{if } t < 0 \\ \dfrac{p}{m}\cos(\omega t) & \text{if } t > 0, \end{cases}$$

$$y_0''(t) = \frac{p}{m}\delta(t) + \begin{cases} 0 & \text{if } t < 0 \\ -\dfrac{\omega p}{m}\sin(\omega t) & \text{if } t > 0, \end{cases}$$

$$m\, y_0''(t) + k\, y_0(t) = p\,\delta(t) + \frac{p}{m}\left[-m\omega + \frac{k}{\omega}\right]\sin(\omega t)h(t) = p\,\delta(t). \qquad \blacksquare$$

The equation $f^{(n)} = 0$

Let f be a generalized function and assume that

$$f'(x) = 0 \quad \text{for} \quad a < x < b,$$

i.e., $f'\{\phi\} = 0$ whenever ϕ is a Schwartz function that vanishes when $x < a$ or $x > b$, *cf.* (22). We will show that

$$f(x) = c \quad \text{for} \quad a < x < b$$

where c is some constant. (This takes some effort because the familiar mean value theorem from calculus cannot be used in the present context!)

Let γ, ϕ be Schwartz functions that vanish when $x < a$ or $x > b$, with

$$\int_{-\infty}^{\infty} \gamma(x) dx = 1.$$

We set

$$c := \int_{-\infty}^{\infty} f(x)\gamma(x) dx, \qquad A := \int_{-\infty}^{\infty} \phi(x) dx,$$

and form the antiderivative

$$\psi(x) := \int_{a}^{x} [\phi(u) - A\gamma(u)] du.$$

By construction, ψ is a Schwartz function that vanishes when $x < a$ or $x > b$, so

$$\int_{-\infty}^{\infty} f(x)[\phi(x) - A\gamma(x)] dx = \int_{-\infty}^{\infty} f(x)\psi'(x) dx =: -\int_{-\infty}^{\infty} f'(x)\psi(x) dx = 0,$$

cf. (17), (18). In this way we see that

$$\int_{-\infty}^{\infty} f(x)\phi(x) dx = Ac = \int_{-\infty}^{\infty} c\phi(x) dx,$$

and thereby show that $f(x) = c$ for $a < x < b$.

It is easy to extend this result to higher derivatives. For example, if

$$f''(x) = 0 \quad \text{for} \quad a < x < b$$

we find in turn

$$\begin{aligned} f'(x) &= c_1 & \text{for} \quad a < x < b, \\ [f(x) - c_1 x]' &= 0 & \text{for} \quad a < x < b, \\ f(x) - c_1 x &= c_0 & \text{for} \quad a < x < b, \\ f(x) &= c_0 + c_1 x & \text{for} \quad a < x < b. \end{aligned}$$

Analogously, if

$$f^{(n)}(x) = 0 \quad \text{for} \quad a < x < b$$

then
$$f(x) = c_0 + c_1 x + \cdots + c_{n-1} x^{n-1} \quad \text{for} \quad a < x < b$$
where $c_0, c_1, \ldots, c_{n-1}$ are suitably chosen constants.

Example: Find all generalized solutions of the differential equation
$$f''(x) = \delta(x + 1) + 2\delta'(x - 1).$$

Solution: We can take antiderivatives and write
$$f'(x) = h(x + 1) + 2\delta(x - 1) + c_1,$$
$$f(x) = (x + 1)h(x + 1) + 2h(x - 1) + c_0 + c_1 x.$$
We can also use the fact that $f''(x)$ vanishes for $-\infty < x < -1$, for $-1 < x < 1$, and for $1 < x < \infty$ to write
$$f(x) = \begin{cases} a_0 + b_0 x & \text{for } -\infty < x < -1 \\ a_1 + b_1 x & \text{for } -1 < x < 1 \\ a_2 + b_2 x & \text{for } 1 < x < \infty, \end{cases}$$

$$f''(x) = (a_1 - a_0 - b_1 + b_0)\delta'(x + 1) + (b_1 - b_0)\delta(x + 1)$$
$$+ (a_2 - a_1 + b_2 - b_1)\delta'(x - 1) + (b_2 - b_1)\delta(x - 1).$$

We force f to satisfy the differential equation by choosing the parameters so that
$$a_1 - a_0 - b_1 + b_0 = 0, \quad b_1 - b_0 = 1, \quad a_2 - a_1 + b_2 - b_1 = 2, \quad b_2 - b_1 = 0,$$
i.e., $a_1 = a_0 + 1$, $b_1 = b_0 + 1$, $a_2 = a_0 + 3$, $b_2 = b_0 + 1$. ∎

Example: Let f be a generalized function and assume that $f(x) = 0$ for $-\infty < x < 0$ and for $0 < x < \infty$. Show that f is a finite linear combination of $\delta, \delta', \delta'', \ldots$.

Solution: We know that $f = g^{(n)}$ for some CSG function g and some nonnegative integer n. Since $g^{(n)}(x) = 0$ for $-\infty < x < 0$ and for $0 < x < \infty$, we can write
$$g(x) = \begin{cases} p_L(x) & \text{for } -\infty < x < 0 \\ p_R(x) & \text{for } 0 < x < \infty \end{cases}$$
where p_L, p_R are polynomials of degree $n - 1$ or less. In this way we see that
$$g'(x) = [p_R(0) - p_L(0)]\delta(x) + \begin{cases} p_L'(x) & \text{for } -\infty < x < 0 \\ p_R'(x) & \text{for } 0 < x < \infty, \end{cases}$$

$$g''(x) = [p_R(0) - p_L(0)]\delta'(x) + [p_R'(0) - p_L'(0)]\delta(x)$$
$$+ \begin{cases} p_L''(x) & \text{for } -\infty < x < 0 \\ p_R''(x) & \text{for } 0 < x < \infty, \end{cases}$$

$\vdots$

and thereby obtain the representation

$$f(x) = g^{(n)}(x) = \sum_{\nu=0}^{n-1} [p_R^{(n-1-\nu)}(0) - p_L^{(n-1-\nu)}(0)] \delta^{(\nu)}(x). \tag{91}$$

∎

Solving differential equations

On occasion we will find it necessary to construct a generalized solution of a differential equation

$$f^{(n)}(x) + c_1 f^{(n-1)}(x) + \cdots + c_n f(x) = d(x)$$

when the (constant) coefficients $c_1, c_2, \ldots, c_n$ and the driving function $d(x)$ are given. In some cases this can be done by splicing together ordinary solutions on certain intervals (a_k, b_k), $k = 1, 2, \ldots, m$. We will give two examples to illustrate the procedure. Additional details can be found in Exs. 7.25 and 7.30, the next section, and Kaplan's text.

Example: Construct a generalized solution of the differential equation

$$f''(x) - f'(x) - 2f(x) = -\delta(x).$$

Solution: The characteristic equation $r^2 - r - 2 = 0$ has the roots $r = -1$, $r = 2$, so any linear combination of the ordinary functions e^{-x}, e^{2x} will satisfy the homogeneous equation

$$y''(x) - y'(x) - 2y(x) = 0.$$

Since f satisfies this homogeneous equation for $-\infty < x < 0$ and for $0 < x < \infty$, we can write

$$f(x) = \begin{cases} a_L e^{-x} + b_L e^{2x} & \text{for } x < 0 \\ a_R e^{-x} + b_R e^{2x} & \text{for } x > 0 \end{cases}$$

where a_L, b_L, a_R, b_R are certain constants. We must set $a_L = 0$, $b_R = 0$ to ensure that f is slowly growing. We then compute

$$f'(x) = (a_R - b_L)\delta(x) + \begin{cases} 2b_L e^{2x} & \text{for } x < 0 \\ -a_R e^{-x} & \text{for } x > 0, \end{cases}$$

$$f''(x) = (a_R - b_L)\delta'(x) - (a_R + 2b_L)\delta(x) + \begin{cases} 4b_L e^{2x} & \text{for } x < 0 \\ a_R e^{-x} & \text{for } x > 0, \end{cases}$$

$$f''(x) - f'(x) - 2f(x) = -(2a_R + b_L)\delta(x) + (a_R - b_L)\delta'(x).$$

The right-hand side reduces to $-\delta(x)$ when $a_R = b_L = 1/3$, so

$$f(x) = \frac{1}{3} \begin{cases} e^{2x} & \text{for } x < 0 \\ e^{-x} & \text{for } x > 0. \end{cases}$$

∎

Example: Construct a generalized solution of the differential equation

$$f''''(x) - f(x) = \delta(x).$$

Solution: The characteristic equation $r^4 - 1 = 0$ has the roots $r = -1$, $r = 1$, $r = -i$, $r = i$, so any linear combination of e^{-x}, e^x, e^{-ix}, e^{ix} will satisfy the homogeneous equation

$$y''''(x) - y(x) = 0.$$

The 8 parameters from

$$f(x) = \begin{cases} a_L e^{-x} + b_L e^x + c_L e^{-ix} + d_L e^{ix} & \text{if } x < 0 \\ a_R e^{-x} + b_R e^x + c_R e^{-ix} + d_R e^{ix} & \text{if } x > 0 \end{cases}$$

can be reduced to 2 by requiring f to be slowly growing ($a_L = 0$, $b_R = 0$) and using the constraints

$$f(0+) - f(0-) = 0,\ f'(0+) - f'(0-) = 0,\ f''(0+) - f''(0-) = 0,\ f'''(0+) - f'''(0-) = 1.$$

This is a bit tedious, however, so we attempt to construct a particular solution by taking a linear combination of the even functions $e^{-|x|}$, $\sin|x|$ (deleting the even solution $\cos x$ of the homogeneous equation). We observe that the function

$$e^{-|x|} + \sin|x| = 1 - |x| + x^2/2 - |x|^3/6 + x^4/24 - \cdots$$
$$+ |x| \qquad\qquad - |x|^3/6 + \cdots$$
$$= 1 + \qquad\quad x^2/2 - |x|^3/3 + x^4/24 + \cdots$$

and its first two derivatives are continuous at $x = 0$, while the third has a jump of -4. In this way we find

$$f(x) = -\tfrac{1}{4}\left(e^{-|x|} + \sin|x|\right) + a\cos x + b\sin x$$

where a, b are arbitrary constants.   ∎

In the next section you will learn how to show that there are no other generalized solutions for the differential equations from these two examples.

7.5 The Fourier Transform Calculus for Generalized Functions

Fourier transform rules

Let f, f_1, f_2 be generalized functions with the generalized Fourier transforms $f^\wedge, f_1^\wedge, f_2^\wedge$. Now that you know how to manipulate generalized functions you can use the following Fourier transform rules.

$c_1 f_1(x) + c_2 f_2(x)$	has the FT	$c_1 f_1^\wedge(s) + c_2 f_2^\wedge(s),$	$c_1, c_2 \in \mathbb{C}$
$f(-x)$	has the FT	$f^\wedge(-s)$	
$f^-(x)$	has the FT	$f^{\wedge-}(-s)$	
$f(x - x_0)$	has the FT	$e^{-2\pi i s x_0} \cdot f^\wedge(s),$	$-\infty < x_0 < \infty$
$e^{2\pi i s_0 x} \cdot f(x)$	has the FT	$f^\wedge(s - s_0),$	$-\infty < s_0 < \infty$
$f^\wedge(x)$	has the FT	$f(-s)$	
$f(ax)$	has the FT	$|a|^{-1} f^\wedge(s/a),$	$a > 0$ or $a < 0$
$f^{(n)}(x)$	has the FT	$(2\pi i s)^n \cdot f^\wedge(s),$	$n = 1, 2, \ldots$
$x^n \cdot f(x)$	has the FT	$(-2\pi i)^{-n} [f^\wedge]^{(n)}(s),$	$n = 1, 2, \ldots$
$\alpha(x) \cdot f(x)$	has the FT	$(\alpha^\wedge * f^\wedge)(s),$	$\alpha, \alpha', \alpha'', \ldots$ are CSG
$(\beta * f)(x)$	has the FT	$\beta^\wedge(s) \cdot f^\wedge(s),$	$\beta^\wedge, \beta^{\wedge\prime}, \beta^{\wedge\prime\prime}, \ldots$ are CSG

Each rule can be established by chasing definitions and using corresponding identities for taking Fourier transforms of Schwartz function. Since you are thoroughly familiar with such patterns from your study of Chapter 3, you will quickly learn to use them within this new setting.

Example: Establish the inversion rule for generalized Fourier transforms.

Solution: Given a generalized function f we write

$$\int_{-\infty}^{\infty} f^{\wedge\wedge}(x)\phi(x)dx := \int_{-\infty}^{\infty} f^\wedge(x)\phi^\wedge(x)dx := \int_{-\infty}^{\infty} f(x)\phi^{\wedge\wedge}(x)dx$$

$$= \int_{-\infty}^{\infty} f(x)\phi(-x)dx =: \int_{-\infty}^{\infty} f(-x)\phi(x)dx, \quad \phi \in \mathbb{S}$$

and thereby conclude that

$$f^{\wedge\wedge}(x) = f(-x). \qquad \blacksquare$$

Example: Let f be a generalized function and let $\alpha, \alpha', \alpha'', \ldots$ be CSG. Establish the product rule

$$(\alpha \cdot f)^\wedge = \alpha^\wedge * f^\wedge. \tag{92}$$

Solution: Let $\psi \in \mathbb{S}$. We chase definitions to find

$$\int_{-\infty}^{\infty} [\psi \cdot \alpha]^\wedge(x)\phi(x)dx := \int_{-\infty}^{\infty} [\psi(x) \cdot \alpha(x)]\phi^\wedge(x)dx$$

$$:= \int_{-\infty}^{\infty} \alpha(x)[\phi^\wedge(x)\psi(x)]dx = \int_{-\infty}^{\infty} \alpha(x)[\phi * \psi^{\wedge\vee}]^\wedge(x)dx$$

$$=: \int_{-\infty}^{\infty} \alpha^\wedge(x)[\phi * \psi^{\wedge\vee}](x)dx =: \int_{-\infty}^{\infty} [\psi^\wedge * \alpha^\wedge](x)\phi(x)dx, \quad \phi \in \mathbb{S},$$

and thereby show that

$$[\psi \cdot \alpha]^\wedge = \psi^\wedge * \alpha^\wedge \quad \text{when} \quad \psi \in \mathbb{S}.$$

We use this special version of the product rule to verify that

$$\phi * \alpha^\wedge \in \mathbb{S} \quad \text{when} \quad \phi \in \mathbb{S}$$

and thereby infer that $\alpha^\wedge * f^\wedge$ is well defined, *cf.* (76) and (78). We then derive the general product rule by writing

$$\int_{-\infty}^{\infty} [\alpha \cdot f]^\wedge(x)\phi(x)dx := \int_{-\infty}^{\infty} [\alpha(x) \cdot f(x)]\phi^\wedge(x)dx$$

$$:= \int_{-\infty}^{\infty} f(x)[\phi^\wedge(x)\alpha(x)]dx = \int_{-\infty}^{\infty} f(x)[\phi * \alpha^{\wedge\vee}]^\wedge(x)dx$$

$$=: \int_{-\infty}^{\infty} f^\wedge(x)[\phi * \alpha^{\wedge\vee}](x)dx =: \int_{-\infty}^{\infty} [\alpha^\wedge * f^\wedge](x)\phi(x)dx, \quad \phi \in \mathbb{S}. \quad \blacksquare$$

Basic Fourier transforms

We know from (66) that

$$\delta(x) \quad \text{has the FT} \quad 1,$$

so we can use the derivative and translation rules to see that

$$\delta^{(n)}(x) \quad \text{has the FT} \quad (2\pi i s)^n, \ n = 0, 1, 2, \ldots \tag{93}$$

$$\delta(x - x_0) \quad \text{has the FT} \quad e^{-2\pi i s x_0}, \ -\infty < x_0 < \infty, \tag{94}$$

and then use the inversion rule to see that

$$x^n \quad \text{has the FT} \quad (-2\pi i)^{-n}\delta^{(n)}(s), \ n = 0, 1, 2, \ldots \tag{95}$$

$$e^{2\pi i s_0 x} \quad \text{has the FT} \quad \delta(s - s_0), \ -\infty < s_0 < \infty. \tag{96}$$

We can use (95)–(96) to find the Fourier transform of any algebraic or trigonometric polynomial.

Example: Find the Fourier transform of the polynomial $f(x) := (x - 1)^3$.

Solution: We can write

$$f(x) = x^3 - 3x^2 + 3x - 1$$

and then use (95) to obtain

$$F(s) = \frac{\delta'''(s)}{(-2\pi i)^3} - \frac{3\delta''(s)}{(-2\pi i)^2} + \frac{3\delta'(s)}{-2\pi i} - \delta(s).$$

We can also use (95) with the translation rule to write

$$F(s) = e^{-2\pi i s} \cdot \frac{\delta'''(s)}{(-2\pi i)^3} .$$

You can use (82) to show that these two expressions are equivalent. ∎

Example: Find the Fourier transforms of

$$f_c(x) := \cos(2\pi x), \qquad f_s(x) := \sin(2\pi x).$$

Solution: We use (96) with the Euler identities

$$f_c(x) = \frac{1}{2}[e^{2\pi i x} + e^{-2\pi i x}], \qquad f_s(x) = \frac{1}{2i}[e^{2\pi i x} - e^{-2\pi i x}]$$

to write

$$F_c(s) = \frac{1}{2}[\delta(s - 1) + \delta(s + 1)], \qquad F_s(s) = \frac{1}{2i}[\delta(s - 1) - \delta(s + 1)],$$

cf. Fig. 7.13. ∎

We know from (69) that the inverse power function

$$p_{-1}(x) \quad \text{has the FT} \quad -\pi i \operatorname{sgn}(s).$$

We use (46) to write

$$p_{-n}(x) = \frac{(-1)^{n-1}}{(n-1)!}p_{-1}^{(n-1)}(x), \quad n = 1, 2, \ldots,$$

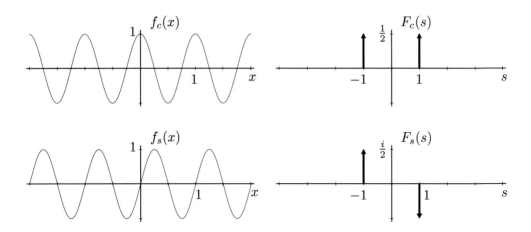

Figure 7.13. The functions $f_c(x) := \cos(2\pi x)$, $f_s(x) := \sin(2\pi x)$ and their Fourier transforms.

and then use the derivative rule to see that

$$p_{-n}(x) \quad \text{has the FT} \quad -\pi i \frac{(-2\pi i s)^{n-1}}{(n-1)!} \operatorname{sgn}(s), \quad n = 1, 2, \ldots . \qquad (97)$$

In Chapter 3 you learned how to find the ordinary Fourier transform of a suitably regular rational function by using (3.22). Now that you have (95) and (97), you can find the generalized Fourier transform of any rational function.

Example: Find the Fourier transform of $f(x) := (x^2 + 1)/(x^2 - 1)$.

Solution: We form the partial fraction expansion

$$f(x) = \frac{x^2 + 1}{x^2 - 1} = 1 + \frac{1}{x-1} - \frac{1}{x+1} = 1 + p_{-1}(x-1) - p_{-1}(x+1)$$

and then Fourier transform term by term to obtain

$$\begin{aligned}
F(s) &= \delta(s) + e^{-2\pi i s}(-\pi i)\operatorname{sgn}(s) - e^{2\pi i s}(-\pi i)\operatorname{sgn}(s) \\
&= \delta(s) - 2\pi \sin(2\pi s)\operatorname{sgn}(s) \\
&= \delta(s) - 2\pi \sin(2\pi|s|).
\end{aligned}$$

    ■

Example: Let y be a generalized solution of the homogeneous differential equation

$$y''''(x) - y(x) = 0.$$

Show that

$$y(x) = c\,e^{ix} + d\,e^{-ix}$$

for suitably chosen constants c, d.

Solution: We Fourier transform the differential equation to obtain

$$[(2\pi i s)^4 - 1] \cdot Y(s) = 0$$

or equivalently,

$$[(2\pi s)^2 + 1][2\pi s + 1][2\pi s - 1] \cdot Y(s) = 0.$$

Since $[(2\pi s)^2 + 1]^{-1}$ and its derivatives are CSG, we can write

$$\left[s + \frac{1}{2\pi}\right]\left[s - \frac{1}{2\pi}\right] Y(s) = 0,$$

$$Y(s) = c\,\delta\left(s - \frac{1}{2\pi}\right) + d\,\delta\left(s + \frac{1}{2\pi}\right),$$

$$y(x) = c\,e^{ix} + d\,e^{-ix}$$

where c, d are constants, *cf.* Ex. 7.25.   ∎

You will often find that it is necessary to determine the Fourier transform of some rational function when you analyze simple physical systems such as those in the following two examples.

Example: Find the generalized solution of the differential equation

$$y''(t) + \omega^2 y(t) = \frac{p}{m}\,\delta(t)$$

that vanishes for $-\infty < t < 0$, *cf.* Figs. 7.1 and 7.2.

Solution: We Fourier transform the differential equation to obtain

$$[(2\pi i s)^2 + \omega^2] \cdot Y(s) = \frac{p}{m}\,.$$

From the equivalent algebraic identity

$$\left(s - \frac{\omega}{2\pi}\right)\left(s + \frac{\omega}{2\pi}\right) \cdot Y(s) = -\frac{p}{4\pi^2 m}$$

we see that

$$Y(s) = c\,\delta\left(s - \frac{\omega}{2\pi}\right) + d\,\delta\left(s + \frac{\omega}{2\pi}\right) - \frac{p}{4\pi m\omega}\left(\frac{1}{s - \omega/2\pi} - \frac{1}{s + \omega/2\pi}\right)$$

where c, d are arbitrary constants, *cf.* Ex. 7.24, so

$$y(t) = c\,e^{i\omega t} + d\,e^{-i\omega t} - \frac{p}{4\pi m\omega}(e^{i\omega t} - e^{-i\omega t})(i\pi)\,\mathrm{sgn}(t)$$

$$= c\,e^{i\omega t} + d\,e^{-i\omega t} + \frac{p}{2m\omega}\,\sin(\omega t)\,\mathrm{sgn}(t).$$

Since we want y to vanish for $-\infty < t < 0$, we determine c, d so that

$$c\,e^{iwt} + d\,e^{-iwt} = \frac{p}{2m\omega}\sin(\omega t),$$

and in this way we obtain the function

$$y(t) = \frac{p}{m\omega}\sin(\omega t)h(t). \qquad\blacksquare$$

Example: A naive automobile suspension system is shown in Fig. 7.14. The mass, spring, and shock absorber process the wheel elevation function $x(t)$ (determined by the shape of the roadbed and the speed of the car) to produce a smoother elevation function $y_0 + y(t)$ for the carriage. The constant y_0 is the elevation of the carriage when $x(t) = 0$ for $-\infty < t < \infty$. The system is governed by the differential equation

$$my''(t) = -k[y(t) - x(t)] - d[y'(t) - x'(t)]$$

where m, k, and d are positive constants. Show that $y = y_\delta * x$ where y_δ is the response of the system to the unit impulse δ, cf. Exs. 4.43, 5.29, and 7.51.

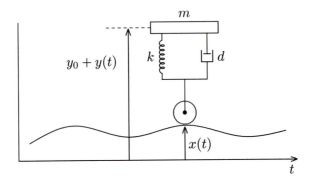

Figure 7.14. An automobile suspension system that links the elevation $x(t)$ of the wheel and the elevation $y(t)$ of the carriage.

Solution: We Fourier transform the differential equation

$$my''(t) + dy'(t) + ky(t) = dx'(t) + kx(t)$$

to obtain

$$[m(2\pi i s)^2 + d(2\pi i s) + k] \cdot Y(s) = [d(2\pi i s) + k] \cdot X(s).$$

The characteristic polynomial

$$\mathcal{P}(r) := mr^2 + dr + k$$

has roots

$$r_\pm := \frac{-d \pm \sqrt{d^2 - 4mk}}{2m}$$

with negative real parts, so

$$\mathcal{P}(2\pi is) = m(2\pi is)^2 + d(2\pi is) + k \neq 0 \;\; \text{for} \;\; -\infty < s < \infty,$$

and $1/\mathcal{P}(2\pi is), [1/\mathcal{P}(2\pi is)]', [1/\mathcal{P}(2\pi is)]'', \ldots$ are CSG. This being the case,

$$Y(s) = Y_\delta(s) \cdot X(s)$$

where

$$Y_\delta(s) := \frac{d(2\pi is) + k}{m(2\pi is)^2 + d(2\pi is) + k}$$

corresponds to the impulsive input $x(t) = \delta(t)$ with $X(s) = 1$. In this way we find the response

$$y(t) = (y_\delta * x)(t)$$

to an arbitrary generalized input x.

The *impulse response*, y_δ, satisfies the homogeneous equation

$$my'' + dy' + ky = 0$$

on $(-\infty, 0)$ and on $(0, +\infty)$. Since r_+, r_- have negative real parts, we must have $y_\delta(t) = 0$ for $t < 0$. We equate the singular parts of

$$my_\delta''(t) + dy_\delta'(t) + ky_\delta(t) = d\,\delta'(t) + k\,\delta(t)$$

to obtain

$$m\{y_\delta(0+)\delta' + y_\delta'(0+)\delta\} + d\{y_\delta(0+)\delta\} = d\,\delta' + k\,\delta$$

and thereby find

$$y_\delta(0+) = \frac{d}{m}, \qquad y_\delta'(0+) = \frac{km - d^2}{m^2}.$$

We can solve the homogeneous differential equation on $(0, +\infty)$ with these initial conditions to find $y_\delta(t)$ for $t \geq 0$. [This is easier than finding the inverse Fourier transform of the rational function $Y_\delta(s)$.] ∎

You can use this procedure to analyze any linear system

$$\mathcal{P}(\mathbf{D})y = \mathcal{Q}(\mathbf{D})x$$

where

$$\mathcal{P}(\mathbf{D}) := a_0 + a_1\mathbf{D} + \cdots + a_n\mathbf{D}^n$$
$$\mathcal{Q}(\mathbf{D}) := b_0 + b_1\mathbf{D} + \cdots + b_m\mathbf{D}^m$$

are polynomials in the derivative operator $\mathbf{D}$ and $\mathcal{P}(z) \neq 0$ when $\text{Re } z = 0$, *cf.* Ex. 7.51.

When we study diffraction in Chapter 9, we will often use the fact that

$$e^{i\pi x^2} \quad \text{has the FT} \quad \frac{1+i}{\sqrt{2}} e^{-i\pi s^2}. \tag{98}$$

It is easy to derive this result within the present context, *cf.* Ex. 3.35. Indeed,

$$f(x) := e^{i\pi x^2}$$

and its ordinary derivatives are CSG, so the generalized functions f, $F := f^\wedge$ satisfy

$$f'(x) = 2\pi i x \cdot f(x),$$
$$2\pi i s \cdot F(s) = -F'(s),$$
$$[e^{i\pi s^2} \cdot F(s)]' = 0,$$
$$e^{i\pi s^2} \cdot F(s) = c$$

where c is a constant. Since $e^{-i\pi s^2}$ and its derivatives are also CSG, we can write

$$F(s) = c\,e^{-i\pi s^2}.$$

The value

$$c = \frac{1+i}{\sqrt{2}} =: \sqrt{i}$$

can be obtained from the Parseval identity

$$\int_{-\infty}^{\infty} e^{i\pi x^2} \cdot e^{-\pi x^2}\, dx = \int_{-\infty}^{\infty} c\,e^{-i\pi s^2} \cdot e^{-\pi s^2}\, ds,$$

cf. Ex. 7.45, or from the analysis leading to (4.71).

Fourier transforms from derivatives

In principle we can find the Fourier transform of any suitably regular function f on $\mathbb{R}$ by evaluating the integral

$$F(s) := \int_{-\infty}^{\infty} f(x)e^{-2\pi i s x}\, dx.$$

The direct evaluation of such Fourier integrals is notoriously difficult, however, and we almost always prefer to use more efficient indirect procedures. The following examples will show you how to find many Fourier transforms by computing generalized *derivatives*. Be prepared to be impressed!

Example: Find the Fourier transform of $f(x) := \Lambda(x)$, cf. (3.25).

Solution: We compute the generalized derivatives

$$f'(x) = \begin{cases} 0 & \text{if} & x < -1 \\ 1 & \text{if} & -1 < x < 0 \\ -1 & \text{if} & 0 < x < 1 \\ 0 & \text{if} & 1 < x, \end{cases}$$

$$f''(x) = \delta(x+1) - 2\delta(x) + \delta(x-1)$$

cf. Fig. 7.15, and then use (94) with the derivative rule to write

$$(2\pi i s)^2 \cdot F(s) = e^{2\pi i s} \cdot 1 - 2 \cdot 1 + e^{-2\pi i s} \cdot 1$$
$$= (e^{\pi i s} - e^{-\pi i s})^2,$$

or equivalently,

$$s^2 \cdot [F(s) - \text{sinc}^2(s)] = 0.$$

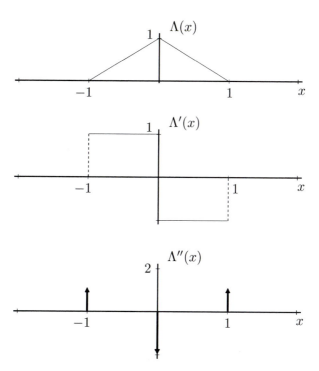

Figure 7.15. The function $\Lambda(x)$ and its generalized derivatives $\Lambda'(x), \Lambda''(x)$.

In view of (89)–(90) this implies that

$$F(s) = \text{sinc}^2(s) + c_0\delta(s) + c_1\delta'(s)$$

for suitably chosen constants c_0, c_1. The piecewise smooth function $f(x)$ vanishes when $|x| \geq 1$, so the Fourier transform F must be an ordinary function on $\mathbb{R}$. This *a priori* knowledge allows us to set $c_0 = c_1 = 0$ and thereby obtain

$$F(s) = \text{sinc}^2(s)$$

(without ever evaluating an integral!). ∎

This procedure, known as *Eagle's method*, can be used to find the Fourier transform of any piecewise polynomial function that vanishes outside some finite interval $[a, b]$, cf. Ex. 7.39.

Example: Find the Fourier series for the 2-periodic function g with $g(x) := x^2$ for $-1 < x < 1$, cf. Fig. 4.2.

Solution: We form the support-limited function

$$f(x) := \begin{cases} x^2 & \text{if } -1 < x < 1 \\ 0 & \text{otherwise,} \end{cases}$$

compute the generalized derivatives

$$f'(x) = \delta(x+1) - \delta(x-1) + \begin{cases} 2x & \text{if } -1 < x < 1 \\ 0 & \text{otherwise,} \end{cases}$$

$$f''(x) = \delta'(x+1) - \delta'(x-1) - 2\delta(x+1) - 2\delta(x-1)$$
$$+ \begin{cases} 2 & \text{if } -1 < x < 1 \\ 0 & \text{otherwise,} \end{cases}$$

$$f'''(x) = \delta''(x+1) - \delta''(x-1) - 2\delta'(x+1) - 2\delta'(x-1)$$
$$+ 2\delta(x+1) - 2\delta(x-1),$$

and then use (93) with the derivative and translation rules to write

$$(2\pi i s)^3 \cdot F(s) = e^{2\pi i s}[(2\pi i s)^2 - 2(2\pi i s) + 2] - e^{-2\pi i s}[(2\pi i s)^2 + 2(2\pi i s) + 2].$$

We know that F is an ordinary function on $\mathbb{R}$, so the quotient

$$F(s) := \frac{e^{2\pi i s}[(2\pi i s)^2 - 2(2\pi i s) + 2] - e^{-2\pi i s}[(2\pi i s)^2 + 2(2\pi i s) + 2]}{(2\pi i s)^3}$$

has a removable singularity at the origin. We use the analysis equation to find

$$F(0) = \int_{-1}^{1} x^2 dx = \frac{2}{3}$$

(since this is much easier than a calculation based on l'Hôpital's rule). Knowing F, we use Poisson's relation (4.18) to obtain the Fourier series

$$g(x) = \sum_{k=-\infty}^{\infty} \frac{1}{2} F\left(\frac{k}{2}\right) e^{2\pi i k x/2}.$$

∎

You can use this procedure to find the Fourier series for any piecewise polynomial p-periodic function on $\mathbb{R}$. You may wish to convince yourself that such a calculation is equivalent to the Bernoulli function expansions that were developed in Section 4.1, *cf.* Exs. 7.73 and 7.75.

Example: Find the Fourier transform of $f(x) := e^{-|x|}$, *cf.* (3.8).

Solution: We write

$$f(x) := \begin{cases} e^x & \text{if } x < 0 \\ e^{-x} & \text{if } x > 0, \end{cases} \qquad f'(x) = \begin{cases} e^x & \text{if } x < 0 \\ -e^{-x} & \text{if } x > 0, \end{cases}$$

$$f''(x) = -2\delta(x) + \begin{cases} e^x & \text{if } x < 0 \\ e^{-x} & \text{if } x > 0, \end{cases}$$

and thereby see that

$$-f''(x) + f(x) = 2\delta(x).$$

We Fourier transform this differential equation and write

$$(4\pi^2 s^2 + 1)F(s) = 2.$$

Since $1/(1 + 4\pi^2 s^2)$ and its derivatives are all CSG, we immediately obtain

$$F(s) = \frac{2}{1 + 4\pi^2 s^2}$$

(without ever evaluating an integral!).

∎

Example: Find the Fourier transforms of the ordinary functions

$$f(x) := \frac{1}{2}\text{sgn}(x) \quad \text{and} \quad h(x) := \begin{cases} 1 & \text{if } x > 0 \\ 0 & \text{if } x < 0. \end{cases}$$

Solution: These functions have the generalized derivatives

$$f'(x) = \delta(x) \quad \text{and} \quad h'(x) = \delta(x)$$

so their Fourier transforms must satisfy

$$2\pi i s \cdot F(s) = 1 \quad \text{and} \quad 2\pi i s \cdot H(s) = 1.$$

Using (88) we see that

$$s \cdot \left[F(s) - \frac{p_{-1}(s)}{2\pi i} \right] = 0 \quad \text{and} \quad s \cdot \left[H(s) - \frac{p_{-1}(s)}{2\pi i} \right] = 0$$

so

$$F(s) = \frac{p_{-1}(s)}{2\pi i} + c\,\delta(s) \quad \text{and} \quad H(s) = \frac{p_{-1}(s)}{2\pi i} + d\,\delta(s)$$

where c, d are constants that we must determine.

The odd function f must have an odd Fourier transform, and since δ is even we must set $c = 0$ and write

$$f(x) = \frac{1}{2}\mathrm{sgn}(x) \quad \text{has the FT} \quad F(s) = \frac{1}{2\pi i s} := \frac{p_{-1}(s)}{2\pi i}\,.$$

[In conjunction with the inversion rule, this gives (69)!] After observing that

$$h(x) = f(x) + \frac{1}{2}$$

we see that

$$h(x) \quad \text{has the FT} \quad H(s) = \frac{1}{2\pi i s} + \frac{1}{2}\delta(s). \tag{99}$$

∎

Example: Find the Fourier transform of $f(x) := |x^2 - 1|$.

Solution: Let

$$g(x) := \begin{cases} 2(1 - x^2) & \text{if } -1 < x < 1 \\ 0 & \text{otherwise.} \end{cases}$$

Since g' has the jumps $4, 4$ and g'' has the jumps $-4, 4$ at the points $x = -1$, $x = 1$, we find

$$g'''(x) = 4\delta'(x + 1) + 4\delta'(x - 1) - 4\delta(x + 1) + 4\delta(x - 1),$$

and thereby obtain

$$G(s) = \frac{4e^{2\pi i s}(2\pi i s - 1) + 4e^{-2\pi i s}(2\pi i s + 1)}{(2\pi i s)^3}$$

(with the singularity at $s = 0$ being removable). We have chosen g to make

$$f(x) = x^2 - 1 + g(x),$$

so we can use (95) to write

$$F(s) = \frac{\delta''(s)}{(-2\pi i)^2} - \delta(s) + G(s).$$

(We cannot obtain the singular part of F from $f''' = g'''$.)   ∎

Such techniques can be used to find the Fourier transform of any piecewise polynomial function with finitely many knots, *cf.* Ex. 7.39.

Example: Find the Fourier transform of $f(x) := 1/(1 + x^4)$.

Solution: We use the power scaling rule as we Fourier transform

$$(1 + x^4) \cdot f(x) = 1$$

to obtain the differential equation

$$F^{(4)}(s) + (2\pi)^4 F(s) = (2\pi)^4 \delta(s).$$

The characteristic polynomial

$$\mathcal{P}(r) = r^4 + (2\pi)^4$$

has the roots

$$r = \sqrt{2}\pi(1 + i), \quad \sqrt{2}\pi(1 - i), \quad \sqrt{2}\pi(-1 + i), \quad \sqrt{2}\pi(-1 - i),$$

and since F must be both even and slowly growing, we can write

$$F(s) = e^{-\sqrt{2}\pi|s|}[c\sin(\sqrt{2}\pi|s|) + d\cos(\sqrt{2}\pi s)]$$

for suitably chosen constants c, d. From the differential equation, we see that

$$F(0+) - F(0-) = 0, \quad F'(0+) - F'(0-) = 0,$$
$$F''(0+) - F''(0-) = 0, \quad F'''(0+) - F'''(0-) = (2\pi)^4,$$

and since

$$F(s) = d + (c - d)|\sqrt{2}\pi s| - c|\sqrt{2}\pi s|^2 + (c + d)|\sqrt{2}\pi s|^3/3 + \cdots$$

we must have

$$c - d = 0, \quad 12(c + d)\frac{(\sqrt{2}\pi)^3}{3} = (2\pi)^4$$

so that

$$c = d = \frac{\pi}{\sqrt{2}}.$$

In this way we obtain the Fourier transform

$$F(s) = \pi\, e^{-\sqrt{2}\pi|s|} \sin\left(\sqrt{2}\pi|s| + \frac{\pi}{4}\right).$$

You used (3.22) (and a lot more algebra) to obtain this result in Ex. 3.13. ∎

You can use this procedure to find the Fourier transform of any rational function f that is defined on all of $\mathbb{R}$, *cf.* Ex. 7.42.

Support- and bandlimited generalized functions

A generalized function β is said to be *σ-support-limited* for some $\sigma > 0$ if $\beta(x) = 0$ for $x < -\sigma$ and for $x > \sigma$, e.g., $\Pi(x)$ is σ-support-limited when $\sigma \geq 1/2$, $\delta(x)$ is σ-support-limited when $\sigma > 0$, and $e^{-\pi x^2}$ is not σ-support-limited for any $\sigma > 0$.

A generalized function α is said to be *σ-bandlimited* for some $\sigma > 0$ if $\alpha^{\wedge}(s) = 0$ for $s < -\sigma$ and for $s > \sigma$, e.g., $\text{sinc}(x)$ is σ-bandlimited when $\sigma \geq 1/2$, x^2 is σ-bandlimited when $\sigma > 0$, and $e^{-\pi x^2}$ is not σ-bandlimited for any $\sigma > 0$. Of course, α is σ-bandlimited if and only if $\alpha^{\wedge}$ is σ-support-limited.

As we develop sampling theory in Chapter 8, we will often have occasion to form products $\alpha \cdot f$, $\beta * f$ where α is σ-bandlimited, β is σ-support-limited, and f is an arbitrary generalized function. We will now verify that such generalized functions α, β satisfy the sufficient conditions (79), (80) and thereby show that these products are well defined.

Let $\beta = g^{(m)}$ where g is CSG and m is a nonnegative integer, and assume that β is σ-support-limited. Then $g^{(m)}(x) = 0$ for $x < -\sigma$ and for $x > \sigma$, so we can write

$$g(x) = p_L(x)h(-\sigma - x) + \gamma(x) + p_R(x)h(x - \sigma) \tag{100}$$

where p_L, p_R are polynomials of degree $m - 1$ or less, where h is the Heaviside function, and where

$$\begin{aligned}&\gamma(x) \text{ is continuous for } -\sigma \leq x \leq \sigma \text{ with} \\ &\gamma(x) = 0 \text{ for } x < -\sigma \text{ or } x > \sigma,\end{aligned} \tag{101}$$

cf. Fig. 7.16. We take m generalized derivatives of (100) to obtain the representation

$$\beta(x) = \gamma^{(m)}(x) + \sum_{\mu=0}^{m-1} [c_\mu \delta^{(\mu)}(x + \sigma) + d_\mu \delta^{(\mu)}(x - \sigma)] \tag{102}$$

where

$$c_\mu := -p_L^{(m-\mu-1)}(-\sigma), \quad d_\mu := p_R^{(m-\mu-1)}(\sigma), \quad \mu = 0, 1, \ldots, m-1.$$

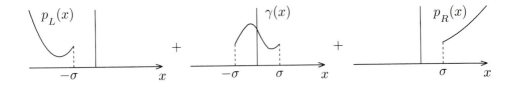

Figure 7.16. The truncated polynomials p_L, p_R and the function γ for the CSG function g that gives a σ-support-limited $\beta = g^{(m)}$.

Now if γ satisfies (101), the Fourier transform

$$\gamma^{\wedge}(s) = \int_{-\sigma}^{\sigma} \gamma(x)e^{-2\pi isx}dx$$

and its derivatives

$$[\gamma^{\wedge}(s)]^{(n)} = \int_{-\sigma}^{\sigma} (-2\pi ix)^n \gamma(x)dx, \quad n = 1, 2, \ldots$$

are continuous and bounded with

$$\left|[\gamma^{\wedge}(s)]^{(n)}\right| \leq (2\pi\sigma)^n \int_{-\sigma}^{\sigma} |\gamma(x)|dx, \quad n = 0, 1, 2, \ldots .$$

This being the case,

$$\beta^{\wedge}(s) = (2\pi is)^m \gamma^{\wedge}(s) + \sum_{\mu=0}^{m-1} (2\pi is)^{\mu} [c_{\mu} e^{2\pi i\sigma s} + d_{\mu} e^{-2\pi i\sigma s}] \tag{103}$$

and all of its derivatives are CSG. You have often observed that the Fourier transform of a function with "small tails" must be "very smooth." In this extreme case where β has "zero tails," the functions $\beta^{\wedge}, \beta^{\wedge\prime}, \beta^{\wedge\prime\prime}, \ldots$ are all CSG.

Finally, when α is σ-bandlimited we apply the above argument to the σ-support-limited function $\beta = \alpha^{\wedge\vee}$ and thereby conclude that $\alpha = \beta^{\wedge}$ and all of its derivatives are CSG. You can now add the σ-bandlimited functions to the list following (79)!

7.6 Limits of Generalized Functions

Introduction

If you dislike the tedious $\delta - \epsilon$ arguments from calculus or chafe a bit at restrictions on the term-by-term differentiation of infinite series, you may not look forward to yet another study of limits. But you cannot do analysis without limits, and you must learn a few elementary concepts before you can make sense of the relations (4) or (5), before you can work with infinite series of generalized functions, or before you can use Fourier analysis to find generalized solutions of partial differential equations. The theory is easy to learn and exceptionally powerful. You will more than double your ability to use generalized functions for solving problems as you master the ideas in this short section!

The limit concept

Let $f, f_1, f_2, \ldots$ be generalized functions. When we feed these functionals some Schwartz function ϕ, we produce a complex number $f\{\phi\}$ and a sequence of complex numbers $f_1\{\phi\}, f_2\{\phi\}, \ldots$. If we have

$$\lim_{n\to\infty} f_n\{\phi\} = f\{\phi\} \quad \text{whenever} \quad \phi \in \mathbb{S},$$

we will write

$$\lim_{n\to\infty} f_n = f.$$

In some situations we will associate a generalized function f_λ with each (real or complex) value of the parameter λ in some neighborhood of a (finite or infinite) point L. If we have

$$\lim_{\lambda\to L} f_\lambda\{\phi\} = f\{\phi\} \quad \text{whenever} \quad \phi \in \mathbb{S},$$

we will write

$$\lim_{\lambda\to L} f_\lambda = f.$$

We will use the unadorned terms *limit* and *converge* in the usual fashion, *e.g.*, we say that $f_1, f_2, \ldots$ converges to f or that f is the limit of $f_1, f_2, \ldots$ when $f = \lim_{n\to\infty} f_n$. In cases where the context admits some competing notion (*e.g.*, pointwise limit, uniform convergence) we will add the modifier *generalized* or *weak* to specify that we are using this new limit concept.

Example: Let $f_n(x) := n\,\Pi(nx)$, $n = 1, 2, \ldots$. Show that $\lim_{n\to\infty} f_n = \delta$.

Solution: Let ϕ be any Schwartz function. We use the integral mean value theorem from calculus to write

$$f_n\{\phi\} = \int_{-\infty}^{\infty} n\,\Pi(nx)\phi(x)dx = n \int_{-1/2n}^{1/2n} \phi(x)dx = \phi(\xi_n)$$

for some choice of ξ_n in the interval $(-1/2n, 1/2n)$. Since ϕ is continuous, we have

$$\lim_{n\to\infty} f_n\{\phi\} = \lim_{n\to\infty} \phi(\xi_n) = \phi(0) = \delta\{\phi\}. \qquad \blacksquare$$

You can use a similar argument to interpret the limit (5), *cf.* Fig. 7.2.

Example: Let $e_s(x) := e^{2\pi i s x}$. Show that $\lim_{s\to\pm\infty} e_s = 0$ even though $|e_s(x)| = 1$ for all real values of s and x.

Solution: Let $\phi \in \mathbb{S}$. Since $\phi^\wedge \in \mathbb{S}$ we can write

$$\lim_{s\to\pm\infty} e_s\{\phi\} = \lim_{s\to\pm\infty} \int_{-\infty}^{\infty} \phi(x)e^{2\pi i s x}dx = \lim_{s\to\pm\infty} \phi^\wedge(-s) = 0. \qquad \blacksquare$$

Example: For $n = 0, 1, \ldots$ and $h > 0$ we define

$$P_{n,h}(x) := \frac{1}{h^n} \sum_{m=0}^{n} (-1)^{n-m} \binom{n}{m} \delta(x - mh), \tag{104}$$

cf. Fig. 7.17. Show that $\lim_{h \to 0} P_{n,h} = (-1)^n \delta^{(n)}$.

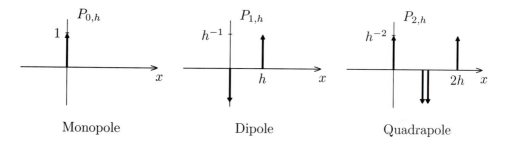

Monopole   Dipole   Quadrapole

Figure 7.17. The monopole $P_{0,h}$, the dipole $P_{1,h}$, and the quadrapole $P_{2,h}$ from (104).

Solution: Let $\phi \in \mathbb{S}$. We observe that

$$P_{1,h}\{\phi\} = \frac{1}{h}[\phi(h) - \phi(0)] = \frac{1}{h} \int_{u_1=0}^{h} \phi'(u_1) du_1,$$

$$P_{2,h}\{\phi\} = \frac{1}{h^2}[\phi(2h) - 2\phi(h) + \phi(0)] = \frac{1}{h^2} \int_{u_2=0}^{h} \int_{u_1=0}^{h} \phi''(u_1 + u_2) du_1\, du_2,$$

$$\vdots$$

and use a multivariate version of the integral mean value theorem to see that

$$\lim_{h \to 0} P_{n,h}\{\phi\} = \phi^{(n)}(0) = (-1)^n \delta^{(n)}\{\phi\}. \qquad \blacksquare$$

Example: For all $\lambda < \mu$ we define

$$f_{\lambda\mu}(x) := \int_{\lambda}^{\mu} e^{2\pi i s x} ds.$$

Show that

$$\lim_{\lambda \to -\infty} \lim_{\mu \to +\infty} f_{\lambda\mu} = \delta,$$

thereby giving meaning to the synthesis equation

$$\delta(x) = \int_{-\infty}^{\infty} 1 \cdot e^{2\pi i s x} ds,$$

which corresponds to (66).

Solution: The CSG function $f_{\lambda\mu}$ is represented by the fundamental functional

$$
\begin{aligned}
f_{\lambda\mu}\{\phi\} &= \int_{x=-\infty}^{\infty} \left\{ \int_{s=\lambda}^{\mu} e^{2\pi i s x} ds \right\} \phi(x) dx \\
&= \int_{s=\lambda}^{\mu} \int_{x=-\infty}^{\infty} \phi(x) e^{2\pi i s x} dx\, ds \\
&= \int_{s=\lambda}^{\mu} \phi^{\wedge}(-s) ds, \quad \phi \in \mathbb{S},
\end{aligned}
$$

so

$$\lim_{\lambda \to -\infty} \lim_{\mu \to +\infty} f_{\lambda\mu}\{\phi\} = \int_{-\infty}^{\infty} \phi^{\wedge}(-s) ds = \phi(0) = \delta\{\phi\}, \quad \phi \in \mathbb{S}$$

(with the order of the iterated limits being of no importance). ∎

Example: Let f be a generalized function with the derivative f'. Show that

$$f'(x) = \lim_{h \to 0} \frac{f(x+h) - f(x)}{h}. \tag{105}$$

Solution: Let $\phi \in \mathbb{S}$ be given. We will define

$$
\begin{aligned}
I_h &:= \int_{-\infty}^{\infty} \frac{f(x+h) - f(x)}{h} \phi(x) dx - \int_{-\infty}^{\infty} f'(x)\phi(x) dx \\
&= h^{-1} \int_{-\infty}^{\infty} f(x)[\phi(x-h) - \phi(x) + h\phi'(x)] dx, \quad h < 0 \text{ or } h > 0,
\end{aligned}
$$

and show that $I_h \to 0$ as $h \to 0$. We choose a CSG function g and nonnegative integer n such that $f = g^{(n)}$ and use Taylor's formula (*cf.* Ex. 2.28) to write

$$
\begin{aligned}
I_h &= h^{-1} \int_{-\infty}^{\infty} g^{(n)}(x)[\phi(x-h) - \phi(x) + h\,\phi'(x)] dx \\
&= (-1)^n h^{-1} \int_{-\infty}^{\infty} g(x)[\phi^{(n)}(x-h) - \phi^{(n)}(x) + h\,\phi^{(n+1)}(x)] dx \\
&= (-1)^n h^{-1} \int_{-\infty}^{\infty} g(x) \int_{u=0}^{h} (h-u)\phi^{(n+2)}(x-u) du\, dx.
\end{aligned}
$$

We choose some $m = 1, 2, \ldots$ so that

$$M := \int_{-\infty}^{\infty} \frac{|g(x)|}{(1 + x^2)^m} dx$$

is finite, and we set

$$B(H) := \max_{-\infty < x < \infty} [1 + (|x| + H)^2]^m \cdot |\phi^{(n+2)}(x)| \quad \text{when} \quad H \geq 0$$

to ensure that

$$\int_{-\infty}^{\infty} \frac{|g(x)|}{(1 + x^2)^m} \cdot (1 + x^2)^m |\phi^{(n+2)}(x - u)| dx \leq M \cdot B(H) \quad \text{when} \quad |u| \leq H.$$

We can now use this bound to write

$$|I_h| \leq |h|^{-1} \int_{u=0}^{|h|} (|h| - u) \int_{-\infty}^{\infty} |g(x)\phi^{(n+2)}(x - u)| dx\, du$$

$$\leq M \cdot B(H) \cdot \frac{|h|}{2} \quad \text{when} \quad 0 < |h| < H,$$

and thereby conclude that $I_h \to 0$ as $h \to 0$. ∎

Example: Let f be a generalized function. Show that

$$\lim_{h \to 0} f(x + h) = f(x). \tag{106}$$

Solution: Let $\phi \in \mathbb{S}$. We use (105) to write

$$\lim_{h \to 0} \int_{-\infty}^{\infty} [f(x + h) - f(x)]\phi(x) dx = \lim_{h \to 0} h \int_{-\infty}^{\infty} \frac{f(x + h) - f(x)}{h} \phi(x) dx$$

$$= 0 \cdot \int_{-\infty}^{\infty} f'(x)\phi(x) dx = 0. \quad ∎$$

Infinite series of generalized functions

Let $f, f_\nu, \nu = 0, \pm 1, \pm 2, \ldots$ be generalized functions. We will write

$$f = \sum_{\nu=0}^{\infty} f_\nu \quad \text{or} \quad f = f_0 + f_1 + f_2 + \cdots$$

provided that

$$f = \lim_{n \to \infty} \sum_{\nu=0}^{n} f_\nu,$$

and we will write

$$f = \sum_{\nu=-\infty}^{\infty} f_\nu \quad \text{or} \quad f = \cdots + f_{-2} + f_{-1} + f_0 + f_1 + f_2 + \cdots$$

provided that

$$f = \lim_{m \to -\infty} \sum_{\nu=m}^{-1} f_\nu + \lim_{n \to +\infty} \sum_{\nu=0}^{n} f_\nu.$$

Example: Show that

$$\text{Ш}(x) = \sum_{\nu=-\infty}^{\infty} \delta(x - \nu). \tag{107}$$

Solution: Let $\phi \in \mathbb{S}$ be given. The series (40) converges absolutely, so we can set $\delta_\nu(x) := \delta(x - \nu)$ and use the sifting relation (64) to write

$$\lim_{\substack{m \to -\infty \\ n \to +\infty}} \left(\sum_{\nu=m}^{n} \delta_\nu \right) \{\phi\} = \lim_{\substack{m \to -\infty \\ n \to +\infty}} \sum_{\nu=m}^{n} \delta_\nu\{\phi\} = \lim_{\substack{m \to -\infty \\ n \to +\infty}} \sum_{\nu=m}^{n} \phi(\nu) = \text{Ш}\{\phi\}. \qquad \blacksquare$$

Example: Let f be the 1-periodic function with

$$f(x) = w_1(x) := -\frac{1}{2}x^2 + \frac{1}{2}x - \frac{1}{12} \quad \text{when } 0 \le x \le 1$$

where w_1 is the Bernoulli function (4.20), *cf.* Fig. 4.5. Show that the corresponding Fourier series

$$f(x) = \sum_{\substack{k=-\infty \\ k \ne 0}}^{\infty} \frac{e^{2\pi i k x}}{(2\pi i k)^2} \tag{108}$$

is weakly convergent

Solution: Let $\phi \in \mathbb{S}$ be chosen, and let

$$M := \max_{-\infty < x < \infty} \left| \sum_{n=-\infty}^{\infty} \phi(x + n) \right|.$$

The Fourier series for the continuous piecewise smooth function f converges uniformly, so we can write

$$\left| \int_{-\infty}^{\infty} f(x)\phi(x)dx - \int_{-\infty}^{\infty} \sum_{\substack{k=m \\ k \neq 0}}^{n} \frac{e^{2\pi i k x}}{(2\pi i k)^2} \phi(x)dx \right|$$

$$= \left| \int_{x=0}^{1} \left(f(x) - \sum_{\substack{k=m \\ k \neq 0}}^{n} \frac{e^{2\pi i k x}}{(2\pi i k)^2} \right) \left(\sum_{n=-\infty}^{\infty} \phi(x+n) \right) dx \right|$$

$$\leq M \left(\sum_{k<m} \frac{1}{|2\pi k|^2} + \sum_{k>n} \frac{1}{|2\pi k|^2} \right)$$

$$\to 0 \quad \text{as} \quad m \to -\infty \quad \text{and} \quad n \to +\infty. \qquad \blacksquare$$

You can use an analogous argument to see that any continuous, piecewise smooth, p-periodic function on $\mathbb{R}$ has a weakly convergent Fourier series.

Example: Let $a_0, a_1, a_2, \ldots$ be complex numbers and assume that the series

$$f(x) = a_0 \delta(x) + a_1 \delta'(x) + a_2 \delta''(x) + \cdots \qquad (109)$$

converges. Show that $a_\nu = 0$ for all but finitely many values of $\nu = 0, 1, 2, \ldots$.

Solution: Since every term of the series vanishes for $-\infty < x < 0$ and for $0 < x < \infty$, the same is true of f, so we can use (91) to write

$$f(x) = \sum_{\nu=0}^{n-1} c_\nu \delta^{(\nu)}(x)$$

for some choice of $n = 1, 2, \ldots$ and some choice of the coefficients $c_0, c_1, \ldots, c_{n-1}$.

Now let $m = n, n+1, \ldots$ be selected. We construct a mesa function $\phi \in \mathbb{S}$ that takes the constant value 1 in some neighborhood of the origin (*cf.* Fig. 7.5) and then set

$$\psi(x) := (x^m/m!)\phi(x).$$

By construction, ψ is a Schwartz function with

$$\psi^{(k)}(0) = \begin{cases} 1 & \text{if } k = m \\ 0 & \text{otherwise,} \end{cases}$$

so

$$(-1)^m a_m = \lim_{k \to \infty} \left(\sum_{\nu=0}^{k} a_\nu \delta^{(\nu)} \right) \{\psi\} = \left(\sum_{\nu=0}^{n-1} c_\nu \delta^{(\nu)} \right) \{\psi\} = 0. \qquad \blacksquare$$

Transformation of limits

Let $f, f_1, f_2, \ldots$ and $g, g_1, g_2, \ldots$ be generalized functions. If you can show that

$$\lim_{n \to \infty} f_n(x) = f(x), \qquad \lim_{n \to \infty} g_n(x) = g(x)$$

then you can use each of the following.

$$\lim_{n \to \infty} [c \, f_n(x) + d \, g_n(x)] = c \, f(x) + d \, g(x), \quad c, d \in \mathbb{C}$$

$$\lim_{n \to \infty} f_n(x - x_0) \qquad = f(x - x_0), \qquad -\infty < x_0 < \infty,$$

$$\lim_{n \to \infty} f_n(ax) \qquad = f(ax), \qquad a < 0 \ \text{ or } \ a > 0,$$

$$\lim_{n \to \infty} f_n^{(k)}(x) \qquad = f^{(k)}(x), \qquad k = 1, 2, \ldots,$$

$$\lim_{n \to \infty} f_n^{\wedge}(s) \qquad = f^{\wedge}(s),$$

$$\lim_{n \to \infty} \alpha(x) \cdot f_n(x) \qquad = \alpha(x) \cdot f(x), \qquad \alpha, \alpha', \alpha'', \ldots \text{ are CSG,}$$

$$\lim_{n \to \infty} (\beta * f_n)(x) \qquad = (\beta * f)(x), \qquad \beta, \beta^{\wedge\prime}, \beta^{\wedge\prime\prime}, \ldots \text{ are CSG.}$$

Corresponding transformations can be used when

$$\lim_{\lambda \to L} f_\lambda(x) = f(x), \qquad \lim_{\lambda \to L} g_\lambda(x) = g(x).$$

You can establish these rules by chasing definitions.

Example: Let $\lim_{n \to \infty} f_n = f$. Show that $\lim_{n \to \infty} f_n' = f'$.

Solution: Given any $\phi \in \mathbb{S}$ we write

$$
\begin{aligned}
f'\{\phi\} &= \int_{-\infty}^{\infty} f'(x)\phi(x)dx && \text{Integral notation} \\[2mm]
&:= \int_{-\infty}^{\infty} f(x)[-\phi'(x)]dx && \text{Definition of } f' \\[2mm]
&= \lim_{n \to \infty} \int_{-\infty}^{\infty} f_n(x)[-\phi'(x)]dx && \text{Since } \lim f_n = f \\[2mm]
&=: \lim_{n \to \infty} \int_{-\infty}^{\infty} f_n'(x)\phi(x)dx && \text{Definition of } f_n' \\[2mm]
&= \lim_{n \to \infty} f_n'\{\phi\}, && \text{Integral notation}
\end{aligned}
$$

and this proves that $\lim_{n \to \infty} f_n' = f'$.   ∎

Example: Let $\lim\limits_{n\to\infty} f_n = f$. Show that $\lim\limits_{n\to\infty} f_n^\wedge = f^\wedge$.

Solution: Given any $\phi \in \mathbb{S}$ we write

$$f^\wedge\{\phi\} = \int_{-\infty}^{\infty} f^\wedge(s)\phi(s)ds \qquad \text{Integral notation}$$

$$:= \int_{-\infty}^{\infty} f(x)[\phi^\wedge(x)]dx \qquad \text{Definition of } f^\wedge$$

$$= \lim_{n\to\infty} \int_{-\infty}^{\infty} f_n(x)[\phi^\wedge(x)]dx \qquad \text{Since } \lim f_n = f$$

$$=: \lim_{n\to\infty} \int_{-\infty}^{\infty} f_n^\wedge(s)\phi(s)ds \qquad \text{Definition of } f_n^\wedge$$

$$= \lim_{n\to\infty} f_n^\wedge\{\phi\}, \qquad \text{Integral notation}$$

and this proves that $\lim\limits_{n\to\infty} f_n^\wedge = f^\wedge$. ∎

The rules for transforming weak limits are exceptionally useful when we work with infinite series. Indeed, if you can somehow show that

$$f(x) = \sum_{\nu=-\infty}^{\infty} f_\nu(x)$$

then you can write

$$f(x-x_0) = \sum_{\nu=-\infty}^{\infty} f_\nu(x-x_0), \qquad -\infty < x_0 < \infty,$$

$$f(ax) = \sum_{\nu=-\infty}^{\infty} f_\nu(ax), \qquad a < 0 \ \text{ or } \ a > 0,$$

$$f^{(k)}(x) = \sum_{\nu=-\infty}^{\infty} f_\nu^{(k)}(x), \qquad k = 1, 2, \ldots,$$

$$f^\wedge(s) = \sum_{\nu=-\infty}^{\infty} f_\nu^\wedge(s),$$

$$\alpha(x) \cdot f(x) = \sum_{\nu=-\infty}^{\infty} \alpha(x) \cdot f_\nu(x), \quad \alpha, \alpha', \alpha'', \ldots \text{ are CSG, and}$$

$$(\beta * f)(x) = \sum_{\nu=-\infty}^{\infty} (\beta * f_\nu)(x), \qquad \beta^\wedge, \beta^{\wedge\prime}, \beta^{\wedge\prime\prime}, \ldots \text{ are CSG.}$$

We will give a few examples to show what can be done with such transformations.

Example: Use the weakly convergent series (107) to show that $\text{III}^\wedge = \text{III}$.

Solution: We Fourier transform (107) term by term to obtain the unusual (but weakly convergent!) infinite series

$$\text{III}^\wedge(s) = \sum_{\nu=-\infty}^{\infty} e^{-2\pi i \nu s}.$$

We use this series with the special Poisson sum formula (67) to write

$$\text{III}^\wedge\{\phi\} = \lim_{\substack{m\to-\infty \\ n\to+\infty}} \sum_{\nu=m}^{n} \int_{-\infty}^{\infty} e^{-2\pi i \nu s} \phi(s) ds = \lim_{\substack{m\to-\infty \\ n\to+\infty}} \sum_{\nu=m}^{n} \phi^\wedge(\nu)$$

$$= \sum_{\nu=-\infty}^{\infty} \phi^\wedge(\nu) = \sum_{\nu=-\infty}^{\infty} \phi(\nu) = \text{III}\{\phi\}, \quad \phi \in \mathbb{S}.$$

This gives a second proof of (68)!   ∎

Example: Use the Fourier series (108) to show that $\text{III}^\wedge = \text{III}$.

Solution: We have shown that (108) converges weakly, so we can differentiate this series term by term to obtain the weakly convergent series

$$f'(x) = \sum_{\substack{k=-\infty \\ k\neq 0}}^{\infty} \frac{e^{2\pi i k x}}{2\pi i k}, \qquad f''(x) = \sum_{\substack{k=-\infty \\ k\neq 0}}^{\infty} e^{2\pi i k x}.$$

After examining the graphs of f, f', f'' as shown in Fig. 7.18, we see that

$$\text{III}(x) = f''(x) + 1 = \sum_{k=-\infty}^{\infty} e^{2\pi i k x}.$$

We take Fourier transforms term by term and use (107) to write

$$\text{III}^\wedge(s) = \sum_{k=-\infty}^{\infty} \delta(s - k) = \text{III}(s).$$

This gives a third proof of (68)!   ∎

Example: Let $p > 0$, so that

$$f(x) := \frac{1}{p} \text{III}\left(\frac{x}{p}\right) \quad \text{has the FT} \quad F(s) = \text{III}(ps).$$

Express f and F in terms of translates of δ.

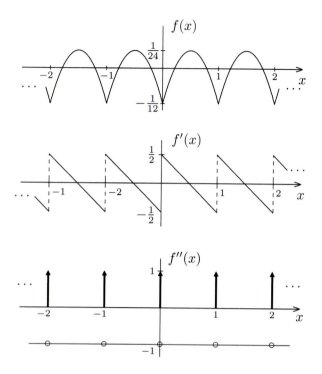

Figure 7.18. The piecewise parabolic Bernoulli function f from (108) and its derivatives f', f''.

Solution: We dilate (107) term by term and use (65) as we write

$$f(x) = \frac{1}{p} \sum_{n=-\infty}^{\infty} \delta\left(\frac{x}{p} - n\right) = \sum_{n=-\infty}^{\infty} \delta(x - np),$$

$$F(s) = \sum_{n=-\infty}^{\infty} \delta(ps - n) = \frac{1}{p} \sum_{n=-\infty}^{\infty} \delta\left(s - \frac{n}{p}\right),$$

(110)

cf. Fig. 7.19. The spacing $1/p$ for the comb F is the reciprocal of the spacing p for the comb f, *cf.* (4.44). It is very easy to derive (110) by using III. ∎

$$f(x) = p^{-1}\text{III}(x/p) \qquad\qquad\qquad F(s) = \text{III}(ps)$$

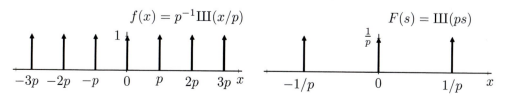

Figure 7.19. The dilated comb and its Fourier transform as given in (110).

Example: Let $c_0, c_1, c_2, \ldots$ be complex numbers and assume that the power series

$$f(x) = c_0 + c_1 x + c_2 x^2 + \cdots$$

converges weakly. Show that $c_\nu = 0$ for all but finitely many values of ν.

Solution: Since the power series converges weakly, we can take Fourier transforms term by term to obtain

$$f^\wedge(s) = c_0 \delta(s) + c_1 \frac{\delta'(s)}{-2\pi i} + c_2 \frac{\delta''(s)}{(-2\pi i)^2} + \cdots .$$

Our previous analysis of (109) shows that $c_\nu = 0$ for all but finitely many ν. ∎

You may be surprised by this result. The familiar Maclaurin series

$$1 - x^2/2! + x^4/4! - x^6/6! + \cdots, \qquad x - x^3/3! + x^5/5! - x^7/7! + \cdots$$

from calculus converge pointwise to the generalized functions $\cos x$, $\sin x$, but these power series do *not* converge weakly. You can freely use the new notion of limit with Fourier series but *not* with power series!

Partial derivatives

Let $u(x, t)$ be a generalized function of x for each choice of the parameter t in some real interval (a, b). We will write

$u_x(x, t)$ for the generalized derivative $\displaystyle\lim_{h \to 0} \frac{u(x + h, t) - u(x, t)}{h}$
of $u(x, t)$ with t being fixed, *cf.* (105);

$u^\wedge(s, t)$ or $U(s, t)$ for the generalized Fourier transform
of $u(x, t)$ with t being fixed; and   (111)

$u_t(x, t)$ for the generalized limit $\displaystyle\lim_{h \to 0} \frac{u(x, t + h) - u(x, t)}{h}$
when this limit exists.

Since the process of taking generalized derivatives and Fourier transforms commutes with that of taking generalized limits we can always write

$$(u_t)_x(x, t) = (u_x)_t(x, t)$$
$$(u_t)^\wedge(s, t) = (u^\wedge)_t(s, t)$$
    (112)

when u_t is defined. We will routinely use such relations as we solve partial differential equations in Chapter 9.

Example: Let f be a generalized function. Show that

$$\lim_{h \to 0} \frac{e^{-2\pi ihx} - 1}{h} f(x) = -2\pi ix \cdot f(x).$$

Solution: We Fourier transform and reflect the limit

$$\lim_{h \to 0} \frac{f^\wedge(s+h) - f^\wedge(s)}{h} = f^{\wedge\prime}(s)$$

from (105) to obtain the desired result. ∎

Example: Let $u(x, t) := \Pi(x - ct)$ where $c > 0$ and where t is a real parameter. Show that $u_t(x, t) = -c\, u_x(x, t)$.

Solution: We will calculate u_x, u_t and verify that $u_x = -c\, u_t$. Since the box $\Pi(x - ct)$ has jumps $+1, -1$ at $x = ct - \frac{1}{2}$, $x = ct + \frac{1}{2}$, we have

$$u_x(x, t) = \delta(x - ct + \tfrac{1}{2}) - \delta(x - ct - \tfrac{1}{2}).$$

Now when $0 < h < 1/c$ we find

$$\Pi(x - ct - ch) - \Pi(x - ct) = \begin{cases} -1 & \text{if } ct - \frac{1}{2} < x < ct - \frac{1}{2} + ch \\ +1 & \text{if } ct + \frac{1}{2} < x < ct + \frac{1}{2} + ch \\ 0 & \text{otherwise} \end{cases}$$

and when $-1/c < h < 0$ we find

$$\Pi(x - ct - ch) - \Pi(x - ct) = \begin{cases} +1 & \text{if } ct - \frac{1}{2} + ch < x < ct - \frac{1}{2} \\ -1 & \text{if } ct + \frac{1}{2} + ch < x < ct + \frac{1}{2} \\ 0 & \text{otherwise.} \end{cases}$$

It follows that

$$\begin{aligned} u_t(x, t) :&= \lim_{h \to 0} h^{-1} [\Pi(x - ct - ch) - \Pi(x - ct] \\ &= c[-\delta(x - ct + \tfrac{1}{2}) + \delta(x - ct - \tfrac{1}{2})] \\ &= -cu_x(x, t). \end{aligned}$$

For a more efficient argument, we can use (105) to write

$$\begin{aligned} u_t(x, t) :&= \quad \lim_{h \to 0} \frac{\Pi(x - ct - ch) - \Pi(x - ct)}{h} \\ &= -c \lim_{h \to 0} \frac{u(x - ch, t) - u(x, t)}{-ch} \\ &= -cu_x(x, t). \end{aligned}$$

∎

7.7 Periodic Generalized Functions

Fourier series

Let $p > 0$, let f be a generalized function, and assume that f is *p-periodic, i.e.,*

$$f(x+p) = f(x) \quad \text{for} \quad -\infty < x < \infty. \tag{113}$$

We will show that f can be represented by a weakly convergent Fourier series

$$f(x) = \sum_{k=-\infty}^{\infty} c_k e^{2\pi i k x/p} \tag{114}$$

with the corresponding weakly convergent series

$$f^{\wedge}(s) = \sum_{k=-\infty}^{\infty} c_k \delta\left(s - \frac{k}{p}\right), \tag{115}$$

giving the generalized Fourier transform. (You will soon learn how to find the Fourier coefficient c_k, $k = 0, \pm 1, \pm 2, \dots$.) Fourier and his contemporaries tried to establish such a representation for an arbitrary p-periodic continuous function, but they were unsuccessful because they did not have a suitable limit concept.

We begin by writing

$$f(x) = g^{(n)}(x)$$

where g is CSG and n is a nonnegative integer. We will replace g by an antiderivative (59), if necessary (and augment n accordingly), to make sure that g is continuously differentiable. Since f is p-periodic, we then have

$$[g(x+p) - g(x)]^{(n)} = 0 \quad \text{for} \quad -\infty < x < \infty,$$

and this implies that

$$g(x+p) - g(x) = q(x) \tag{116}$$

where q is a polynomial of degree $n - 1$ or less.

We determine coefficients $a_0, a_1, \dots, a_{n-1}$ such that

$$q(x) = a_0 x^{[0]} + a_1 x^{[1]} + a_2 x^{[2]} + \cdots + a_{n-1} x^{[n-1]}$$

where

$$x^{[0]} := 1, \quad x^{[1]} := x, \quad x^{[2]} := x(x-p), \quad x^{[3]} := x(x-p)(x-2p), \dots .$$

These *factorial powers* have the forward differences

$$(x+p)^{[k]} - x^{[k]} = kp \cdot x^{[k-1]}, \quad k = 1, 2, \dots ,$$

so the polynomial

$$q_0(x) := \sum_{k=1}^{n} a_{k-1} \frac{x^{[k]}}{kp}$$

of degree n or less has the forward difference

$$q_0(x+p) - q_0(x) = \sum_{k=1}^{n} a_{k-1} \frac{(x+p)^{[k]} - x^{[k]}}{kp} = q(x). \tag{117}$$

We now set

$$g_0(x) := g(x) - q_0(x)$$

and use (116), (117) to see that

$$
\begin{aligned}
g_0(x+p) &= g(x+p) - q_0(x+p) \\
&= [g(x) + q(x)] - [q_0(x) + q(x)] \\
&= g_0(x) \quad \text{for} \quad -\infty < x < \infty.
\end{aligned}
$$

Since g_0 is a continuously differentiable p-periodic function on $\mathbb{R}$, the analysis from Section 1.5 shows that the Fourier coefficients

$$G_0[k] := \frac{1}{p} \int_0^p g_0(x) e^{-2\pi i k x/p} dx, \quad k = 0, \pm 1, \pm 2, \dots$$

are absolutely summable. This guarantees that the ordinary Fourier series

$$g_0(x) = \sum_{k=-\infty}^{\infty} G_0[k] e^{2\pi i k x/p}$$

converges weakly. We repeatedly differentiate this series term by term and write

$$f(x) = g^{(n)}(x) = g_0^{(n)}(x) + q_0^{(n)}(x) = \sum_{k=-\infty}^{\infty} \left(\frac{2\pi i k}{p}\right)^n G_0[k] e^{2\pi i k x/p} + q_0^{(n)}(0).$$

(Since q_0 is a polynomial of degree n or less, $q_0^{(n)}$ is a constant.) In this way we show that f has the representation (114) with

$$c_k = \begin{cases} q_0^{(n)}(0) & \text{if } k = 0 \\ \left(\dfrac{2\pi i k}{p}\right)^n G_0[k] & \text{if } k = \pm 1, \pm 2, \dots \end{cases} \tag{118}$$

and

$$|c_k| \le \left\{ \left(\frac{2\pi}{p}\right)^n \cdot \frac{1}{p} \int_0^p |g_0(x)| dx \right\} |k|^n, \quad k = \pm 1, \pm 2, \dots . \tag{119}$$

Example: Find the Fourier transform of the 5-periodic generalized function

$$f(x) := \sum_{n=-\infty}^{\infty} \Pi(x - 5n). \tag{120}$$

Solution: We use Poisson's relation (4.18) to obtain the Fourier series

$$f(x) = \sum_{k=-\infty}^{\infty} \frac{1}{5} \operatorname{sinc}\left(\frac{k}{5}\right) e^{2\pi i k x/5}.$$

This series converges weakly (differentiate the antiderivative!), so we can take Fourier transforms term by term to obtain

$$f^{\wedge}(s) = \sum_{k=-\infty}^{\infty} \frac{1}{5} \operatorname{sinc}\left(\frac{k}{5}\right) \delta\left(s - \frac{k}{5}\right), \tag{121}$$

cf. Fig. 7.20.   ∎

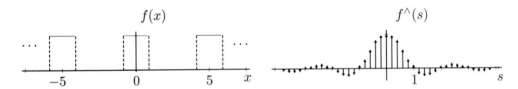

$f(x)$ $f^{\wedge}(s)$

Figure 7.20. The 5-periodic function f from (120) and the corresponding generalized Fourier transform $f^{\wedge}$ from (121).

Example: Use (107) to infer that III, $\text{III}^{\wedge}$ are 1-periodic and thereby show that $\text{III}^{\wedge} = \text{III}$.

Solution: We translate (107) term by term to infer that

$$\text{III}(x + 1) = \sum_{\nu=-\infty}^{\infty} \delta(x + 1 - \nu) = \sum_{\nu=-\infty}^{\infty} \delta(x - \nu) = \text{III}(x).$$

Knowing that III is 1-periodic, we use (115) (with $p = 1$) to write

$$\text{III}^{\wedge}(s) = \sum_{k=-\infty}^{\infty} c_k \delta(s - k)$$

for suitably chosen c_k's. Since $e^{-2\pi i x}$ and each of its derivatives is CSG, we can apply this factor to the terms of (107) and then use (82) to see that

$$e^{-2\pi i x} \cdot \text{III}(x) = \sum_{\nu=-\infty}^{\infty} e^{-2\pi i x} \cdot \delta(x - \nu) = \sum_{\nu=-\infty}^{\infty} \delta(x - \nu) = \text{III}(x).$$

We then use the modulation rule to show that $\text{III}^{\wedge}$ is 1-periodic,

$$\text{III}^{\wedge}(s + 1) = \text{III}^{\wedge}(s),$$

and thereby infer that c_k is independent of k, *i.e.*,

$$\text{III}^{\wedge}(s) = c \, \text{III}(s)$$

for some constant c. Finally, we use the Parseval identity as we write

$$\sum_{n=-\infty}^{\infty} e^{-\pi n^2} = \int_{-\infty}^{\infty} \text{III}(x) e^{-\pi x^2} dx$$

$$= \int_{-\infty}^{\infty} \text{III}^{\wedge}(s) e^{-\pi s^2} ds = c \int_{-\infty}^{\infty} \text{III}(s) e^{-\pi s^2} ds = c \sum_{n=-\infty}^{\infty} e^{-\pi n^2}$$

and thereby show that $c = 1$. This gives a fourth proof of (68)! ∎

Example: Let $p > 0$ and assume that the zero function $z(x) = 0$ is given by the weakly convergent series

$$z(x) = \sum_{k=-\infty}^{\infty} d_k e^{2\pi i k x / p}.$$

Show that $d_k = 0$ for each k [and thereby prove that the representation (114) is unique].

Solution: Let $k = 0, \pm 1, \pm 2, \ldots$ be selected. We construct a mesa function $\phi_k \in \mathbb{S}$ such that

$$\phi_k \left(\frac{k}{p} \right) = 1 \text{ and } \phi_k(x) = 0 \text{ when } \left| x - \frac{k}{p} \right| \geq \frac{1}{p},$$

cf. Fig. 7.5. Since

$$z^{\wedge}(s) = \sum_{\nu=-\infty}^{\infty} d_\nu \delta \left(s - \frac{\nu}{p} \right)$$

we can then write

$$d_k = \lim_{\substack{m \to -\infty \\ n \to +\infty}} \sum_{\nu=m}^{n} d_\nu \phi_k \left(\frac{\nu}{p}\right) = z^\wedge\{\phi_k\} = 0. \qquad \blacksquare$$

If the series (114) converges weakly, then the limit function f is p-periodic and (119) shows that the coefficients are slowly growing in the sense that

$$|c_k| \le B \cdot |k|^n \quad \text{when} \quad k = \pm 1, \pm 2, \dots \qquad (122)$$

for some choice of $B > 0$ and some nonnegative integer n. We will now establish the converse.

Let c_k, $k = 0, \pm 1, \pm 2, \dots$ satisfy (122) and let $p > 0$ be selected. Given $\phi \in \mathbb{S}$ we set

$$C := \max_{|x| \ge 1} \left| x^{n+2} \cdot \phi \left(\frac{x}{p}\right) \right|$$

and use this bound with (122) to see that

$$\sum_{k \ne 0} \left| c_k \phi \left(\frac{k}{p}\right) \right| \le \sum_{k \ne 0} \left| B k^n \cdot k^2 \phi \left(\frac{k}{p}\right) \cdot k^{-2} \right| \le BC \sum_{k \ne 0} k^{-2} < \infty.$$

In this way we see in turn that

$$\sum_{k=-\infty}^{\infty} c_k \phi \left(\frac{k}{p}\right), \quad \phi \in \mathbb{S}$$

converges absolutely, that

$$\sum_{k=-\infty}^{\infty} c_k \delta \left(s - \frac{k}{p}\right)$$

converges weakly, and that

$$\sum_{k=-\infty}^{\infty} c_k e^{2\pi i k x / p}$$

converges weakly to a generalized function f that is p-periodic.

Example: Let $p > 0$. For which values of the complex parameter z does the series

$$\sum_{k=-\infty}^{\infty} z^{|k|} e^{2\pi i k x / p}$$

converge weakly?

Solution: The series converges weakly if and only if the coefficients $z^{|k|}$ are slowly growing, *i.e.*, if and only if $|z| \le 1$. $\qquad \blacksquare$

The analysis equation

A p-periodic generalized function f can be synthesized by the Fourier series (114), but the expression (118) for the c_k's is a bit unwieldly for routine use. We will suitably modify the familiar formula

$$c_k = \frac{1}{p}\int_0^p f(x)e^{-2\pi ikx/p}dx, \quad k = 0, \pm1, \pm2, \ldots$$

to obtain an analysis equation that makes sense within the present context.

When we use $a = -1$, $b = c = 0$, $d = 1$ for the mesa function of Fig. 7.5, we obtain the *tapered box*

$$b(x) := \frac{\int_{|x|}^1 \beta(u)du}{\int_0^1 \beta(u)du} \tag{123}$$

where

$$\beta(u) := \begin{cases} e^{-1/[x(1-x)]} & \text{if } 0 < x < 1 \\ 0 & \text{otherwise.} \end{cases} \tag{124}$$

From (123) we see that b is a Schwartz function with

$$\begin{aligned} b(x) &= b(-x) && \text{for } -1 < x < 1, \\ b(x) &> 0 && \text{for } -1 < x < 1 \\ b(x) &= 0 && \text{for } -\infty < x \le -1 \text{ or } 1 \le x < \infty, \end{aligned} \tag{125}$$

cf. Fig. 7.21. We use (123) with the symmetry

$$\beta(u) = \beta(1-u), \quad -\infty < u < \infty$$

to see that

$$b(x) + b(1-x) = 1 \quad \text{for } 0 < x < 1$$

or equivalently (since b is even),

$$b(x) + b(x-1) = 1 \quad \text{for } 0 < x < 1. \tag{126}$$

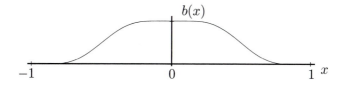

Figure 7.21. The tapered box b from (123).

By using (125)–(126) we easily compute the samples

$$
\begin{aligned}
B(k) : = \int_{-\infty}^{\infty} b(x)e^{-2\pi ikx}dx &= \sum_{n=-\infty}^{\infty} \int_{n}^{n+1} b(x)e^{-2\pi ikx}dx \\
&= \sum_{n=-\infty}^{\infty} \int_{0}^{1} b(x+n)e^{-2\pi ikx}dx = \int_{0}^{1} [b(x)+b(x-1)]e^{-2\pi ikx}dx \quad (127) \\
&= \int_{0}^{1} 1 \cdot e^{2\pi ikx}dx = \begin{cases} 1 & \text{if } k = 0 \\ 0 & \text{if } k = \pm 1, \pm 2, \dots \end{cases}
\end{aligned}
$$

of the Fourier transform. Since b and B are Schwartz functions, we can use (107), (82), and (127) to write

$$
B(s) \cdot \text{III}(s) = \sum_{k=-\infty}^{\infty} B(s) \cdot \delta(s-k) = \sum_{k=-\infty}^{\infty} B(k) \cdot \delta(s-k) = \delta(s)
$$

and thereby obtain the weakly converging series

$$
\sum_{k=-\infty}^{\infty} b(x-k) = (b * \text{III})(x) = 1. \quad (128)
$$

Now let f be a p-periodic generalized function with the Fourier series (114). We evaluate f at the Schwartz function $e^{-2\pi ikx/p}b(x/p)$ and use (127) to write

$$
\begin{aligned}
\int_{-\infty}^{\infty} f(x) \left[e^{-2\pi ikx/p}\, b\left(\frac{x}{p}\right) \right] dx &= \int_{-\infty}^{\infty} \left(\sum_{\ell=-\infty}^{\infty} c_\ell e^{-2\pi i(k-\ell)x/p} \right) b\left(\frac{x}{p}\right) dx \\
&= \sum_{\ell=-\infty}^{\infty} c_\ell \int_{-\infty}^{\infty} e^{-2\pi i(k-\ell)x/p}\, b\left(\frac{x}{p}\right) dx \\
&= p \sum_{\ell=-\infty}^{\infty} c_\ell B(k-\ell) \\
&= p\, c_k, \quad k = 0, \pm 1, \pm 2, \dots .
\end{aligned}
$$

In this way we obtain the desired *analysis equation*

$$
c_k = \frac{1}{p} \int_{-\infty}^{\infty} f(x)e^{-2\pi ikx/p}\, b\left(\frac{x}{p}\right) dx, \quad k = 0, \pm 1, \pm 2, \dots \quad (129)
$$

for the Fourier coefficients of a p-periodic generalized function f.

Convolution of p-periodic generalized functions

When f, g are piecewise smooth ordinary functions on $\mathbb{T}_p$, we use the familiar integral

$$\int_0^p f(u)g(x-u)du$$

from (2.1) to define the convolution product. We will now introduce a corresponding definition that allows us to work with generalized functions on $\mathbb{T}_p$. [In essence, we will formally replace du by $b(u/p)du$ and change the limits in the above integral to $\pm\infty$, cf. (129).]

Let $p > 0$, let

$$f(x) = \sum_{k=-\infty}^{\infty} c_k e^{2\pi ikx/p}, \quad g(x) = \sum_{k=-\infty}^{\infty} d_k e^{2\pi ikx/p} \tag{130}$$

be arbitrary p-periodic generalized functions, and let

$$b_p(x) := b\left(\frac{x}{p}\right).$$

Since b_p is a Schwartz function, we can form the product $b_p \cdot f$, and since $b_p \cdot f$ is p-support limited we can form the convolution product $(b_p \cdot f) * g$. We will define

$$f \circledast g := (b_p \cdot f) * g \tag{131}$$

and show that

$$(f \circledast g)(x) = \sum_{k=-\infty}^{\infty} p\, c_k d_k e^{2\pi ikx/p}, \tag{132}$$

cf. (2.18). Here $*$ is the convolution product for generalized functions on $\mathbb{R}$ and $\circledast$ is the new convolution product for generalized functions on $\mathbb{T}_p$. Indeed, we first convolve the p-bandlimited function $b_p^\wedge$ with

$$f^\wedge(s) = \sum_{\ell=-\infty}^{\infty} c_\ell \delta\left(s - \frac{\ell}{p}\right)$$

to obtain the ordinary function

$$(b_p^\wedge * f^\wedge)(s) = \sum_{\ell=-\infty}^{\infty} c_\ell b_p^\wedge\left(s - \frac{\ell}{p}\right) = \sum_{\ell=-\infty}^{\infty} p\, c_\ell B(ps - \ell),$$

and then use (127) to evaluate

$$(b_p \cdot f)^\wedge\left(\frac{k}{p}\right) = (b_p^\wedge * f^\wedge)\left(\frac{k}{p}\right) = p\, c_k, \quad k = 0, \pm1, \pm2, \dots. \tag{133}$$

We multiply

$$g^{\wedge}(s) = \sum_{k=-\infty}^{\infty} d_k \delta\left(s - \frac{k}{p}\right)$$

by the bandlimited function $(b_p \cdot f)^{\wedge}$, use (82) and (133) to write

$$[(b_p \cdot f)^{\wedge} \cdot g^{\wedge}](s) = \sum_{k=-\infty}^{\infty} d_k (b_p \cdot f)^{\wedge}(s)\delta\left(s - \frac{k}{p}\right)$$

$$= \sum_{k=-\infty}^{\infty} d_k (b_p \cdot f)^{\wedge}\left(\frac{k}{p}\right) \delta\left(s - \frac{k}{p}\right) = \sum_{k=-\infty}^{\infty} p\, c_k d_k \delta\left(s - \frac{k}{p}\right),$$

and thereby obtain (132).

You can use (132) to verify that ⊛ has all of the algebraic properties that we would demand of a convolution product on $\mathbb{T}_p$. Indeed, the set of p-periodic generalized functions is closed under ⊛, ⊛ is commutative and associative, ⊛ distributes over addition, and the dilated comb

$$\delta_p(x) := \frac{1}{p}\text{III}\left(\frac{x}{p}\right) = \sum_{\ell=-\infty}^{\infty} \delta\left(x - \ell p\right) = \sum_{k=-\infty}^{\infty} \frac{1}{p}e^{2\pi i k x/p} \tag{134}$$

is the multiplicative identity for ⊛ with

$$\delta_p \circledast f = f \circledast \delta_p = f,$$

when f is p-periodic.

Discrete Fourier transforms

Let f be a function on $\mathbb{P}_N$ and let F be the corresponding discrete Fourier transform. The N-periodic sequence $f[n]$, $n = 0, \pm 1, \pm 2, \ldots$ is slowly growing, so we can define the generalized function

$$g(x) := \sum_{n=-\infty}^{\infty} f[n]\delta(x - n).$$

Since f is N-periodic, we can use (65) and (107) to see that

$$g(x) = \sum_{n=0}^{N-1} f[n] \sum_{m=-\infty}^{\infty} \delta(x - n - mN) = \sum_{n=0}^{N-1} f[n]\frac{1}{N} \sum_{m=-\infty}^{\infty} \delta\left(\frac{x-n}{N} - m\right)$$

$$= \sum_{n=0}^{N-1} f[n]\frac{1}{N}\text{III}\left(\frac{x-n}{N}\right).$$

We Fourier transform this identity and use (107), (65), (82) in turn to write

$$g^{\wedge}(s) = \sum_{n=0}^{N-1} f[n]e^{-2\pi i n s}\, \text{III}(Ns) = \sum_{n=0}^{N-1} f[n]e^{-2\pi i n s} \sum_{k=-\infty}^{\infty} \delta(Ns - k)$$

$$= \sum_{n=0}^{N-1} f[n]e^{-2\pi i n s}\, \frac{1}{N} \sum_{k=-\infty}^{\infty} \delta\left(s - \frac{k}{N}\right)$$

$$= \sum_{k=-\infty}^{\infty} \left(\frac{1}{N} \sum_{n=0}^{N-1} f[n]e^{-2\pi i k n/N}\right) \delta\left(s - \frac{k}{N}\right)$$

$$= \sum_{k=-\infty}^{\infty} F[k]\left(s - \frac{k}{N}\right).$$

Example: Find the Fourier transform of

$$g(x) = \sum_{m=-\infty}^{\infty} f[n]\delta(x - n) \tag{135}$$

when f is the 12-periodic function on $\mathbb{Z}$ with

$$f[n] = \Lambda\left(\frac{n}{3}\right) \quad \text{when} \quad n = 0, \pm 1, \ldots, \pm 6. \tag{136}$$

Solution: We use the table from Appendix 1 to see that the 12-periodic DFT of f has the components

$$F[k] = \frac{1}{4}\frac{\text{sinc}^2(k/4)}{\text{sinc}^2(k/12)} \quad \text{when} \quad k = \pm 1, \ldots, \pm 6. \tag{137}$$

We then use the above analysis to write

$$g^{\wedge}(s) = \sum_{k=-\infty}^{\infty} F[k]\delta\left(s - \frac{k}{12}\right), \tag{138}$$

cf. Fig. 7.22. ∎

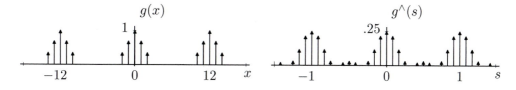

Figure 7.22. The 12-periodic generalized function g from (135)–(136) and the corresponding generalized Fourier transform $g^{\wedge}$ from (137)–(138).

Connections

In Chapter 1 we showed how to construct the Fourier transform of a suitably regular function on $\mathbb{T}_p$, $\mathbb{Z}$, or $\mathbb{P}_N$ from the Fourier transform of some corresponding function on $\mathbb{R}$ by using p-summation or h-sampling, *cf.* Fig. 1.22. The generalized Fourier transform (GFT) that we have studied in this chapter allows us to accomplish the same objective. Indeed, by using (114)–(115) and the above analysis of the DFT we see that:

If f on $\mathbb{T}_p$ has the FT F on $\mathbb{Z}$, then

$$f(x) \text{ has the GFT } \sum_{k=-\infty}^{\infty} F[k]\delta\left(s - \frac{k}{p}\right); \tag{139}$$

If f on $\mathbb{Z}$ has the FT F on $\mathbb{T}_p$, then

$$\sum_{n=-\infty}^{\infty} f[n]\frac{1}{p}\delta\left(x - \frac{n}{p}\right) \text{ has the GFT } F(s); \tag{140}$$

If f on $\mathbb{P}_N$ has the FT F on $\mathbb{P}_N$, then

$$\sum_{n=-\infty}^{\infty} f[n]\delta(x - n) \text{ has the GFT } \sum_{k=-\infty}^{\infty} F[k]\delta\left(s - \frac{k}{N}\right). \tag{141}$$

In principle, you can use these identities to solve the various exercises from Chapter 4 (but in practice you will find that it is usually much easier to use more elementary direct methods!).

7.8 Alternative Definitions for Generalized Functions

Functionals on $\mathbb{S}$

After reading this chapter and working some of the exercises, you may wish to consult other references to enhance your understanding of generalized functions. You will quickly discover that there are three ways to specify the corresponding functionals on $\mathbb{S}$.

(i) The functional f has the representation

$$f\{\phi\} = (-1)^n \int_{-\infty}^{\infty} g(x)\phi^{(n)}(x)dx, \quad \phi \in \mathbb{S}$$

of (20) where g is CSG and n is some nonnegative integer.

(ii) The functional f is given by

$$f\{\phi\} = \lim_{\nu \to \infty} \int_{-\infty}^{\infty} f_\nu(x)\phi(x)dx, \quad \phi \in \mathbb{S}$$

where $f_1, f_2, \ldots$ are Schwartz functions that have been chosen to ensure that the limit exists (for every choice of $\phi \in \mathbb{S}$).

(iii) The functional f is *linear*, i.e.,

$$f\{c\phi\} = c\,f\{\phi\} \qquad\qquad \text{when} \quad c \in \mathbb{C}, \ \phi \in \mathbb{S}$$

$$f\{\phi_1 + \phi_2\} = f\{\phi_1\} + f\{\phi_2\} \quad \text{when} \quad \phi_1, \phi_2 \in \mathbb{S},$$

and *continuous* in the sense that

$$\lim_{\nu \to \infty} f\{\phi_\nu\} = 0$$

whenever $\phi_1, \phi_2, \ldots$ is a sequence from $\mathbb{S}$ with the property

$$\lim_{\nu \to \infty} \ \max_{-\infty < x < \infty} \ |x^n \phi_\nu^{(m)}(x)| = 0$$

for every choice of $n = 0, 1, \ldots$ and $m = 0, 1, \ldots$.

The properties of generalized functions can be developed from (i) as done in this text, from (ii) as done in the well-known monograph of M.J. Lighthill, or from (iii) as done in the treatise of L. Schwartz (and almost all advanced mathematics books that deal with this portion of functional analysis). You can use elementary arguments to show that (i) implies (ii) and that (i) implies (iii), *cf.* Exs. 7.81 and 7.82. More sophisticated concepts (which lie beyond the scope of this text) are needed to show that (ii) implies (i) and that (iii) implies (i). You can find proofs for these *structure theorems* in Jones, pp. 81–90, or in Richards and Youn, pp. 137–140.

Other test functions

L. Schwartz used (iii) with the *test functions*

$\mathbb{E} := \{\phi : \phi, \phi', \phi'', \ldots \text{ are continuous on } \mathbb{R}\}$,

$\mathbb{S} := \{\phi \in \mathbb{E} : x^n \phi^{(m)}(x) \to 0 \text{ as } x \to \pm\infty \text{ for each } m, n = 0, 1, \ldots\}$ and

$\mathbb{D} := \{\phi \in \mathbb{E} : \phi(x) = 0 \text{ for all sufficiently large values of } |x|\}$

to construct corresponding sets of generalized functions (or distributions) on $\mathbb{R}$. When we replace $\mathbb{S}$ by $\mathbb{D}$, $\mathbb{E}$ in (iii) we get a larger, smaller class of functionals because $\mathbb{D} \subset \mathbb{S}$, $\mathbb{E} \supset \mathbb{S}$, respectively. The test functions $\mathbb{E}$, $\mathbb{S}$, $\mathbb{D}$ are closed under differentiation, so we can always differentiate the corresponding generalized functions. (The generalized functions that correspond to $\mathbb{D}$ contain ordinary functions of rapid growth, *e.g.*, e^x, so they are particularly useful for solving differential equations.) Unfortunately, the test functions $\mathbb{E}$, $\mathbb{D}$ are not closed under Fourier transformation, so we cannot Fourier transform the corresponding generalized functions. We must use the test functions $\mathbb{S}$ when we do Fourier analysis!

References

R.N. Bracewell, *The Fourier Transform and Its Applications, 2nd ed.*, McGraw-Hill, New York, 1986.

Chapter 5 gives an informal introduction to the generalized functions most commonly used in electrical engineering. The subsequent chapters show how such generalized functions are used in practice.

L. Hörmander, *The Analysis of Linear Partial Differential Equations I*, Springer-Verlag, New York, 1983.

A tightly written treatise on distribution theory and Fourier analysis.

D.S. Jones, *The Theory of Generalized Functions, 2nd ed.*, Cambridge University Press, Cambridge, 1982.

A comprehensive intermediate-level mathematical monograph that builds on Lighthill's introduction.

W. Kaplan, *Operational Methods for Linear Systems*, Addison-Wesley, Reading, MA, 1962.

Dirac's δ is introduced informally in Chapter 2 and used to solve various ordinary differential equations (with constant coefficients).

M.J. Lighthill, *An Introduction to Fourier Analysis and Generalized Functions*, Cambridge University Press, Cambridge, 1958.

A mathematically correct elementary introduction to generalized functions.

J. Lützen, *The Prehistory of the Theory of Distributions*, Springer-Verlag, New York, 1982.

An account of the origins of distribution theory with selected quotations from the work of Fourier, Heaviside, Dirac, Schwartz, ...

J.I. Richards and H.K. Youn, *Theory of Distributions*, Cambridge University Press, Cambridge, 1990.

A modern nontechnical introduction to the theory of distributions.

L. Schwartz, *Mathematics for the Physical Sciences*, Addison-Wesley, Reading, MA, 1966.

An introduction to distribution theory by its creator!

R. Strichartz, *A Guide to Distribution Theory and Fourier Transforms*, CRC Press, Boca Raton, FL, 1994.

A very readable exposition of the elements of distribution theory with selected applications.

Exercise Set 7

▶ **EXERCISE 7.1:**   Let f_ϵ, y_ϵ be given by (1), (2).

(a) Verify that $y_\epsilon(0-) = y_\epsilon(0+)$, $y_\epsilon(\epsilon-) = y_\epsilon(\epsilon+)$, $y'_\epsilon(0-) = y'_\epsilon(0+)$, $y'_\epsilon(\epsilon-) = y'_\epsilon(\epsilon+)$, and thereby show that y_ϵ, y'_ϵ are continuous functions on $\mathbb{R}$.

(b) Verify that $y_\epsilon(t)$ satisfies the differential equation (3) at each point $t \neq 0, \epsilon$.

(c) What feature of the model allows us to distinguish between the tap of a tack hammer and a blow from a sledgehammer?

▶ **EXERCISE 7.2:**   In this exercise you will sort out the details for the construction (8) that allows us to recover the values of a suitably regular function from the corresponding fundamental functional.

(a) Let f, ϕ be continuous real-valued functions with ϕ being nonnegative for $a \leq x \leq b$. Show that there is some point ξ with $a \leq \xi \leq b$ such that

$$\int_a^b f(x)\phi(x)dx = f(\xi) \cdot \int_a^b \phi(x)dx.$$

Hint. Begin with the inequality

$$m \leq f(x) \leq M \quad \text{for} \ \ a \leq x \leq b$$

where m, M are the minimum, maximum values taken by f, and then make suitable use of the intermediate value theorem.

(b) Let f be continuous for $a \leq x \leq b$ and let ϕ be any nonnegative Schwartz function such that

$$\phi(x) = 0 \ \ \text{for} \ \ |x| \geq 1 \ \ \text{and} \ \ \int_{-1}^1 \phi(x)dx = 1.$$

[You can use the construction of (14)–(15) to produce such a function.] Use (a) to show that

$$\lim_{n\to\infty} \int_a^b f(x)\left[n\,\phi(n(x-x_0))\right]dx = f(x_0), \quad a < x_0 < b.$$

▶ **EXERCISE 7.3:**   Show that the Dirac functional (9) is not the fundamental functional (7) of some piecewise smooth slowly growing function on $\mathbb{R}$.

Hint. Apply (7) to the sequence of dilates $\phi_n(x) = \phi(nx)$, $n = 1, 2, \ldots$, of a Schwartz function ϕ that has been constructed in such a manner that $\phi(x) > 0$ for $-1 < x < 1$ and $\phi(x) = 0$ otherwise.

▶ **EXERCISE 7.4:**   Use the mesa function from Fig. 7.5 to construct a Schwartz function ϕ with the following properties.

(a) Let $-\infty < a < b < c < d < \infty$ and let p be a polynomial. Construct $\phi \in \mathbb{S}$ such that

$$\phi(x) = 0 \qquad\qquad\quad \text{for} \quad x \le a \ \text{ or } \ x \ge d;$$
$$\phi(x) \text{ lies between } p(x) \text{ and } 0 \quad \text{for} \quad a \le x \le b \ \text{ or } \ c \le x \le d; \ \text{ and}$$
$$\phi(x) = p(x) \qquad\qquad\quad \text{for} \quad b \le x \le c.$$

(b) Let $a_0, a_1, \dots, a_n$ be constants and let $\epsilon > 0$. Construct $\phi \in \mathbb{S}$ such that

$$\phi(0) = a_0, \ \phi'(0) = a_1, \ \dots, \ \phi^{(n)}(0) = a_n; \ \text{ and}$$
$$\phi(x) = 0 \ \text{ for } \ |x| \ge \epsilon.$$

▶ **EXERCISE 7.5:**   Show that if $\phi \in \mathbb{S}$, then $\phi^\wedge \in \mathbb{S}$.

Hint. Use the results of Ex. 3.42 and the fact that

$$\mathbf{D}^n\{(-2\pi i x)^m f(x)\} \text{ has the FT } (2\pi i s)^n (\mathbf{D}^m f^\wedge)(s)$$

when f is a suitably regular function on $\mathbb{R}$ and $\mathbf{D}$ is the derivative operator.

▶ **EXERCISE 7.6:**   Let $\phi_1, \phi_2 \in \mathbb{S}$. Show that $\phi_1 * \phi_2 \in \mathbb{S}$.

Hint. You can work with $\phi_1^\wedge \cdot \phi_2^\wedge$ and use Ex. 7.5, or you can use the defining integral (2.1).

▶ **EXERCISE 7.7:**   Find the value of each of the following "integrals."

(a) $\displaystyle\int_{-\infty}^{\infty} \delta(3x)e^{-\pi x^2} dx$ (b) $\displaystyle\int_{-\infty}^{\infty} \delta(x-2)e^{-\pi x^2} dx$ (c) $\displaystyle\int_{-\infty}^{\infty} \delta'(x)e^{-\pi x^2} dx$

(d) $\displaystyle\int_{-\infty}^{\infty} \delta'(2x)e^{-\pi x^2} dx$ (e) $\displaystyle\int_{-\infty}^{\infty} \delta'(x-1)e^{-\pi x^2} dx$ (f) $\displaystyle\int_{-\infty}^{\infty} \delta''(x)e^{-\pi x^2} dx$

(g) $\displaystyle\int_{-\infty}^{\infty} [\cos(\pi x)\delta(x)]e^{-\pi x^2} dx$ (h) $\displaystyle\int_{-\infty}^{\infty} [\sin(\pi x)\delta'(x)]e^{-\pi x^2} dx$ (i) $\displaystyle\int_{-\infty}^{\infty} (\delta'*\delta')(x)e^{-\pi x^2} dx$

▶ **EXERCISE 7.8:**   Find and simplify the functional $f\{\phi\}$, $\phi \in \mathbb{S}$, that is used to represent the generalized function f when:

(a) $f(x) = x^2$;   (b) $f(x) = \delta(3x)$;   (c) $f(x) = \lfloor x \rfloor$;

(d) $f(x) = \delta'(2x)$;   (e) $f(x) = \log|x|$;   (f) $f(x) = \Pi'(x)$;

(g) $f(x) = \sin(e^x)$;   (h) $f(x) = \delta''(x-5)$;   (i) $f(x) = \lfloor x - (1/2) \rfloor'$;

(j) $\displaystyle f(x) := \frac{27}{(x^2 + 5x + 4)^2} = \frac{2}{x+4} + \frac{3}{(x+4)^2} - \frac{2}{x+1} + \frac{3}{(x+1)^2}.$

▶ **EXERCISE 7.9:** Let $g(x) := h(ax + b)$ where a, b are real, $a \neq 0$, and h is the Heaviside function.

(a) Find a simple expression for the functional

$$g^{(n)}\{\phi\} = \int_{-\infty}^{\infty} g^{(n)}(x)\phi(x)dx, \quad \phi \in \mathbb{S}, \quad n = 1, 2, \dots .$$

Hint. Sketch the graph of $h(ax + b)$ when $a > 0$ and when $a < 0$.

(b) Using (a), explain why we write

$$g^{(n)}(x) = \text{sgn}(a)\, \delta^{(n-1)}\left(x + \frac{b}{a}\right), \quad n = 1, 2, \dots .$$

▶ **EXERCISE 7.10:** A generalized function f is even, odd when $f^{\vee} = f$, $f^{\vee} = -f$, respectively.

(a) Show that f is even, odd if and only if $f\{\phi\} = 0$ whenever $\phi \in \mathbb{S}$ is odd, even, respectively.

(b) Show that $\delta^{(n)}$ is even, odd when the nonnegative integer n is even, odd, respectively.

(c) Show that p_n is even, odd when the integer n is even, odd, respectively. The power function p_n is defined by (24) when n is nonnegative and by (46) when n is negative.

▶ **EXERCISE 7.11:** Find and simplify the functional $f\{\phi\}$, $\phi \in \mathbb{S}$, that is used to represent the generalized function f when:

(a) $f(x) = p_1(x) \cdot \delta'(2x)$;   (b) $f(x) = (\delta' * p_{-1})(x)$;   (c) $f(x) = x^2 \cdot \text{Ш}(x)$;

(d) $f^{\wedge}(s) = \delta''(s)$;   (e) $f^{\wedge}(s) = 2\Pi(2s) * \text{Ш}(s)$;   (f) $f^{\wedge}(s) = |s|$.

▶ **EXERCISE 7.12:** Let $\mathcal{P}, \mathcal{Q}$ be polynomials and let $\mathbf{D}$ be the derivative operator. Show that

$$\int_{-\infty}^{\infty} \{\mathcal{P}(x) \cdot [\mathcal{Q}(\mathbf{D})\delta](x)\}\phi(x)dx = \{[\mathcal{P}(-\mathbf{D})\mathcal{Q}](-\mathbf{D})\}\phi(0), \quad \phi \in \mathbb{S}.$$

Hint. Begin with the special case where $\mathcal{P}(x) = x^m$ and $\mathcal{Q}(x) = x^n$.

▶ **EXERCISE 7.13:** Let $\alpha, \alpha', \alpha'', \dots$ be CSG, let f be a generalized function, and let $\delta_{x_0}^{(k)}$ be given by (81).

(a) Derive the identity (82) for the product $\alpha \cdot \delta_{x_0}^{(k)}$.

Hint. Use the Leibnitz rule (2.29) and the sifting relation (64) to simplify

$$\int_{-\infty}^{\infty} [\alpha(x) \cdot \delta^{(k)}(x - x_0)]\phi(x)dx := (-1)^k \int_{-\infty}^{\infty} \delta(x - x_0)[\phi(x) \cdot \alpha(x)]^{(k)}dx, \quad \phi \in \mathbb{S}.$$

(b) Derive the identity (85) for the convolution product $\delta_{x_0}^{(k)} * f$.

▶ **EXERCISE 7.14:**   Let a, b, c, d be real numbers with $a > 0$ and $c > 0$ and let $m, n = 0, 1, 2, \ldots$. Find a simple representation for the convolution product $\beta * f$ when:

(a) $\beta(x) = \delta(ax + b)$, $f(x) = \cos(cx + d)$;   (b) $\beta(x) = e^{-ax^2}$, $f(x) = \cos(cx + d)$;

(c) $\beta(x) = \Pi(x)$, $f(x) = \text{III}^{(n)}(x)$;   (d) $\beta(x) = \delta^{(n)}(x)$, $f(x) = \delta^{(m)}(x)$;

(e) $\beta(x) = \delta(ax + b)$, $f(x) = \text{III}(x)$;   (f) $\beta(x) = \delta'(ax+b)$, $f(x) = \text{sgn}(x)$.

▶ **EXERCISE 7.15:**   Find the following convolution products.

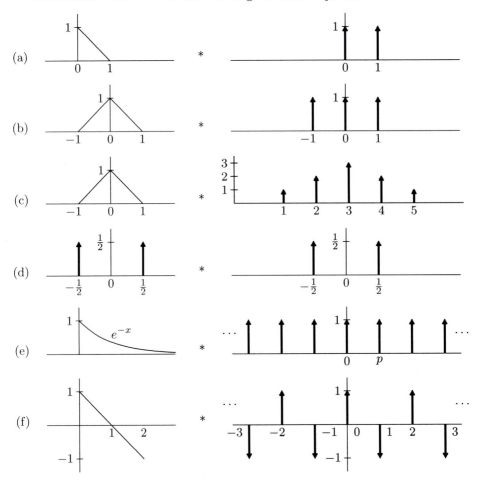

▶ **EXERCISE 7.16:**   In this exercise you will establish the power rule (88) for the generalized power functions of (24) and (46).

(a) Explain why $x^m \cdot p_n(x) = p_{m+n}(x)$ when m, n are nonnegative integers.

(b) Show that $x^m \cdot p_{-n}(x) = p_{m-n}(x)$ when m is a nonnegative integer and $n = 1, 2, \ldots$.

Hint. Use the functionals of (53)–(54) and the identity

$$\left.\frac{[x^m \phi(x)]^{(k)}}{k!}\right|_{x=0} = \begin{cases} 0 & \text{if } k = 0, 1, \ldots, m-1 \\ \dfrac{\phi^{(k-m)}(0)}{(k-m)!} & \text{if } k = m, m+1, \ldots \end{cases}$$

that follows from the Leibnitz rule (2.29).

Note. In view of (a)–(b), you can write $p_{m_1} \cdot (p_{m_2} \cdot p_n) = (p_{m_1} \cdot p_{m_2}) \cdot p_n$ when m_1, m_2 are nonnegative integers and n is any integer.

▶ **EXERCISE 7.17:**   In this exercise you will show that

$$\int_{s=-\infty}^{\infty} \int_{x=-\infty}^{\infty} \frac{\phi(s)\sin(sx)}{x} \, dx \, ds = \int_{x=-\infty}^{\infty} \int_{s=-\infty}^{\infty} \frac{\phi(s)\sin(sx)}{x} \, ds \, dx$$

when ϕ is a Schwartz function. This identity (with x replaced by $2\pi x$) is used to obtain the Fourier transform (69) for p_{-1}.

(a) Explain why it is sufficient to observe that

$$\int_{s=0}^{\infty} \int_{x=0}^{\infty} \frac{\phi(s)\sin(sx)}{x} \, dx \, ds$$

is well defined and then show that each of the integrals

$$\int_{x=0}^{N} \int_{s=N}^{\infty} \left| \frac{\phi(s)\sin(sx)}{x} \right| ds \, dx, \qquad \int_{x=N}^{\infty} \int_{s=0}^{\infty} \frac{\phi(s)\sin(sx)}{x} ds \, dx,$$

$$\int_{s=N}^{\infty} \int_{x=N}^{\infty} \frac{\phi(s)\sin(sx)}{x} dx \, ds, \qquad \int_{s=0}^{N} \int_{x=N}^{\infty} \frac{\phi(s)\sin(sx)}{x} dx \, ds,$$

has the limit 0 as $N \to +\infty$ when $\phi \in \mathbb{S}$.

(b) Verify that

$$\int_{x=0}^{N} \int_{s=N}^{\infty} \left| \frac{\phi(s)\sin(sx)}{x} \right| ds \, dx \leq N \int_{s=N}^{\infty} |s\phi(s)| ds \to 0 \text{ as } N \to \infty.$$

(The same argument works when the limits are reversed.)

(c) Use an integration by parts to verify that

$$\left| \int_{x=N}^{\infty} \frac{1}{x^2} \int_{s=0}^{\infty} \phi(s) \frac{\partial}{\partial s} [\cos(sx)] ds \, dx \right| \to 0 \text{ as } N \to \infty.$$

(d) Verify that

$$\left| \int_{s=N}^{\infty} \phi(s) \int_{x=N}^{\infty} \frac{\sin(sx)}{x} dx \, ds \right| \leq \int_{s=N}^{\infty} |\phi(s)| \left| \int_{u=Ns}^{\infty} \frac{\sin u}{u} du \right| ds \to 0 \text{ as } N \to \infty.$$

Hint. The areas associated with $(\sin u)/u$ form an alternating series.

(e) Use an integration by parts (with Ex. 1.38) to verify that

$$\left| \int_{s=0}^{N} \phi(s) \int_{x=N}^{\infty} \frac{\sin(sx)}{x} dx\, ds \right| = \left| \int_{s=0}^{N} \Phi'(s) g(Ns) ds \right|$$

$$\leq |\Phi(N) g(N^2)| + \left| \int_{0}^{N} \frac{\Phi(s)}{s} \sin(Ns) ds \right| \to 0 \quad \text{as} \quad N \to \infty$$

where $\Phi(s) := \int_{0}^{s} \phi(u) du$, $g(s) := \int_{x=s}^{\infty} \frac{\sin u}{u} du$.

▶ **EXERCISE 7.18:**   Let g be a CSG function and let $\phi \in \mathbb{S}$. Show that

$$g(x) := \int_{-\infty}^{\infty} g(u) \phi(x - u) du$$

is well defined, continuous, and slowly growing.

Hint. Choose $m = 0, 1, 2, \ldots$ to make $g(u)[1+u^2]^{-m}$ absolutely integrable, and then write

$$g(u)\phi(x - u) = \{g(u)[1 + u^2]^{-m}\}[1 + (x - u - x)^2]^m \phi(x - u)$$

$$= \sum_{k,\ell} c_{k\ell} x^\ell \cdot \{g(u)[1 + u^2]^{-m}\} \cdot \{(x - u)^k \phi(x - u)\}$$

for suitably chosen constants $c_{k\ell}$.

▶ **EXERCISE 7.19:**   Let $f, \beta, \beta_1, \beta_2, \ldots$ be generalized functions and assume that $\beta^\wedge, \beta_1^\wedge, \beta_2^\wedge, \ldots$ and all of their derivatives are CSG.

(a) Explain why you can freely associate and commute the factors from the convolution product of $f, \beta_1, \beta_2, \ldots, \beta_n$, e.g., $(\beta_1 * \beta_2) * f = \beta_2 * (\beta_1 * f)$.

(b) Use (a) with the identity $(\beta * f)' = \delta' * (\beta * f)$ from (85) to derive the differentiation rule $(\beta * f)' = \beta' * f = \beta * f'$.

(c) Let $-\infty < x_0 < \infty$ and let $f_{x_0}(x) := f(x + x_0)$. Use (a) with the identity $(\beta * f)_{x_0} = \delta_{x_0} * (\beta * f)$ from (85) to derive the translation rule $(\beta * f)_{x_0} = \beta_{x_0} * f = \beta * f_{x_0}$.

Note. You can use this analysis to solve Ex. 2.21.

▶ **EXERCISE 7.20:**   Let f be a generalized function and assume that

$$f = g_1^{(m_1)} \quad \text{and} \quad f = g_2^{(m_2)}$$

where g_1, g_2 are CSG and m_1, m_2 are nonnegative integers, e.g.,

$$\delta(x) = [xh(x)]'' \quad \text{and} \quad \delta(x) = \frac{1}{2}[x^2 h(x) - x]'''.$$

(a) Use (60) to show that $f = g_1^{(m_1)}$ and $f = g_2^{(m_2)}$ produce the same functional for the translate $f(x - x_0)$, $-\infty < x_0 < \infty$.

(b) Use (63) to show that $f = g_1^{(m_1)}$ and $f = g_2^{(m_2)}$ produce the same functional for the Fourier transform $f^\wedge(s)$.

Note. Similar arguments are used for the dilate, derivative, α-product, and β-convolution product.

▶ **EXERCISE 7.21:**   In this exercise you will show that the linear space

$$\mathbb{G} = \{g^{(m)} : g \text{ is a CSG function on } \mathbb{R}, \quad m = 0, 1, \dots\}$$

of generalized functions is closed under the operations of translation, dilation, differentiation, α-multiplication, and β-convolution as defined by (60)–(62) and (75)–(76). Let $f = \gamma^{(\mu)}$ where γ is CSG and μ is a nonnegative integer. Show that it is possible to express each of the following in the form $g^{(m)}$ where g is a CSG function and m is a nonnegative integer (which depend on γ and μ).

(a) $f(x + x_0)$ where $-\infty < x_0 < \infty$ (b) $f(ax)$ where $a < 0$ or $a > 0$ (c) $f'(x)$

(d) $\alpha(x) \cdot f(x)$ where $\alpha, \alpha', \alpha'', \dots$ are CSG

Hint. Use the Leibnitz rule (2.29) and (59) to write

$$\alpha \cdot \gamma^{(\mu)} = (\alpha \cdot \gamma)^{(\mu)} - \binom{\mu}{1}(\alpha' \cdot \gamma)^{(\mu-1)} + \binom{\mu}{2}(\alpha'' \cdot \gamma)^{(\mu-2)} - \dots = g^{(\mu)}.$$

(e) $(\beta * f)(x)$ where $\beta^{\wedge}, \beta^{\wedge\prime}, \beta^{\wedge\prime\prime}, \dots$ are CSG.

Hint. Use (d), the convolution rule, and the closure of $\mathbb{G}$ under Fourier transformation (which was established in the text).

▶ **EXERCISE 7.22:**   Let $x_1 < x_2 < \dots < x_m$, let f be a generalized function, and assume that $f(x) = 0$ for $x < x_1$, $x_1 < x < x_2$, $\dots$, $x_{m-1} < x < x_m$, and $x_m < x$. Show that there is some nonnegative integer n and constants $c_{\mu k}$ such that

$$f(x) = \sum_{\mu=0}^{m} \sum_{\nu=0}^{n} c_{\mu\nu} \delta^{(\nu)}(x - x_\mu).$$

Hint. Modify the analysis that leads to (91).

▶ **EXERCISE 7.23:**   Let f be a generalized function, let $-\infty < x_0 < \infty$, let $n = 0, 1, \dots$, and assume that $(x - x_0)^n \cdot f(x) = 0$. Show that f has the representation (90).

Hint. Set $g(x) := f(x+x_0)$, observe that $(g^{\wedge})^{(n)} = 0$, and use the analysis from Section 7.4.

▶ **EXERCISE 7.24:**   Let the polynomial $\mathcal{P}$ have the real roots $\alpha_1 < \dots < \alpha_r$ with multiplicities $n_1, \dots, n_r$, respectively, and assume that the other roots of $\mathcal{P}$, if any, have nonzero imaginary parts. Let f be a generalized function and assume that

$$\mathcal{P}(x) \cdot f(x) = 0.$$

Show that f has the representation

$$f(x) = \sum_{\mu=1}^{r} \sum_{\nu=0}^{n_\mu-1} c_{\mu\nu} \delta^{(\nu)}(x - \alpha_\mu)$$

for some choice of the constants $c_{\mu\nu}$. You can organize the demonstration as follows.

(a) Show that

$$\{(x - \alpha_1)^{n_1} (x - \alpha_2)^{n_2} \cdots (x - \alpha_r)^{n_r}\} \cdot f(x) = 0.$$

Hint. You can divide by a polynomial that has no real roots.

(b) Let α, β be distinct real numbers, let $n = 1, 2, \ldots$, and let $\nu = 0, 1, \ldots, n - 1$. Show that the inhomogeneous equation

$$(x - \alpha)^n \cdot g(x) = \delta^{(\nu)}(x - \beta)$$

has a generalized solution g of the form

$$g(x) = c_0 \delta(x - \beta) + c_1 \delta'(x - \beta) + \cdots + c_\nu \delta^{(\nu)}(x - \beta)$$

where $c_0, c_1, \ldots, c_\nu$ are constants.

(c) Use the result of Ex. 7.23 with (b) to determine in turn the structure of the generalized solutions $g_1(x), g_2(x), \ldots, g_{r-1}(x), f(x)$ of $(x - \alpha_1)^{n_1} \cdot g_1(x) = 0$, $(x - \alpha_2)^{n_2} \cdot g_2(x) = g_1(x)$, $\ldots$, $(x - \alpha_{r-1})^{n_{r-1}} \cdot g_{r-1}(x) = g_{r-2}(x)$, $(x - \alpha_r)^{n_r} \cdot f(x) = g_{r-1}(x)$.

▶ **EXERCISE 7.25:**   Let the polynomial $\mathcal{P}$ have the distinct pure imaginary roots $2\pi i \beta_1, \ldots, 2\pi i \beta_r$ with multiplicities $n_1, \ldots, n_r$, respectively, and assume that the other roots of $\mathcal{P}$, if any, have nonzero real parts. Let f be a generalized function and assume that

$$\mathcal{P}(\mathbf{D}) f(x) = 0,$$

where $\mathbf{D}$ is the derivative operator. Show that f has the representation

$$f(x) = \sum_{\mu=1}^{r} \sum_{\nu=0}^{n_\mu - 1} c_{\mu\nu} x^\nu e^{2\pi i \beta_\mu x}$$

for some choice of the constants $c_{\mu\nu}$. Thus every generalized solution of the differential equation is a slowly growing ordinary solution.

Hint. Use Ex. 7.24.

▶ **EXERCISE 7.26:**   Find all generalized functions f that satisfy each of the following homogeneous equations.

(a) $(x^2 - 1) \cdot f(x) = 0$ (b) $(x^4 - 1) \cdot f(x) = 0$ (c) $(x^4 - 1)^2 \cdot f(x) = 0$

(d) $(\mathbf{D}^2 - 1) f(x) = 0$ (e) $(\mathbf{D}^4 - 1) f(x) = 0$ (f) $(\mathbf{D}^4 - 1)^2 f(x) = 0$

Hint. Use the representations from Exs. 7.24 and 7.25.

▶ **EXERCISE 7.27:**   Find all generalized solutions for each of the following inhomogeneous equations.

(a) $(x - 1) \cdot f(x) = \delta(x)$ (b) $(x + 1) \cdot f(x) = \delta(x)$ (c) $(x^2 - 1) \cdot f(x) = \delta(x)$

(d) $(x^2 + 1) \cdot f(x) = \delta(x)$ (e) $(x^4 - 1) \cdot f(x) = \delta(x)$ (f) $(x^4 + 1) \cdot f(x) = \delta(x)$

(g) $(\mathbf{D} - 1)f(x) = \delta(x)$ (h) $(\mathbf{D} + 1)f(x) = \delta(x)$ (i) $(\mathbf{D}^2 - 1)f(x) = \delta(x)$

(j) $(\mathbf{D}^2 + 1)f(x) = \delta(x)$ (k) $(\mathbf{D}^4 - 1)f(x) = \delta(x)$ (l) $(\mathbf{D}^4 + 1)f(x) = \delta(x)$

Hint. Construct a particular solution to combine with solutions of the corresponding homogeneous equation, and *cf.* Exs. 7.24 and 7.25.

▶ **EXERCISE 7.28:**   Let β be a σ-support limited generalized function, let $g(x) := |x|$, and let $f := \beta * g$. Show that $f'' = 2\beta$.

▶ **EXERCISE 7.29:**   Let f be a generalized function, let g be an ordinary CSG function, and assume that $x \cdot f(x) = g(x)$ for $-\infty < x < \infty$.

(a) Show that

$$f\{\phi\} = \int_{-\infty}^{\infty} \frac{g(x)}{x} \cdot \phi(x)dx$$

when $\phi \in \mathbb{S}$ and $\phi(0) = 0$.

Hint. Use (16) to see that there is a Schwartz function ψ such that $x\psi(x) = \phi(x)$.

(b) Show that

$$f(x) = \frac{g(x)}{x} \quad \text{for} \quad -\infty < x < 0 \quad \text{and for} \quad 0 < x < \infty.$$

(c) Give an example to show that f is not determined by the ratios from (b).

▶ **EXERCISE 7.30:**   Let $\mathcal{P}(\mathbf{D})$ be a polynomial in the derivative operator $\mathbf{D}$, let f be a generalized function, let $-\infty \leq a < b \leq \infty$, and assume that

$$[\mathcal{P}(\mathbf{D})f](x) = 0 \quad \text{for} \quad a < x < b.$$

Show that $f(x) = y(x)$ for $a < x < b$ where y is some ordinary solution of

$$[\mathcal{P}(\mathbf{D})y](x) = 0 \quad \text{for} \quad -\infty < x < \infty$$

(which is slowly growing if $a = -\infty$ or $b = +\infty$).

Hint. Begin with a representation $f = \mathbf{D}^m g$ where $g, g', \dots, g^{(n)}$ are CSG and $n := \deg \mathcal{P}$. The function g must then be a solution of the homogeneous equation $\mathbf{D}^m[\mathcal{P}(\mathbf{D})g] = 0$.

▶ **EXERCISE 7.31:**   Let f, β be generalized functions and assume that $\beta^{\wedge}, \beta^{\wedge\prime}, \beta^{\wedge\prime\prime}, \dots$ are CSG. Establish the convolution rule $(\beta * f)^{\wedge} = \beta^{\wedge} \cdot f^{\wedge}$.

▶ **EXERCISE 7.32:**   Let f be a generalized function and let

$$f_1 := \Pi * f, \quad f_2 := \Pi * \Pi * f, \quad f_3 := \Pi * \Pi * \Pi * f, \quad \ldots .$$

Show that f_n is an ordinary CSG function when n is sufficiently large.

Hint. Begin with $f = g^{(m)}$ where g is CSG and m is a nonnegative integer.

▶ **EXERCISE 7.33:**   Find the Fourier transform of each of the following generalized functions. In so doing, freely use the rules from the Fourier transform calculus together with (93)–(96).

(a) $f(x) = \delta(x-1) + \delta(x+1)$ (b) $f(x) = e^{2\pi i x} \cdot \delta(x)$ (c) $f(x) = \delta'(5x)$

(d) $f(x) = 3x^2 + 2x + 1$ (e) $f(x) = \sin(4\pi x)$ (f) $f(x) = \cos^2(\pi x)$

(g) $f(x) = \delta'(x) * x$ (h) $f(x) = x \cdot \delta'(x)$ (i) $f(x) = 2\pi x \cdot \cos(2\pi x)$

(j) $f(x) = \delta'(x+1) * \delta''(x-1)$ (k) $f(x) = e^{-\pi x^2} \cdot 2\pi x$ (l) $f(x) = e^{-\pi x^2} * 2\pi x$

(m) $f(x) = \delta(2x+1) * \delta(4x+2)$ (n) $f(x) = \delta''(x) - 4\pi^2 x^2$ (o) $f(x) = \cos(x) \cdot \delta''(x)$

▶ **EXERCISE 7.34:**   Verify each of the following identities by showing that the corresponding Fourier transforms are equal. In so doing, freely use the rules from the Fourier transform calculus together with (93)–(96). The parameters a, b are real and $m, n = 0, 1, 2, \ldots$.

(a) $\delta(ax) = |a|^{-1}\delta(x), \ a \neq 0$ (b) $\delta(ax+b) = |a|^{-1}\delta(x+(b/a)), \ a \neq 0$

(c) $x\,\delta(x) = 0$ (d) $x\,\delta'(x) = -\delta(x)$

(e) $[\delta(ax)]' = a\delta'(ax), \ a \neq 0$ (f) $[\delta(ax)]'' = a^2\delta''(ax)$

(g) $\delta(x-a) * \delta(x-b) = \delta(x-a-b)$ (h) $\delta^{(m)}(x) * \delta^{(n)}(x) = \delta^{(m+n)}(x)$

(i) $x^n \delta^{(m)}(x) = \begin{cases} 0 & \text{if } n > m \\ (-1)^n [m!/(m-n)!]\delta^{(m-n)}(x) & \text{if } n \leq m \end{cases}$

▶ **EXERCISE 7.35:**   Derive the addition formula for the cosine as follows.

(a) Find the Fourier transform of $\cos(px)$, $-\infty < p < \infty$.

(b) Using (a) and the product rule, find the Fourier transform of

$$f(x) = \cos(\alpha x)\cos(\beta x), \quad -\infty < \alpha < \infty, \quad -\infty < \beta < \infty.$$

(c) Using (a) with $p = \alpha \pm \beta$, find the Fourier transform of

$$g(x) = \tfrac{1}{2}\{\cos(\alpha + \beta)x + \cos(\alpha - \beta)x]\}.$$

(d) Using (b), (c) derive the familiar trigonometric identity

$$\cos\alpha\,\cos\beta = \tfrac{1}{2}\{\cos(\alpha + \beta) + \cos(\alpha - \beta)\}.$$

▶ **EXERCISE 7.36:** Find the Fourier transform for each of the following generalized functions. In so doing, freely use the rules from the Fourier transform calculus together with the identities $p^\wedge_{-1}(s) = -\pi i \, \text{sgn}(s)$ and $\text{sgn}^\wedge(s) = (\pi i)^{-1} p_{-1}(s)$ from the text.

(a) $f(x) = \text{sgn}(3x)$

(b) $f(x) = e^{4\pi i x} \cdot \text{sgn}(x)$

(c) $f(x) = e^{2\pi i x} \cdot h(x - 1)$

(d) $f(x) = \sin(x) \cdot \text{sgn}(x)$

(e) $f(x) = 1/(x - 1)$

(f) $f(x) = 1/(x - 1)^2$

(g) $f(x) = 1/(x^2 - 4)$

(h) $f(x) = \delta'(x) * \log|x|$

(i) $f(x) = x \cdot (1/x)$

(j) $f(x) = \sin(\pi x) \cdot (1/\pi x)$

(k) $f(x) = \{\Pi(x + 1) - \Pi(x - 1)\} * \text{sgn}(x)$

(l) $f(x) = \cos(\pi x) \cdot (1/\pi i x)$

▶ **EXERCISE 7.37:** Find the δ identity in the Fourier transform domain that corresponds to each of the following.

(a) $1 \cdot f(x) = f(x)$

(b) $e^{2\pi i a x} \cdot e^{2\pi i b x} = e^{2\pi i (a+b)x}, \quad -\infty < a < \infty, \quad -\infty < b < \infty$

(c) $(x^n)' = n\, x^{n-1}, \quad n = 1, 2, \ldots$

(d) $x^n \cdot x^m = x^{n+m}, \quad n, m = 0, 1, \ldots$

(e) $(1 - 2\pi i x)^2 = 1 + 2 \cdot (-2\pi i x) + (-2\pi i x)^2$

(f) $x \cdot (1/x) = 1$

Note. cf. Ex. 7.16 for the law of exponents as used in (d), (e), (f).

▶ **EXERCISE 7.38:** Find the Fourier transform of each of the following generalized functions assuming that $a > 0$, $-\infty < b < \infty$, and $n = 1, 2, \ldots$.

(a) $f(x) = \delta^{(n)}(ax + b)$

(b) $f(x) = x^n \cdot \delta^{(n)}(ax+b)$

(c) $f(x) = \cos^n(ax + b)$

(d) $f(x) = (x+b)^n \cdot (x+b)^{-n}$

(e) $f(x) = (x + a)^{-n}$

(f) $f(x) = (x + ia)^{-n}$

▶ **EXERCISE 7.39:** Let $x_1 < x_2 < \cdots < x_m$ and let

$$f(x) := \begin{cases} f_0(x) & \text{for } x < x_1 \\ f_\mu(x) & \text{for } x_\mu < x < x_{\mu+1}, \quad \mu = 1, 2, \ldots, m-1 \\ f_m(x) & \text{for } x_m < x \end{cases}$$

where $f_0, f_1, \ldots, f_m$ are all polynomials of degree n or less. In this exercise you will show that f has the Fourier transform

$$F(s) = \frac{1}{2}\left\{ f_0\left(\frac{-\mathbf{D}}{2\pi i}\right) + f_m\left(\frac{-\mathbf{D}}{2\pi i}\right) \right\} \delta(s) + \sum_{\mu=1}^{m} \sum_{\nu=0}^{n} \left[f^{(\nu)}(x_\mu+) - f^{(\nu)}(x_\mu-) \right] \frac{e^{-2\pi i s x_\mu}}{(2\pi i s)^{\nu+1}}$$

where $\mathbf{D}$ is the derivative operator

(a) Verify that

$$f^{(n+1)}(x) = \sum_{\mu=1}^{m} \sum_{\nu=0}^{n} [f^{(\nu)}(x_\mu+) - f^{(\nu)}(x_\mu-)]\delta^{(n-\nu)}(x - x_\mu).$$

(b) Show that there are constants $c_0, c_1, \ldots, c_n$ such that

$$F(s) = c_0\delta(s)+c_1\delta'(s)+\cdots+c_n\delta^{(n)}(s)+\sum_{\mu=1}^{m}\sum_{\nu=0}^{n}\left[f^{(\nu)}(x_\mu+)-f^{(\nu)}(x_\mu-)\right]\frac{e^{-2\pi i x_\mu s}}{(2\pi i s)^{\nu+1}}.$$

(c) Explain why the Fourier transform of

$$g(x) := f(x) - \tfrac{1}{2}[f_m(x) + f_o(x)] - \tfrac{1}{2}[f_m(x) - f_o(x)]\mathrm{sgn}(x)$$

is a smooth ordinary function on $\mathbb{R}$.

(d) Show that the Fourier transform of $[f_m(x) - f_o(x)]\mathrm{sgn}(x)$ is a linear combination of the inverse power functions $p_{-1}, p_{-2}, \ldots, p_{-n-1}$.

(e) Show that the $\delta, \delta', \ldots$ terms in the Fourier transform F can be obtained from the formula

$$\frac{1}{2}\{F_m(s) + F_o(s)\} = \frac{1}{2}\left\{ f_o\left(\frac{-\mathbf{D}}{2\pi i}\right) + f_m\left(\frac{-\mathbf{D}}{2\pi i}\right)\right\}\delta(s).$$

Note. The Fourier transform F has no $\delta, \delta', \ldots, \delta^{(n)}$ terms when $f_m = -f_o$ and no $p_{-1}, p_{-2}, \ldots, p_{-n-1}$ terms when $f_m = f_o$.

▶ **EXERCISE 7.40:**   Find the Fourier transform of each of the following functions. You may wish to use the analysis of Ex. 7.39 to expedite the process.

(a)
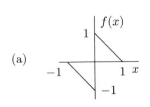

(b)
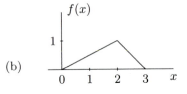

(c)
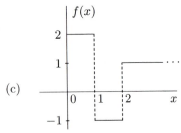

(d)
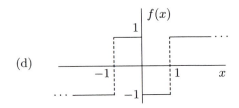

(e) $f(x) = |x|$

(f) $f(x) = x \cdot |x|$

(g) $f(x) = \max\{1 - x^2, 0\}$

(h) $f(x) = \max\{1, x^2\}$

(i) $f(x) = \begin{cases} x^2 & \text{if } 0 < x < 1 \\ 0 & \text{otherwise} \end{cases}$

(j) $f(x) = \begin{cases} x^2 & \text{if } x \geq 0 \\ 0 & \text{if } x \leq 0 \end{cases}$

▶ **EXERCISE 7.41:**   You can find the Fourier transform of any rational function by using (3.22), (95), and (97) with a partial fraction decomposition. Use this method to find the Fourier transform of each of the following rational functions.

Note. You cannot obtain the Fourier transform of $(x - \alpha)^{-n}$ by letting $\beta \to 0+$ in the Fourier transform for $(x - \alpha - i\beta)^{-n}$, and you cannot obtain the Fourier transform for $(x - \alpha - i\beta)^{-n}$, by simply making the substitution $\gamma = \alpha + i\beta$ in the Fourier transform for $(x - \gamma)^{-n}$. You can use the conjugation rule to obtain the Fourier transform of $(x - \alpha + i\beta)^{-n}$ from that of $(x - \alpha - i\beta)^{-n}$.

(a) $f(x) = \dfrac{1}{1 + x^2} = \dfrac{1}{2i} \left\{ \dfrac{1}{x - i} - \dfrac{1}{x + i} \right\}$

(b) $f(x) = \dfrac{4x^5}{x^4 - 1} = 4x + \dfrac{1}{x + 1} + \dfrac{1}{x - 1} - \dfrac{1}{x + i} - \dfrac{1}{x - i}$

(c) $f(x) = \dfrac{27}{(x^2 + 5x + 4)^2} = \dfrac{2}{x + 4} + \dfrac{3}{(x + 4)^2} - \dfrac{2}{x + 1} + \dfrac{3}{(x + 1)^2}$

(d) $f(x) = \dfrac{32}{(x^2 + 2x + 5)^2} = \dfrac{i}{x + 1 + 2i} - \dfrac{2}{(x + 1 + 2i)^2} - \dfrac{i}{x + 1 - 2i} - \dfrac{2}{(x + 1 - 2i)^2}$

▶ **EXERCISE 7.42:**   Suppose that we want to find the Fourier transform of the rational function $f(x) := \mathcal{P}(x)/\mathcal{Q}(x)$ where $\mathcal{P}, \mathcal{Q}$ are polynomials. We write

$$\mathcal{Q}(x) \cdot f(x) = \mathcal{P}(x)$$

and use the power scaling rule to obtain the differential equation

$$\mathcal{Q}\left(\dfrac{\mathbf{D}}{-2\pi i}\right) F(s) = \mathcal{P}\left(\dfrac{\mathbf{D}}{-2\pi i}\right) \delta(s)$$

where $\mathbf{D}$ is the derivative operator. In this way we see that

$$F(s) = h(s)F_+(s) + h(-s)F_-(s) + c_0\delta(s) + c_1\delta'(s) + \cdots$$

where F_+, F_- are solutions of the homogeneous equation

$$\mathcal{Q}\left(\dfrac{\mathbf{D}}{-2\pi i}\right) F(s) = 0$$

which are slowly growing on $(0, +\infty)$, $(-\infty, 0)$, respectively, and where $c_0, c_1, \ldots$ are suitably chosen constants. Use this representation to find the Fourier transform of each of the following rational functions.

(a) $f(x) = \dfrac{1}{1 - 2\pi i x}$ (b) $f(x) = \dfrac{1 - 2\pi i x}{1 + 2\pi i x}$ (c) $f(x) = \dfrac{1}{(2\pi i x)^2}$

(d) $f(x) = \dfrac{1}{1 + (2\pi x)^2}$ (e) $f(x) = \dfrac{1}{1 - (2\pi x)^2}$ (f) $f(x) = \dfrac{(2\pi x)^2}{1 - (2\pi x)^2}$

Hint. The functions (c)–(f) are all even.

▶ **EXERCISE 7.43:**   Find the simplified form of the Parseval identity

$$\int_{-\infty}^{\infty} f^{\wedge}(x)\phi(x)dx = \int_{-\infty}^{\infty} f(s)\phi^{\wedge}(s)ds$$

which results when ϕ is an arbitrary Schwartz function and f is specified as follows. Assume that $-\infty < x_0 < \infty$, $-\infty < s_0 < \infty$, $p > 0$, and $n = 0, 1, 2, \ldots$.

(a) $f(s) = e^{2\pi i x_0 s}$

(b) $f(s) = \delta(s - s_0)$

(c) $f(s) = (2\pi i)^{-n}\delta^{(n)}(s)$

(d) $f(s) = (-2\pi i s)^n$

(e) $f(s) = e^{2\pi i x_0 s}III(ps)$

(f) $f(s) = -\dfrac{1}{\pi i}p_{-1}(s)$

▶ **EXERCISE 7.44:**   Find the Fourier transform of each of the following generalized functions. In so doing, freely use the rules from the Fourier transform calculus together with (98).

(a) $f(x) = e^{ix^2}$

(b) $f(x) = e^{i\pi(x^2 - 2x)}$

(c) $f(x) = 2\pi i x\, e^{i\pi x^2}$

(d) $f(x) = \cos(\pi x^2)$

(e) $f(x) = \cos(\pi x)\cos(\pi x^2)$

(f) $f(x) = \cos^2(\pi x^2)$

▶ **EXERCISE 7.45:**   In this exercise you will use the Parseval identity

$$\int_{-\infty}^{\infty} e^{i\pi x^2} e^{-\pi x^2} dx = \int_{-\infty}^{\infty} c\, e^{-i\pi s^2} e^{-\pi s^2} ds$$

to determine the constant c for the Fourier transform (98).

(a) Show that $c = (a + ib)/(a - ib)$ where

$$a + ib := \int_{-\infty}^{\infty} e^{-\pi(1-i)x^2} dx.$$

(b) Use polar coordinates to show that

$$(a + ib)^2 = \int_{-\infty}^{\infty}\int_{-\infty}^{\infty} e^{-\pi(1-i)(x^2+y^2)} dx\, dy = \frac{1+i}{2}.$$

(c) Use (a), (b) to obtain $c = (1+i)/\sqrt{2}$.

▶ **EXERCISE 7.46:**   The *diffraction kernel* (with wavelength $\lambda = 1$) is defined by

$$\gamma_d(x) := \begin{cases} \dfrac{1-i}{\sqrt{2d}}e^{i\pi x^2/d} & \text{if } d > 0 \\[2mm] \delta(x) & \text{if } d = 0 \\[2mm] \dfrac{1+i}{\sqrt{-2d}}e^{i\pi x^2/d} & \text{if } d < 0. \end{cases}$$

(a) Use (98) to find the Fourier transform Γ_d of γ_d.

(b) Explain why the convolution product of γ_d and a generalized function f is well defined.

(c) Show that $\gamma_{d_1} * \gamma_{d_2} = \gamma_{d_1+d_2}$.

(d) Find a simple expression for the inverse of the Fresnel operator $\mathcal{A}_d f := \gamma_d * f$.

▶ **EXERCISE 7.47:**   In this exercise you will show that the floor function

$$f(x) := \lfloor x \rfloor := n \quad \text{for} \quad n \le x < n+1, \quad n = 0, \pm 1, \pm 2, \ldots$$

has the Fourier transform

$$F(s) = \sum_{k=1}^{\infty} \frac{\delta(s-k) - \delta(s+k)}{2\pi i k} - \frac{\delta(s)}{2} - \frac{\delta'(s)}{2\pi i}.$$

(a) Verify that $f(x) - x + \frac{1}{2}$ is the derivative of the weakly convergent Fourier series (108) and thereby obtain the above expression for F.

(b) Observe that $g(x) := f(x) + \frac{1}{2}$ is an odd function with the derivative $g'(x) = f'(x) = \text{III}(x)$. Find the odd solution of $2\pi i s \cdot G(s) = \text{III}(s)$ and thereby obtain the above expression for F.

▶ **EXERCISE 7.48:**   The function $f(x) := |x|^{-1/2}$ is locally integrable and slowly growing, so we can use the fundamental functional

$$f\{\phi\} := \int_{-\infty}^{\infty} |x|^{-1/2} \phi(x) dx, \quad \phi \in \mathbb{S}$$

to define a generalized function.

(a) Show that the generalized Fourier transform of f is represented by the fundamental functional of the ordinary function

$$F(s) = 2 \int_0^{\infty} \frac{\cos(2\pi|s|x)}{\sqrt{x}} dx, \quad s \ne 0.$$

(b) Use the transformation $u^2 = 2|s|x$ to show that $F(s) = |s|^{-1/2}$.

   Hint. Use (98) with $s = 0$ to obtain the integral of $\cos(\pi u^2)$.

▶ **EXERCISE 7.49:**   In this exercise you will use the even CSG function

$$g(x) := \int_0^x \text{sgn}(u) \log|u| du = |x| \log|x| - |x|$$

to construct generalized functions

$$\frac{1}{|x|}, \quad \frac{1}{x|x|}, \quad \frac{1}{x^2|x|}, \quad \cdots .$$

(a) Explain why we might identify the generalized function

$$f_n(x) := (-1)^n \frac{g^{(n+2)}(x)}{n!}$$

with the ordinary function $\text{sgn}(x)/x^{n+1}$, $n = 0, 1, \ldots$.

(b) Show that f_0 is represented by the functional

$$f_0\{\phi\} = \int_{|x|<1} \frac{\phi(x) - \phi(0)}{|x|} \, dx + \int_{|x|>1} \frac{\phi(x)}{|x|} \, dx, \quad \phi \in \mathbb{S}.$$

(c) Use the functional of (b) to show that $x \cdot f_0(x) = \text{sgn}(x)$.

(d) Find all even generalized functions f that satisfy $x \cdot f(x) = \text{sgn}(x)$.

Note. The function p_{-1} is the unique odd solution of $x \cdot f(x) = 1$, so there is only one sensible choice for "$1/x$". Unfortunately, "$1/|x|$" is not unique.

(e) Show that f_0 does *not* satisfy the dilation relation

$$|a| \cdot f_0(ax) = f_0(x), \quad a < 0 \text{ or } a > 0$$

that corresponds to the ordinary algebraic identity $|a|/|ax| = 1/|x|$.

(f) Use (c) to show that

$$f_0(x) \text{ has the FT } F_0(s) = -2 \log |s| + C$$

where C is a constant that can be determined from the Parseval identity

$$f_0\{\phi\} = F_0\{\phi\}, \quad \phi(x) := e^{-\pi x^2}.$$

Note. It can be shown that $C = -2\gamma - 2 \log(2\pi) = -4.83018\ldots$ where $\gamma := .57721\ldots$ is Euler's constant, *cf.* Richards and Youn, pp. 73–74.

(g) Use (f) to show that

$$f(x) := \log |x| \text{ has the FT } F(s) = \tfrac{1}{2}\{C\delta(s) - f_0(s)\}.$$

(h) Use (f) to show that

$$f_n(x) \quad \text{has the FT} \quad F_n(s) = \frac{(-2\pi i s)^n}{n!} [-2 \log |s| + C], \quad n = 0, 1, 2, \ldots .$$

▶ **EXERCISE 7.50:**   Find and correct the flaw(s) in the following arguments.

(a) "We know that

$$\delta(x) = 0 \quad \text{for} \quad x < 0 \quad \text{and for} \quad x > 0$$
$$\delta(2x) = 0 \quad \text{for} \quad x < 0 \quad \text{and for} \quad x > 0.$$

Since $2 \cdot 0 = 0$ we must have

$$\delta(x) = \delta(2x) \quad \text{for} \quad -\infty < x < \infty."$$

(b) "Let $f(x) := e^{-x} h(x)$ where h is the Heaviside function. Since

$$f'(x) = -e^{-x} h(x) = -f(x)$$

we can use the derivative rule to write

$$2\pi i s\, F(s) = -F(s).$$

From the equivalent identity

$$(2\pi i s + 1)F(s) = 0, \quad -\infty < s < \infty$$

we see that f has the Fourier transform $F(s) \equiv 0$."

(c) "Let $x \cdot f(x) = g(x)$ where g is CSG. We have

$$f\{\phi_R\} = \int_{-\infty}^{\infty} \frac{g(x)}{x} \phi_R(x) \quad \text{if } \phi_R \in \mathbb{S} \text{ and } \quad \phi_R(x) = 0 \text{ when } x < 0$$

$$f\{\phi_L\} = \int_{-\infty}^{\infty} \frac{g(x)}{x} \phi_L(x) \quad \text{if } \phi_L \in \mathbb{S} \text{ and } \quad \phi_L(x) = 0 \text{ when } x > 0.$$

Given $\phi \in \mathbb{S}$, we write $\phi = \phi_L + \phi_R$ where $\phi_R(x) = 0$ for $x < 0$, $\phi_L(x) = 0$ for $x > 0$ and use the linearity of f to conclude that

$$f\{\phi\} = \int_{-\infty}^{\infty} \frac{g(x)}{x} \phi(x)dx, \quad \phi \in \mathbb{S}."$$

(d) "Let $f(x) := x^2 \cdot h(x)$. Since

$$f''(x) = 2h(x)$$

we can use the derivative rule to write

$$(2\pi i s)^2 F(s) = \delta(s) + \frac{2}{2\pi i s}$$

and thereby obtain

$$F(s) = \frac{\delta(s)}{(2\pi i s)^2} + \frac{2}{(2\pi i s)^3} ."$$

(e) "Let $f(x) := \cos(2\pi x)$. Since

$$f''(x) + 4\pi^2 f(x) = 0$$

we write

$$(2\pi i s)^2 F(s) + 4\pi^2 F(s) = 0,$$

i.e.,

$$(s^2 - 1)F(s) = 0,$$

and thereby conclude that

$$F(s) = c_+\delta(x + 1) + c_-\delta(x - 1).$$

Since c_+, c_- are arbitrary constants, we can take $c_+ = 0$, $c_- = 1$ and write

$$F(s) = \delta(s - 1)."$$

▶ **EXERCISE 7.51:**   Let $\mathcal{P}, \mathcal{Q}$ be polynomials, and assume that $\mathcal{P}$ has no pure imaginary roots (including 0).

(a) Let x be a generalized function. Show that there is a unique generalized solution of the differential equation
$$[\mathcal{P}(\mathbf{D})y](t) = [\mathcal{Q}(\mathbf{D})x](t)$$
(where $\mathbf{D}$ is the derivative operator).

Note. Such differential equations do not have unique solutions when we work with ordinary functions.

(b) Let the operator $\mathcal{A}$ be defined by writing $\mathcal{A}x := y$ when y is obtained from x as in (a). Show that $\mathcal{A}$ is linear and translation invariant.

Hint. cf. Ex. 5.28.

(c) Let $y_\delta := \mathcal{A}\delta$ be the *impulse response* of $\mathcal{A}$. Show that $\mathcal{A}x = y_\delta * x$.

Note. This convolution product is well defined because $y_\delta^\wedge$ and its derivatives are CSG.

(d) Let h be the Heaviside function and let $y_h := \mathcal{A}h$ be the *step response* of $\mathcal{A}$. Show that $y_\delta = y_h'$.

(e) Let $e_\sigma(t) := e^{2\pi i \sigma t}$ with $-\infty < \sigma < \infty$. Show that $\mathcal{A}e_\sigma = \lambda_\sigma \cdot e_\sigma$ where λ_σ is a complex constant that you should express in terms of $\mathcal{P}$, $\mathcal{Q}$, and σ.

▶ **EXERCISE 7.52:**   In this exercise you will learn an *algebraic* approach for solving certain differential equations.

(a) Given a complex number λ with Re $\lambda \neq 0$, we define
$$g_\lambda(x) := \begin{cases} e^{\lambda x} h(x) & \text{if Re } \lambda < 0 \\ -e^{\lambda x} h(-x) & \text{if Re } \lambda > 0 \end{cases}$$
where h is the Heaviside function. Show that
$$(\delta' - \lambda\delta) * f = \delta$$
if and only if $f = g_\lambda$.

(b) Let the polynomial $\mathcal{P}$ have roots $\lambda_1, \lambda_2, \ldots, \lambda_n$ with nonzero real parts and let g be a generalized function. Show how to find the generalized function f that satisfies the forced differential equation $\mathcal{P}(\mathbf{D})f = g$ or equivalently,
$$(\delta' - \lambda_1\delta) * (\delta' - \lambda_2\delta) * \cdots * (\delta' - \lambda_n f) * f = g.$$

Hint. From (a) you know that convolution with g_λ undoes convolution with $\delta' - \lambda\delta$.

(c) Use (b) to solve
$$f''(x) + 2f'(x) + f(x) = e^{-x}h(x) + \delta'(x).$$

(d) Why can't you use the procedure of (b) to solve $\delta' * f = g$ by using the fact that $\delta' * h = \delta$?

▶ **EXERCISE 7.53:**   The following systems map a generalized function $x(t)$ to a generalized function $y(t)$. Find and sketch the step response y_h (to the Heaviside step $x = h$), the impulse response y_δ (to $x = \delta$), and the frequency response $Y_\delta := y_\delta^\wedge$ for each of them.

(a) $y' + y = x$

(b) $y' - y = x$

(c) $y'' + 2y' + 10y = x$

(d) $y'' + 2y' + y = 2x' + x$

Hint. Use the analysis of Exs. 7.51 and 7.52.

▶ **EXERCISE 7.54:**   Let $a_{00}, a_{01}, a_{10}, a_{11}$ be complex constants with $a_{00}a_{11} \neq a_{01}a_{10}$. Find generalized functions $f_{00}, f_{01}, f_{10}, f_{11}$ such that

$$\begin{bmatrix} a_{00}\delta & a_{01}\delta' \\ a_{10}\delta' & a_{11}\delta'' \end{bmatrix} * \begin{bmatrix} f_{00} & f_{01} \\ f_{10} & f_{11} \end{bmatrix} = \begin{bmatrix} \delta & 0 \\ 0 & \delta \end{bmatrix}.$$

▶ **EXERCISE 7.55:**   The *Hilbert transform* of a suitably regular function f can be defined by writing

$$(\mathcal{H}f)(x) := -i \int_{s=-\infty}^{\infty} f^\wedge(s) \cdot \operatorname{sgn}(s) e^{2\pi i s x} \, ds$$

[as done in (5.83)], by writing

$$(\mathcal{H}f)(x) := \frac{1}{\pi} \int_{u=-\infty}^{\infty} \frac{f(u)}{x - u} \, du \ ,$$

or by writing

$$\mathcal{H}f := f * \pi^{-1} p_{-1}$$

where p_{-1} is the generalized inverse power function of (46) and (52). Reconcile these definitions.

▶ **EXERCISE 7.56:**   Solve the following convolution equations.

(a) $\displaystyle\int_{-\infty}^{\infty} e^{-|x-u|} f(u) \, du = x^4$

(b) $\displaystyle -2f''(x) + \int_{-\infty}^{\infty} e^{-|x-u|} f(u) \, du = x^4$

▶ **EXERCISE 7.57:**   The rules for taking Fourier transforms of generalized functions can be deduced from the corresponding rules for taking Fourier transforms of Schwartz functions by "chasing definitions."

(a) Let f be a generalized function and let $g(x) := f(x - x_0)$ where $-\infty < x_0 < \infty$. Justify each step in the following proof of the translation rule.

$$\int_{-\infty}^{\infty} g^\wedge(s)\phi(s) \, ds = \int_{-\infty}^{\infty} g(x)\phi^\wedge(x) \, dx = \int_{-\infty}^{\infty} f(x - x_0)\phi^\wedge(x) \, dx = \int_{-\infty}^{\infty} f(x)\phi^\wedge(x + x_0) \, dx$$

$$= \int_{-\infty}^{\infty} f^\wedge(s)\{e^{-2\pi i x_0 s} \cdot \phi(s)\} \, ds = \int_{-\infty}^{\infty} \{e^{-2\pi i x_0 s} \cdot f^\wedge(s)\}\phi(s) \, ds, \quad \phi \in \mathbb{S}$$

$$\therefore \ g^\wedge(s) = e^{-2\pi i x_0 s} \cdot f^\wedge(s)$$

(b) Give an analogous proof of the dilation rule.

(c) Give an analogous proof of the derivative rule.

▶ **EXERCISE 7.58:**   A generalized function f is said to be *nonnegative* provided that $f\{\phi\} \geq 0$ whenever ϕ is a nonnegative Schwartz function. Which of the following generalized functions, if any, are nonnegative?

(a) $f(x) = e^{-\pi x^2}$

(b) $f(x) = \log|x|$

(c) $f(x) = 1/x^2$

(d) $f(x) = \delta(x)$

(e) $f(x) = \text{III}(x)$

(f) $f(x) = \delta(x) - \delta''(x)$

▶ **EXERCISE 7.59:**   In this exercise you will find a power series for the Fourier transform $F(s)$ of the function

$$f(x) := e^{-4\pi^4 x^4}.$$

(a) Show that $F'''(s) = sF(s)$. (The constant $4\pi^4$ was chosen to simplify this differential equation.)

(b) Show that $F(0) = I_0$, $F'(0) = 0$, $F''(0) = -I_2$ where

$$I_0 := \int_{-\infty}^{\infty} e^{-4\pi^4 x^4}\, dx, \qquad I_2 := 4\pi^2 \int_{-\infty}^{\infty} x^2 e^{-4\pi^4 x^4}\, dx.$$

(c) Using (a), (b), obtain the rapidly converging series

$$F(s) = I_0\left(1 + \frac{1 \cdot s^4}{4!} + \frac{1 \cdot 5 \cdot s^8}{8!} + \frac{1 \cdot 5 \cdot 9 \cdot s^{12}}{12!} + \cdots\right)$$
$$- I_2\left(\frac{s^2}{2!} + \frac{3 \cdot s^6}{6!} + \frac{3 \cdot 7 \cdot s^{10}}{10!} + \frac{3 \cdot 7 \cdot 11 \cdot s^{14}}{14!} + \cdots\right).$$

▶ **EXERCISE 7.60:**   Verify that each of the following limits is δ.

(a) $\displaystyle\lim_{\sigma \to 0+} \frac{e^{-x^2/2\sigma^2}}{\sqrt{2\pi}\,\sigma}$

(b) $\displaystyle\lim_{a \to +\infty} \frac{\sin(ax)}{\pi x}$

(c) $\displaystyle\lim_{\lambda \to +\infty} \lambda e^{-2\lambda|x|}$

(d) $\displaystyle\lim_{d \to 0+} \frac{(1-i)e^{i\pi x^2/d}}{\sqrt{2d}}$

(e) $\displaystyle\lim_{\lambda \to +\infty} \lambda b(\lambda x)$

(f) $\displaystyle\lim_{N \to \infty} b(x) \sum_{n=-N}^{N} e^{2\pi inx}$

Note. The tapered box (125) is used in (e), (f).

▶ **EXERCISE 7.61:**   Verify each of the following III identities.

(a) $\text{III}(x - n) = \text{III}(x)$, $n = 0, \pm 1, \pm 2, \ldots$

(b) $(\Pi * \text{III})(x) = 1$

(c) $\text{sinc}(x) \cdot \text{III}(x) = \delta(x)$

(d) $\cos(n\pi x) \cdot \text{III}(x) = \dfrac{1}{2}\left\{\text{III}\left(\dfrac{x}{2}\right) + (-1)^n \text{III}\left(\dfrac{x-1}{2}\right)\right\}$, $n = 0, \pm 1, \ldots$

(e) $\text{III}(x) + \text{III}\left(x - \dfrac{1}{n}\right) + \text{III}\left(x - \dfrac{2}{n}\right) + \cdots + \text{III}\left(x - \dfrac{n-1}{n}\right) = n\,\text{III}(nx)$

(f) $e^{i\pi x^2} \cdot \text{III}(x) = \frac{1}{2}\text{III}\left(\frac{x}{2}\right) - \frac{1}{2}\text{III}\left(\frac{x-1}{2}\right)$

Hint. You can use a direct argument based on (40), you can use the series (107) and known properties of δ, or you can use (68) to establish the identity for the Fourier transforms.

▶ **EXERCISE 7.62:** Find each of the following weak limits.

(a) $\displaystyle\lim_{\sigma\to 0+}\int_{u=-\infty}^{x}\frac{e^{-u^2/2\sigma^2}}{\sqrt{2\pi}\,\sigma}\,du$

(b) $\displaystyle\lim_{\sigma\to 0+}\int_{u=-\infty}^{\infty}|x-u|\frac{e^{-u^2/2\sigma^2}}{\sqrt{2\pi}\,\sigma}\,du$

(c) $\displaystyle\lim_{\sigma\to 0+}\int_{u=-\infty}^{\infty}|x-u|\frac{d}{du}\left\{\frac{e^{-u^2/2\sigma^2}}{\sqrt{2\pi}\,\sigma}\right\}du$

(d) $\displaystyle\lim_{\sigma\to 0+}e^{-\sigma^2 x^2}\sum_{n=-\infty}^{\infty}\frac{e^{-(x-n)^2/2\sigma^2}}{\sqrt{2\pi}\,\sigma}$

(e) $\displaystyle\lim_{a\to+\infty}\delta(x-a)$

(f) $\displaystyle\lim_{N\to+\infty}\sum_{n=-N}^{N}e^{2\pi inx}$

(g) $\displaystyle\lim_{h\to 0+}\frac{\delta(x+h)-2\delta(x)+\delta(x-h)}{h^2}$

(h) $\displaystyle\lim_{h\to 0+}\frac{\lfloor x+h\rfloor - \lfloor x\rfloor}{h}$

(i) $\displaystyle\lim_{p\to 0+}p\,\text{III}(px)$

(j) $\displaystyle\lim_{p\to+\infty}\text{III}(px)$

▶ **EXERCISE 7.63:** Let $h_\alpha(x) := e^{-\alpha x}h(x)$, $\alpha \geq 0$, be an approximation to the Heaviside function h. In this exercise you will determine the Fourier transform of h by using the weak limit $\displaystyle\lim_{\alpha\to 0+}H_\alpha(s)$.

(a) Show that

$$\lim_{a\to 0+}\int_{-\infty}^{\infty}h_\alpha(x)\phi(x)dx = \int_{-\infty}^{\infty}h(x)\phi(x)dx$$

when $\phi \in \mathbb{S}$ and thereby prove that h_α converges weakly to h as $\alpha \to 0+$.

(b) Verify that the Fourier transform H_α of h_α has the real, imaginary parts

$$H_{R,\alpha}(s) := \frac{\alpha}{\alpha^2 + 4\pi^2 s^2}, \qquad H_{I,\alpha}(s) := \frac{-2\pi s}{\alpha^2 + 4\pi^2 s^2}.$$

(c) Show that

$$\lim_{\alpha\to 0+}\int_{-\infty}^{\infty}H_{R,\alpha}(s)\Phi(s)ds = \tfrac{1}{2}\Phi(0)$$

when $\phi \in \mathbb{S}$ and thereby prove that $H_{R,\alpha}$ converges weakly to $\frac{1}{2}\delta$ as $\alpha \to 0+$.

Hint. $\displaystyle\left|\int_{-\infty}^{\infty}H_{R,\alpha}(s)[\Phi(s)-\Phi(0)]ds\right| \leq \max_{|s|\leq\epsilon}|\Phi(s)-\Phi(0)|\cdot\int_{|s|\leq\epsilon}H_{R,\alpha}(s)ds$

$$+2\cdot\max_{-\infty<s<\infty}|\Phi(s)|\cdot\int_{|s|\geq\epsilon}H_{R,\alpha}(s)ds.$$

(d) Show that

$$\lim_{\alpha \to 0+} \int_{-\infty}^{\infty} H_{I,\alpha}(s)\Phi(s)\,ds = \int_{-\infty}^{\infty} \frac{\Phi(s) - \Phi(0)}{-2\pi s}$$

when $\phi \in \mathbb{S}$ and thereby prove that $H_{I,\alpha}$ converges weakly to $(-1/2\pi)p_{-1}$.

(e) Use (b)–(d) to obtain the Fourier transform (99) of the Heaviside function h. [The *pointwise* limit of $H_\alpha(s)$ gives the first (but not the second) term of this sum.]

▶ **EXERCISE 7.64:**   Let f be a 1-periodic generalized function on $\mathbb{R}$ and let

$$\delta_{D,n}(x) := \sum_{k=-n}^{n} e^{2\pi i k x}, \quad \delta_{F,n}(x) := \frac{1}{n+1}\sum_{\ell=0}^{n} \delta_{D,\ell}(x), \quad n = 0, 1, 2, \ldots$$

be the 1-periodic Dirichlet and Fejer kernel functions from Exs. 1.31 and 1.32. Establish each of the following weak limits.

(a) $\displaystyle\lim_{n \to +\infty} \delta_{D,n} = \text{III}$

(b) $\displaystyle\lim_{n \to +\infty} \delta_{F,n} = \text{III}$

(c) $\displaystyle\lim_{n \to +\infty} \delta_{D,n} \circledast f = f$

(d) $\displaystyle\lim_{n \to +\infty} \delta_{F,n} \circledast f = f$

Note. The convolution product $\circledast$ is defined by (131).

▶ **EXERCISE 7.65:**   In this exercise you will show that

$$f(x) := \text{sech}(\pi x) \quad \text{has the FT} \quad F(s) = \text{sech}(\pi s).$$

(a) Verify the pointwise limits

$$\text{sech}(\pi x) = 2\sum_{\nu=0}^{\infty} (-1)^{\nu} e^{-(2\nu+1)\pi|x|}, \quad x < 0 \text{ or } x > 0$$

$$\text{sech}(\pi s) = \frac{4}{\pi}\sum_{\nu=0}^{\infty} \frac{(-1)^{\nu}(2\nu+1)}{4s^2 + (2\nu+1)^2}, \quad -\infty < s < \infty.$$

Hint. Use the identity

$$\text{sech}(u) = \frac{2e^{-|u|}}{1 + e^{-2|u|}}$$

to establish the first and use the result of Ex. 4.6(b) for the second.

(b) Establish the bounds

$$\left| 2\sum_{\nu=n}^{\infty} (-1)^{\nu} e^{-(2\nu+1)\pi|x|} \right| \le 2e^{-(2n+1)\pi|x|}, \quad \left| \frac{4}{\pi}\sum_{\nu=n}^{\infty} \frac{(-1)^{\nu}(2\nu+1)}{4s^2 + (2\nu+1)^2} \right| \le \frac{4}{2n+1}$$

for the tails of the series in (a).

Hint. Use the alternating series test.

(c) Using (b) show that the series from (a) both converge weakly.

(d) Show that the term-by-term Fourier transform of one of the series from (a) gives the other.

▶ **EXERCISE 7.66:**   Find and correct the flaw(s) in the following arguments.

(a) "Let $f_n(x) := e^{2\pi inx} \cdot n\Pi(nx)$, $n = 1, 2, \dots$. Since $\lim\limits_{n\to\infty} n\,\Pi(nx) = \delta(x)$ and since

$\alpha(x) \cdot \delta(x) = \alpha(0) \cdot \delta(x)$ when $\alpha(x) := e^{2\pi inx}$ we can write

$$\lim_{n\to\infty} f_n(x) = 1 \cdot \delta(x) = \delta(x)."$$

(b) "Let $f_\alpha(x) := \frac{1}{2}\delta(x) + \sum_{m=1}^{\infty} e^{-\alpha m}\delta(x - m)$, $\alpha \geq 0$. Since $f_0 = \lim\limits_{\alpha\to 0+} f_\alpha$, we can write

$$F_0(s) = \lim_{\alpha\to 0+} F_\alpha(s) = \lim_{\alpha\to 0+} \left\{ \frac{1}{2} + \sum_{m=1}^{\infty} e^{-(\alpha+2\pi is)m} \right\} = \frac{1}{2} + \frac{e^{-2\pi is}}{1 - e^{-2\pi is}} = -\frac{i}{2}\cot(\pi s),$$

and thereby obtain the functional

$$F_0\{\phi\} = -\frac{i}{2}\int_{s=-\infty}^{\infty} \cot(\pi s)\phi(s)ds, \quad \phi \in \mathbb{S}."$$

▶ **EXERCISE 7.67:**   Two Fourier analysis students are trying to prove that

$$\delta(x - a) = \sum_{n=0}^{\infty} \frac{(-a)^n}{n!}\delta^{(n)}(x)$$

where $-\infty < a < \infty$.

"It's easy if you work in the transform domain since the series

$$e^{-2\pi ias} = \sum_{n=0}^{\infty} (-1)^n (2\pi is)^n \cdot \frac{a^n}{n!}$$

is certainly correct," says one of the students.

"I think you have to use Schwartz functions," says the second. "Use Maclaurin's formula from calculus to write

$$\int_{-\infty}^{\infty} \delta(x - a)\phi(x)dx = \phi(a) = \sum_{n=0}^{\infty} \phi^{(n)}(0)\frac{a^n}{n!}$$

$$= \sum_{n=0}^{\infty} \left\{ \int_{-\infty}^{\infty} (-1)^n \delta^{(n)}(x)\phi(x)dx \right\} \frac{a^n}{n!} = \int_{-\infty}^{\infty} \left\{ \sum_{n=0}^{\infty} \frac{(-a)^n}{n!}\delta^{(n)}(x) \right\} \phi(x)dx$$

when $\phi \in \mathbb{S}."$

(a) Find the flaw in the first argument.

(b) Find the flaw in the second argument.

(c) Explain why the $\delta^{(n)}$ series does *not* converge weakly.

▶ **EXERCISE 7.68:**   Use a suitable weakly converging series to find the Fourier transform of each of the following generalized functions.

(a) $f(x) = e^{\Pi(x)}$

(b) $f(x) = 1/(2 - e^{-\pi x^2})$

(c) $f(x) = 1 + \dfrac{e^{-\pi x^2}}{1!} + \dfrac{e^{-\pi x^2} * e^{-\pi x^2}}{2!} + \dfrac{e^{-\pi x^2} * e^{-\pi x^2} * e^{-\pi x^2}}{3!} + \cdots$

(d) $f(x) = \cos[\sin(2\pi x)]$

   Hint. Use Ex. 4.19(c).

▶ **EXERCISE 7.69:**   In this exercise you will study transformations of the limit

$$\frac{\partial}{\partial \lambda} \delta(x - \lambda) := \lim_{h \to 0} \frac{\delta(x - \lambda - h) - \delta(x - \lambda)}{h}$$

where λ (like x) is a real parameter).

(a) Show that the partial derivative has the expected form

$$\frac{\partial}{\partial \lambda} \delta(x - \lambda) = -\delta'(x - \lambda).$$

(b) What identity results when you transform the limit by replacing δ with the derivative δ'?

(c) What identity results when you transform the limit by replacing δ with the Fourier transform $\delta^\wedge$?

(d) What identity results when you transform the limit by replacing δ with $\alpha \cdot \delta$? Assume that $\alpha, \alpha', \alpha'', \ldots$ are CSG functions on $\mathbb{R}$.

▶ **EXERCISE 7.70:**   Let f, g be p-periodic generalized functions with the Fourier series

$$f(x) = \sum_{k=-\infty}^{\infty} c_k e^{2\pi i k x/p}, \qquad g(x) = \sum_{k=-\infty}^{\infty} d_k e^{2\pi i k x/p}.$$

Find the Fourier series for each of the following.

(a) $af(x) + bg(x)$ where $a, b \in \mathbb{C}$

(b) $f(-x)$

(c) $f(x - x_0)$ where $-\infty < x_0 < \infty$

(d) $\overline{f(x)}$

(e) $e^{2\pi i k_0 x/p} \cdot f(x)$ where $k_0 = 0, \pm 1, \pm 2, \ldots$

(f) $(f \circledast g)(x)$ where $\circledast$ is given by (131)

(g) $(f \cdot g)(x)$ in the case where f is σ-bandlimited for some $\sigma > 0$

(h) $f^{(m)}(x)$ where $m = 0, 1, 2, \ldots$

(i) $f(mx)$ where $m = 1, 2, \ldots$

(j) $\displaystyle\sum_{\ell=0}^{m-1} f(x/m - \ell p/m)$ where $m = 1, 2, \ldots$

Note. You can freely manipulate the Fourier series for p-periodic generalized functions!

▶ **EXERCISE 7.71:**   Let f be a p-periodic generalized function with the Fourier coefficients $\ldots c_{-2}, c_{-1}, c_0, c_1, c_2, \ldots$. What can you infer about these coefficients if you know that:

(a) $f(-x) = f(x)$ (reflection symmetry)?

(b) $f\left(\dfrac{p}{2} - x\right) = f\left(\dfrac{p}{2} + x\right)$ (reflection symmetry)?

(c) $f\left(x + \dfrac{p}{2}\right) = f(x)$ (translation symmetry)?

(d) $f\left(x + \dfrac{p}{2}\right) = -f(x)$ (glide symmetry)?

Hint. The function $g(x) := e^{2\pi i x/p} \cdot f(x)$ is $p/2$-periodic.

▶ **EXERCISE 7.72:**   In this exercise you will use three methods to solve the differential equation

$$y'(t) + \mu\, y(t) = x(t)$$

when the forcing function is the periodic δ train

$$x(t) := A \sum_{m=-\infty}^{\infty} \delta(t - mp) = \frac{A}{p} \mathrm{III}\left(\frac{t}{p}\right).$$

Here μ, A, p are positive constants. (For an interesting interpretation, regard y as the blood concentration of a medicinal drug that is injected in equal doses at intervals of length p.)

(a) Show that the solution y, if any, is unique.

(b) Show that the solution y, if any, is p-periodic.

(c) Use the fact that $y'(t) + \mu\, y(t) = 0$ for $0 < t < p$ to infer that the p-periodic extension of the ordinary function

$$y(t) = \frac{A e^{-\mu t}}{1 - e^{-\mu p}}, \qquad 0 < t < p$$

satisfies the differential equation

(d) Use the differential equation and the Fourier series for $x(t)$ to determine the Fourier coefficients for the Fourier series

$$y(t) = \sum_{k=-\infty}^{\infty} c_k e^{2\pi i k x/p}.$$

[You can then check your work by using the analysis equation (1.6) with the p-periodic function from (c).]

(e) Find the generalized function g that satisfies the differential equation

$$g'(t) + \mu\, g(t) = \delta(t),$$

observe that g satisfies the condition (80), and construct

$$y(t) = (g * x)(t)$$

by summing a certain geometric series. [Since

$$(g * x)' + \mu(g * x) = (g' + \mu\, g) * x = \delta * x = x,$$

this procedure must give the same solution found in (c)!]

(f) Show that the ordinary function y from (c)–(e) will satisfy

$$\frac{1}{2}A \le y(t) \le 2A \text{ for } -\infty < t < \infty$$

provided that $\log 2 \le \mu\, p \le \log 3$. (We can control the concentration of the drug by our choice of p and A!)

▶ **EXERCISE 7.73:**   Let β be a σ-support-limited generalized function with the Fourier transform B and let $p > 0$.

(a) Show that we can define a p-periodic generalized function f by writing

$$f(x) = \sum_{k=-\infty}^{\infty} \beta(x - kp).$$

Hint. Convolve β with $p^{-1}\text{III}(x/p)$.

(b) Show that the Fourier transform is given by the weakly converging series

$$f^\wedge(s) = \sum_{k=-\infty}^{\infty} \frac{1}{p}B\left(\frac{k}{p}\right)\delta\left(s - \frac{k}{p}\right).$$

(c) Show that f has the Fourier series

$$f(x) = \sum_{k=-\infty}^{\infty} \frac{1}{p}B\left(\frac{k}{p}\right)e^{2\pi ikx/p},$$

i.e., show that Poisson's relation can be used in this context.

(d) Find the Fourier series when $\beta(x) := \delta^{(m)}(x)$ and m is a nonnegative integer.

(e) Find the Fourier series when

$$\beta(x) := \begin{cases} \beta_1(x) & \text{for } x_1 < x < x_2 \\ \beta_2(x) & \text{for } x_2 < x < x_3 \\ \vdots \\ \beta_{m-1}(x) & \text{for } x_{m-1} < x < x_m \\ \beta_m(x) & \text{for } x_m < x < x_1 + p \end{cases}$$

where $\beta_1, \beta_2, \ldots, \beta_m$ are polynomials of degree n or less and $x_1 < x_2 < \cdots < x_m < x_1 + p$.

Hint. Use the analysis from Ex. 7.39 and *cf.* Ex. 7.75.

▶ **EXERCISE 7.74:** Use Poisson's relation to find the Fourier series for the 2-periodic function

$$g(x) := \sum_{m=-\infty}^{\infty} f(x - 2m)$$

where

$$f(x) := \begin{cases} -\sin^2(\pi x) & \text{if } -1 < x < 0 \\ \sin^2(\pi x) & \text{if } \ 0 < x < 1 \\ 0 & \text{otherwise.} \end{cases}$$

Hint. First show that

$$f'''(x) + 4\pi^2 f'(x) = 2\pi^2[-\delta(x + 1) + 2\delta(x) - \delta(x - 1)].$$

▶ **EXERCISE 7.75:** Let f be an ordinary p-periodic function on $\mathbb{R}$, let $m = 1, 2, \ldots$, let $0 \le x_1 < x_2 < \cdots < x_m < p$, and assume that $f^{(n+1)}(x) = 0$ at all points $0 \le x < p$ except for $x_1, x_2, \ldots, x_m$ (*i.e.*, assume that f is a piecewise polynomial function). Let $f, f', f'', \ldots$ have the jumps

$$J_\mu := f(x_\mu+) - f(x_\mu-), \ \ J'_\mu := f'(x_\mu+) - f(x_\mu-), \ \ J''_\mu := f''(x_\mu+) - f''(x_\mu-), \ldots$$

at the point x_μ, $\mu = 1, 2, \ldots, m$, and let

$$A := \frac{1}{p} \int_0^p f(x)\,dx$$

be the average value of f.

(a) Express the coefficients ( ) for

$$f^{(n+1)}(x) = \sum_{\mu=1}^{m} \{J_\mu \cdot (\ \ ) + J'_\mu \cdot (\ \ ) + J''_\mu \cdot (\ \ ) + \cdots \}$$

in terms of dilates of derivatives of III.

(b) Express the coefficients ( ) for the Fourier transform

$$f^\wedge(s) = A\,\delta(s) + \sum_{\mu=1}^{m} \{J_\mu \cdot (\ \ ) + J'_\mu \cdot (\ \ ) + J''_\mu \cdot (\ \ ) + \cdots \}$$

of f in terms of III.

(c) Find the coefficients ( ) for the Fourier series

$$f(x) = A + \sum_{\substack{k=-\infty \\ k \ne 0}}^{\infty} \sum_{\mu=1}^{m} \{J_\mu \cdot (\ \ ) + J'_\mu \cdot (\ \ ) + J''_\mu \cdot (\ \ ) + \cdots \} e^{2\pi i k x/p}.$$

(d) Express $f(x)$ as a linear combination of translates of dilates of the Bernoulli functions (4.23) plus the constant A.

Note. Such formulas were developed by A. Eagle, *A Practical Treatise on Fourier's Theorem and Harmonic Analysis*, Longmans, Green & Co., London, 1925, pp. 57–59.

▶ **EXERCISE 7.76:**   Let $g(x) = \alpha(x) \cdot f(x)$ where f is a p-periodic generalized function having the Fourier series

$$f(x) = \sum_{k=-\infty}^{\infty} c_k e^{2\pi i k x / p},$$

and where $\alpha, \alpha', \alpha'', \ldots$ are CSG. Express the Fourier transform of g in terms of $A := \alpha^{\wedge}$ and the c_k's, and then specialize this general formula to the case where:

(a) $\alpha(x) = e^{-\pi x^2}$, $f(x) = \cos(2\pi x)$;   (b) $\alpha(x) = \mathrm{sinc}^2(x)$, $f(x) = \mathrm{III}(x)$;
(c) $\alpha(x) = (1 - 2\pi i x)^{-1}$, $f(x) = \mathrm{III}(x)$.

▶ **EXERCISE 7.77:**   What can you infer about the Fourier transform $f^{\wedge}$ of the generalized function f on $\mathbb{R}$ if you know that:

(a) $\sin(2\pi x) \cdot f(x) = 0$?   (b) $\cos(2\pi x) \cdot f(x) = 0$?

(c) $f'(x + 1) = f'(x)$?   (d) $f(x+1) = f(x)$ and $f(x+\sqrt{2}) = f(x)$?

(e) $f'(x + 1) = f(x)$?   (f) $f(x + 1) = f(x) * e^{-\pi x^2}$?

▶ **EXERCISE 7.78:**   Let f be a generalized function, let $p > 0$, and assume that $\boldsymbol{\Delta}_p^n f = 0$ where $(\boldsymbol{\Delta}_p f)(x) := p^{-1}[f(x + p) - f(x)]$.

(a) Show that f has the representation

$$f(x) = \sum_{k=-\infty}^{\infty} (c_{0k} + c_{1k}x + \cdots + c_{n-1,k}x^{n-1}) e^{2\pi i k x / p}$$

where $c_{0k}, c_{1k}, \ldots, c_{n-1,k}$ are slowly growing complex coefficients. (It is easy to see that the converse also holds.)

Hint. Begin by observing that $\boldsymbol{\Delta}_p^{n-1} f$ is a p-periodic generalized function with the weakly convergent Fourier series (114).

(b) Show that f has the representation

$$f(x) = f_0(x) + x f_1(x) + \cdots + x^{n-1} f_{n-1}(x)$$

where $f_0, f_1, \ldots, f_{n-1}$ are suitably chosen p-periodic generalized functions.

▶ **EXERCISE 7.79:**   Let $a(x)$, $d(x)$ be p-periodic generalized functions with the Fourier series

$$a(x) = \sum_{k=-\infty}^{\infty} \alpha_k e^{2\pi i k x / p}, \qquad d(x) = \sum_{k=-\infty}^{\infty} \delta_k e^{2\pi i k x / p}.$$

What conditions must we impose on the coefficients to ensure that the convolution equation

$$a \circledast f = d$$

has a unique p-periodic generalized solution

$$f(x) = \sum_{k=-\infty}^{\infty} c_k e^{2\pi i k x / p} ?$$

[Here $\circledast$ is given by (131).]

▶ **EXERCISE 7.80:**   Let the polynomials $\mathcal{P}, \mathcal{Q}$ and the LTI operator $\mathcal{A}$ be as defined in Ex. 7.51.

(a) Find the Fourier series for $y_{\text{III}} := \mathcal{A}\,\text{III}$.

   Hint. First find $y_{\text{III}}^{\wedge}$.

(b) Let x be a 1-periodic generalized function having the Fourier series

$$x(t) = \sum_{k=-\infty}^{\infty} c_k e^{2\pi i k t}.$$

   Show that $\mathcal{A}x$ is 1-periodic and find the corresponding Fourier series.

(c) Let x be a 1-periodic generalized function. Show that

$$\mathcal{A}x = y_{\text{III}} \circledast x.$$

   Note. The function y_{III} serves as the *impulse response* for $\mathcal{A}$ when we work with generalized functions on $\mathbb{T}_1$.

▶ **EXERCISE 7.81:**   We use the tapered box (123) to construct the Schwartz functions

$$b_n(x) := nb(nx), \qquad m_n(x) := \Pi\left(\frac{x}{2n}\right) * b(x), \qquad n = 1, 2, \ldots$$

that serve as approximate convolution and multiplication identities when n is large.

(a) Sketch the graphs of b_n and m_n.

(b) Sketch the graphs of $m_n \cdot f$ and $b_n * (m_n \cdot f)$ when $f(x) := |x|$ and n is large.

(c) Let g be CSG and let $\phi \in \mathbb{S}$. Show that $\lim\limits_{n \to \infty} \int_{-\infty}^{\infty} |g(x)|\,|(\phi * b_n)(x) - \phi(x)|dx = 0$.

   Note. Observe that $\min\limits_{|x-u| \leq 1/n} \phi(u) \leq (\phi * b_n)(x) \leq \max\limits_{|x-u| \leq 1/n} \phi(u)$.

(d) Let f be a generalized function. Show that $\lim\limits_{n \to \infty} b_n * f = f$.

(e) Let g be CSG and let $\phi \in \mathbb{S}$. Show that $\lim\limits_{n \to \infty} \int_{|x| \geq n-2} |g(x)(\phi * b_n)(x)|dx = 0$.

(f) Let f be a generalized function. Show that $\lim\limits_{n \to \infty} m_n \cdot f = f$.

(g) Let f be a generalized function. Show that $\lim\limits_{n \to \infty} b_n * (m_n \cdot f) = f$.

Note. In this way you show that every generalized function is the weak limit of some sequence of Schwartz functions.

Hint.

$$\left| \int_{-\infty}^{\infty} g(x)[m_n(x) \cdot (\phi * b_n)(x) - \phi(x)]^{(m)} \, dx \right|$$

$$\leq \int_{-\infty}^{\infty} |g(x)| \cdot |(\phi^{(m)} * b_n)(x) - \phi^{(m)}(x)| \, dx$$

$$+ \int_{-\infty}^{\infty} |g(x)| \cdot [1 - m_n(x)] \cdot |(\phi^{(m)} * b_n)(x) - \phi^{(m)}(x)| \, dx$$

$$+ \sum_{\ell=1}^{m} \binom{m}{\ell} \int_{-\infty}^{\infty} |g(x)| \cdot |m_n^{(\ell)}(x)| \cdot |(\phi^{(m-\ell)} * b_n)(x)| \, dx.$$

▶ **EXERCISE 7.82:**   Show that the linear functional

$$f\{\phi\} := (-1)^{\mu} \int_{-\infty}^{\infty} g(x) \phi^{(\mu)}(x) \, dx, \qquad \phi \in \mathbb{S}$$

that is constructed from a CSG function g and a nonnegative integer μ is *Schwartz continuous, i.e.,*

$$\lim f\{\phi_{\nu}\} = 0$$

whenever $\phi_1, \phi_2, \ldots \in \mathbb{S}$ are chosen in such a manner that

$$\lim_{\nu \to \infty} \max_{-\infty < x < \infty} |x^n \phi_{\nu}^{(m)}(x)| = 0$$

for every choice of $n = 0, 1, \ldots$ and $m = 0, 1, \ldots$.

Note. The Hahn-Banach theorem from functional analysis can be used to prove that every Schwartz continuous linear functional has such a representation.

Chapter 8

Sampling

8.1 Sampling and Interpolation

Introduction

As I speak the word *Fourier*, a microphone converts the pressure wave from my voice into an electrical voltage $f(t)$, $0 \leq t \leq .5$ sec. The sound card in my computer discretizes this signal by producing the samples

$$f(n\mathrm{T}), \qquad n = 0, 1, \ldots, 4000, \qquad \mathrm{T} := 1/8000 \text{ sec.}$$

Figure 8.1 shows the (overlapping) line segments joining

$$(n\mathrm{T}, 0) \text{ to } (n\mathrm{T}, f(n\mathrm{T})), \quad n = 0, 1, \ldots, 4000.$$

You can identify the hissy initial consonant "f", the three long vowels "o", "e", "a" (as in *boat, beet, bait*), and the semivowel "r" (as in *burr*).

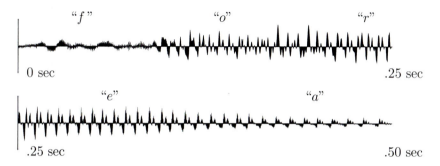

Figure 8.1. An audio signal for the spoken word *Fourier*.

An expanded 80-sample segment of the 10-ms interval $.33$ sec $\le t \le .34$ sec (a part of the "e" sound) is shown in Fig. 8.2. It would appear that we have enough sample points to construct a good approximation to the original audio signal.

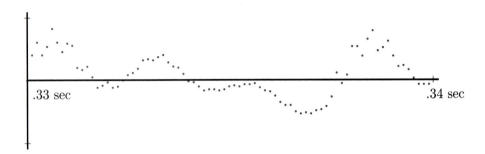

Figure 8.2. An 80-sample segment of the "e" sound from the spoken word *Fourier*.

In practice, we work with a quantized approximation of the sample $f(n\text{T})$. For example, the *Fourier* recording uses 1 byte $:= 8$ bits per sample, with one bit specifying the sign of $f(n\text{T})$ and with seven bits specifying the modulus $|f(n\text{T})|$ to within 1 part in 127 (of some fixed maximum modulus). It takes 4000 bytes of storage for the *Fourier* recording, more than we need for a page of this text with

$$\frac{1 \text{ byte}}{\text{character}} \cdot \frac{80 \text{ characters}}{\text{line}} \cdot \frac{45 \text{ lines}}{\text{page}} = \frac{3600 \text{ bytes}}{\text{page}}.$$

A digitized sound file that uses 8000 8-bit samples/sec is of telephone quality. For high fidelity we must increase the sampling rate and reduce the quantization error. A compact disk recording uses 44100 16-bit samples/sec or

$$\frac{44100 \text{ samples}}{\text{sec}} \cdot \frac{3600 \text{ sec}}{\text{hour}} \cdot \frac{2 \text{ bytes}}{\text{sample}} = \frac{318 \text{ Mbytes}}{\text{hour}}.$$

Digitized sound files are very big!

Shannon's hypothesis

Let f be a function on $\mathbb{R}$, let $\text{T} > 0$, and let

$$f_n := f(n\text{T}), \qquad n = 0, \pm 1, \pm 2, \dots . \tag{1}$$

It is easy to produce a function y on $\mathbb{R}$ that *interpolates* f at the sample points, *i.e.*,

$$y(n\text{T}) = f_n, \qquad n = 0, \pm 1, \pm 2, \dots . \tag{2}$$

For example, the broken line

$$y(t) := \sum_{n=-\infty}^{\infty} f_n \Lambda \left(\frac{t - n\mathrm{T}}{\mathrm{T}} \right), \qquad -\infty < t < \infty \tag{3}$$

satisfies (2). When T is small, this approximation will be close to f. Figure 8.3 shows the broken-line approximation (3) for the sample points from Fig. 8.2.

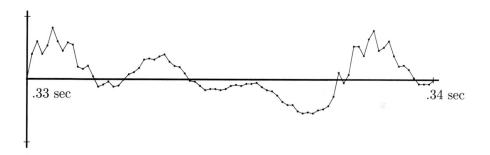

Figure 8.3. The broken-line interpolant (3) for the sample points from Fig. 8.2.

In this chapter, however, we have a more ambitious goal. A good approximation is not good enough! We want to reconstruct f precisely from the samples (1), *i.e.*, we want to produce an interpolating function y with $y(t) = f(t)$ for all t, not just for $t = 0, \pm\mathrm{T}, \pm2\mathrm{T}, \dots$. Of course, this is possible only if we have certain *a priori* knowledge of the function f. For example, if we know that $f'' \equiv 0$, then we can use (3) to recover f with no error.

In sampling theory, we follow Claude Shannon and assume that f is σ-*bandlimited*, *i.e.*, we assume that f can be synthesized from the complex exponentials $e^{2\pi i s t}$ with $-\sigma \le t \le \sigma$. The fact that (3) reproduces f when $f'' \equiv 0$ is an odd curiosity, but Shannon's hypothesis is of enormous practical importance. After all, the sounds made by the human voice are contained well within the 27.5–4186 Hz range of the piano keyboard (*cf.* Appendix 7), and the human ear does not respond to frequencies above 20000 Hz. Thus speech, music signals are (essentially) σ-bandlimited with $\sigma = 4000$ Hz, $\sigma = 20000$ Hz, respectively.

For purposes of mathematical analysis, we will allow the Fourier transform F of our function f to be an arbitrary σ-support-limited generalized function on $\mathbb{R}$. This forces f to have the representation

$$f(t) = t^m \int_{-\sigma}^{\sigma} \Gamma(s) e^{2\pi i s t} ds + \sum_{\mu=0}^{m-1} t^\mu [c_\mu e^{2\pi i \sigma t} + d_\mu e^{-2\pi i \sigma t}] \tag{4}$$

where m is a nonnegative integer, $\Gamma(s)$ is a continuous complex-valued function for $-\sigma \le s \le \sigma$, and $c_\mu, d_\mu \in \mathbb{C}$ for $\mu = 0, 1, \ldots, m$, cf. (7.103). Using (4), we see that $f, f', f'', \ldots$ are all well defined, and it is a fairly simple matter to verify that

$$|f^{(k)}(t)| \le B \cdot (2\pi\sigma)^k \cdot (1 + |t|^m) \cdot (1 + k^m), \quad -\infty < t < \infty, \quad k = 0, 1, \ldots \quad (5)$$

where B is a constant that depends on m, Γ, and the constants c_μ, d_μ, cf. Ex. 8.19. You can then use this bound to show that f is represented by the pointwise convergent Maclaurin series

$$f(t) = \sum_{k=0}^{\infty} f^{(k)}(0) \frac{t^k}{k!}, \quad -\infty < t < \infty,$$

with the convergence being uniform on every finite interval, cf. Ex. 8.20.

The Nyquist condition

We will use pure complex exponentials to determine the appropriate size for the sampling interval $\text{T} > 0$. You will observe that

$$f_1(t) := e^{2\pi i s_1 t}, \quad f_2(t) := e^{2\pi i s_2 t}, \quad s_1 \ne s_2,$$

have precisely the same T-samples

$$e^{2\pi i s_1 n \text{T}} = e^{2\pi i s_2 n \text{T}} \quad \text{for all} \quad n = 0, \pm 1, \pm 2, \ldots$$

when $(s_2 - s_1)\text{T}$ is an integer. We eliminate this possibility if we choose T so that $|s_2 - s_1|\text{T} < 1$. In particular, we can distinguish between the T-samples of the σ-bandlimited functions

$$f_1(t) := e^{-2\pi i \sigma t}, \quad f_2(t) := e^{2\pi i \sigma t}$$

if the sampling interval T satisfies the *Nyquist condition*

$$2\sigma\text{T} < 1. \quad (6)$$

We often observe *aliasing* when we violate (6), *i.e.*, when we do not take at least two samples per cycle of the band frequency σ. For example, the *chirp*

$$f(t) = \sin[2\pi(5t + 27t^2)], \quad 0 \le t \le 1 \quad (7)$$

is shown in Fig. 8.4 along with samples taken with $\text{T} := 1/64$. The local frequency increases linearly from 5 to 59 as t increases from 0 to 1, so we have at least two samples/cycle on the left half of the curve but not on the right. Indeed, the samples appear to come from the alias

$$f_a(t) := \begin{cases} f(t) & \text{if } 0 \le t \le 1/2 \\ f(1-t) & \text{if } 1/2 \le t \le 1, \end{cases} \tag{8}$$

which is also shown in Fig. 8.4. A less exotic example of aliasing appears in Fig. 1.6.

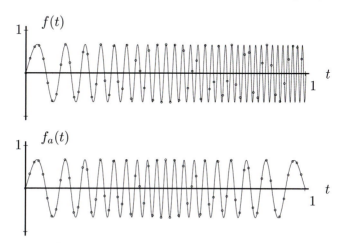

Figure 8.4. The chirp (7) and a corresponding alias (8) that results when the Nyquist condition (6) is violated.

8.2 Reconstruction of f from Its Samples

A weakly convergent series

Let f be σ-bandlimited and let $\text{T} > 0$. Since $f, f', f'', \ldots$ are CSG, we can write

$$f(t) \cdot \text{III}\left(\frac{t}{\text{T}}\right) = f(t) \cdot \sum_{n=-\infty}^{\infty} \delta\left(\frac{t}{\text{T}} - n\right) = f(t) \cdot \sum_{n=-\infty}^{\infty} \text{T}\delta(t - n\text{T})$$

$$= \sum_{n=-\infty}^{\infty} \text{T}f(t) \cdot \delta(t - n\text{T}) = \sum_{n=-\infty}^{\infty} \text{T}f(n\text{T})\delta(t - n\text{T}),$$

$$F(s) * \text{T}\text{III}(\text{T}s) = F(s) * \sum_{m=-\infty}^{\infty} \text{T}\delta(\text{T}s - m) = F(s) * \sum_{m=-\infty}^{\infty} \delta\left(s - \frac{m}{\text{T}}\right)$$

$$= \sum_{m=-\infty}^{\infty} F(s) * \delta\left(s - \frac{m}{\text{T}}\right) = \sum_{m=-\infty}^{\infty} F\left(s - \frac{m}{\text{T}}\right),$$

with all of the infinite series converging weakly. From these identities we obtain the *Poisson relation*

$$\sum_{m=-\infty}^{\infty} F\left(s - \frac{m}{T}\right) = \sum_{k=-\infty}^{\infty} T f(nT)e^{-2\pi i n Ts}, \tag{9}$$

which we will use to reconstruct f from its T-samples.

Let $0 < \alpha < 1$ and let b be the tapered box of (7.123) and Fig. 7.21. We mollify the unit box by defining

$$R_\alpha(s) := \frac{2}{\alpha} b\left(\frac{2s}{\alpha}\right) * \Pi(s). \tag{10}$$

The scaled dilate of b is an approximate delta that vanishes when $|s| \geq \alpha/2$, so R_α is a mesa function with

$$R_\alpha(s) = 1 \quad \text{when} \quad |s| \leq \frac{1-\alpha}{2}, \qquad R_\alpha(s) = 0 \quad \text{when} \quad |s| \geq \frac{1+\alpha}{2},$$

as shown in Fig. 8.5. The Fourier transform of (10) is the Schwartz function

$$r_\alpha(t) = B(\alpha t/2) \cdot \text{sinc}(t). \tag{11}$$

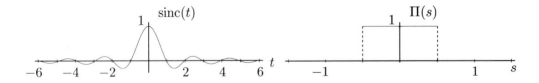

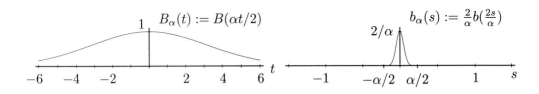

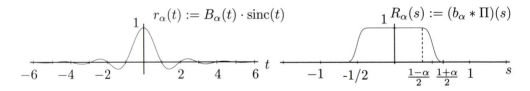

Figure 8.5. Construction of the damped sinc function r_α and the mollified box R_α via (11) and (10).

Let σ, T satisfy the Nyquist condition (6). We choose $0 < \alpha < 1$ so that

$$0 < \sigma < \frac{1 - \alpha}{2\text{T}} < \frac{1}{2\text{T}}. \tag{12}$$

The T-dilate of R_α will then be a mesa function with

$$R_\alpha(\text{T}s) = 1 \quad \text{if} \quad |s| \le \frac{1 - \alpha}{2\text{T}}, \qquad R_\alpha(\text{T}s) = 0 \quad \text{if} \quad |s| \ge \frac{1 + \alpha}{2\text{T}},$$

so that

$$R_\alpha(\text{T}s) \cdot F\left(s - \frac{m}{\text{T}}\right) = \begin{cases} F(s) & \text{if } m = 0 \\ 0 & \text{if } m = \pm 1, \pm 2, \dots \end{cases} \tag{13}$$

(This product is well defined since $R_\alpha \in \mathbb{S}$.) We use (13) and (9) to write

$$F(s) = \sum_{m=-\infty}^{\infty} R_\alpha(\text{T}s) \cdot F\left(s - \frac{m}{\text{T}}\right) = R_\alpha(\text{T}s) \cdot \sum_{m=-\infty}^{\infty} F\left(s - \frac{m}{\text{T}}\right)$$

$$= R_\alpha(\text{T}s) \cdot \sum_{n=-\infty}^{\infty} \text{T} f(n\text{T}) e^{-2\pi i n \text{T} s} = \sum_{n=-\infty}^{\infty} f(n\text{T}) e^{-2\pi i n \text{T} s} \cdot \text{T} R_\alpha(\text{T}s),$$

and thereby obtain

$$f(t) = \sum_{n=-\infty}^{\infty} f(n\text{T}) r_\alpha\left(\frac{t - n\text{T}}{\text{T}}\right). \tag{14}$$

The weakly convergent series (14) allows us to reconstruct the σ-bandlimited function f from its T-samples! Figure 8.6 illustrates the steps that we used to produce this remarkable interpolation formula.

Example: Let f, r_α, T be as in the above discussion. Show that the series (14) converges uniformly on every finite interval.

Solution: The σ-bandlimited function f is slowly growing, so we can write

$$|f(n\text{T})| \le B|n|^m, \qquad n = \pm 1, \pm 2, \dots$$

for a suitable choice of $B > 0$ and the nonnegative integer m. Since $r_\alpha \in \mathbb{S}$ we can also write

$$|r_a(u)| \le \frac{C}{|u|^{m+2}} \quad \text{when} \quad |u| \ge 1$$

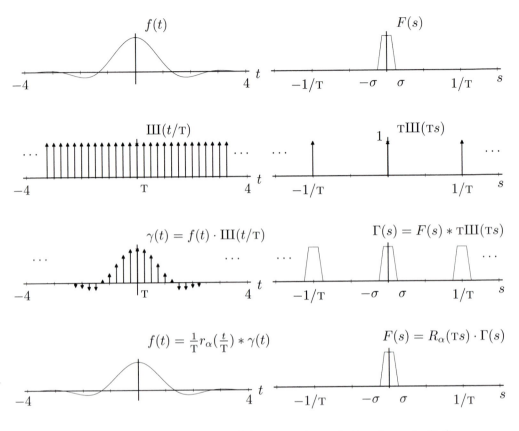

Figure 8.6. Development of the exact interplation formula (14) for a σ-bandlimited function f.

when $C > 0$ is suitably chosen. Let M be any positive integer and let t be any point in the interval $-M\mathrm{T} \leq t \leq M\mathrm{T}$. Given $\epsilon > 0$ we will then have

$$\sum_{|n|>N} \left| f(n\mathrm{T}) r_\alpha \left(\frac{t - n\mathrm{T}}{\mathrm{T}} \right) \right| \leq 2 \sum_{n=N+1}^{\infty} B\, n^m \cdot \frac{C}{(n-M)^{m+2}}$$

$$\leq 2BC(M+1)^m \sum_{n=N+1}^{\infty} \frac{1}{(n-M)^2}$$

$$\leq \frac{2BC(M+1)^m}{N-M}$$

$$< \epsilon$$

when N is sufficiently large.

The cardinal series

In some situations we can eliminate the mollification parameter α from (14) by replacing $r_\alpha(u)$ with

$$\text{sinc}(u) = \lim_{\alpha \to 0+} r_\alpha(u),$$

cf. (11). The resulting *cardinal series* representation

$$f(t) = \sum_{n=-\infty}^{\infty} f(n\text{T}) \, \text{sinc}\left(\frac{t - n\text{T}}{\text{T}}\right) \tag{15}$$

for a suitably regular σ-bandlimited function f is the best known identity from sampling theory. This remarkable formula was discovered independently by the mathematicians E. Borel and J.M. Whittaker as well as the electrical engineers V. Kotel'nikov and C.E. Shannon, who first recognized its significance in communication theory.

We cannot establish (15) by simply replacing R_α with $R_0 := \Pi$ in the above argument because we cannot justify term-by-term multiplication by Π in the steps leading to (14) without imposing additional regularity conditions on f. It seems best to use a completely different approach.

Let f be a σ-bandlimited function with an ordinary Fourier transform F, let $0 < 2\sigma\text{T} \leq 1$, and assume that F, F' are continuous except at finitely many points where jumps can occur. We will use the validity of Fourier's representation for f and for the T^{-1}-periodization (9) of F (as established in Section 1.5) to show that the cardinal series (15) converges uniformly (and weakly) on $\mathbb{R}$.

As a first step, we use the Plancherel identity for (9) to write

$$\text{T} \sum_{n=-\infty}^{\infty} |f(n\text{T})|^2 = \int_{-1/2\text{T}}^{1/2\text{T}} |F(s)|^2 ds < \infty. \tag{16}$$

We choose integers $M < N$ for the summation limits. Using the synthesis equation for functions on $\mathbb{R}$, the Cauchy-Schwartz inequality

$$\left| \int_a^b \alpha(s)\beta(s)ds \right|^2 \leq \int_a^b |\alpha(s)|^2 ds \cdot \int_a^b |\beta(s)|^2 ds,$$

the synthesis equation for functions on $\mathbb{T}_{1/\text{T}}$, and Plancherel's identity for functions on $\mathbb{T}_{1/\text{T}}$ in turn, we write

$$
\left| f(t) - \sum_{n=M}^{N} f(n\text{T}) \operatorname{sinc}\left(\frac{t-n\text{T}}{\text{T}}\right) \right|^2
$$

$$
= \left| \int_{-1/2\text{T}}^{1/2\text{T}} \left\{ F(s) - \sum_{n=M}^{N} f(n\text{T}) e^{-2\pi i n \text{T}s}\, \text{T}\,\Pi(\text{T}s) \right\} e^{2\pi i s t} ds \right|^2
$$

$$
= \left| \int_{-1/2\text{T}}^{1/2\text{T}} \left\{ F(s) - \sum_{n=M}^{N} \text{T}\, f(n\text{T}) e^{-2\pi i n \text{T}s} \right\} \left\{ \Pi(\text{T}s) e^{2\pi i s t} \right\} ds \right|^2 \qquad (17)
$$

$$
\leq \int_{-1/2\text{T}}^{1/2\text{T}} \left| F(s) - \sum_{n=M}^{N} \text{T}\, f(n\text{T}) e^{-2\pi i n \text{T}s} \right|^2 ds \cdot \int_{-1/2\text{T}}^{1/2\text{T}} |e^{2\pi i s t}|^2 ds
$$

$$
= \frac{1}{\text{T}} \int_{-1/2\text{T}}^{1/2\text{T}} \left| \sum_{n<M} \text{T}\, f(n\text{T}) e^{-2\pi i n \text{T}s} + \sum_{n>N} \text{T} f(n\text{T}) e^{-2\pi i n \text{T}s} \right|^2 ds
$$

$$
= \sum_{n<M} |f(n\text{T})|^2 + \sum_{n>N} |f(n\text{T})|^2.
$$

In conjunction with (16), this inequality establishes the uniform (and the weak) convergence of (15). Related identities for $f', f'', \ldots$ are given in Ex. 8.17.

A σ-bandlimited function f is also σ'-bandlimited when $\sigma' \geq \sigma$, and this simple observation leads to a useful generalization of (15). When

$$
\sigma \leq \sigma' \leq \frac{1}{2\text{T}}
$$

we can replace

$$
\operatorname{sinc}\left(\frac{t-n\text{T}}{\text{T}}\right) \quad \text{by} \quad 2\sigma'\text{T} \operatorname{sinc}[2\sigma'(t-n\text{T})], \quad \text{and}
$$

$$
\text{T}\,\Pi(\text{T}s) \qquad \text{by} \quad \text{T}\,\Pi\left(\frac{s}{2\sigma'}\right)
$$

in the first three lines of (17), and thereby show that

$$
f(t) = (2\sigma'\text{T}) \sum_{n=-\infty}^{\infty} f(n\text{T}) \operatorname{sinc}[2\sigma'(t-n\text{T})].
$$

Equivalently, we can write

$$f(t) = a \sum_{n=-\infty}^{\infty} f(n\mathrm{T}) \operatorname{sinc}\left[a\left(\frac{t - n\mathrm{T}}{\mathrm{T}}\right)\right] \quad \text{when } 2\sigma\mathrm{T} \le a \le 1, \tag{18}$$

with the series converging uniformly (and weakly) on $\mathbb{R}$.

Example: Let a_1, a_2, T, t_1, t_2 be real numbers with $0 < a_1 \le a_2 \le 1$ and $\mathrm{T} > 0$. Show that

$$\sum_{n=-\infty}^{\infty} \operatorname{sinc}\left\{\frac{a_1(t_1 - n\mathrm{T})}{\mathrm{T}}\right\} \operatorname{sinc}\left\{\frac{a_2(t_2 - n\mathrm{T})}{\mathrm{T}}\right\} = \frac{1}{a_2} \operatorname{sinc}\left\{\frac{a_1(t_2 - t_1)}{\mathrm{T}}\right\}. \tag{19}$$

Solution: The function

$$f(t) := \frac{1}{a_2} \operatorname{sinc}\left\{\frac{a_1(t - t_1)}{\mathrm{T}}\right\}$$

is σ-bandlimited with $\sigma = a_1/2\mathrm{T}$. We use this function with $t := t_2$, $a := a_2$ in (18) to obtain (19). Related identities are developed in Ex. 8.13 and Ex. 8.14. ∎

You can use (14) with any σ-bandlimited generalized function f, but our proof of (15) only covers the case where F is a piecewise smooth ordinary function. The following examples show what can happen to the cardinal series when we allow F to be a generalized function.

Example: Let $f(t) := 2\pi it$ [so that $F(s) = -\delta'(s)$] and let $\mathrm{T} := 1$. Show that the corresponding cardinal series

$$\sum_{n=-\infty}^{\infty} 2\pi i n \cdot \operatorname{sinc}(t - n)$$

does not converge weakly.

Solution: Let the Schwartz function ϕ be constructed so that

$$\phi^{\wedge}(s) = s \quad \text{when} \quad -\tfrac{1}{2} \le s \le \tfrac{1}{2},$$

cf. Ex. 7.4. We use Parseval's identity to write

$$2\pi i n \int_{-\infty}^{\infty} \operatorname{sinc}(t - n)\phi(t)dt = 2\pi i n \int_{-\infty}^{\infty} e^{2\pi i n s}\Pi(s)\phi^{\wedge}(s)ds$$

$$= \int_{-1/2}^{1/2} 2\pi i n s \cdot e^{2\pi i n s}ds = \begin{cases} 0 & \text{if } n = 0 \\ (-1)^n & \text{if } n = \pm 1, \pm 2, \ldots, \end{cases}$$

and observe that the series

$$\sum_{n=-\infty}^{\infty} \int_{-\infty}^{\infty} 2\pi i n \cdot \mathrm{sinc}(t-n)\phi(t)dt = \sum_{\substack{n=-\infty \\ n\neq 0}}^{\infty} (-1)^n$$

does not converge.   ∎

Example: Let $f(t) := e^{2\pi i s_0 t}$ [so that $F(s) = \delta(s - s_0)$] and let $\mathrm{T} > 0$ be selected so that $-1 < 2s_0\mathrm{T} < 1$. Show that f is given by the weak limit

$$f(t) = \lim_{N\to\infty} \sum_{n=-N}^{N} f(n\mathrm{T})\, \mathrm{sinc}\left(\frac{t-n\mathrm{T}}{\mathrm{T}}\right). \tag{20}$$

Solution: Let $\phi \in \mathbb{S}$ be selected. The piecewise smooth T^{-1}-periodic function

$$\sum_{m=-\infty}^{\infty} \phi^{\wedge}\left(s - \frac{m}{\mathrm{T}}\right) \Pi\left\{\mathrm{T}\left(s - \frac{m}{\mathrm{T}}\right)\right\}$$

and its derivative are continuous except at the points $s = \pm 1/2\mathrm{T}, \pm 3/2\mathrm{T}, \ldots$ where there can be jump discontinuities. This being the case, the analysis from Section 1.5 shows that the (symmetrized) Fourier series converges to $\phi^{\wedge}(-s_0)$ at the point $s = -s_0$. In this way we see that

$$\int_{-\infty}^{\infty} e^{2\pi i s_0 t}\phi(t)dt = \phi^{\wedge}(-s_0)$$

$$= \lim_{N\to\infty} \sum_{n=-N}^{N} \left\{\mathrm{T}\int_{-1/2\mathrm{T}}^{1/2\mathrm{T}} \phi^{\wedge}(s)e^{-2\pi i n\mathrm{T}s}ds\right\}e^{-2\pi i n\mathrm{T}s_0}$$

$$= \lim_{N\to\infty} \sum_{n=-N}^{N} e^{2\pi i n\mathrm{T}s_0}\int_{-\infty}^{\infty} e^{2\pi i n\mathrm{T}s}\cdot\mathrm{T}\,\Pi(\mathrm{T}s)\phi^{\wedge}(s)ds$$

$$= \lim_{N\to\infty} \sum_{n=-N}^{N} e^{2\pi i n\mathrm{T}s_0}\int_{-\infty}^{\infty} \mathrm{sinc}\left(\frac{t-n\mathrm{T}}{\mathrm{T}}\right)\phi(t)dt$$

$$= \lim_{N\to\infty} \int_{-\infty}^{\infty}\left\{\sum_{n=-N}^{N} e^{2\pi i n\mathrm{T}s_0}\,\mathrm{sinc}\left(\frac{t-n\mathrm{T}}{\mathrm{T}}\right)\right\}\phi(t)dt. \quad\blacksquare$$

Recovery of an alias

Let f be σ-bandlimited and let

$$f_m(t) := e^{2\pi i m t/\mathrm{T}} \cdot f(t) \tag{21}$$

where $\mathrm{T} > 0$ and m is a nonzero integer. By construction,

$$f_m(n\mathrm{T}) = f(n\mathrm{T}) \quad \text{for all} \quad n = 0, \pm 1, \pm 2, \dots \, .$$

Now if $2\sigma\mathrm{T} < 1$, we can use (14) and (21) to write

$$f_m(t) = e^{2\pi i m t/\mathrm{T}} \cdot \sum_{n=-\infty}^{\infty} f_m(n\mathrm{T}) r_\alpha \left(\frac{t - n\mathrm{T}}{\mathrm{T}} \right) \tag{22}$$

with the series converging weakly on $\mathbb{R}$. Similarly, if $2\sigma\mathrm{T} \leq 1$ and F is piecewise smooth, we can use (15) and (21) to write

$$f_m(t) = e^{2\pi i m t/\mathrm{T}} \cdot \sum_{n=-\infty}^{\infty} f_m(n\mathrm{T}) \operatorname{sinc} \left(\frac{t - n\mathrm{T}}{\mathrm{T}} \right) \tag{23}$$

with the series converging uniformly (and weakly) on $\mathbb{R}$. If we know the phase shift parameter m/T, we can reconstruct f_m from its T-samples, even when T is much larger than the Nyquist sample size $\mathrm{T}/(2\sigma\mathrm{T} + 2|m|)$ for this function.

Fragmentation of Π

Let $k_1, k_2, \dots, k_M$ be integers and let $-\frac{1}{2} = s_0 < s_1 < \cdots < s_M = \frac{1}{2}$. We split Π into

$$\Pi_\mu(s) := \begin{cases} 1 & \text{if } s_{\mu-1} \leq s < s_\mu \\ 0 & \text{otherwise,} \end{cases} \qquad \mu = 1, 2, \dots, M,$$

sum translates of these fragments to produce

$$P(s) := \sum_{\mu=1}^{M} \Pi_\mu(s - k_\mu), \tag{24}$$

and synthesize the corresponding basis function

$$p(t) := \int_{-\infty}^{\infty} P(s) e^{2\pi i s t} ds, \tag{25}$$

as illustrated in Fig. 8.7.

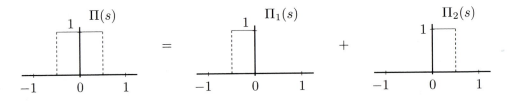

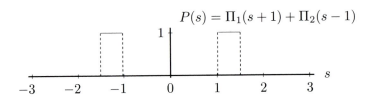

$$P(s) = \Pi_1(s+1) + \Pi_2(s-1)$$

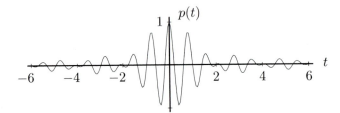

Figure 8.7. Construction of the functions P, p by using translates of fragments of Π.

Now let F be a piecewise smooth support-limited function, let

$$f(t) = \int_{-\infty}^{\infty} F(s)e^{2\pi i s t} ds,$$

and let $\text{T} > 0$. We will not require f to be $1/2\text{T}$-bandlimited [as we did for the proof of (15)], but we will insist that

$$F(s) = P(\text{T}s) \cdot F(s) \tag{26}$$

so the points where $F(s) \neq 0$ can be covered by M intervals having the collective length $1/\text{T}$. Since the 1-translates of $[-\frac{1}{2}, \frac{1}{2})$ exactly cover $\mathbb{R}$, we must have

$$\sum_{m=-\infty}^{\infty} P(s-m) = 1$$

and

$$P(s)P(s-m) = \begin{cases} P(s) & \text{if } m = 0 \\ 0 & \text{if } m = \pm 1, \pm 2, \ldots . \end{cases}$$

We use this identity with (26) to see that

$$P(\mathrm{T}s) \cdot F\left(s - \frac{m}{\mathrm{T}}\right) = P(\mathrm{T}s) \cdot \left\{ P\left[\mathrm{T}\left(s - \frac{m}{\mathrm{T}}\right)\right] F\left(s - \frac{m}{\mathrm{T}}\right) \right\}$$

$$= \{P(\mathrm{T}s)P(\mathrm{T}s - m)\} F\left(s - \frac{m}{\mathrm{T}}\right)$$

$$= \begin{cases} F(s) & \text{if } m = 0 \\ 0 & \text{if } m = \pm 1, \pm 2, \dots, \end{cases}$$

at each point s. In conjunction with (9) this leads us to conjecture that

$$F(s) = \sum_{n=-\infty}^{\infty} \mathrm{T} f(n\mathrm{T}) e^{-2\pi i n \mathrm{T}s} P(\mathrm{T}s)$$

and

$$f(t) = \sum_{n=-\infty}^{\infty} f(n\mathrm{T}) \, p\left(\frac{t - n\mathrm{T}}{\mathrm{T}}\right). \tag{27}$$

A minor variation of the argument from (17) gives the bound

$$\left| f(t) - \sum_{n=M}^{N} f(n\mathrm{T}) \, p\left(\frac{t - n\mathrm{T}}{\mathrm{T}}\right) \right|^2 \leq \sum_{n<M} |f(n\mathrm{T})|^2 + \sum_{n>N} |f(n\mathrm{T})|^2, \tag{28}$$

cf. Ex. 8.21. In this way we establish (27) and verify that the series converges uniformly (and weakly) on $\mathbb{R}$. Both (15) and (23) are special cases of (27).

8.3 Reconstruction of f from Samples of $a_1*f, a_2*f, \ldots$

Filters

Let a be a support-limited generalized function with the Fourier transform A, and let f be σ-bandlimited. For simplicity, we will again assume that F, F' are continuous except for finitely many points where finite jumps can occur. We will use the function a to *filter* f by writing

$$g := a * f, \qquad G = A \cdot F. \tag{29}$$

For example, when

$$a(t) = \delta(t - t_0), \quad \delta^{(k)}(t), \quad \frac{1}{2h} \Pi\left(\frac{t}{2h}\right)$$

with $-\infty < t_0 < \infty$, $k = 0, 1, 2, \ldots$, $h > 0$, we find

$$g(t) = f(t - t_0), \quad f^{(k)}(t), \quad \frac{1}{2h} \int_{u=t-h}^{t+h} f(u)\,du,$$

respectively.

From (29) we see that g, like f, is σ-bandlimited, so we can use (16) to show that the T-samples of g are square summable with

$$\text{T} \sum_{n=-\infty}^{\infty} |(a * f)(n\text{T})|^2 \leq \int_{-1/2\text{T}}^{1/2\text{T}} |A(s) \cdot F(s)|^2 ds \tag{30}$$

when $0 < 2\sigma\text{T} \leq 1$.

Samples from one filter

Let a, f, g be as above, let $0 < 2\sigma\text{T} \leq 1$, and assume that

$$A(s) \neq 0 \quad \text{for} \quad -\frac{1}{2\text{T}} \leq s \leq \frac{1}{2\text{T}}. \tag{31}$$

We will show that it is possible to reconstruct f from the T-samples of g.

For this purpose we define

$$P(s) := \begin{cases} \dfrac{1}{A(s)} & \text{if } -\dfrac{1}{2\text{T}} \leq s \leq \dfrac{1}{2\text{T}} \\ 0 & \text{otherwise,} \end{cases} \tag{32}$$

$$p(t) := \int_{-1/2\text{T}}^{1/2\text{T}} P(s) e^{2\pi i s t}\,ds, \tag{33}$$

$$\gamma(t) := g(t) \cdot \text{III}\left(\frac{t}{\text{T}}\right) = \sum_{n=-\infty}^{\infty} \text{T}\, g(n\text{T})\delta(t - n\text{T}), \tag{34}$$

$$\Gamma(s) = G(s) * \text{T}\,\text{III}(\text{T}s) = \sum_{m=-\infty}^{\infty} G\left(s - \frac{m}{\text{T}}\right). \tag{35}$$

We use (35) with (32) and (29) to show that

$$P(s) \cdot \Gamma(s) = \sum_{m=-\infty}^{\infty} P(s) \cdot G\left(s - \frac{m}{\text{T}}\right) = P(s) \cdot G(s) = F(s)$$

at each point s, and then informally use (34) to write

$$f(t) = \sum_{n=-\infty}^{\infty} g(n\mathrm{T}) \cdot \mathrm{T}\, p(t - n\mathrm{T}). \tag{36}$$

If we replace P by a suitable mollification P_α (cf. Fig. 8.5), we can justify the term-by-term multiplication and convolution that we use in this derivation.

It is a bit easier to modify the argument of (17) by writing

$$\left| f(t) - \sum_{n=M}^{N} \mathrm{T}\, g(n\mathrm{T}) p(t - n\mathrm{T}) \right|^2$$

$$= \left| \int_{-1/2\mathrm{T}}^{1/2\mathrm{T}} \left\{ F(s) - \sum_{n=M}^{N} \mathrm{T}\, g(n\mathrm{T}) e^{-2\pi i n \mathrm{T}s} P(s) \right\} e^{2\pi i s t} ds \right|^2$$

$$= \left| \int_{-1/2\mathrm{T}}^{1/2\mathrm{T}} \left\{ G(s) - \sum_{n=M}^{N} \mathrm{T}\, g(n\mathrm{T}) e^{-2\pi i n \mathrm{T}s} \right\} \{ P(s) e^{2\pi i s t} \} ds \right|^2 \tag{37}$$

$$\leq \int_{-1/2\mathrm{T}}^{1/2\mathrm{T}} \left| G(s) - \sum_{n=M}^{N} \mathrm{T}\, g(n\mathrm{T}) e^{-2\pi i n \mathrm{T}s} \right|^2 ds \cdot \int_{-1/2\mathrm{T}}^{1/2\mathrm{T}} |P(s)|^2 ds$$

$$= \mathrm{T} \left\{ \sum_{n<M} |g(n\mathrm{T})|^2 + \sum_{n>N} |g(n\mathrm{T})|^2 \right\} \cdot \int_{-1/2\mathrm{T}}^{1/2\mathrm{T}} \frac{ds}{|A(s)|^2}.$$

In conjunction with (30) this shows that the series (36) converges uniformly (and weakly) on $\mathbb{R}$.

The Papoulis generalization

When the σ-bandlimited function f and the filter a are suitably regular and $0 < 2\sigma\mathrm{T} \leq 1$, we can reconstruct f from the samples

$$(a * f)(n\mathrm{T}), \qquad n = 0, \pm 1, \pm 2, \dots .$$

If we are given two, three, ... times as much data in each sampling interval, we might reasonably expect to use steps that are two, three, ... times as large. For example, we might reasonably expect to reconstruct f from the samples

$$f(n{\cdot}2\mathrm{T}), \quad f'(n{\cdot}2\mathrm{T}), \quad n = 0, \pm 1, \pm 2, \dots$$

or

$$f(n{\cdot}3\mathrm{T}), \quad f'(n{\cdot}3\mathrm{T}), \quad f''(n{\cdot}3\mathrm{T}), \quad n = 0, \pm 1, \pm 2, \dots$$

when $0 < 2\sigma\mathrm{T} \leq 1$. We will now present the clever analysis that A. Papoulis created to produce the corresponding interpolation formulas.

Let $a_1, a_2, \dots, a_K$ be support-limited generalized functions, and let f be a σ-bandlimited function with a piecewise smooth Fourier transform F. The filtered function

$$g_\kappa := a_\kappa * f \quad \text{with FT} \quad G_\kappa = A_\kappa \cdot F \tag{38}$$

is then σ-bandlimited, so we can define

$$\gamma_\kappa(t) := g_\kappa(t) \cdot \mathrm{III}\left(\frac{t}{K\mathrm{T}}\right) = \sum_{n=-\infty}^{\infty} K\mathrm{T}\, g_\kappa(n \cdot K\mathrm{T}) \delta(t - n \cdot K\mathrm{T}) \tag{39}$$

and write

$$\Gamma_\kappa(s) = G_\kappa(s) * K\mathrm{T}\, \mathrm{III}(K\mathrm{T}s) = \sum_{m=-\infty}^{\infty} G_\kappa\left(s + \frac{m}{K\mathrm{T}}\right)$$

$$= \sum_{m=-\infty}^{\infty} A_\kappa\left(s + \frac{m}{K\mathrm{T}}\right) F\left(s + \frac{m}{K\mathrm{T}}\right) \tag{40}$$

for each $\kappa = 1, 2, \dots, K$. Our goal is to reconstruct f from the samples

$$g_\kappa(n \cdot K\mathrm{T}), \quad n = 0, \pm 1, \pm 2, \dots, \quad \kappa = 1, 2, \dots, K$$

or equivalently from the periodic functions $\Gamma_1, \Gamma_2, \dots, \Gamma_K$, cf. Fig. 8.8.

We will assume that $0 < 2\sigma\mathrm{T} \leq 1$ [but insist that $F(\pm\sigma) = 0$ when $2\sigma\mathrm{T} = 1$]. For notational convenience, we will use

$$S := \frac{1}{2\mathrm{T}}$$

for the upper bound on σ and

$$H := \frac{1}{K\mathrm{T}} = \frac{2S}{K} \tag{41}$$

for the periodization parameter from (40). Since the sampling step, $K\mathrm{T}$, in (39) is K times larger than the Nyquist step T, the periodization parameter, H, in (40) is K times smaller than the parameter $1/\mathrm{T}$, which is guaranteed to separate the translates of F, cf. Figs. 8.6 and 8.9. As many as K terms of (40) can contribute to the value of Γ_κ at a given point s. In particular,

$$\sum_{\lambda=1}^{K} A_\kappa(s + (\lambda - 1)H) \cdot F(s + (\lambda - 1)H) = \Gamma_\kappa(s), \quad \kappa = 1, 2, \dots, K \tag{42}$$

$$\text{when} \quad -S \leq s < -S + H.$$

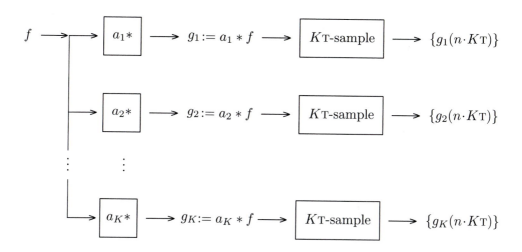

Figure 8.8. Comparison of the Shannon samples and the Papoulis samples that are used to reconstruct a σ-bandlimited function f when $2\sigma\mathrm{T} \leq 1$.

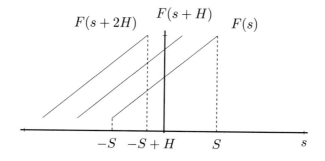

Figure 8.9. The functions $F(s)$, $F(s+H)$, $F(s+2H)$ can be nonzero on $[-S, -S+H)$ when $H = 2S/3$.

If $A_1, A_2, \ldots, A_K$ are suitably regular, we can solve this system of K linear equations to find the K unknowns

$$F(s), F(s+H), \ldots, F(s+(K-1)H) \quad \text{when} \quad -S \le s < -S+H.$$

In this way we obtain $F(s)$ for $-S \le s < S$, so we can synthesize

$$f(t) = \int_{-S}^{S} F(s)e^{2\pi i s t} dt.$$

It is one thing to see that such a reconstruction is feasible, but quite another to work out the details, especially in the way that Papoulis did! We will describe the procedure he used to find the interpolation formula.

We abandon (42) and use the system of K linear equations

$$\sum_{\kappa=1}^{K} A_\kappa(s+(\lambda-1)H)P_\kappa(s,t) = e^{2\pi i(\lambda-1)Ht}, \quad \lambda = 1, 2, \ldots, K \qquad (43)$$

to find the K unknown functions

$$P_1(s,t), P_2(s,t), \ldots, P_K(s,t) \quad \text{when} \quad -S \le s < -S+H, \quad -\infty < t < \infty.$$

Of course, we must impose a suitable regularity condition on the filters $a_1, a_2, \ldots, a_K$ [the same as that required for (42)] to ensure that this can be done for each s.

We now define

$$p_\kappa(t) := \frac{1}{H} \int_{-S}^{-S+H} P_\kappa(s,t)e^{2\pi i s t} ds, \quad -\infty < t < \infty, \qquad (44)$$

and let

$$\langle P_\kappa(s,t)e^{2\pi i s t} \rangle_H$$

denote the H-periodic extension of

$$P_\kappa(s,t)e^{2\pi i s t}, \quad -S \le s < S+H$$

to $-\infty < s < \infty$. We use the analysis equation integral

$$\frac{1}{H} \int_{-S}^{-S+H} P_\kappa(s,t)e^{2\pi i s t} \cdot e^{-2\pi i n s/H} ds = p_\kappa\left(t - \frac{n}{H}\right)$$

with (41) to obtain the Fourier series

$$\langle P_\kappa(s,t)e^{2\pi i s t} \rangle_H = \sum_{n=-\infty}^{\infty} p_\kappa(t - n\cdot K_T)e^{2\pi i n\,K_T s}. \qquad (45)$$

We also rewrite (43) in the form

$$\sum_{\kappa=1}^{K} A_\kappa(s + (\lambda - 1)H)P_\kappa(s,t)e^{2\pi i s t} = e^{2\pi i (s + (\lambda-1)H)t}$$

$$\lambda = 1, 2, \ldots, K, \qquad -S \le s < -S + H,$$

and thereby see that

$$\sum_{\kappa=1}^{K} A_\kappa(s)\langle P_\kappa(s,t)e^{2\pi i s t}\rangle_H = e^{2\pi i s t}, \qquad -S \le s < S. \tag{46}$$

The hard work is done! We formally use (46), (45), and (38) in turn to write

$$f(t) = \int_{-S}^{S} F(s)e^{2\pi i s t}ds$$

$$= \int_{-S}^{S} F(s) \sum_{\kappa=1}^{K} A_\kappa(s)\langle P_\kappa(s,t)e^{2\pi i s t}\rangle_H ds$$

$$= \int_{-S}^{S} F(s) \sum_{\kappa=1}^{K} A_\kappa(s) \sum_{n=-\infty}^{\infty} p_\kappa(t - n \cdot K_{\mathrm{T}})e^{2\pi i n K_{\mathrm{T}} s}ds$$

$$\overset{?}{=} \sum_{\kappa=1}^{K} \sum_{n=-\infty}^{\infty} p_\kappa(t - n \cdot K_{\mathrm{T}}) \int_{-S}^{S} F(s)A_\kappa(s)e^{2\pi i n K_{\mathrm{T}} s}ds$$

$$= \sum_{\kappa=1}^{K} \sum_{n=-\infty}^{\infty} p_\kappa(t - n \cdot K_{\mathrm{T}}) \int_{-S}^{S} G_\kappa(s)e^{2\pi i n K_{\mathrm{T}} s}ds$$

$$= \sum_{\kappa=1}^{K} \sum_{n=-\infty}^{\infty} g_\kappa(n \cdot K_{\mathrm{T}})p_\kappa(t - n \cdot K_{\mathrm{T}}),$$

and thereby obtain the Papoulis formula

$$f(t) = \sum_{\kappa=1}^{K} \sum_{n=-\infty}^{\infty} g_\kappa(n \cdot K_{\mathrm{T}})p_\kappa(t - n \cdot K_{\mathrm{T}}) \tag{47}$$

for reconstructing f from the K_{T}-samples of $g_1, g_2, \ldots, g_K$.

You can use this procedure to find a multitude of useful interpolation formulas, *cf.* Ex. 8.24. As you work through the details with a particular set of filters $a_1, a_2, \ldots, a_K$ you can check to make sure that the system (43) is nonsingular, that the Fourier series (45) makes good sense, etc. In most cases it is easier to do an *a posteriori* analysis to establish the formula (47).

Example: Show how to recover the σ-bandlimited function f from the samples $f(n \cdot 2\text{T})$, $f'(n \cdot 2\text{T})$, $n = 0, \pm 1, \pm 2, \ldots$ when $2\sigma\text{T} \le 1$ and F is piecewise smooth.

Solution: We define $S := 1/2\text{T}$, $H := 1/2\text{T}$, and use the filters

$$a_1(t) := \delta(t) \quad \text{with FT} \quad A_1(s) = 1,$$
$$a_2(t) := \delta'(t) \quad \text{with FT} \quad A_2(s) = 2\pi i s.$$

We solve the linear equations

$$1 \cdot P_1(s,t) + 2\pi i s \quad\quad\quad \cdot P_2(s,t) = 1$$
$$1 \cdot P_1(s,t) + 2\pi i (s + H) \cdot P_2(s,t) = e^{2\pi i H t}$$

from (43) to obtain

$$P_1(s,t) = \frac{H + s - s\,e^{2\pi i H t}}{H}, \quad\quad P_2(s,t) = \frac{e^{2\pi i H t} - 1}{2\pi i H}$$

(when $-S \le s < 0$ and $-\infty < t < \infty$), and then compute

$$p_1(t) := \frac{1}{H} \int_{-H}^{0} P_1(s,t) e^{2\pi i s t}\,ds = \frac{1}{H^2} \int_{-H}^{0} \{(H + s)e^{2\pi i s t} - s\,e^{2\pi i (H+s)t}\}\,ds$$

$$= \frac{1}{H^2} \int_{-H}^{H} (H - |s|)e^{2\pi i s t}\,ds = \text{sinc}^2(Ht),$$

$$p_2(t) := \frac{1}{H} \int_{-H}^{0} P_2(s,t) e^{2\pi i s t}\,ds = \frac{e^{2\pi i H t} - 1}{2\pi i H^2} \int_{-H}^{0} e^{2\pi i s t}\,ds$$

$$= \frac{t(e^{2\pi i H t} - 1)(1 - e^{-2\pi i H t})}{(2\pi i H t)^2} = t\,\text{sinc}^2(Ht).$$

We use these expressions with (47) to obtain the interpolation formula

$$f(t) = \sum_{n=-\infty}^{\infty} \{f(n \cdot 2\text{T}) + (t - n \cdot 2\text{T})f'(n \cdot 2\text{T})\} \text{sinc}^2\left(\frac{t - n \cdot 2\text{T}}{2\text{T}}\right) \quad\quad (48)$$

for recovering f from the samples $f(n \cdot 2\text{T})$, $f'(n \cdot 2\text{T})$. [The analogous formula

$$f(t) = \sum_{n=-\infty}^{\infty} \left\{f(n \cdot 3\text{T}) + (t - n \cdot 3\text{T}) + \frac{(t - n \cdot 3\text{T})^2}{2!} f''(n \cdot 3\text{T})\right\} \text{sinc}^3\left(\frac{t - n \cdot 3\text{T}}{3\text{T}}\right)$$

for recovering f from the samples $f(n \cdot 3\text{T})$, $f'(n \cdot 3\text{T})$, $f''(n \cdot 3\text{T})$ reveals the general pattern.] ∎

8.4 Approximation of Almost Bandlimited Functions

Let f have the piecewise smooth Fourier transform F, and assume that the tails of F are small in the sense that

$$|F(s)| \le G(|s|), \qquad -\infty < s < \infty$$

where G is a continuous, decreasing, integrable function on $0 \le s < \infty$. We will show that any such *almost bandlimited* function f is well approximated by the cardinal series when the sampling interval $\text{T} > 0$ is sufficiently small.

With this in mind we split F into the segments

$$F_m(s) := \begin{cases} F(s) & \text{if } -\dfrac{1}{2\text{T}} \le s - \dfrac{m}{\text{T}} < \dfrac{1}{2\text{T}} \\ 0 & \text{otherwise,} \end{cases} \qquad m = 0, \pm 1, \pm 2, \dots, \tag{49}$$

and set

$$f_m(t) := \int_{-\infty}^{\infty} F_m(s) e^{2\pi i s t} ds, \qquad m = 0, \pm 1, \pm 2, \dots . \tag{50}$$

We show that

$$f(t) = \sum_{m=-\infty}^{\infty} f_m(t) \tag{51}$$

(with the series converging uniformly for $-\infty < t < \infty$) by using (50), (49), and the tail hypothesis to write

$$\sum_{m=-\infty}^{\infty} |f_m(t)| \le \sum_{m=-\infty}^{\infty} \int_{-\infty}^{\infty} |F_m(s)| ds = \int_{-\infty}^{\infty} |F(s)| ds < \infty.$$

We also use the tail hypothesis to verify that

$$\sum_{m=-\infty}^{\infty} \sum_{n=-\infty}^{\infty} f_m(n\text{T}) \operatorname{sinc}\left(\frac{t - n\text{T}}{\text{T}}\right)$$

converges absolutely (so that we can switch the order of the m, n summations), *cf.* Ex. 8.26. We can then use (51) with the alias lemma (23) to write

$$
\left| f(t) - \sum_{n=-\infty}^{\infty} f(n\mathrm{T}) \operatorname{sinc}\left(\frac{t - n\mathrm{T}}{\mathrm{T}}\right) \right|
$$

$$
= \left| \sum_{m=-\infty}^{\infty} f_m(t) - \sum_{n=-\infty}^{\infty} \sum_{m=-\infty}^{\infty} f_m(n\mathrm{T}) \operatorname{sinc}\left(\frac{t - n\mathrm{T}}{\mathrm{T}}\right) \right|
$$

$$
= \left| \sum_{m=-\infty}^{\infty} \left\{ f_m(t) - \sum_{n=-\infty}^{\infty} f_m(n\mathrm{T}) \operatorname{sinc}\left(\frac{t - n\mathrm{T}}{\mathrm{T}}\right) \right\} \right| \tag{52}
$$

$$
= \left| \sum_{m=-\infty}^{\infty} \{1 - e^{-2\pi i m t/\mathrm{T}}\} f_m(t) \right|
$$

$$
\leq 2 \sum_{m \neq 0} |f_m(t)| \leq 2 \int_{|s| > 1/2\mathrm{T}} |F(s)| ds.
$$

Given $\epsilon > 0$, we will have

$$
\left| f(t) - \sum_{n=-\infty}^{\infty} f(n\mathrm{T}) \operatorname{sinc}\left(\frac{t - n\mathrm{T}}{\mathrm{T}}\right) \right| < \epsilon, \quad -\infty < t < \infty,
$$

if we choose $\mathrm{T} > 0$ so that

$$
\int_{|s| > 1/2\mathrm{T}} |F(s)| ds < \frac{\epsilon}{2}.
$$

Each term of the cardinal series is $1/2\mathrm{T}$-bandlimited, so the same is true of the approximating sum, *cf.* Ex. 8.6.

Example: Apply the above analysis to

$$
f(t) := \operatorname{sinc}(2t - 1),
$$

and thereby show that the bound (52) is sharp.

Solution: The Fourier transform

$$
F(s) = e^{-\pi i s} \cdot \frac{1}{2} \Pi\left(\frac{s}{2}\right)
$$

vanishes when $|s| > 1$, so f is σ-bandlimited with $\sigma = 1$. If we use the sample spacing $\mathrm{T} = 1$ (twice as large as the Nyquist sample spacing $1/2\sigma = 1/2$), we find

$$
f(n\mathrm{T}) = \operatorname{sinc}(2n - 1) = 0 \quad \text{for} \quad n = 0, \pm 1, \pm 2, \dots .
$$

The error bound

$$\left| \text{sinc}(2t - 1) \right| = \left| f(t) - \sum_{n=-\infty}^{\infty} f(n\text{T}) \, \text{sinc} \left(\frac{t - n\text{T}}{\text{T}} \right) \right| \leq 2 \int_{|s| > 1/2\text{T}} F(s) ds = 1$$

of (52) is sharp when $t = 1/2$.   ∎

Example: Use (52) to estimate the error in the approximation

$$\frac{1}{1 + t^2} \approx \sum_{n=-\infty}^{\infty} \frac{1}{1 + n^2 \text{T}^2} \, \text{sinc} \left(\frac{t - n\text{T}}{\text{T}} \right)$$

when $\text{T} = 1, 1/2, 1/4, 1/8$.

Solution: Since

$$f(t) = \frac{1}{1 + t^2} \quad \text{has the FT} \quad F(s) = \pi \, e^{-2\pi |s|},$$

the bound from (52) is given by

$$2 \int_{|s| > 1/2\text{T}}^{\infty} |F(s)| ds = 4 \int_{s = 1/2\text{T}}^{\infty} \pi \, e^{-2\pi s} ds = 2 \, e^{-\pi/\text{T}}.$$

When $\text{T} = 1, 1/2, 1/4, 1/8$ the error does not exceed $2e^{-\pi} \approx 8.64 \cdot 10^{-2}$, $2e^{-2\pi} \approx 3.73 \cdot 10^{-3}$, $2e^{-4\pi} \approx 6.98 \cdot 10^{-6}$, $2e^{-8\pi} \approx 2.43 \cdot 10^{-11}$, respectively.   ∎

We will now generalize (17). The resulting *real-world* sampling theorem shows that we can produce an arbitrarily good uniform approximation to an almost bandlimited function f by truncating a cardinal series.

As a first step, we use the Cauchy-Schwartz inequality

$$\left| \sum_{n=-\infty}^{\infty} \alpha_n \beta_n \right|^2 \leq \sum_{n=-\infty}^{\infty} |\alpha_n|^2 \sum_{n=-\infty}^{\infty} |\beta_n|^2 \tag{53}$$

and the identity

$$\sum_{n=-\infty}^{\infty} \text{sinc}^2(\tau - n) = 1, \qquad -\infty < \tau < \infty$$

(*cf.* Ex. 8.14) to write

$$\left| \sum_{n < M} f(n\text{T}) \, \text{sinc} \left(\frac{t - n\text{T}}{\text{T}} \right) \right|^2 \leq \sum_{n < M} |f(n\text{T})|^2,$$

$$\left| \sum_{n > N} f(n\text{T}) \, \text{sinc} \left(\frac{t - n\text{T}}{\text{T}} \right) \right|^2 \leq \sum_{n > N} |f(n\text{T})|^2. \tag{54}$$

We use the tail hypothesis [*cf.* Ex. 8.26(c)] to show that the series

$$\sum_{m=-\infty}^{\infty} F\left(s - \frac{m}{\mathrm{T}}\right)$$

converges uniformly on $\mathbb{R}$ to a bounded T^{-1}-periodic function with the Fourier coefficients $\mathrm{T}\, f(n\mathrm{T})$, $n = 0, \pm 1, \pm 2, \dots$. We then use Bessel's inequality to write

$$\sum_{n=-\infty}^{\infty} |f(n\mathrm{T})|^2 \le \frac{1}{\mathrm{T}} \int_{-1/2\mathrm{T}}^{1/2\mathrm{T}} \left| \sum_{m=-\infty}^{\infty} F\left(s - \frac{m}{\mathrm{T}}\right) \right|^2 ds < \infty.$$

Knowing that F is absolutely integrable and that the samples $f(n\mathrm{T})$ are square summable, we use (52) and (54) to obtain the bound

$$\left| f(t) - \sum_{n=M}^{N} f(n\mathrm{T}) \operatorname{sinc}\left(\frac{t - n\mathrm{T}}{\mathrm{T}}\right) \right| \le \left| f(t) - \sum_{n=-\infty}^{\infty} f(n\mathrm{T}) \operatorname{sinc}\left(\frac{t - n\mathrm{T}}{\mathrm{T}}\right) \right|$$

$$+ \left| \sum_{n<M} f(n\mathrm{T}) \operatorname{sinc}\left(\frac{t - n\mathrm{T}}{\mathrm{T}}\right) \right| + \left| \sum_{n>N} f(n\mathrm{T}) \operatorname{sinc}\left(\frac{t - n\mathrm{T}}{\mathrm{T}}\right) \right| \qquad (55)$$

$$\le 2 \int_{s>1/2\mathrm{T}} |F(s)|\,ds + \sum_{n<M} |f(n\mathrm{T})|^2 + \sum_{n>N} |f(n\mathrm{T})|^2.$$

If we are given some $\epsilon > 0$, we can make the integral less than $\epsilon/3$ by choosing a sufficiently small sampling interval $\mathrm{T} > 0$. We can then make each of the sums less than $\epsilon/3$ by suitably choosing M, N. In this way we produce a truncated cardinal series approximation to f with

$$\left| f(t) - \sum_{n=M}^{N} f(n\mathrm{T}) \operatorname{sinc}\left(\frac{t - n\mathrm{T}}{\mathrm{T}}\right) \right| < \epsilon \quad \text{for} \quad -\infty < t < \infty. \qquad (56)$$

We can almost reconstruct any almost bandlimited function f from finitely many T-samples. This remarkable inequality is the mathematical foundation for digital signal processing!

References

J.R. Higgins, *Sampling Theory in Fourier and Signal Analysis*, Clarendon Press, Oxford, 1996.

  A rigorous graduate-level introduction to the mathematics of sampling theory.

J.R. Higgins, Five short stories about the cardinal series, *Bull. Amer. Math. Soc.* **12**(1985), 45–89.

  A definitive history of the sampling theorem and its generalizations.

R.J. Marks II, *Introduction to Shannon Sampling and Interpolation Theory*, Springer-Verlag, New York, 1991.

A graduate-level text in sampling theory for electrical engineers.

A.V. Oppenheim, A.S. Willsky, and I.T. Young, *Signals and Systems*, Prentice Hall, Englewood Cliffs, NJ, 1983.

Chapter 8 provides an exceptionally well written elementary introduction to the sampling theorem for scientists and engineers.

A. Papoulis, *Signal Analysis*, McGraw-Hill, New York, 1977.

Chapter 6 contains Papoulis's own exposition of his sampling theorem.

A.I. Zayed, *Advances in Shannon's Sampling Theory*, CRC Press, Boca Raton, FL, 1993.

A mathematical monograph dealing with the sampling theorem, its generalizations, and its connections to other branches of mathematics.

Exercise Set 8

▶ **EXERCISE 8.1:** Let f_k be σ_k-bandlimited, $k = 1, 2, \ldots, K$. How must $\mathrm{T} > 0$ be chosen to ensure that we can recover f from the samples $f(n\mathrm{T})$, $n = 0, \pm 1, \pm 2, \ldots$ when:

(a) $f := f_1 \cdot f_2 \cdot \cdots \cdot f_K$? (b) $f := f_1 * f_2 * \cdots * f_K$?

Note. An additional regularity is needed to ensure that the convolution product is well defined, *e.g.*, you may assume that $F_1, F_2, \ldots, F_K$ are piecewise smooth ordinary functions on $\mathbb{R}$.

▶ **EXERCISE 8.2:** A certain computer sound card records and plays at a rate of 20,000 samples/sec. The card is used to generate samples $f_0, f_1, f_2, \ldots, f_{39999}$ of the audio waveform produced when a human voice slowly reads the words *Joseph Fourier*. Describe what you would hear when you play the samples:

(a) $f_0, f_1, f_2, \ldots, f_{39999}$? (b) $f_{39999}, f_{39998}, f_{39997}, \ldots, f_0$?

(c) $f_0, f_2, f_4, f_6, \ldots, f_{39998}$? (d) $f_0, f_0, f_1, f_1, f_2, f_2, \ldots, f_{39999}, f_{39999}$?

(e) $\dfrac{f_0 + f_1 + f_2}{3}, \dfrac{f_1 + f_2 + f_3}{3}, \dfrac{f_2 + f_3 + f_4}{3}, \ldots, \dfrac{f_{39997} + f_{39998} + f_{39999}}{3}$?

(f) $\dfrac{f_0 + f_{20000}}{2}, \dfrac{f_1 + f_{20001}}{2}, \dfrac{f_2 + f_{20002}}{2}, \ldots, \dfrac{f_{19999} + f_{39999}}{2}$?

Note. Regard $f_0, f_1, \ldots$ as samples of the audio waveform. In practice such samples are scaled, quantized, and shifted to produce nonnegative integers (*e.g.*, $0, 1, 2, \ldots, 255$ for an 8-bit sound card).

▶ **EXERCISE 8.3:**   When you are at reading distance from the following grating your eye will see 101 black bars separated by 100 white bars.

When you are far away, the whole grating will appear to be a uniform shade of gray.

(a) Experimentally determine the maximum distance D where your eye can resolve these bars when they are horizontal and when they are vertical.

(b) The images your eye can see are σ-bandlimited for some choice of σ. Use the result of (a) to estimate σ (with the unit cycles/radian).

(c) When your are at reading distance from the graytone image of Fig. 8.10 your eye will see the individual pixels, but when your are far away your eye will perceive the image as a fine glossy photograph. Use the result from (a) to predict the distance where this transition occurs ... and then confirm (or disprove) your prediction by actually observing the phenomenon.

Figure 8.10. An 80 pixel by 100 pixel graytone image of *Joseph Fourier*.

▶ **EXERCISE 8.4:**   Let $N = 3, 4, 5, \ldots$, let $T > 0$, and let

$$\gamma(t) := \sum_{n=-\infty}^{\infty} \cos\left(\frac{2\pi n}{N}\right) \cdot T\, \delta(t - nT).$$

(a) Show that $\gamma(t) = \cos(2\pi\sigma_m t) \cdot \text{III}(t/T)$ when

$$\sigma_m := \frac{1}{NT} + \frac{m}{T}, \quad m = 0, \pm 1, \pm 2, \ldots .$$

(b) Find and sketch the Fourier transform Γ of γ, showing the effects of the parameters N, T, and m.

(c) Find a simple formula for the function $f(t)$ that has the Fourier transform

$$F(s) := R_\alpha(Ts) \cdot \Gamma(s)$$

where R_α is given by (10) (*cf.* Fig. 8.5) and $0 < \alpha \leq 1/3$.

Note. We can generate a pure sinusoidal tone by using N samples of the cosine function, a suitable choice of the sampling interval $T > 0$, and a suitable low-pass filter R_α.

▶ **EXERCISE 8.5:**   Let f, g be σ-bandlimited functions with $G, G', G'', \ldots$ being continuous. Let $T > 0$ and assume that $0 < 2\sigma T < 1$. Show that

$$(g * f)(t) = \sum_{n=-\infty}^{\infty} T\, f(nT) g(t - nT).$$

▶ **EXERCISE 8.6:**   Let $f_1, f_2, \ldots$ be a sequence of σ-bandlimited generalized functions that has the weak limit f. Show that f is σ-bandlimited.

▶ **EXERCISE 8.7:**   Let f be a σ-bandlimited generalized function on $\mathbb{R}$. What can you infer about f if you also know that:

(a) f is p-periodic?   (b) f is support limited?

Hint. $\sum f(t - mp)$ is a p-periodic function that vanishes on an interval centered at $t = p/2$ when p is large.

▶ **EXERCISE 8.8:**   Let f be a σ-bandlimited generalized function, let $0 < \beta < 1$, let $T > 0$, and assume that $0 < 2\sigma T < 1$. The generalized function

$$q_{\beta,T}(t) := \sum_{m=-\infty}^{\infty} \frac{1}{\beta}\Pi\left(\frac{t - nT}{\beta T}\right)$$

is said to be a *pulse amplitude carrier*, and the product

$$g_{\beta,T}(t) := f(t) \cdot q_{\beta,T}(t)$$

is said to be a *pulse amplitude modulation* of f.

(a) Explain why $g_{\beta,\mathrm{T}}$ is a well-defined generalized function.

(b) Prepare representative sketches of f, F and of $g_{\beta,\mathrm{T}}, G_{\beta,\mathrm{T}}$, showing the effects of the parameters σ, β, and T.

(c) Show how to recover f from $g_{\beta,\mathrm{T}}$.

   Hint. Multiply $G_{\beta,\mathrm{T}}$ by $R_\alpha(\mathrm{T}s)$ using an α that satisfies (12).

   Note. In view of this result you can periodically eavesdrop on N concurrent conversations and then perfectly reconstruct each of them!

▶ **EXERCISE 8.9:**   Let $M, \mathrm{T} = 1, 2, \ldots$ and let $N := M \cdot \mathrm{T}$. In this exercise you will develop an interpolation formula for recovering a suitably bandlimited function f on $\mathbb{P}_N$ from the samples $f[m\mathrm{T}]$, $m = 0, 1, 2, \ldots, M - 1$.

(a) Show that $\gamma_{\mathrm{T}} := f \cdot \mathrm{T} \, c_{\mathrm{T}}$ has the Fourier transform $\Gamma_{\mathrm{T}} = F * c_M$. Here c_{T}, c_M are the discrete combs on $\mathbb{P}_N$ with tooth spacing T, M, respectively, *cf.* (4.43) and (4.44).

(b) Let p be a function on $\mathbb{P}_N$ and assume that the discrete Fourier transforms of p, f are related in such a manner that

$$P[k] \cdot F[k] = F[k], \quad k = 0, 1, \ldots, N - 1$$
$$P[k - t \cdot M] \cdot F[k] = 0, \quad k = 0, 1, \ldots, N - 1, \quad t = 1, 2, \ldots, \mathrm{T} - 1.$$

Show that $F = \Gamma_{\mathrm{T}} \cdot P$ and thereby obtain the interpolation formula

$$f[n] = \frac{1}{M} \sum_{m=0}^{M-1} f[m\mathrm{T}] \cdot p[n - m\mathrm{T}].$$

(c) Find the p that corresponds to the box

$$P[k] := \begin{cases} 1 & \text{if } k \equiv 0, 1, \ldots, M - 1 (\mathrm{mod}\ N) \\ 0 & \text{otherwise.} \end{cases}$$

(d) Find the p that corresponds to the shifted box

$$P[k] := \begin{cases} 1 & \text{if } k \equiv L, L + 1, \ldots, L + M - 1 (\mathrm{mod}\ N) \\ 0 & \text{otherwise.} \end{cases}$$

▶ **EXERCISE 8.10:**   In this exercise you will develop a sampling theorem for a suitably bandlimited function on $\mathbb{Z}$.

(a) Let $\mathrm{T} = 1, 2, 3, \ldots$. Find a simple expression for the (1-periodic) generalized Fourier transform of the comb

$$c_{\mathrm{T}}[n] := \begin{cases} 1 & \text{if } n = 0, \pm\mathrm{T}, \pm 2\mathrm{T}, \ldots \\ 0 & \text{otherwise} \end{cases}$$

on $\mathbb{Z}$.

(b) Let $F_\sigma(s)$ be a generalized function on $\mathbb{R}$ that vanishes when $|s| > \sigma$ and let

$$f[n] := F_\sigma^{\wedge\vee}(n), \quad n = 0, \pm 1, \pm 2, \dots$$

be the corresponding slowly growing function on $\mathbb{Z}$. Show that

$$\gamma[n] := c_{\mathrm{T}}[n]f[n], \quad n = 0, \pm 1, \pm 2, \dots$$

has the (1-periodic) generalized Fourier transform

$$\Gamma(s) = \frac{1}{\mathrm{T}} \sum_{n=-\infty}^{\infty} F_\sigma\left(s - \frac{n}{\mathrm{T}}\right).$$

(c) Show how to recover f from the samples $f[n\mathrm{T}]$, $n = 0, \pm 1, \pm 2, \dots$ when $2\sigma\mathrm{T} < 1$.

Hint. Multiply $\Gamma(s)$ by $\mathrm{T}\,R_\alpha(\mathrm{T}s)$ using an α that satisfies (12).

▶ **EXERCISE 8.11:**   Several Fourier analysis students are trying to find a simple expression for the cardinal series

$$f(t) := \sum_{m=-\infty}^{\infty} \mathrm{sinc}(t - mN)$$

where N is a positive integer.

"Each partial sum is σ-bandlimited with $\sigma = 1/2$ so the same is true of f," says the first.

"Yes, and since the limit is obviously N-periodic this forces f to be a trigonometric polynomial," adds the second.

"This is going to be easy," says a third, "since f is a real function that takes the values $1, 0, 0, \dots, 0$ when $t = 0, 1, 2, \dots, N - 1$."

"Wait a minute," exclaims a fourth student. "I'm not even convinced that the sum is properly defined."

(a) Use the observations of the first three students to show that

$$f(t) = \begin{cases} 1 & \text{if } t = 0, \pm N, \pm 2N, \dots \\[2mm] \dfrac{\sin(\pi t)}{N\sin(\pi t/N)} & \text{if } N = 1, 3, 5, \dots \text{ and } t \neq 0, \pm N, \pm 2N, \dots \\[2mm] \dfrac{\sin(\pi t)}{N\tan(\pi t/N)} & \text{if } N = 2, 4, 6, \dots \text{ and } t \neq 0, \pm N, \pm 2N, \dots \ . \end{cases}$$

(b) Explain why the above sum converges weakly as well as uniformly on every finite interval.

▶ **EXERCISE 8.12:**   Let f be a trigonometric polynomial

$$f(t) := \sum_{k=-K}^{K} c_k e^{2\pi i k t/p}$$

and let $T := p/(2N+1)$ where $N \geq K$. Use the cardinal series

$$f(t) = \sum_{n=-\infty}^{\infty} f(nT) \operatorname{sinc}\left(\frac{t - nT}{T}\right)$$

to derive the interpolation formula

$$f(t) = \sum_{n=0}^{2N} f(nT) \frac{\operatorname{sinc}\{(t - nT)/T\}}{\operatorname{sinc}\{(t - nT)/p\}}, \qquad 0 \leq t < p.$$

Hint. Use the identity from Ex. 8.11(a).

▶ **EXERCISE 8.13:**   Let $M = 2, 3, \ldots$ and let $m = 1, 2, \ldots, M - 1$. Show that

$$\sum_{n=-\infty}^{\infty} \frac{1}{(m + nM)^2} = \left\{ \frac{\pi}{M} \csc\left(\frac{m\pi}{M}\right) \right\}^2$$

and thereby generalize the identity of Ex. 4.14.

Hint. Use (19) with $a_1 = a_2 = 1$, $T = \pi$ and $t_1 = t_2 = m\pi/M$.

▶ **EXERCISE 8.14:**   In this exercise you will establish several properties of the basis functions for the cardinal series.

(a) Let $M < N$ be positive integers and let $t \geq 0$. Show that

$$\left| \sum_{n=M}^{N} \operatorname{sinc}(t + n) \right| \leq \frac{2}{\pi M}.$$

Hint. The identity

$$\operatorname{sinc}(t + n) = \frac{\sin(\pi t)}{\pi} \frac{(-1)}{t + n}$$

shows that the terms alternate in sign and decrease in modulus.

(b) Let $M < N$ be integers and let $-\infty < t < \infty$. Show that

$$\left| \sum_{n=M}^{N} \operatorname{sinc}(t - n) \right| \leq 3.$$

(c) Let $-\infty < t < \infty$. Show that

$$\sum_{n=-\infty}^{\infty} \operatorname{sinc}(t - n) = 1.$$

Hint. Work with the Fourier series for the piecewise smooth 1-periodic function f with

$$f(x) := e^{2\pi i t x} \quad \text{when} \quad -\frac{1}{2} < x < \frac{1}{2}.$$

(d) Let $-\infty < t < \infty$. Show that

$$\sum_{n=-\infty}^{\infty} \text{sinc}^2(t-n) = 1 \quad \text{when} \quad -\infty < t < \infty.$$

Hint. Use Plancherel's identity with the Fourier series from (c) or use (19).

▶ **EXERCISE 8.15:** In this exercise you will establish a minimum bandwidth characterization of the sinc function.

(a) Find a $(1/2\text{T})$-bandlimited function on $\mathbb{R}$ that interpolates the points $(0,1)$ and $(n\text{T}, 0)$, $n = \pm 1, \pm 2, \ldots$.

(b) Show that there is no σ-bandlimited function f that interpolates the points of (a) when $2\sigma\text{T} < 1$.

▶ **EXERCISE 8.16:** Let $N := M \cdot \text{T}$ where M, T are positive integers and let the complex numbers $y_0, y_1, \ldots, y_{M-1}$ be given. There are many ways to produce a function f on $\mathbb{P}_N$ that interpolates the data in the sense that

$$f[m\text{T}] = y_m, \quad m = 0, 1, \ldots, M-1.$$

Show how to construct such a function f that has the "smallest" possible bandwidth.

Hint. cf. Ex. 4.47.

▶ **EXERCISE 8.17:** Let f be a σ-bandlimited function with a piecewise smooth Fourier transform F, let $\text{T} > 0$ with $2\sigma\text{T} \leq 1$, and let $k = 1, 2, \ldots$. In this exercise you will generalize (15) by showing that

$$f^{(k)}(t) = \sum_{n=-\infty}^{\infty} \frac{f(n\text{T})}{\text{T}^k} \text{sinc}^{(k)} \left(\frac{t - n\text{T}}{\text{T}} \right)$$

with the series converging uniformly (and weakly) on $\mathbb{R}$.

(a) Let M, N be integers with $M < N$. Show that

$$f^{(k)}(t) - \sum_{n=M}^{N} \frac{f(n\text{T})}{\text{T}^k} \text{sinc}^{(k)} \left(\frac{t - n\text{T}}{\text{T}} \right)$$

$$= \int_{-1/2\text{T}}^{1/2\text{T}} (2\pi i s)^k \left\{ F(s) - \sum_{n=M}^{N} \text{T} f(n\text{T}) e^{-2\pi i n\text{T} s} \right\} e^{2\pi i s t} ds.$$

(b) Show that

$$\left| f^{(k)}(t) - \sum_{n=M}^{N} \frac{f(n\mathrm{T})}{\mathrm{T}^k} \mathrm{sinc}^{(k)} \left(\frac{t - n\mathrm{T}}{\mathrm{T}} \right) \right|^2$$

$$\leq \frac{1}{2k+1} \left(\frac{\pi}{\mathrm{T}} \right)^{2k} \left\{ \sum_{n<M} |f(n\mathrm{T})|^2 + \sum_{n>N} |f(n\mathrm{T})|^2 \right\}.$$

Hint. Suitably modify the proof of (17).

▶ **EXERCISE 8.18:**   Let f be a σ-bandlimited function with a piecewise smooth Fourier transform F, and let

$$M := \max_{-\infty < t < \infty} |f(t)|.$$

In this exercise you will derive Bernstein's bound

$$|f^{(k)}(t)| \leq (2\pi\sigma)^k M, \quad -\infty < t < \infty, \quad k = 1, 2, \ldots$$

for the derivatives of f.

(a) Let $-\infty < u < \infty$, $-\infty < \tau < \infty$, and let $\mathrm{T} := 1/2\sigma$. Use Ex. 8.17 to show that

$$f'(u + \tau) = \sum_{n=-\infty}^{\infty} f(n\mathrm{T} + \tau) \frac{\cos\{\pi(u - n\mathrm{T})/\mathrm{T}\} - \mathrm{sinc}\{(u - n\mathrm{T})/\mathrm{T}\}}{u - n\mathrm{T}},$$

set $u = \mathrm{T}/2$, and thereby obtain

$$f'\left(\tau + \frac{\mathrm{T}}{2}\right) = \frac{8\sigma}{\pi} \sum_{n=-\infty}^{\infty} \frac{(-1)^{n+1} f(n\mathrm{T} + \tau)}{(2n-1)^2}.$$

(b) Use (a) with the identity

$$\sum_{n=-\infty}^{\infty} \frac{1}{(2n-1)^2} = \frac{\pi^2}{4}$$

from Ex. 8.13 to show that

$$|f'(t)| \leq 2\pi\sigma M, \quad -\infty < t < \infty.$$

(c) Use (b) to derive the above Bernstein bound for $f^{(k)}(t)$.

▶ **EXERCISE 8.19:**   In this exercise you will establish the bound

$$|f^{(k)}(t)| \le B \cdot (2\pi\sigma)^k \cdot (1 + |t|^m) \cdot (1 + k^m), \quad -\infty < t < \infty, \quad k = 0, 1, 2, \dots$$

for the kth derivative of a σ-bandlimited function f. Here $m > 0$, $\sigma > 0$, and the constant B depend on f (but not on t or k).

(a) Derive such a bound for the function $g(t) := t^2 \cos(2\pi\sigma t)$.

   Hint. Begin by using the Leibnitz rule (2.29) to write

$$g^{(k)}(t) = [\cos(2\pi\sigma t)]^{(k)} \cdot t^2 + \binom{k}{1} [\cos(2\pi\sigma t)]^{(k-1)} \cdot 2t + \binom{k}{2} [\cos(2\pi\sigma t)]^{(k-2)} \cdot 2.$$

(b) Let $\Gamma(s)$ be continuous for $-\sigma \le s \le \sigma$ and let

$$f_0(t) := \int_{-\sigma}^{\sigma} \Gamma(s) e^{2\pi i s t} ds, \quad -\infty < t < \infty.$$

   Show that

$$|f_0^{(k)}(t)| \le (2\pi\sigma)^k \int_{-\sigma}^{\sigma} |\Gamma(s)| ds, \quad k = 0, 1, \dots \ .$$

(c) Use the representation (4) together with the Leibnitz rule (2.29) and (b) to establish the above bound for $f^{(k)}$.

▶ **EXERCISE 8.20:**   Let f be a σ-bandlimited function on $\mathbb{R}$.

(a) Show that f is *entire*, i.e., show that the Maclaurin series (4) converges to $f(t)$ for every choice of the point $-\infty < t < \infty$.

   Hint. Use Taylor's formula from Ex. 2.28 with the bound from Ex. 8.19.

(b) Show that f is of *exponential type* $2\pi\sigma$, i.e., show that for every choice of $\epsilon > 0$ there is a constant $B(\epsilon)$ (that depends on f and ϵ) such that

$$\left| f(0) + f'(0)\frac{z}{1!} + f''(0)\frac{z^2}{2!} + \cdots \right| \le B(\epsilon) e^{(2\pi\sigma + \epsilon)|z|}$$

   for every choice of the point $z \in \mathbb{C}$.

Note. The Paley-Wiener theorem (from complex analysis) asserts that a square integrable function f on $\mathbb{R}$ is σ-bandlimited if it has the properties (a) and (b).

▶ **EXERCISE 8.21:**   Let the basis function (25) be constructed from the fragmentation (24) of Π, let $T > 0$, and let $C := \{s : P(Ts) = 1\}$. Let f have a piecewise smooth
Fourier transform F that satisfies (26).

(a) Show that the piecewise smooth function

$$G(s) := \sum_{n=-\infty}^{\infty} F\left(s - \frac{n}{T}\right)$$

is represented by the Fourier series

$$G(s) = \sum_{n=-\infty}^{\infty} T f(nT) e^{-2\pi i n T s}$$

at each point s where G is continuous.

Hint. Use the identity

$$\int_0^{1/T} G(s) e^{2\pi i n T s} ds = \int_C F(s) e^{2\pi i n T s} ds$$

to evaluate the Fourier coefficients and then use the representation theorem from
Chapter 1.

(b) Show that (28) holds and thereby establish the uniform (and weak) convergence of
the series (27).

Hint. Suitably modify the argument of (17), using (26) and the identities

$$\int_C ds = \frac{1}{T},$$

$$\int_C \left| F(s) - \sum_{n=M}^{N} T f(nT) e^{-2\pi i n T s} \right|^2 ds = \int_0^{1/T} \left| G(s) - \sum_{n=M}^{N} T f(nT) e^{-2\pi i n T s} \right|^2 ds.$$

▶ **EXERCISE 8.22:**   Let $m, n = 0, \pm 1, \pm 2, \ldots$.

(a) Show that

$$\int_{-\infty}^{\infty} \text{sinc}(t - m)\text{sinc}(t - n)dt = \begin{cases} 1 & \text{if } m = n \\ 0 & \text{if } m \neq n. \end{cases}$$

Hint. Use Parseval's identity.

(b) Show that

$$\int_{-\infty}^{\infty} p(t - m)\overline{p(t - n)}dt = \begin{cases} 1 & \text{if } m = n \\ 0 & \text{if } m \neq n \end{cases}$$

when p is constructed from a fragmentation of Π by using (24)–(25).

▶ **EXERCISE 8.23:**   Let f be a σ-bandlimited function with a piecewise smooth Fourier transform F, and let

$$g(t) := (\mathcal{H} f)(t) := -i \int_{-\sigma}^{\sigma} \text{sgn}(s) F(s) e^{2\pi i s t} ds$$

be the corresponding Hilbert transform, *cf.* (5.83) and Ex. 7.55. Let $\text{T} > 0$ and assume that $2\sigma\text{T} \leq 1$.

(a) Use an informal argument to derive the sampling series

$$f(t) = - \sum_{n=-\infty}^{\infty} g(n\text{T}) \sin\left\{\frac{\pi(t - n\text{T})}{2\text{T}}\right\} \text{sinc}\left\{\frac{t - n\text{T}}{2\text{T}}\right\}.$$

Hint. Observe that $G(s) = A(s) \cdot F(s)$ when

$$A(s) := \begin{cases} -i\,\text{sgn}(s) & \text{if } |s| < 1/2\text{T} \\ 0 & \text{otherwise.} \end{cases}$$

(b) Show that

$$\text{T} \sum_{n=-\infty}^{\infty} |g(n\text{T})|^2 = \int_{-1/2\text{T}}^{1/2\text{T}} |F(s)|^2 ds.$$

(c) Let M, N be integers with $M < N$. Show that

$$\left| f(t) + \sum_{n=M}^{N} g(n\text{T}) \sin\left\{\frac{\pi(t - n\text{T})}{2\text{T}}\right\} \text{sinc}\left\{\frac{t - n\text{T}}{2\text{T}}\right\} \right|^2 \leq \sum_{n<M} |g(n\text{T})|^2 + \sum_{n>N} |g(n\text{T})|^2.$$

In conjunction with (b) this shows that the series of (a) converges uniformly (and weakly) on $\mathbb{R}$.

Hint. Suitably adapt the argument of (17) after writing

$$f(t) + \sum_{n=M}^{N} \text{T}\, g(n\text{T}) a(t - n\text{T}) = \int_{-1/2\text{T}}^{1/2\text{T}} A(s)\left\{ -G(s) + \sum_{n=M}^{N} \text{T}\, g(n\text{T}) e^{-2\pi i n \text{T} s} \right\} e^{2\pi i s t} ds.$$

▶ **EXERCISE 8.24:**   Let f be a σ-bandlimited function with a piecewise smooth Fourier transform F, let $\text{T} > 0$ with $2\sigma\text{T} \leq 1$, and let $0 < \alpha < 1$. In this exercise you will use the Papoulis analysis to recover $f(t)$, $-\infty < t < \infty$, from the bunched samples

$$f(n\cdot 2\text{T} - \alpha\text{T}), \quad f(n\cdot 2\text{T} + \alpha\text{T}), \quad n = 0, \pm 1, \pm 2, \ldots .$$

(a) Verify that

$$f(n\cdot 2\text{T} - \alpha\text{T}) = (a_1 * f)(n\cdot 2\text{T}), \quad f(n\cdot 2\text{T} + \alpha\text{T}) = (a_2 * f)(n\cdot 2\text{T})$$

when

$$a_1(t) := \delta(t - \alpha\text{T}), \quad a_2(t) := \delta(t + \alpha\text{T}).$$

(b) Let $g_\kappa(t) := (a_\kappa * f)(t)$, $\gamma_\kappa(t) := g_\kappa(t) \cdot \text{III}(t/2\text{T})$, $\kappa = 1, 2$. Show that

$$\det \begin{vmatrix} A_1(s) & A_1(s + 1/2\text{T}) \\ A_2(s) & A_2(s + 1/2\text{T}) \end{vmatrix} = 2i\sin(\alpha\pi) \neq 0$$

and thereby prove that $F(s)$, $F(s+1/2\text{T})$ can be uniquely determined from the linear equations

$$A_1(s)F(s) + A_1(s + 1/2\text{T})F(s + 1/2\text{T}) = \Gamma_1(s)$$
$$A_2(s)F(s) + A_2(s + 1/2\text{T})F(s + 1/2\text{T}) = \Gamma_2(s)$$

of (42) when $-1/2\text{T} \leq s \leq 0$.

(c) Show that the linear equations

$$A_1(s)P_1(s,t) + \qquad A_2(s)P_2(s,t) = 1$$
$$A_1(s + 1/2\text{T})P_1(s,t) + A_2(s + 1/2\text{T})P_2(s,t) = e^{2\pi it/2\text{T}}$$

from (43) have the solution

$$P_1(s,t) = \quad e^{2\pi i\alpha\text{T}s}(e^{\pi i\alpha} - e^{\pi it/\text{T}})/(e^{\pi i\alpha} - e^{-\pi i\alpha})$$
$$P_2(s,t) = -e^{-2\pi i\alpha\text{T}s}(e^{-\pi i\alpha} - e^{\pi it/\text{T}})/(e^{\pi i\alpha} - e^{-\pi i\alpha})$$

when $-1/2\text{T} \leq s \leq 0$.

(d) Show that the corresponding basis functions

$$p_k(t) := 2\text{T} \int_{-1/2\text{T}}^{0} P_k(s,t)e^{2\pi ist}ds, \quad k = 1, 2$$

from (44) are given by

$$p_1(t) = \frac{-\text{T}\{\cos(\pi\alpha) - \cos(\pi t/\text{T})\}}{\pi(t + \alpha\text{T})\sin(\pi\alpha)}, \qquad p_2(t) = \frac{\text{T}\{\cos(\pi\alpha) - \cos(\pi t/\text{T})\}}{\pi(t - \alpha\text{T})\sin(\pi\alpha)}.$$

(e) Use (a), (d), and (47) to obtain the interpolation formula

$$f(t) = \frac{\text{T}\{\cos(\pi\alpha) - \cos(\pi t/\text{T})\}}{\pi\sin\pi\alpha} \sum_{n=-\infty}^{\infty} \left\{ \frac{f(n\cdot2\text{T} + \alpha\text{T})}{t - n\cdot2\text{T} - \alpha\text{T}} - \frac{f(n\cdot2\text{T} - \alpha\text{T})}{t - n\cdot2\text{T} + \alpha\text{T}} \right\}.$$

▶ **EXERCISE 8.25:**   Let f_1, f_2 be σ-bandlimited functions with piecewise smooth Fourier transforms F_1, F_2, and let $\text{T} := 1/2\sigma$. In this exercise you will show how to recover f_1, f_2 from the T/2-samples of

$$f(t) := f_1(t) \cdot e^{-2\pi i\sigma t} + f_2(t) \cdot e^{2\pi i\sigma t}.$$

(You can combine two conversations f_1, f_2 to form f, transmit samples of f over a phone line, and reconstruct f_1, f_2 at the other end!)

(a) Show how to obtain F_1, F_2 from F.

(b) Informally derive the sampling series

$$f_1(t) = \frac{1}{2} \sum_{n=-\infty}^{\infty} i^n f\left(\frac{n\mathrm{T}}{2}\right) \operatorname{sinc}\left(\frac{2t - n\mathrm{T}}{2\mathrm{T}}\right), \quad f_2(t) = \frac{1}{2} \sum_{n=-\infty}^{\infty} (-i)^n f\left(\frac{n\mathrm{T}}{2}\right) \operatorname{sinc}\left(\frac{2t - n\mathrm{T}}{2\mathrm{T}}\right).$$

Hint. Begin with $\gamma(t) := f(t) \cdot \mathrm{III}(2t/\mathrm{T})$.

(c) Use (17) with the σ-bandlimited functions $f_k(t)$, $f_k(t + \mathrm{T}/2)$ to show that

$$f_k(t) = \sum_{n=\text{even}} f_k\left(\frac{n\mathrm{T}}{2}\right) \operatorname{sinc}\left(\frac{2t - n\mathrm{T}}{2\mathrm{T}}\right) = \sum_{n=\text{odd}} f_k\left(\frac{n\mathrm{T}}{2}\right) \operatorname{sinc}\left(\frac{2t - n\mathrm{T}}{2\mathrm{T}}\right), \quad k = 1, 2$$

with each of these series converging uniformly for $-\infty < t < \infty$.

(d) Use (c) with the identities

$$i^n f\left(\frac{n\mathrm{T}}{2}\right) = f_1\left(\frac{n\mathrm{T}}{2}\right) + (-1)^n f_2\left(\frac{n\mathrm{T}}{2}\right), \quad (-i)^n f\left(\frac{n\mathrm{T}}{2}\right) = (-1)^n f_1\left(\frac{n\mathrm{T}}{2}\right) + f_2\left(\frac{n\mathrm{T}}{2}\right)$$

to show that the modified cardinal series of (b) converge uniformly (and weakly) for $-\infty < t < \infty$, and provide bounds for the truncation errors.

Note. Analogously we can recover the σ-bandlimited functions $f_1, f_2, \ldots, f_K$ from the T/K-samples of

$$f(t) := f_1(t) \cdot e^{2\pi i(-K+1)\sigma t} + f_2(t) \cdot e^{2\pi i(-K+3)\sigma t} + \cdots + f_K(t) \cdot e^{2\pi i(K-1)\sigma t}.$$

▶ **EXERCISE 8.26:** Let f have the piecewise smooth Fourier transform F and assume that

$$|F(s)| \leq G(|s|), \quad -\infty < s < \infty$$

where G is continuous, decreasing, and integrable on $0 \leq s < \infty$. In this exercise you will show that

$$\sum_{m=-\infty}^{\infty} \sum_{n=-\infty}^{\infty} \left| f_m(n) \operatorname{sinc}\left(\frac{t - n\mathrm{T}}{\mathrm{T}}\right) \right| < M, \quad -\infty < t < \infty$$

where f_m is given by (49)–(50), $\mathrm{T} > 0$, and M is a constant (that depends on T and G.) This absolute convergence allows us to exchange the order of the m, n summations in (52).

(a) Use the Cauchy-Schwartz inequality (53) and the identity of Ex. 8.14(d) to show that

$$\sum_{n=-\infty}^{\infty} \left| f_m(n) \operatorname{sinc}\left(\frac{t - n\mathrm{T}}{\mathrm{T}}\right) \right| \leq \left\{ \sum_{n=-\infty}^{\infty} |f_m(n)|^2 \right\}^{1/2}.$$

(b) Use (16) with the tail hypothesis to show that

$$\sum_{n=-\infty}^{\infty} |f_m(n)|^2 = \frac{1}{\mathrm{T}} \int_{-\infty}^{\infty} |F_m(s)|^2 ds \leq \frac{1}{\mathrm{T}^2} \max\left\{ |G(s)|^2 : \left| s - \frac{m}{\mathrm{T}} \right| \leq \frac{1}{2\mathrm{T}} \right\}.$$

(c) Use (a), (b) with the tail hypothesis to show that

$$\sum_{m=-\infty}^{\infty} \sum_{n=-\infty}^{\infty} \left| f_m(n) \operatorname{sinc}\left(\frac{t-n\mathrm{T}}{\mathrm{T}}\right) \right| \leq G(0) + 2G\left(\frac{1}{2\mathrm{T}}\right) + 2G\left(\frac{3}{2\mathrm{T}}\right) + 2G\left(\frac{5}{2\mathrm{T}}\right) + \cdots$$

$$\leq 3G(0) + 2\int_{1/2\mathrm{T}}^{\infty} G(s)ds.$$

▶ **EXERCISE 8.27:**   Two Fourier analysis students are trying to construct a good σ-bandlimited approximation g to a suitably regular function f when $\sigma > 0$ is specified.

"I think I'll go with a least squares approximation and choose the σ-bandlimited function g that minimizes the integral

$$\int_{-\infty}^{\infty} |f(t) - g(t)|^2 dt"$$

says the first.

"I like the idea of using least squares," says the second, "but I would rather minimize the sum

$$\sum_{n=-\infty}^{\infty} |f(n\mathrm{T}) - g(n\mathrm{T})|^2$$

with $\mathrm{T} := 1/2\sigma$.

(a) Show how to synthesize the σ-bandlimited function g_1 that minimizes the integral.

Hint. Use Plancherel's identity.

(b) Show how to synthesize the σ-bandlimited function g_2 that minimizes the sum.

Hint. The minimum is 0.

Note. The mapping that takes f to g_1 is LTI, but the mapping that takes f to g_2 is not.

▶ **EXERCISE 8.28:**   Let $\epsilon > 0$, $\sigma > 0$, $B > 0$ be given. Construct a continuously differentiable function f with a piecewise smooth absolutely integrable Fourier transform F such that

$$\int_{-\infty}^{\infty} |f(t)|^2 dt < \epsilon,$$

$$F(s) = 0 \quad \text{for} \quad -\sigma < s < \sigma, \quad \text{and}$$

$$|f(0)| > B.$$

Hint. Suitably choose $A > 0$, $\alpha > 0$, $\beta > \alpha/2$ and set $f(t) := A\cos(2\pi\beta t)\operatorname{sinc}(\alpha t)$.

Chapter 9

Partial Differential Equations

9.1 Introduction

Vibrating strings on bows, lyres, dulcimers, ... are as old as civilization. The plane vibration of such a string can be specified by a displacement function $u(x, t)$ which depends on a space variable, x, and a time variable t. Fix $t = t_0$ and $u(x, t_0)$ gives a snapshot of the shape of the whole string at that time. Fix $x = x_0$ and $u(x_0, t)$ specifies the trajectory of the corresponding spot on the string as time evolves. From personal observation, you undoubtedly know that when the string emits an audible tone it moves much too fast for you to see such time slices or follow such point trajectories without using a stroboscope, high-speed camera, etc. (You can generate traveling waves, stationary waves, ... that you *can* see by shaking one end of a long suitably tensioned telephone extension cord or Slinky!)

In 1747 d'Alembert used a bit of elementary physics to derive a new mathematical construct, the *partial differential equation* (PDE),

$$\frac{\partial^2 u(x, t)}{\partial t^2} = c^2 \frac{\partial^2 u(x, t)}{\partial x^2}$$

which specifies the local behavior of the displacement function, *cf.* Section 9.2. Here $c > 0$ is a constant. We will use the subscripts x, t as tags [*cf.* (5.35)] to write this equation in the compact form

$$u_{tt} = c^2 u_{xx}. \tag{1}$$

By means of a clever change of variables, *cf.* Ex. 9.1, d'Alembert managed to solve this equation and find a general form for the displacement function.

From that time partial differential equations have been used to model physical processes that take place on a continuum of space. There are dozens of named PDEs that succinctly summarize corresponding laws of nature, and you will certainly encounter them as you study physics, chemistry, geology, engineering, physiology, ...

Perhaps you have already learned to construct solutions by using the elementary but powerful *separation of variables* technique, *e.g.*, as described in W.E. Boyce and R.C. DiPrima, *Ordinary Differential Equations and Boundary Value Problems*, *6th ed.*, John Wiley & Sons, New York, 1997, pp. 543–619.

In this chapter we will use Fourier analysis to study three linear constant co-efficient PDEs from mathematical physics, the *wave equation* (1), the *diffusion equation*

$$u_t = a^2 u_{xx}, \tag{2}$$

and the *diffraction equation*

$$u_t = (i\lambda/4\pi)u_{xx} \tag{3}$$

where $a > 0$ and $\lambda > 0$ are constants. It is easy to see that these PDEs have the particular solutions

$$e^{2\pi i[sx - \nu(s)t]}, \qquad -\infty < s < \infty, \tag{4}$$

with

$$\nu(s) = \pm cs, \quad \nu(s) = -2\pi i a^2 s^2, \quad \nu(s) = \lambda s^2/2,$$

respectively, *cf.* Ex. 9.21. We can synthesize the corresponding general solutions

$$u(x,t) = \int_{-\infty}^{\infty} \left\{ A_+(s)e^{2\pi isct} + A_-(s)e^{-2\pi isct} \right\} e^{2\pi isx} ds, \tag{5}$$

$$u(x,t) = \int_{-\infty}^{\infty} \left\{ A(s)e^{-4\pi^2 a^2 s^2 t} \right\} e^{2\pi isx} ds, \tag{6}$$

$$u(x,t) = \int_{-\infty}^{\infty} \left\{ A(s)e^{-\pi i\lambda s^2 t} \right\} e^{2\pi isx} ds \tag{7}$$

from these particular solutions and use the Fourier transform calculus to express the functions A_+, A_-, A in terms of certain initial values of u, u_t. As you work through the details in the following sections of the chapter you will see the extraordinary power and elegance of this procedure.

A PDE is best studied within some definite physical context. We will show how (1), (2), (3) arise within settings where the names *wave equation, diffusion equation,* and *diffraction equation* are most natural. Simple concepts (Newton's second law of motion, conservation of energy, Huygens' principle) allow us to formulate mathematical models that use PDEs with initial conditions (and boundary conditions) to capture the essence of the physics without getting bogged down with irrelevant details. A mathematical model allows us to make formal deductions that we can interpret within a meaningful physical setting.

As we derive the PDEs we will tacitly assume that each $u(x,t)$, $u_x(x,t)$, $u_t(x,t), \ldots$ we encounter is a suitably smooth ordinary function of the variables x and t. Once we formulate the mathematical model, however, we will routinely extend the class of admissible u's. We will allow $u(x,t)$ to be a generalized function of x (as described in Chapter 7) for each value of the parameter t. The partial derivatives of u will be defined in terms of weak limits

$$u_x(x,t) := \lim_{h\to 0} \frac{u(x+h,t) - u(x,t)}{h}, \quad u_{xx}(x,t) := \lim_{h\to 0} \frac{u_x(x+h,t) - u_x(x,t)}{h}, \ \ldots$$

$$u_t(x,t) := \lim_{h\to 0} \frac{u(x,t+h) - u(x,t)}{h}, \quad u_{tt}(x,t) := \lim_{h\to 0} \frac{u_t(x,t+h) - u_t(x,t)}{h}, \ \ldots \ .$$

You will quickly discover that this convention enables us to handle various limits with minimal effort, *e.g.*, as done in our study of bowed, plucked, and struck strings.

Example: Show that $u(x,t) := \Pi(x - ct)$ satisfies the wave equation (1) with

$$\lim_{t\to 0+} u(x,t) = u(x,0) = \Pi(x)$$

$$\lim_{t\to 0+} u_t(x,t) = u_t(x,0) = -c\{\delta(x+1/2) - \delta(x-1/2)\}.$$

Solution: A routine calculation (given at the end of Section 7.6) shows that

$$u_x(x,t) = \delta(x - ct + 1/2) - \delta(x - ct - 1/2) = -c^{-1} u_t(x,t)$$
$$u_{xx}(x,t) = \delta'(x - ct + 1/2) - \delta'(x - ct - 1/2) = c^{-2} u_{tt}(x,t),$$

so this u satisfies (1). The weak continuity relation (7.106) then gives the weak limits for $u(x,0+)$ and $u_t(x,0+)$. ∎

The equations (1), (2), (3) can be extended to a multivariate setting, *e.g.*, (2) takes the form

$$u_t = a^2\{u_{xx} + u_{yy}\}, \quad u_t = a^2\{u_{xx} + u_{yy} + u_{zz}\} \tag{8}$$

when we work with 2,3-space variables. We obtain *Laplace's equation*

$$u_{xx} + u_{yy} = 0, \quad u_{xx} + u_{yy} + u_{zz} = 0 \tag{9}$$

when we set $u_t = 0$ in (8) so that we can find steady-state (*i.e.*, time invariant) solutions. Fourier analysis can be used to solve such PDEs on special domains, *cf.* Ex. 9.48. A meaningful discussion of Laplace's equation and the 2,3-dimensional versions of (1), (2) is best done within a context that admits space domains bounded by suitably regular curves in $\mathbb{R}^2$ or suitably regular surfaces in $\mathbb{R}^3$. You can find such a treatment in the PDE texts given in the references.

9.2 The Wave Equation

A physical context: Plane vibration of a taut string

We consider the plane motion of a taut string that has a linear mass density ρ. We assume that a smooth function $u(x,t)$ gives the displacement of the string from the x-axis at coordinate x and time t. We assume that the string is perfectly flexible, *i.e.* (unlike a rod), it offers no resistance to bending. This being the case, the sections of the string that lie to the left and to the right of the point x supply balancing tensile forces directed along the local tangent line which makes an angle

$$\theta(x,t) := \arctan u_x(x,t)$$

with the x-axis, *cf.* Fig. 9.1.

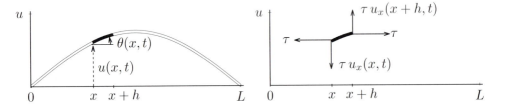

Figure 9.1. A segment of a string under tension and the forces used with Newton's second law in (10).

From trigonometry,

$$\cos\theta(x,t) = \frac{1}{\sqrt{1+\tan^2\theta(x,t)}} = \frac{1}{\sqrt{1+u_x^2(x,t)}},$$

$$\sin\theta(x,t) = \frac{\tan\theta(x,t)}{\sqrt{1+\tan^2\theta(x,t)}} = \frac{u_x(x,t)}{\sqrt{1+u_x^2(x,t)}}$$

when $-\pi/2 < \theta(x,t) < \pi/2$. We will assume that u_x is uniformly small, with u_x^2 being negligible in comparison to 1. We can then use the approximation

$$\cos\theta(x,t) \approx 1$$

to identify the magnitude of the tensile force with that of its horizontal component, replacing both by the constant τ. To the same degree of approximation

$$\sin\theta(x,t) \approx u_x(x,t),$$

and (with an appropriate sign) we use $\tau\, u_x(x,t)$ for the vertical component of the tensile force at coordinate x.

The segment of the string that we associate with the small interval $[x, x + h]$ has mass ρh. We neglect the effects of gravity, air resistence, etc. in comparison to the vertical components of the large tensile forces that act at the ends of the segment as shown in Fig. 9.1. We can then use Newton's second law to write

$$\rho h\, u_{tt}(x + h/2, t) = \text{mass} \cdot \text{acceleration at center of mass}$$
$$= \text{force} \tag{10}$$
$$= \tau\, u_x(x + h, t) - \tau\, u_x(x, t).$$

We divide both sides of this equation by the mass, ρh, and then let $h \to 0$ to obtain

$$u_{tt}(x, t) = (\tau/\rho)u_{xx}(x, t).$$

In this way we obtain the wave equation (1) with

$$c := (\tau/\rho)^{1/2}. \tag{11}$$

The above derivation is quite informal, and you should understand that we cannot use (1) to determine the *exact* motion of any real string. Real strings stretch, they resist bending (if ever so slightly), they move horizontally as well as vertically, they are damped by air resistance, etc., so a precise analysis requires suitably framed arguments from continuum mechanics, *cf.* S. Antman, The equations for large vibrations of strings, *Amer. Math. Monthly* **87**(1980), 359–370. Nevertheless, (1) does capture the essence of the taut string and serves as a good mathematical model. We will now use Fourier analysis to construct solutions of this PDE.

The wave equation on $\mathbb{R}$

Let f, g be generalized functions on $\mathbb{R}$. We will construct the unique generalized solution of the wave equation (1) that has the initial position

$$u(x, 0) := f(x) \tag{12}$$

and the initial velocity

$$u_t(x, 0) := g(x). \tag{13}$$

If such a solution exists, we can Fourier transform (1), (12), and (13) to obtain

$$U_{tt}(s, t) = c^2(2\pi i s)^2 U(s, t), \qquad U(s, 0) = F(s), \quad U_t(s, 0) = G(s).$$

Using the fact that the initial value problem

$$y''(t) + \omega^2 y(t) = 0, \quad y(0) = A, \quad y'(0) = B$$

from ordinary differential equations has the solution

$$y(t) = A\cos(\omega t) + (B/\omega)\sin(\omega t)$$

(when $\omega > 0$), we then write

$$U(s,t) = \begin{cases} F(0) + t\,G(0) & \text{if } s = 0 \\ F(s)\cos(2\pi cst) + \{G(s)/2\pi cs\}\sin(2\pi cst) & \text{otherwise.} \end{cases} \tag{14}$$

You will recognize the equivalent

$$U(s,t) = \frac{1}{2}e^{2\pi icts}F(s) + \frac{1}{2}e^{-2\pi icts}F(s) + t\,\text{sinc}(2cts)\cdot G(s)$$

as the Fourier transform of the *d'Alembert formula*

$$u(x,t) = \frac{1}{2}f(x+ct) + \frac{1}{2}f(x-ct) + \frac{1}{2c}\Pi\left(\frac{x}{2ct}\right) * g(x). \tag{15}$$

[Of course, d'Alembert worked with ordinary functions and wrote (15) in the form

$$u(x,t) = \frac{1}{2}f(x+ct) + \frac{1}{2}f(x-ct) + \frac{1}{2c}\int_{\xi=-ct}^{ct} g(x-\xi)d\xi$$

when he derived this result by other means some 250 years ago!] We can use the *wave kernel*

$$r(x,t) := \begin{cases} 0 & \text{if } t = 0 \\ \frac{1}{2c}\Pi\left(\frac{x}{2ct}\right) & \text{otherwise} \end{cases} \tag{16}$$

that has the partial derivative

$$r_t(x,t) = \frac{1}{2}\delta(x+ct) + \frac{1}{2}\delta(x-ct), \tag{17}$$

cf. Ex. 9.8, to write (15) in the compact form

$$u(x,t) = r_t(x,t) * f(x) + r(x,t) * g(x). \tag{18}$$

At this point it is easy to see that our initial value problem always has a solution. Given f, g we can use (15) to construct u. A straightforward calculation then shows that this u satisfies (1), (12), and (13).

Example: Solve $u_{tt} = c^2\,u_{xx}$ with the initial position and velocity

$$u(x,0) := \frac{e^{-x^2/2\sigma^2}}{\sqrt{2\pi}\,\sigma}, \qquad u_t(x,0) = 0$$

where $\sigma > 0$.

Solution: The d'Alembert formula (15) gives

$$u(x,t) = \frac{e^{-(x+ct)^2/2\sigma^2}}{2\sqrt{2\pi}\,\sigma} + \frac{e^{-(x-ct)^2/2\sigma^2}}{2\sqrt{2\pi}\,\sigma}$$

with u and its partial derivatives being ordinary functions that satisfy the PDE in the ordinary fashion. You can use the physical context to interpret this result. If a motionless taut string is forced to assume the shape of a gaussian and then suddenly released, two half-sized gaussians emerge from the initial distortion and travel to the left and to the right with the *wave velocity c.* ∎

Example: Solve the two initial value problems

$$u_{tt} = c^2\,u_{xx} \text{ with } u(x,0) = \Pi(x),\ u_t(x,0) = 0 \text{ and} \tag{19}$$
$$v_{tt} = c^2\,v_{xx} \text{ with } v(x,0) = 0,\qquad v_t(x,0) = \Pi(x). \tag{20}$$

Solution: We use the d'Alembert formula (15) to write

$$u(x,t) = \frac{1}{2}\Pi(x+ct) + \frac{1}{2}\Pi(x-ct), \tag{21}$$

$$v(x,t) = \frac{1}{2c}\Pi\left(\frac{x}{2ct}\right) * \Pi(x)$$
$$= \begin{cases} \dfrac{1}{2c}\min\{1, 2ct, \tfrac{1}{2} + ct - |x|\} & \text{if } |x| \le \tfrac{1}{2} + ct \\ 0 & \text{if } |x| > \tfrac{1}{2} + ct, \end{cases} \tag{22}$$

cf. Ex. 9.2. Although u, v are ordinary functions with the time slices shown in Fig. 9.2, the partial derivatives u_{tt}, u_{xx}, v_{tt}, v_{xx} are all generalized functions of x for each value of t. The functions u, v are linked, and by using (16)–(18) we write

$$u = r_t * \Pi = (r * \Pi)_t = v_t.$$

This relation is consistent with the time slices from Fig. 9.2.

You can again use the physical context to interpret the solutions. For example, the spreading plateau (22) is the response of a taut quiescent string to a blow from a broad hammer that instantly imparts momentum to the segment $-\tfrac{1}{2} < x < \tfrac{1}{2}$. If we sit at some point x_0 with $|x_0| > \tfrac{1}{2}$, we do not feel the effect of the hammer blow until the time interval

$$\frac{|x_0| - \tfrac{1}{2}}{c} \le t \le \frac{|x_0| + \tfrac{1}{2}}{c}$$

when the string rises linearly from $u = 0$ to $u = 1/2c$ (where it remains). ∎

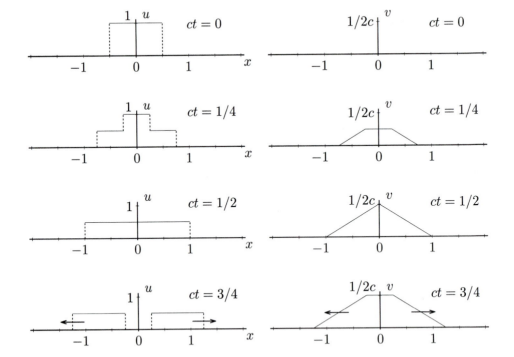

Figure 9.2. Time slices for the functions (21), (22) that satisfy (19) and (20), respectively.

Let w be a generalized function with the Fourier transform W. We can construct *traveling* solutions

$$w(x + ct), \qquad w(x - ct)$$

of the wave equation (1) that move left, right with the wave velocity c. Each time slice of such a solution is a translate of w, so these waves do not change shape as they advance. We can formally synthesize such traveling waves from the complex exponential solutions

$$e^{2\pi is(x+ct)}, \qquad e^{2\pi is(x-ct)}, \qquad -\infty < s < \infty$$

of (1) by writing

$$\int_{-\infty}^{\infty} W(s)e^{2\pi is(x+ct)}\,ds, \qquad \int_{-\infty}^{\infty} W(s)e^{2\pi is(x-ct)}\,ds. \tag{23}$$

We obtain the same functions when we use d'Alembert's formula (15) with the initial position $w(x)$ and the initial velocity $\pm c\,w'(x)$.

We can use a traveling wave to transmit a message along a taut string. For example, if we want to transmit a message encoded with the bit string $b_1, b_2, \ldots, b_N$ (where $b_n = 0$ or 1 for each n) we can launch a traveling wave constructed from a suitable mollification of

$$w(x) = \sum_{n=1}^{N} b_n \Pi(x + n).$$

The wave will travel along an ideal taut string with no attenuation, carrying our message at the wave velocity c. Of course, we cannot send a nondissipative wave along a real string, but we can produce comparable electromagnetic waves. When you see a star in the night sky, the retina of your eye has just processed a traveling light wave that has completed a journey of some $10^{13} - 10^{16}$ km!

Nondissipative traveling waves are possible because of a remarkable *conservation law* associated with most ordinary solutions u of the wave equation (1). The integral

$$\mathcal{E}_t\{u\} := \int_{-\infty}^{\infty} \{c^2 |u_x(x,t)|^2 + |u_t(x,t)|^2\} dx \qquad (24)$$

is a constant of the motion! [When scaled by $\rho/2$, cf. (11), the $c^2 u_x^2$ integral gives the potential energy and the u_t^2 integral gives the kinetic energy of the string at time t.] Indeed, when the initial position (12) and the initial velocity (13) are piecewise smooth square integrable functions on $\mathbb{R}$, we can use Plancherel's identity (1.15) with (14) (and a bit of algebra) to write

$$\mathcal{E}_t\{u\} = \int_{-\infty}^{\infty} \{|2\pi isc\, U(s,t)|^2 + |U_t(s,t)|^2\} ds$$

$$= \int_{-\infty}^{\infty} \{|2\pi sc\, F(s)\cos(2\pi sct) + G(s)\sin(2\pi sct)|^2$$

$$+ |-2\pi sc\, F(s)\sin(2\pi sct) + G(s)\cos(2\pi sct)|^2\} ds$$

$$= \int_{-\infty}^{\infty} \{|2\pi isc\, F(s)|^2 + |G(s)|^2\} ds$$

$$= \mathcal{E}_0\{u\}.$$

The wave equation on $\mathbb{T}_p$

We will specialize the results of the preceding section to the case where the initial position f and the initial velocity g are both p-periodic with $p > 0$. The d'Alembert formula (15) then gives a generalized solution $u(x,t)$ of the wave equation that is p-periodic in x.

Example: Solve the two initial value problems

$$u_{tt} = c^2\, u_{xx}, \quad u(x,0) = \frac{1}{p}\text{III}\left(\frac{x}{p}\right), \qquad u_t(x,0) = 0, \text{ and} \tag{25}$$

$$v_{tt} = c^2\, v_{xx}, \quad v(x,0) = 0, \qquad\qquad v_t(x,0) = \frac{1}{p}\text{III}\left(\frac{x}{p}\right). \tag{26}$$

Solution: We will define

$$r^p(x,t) := r(x,t) * \frac{1}{p}\text{III}\left(\frac{x}{p}\right) = \sum_{m=-\infty}^{\infty} r(x - mp, t) \tag{27}$$

so that we can use (18) with (16)–(17) to write

$$v(x,t) = r(x,t) * \frac{1}{p}\text{III}\left(\frac{x}{p}\right) = r^p(x,t), \tag{28}$$

$$
\begin{aligned}
u(x,t) &= r_t(x,t) * \frac{1}{p}\text{III}\left(\frac{x}{p}\right) = r_t^p(x,t) \\
&= \frac{1}{2p}\text{III}\left(\frac{x+ct}{p}\right) + \frac{1}{2p}\text{III}\left(\frac{x-ct}{p}\right).
\end{aligned}
\tag{29}
$$

You can visualize the evolution of v as follows. At time $t = 0$ a string at rest is subjected to impulsive blows at the points $x = 0, \pm p, \pm 2p, \dots$. The blow at $x = kp$ produces a spreading box with height $1/2c$, center $x = kp$, and width $2ct$ at time t. When these spreading boxes collide at times $p/2c, 2p/2c, 3p/2c, \dots$ the string undergoes additional jumps of size $1/2c$, cf. Fig. 9.3.

The function u is a sum of traveling combs. You can easily verify that this function is periodic in time with the period $T = p/c$. ∎

When f, g are p-periodic we usually find it preferable to work with the weakly convergent Fourier series

$$f(x) = \sum_{k=-\infty}^{\infty} F[k]e^{2\pi i k x/p}, \qquad g(x) = \sum_{k=-\infty}^{\infty} G[k]e^{2\pi i k x/p}. \tag{30}$$

Since the solution of

$$u_{tt} = c^2\, u_{xx}, \quad u(x,0) = f(x), \quad u_t(x,0) = g(x) \tag{31}$$

is a p-periodic generalized function of x for each choice of t, we can also write

$$u(x,t) = \sum_{k=-\infty}^{\infty} U[k,t]e^{2\pi i k x/p}.$$

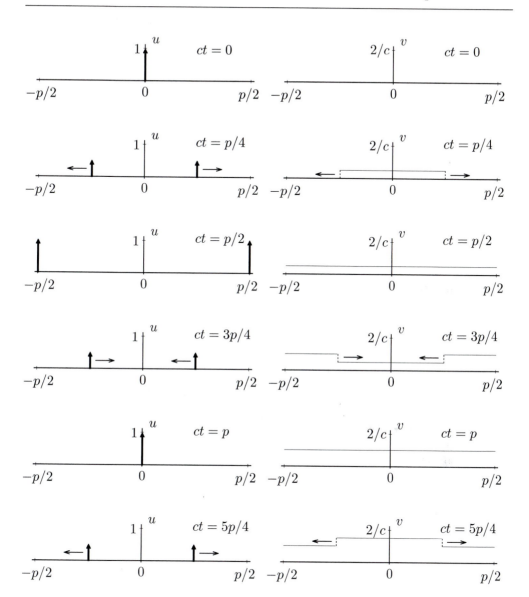

Figure 9.3. Time slices for the functions $u = r_t^p$ and $v = r^p$ that satisfy (25) and (26), respectively.

The kth Fourier coefficient is a function of time, and by using (31) we see that

$$U_{tt}[k,t] = c^2 \left(\frac{2\pi i k}{p} \right)^2 U[k,t], \quad U[k,0] = F[k], \quad U_t[k,0] = G[k]. \tag{32}$$

We solve this initial value problem [*cf.* (14)] to find

$$U[k,t] = F[k]\cos\left(\frac{2\pi kct}{p}\right) + t\,G[k]\,\text{sinc}\left(\frac{2kct}{p}\right). \tag{33}$$

We can use (33) to express u in terms of the Fourier coefficients from (30), *i.e.*,

$$u(x,t) = \sum_{k=-\infty}^{\infty} \left\{F[k]\cos\left(\frac{2\pi kct}{p}\right) + t\,G[k]\,\text{sinc}\left(\frac{2kct}{p}\right)\right\} e^{2\pi ikx/p}. \tag{34}$$

We can also use (33) with the modulation rule, the convolution rule, and the Fourier series

$$r^p(x,t) = \frac{t}{p}\sum_{k=-\infty}^{\infty} \text{sinc}\left(\frac{2kct}{p}\right) e^{2\pi ikx/p}$$

for (27) to obtain the d'Alembert formula

$$u(x,t) = \frac{1}{2}f(x+ct) + \frac{1}{2}f(x-ct) + r^p(x,t) \circledast g(x)$$

that corresponds to (15). Here $\circledast$ is the convolution product (7.131) for generalized functions on $\mathbb{T}_p$. [We can use (2.2) when g is suitably regular.]

Example: Find the Fourier series for a traveling Λ that goes around $\mathbb{T}_p$, $p > 2$, with velocity c as shown in Fig. 9.4.

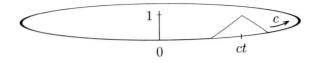

Figure 9.4. A traveling triangle wave on $\mathbb{T}_p$, $p > 2$.

Solution: We use Poisson's relation (4.18) as we write

$$u(x,t) = \sum_{m=-\infty}^{\infty} \Lambda(x - ct - mp) = \sum_{k=-\infty}^{\infty} \frac{1}{p}\,\text{sinc}^2\left(\frac{k}{p}\right) e^{2\pi ik(x-ct)/p}.$$

Of course, you can get the same result by using (34) with

$$f(x) = \sum_{m=-\infty}^{\infty} \Lambda(x - mp) = \sum_{k=-\infty}^{\infty} \frac{1}{p} \operatorname{sinc}^2\left(\frac{k}{p}\right) e^{2\pi i kx/p}$$

$$g(x) = -c \sum_{m=-\infty}^{\infty} \Lambda'(x - mp) = \sum_{k=-\infty}^{\infty} -\frac{c}{p}\left(\frac{2\pi i k}{p}\right) \operatorname{sinc}^2\left(\frac{k}{p}\right) e^{2\pi i kx/p}. \qquad \blacksquare$$

A taut string of infinite length may be only a figment of our mathematical imagination, but we can use the above analysis to study a string of length $L > 0$. We will assume that the functions f, g from (31) are odd and periodic with period $p := 2L$. This being the case, the Fourier coefficients $F[k], G[k]$ are odd, $U[k, t]$ is an odd function of k, and (34) can be rewritten in the form

$$u(x, t) = \sum_{k=1}^{\infty} \left\{ a_k \cos\left(\frac{\pi k ct}{L}\right) + b_k \sin\left(\frac{\pi k ct}{L}\right) \right\} \sin\left(\frac{\pi k x}{L}\right) \qquad (35)$$

where

$$a_k := 2i\, F[k], \quad b_k := 2i\, G[k]\left(\frac{L}{\pi k c}\right), \quad k = 1, 2, \ldots\,. \qquad (36)$$

In view of the parity and periodicity, we have

$$u(-x, t) = -u(x, t), \qquad u(L - x, t) = -u(x - L, t) = -u(L + x, t)$$

for all values of t, and we will say that u satisfies the *boundary conditions*

$$u(0, t) = 0, \qquad u(L, t) = 0 \qquad (37)$$

for a finite string with fixed endpoints even in cases where it makes no sense to evaluate u at the points $x = 0$, $x = L$.

Of course, when f, g are suitably regular ordinary functions, (35) produces an ordinary solution of (31) that satisfies (37). We can then use the analysis equations for f, g to write

$$a_k = \frac{2}{L} \int_0^L f(x) \sin\left(\frac{\pi k x}{L}\right) dx, \quad b_k = \frac{2}{\pi k c} \int_0^L g(x) \sin\left(\frac{\pi k x}{L}\right), \qquad (38)$$

$$k = 1, 2, \ldots\,,$$

and thereby express the coefficients (36) in terms of the initial position and velocity of the string on the interval $0 \le x \le L$.

Example: Find all *separable* solutions $u(x,t) = v(x) \cdot w(t)$ of the wave equation $u_{tt} = c^2 u_{xx}$ that satisfy the boundary conditions (37).

Solution: In view of the previous discussion, we can assume that v is an odd $2L$-periodic generalized function of x. Since u satisfies the wave equation we have

$$v(x)w''(t) = c^2 v''(x)w(t)$$

and

$$w''(t)V[k] = c^2 w(t) \left(\frac{2\pi i k}{2L} \right)^2 V[k], \quad k = 0, \pm 1, \pm 2, \dots .$$

If $V[k] \neq 0$ for some $k = 1, 2, \dots$ we see in turn that

$$w''(t) + \left(\frac{\pi k c}{L} \right)^2 w(t) = 0$$

and

$$w(t) = a_k \cos\left(\frac{\pi k c t}{L} \right) + b_k \sin\left(\frac{\pi k c t}{L} \right)$$

for some coefficients a_k, b_k. Since V is odd, the function u must have the form

$$u(x,t) = \left\{ a_k \cos\left(\frac{\pi k c t}{L} \right) + b_k \sin\left(\frac{\pi k c t}{L} \right) \right\} \sin\left(\frac{\pi k x}{L} \right). \tag{39}$$

Each point on the string vibrates with the *frequency*

$$f_k = \frac{kc}{2L} = \frac{k}{2L} \left(\frac{\tau}{\rho} \right)^{1/2} \tag{40}$$

(a formula found by Mersenne in 1648) and with the *shape* $\sin(\pi k x/L)$ as shown in Fig. 9.5. Any generalized solution of the wave equation that satisfies the boundary conditions (37) can be synthesized from such *normal vibrational modes* by using the weakly converging series (35)!

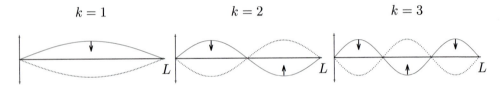

Figure 9.5. The normal vibrational modes (39) corresponding to $k = 1, 2, 3$ for a taut string with fixed endpoints.

Example: Find the series (35) for a vibrating string with fixed ends when the initial position and velocity are given by $u(x,0) := x(L-x)$, $u_t(x,0) := 0$.

Solution: We use Kronecker's rule (4.5) with (38) to write

$$a_k = \frac{2}{L} \int_0^L x(L-x) \sin\left(\frac{\pi kx}{L}\right) dx$$

$$= \frac{2}{L} \left\{ x(L-x) \left[-\left(\frac{L}{\pi k}\right) \cos\left(\frac{\pi kx}{L}\right) \right] - (L-2x) \left[-\left(\frac{L}{\pi k}\right)^2 \sin\left(\frac{\pi kx}{L}\right) \right] \right.$$

$$\left. + (-2) \left[\left(\frac{L}{\pi k}\right)^3 \cos\left(\frac{\pi kx}{L}\right) \right] \right\} \Big|_0^L$$

$$= \begin{cases} \dfrac{8L^2}{\pi^3 k^3} & \text{if } k = 1,3,5,\dots \\ 0 & \text{if } k = 2,4,6,\dots, \end{cases}$$

and thereby find

$$u(x,t) = \frac{8L^2}{\pi^3} \left\{ \frac{\cos(\pi ct/L)\sin(\pi x/L)}{1^3} + \frac{\cos(3\pi ct/L)\sin(3\pi x/L)}{3^3} \right.$$

$$\left. + \frac{\cos(5\pi ct/L)\sin(5\pi x/L)}{5^3} + \cdots \right\}. \qquad \blacksquare$$

We can make a taut string vibrate by *bowing* it (violin, cello, ...), by *plucking* it (guitar, harp, ...), or by *striking* it (piano, dulcimer, ...). We will use generalized functions [not the integrals (38)] to analyze the corresponding motions.

Example: Let $0 < x_0 < L$. When we bow, pluck, strike a string at the point $x = x_0$, we cause it to vibrate with the initial position and velocity functions

$$u^b(x,0) = 0, \qquad\qquad u_t^b(x,0) = \min\left\{ \frac{x}{x_0}, \frac{L-x}{L-x_0} \right\}, \qquad (41)$$

$$u^p(x,0) = \min\left\{ \frac{x}{x_0}, \frac{L-x}{L-x_0} \right\}, \qquad u_t^p(x,0) = 0, \qquad (42)$$

$$u^s(x,0) = 0, \qquad\qquad u_t^s(x,0) = -\frac{Lc^2}{x_0(L-x_0)} \delta(x-x_0), \qquad (43)$$

on the interval $0 \le x \le L$. Find the corresponding series (35) for the displacement functions.

Solution: We use the wave equation $u_{tt}^b(x,t) = c^2 u_{xx}^b(x,t)$ with (41) to find

$$u_{tt}^b(x,0) = c^2 u_{xx}^b(x,0) = 0$$

$$u_{ttt}^b(x,0) = c^2 u_{txx}^b(x,0) = -\frac{Lc^2}{x_0(L-x_0)} \delta(x-x_0) \text{ when } 0 < x < L.$$

In conjunction with (42), (43) this gives

$$u^p(x,0) = u_t^b(x,0), \qquad u_t^p(x,0) = u_{tt}^b(x,0),$$
$$u^s(x,0) = u_{tt}^b(x,0), \qquad u_t^s(x,0) = u_{ttt}^b(x,0),$$

and it follows that

$$u^p(x,t) = u_t^b(x,t),$$
$$u^s(x,t) = u_t^p(x,t) = u_{tt}^b(x,t).$$

We produce an odd $2L$-periodic function by suitably extending the definition of $u_t^s(x,0)$, and use the series (7.107) for III to write

$$u_t^s(x,0) = \frac{-Lc^2}{x_0(L-x_0)} \sum_{m=-\infty}^{\infty} \{\delta(x - x_0 - m\,2L) - \delta(x + x_0 - m\,2L)\}$$

$$= \frac{-c^2}{2x_0(L-x_0)} \left\{ \text{III}\left(\frac{x-x_0}{2L}\right) - \text{III}\left(\frac{x+x_0}{2L}\right) \right\}$$

$$= \frac{-c^2}{2x_0(L-x_0)} \sum_{k=-\infty}^{\infty} \left\{ e^{\pi ik(x-x_0)/L} - e^{\pi ik(x+x_0)/L} \right\}$$

$$= \frac{-2c^2}{x_0(L-x_0)} \sum_{k=1}^{\infty} \sin\left(\frac{\pi k x_0}{L}\right) \sin\left(\frac{\pi k x}{L}\right).$$

We can then write

$$u_t^s(x,t) = \frac{-2c^2}{x_0(L-x_0)} \sum_{k=1}^{\infty} \sin\left(\frac{\pi k x_0}{L}\right) \sin\left(\frac{\pi k x}{L}\right) \cos\left(\frac{2\pi kt}{T}\right)$$

and take antiderivatives term by term to obtain

$$u^s(x,t) = \frac{-2c^2}{x_0(L-x_0)} \sum_{k=1}^{\infty} \left(\frac{L}{\pi ck}\right) \sin\left(\frac{\pi k x_0}{L}\right) \sin\left(\frac{\pi k x}{L}\right) \sin\left(\frac{2\pi kt}{T}\right), \qquad (44)$$

$$u^p(x,t) = \frac{2c^2}{x_0(L-x_0)} \sum_{k=1}^{\infty} \left(\frac{L}{\pi ck}\right)^2 \sin\left(\frac{\pi k x_0}{L}\right) \sin\left(\frac{\pi k x}{L}\right) \cos\left(\frac{2\pi kt}{T}\right), \qquad (45)$$

$$u^b(x,t) = \frac{2c^2}{x_0(L-x_0)} \sum_{k=1}^{\infty} \left(\frac{L}{\pi ck}\right)^3 \sin\left(\frac{\pi k x_0}{L}\right) \sin\left(\frac{\pi k x}{L}\right) \sin\left(\frac{2\pi kt}{T}\right), \qquad (46)$$

where

$$T := 2L/c \qquad (47)$$

is the period of the fundamental vibrational mode. The musical tone emitted by the string depends on the *point* of excitation [through the common factor $\sin(\pi k x_0/L)$] and on the *mode* of excitation (through the exponent 1, 2, or 3 that we place on k). You can use these series to determine the shape of the string at any time t. The time slices shown in Fig. 9.6 will help you to visualize the motion. ∎

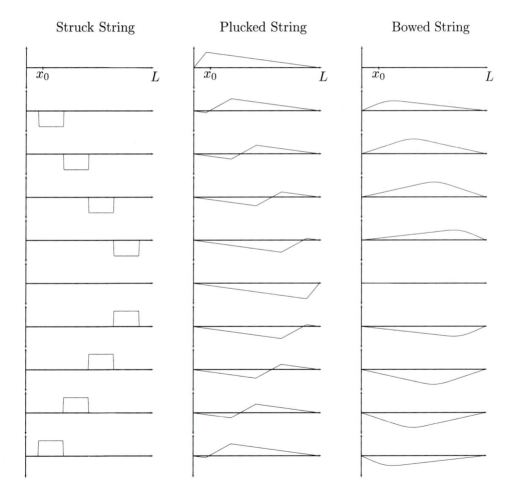

Figure 9.6. Time slices $u(x, kT/10)$, $k = 0, 1, \ldots, 9$, for the struck, plucked, and bowed strings as given by (44), (45), and (46) when $x_0 = L/10$.

9.3 The Diffusion Equation

A physical context: Heat flow along a long rod

We consider the flow of heat along a long, perfectly insulated, homogeneous rod that has a uniform cross-sectional area A. The insulation ensures that no heat energy is conducted or radiated from the sides of the rod into the surrounding environment. We assume that a smooth function $u(x,t)$ gives the temperature of all points having the coordinate x (measured along the axis of the rod) at time t, *cf.* Fig. 9.7.

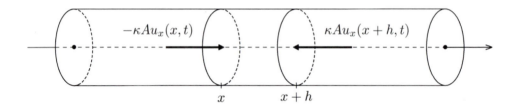

Figure 9.7. A segment of a rod and the heat flows that are used with (48) to derive the diffusion equation (49).

Heat energy will flow from a hot segment of the rod into cooler adjacent regions, with the flow being greatest in regions where u changes most rapidly with respect to x. Following Fourier, we will assume that the rate at which heat flows through the cross section of the rod at coordinate x is given by $-\kappa A u_x(x,t)$. The *thermal conductivity* κ is a constant that depends on the composition of the rod (*e.g.*, $\kappa = 420$ joules/(sec $\cdot$ m $\cdot^\circ$C) for silver, a good conductor, and $\kappa = 1.2$ joules/(sec $\cdot$ m $\cdot^\circ$C) for glass, a poor conductor.)

Suppose we are given an isolated segment of the rod that has length h and a uniform temperature. When we add heat energy to this segment, the temperature will rise. We will assume that the amount of additional heat is proportional to the change in temperature. For the constant of proportionality we will use the product of the volume, Ah, the mass density ρ, and the specific *heat capacity $\mathcal{C}$*, another constant that depends on the composition of the rod. Of course, the same proportionality constant allows us to determine the rate at which heat is added to the segment when we know the rate at which the temperature changes with respect to time.

We will now focus our attention on the segment of the rod between the coordinates x and $x + h$ as shown in Fig. 9.7. The heat that flows into this segment through its left and right faces causes the temperature to rise, so we can write

$$-\kappa A u_x(x,t) + \kappa A u_x(x+h,t) = \text{rate of heat influx}$$
$$= \text{rate of heat storage} \qquad (48)$$
$$= \mathcal{C}Ah\rho u_t(x',t)$$

where x' is a suitably chosen point within the segment. We divide both sides by $\mathcal{C}Ah\rho$, and then let $h \to 0$ to obtain the PDE

$$u_t(x,t) = \left(\frac{\kappa}{\mathcal{C}\rho}\right) u_{xx}(x,t). \qquad (49)$$

In this way we obtain the diffusion equation (2) with

$$a^2 := \frac{\kappa}{\mathcal{C}\rho}. \qquad (50)$$

Like κ, $\mathcal{C}$, and ρ, the *thermal diffusivity* a^2 is a constant that depends on the composition of the rod (*e.g.*, $a^2 = 1.7 \cdot 10^{-4}$ m^2/sec for silver and $a^2 = 5.8 \cdot 10^{-7}$ m^2/sec for glass).

If you regard temperature as a measure of the *concentration* of heat energy, you will find it very easy to see that the PDE (2) can be used to study other diffusion processes. For example, we can replace the rod with a long tube filled with water, air, ... and let $u(x,t)$ be the concentration of some dye, perfume, ... at coordinate x and time t. You can interpret the above derivation of (49) within such contexts.

Heat flow is a far more complicated process than the above derivation suggests, so you should consult a treatise, *e.g.*, the one by Carslaw and Jaeger, if you must perform a precise analysis of this phenomenon. (In practice the "constants" κ, ρ, and $\mathcal{C}$ depend on both u and x.) Fourier's simple model (49) is sufficiently accurate for our purposes, however, and we will now construct solutions of this PDE.

The diffusion equation on $\mathbb{R}$

Let f be a generalized function on $\mathbb{R}$. We will construct the unique generalized solution of the diffusion equation (2) that has the initial temperature

$$u(x,0) = f(x). \qquad (51)$$

If such a solution exists, we can Fourier transform (2) and (51) to obtain

$$U_t(s,t) = a^2(2\pi i s)^2 U(s,t), \qquad U(s,0) = F(s).$$

Using the fact that the initial value problem

$$y'(t) + \alpha y(t) = 0, \qquad y(0) = A$$

has the solution

$$y(t) = A e^{-\alpha t}, \qquad t \geq 0,$$

we then write

$$U(s,t) = e^{-4\pi^2 a^2 s^2 t} F(s), \qquad t \geq 0. \tag{52}$$

The gaussian factor

$$K(s,t) := e^{-4\pi^2 a^2 s^2 t}, \qquad t \geq 0$$

is the Fourier transform of the *diffusion kernel*

$$k(x,t) := \begin{cases} \delta(x) & \text{if } t = 0, \\[2mm] \dfrac{e^{-x^2/4a^2 t}}{\sqrt{4\pi a^2 t}} & \text{if } t > 0, \end{cases} \tag{53}$$

a normal density with mean $\mu = 0$ and variance $\sigma^2 := 2a^2 t$. We recognize

$$U(s,t) = K(s,t) \cdot F(s)$$

as the Fourier transform of

$$u(x,t) = k(x,t) * f(x). \tag{54}$$

The convolution product (54) is defined for every choice of f and for every $t \geq 0$, and you can verify that the resulting u satisfies both (2) and (51), *cf.* Ex. 9.35. Of course, when f is a suitably regular ordinary function and $t > 0$, we can also write

$$u(x,t) = \frac{1}{\sqrt{4\pi a^2 t}} \int_{-\infty}^{\infty} f(\xi) e^{-(x-\xi)^2/4a^2 t} d\xi. \tag{55}$$

The gaussian kernel is a Schwartz function, so we can use (54) to infer that $u, u_x, u_{xx}, \ldots$ are CSG functions of x when $t > 0$ is fixed. You can begin with any generalized temperature f (no matter how badly behaved) and an instant $t > 0$ later the diffusion process will give you a supersmooth temperature profile $u(x,t)$! [In cases where the generalized function f is support limited, we can use (53) to see that u is a Schwartz function of x when $t > 0$.]

Example: Solve $u_t = a^2 u_{xx}$ with the initial temperature

$$u(x,0) = \frac{e^{-x^2/2\sigma^2}}{\sqrt{2\pi}\,\sigma}, \qquad \sigma > 0.$$

Solution: The convolution product (54) of (zero mean) normal densities having the variances $2a^2 t$, σ^2 is the normal density

$$u(x,t) = \frac{e^{-x^2/2\sigma^2(t)}}{\sqrt{2\pi}\,\sigma(t)}, \qquad t \geq 0$$

with

$$\sigma(t) := (\sigma^2 + 2a^2 t)^{1/2}.$$

You can use the context to interpret this result. For example, when $\sigma \ll 1$ the rod has a tightly focused hot spot that spreads smoothly as the rod cools. ∎

Example: Solve $u_t = a^2 u_{xx}$ with the initial temperature $u(x, 0) = \Pi(x)$.

Solution: We can use (55) to write

$$u(x, t) = \frac{1}{\sqrt{4\pi a^2 t}} \int_{\xi=-1/2}^{1/2} e^{-(x-\xi)^2/4a^2 t} d\xi, \qquad (56)$$

and express the integral in terms of the error function, *cf.* Ex. 9.22. We can also use (52) to write

$$U(s, t) = e^{-4\pi^2 a^2 s^2 t} \operatorname{sinc}(s)$$

and synthesize

$$u(x, t) = \int_{s=-\infty}^{\infty} U(s, t) e^{2\pi i s x} ds = \int_{-\infty}^{\infty} e^{-4\pi^2 a^2 s^2 t} \operatorname{sinc}(s) \cos(2\pi s x) dx,$$

cf. Ex. 1.20. Figure 9.8 shows a few selected time slices of the temperature u. The discontinuous $u(x, 0) = \Pi(x)$ [with the decidedly unphysical initial temperature gradient $u_x(x, 0) = \delta(x + 1/2) - \delta(x - 1/2)$] is instantly mollified. The box rapidly evolves into an approximate gaussian (*cf.* Ex. 9.28) that spreads in accordance with the analysis from the previous example. ∎

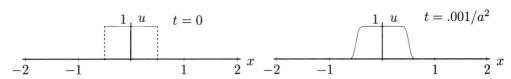

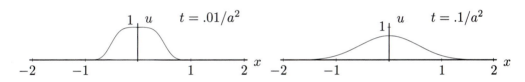

Figure 9.8. Time slices for the solution (56) of $u_t = a^2 u_{xx}$ with $u(x, 0) = \Pi(x)$.

Example: Solve $u_t = a^2 u_{xx}$ with the initial temperature $u(x, 0) = x^2$.

Solution: We use (52) to obtain

$$U(s, t) = e^{-4\pi^2 a^2 s^2 t} \frac{\delta''(s)}{(-2\pi i)^2},$$

and then simplify the synthesis integral by writing

$$u(x, t) = -\frac{1}{4\pi^2} \int_{-\infty}^{\infty} e^{-4\pi^2 a^2 s^2 t} \cos(2\pi s x) \delta''(s) ds$$

$$= -\frac{1}{4\pi^2} \frac{\partial^2}{\partial s^2} \left\{ e^{-4\pi^2 a^2 s^2 t} \cos(2\pi s x) \right\} \Big|_{s=0} = x^2 + 2a^2 t.$$

An alternative analysis is given in Ex. 9.26. ∎

When $u(x, t)$ is a suitably regular ordinary solution of (2), we can scale

$$\mathcal{E}_t\{u\} := \int_{-\infty}^{\infty} u(x, t) dx \tag{57}$$

(by $C\rho A$) to obtain the total heat energy stored in the rod at time t. Indeed, if the initial temperature f is piecewise smooth and absolutely integrable, we can use (54) with the fact the diffusion kernel (53) has unit area to write

$$\mathcal{E}_t\{u\} = \int_{x=-\infty}^{\infty} \int_{\xi=-\infty}^{\infty} k(x - \xi, t) f(\xi) d\xi \, dx = \int_{x=-\infty}^{\infty} f(x) dx = \mathcal{E}_0\{u\}, \quad t > 0,$$

cf. Ex. 2.26. The energy integral (57) is conserved by the diffusion process!

There are many ways to rearrange the heat energy in a rod without changing the integral (57). You know from experience, however, that heat never flows up a temperature gradient: the hottest spot on the rod (if any) does not get even hotter at some later time. A simple proof of this observation follows from the fact that (53) is a positive kernel with unit area. Indeed, if the initial temperature is a piecewise smooth absolutely integrable function with

$$m \le f(\xi) \le M, \qquad -\infty < \xi < \infty \tag{58}$$

for some choice of the constants m, M, then for each choice of x and $t > 0$ we must also have

$$m \, k(x - \xi, t) \le f(\xi) k(x - \xi, t) \le M \, k(x - \xi, t).$$

We integrate these inequalities from $\xi = -\infty$ to $\xi = +\infty$ and use (55) to see that

$$m \le u(x, t) \le M, \quad -\infty < x < \infty, \quad t > 0. \tag{59}$$

The values of $u(x,t)$ at time $t > 0$ never lie beyond the extremes m, M that bound $u(x,0)$ when $t = 0$! This is known as the *extreme value principle* for solutions of the diffusion equation (2). We can use this principle to show that a small change in the initial temperature f produces a small change in u, *cf.* Ex. 9.25.

On occasion, we wish to study the flow of heat in a semi-infinite rod corresponding to the interval $0 \leq x < \infty$. We must then specify some *boundary condition* at the end, $x = 0$. For example, we set $u(0,t) = c$, $t \geq 0$, if the end is kept at some constant temperature, and we set $u_x(0,t) = d$, $t \geq 0$, if there is a constant flow of heat into the rod at $x = 0$. In many cases we can suitably extend the domain of u to $-\infty < x < \infty$ and solve the problem by using (54). The following examples will show you how this is done.

Example: Find the solution of $u_t = a^2 u_{xx}$ on the half line $x \geq 0$ with the boundary condition

$$u(0,t) = 1, \qquad t \geq 0 \tag{60}$$

and the initial temperature

$$u(x,0) = \begin{cases} 1 & \text{if } 0 \leq x \leq 1 \\ 0 & \text{otherwise.} \end{cases} \tag{61}$$

Solution: We can find u if we can find the function

$$v(x,t) := u(x,t) - 1$$

that satisfies the diffusion equation with the homogeneous boundary condition

$$v(0,t) = 0, \qquad t \geq 0$$

and the initial condition

$$v(x,0) = \begin{cases} 0 & \text{if } 0 \leq x \leq 1 \\ -1 & \text{if } 1 < x. \end{cases}$$

We can construct such a function v by using (55) with the odd initial temperature

$$f(x) = \begin{cases} 1 & \text{if} & x < -1 \\ 0 & \text{if} & -1 \leq x \leq 1 \\ -1 & \text{if} & 1 < x \end{cases}$$

on all of $\mathbb{R}$. In this way we obtain

$$u(x,t) = 1 + \frac{1}{\sqrt{4\pi a^2 t}} \int_{\xi=1}^{\infty} \left\{ e^{-(x+\xi)^2/4a^2 t} - e^{-(x-\xi)^2/4a^2 t} \right\} d\xi$$

$$= 1 - \frac{1}{\sqrt{4\pi a^2 t}} \int_{1-x}^{1+x} e^{-\xi^2/4a^2 t} d\xi, \tag{62}$$

cf. Ex. 9.22. Figure 9.9 shows a few selected time slices of the temperature u. ∎

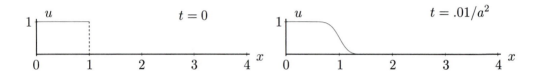

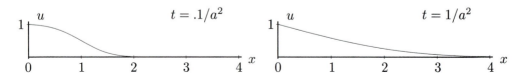

Figure 9.9. Time slices for the solution (62) of $u_t = a^2 u_{xx}$ on $0 \le x < \infty$ with the boundary condition (60) and the initial temperature (61).

Example: Find the steady-state solution of $u_t = a^2 u_{xx}$ on the half line $x \ge 0$ when u satisfies the time-varying boundary condition

$$u(0,t) = c_0 + c_1 \cos\left(\frac{2\pi t}{T}\right), \qquad t \ge 0. \tag{63}$$

Here c_0, c_1, and $T > 0$ are constants.

Solution: We consider the forced diffusion equation

$$v_t(x,t) = a^2 v_{xx}(x,t) + \delta(x) e^{2\pi i t / T} \tag{64}$$

on the line $-\infty < x < \infty$. (The forcing term allows us to inject or remove heat at the point $x = 0$ so we can control the temperature at that point.) We Fourier transform the PDE and the initial condition

$$v(x,0) = f(x)$$

to obtain

$$V_t(s,t) + 4\pi^2 a^2 s^2 V(s,t) = e^{2\pi i t / T},$$
$$V(s,0) = F(s).$$

It is easy to verify that this forced ordinary differential equation has the solution

$$V(s,t) = [F(s) - A(s)] e^{-4\pi^2 a^2 s^2 t} + A(s) e^{2\pi i t / T} \tag{65}$$

where

$$A(s) := \frac{1}{4\pi^2 a^2 s^2 + 2\pi i / T} = \frac{1}{\pi} \frac{T}{L^2 s^2 + 2i} \tag{66}$$

and

$$L := (4\pi a^2 T)^{1/2}. \tag{67}$$

The first term on the right side of (65) is a transient that converges weakly to the zero function as $t \to +\infty$. We discard this transient, e.g., by choosing $F(s) = A(s)$, and focus our attention on the steady-state portion

$$V(s,t) = A(s)e^{2\pi i t/T}.$$

We use the dilation rule and the fact that

$$e^{-2\pi(1+i)|x|} \quad \text{has the FT} \quad \frac{1}{\pi}\frac{1+i}{s^2+2i}, \tag{68}$$

cf. Ex. 9.31, to obtain the steady-state solution

$$v(x,t) = \frac{T}{(1+i)L}e^{-2\pi(1+i)|x|/L}e^{2\pi i t/T} \tag{69}$$

of (64). Since (69) is a solution of the diffusion equation at each point where $t > 0$ and $x \neq 0$, we see that the same is true of

$$u(x,t) = c_0 + c_1 e^{-2\pi|x|/L}\cos\left[2\pi\left(\frac{t}{T} - \frac{|x|}{L}\right)\right]. \tag{70}$$

By design, (70) satisfies the forced boundary condition (63).

For Fourier's classical interpretation of this result, we let $u(x,t)$ be the temperature at depth $x \geq 0$ and time t in a large slab of soil that is subjected to a seasonal variation of temperature (63) at all points on the surface. Using the approximate diffusivity $a^2 \approx 4.8 \cdot 10^{-7}$ m^2/sec for soil and $T = 1$ year with (67), we find $L \approx 13.8$ m. At half that depth, the phase lag in (70) causes the soil to be cold in summer and warm in winter, cf. Fig. 9.10. The temperature fluctuations decrease as x increases and you can use this observation to determine how deep you must bury a water pipe to keep it from freezing, cf. Ex. 9.32. ∎

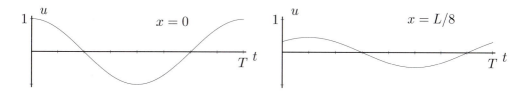

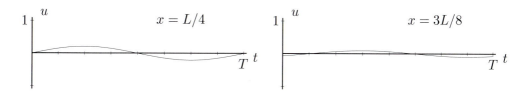

Figure 9.10. Slices of $u(x,t)$, $0 \leq t \leq T$, at selected choices of x when u is given by (70) and $c_0 = 0$, $c_1 = 1$.

The diffusion equation on $\mathbb{T}_p$

We will specialize the above analysis to the case where the initial temperature f is p-periodic with $p > 0$. The convolution product (54) then gives a generalized solution of (2) that is p-periodic in x.

Example: Solve the initial value problem

$$u_t = a^2 u_{xx}, \qquad u(x,0) = \frac{1}{p}\text{III}\left(\frac{x}{p}\right). \tag{71}$$

Solution: We use (54) to see that

$$u(x,t) = k^p(x,t)$$

where

$$k^p(x,t) := k(x,t) * \frac{1}{p}\text{III}\left(\frac{x}{p}\right) = \sum_{m=-\infty}^{\infty} k(x - mp, t). \tag{72}$$

In particular,

$$k^p(x,t) = \frac{1}{\sqrt{4\pi a^2 t}} \sum_{m=-\infty}^{\infty} e^{-(x-mp)^2/4a^2 t} \quad \text{when } t > 0. \tag{73}$$

You can interpret this formula as follows. At time $t = 0$, a heat energy impulse (of size $C\rho A$) is injected at each point $x = 0, \pm p, \pm 2p, \ldots$ in a rod that is everywhere at temperature $u = 0$. Each impulse gives rise to one of the spreading gaussian hot spots from (73). When t is large, these gaussians overlap and it is not so easy to compute u by using (73). We can use the Poisson relation (4.18) to write

$$k^p(x,t) = \frac{1}{p} \sum_{k=-\infty}^{\infty} e^{-(2\pi ak/p)^2 t} e^{2\pi ikx/p}, \tag{74}$$

and thereby see that u converges to the constant $1/p$ as $t \to +\infty$, *cf.* Fig. 9.11. ∎

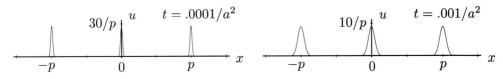

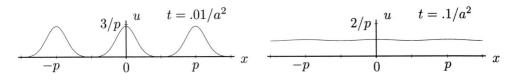

Figure 9.11. Time slices for the function $u = k^p$ from (72) that satisfies (71).

Example: Solve $u_t = a^2 u_{xx}$ with the initial temperature $u(x,0) = \cos(2\pi x/p)$.

Solution: Using (52) and (7.82), we write

$$U(s,t) = e^{-4\pi^2 a^2 s^2 t} \frac{1}{2}\left\{\delta\left(s - \frac{1}{p}\right) + \delta\left(s + \frac{1}{p}\right)\right\}$$

$$= e^{-4\pi^2 a^2 t/p^2} \frac{1}{2}\left\{\delta\left(s - \frac{1}{p}\right) + \delta\left(s + \frac{1}{p}\right)\right\},$$

and thereby find

$$u(x,t) = e^{-4\pi^2 a^2 t/p^2} \cos\left(\frac{2\pi x}{p}\right).$$ ∎

When we work with p-periodic functions, we usually find it advantageous to use Fourier series. If

$$u(x,t) = \sum_{k=-\infty}^{\infty} U[k,t] e^{2\pi i k x/p}$$

is the solution of $u_t = a^2 u_{xx}$ with the initial temperature

$$f(x) = \sum_{k=-\infty}^{\infty} F[k] e^{2\pi i k x/p},$$

then the kth Fourier coefficient must satisfy

$$U_t[k,t] + \left(\frac{2\pi a k}{p}\right)^2 U[k,t] = 0, \quad U[k,0] = F[k].$$

We solve this initial value problem to find

$$U[k,t] = F[k] e^{-(2\pi a k/p)^2 t}, \quad k = 0, \pm 1, \pm 2, \ldots,$$

and synthesize

$$u(x,t) = \sum_{k=-\infty}^{\infty} F[k] e^{-(2\pi a k/p)^2 t} e^{2\pi i k x/p}. \tag{75}$$

You will observe that $U[k,t]$ is the product of the Fourier coefficient $F[k]$ for f and the corresponding p scaled Fourier coefficient from the p-periodic diffusion kernel (74). We can use the convolution rule for p-periodic functions to see that

$$u(x,t) = k^p(x,t) \circledast f(x). \tag{76}$$

Example: A circle with circumference $p > 0$ is made from a thin rod. At time $t = 0$ the right half of the circle has temperature $u = +1$ and the left half has temperature $u = -1$. Find the temperature function at time $t > 0$.

Solution: We use the Fourier series

$$f(x) = \sum_{k=\text{odd}} \frac{2e^{2\pi i k x/p}}{\pi i k} = \begin{cases} 1 & \text{if } 0 < x < \dfrac{p}{2} \\ -1 & \text{if } \dfrac{p}{2} < x < p \end{cases}$$

with (75) to write

$$u(x,t) = \frac{4}{\pi} \left\{ e^{-(2\pi a/p)^2 t} \sin\left(\frac{2\pi x}{p}\right) + \frac{1}{3} e^{-(6\pi a/p)^2 t} \sin\left(\frac{6\pi x}{p}\right) \right.$$
$$\left. + \frac{1}{5} e^{-(10\pi a/p)^2 t} \sin\left(\frac{10\pi x}{p}\right) + \cdots \right\}. \qquad \blacksquare$$

On occasion we want to study the flow of heat in a finite rod with the space coordinates $0 \le x \le L$. We must then introduce a suitable boundary condition at each end. For example, we might specify the temperature u or the (scaled) heat flux u_x. It is sometimes possible to eliminate such boundary conditions by using symmetry. We will give three examples to show how this is done.

Example: Let $0 < x_0 < L$. A rod of length L initially has the temperature

$$u(x,0) = f(x) := \begin{cases} 1 & \text{if } 0 \le x \le x_0 \\ 0 & \text{if } x_0 < x \le L. \end{cases}$$

The ends of the rod are insulated so that

$$u_x(0,t) = 0, \quad u_x(L,t) = 0 \quad \text{for} \quad t > 0.$$

Find the temperature in the rod at time $t > 0$ by solving the diffusion equation $u_t = a^2 u_{xx}$ subject to these constraints.

Solution: Let $p := 2L$. We extend f from $[0, L]$ to $\mathbb{R}$ by writing

$$f(x) = \sum_{m=-\infty}^{\infty} \Pi\left(\frac{x - m\,2L}{2x_0}\right) = \sum_{k=-\infty}^{\infty} \frac{x_0}{L} \operatorname{sinc}\left(\frac{kx_0}{L}\right) e^{2\pi i k x/2L}.$$

By construction, this p-periodic function has the even symmetries

$$f(-x) = f(x), \qquad f(L - x) = f(L + x).$$

We produce a solution of the diffusion equation that has the initial temperature f by using (75) to write

$$u(x,t) = \sum_{k=-\infty}^{\infty} \frac{x_0}{L} \, \text{sinc}\left(\frac{kx_0}{L}\right) e^{-(2\pi ak/2L)^2 t} e^{2\pi ikx/2L}$$

$$= \frac{x_0}{L} + \frac{2x_0}{L} \sum_{k=1}^{\infty} \text{sinc}\left(\frac{kx_0}{L}\right) e^{-(\pi ak/L)^2 t} \cos(\pi kx/L).$$

Since u is a smooth function of x that inherits the symmetries

$$u(-x,t) = u(x,t), \qquad u(L-x,t) = u(L+x,t)$$

of f, u_x must vanish at $x = 0$ and $x = L$. ∎

Example: A rod of length L at the uniform temperature $u = 0$ is given an impulse of heat to produce an initial temperature $u(x,0) = f(x) := \delta(x - x_0)$. Find the temperature at time $t > 0$ if the ends of the rod are held at the temperature $u(0,t) = 0$, $u(L,t) = 0$ for $t \geq 0$.

Solution: Let $p = 2L$. We extend the initial temperature from $[0, L]$ to $\mathbb{R}$ by writing

$$f(x) = \sum_{m=-\infty}^{\infty} \delta(x - x_0 - m\,2L) - \sum_{m=-\infty}^{\infty} \delta(x + x_0 - m\,2L)$$

$$= \frac{1}{2L} \left\{ \text{III}\left(\frac{x - x_0}{2L}\right) - \text{III}\left(\frac{x + x_0}{2L}\right) \right\}$$

$$= \frac{1}{2L} \sum_{k=-\infty}^{\infty} \left\{ e^{2\pi ik(x-x_0)/2L} - e^{2\pi ik(x+x_0)/2L} \right\}$$

$$= -\frac{i}{L} \sum_{k=-\infty}^{\infty} \sin\left(\frac{\pi kx_0}{L}\right) e^{2\pi ikx/2L}.$$

By construction, this p-periodic function has the odd symmetries

$$f(-x) = -f(x), \qquad f(L-x) = -f(L+x).$$

We produce a solution of the diffusion equation that has the initial temperature f by using (75) to write

$$u(x,t) = -\frac{i}{L} \sum_{k=-\infty}^{\infty} \sin\left(\frac{\pi kx_0}{L}\right) e^{-(2\pi ak/2L)^2 t} e^{2\pi ikx/2L}$$

$$= \frac{2}{L} \sum_{k=1}^{\infty} \sin\left(\frac{\pi kx_0}{L}\right) e^{-(\pi ak/L)^2 t} \sin(\pi kx/L).$$

The smooth function u inherits the symmetries

$$u(-x, t) = -u(x, t), \qquad u(L - x, t) = -u(L + x, t)$$

of f, so u must vanish at $x = 0$ and $x = L$.   ∎

Example: The ends of a rod of length L are kept in water baths with temperatures $u = 0$, $u = 100$ until thermal equilibrium is reached. The water baths are then reversed, *i.e.*, the boundary conditions are switched. Find the temperature at all points of the rod at time $t > 0$ after the switch.

Solution: We must solve the diffusion equation $u_t = a^2 u_{xx}$ for $0 < x < L$ subject to the initial condition

$$u(x, 0) = 100 \left(1 - \frac{x}{L} \right), \quad 0 < x < L, \tag{77}$$

and the boundary conditions

$$u(0, t) = 0, \quad u(L, t) = 100 \quad \text{for} \quad t > 0. \tag{78}$$

We know that the rod will eventually reach thermal equilibrium at the temperature

$$u^\infty(x, t) := 100 \frac{x}{L}, \qquad 0 < x < L.$$

The difference

$$v(x, t) := u(x, t) - u^\infty(x, t)$$

satisfies the diffusion equation for $0 < x < L$ with the initial temperature

$$v(x, 0) = u(x, 0) - u^\infty(x, 0) = 200 \left(\frac{1}{2} - \frac{x}{L} \right), \quad 0 < x < L,$$

and the boundary conditions

$$v(0, t) = 0, \qquad v(L, t) = 0, \quad t \geq 0.$$

We L-periodically extend $v(x, 0)$ to obtain the odd scaled Bernoulli function

$$f(x) = 200 \sum_{\substack{k=-\infty \\ k \neq 0}}^{\infty} \frac{e^{2\pi i k x / L}}{2\pi i k}$$

of (4.23), and then use (75) to write

$$v(x, t) = 200 \sum_{\substack{k=-\infty \\ k \neq 0}}^{\infty} \frac{e^{-(2\pi a k / L)^2 t} e^{2\pi i k x / L}}{2\pi i k}$$

$$= \frac{200}{\pi} \sum_{k=1}^{\infty} \frac{e^{-(2\pi a k / L)^2 t} \sin(2\pi k x / L)}{k}.$$

In this way we find

$$u(x,t) = v(x,t) + 100\frac{x}{L}$$

when $0 \le x \le L$ and $t > 0$, *cf.* Fig. 9.12. ∎

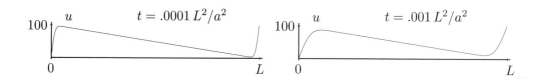

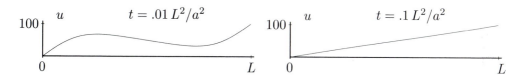

Figure 9.12. Time slices for the solution of the diffusion equation $u_t = a^2 u_{xx}$ with the initial condition (77) and the boundary conditions (78).

9.4 The Diffraction Equation

A physical context: Diffraction of a laser beam

When a laser beam (with a diameter of a few millimeters) passes through a suitably prepared 35-mm slide, there is a point-by-point modification of the phase and amplitude of the light wave. The emergent beam spreads to form a spot (with a diameter of a few centimeters) on a screen placed a few meters from the slide. The intensity of the light varies from point to point within the spot, and in this way we obtain a *diffraction pattern* that corresponds to the image on the slide. (A striking collection of photographs of such diffraction patterns can be found in G. Harburn, C.A. Taylor, and T.R. Welberry, *Atlas of Optical Transforms*, G. Bell & Sons, London, 1975.)

A precise analysis of this phenomenon must be based on Maxwell's equations, *e.g.*, as done in the classic treatise of Born and Wolf. We will give a simplified derivation that can be used when the light is monochromatic with wavelength λ, *e.g.*, $\lambda = .6328 \cdot 10^{-3}$ mm for the red light from a He-Ne laser. The same derivation can be used for sound waves, water waves, etc. provided that all of the waves have a common wavelength and the dimensions (as measured in wavelengths) are comparable to those for the diffraction of a laser beam.

We will use the coordinate z to measure distance along the axis of the laser beam. The coordinates x, y will locate points in a plane perpendicular to the beam. In this chapter we will suppress the y-coordinate and analyze the way a univariate diffraction pattern evolves as we vary the parameter z, *i.e.*, as we position the viewing screen at various points along the z-axis.

Suppose that we have a *source* of waves at the coordinate ξ on the line perpendicular to the z-axis at the origin as shown in Fig. 9.13. Circular waves emanate from this point and travel to an observation point with coordinates (x, z). (You can produce such water waves with $\lambda = 1$ cm by touching the surface of a still lake with the tip of a pencil that oscillates up and down along its axis with a frequency of 12.5 Hz, *cf.* Ex. 9.47.) Let $\mathcal{U}(x, z, t)$ be a real measure of the size of the disturbance at the point (x, z) and time t. For example, when we work with water, sound, or light waves we can let $\mathcal{U}$ represent the vertical displacement of the surface of the water from its equilibrium position, the departure of air pressure from its equilibrium value, or the x-component of the electric field vector at the point in question.

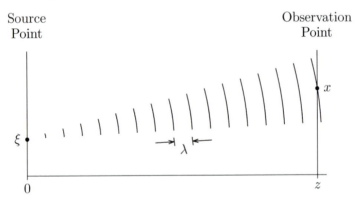

Figure 9.13. The parameters ξ, x, z, and λ that are used for (79)–(80).

At the source, we expect $\mathcal{U}$ to be proportional to some sinusoid

$$\cos\left\{-2\pi\frac{t}{T} + \phi\right\}$$

that has the period $T > 0$ and the initial phase ϕ. As we travel away from the source along a ray, we pass over one complete wave each time we move a distance λ, so we expect the disturbance $\mathcal{U}$ to be proportional to

$$\cos\left\{2\pi\left(\frac{r}{\lambda} - \frac{t}{T}\right) + \phi\right\}$$

when we are at a distance r from the source. We expect the amplitude of the disturbance to diminish as r increases, and we must include some factor to specify the strength of the source. These considerations lead us to assume that $\mathcal{U}$ has the form

$$\mathcal{U}(x, z, t) = A\, a(r) \cos\left\{ 2\pi \left(\frac{r}{\lambda} - \frac{t}{T} \right) + \phi \right\} \tag{79}$$

where

$$r := \left[(x - \xi)^2 + z^2 \right]^{1/2}, \tag{80}$$

$A > 0$ is a scalar, and $a(r)$ is a well-behaved function of r (which we will determine presently). Huygens, a contemporary of Newton, introduced such wave forms for his study of diffraction three centuries ago!

We will now describe two approximations that greatly simplify the subsequent analysis. We first use the binomial series with (80) to write

$$
\begin{aligned}
r &= z \left\{ 1 + \frac{(x - \xi)^2}{z^2} \right\}^{1/2} \\
&= z \left\{ 1 + \frac{(x - \xi)^2}{2z^2} - \frac{(x - \xi)^4}{8z^4} + \frac{(x - \xi)^6}{16z^6} - \cdots \right\}, \quad |x - \xi| < z.
\end{aligned}
\tag{81}
$$

For example, when $|x - \xi| = 1$ cm and $z = 1$ m (reasonable choices for the laser beam diffraction analysis) the terms of this series have the numerical values

$$r = 1 \text{ m } \cdot \left\{ 1 + \frac{1}{2 \cdot 10^4} - \frac{1}{8 \cdot 10^8} + \frac{1}{16 \cdot 10^{12}} - \cdots \right\},$$

and you will readily see why we choose to use the approximation

$$a(r) \approx a(z)$$

for the amplitude reduction factor that appears in (79). On the other hand, when $|x - \xi| = 1$ cm, $z = 1$ m, and $\lambda = .6328 \cdot 10^{-3}$ mm, the same series gives

$$
\begin{aligned}
\frac{r}{\lambda} &= \frac{1 \text{ m}}{.6328 \cdot 10^{-6} \text{ m}} \cdot \left\{ 1 + \frac{1}{2 \cdot 10^4} - \frac{1}{8 \cdot 10^8} + \frac{1}{16 \cdot 10^{12}} - \cdots \right\} \\
&= 1.58 \ldots \cdot 10^6 + 79.0 \ldots - .00197 \ldots + \cdots .
\end{aligned}
$$

The very large z/λ ratio forces us to use at least two terms of the series if we want to compute the cosine factor from (79) with reasonable accuracy. Such considerations lead us to replace (79) with the *Fresnel approximation*

$$\mathcal{U}(x, z, t) = A\, a(z) \cos\left\{ 2\pi \left(\frac{z}{\lambda} - \frac{t}{T} \right) + \frac{\pi (x - \xi)^2}{\lambda z} + \phi \right\}. \tag{82}$$

You can safely use this approximation when x, ξ, z, and λ satisfy both

$$|x - \xi|^2 \ll z^2 \quad \text{and} \quad |x - \xi|^4 \ll \lambda z^3. \tag{83}$$

Suppose now that we have two point sources on the line $z = 0$ at coordinates ξ_1, ξ_2 with the amplitudes A_1, A_2 and the initial phases ϕ_1, ϕ_2. We will assume that we can find the resulting disturbance at (x, z) by adding the two corresponding functions (82). To facilitate the analysis we use complex arithmetic and write

$$
\begin{aligned}
\mathcal{U}(x, z, t) &= A_1\, a(z) \cos\left\{ 2\pi \left(\frac{z}{\lambda} - \frac{t}{T} \right) + \frac{\pi(x - \xi_1)^2}{\lambda z} + \phi_1 \right\} \\
&\quad + A_2\, a(z) \cos\left\{ 2\pi \left(\frac{z}{\lambda} - \frac{t}{T} \right) + \frac{\pi(x - \xi_2)^2}{\lambda z} + \phi_2 \right\} \\
&= \operatorname{Re}\left\{ \left[A_1 e^{i\phi_1} \cdot a(z) e^{i\pi(z - \xi_1)^2/\lambda z} + A_2 e^{i\phi_2} \cdot a(z) e^{i\pi(x - \xi_2)^2/\lambda z} \right] \right. \\
&\qquad\qquad\qquad\qquad\qquad\qquad\qquad \left. \cdot e^{2\pi i(z/\lambda - t/T)} \right\}.
\end{aligned}
\tag{84}
$$

We will introduce a *reduced wave function*

$$
u(x, z) := A_1 e^{i\phi_1} \cdot a(z) e^{i\pi(x - \xi_1)^2/\lambda z} + A_2 e^{i\phi_2} \cdot a(z) e^{i\pi(x - \xi_2)^2/\lambda z}
$$

with modulus

$$
A(x, z) := |u(x, z)|,
$$

and choose $0 \leq \phi(x, z) < 2\pi$ so that

$$
u(x, z) = A(x, z) e^{i\phi(x, z)}.
$$

We can use this expression with (84) to write

$$
\begin{aligned}
\mathcal{U}(x, z, t) &= \operatorname{Re}\{ u(x, z) \cdot e^{2\pi i(z/\lambda - t/T)} \} \\
&= \operatorname{Re}\{ A(x, z) e^{i\phi(x, z)} \cdot e^{2\pi i(z/\lambda - t/T)} \} \\
&= A(x, z) \cos\left\{ 2\pi \left(\frac{z}{\lambda} - \frac{t}{T} \right) + \phi(x, z) \right\} \\
&= |u(x, z)| \cos\left\{ 2\pi \left(\frac{z}{\lambda} - \frac{t}{T} \right) + \phi(x, z) \right\}.
\end{aligned}
$$

When we add the wave functions (82) for two point sources, we obtain a T-periodic disturbance $\mathcal{U}$ at (x, z) that has the amplitude $|u(x, z)|$. The energy density of a wave is always proportional to the *square* of its amplitude, so the diffraction pattern will be bright, dark at points where $|u(x, z)|^2$ is large, small, respectively.

It is easy to generalize this analysis. If we have point sources at $\xi_1, \xi_2, \ldots, \xi_M$ with amplitudes $A_1, A_2, \ldots, A_M$ and initial phases $\phi_1, \phi_2, \ldots, \phi_M$, our reduced wave function takes the form

$$
u(x, z) = \sum_{m=1}^{M} A_m e^{i\phi_m} \cdot a(z) e^{i\pi(x - \xi_m)^2/\lambda z},
$$

and $|u(x, z)|$ is the amplitude of the resulting disturbance $\mathcal{U}$ at (x, z). Of course, we can replace such a sum by an integral. Indeed, suppose that we are given a suitably regular complex-valued source function $u(\xi, 0)$, $-\infty < \xi < \infty$. We will regard $|u(\xi, 0)| d\xi$ as the amplitude of an infinitesimal point source at ξ and use the phase of $u(\xi, 0)$ for the initial phase of this source. The reduced wave function is then given by

$$u(x, z) = \int_{-\infty}^{\infty} u(\xi, 0) \cdot a(z) e^{i\pi(x-\xi)^2/\lambda z} d\xi, \qquad (85)$$

and $|u(x, z)|$ is the amplitude of the resulting disturbance $\mathcal{U}$ at (x, z). For simplicity, we use the limits $-\infty, \infty$ for the integral. [The functions $u(\xi, 0)$, $-\infty < \xi < \infty$, and $u(x, z)$, $-\infty < x < \infty$, will be suitably localized near the axis of the laser beam when (85) really does correspond to a diffraction pattern.]

We will determine the function $a(z)$ by forcing (85) to satisfy a natural *consistency relation*. By using the convolution rule, the dilation rule, and the fact that

$$e^{i\pi x^2} \quad \text{has the FT} \quad e^{i\pi/4} \cdot e^{-i\pi s^2}$$

[*cf.* (7.98)], we obtain the Fourier transform

$$U(s, z) = e^{i\pi/4} a(z) \sqrt{\lambda z} \, e^{-i\pi\lambda z s^2} \cdot U(s, 0) \qquad (86)$$

of (85). Now if we use (85) to advance some initial source function $u(x, 0)$ from $z = 0$ to $z = z_1$, (86) shows that the Fourier transform changes from

$$U(s, 0) \quad \text{to} \quad e^{i\pi/4} a(z_1) \sqrt{\lambda z_1} \, e^{-i\pi\lambda z_1 s^2} \cdot U(s, 0).$$

If we then use (85) to advance the new source function $u(x, z_1)$ from $z = z_1$ to $z = z_1 + z_2$, (86) shows that the Fourier transform changes from

$$e^{i\pi/4} a(z_1) \sqrt{\lambda z_1} \, e^{-i\pi\lambda z_1 s^2} \cdot U(s, 0) \qquad \text{to}$$
$$e^{i\pi/4} a(z_2) \sqrt{\lambda z_2} \, e^{-i\pi\lambda z_2 s^2} \cdot e^{i\pi/4} a(z_1) \sqrt{\lambda z_1} \, e^{-i\pi\lambda z_1 s^2} \cdot U(s, 0).$$

Of course, if we use (85) to advance $u(x, 0)$ from $z = 0$ to $z = z_1 + z_2$ in a single step, we can use (86) to see that the Fourier transform changes from

$$U(s, 0) \text{ to } e^{i\pi/4} a(z_1 + z_2) \sqrt{\lambda(z_1 + z_2)} \, e^{-i\pi\lambda(z_1+z_2)s^2} \cdot U(s, 0).$$

We insist that two successive steps of size z_1, z_2 should produce the same effect as a single step of size $z_1 + z_2$, and this will be the case when we define

$$a(z) := \frac{e^{-i\pi/4}}{\sqrt{\lambda z}}, \qquad (87)$$

cf. Ex. 9.36. In this way we complete the derivation of the *Fresnel convolution equation*

$$u(x, z) = \int_{-\infty}^{\infty} u(\xi, 0) \frac{e^{-i\pi/4}}{\sqrt{\lambda z}} e^{i\pi(x-\xi)^2/\lambda z} d\xi \tag{88}$$

that will serve as our univariate mathematical model for diffraction. Since we have chosen $a(z)$ to make

$$U(s, z) = U(s, 0) \cdot e^{-i\pi\lambda z s^2},$$

it is easy to compute the partial derivative

$$U_z(s, z) = -i\pi\lambda s^2 U(s, z) = \frac{i\lambda}{4\pi}(2\pi i s)^2 U(s, z),$$

and thereby see that the reduced wave function must satisfy the partial differential equation

$$u_z(x, z) = \frac{i\lambda}{4\pi} u_{xx}(x, z). \tag{89}$$

This is the diffraction equation (3) with z (position along the laser beam) in place of t.

When you take a course in quantum mechanics, you will learn to use a complex wave function $\Psi(x, t)$ that evolves in accordance with *Schrödinger's equation*

$$\frac{1}{2m}\left(\frac{h}{2\pi i}\right)^2 \frac{\partial^2}{\partial x^2}\Psi(x, t) + V(x, t) \cdot \Psi(x, t) = -\frac{h}{2\pi i}\frac{\partial}{\partial t}\Psi(x, t)$$

to predict the behavior of a particle with a small mass m in the presence of a potential $V(x, t)$. (The parameter $h = 6.626 \cdot 10^{-34}$ joule sec is *Planck's constant*.) We say that the particle is *free* when $V(x, t) \equiv 0$, in which case $\Psi(x, t)$ evolves in accordance with the diffraction equation

$$\Psi_t = \frac{ih/m}{4\pi}\Psi_{xx}.$$

In the rest of this section we will analyze the way the reduced wave function $u(x, z)$ for a diffracting laser beam evolves as we move away from the source function along the axis of the beam. Of course, if we move with velocity $v = 1$, we can simplify identity t with z in (89) and we will do so. This will make it easier for you to compare solutions of the diffraction equation with solutions of the wave and diffusion equations. [And you will already know how to visualize the time evolution of $\Psi(x, t)$ for a free particle when you study quantum mechanics!]

The diffraction equation on $\mathbb{R}$

Let f be a generalized function on $\mathbb{R}$. We will construct the unique generalized solution of the diffraction equation (3) that has the initial source

$$u(x,0) := f(x). \tag{90}$$

In the Fourier transform domain we have

$$U_t(s,t) = \frac{i\lambda}{4\pi}(2\pi is)^2 U(s,t) = -i\pi s^2 \lambda\, U(s,t), \qquad U(s,t) = F(s).$$

We solve this initial value problem to obtain

$$U(s,t) = F(s)e^{-i\pi s^2 \lambda t}, \tag{91}$$

observing that the product is well defined since the exponential

$$\Gamma(s,t) := e^{-i\pi s^2 \lambda t}$$

and its partial derivatives $\Gamma_s, \Gamma_{ss}, \ldots$ are all CSG. The function Γ is the Fourier transform of the *diffraction kernel*

$$\gamma(x,t) := \begin{cases} \dfrac{e^{-i\pi/4}}{\sqrt{\lambda t}}\, e^{i\pi x^2/\lambda t} & \text{if } t > 0 \\[2mm] \delta(x) & \text{if } t = 0 \\[2mm] \dfrac{e^{i\pi/4}}{\sqrt{-\lambda t}}\, e^{i\pi x^2/\lambda t} & \text{if } t < 0, \end{cases} \tag{92}$$

cf. Ex. 7.46, so we can use the convolution rule to write

$$u(x,t) = \gamma(x,t) * f(x), \tag{93}$$

cf. (18) and (54). Of course, when the source f is a suitably regular ordinary function we can use the corresponding integral

$$u(x,t) = \frac{e^{-i\pi/4}}{\sqrt{\lambda t}} \int_{\xi=-\infty}^{\infty} f(\xi)e^{i\pi(x-\xi)^2/\lambda t}\,d\xi, \qquad t > 0. \tag{94}$$

Example: Solve $u_t = (i\lambda/4\pi)u_{xx}$ with the initial source

$$f(x) = \delta(x - a) + \delta(x + a), \qquad a > 0.$$

Solution: Using (93) and the sifting relation (7.64) for δ, we find

$$u(x,t) = \frac{e^{-i\pi/4}}{\sqrt{\lambda t}} \left\{ e^{i\pi(x-a)^2/\lambda t} + e^{i\pi(x+a)^2/\lambda t} \right\}$$

$$= \frac{2e^{-i\pi/4}e^{i\pi(x^2+a^2)/\lambda t}}{\sqrt{\lambda t}} \cos(2\pi ax/\lambda t).$$

The Fresnel factor $e^{i\pi x^2/\lambda t}$ causes the real and imaginary parts of the t-slices of u to oscillate wildly, but the *intensity*

$$|u(x,t)|^2 = \frac{4}{\lambda t} \cos^2\left(\pi \frac{2ax}{\lambda t} \right)$$

is well behaved. You will find it instructive to interpret this result within the context of Thomas Young's famous 1801 experiment that demonstrated the wave nature of light. The δ's from $u(x,0)$ correspond to two *coherent* (i.e., in phase) sources of light passing through infinitesimally small slits separated by the distance $2a$. Our expression for $|u(x,t)|^2$ shows that the diffraction pattern at distance t consists of a series of alternating bright and dark fringes with period $\lambda t/2a$. ∎

Example: Solve $u_t = (i\lambda/4\pi)u_{xx}$ with the initial source $u(x,0) = \Pi(x)$.

Solution: We can express the solution in terms of the Fresnel function

$$\mathcal{F}(r) := \int_{\rho=0}^{r} e^{i\pi\rho^2/2}d\rho. \tag{95}$$

(Many mathematical handbooks have tables for the real and imaginary parts of this function, and most mathematical software packages include code for evaluating $\mathcal{F}$ numerically.) By using (94) with a routine change of variable we find

$$u(x,t) = \frac{e^{-i\pi/4}}{\sqrt{\lambda t}} \int_{\xi=-1/2}^{1/2} e^{i\pi(x-\xi)^2/\lambda t}d\xi$$

$$= \frac{e^{-i\pi/4}}{\sqrt{2}} \int_{(x-1/2)/\sqrt{\lambda t/2}}^{(x+1/2)/\sqrt{\lambda t/2}} e^{i\pi\rho^2/2}d\rho \tag{96}$$

$$= \frac{1-i}{2} \left\{ \mathcal{F}\left(\frac{x+1/2}{\sqrt{\lambda t/2}} \right) - \mathcal{F}\left(\frac{x-1/2}{\sqrt{\lambda t/2}} \right) \right\}.$$

Figure 9.14 shows a few selected t-slices of the intensity $|u|^2$. This function corresponds to the diffraction pattern produced by shining a laser beam through a slit. When the viewing screen is close to the slit, we see a (slightly fuzzy) image of the slit. When we are far away, the diffraction pattern consists of a system of uniformly spaced bright fringes that diminish in intensity as we move away from the center. ∎

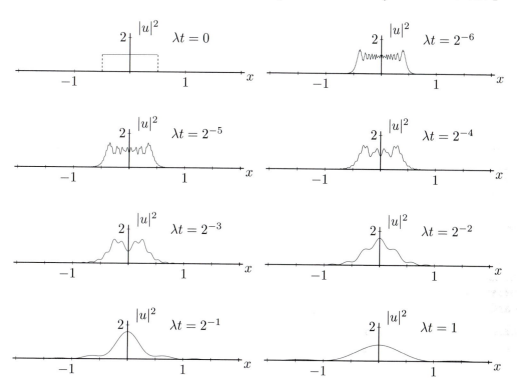

Figure 9.14. Selected t-slices for the intensity $|u(x,t)|^2$ of the reduced wave function (96) that corresponds to the slit $u(x,0) = \Pi(x)$.

Example: Solve $u_t = (i\lambda/4\pi)u_{xx}$ with the initial source

$$u(x,0) = \frac{e^{-x^2/4\sigma^2}}{(2\pi\sigma^2)^{1/4}}, \qquad \sigma > 0. \tag{97}$$

[The function $u(x,0)$ is real, *i.e.*, the light is coherent, and $|u(x,0)|^2$ is a normal density with the standard deviation σ.]

Solution: We use the dilation rule to find the Fourier transform of the gaussian (97), and then use (91) to see that

$$U(s,t) = (8\pi\sigma^2)^{1/4}e^{-\pi(4\pi\sigma^2+i\lambda t)s^2}.$$

When α, β are real numbers with $\alpha > |\beta|$,

$$e^{-\pi[x/(\alpha+i\beta)]^2} \quad \text{has the FT} \quad (\alpha+i\beta)e^{-\pi[(\alpha+i\beta)s]^2},$$

cf. Ex. 9.39, so we can write

$$u(x,t) = \frac{(8\pi\sigma^2)^{1/4}}{\sqrt{4\pi\sigma^2 + i\lambda t}} e^{-\pi x^2/(4\pi\sigma^2 + i\lambda t)} \tag{98}$$

provided we use the square root that has a positive real part. The real and imaginary parts of the t-slices of u oscillate wildly when $t > 0$, but we can use a bit of algebra to verify that the intensity

$$|u(x,t)|^2 = \left\{ \frac{8\pi\sigma^2}{|4\pi\sigma^2 + i\lambda t|^2} \right\}^{1/2} e^{-\frac{2\pi x^2 \cdot 4\pi\sigma^2}{|4\pi\sigma^2 + i\lambda t|^2}} = \frac{e^{-x^2/2\sigma^2(t)}}{\sqrt{2\pi}\,\sigma(t)} \tag{99}$$

is a normal density with the standard deviation

$$\sigma(t) := \left\{ \sigma^2 + \left(\frac{\lambda t}{4\pi\sigma} \right)^2 \right\}^{1/2}. \tag{100}$$

When the laser beam is coherent and the intensity is gaussian at the source plane $t = 0$, the diffracted beam will have a gaussian intensity for each $t > 0$. As the beam advances, the diffraction process causes it to spread in accordance with (100). For example, a He-Ne gaussian laser beam with $\lambda = .6328 \cdot 10^{-3}$ mm and $\sigma = 1$ mm will form a spot with $\sigma(t) \approx 19$ km after traveling the $t = 384 \cdot 10^3$ km from the earth to the moon! ∎

We will show that it is possible to *translate* a diffraction pattern by suitably adjusting the phase of the source function. Indeed, let u_0, u be solutions of the diffraction equation (3) with the source functions

$$u_0(x,0) = f_0(x), \quad u(x,0) = e^{2\pi i\beta x} f_0(x) \tag{101}$$

where β is a real parameter. We use (91) and the modulation rule to write

$$\begin{aligned} U(s,t) &= F_0(s - \beta)e^{-i\pi s^2 \lambda t} \\ &= F_0(s - \beta)e^{-i\pi[(s-\beta)^2 + 2\beta s - \beta^2]\lambda t} \\ &= e^{i\pi\beta^2 \lambda t} e^{-2\pi i(\beta\lambda t)s} U_0(s - \beta, t). \end{aligned}$$

We easily verify that this is the Fourier transform of

$$u(x,t) = e^{i\pi\beta^2 \lambda t}\, e^{2\pi i\beta(x - \beta\lambda t)}\, u_0(x - \beta\lambda t, t) \tag{102}$$

by using the translation and modulation rules. The phase factor that we apply in (101) translates the intensity

$$|u(x,t)|^2 = |u_0(x - \beta\lambda t, t)|^2$$

of the t-slice by $\beta\lambda t$.

Example: Solve $u_t = (i\lambda/4\pi)u_{xx}$ with the initial source

$$u(x,0) = e^{2\pi i\beta x}\frac{e^{-x^2/4\sigma^2}}{(2\pi\sigma^2)^{1/4}}, \quad \sigma > 0, \quad -\infty < \beta < \infty. \tag{103}$$

Solution: We use (102) with the expressions (98)–(100) for u_0 to write

$$u(x,t) = e^{i\pi\beta^2\lambda t}e^{2\pi i\beta(x-\beta\lambda t)}\frac{(8\pi\sigma^2)^{1/4}}{\sqrt{4\pi\sigma^2 + i\lambda t}}e^{-\pi(x-\beta\lambda t)^2/(4\pi\sigma^2 + i\lambda t)} \tag{104}$$

and

$$|u(x,t)|^2 = \frac{e^{-(x-\beta\lambda t)^2/\sigma^2(t)}}{\sqrt{2\pi}\,\sigma(t)}. \tag{105}$$

When we apply the phase function $e^{2\pi i\beta t}$ to the coherent (real) source (97), the diffracted gaussian beam swings through the angle arctan $(\beta\lambda)$. We can *aim* the beam by properly choosing the parameter β! [You will find the formulas (104), (105) in most quantum mechanics texts, where they are used for the analysis of a traveling gaussian wave packet.] ∎

Example: Solve $u_t = (i\lambda/4\pi)u_{xx}$ with the initial source

$$u(x,0) = e^{2\pi i\beta(x+a)}\frac{e^{-(x+a)^2/4\sigma^2}}{(2\pi\sigma^2)^{1/4}} + e^{-2\pi i\beta(x-a)}\frac{e^{-(x-a)^2/4\sigma^2}}{(2\pi\sigma^2)^{1/4}},$$

$$\sigma > 0, \quad \beta > 0, \quad a > 0.$$

Solution: We obtain the left term of the source by shifting (103) a units to the left, and we obtain the right term by shifting (103) a units to the right and changing the sign of β. Since the diffraction equation is linear and translation invariant, we can use (104) to write

$$u(x,t) = e^{i\pi\beta^2\lambda t}\frac{(8\pi\sigma^2)^{1/4}}{\sqrt{4\pi\sigma^2 + i\lambda t}}\left\{e^{2\pi i\beta(x+a-\beta\lambda t)}e^{-\pi(x+a-\beta\lambda t)^2/(4\pi\sigma^2+i\lambda t)}\right.$$
$$\left. + e^{-2\pi i\beta(x-a+\beta\lambda t)}e^{-\pi(x-a+\beta\lambda t)^2/(4\pi\sigma^2+i\lambda t)}\right\}. \tag{106}$$

The left term gives a spreading gaussian beam with an axis along the ray from $x = -a$, $t = 0$ to $x = 0$, $t = a/\beta\lambda$. The right term gives a spreading gaussian beam with an axis along the ray from $x = a$, $t = 0$ to $x = 0$, $t = a/\beta\lambda$. In the region where the beams overlap, there are interference fringes, *e.g.*, at the crossover point

$$|u(x, a/\beta\lambda)|^2 = \frac{e^{-x^2/2\sigma^2(t)}}{\sqrt{2\pi}\,\sigma(t)}\,|2\cos(2\pi\beta x)|^2,$$

cf. Fig. 9.15. The cover illustration of this book shows $|u(x,t)|^2$ as a color density with x increasing from left to right and with t increasing from top to bottom. ∎

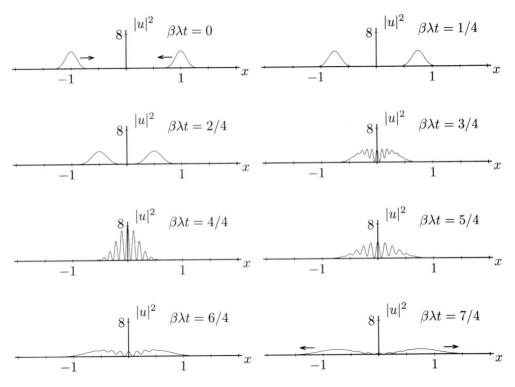

Figure 9.15. Selected t-slices for the intensity $|u(x,t)|^2$ of the colliding gaussian beams (106) when $a = 1$, $\sigma = .1$, and $\beta = 4.5$.

When $u(x,t)$ is a suitably regular ordinary solution of (3), we can scale the integral

$$\mathcal{E}_t\{u\} := \int_{x=-\infty}^{\infty} |u(x,t)|^2 dx \tag{107}$$

to obtain the energy the diffracted beam deposits on a screen at distance t from the source (during a unit interval of time). Indeed, if the source function f is piecewise smooth and square integrable, we can use Plancherel's identity (1.15) with (91) to write

$$\mathcal{E}_t\{u\} = \int_{-\infty}^{\infty} |F(s)e^{-i\pi s^2 \lambda t}|^2 ds = \int_{-\infty}^{\infty} |F(s)|^2 ds = \mathcal{E}_0\{u\}, \qquad t > 0.$$

In this way we see that the energy integral (107) is conserved.

In practice, it is almost never possible to evaluate the integral (94) by using the techniques from calculus and elementary Fourier analysis. We must use numerical methods (as we did for the graphs in Fig. 9.14) or we must find some other simplification. Fraunhofer, a contemporary of Fourier and Fresnel, found an exceptionally useful way to approximate $u(x, t)$ and $|u(x, t)|^2$ when the diffraction screen is far from the source. Let f be a piecewise smooth function and assume that f is essentially a-support limited in the sense that

$$f(x) \approx 0 \quad \text{when} \quad |x| > a$$

for some $a > 0$. Let $t \gg a^2/\lambda$ so that

$$\frac{\xi^2}{\lambda t} \approx 0 \quad \text{when} \quad -a \leq \xi \leq a.$$

We can then approximate the integral (94) by writing

$$\begin{aligned}
u(x, t) &\approx \frac{e^{-i\pi/4}}{\sqrt{\lambda t}} \int_{-a}^{a} f(\xi) e^{i\pi(x-\xi)^2/\lambda t} d\xi \\
&= \frac{e^{-i\pi/4} e^{i\pi x^2/\lambda t}}{\sqrt{\lambda t}} \int_{-a}^{a} f(\xi) e^{-2\pi i \xi x/\lambda t} e^{i\pi \xi^2/\lambda t} d\xi \\
&\approx \frac{e^{-i\pi/4} e^{i\pi x^2/\lambda t}}{\sqrt{\lambda t}} \int_{-a}^{a} f(\xi) e^{-2\pi i \xi x/\lambda t} d\xi \\
&\approx \frac{e^{-i\pi/4} e^{i\pi x^2/\lambda t}}{\sqrt{\lambda t}} F\left(\frac{x}{\lambda t}\right)
\end{aligned} \qquad (108)$$

where F is the Fourier transform of f. In this way we see that the initial intensity

$$|u(x, 0)|^2 = |f(x)|^2 \quad \text{evolves into} \quad |u(x, t)|^2 \approx \frac{1}{\lambda t}\left|F\left(\frac{x}{\lambda t}\right)\right|^2 \qquad (109)$$

when t is large. You will observe this phenomenon as you examine the t-slices from Fig. 9.14 [where $\Pi^2(x)$ evolves into $|\lambda t|^{-1} \operatorname{sinc}^2(x/\lambda t)$]. With a bit of analysis, you can verify that the *Fraunhofer approximation* (109) can also be used for the gaussian beams (98) and (104) when $t \gg \sigma^2/\lambda$.

We can use (23) to construct a localized solution of the wave equation that travels without changing its shape. The diffraction equation has no such solutions. When the source $u(x, 0)$ is localized, the intensity $|u(x, t)|^2$ spreads out as t increases, e.g., as seen in Fig. 9.14, (99)–(100), and (109). There is a simple explanation for this phenomenon. It is easy to verify that

$$e^{2\pi i s[x-(\lambda s/2)t]}$$

is a solution of the diffraction equation that travels with the *phase velocity*

$$c(s) := \frac{\lambda s}{2}, \qquad -\infty < s < \infty. \tag{110}$$

This nonlocalized wave moves left, right when $s < 0$, $s > 0$, and it moves fast, slow when $|s|$ is large, small. We combine such traveling sinusoids when we synthesize

$$u(x, t) = \int_{-\infty}^{\infty} F(s) e^{2\pi i s[x-c(s)t]} ds$$

from the Fourier transform (91). The integral has components that travel with different velocities (unless F is a translate of δ), so $|u(x, t)|^2$ disperses as t increases. (A caravan on an interstate highway will maintain its form when every vehicle has the same velocity, but it will disperse if the cars, trucks, buses, motorcycles, ... travel at different speeds!)

The diffraction equation on $\mathbb{T}_p$

We will specialize the above analysis to the case where the source function f is p-periodic with $p > 0$. The convolution product (93) then gives a generalized solution of (3) that is p-periodic in x.

Example: Solve the initial value problem

$$u_t = \left(\frac{i\lambda}{4\pi}\right) u_{xx}, \qquad u(x, 0) = \frac{1}{p} \text{III}\left(\frac{x}{p}\right). \tag{111}$$

Solution: We use (92)–(93) to see that $u(x, t) = \gamma^p(x, t)$ where

$$\gamma^p(x, t) := \gamma(x, t) * \frac{1}{p} \text{III}\left(\frac{x}{p}\right),$$

cf. (27) and (72). We can then use Poisson's relation (4.18) to see that γ^p is given by the weakly convergent series

$$\gamma^p(x, t) = \frac{e^{-i\pi/4}}{\sqrt{\lambda t}} \sum_{m=-\infty}^{\infty} e^{i\pi(x-mp)^2/\lambda t}$$

$$= \frac{1}{p} \sum_{k=-\infty}^{\infty} e^{-i\pi k^2 \lambda t/p^2} e^{2\pi i k x/p} \tag{112}$$

when $t > 0$. Every summand has modulus 1, so both of these series diverge at each point x. We cannot use such series to produce t-slices of γ^p that correspond to the time slices of r^p, k^p shown in Figs. 9.3 and 9.11! ∎

Example: Solve $u_t = (i\lambda/4\pi)u_{xx}$ with the initial source $u(x, 0) = \cos(2\pi x/p)$, $p > 0$.

Solution: Using (91) and (7.82) we write

$$U(s,t) = e^{-i\pi s^2 \lambda t} \frac{1}{2}\left\{\delta\left(s + \frac{1}{p}\right) + \delta\left(s - \frac{1}{p}\right)\right\}$$

$$= e^{-i\pi\lambda t/p^2}\frac{1}{2}\left\{\delta\left(s + \frac{1}{p}\right) + \delta\left(s - \frac{1}{p}\right)\right\},$$

and thereby find

$$u(x,t) = e^{-i\pi\lambda t/p^2}\cos\left(\frac{2\pi x}{p}\right).$$

∎

When we work with p-periodic functions we often use Fourier series. If

$$u(x,t) = \sum_{k=-\infty}^{\infty} U[k,t]e^{2\pi ikx/p}$$

is the solution of $u_t = (i\lambda/4\pi)u_{xx}$ with the initial source

$$f(x) = \sum_{k=-\infty}^{\infty} F[k]e^{2\pi ikx/p},$$

then the kth Fourier coefficient must satisfy

$$U_t[k,t] + \frac{i\pi\lambda k^2}{p^2}U[k,t] = 0, \quad U[k,0] = F[k].$$

We solve this initial value problem to find

$$U[k,t] = F[k]e^{-i\pi\lambda tk^2/p^2}, \quad k = 0, \pm 1, \pm 2, \ldots,$$

and synthesize

$$u(x,t) = \sum_{k=-\infty}^{\infty} F[k]e^{-i\pi\lambda tk^2/p^2}e^{2\pi ikx/p}. \tag{113}$$

We use (112) with the convolution product (7.131) (for functions on $\mathbb{T}_p$) to write

$$u(x,t) = \gamma^p(x,t) \circledast f(x). \tag{114}$$

The kernel $\gamma^p(x,t)$ is $2p^2/\lambda$-periodic in t (as well as p-periodic in x), and the same is true of $u(x,t)$.

Example: Solve $u_t = (i\lambda/4\pi)u_{xx}$ when the initial source is a periodic gaussian array

$$u(x,0) = \sum_{m=-\infty}^{\infty} \frac{e^{-(x-mp)^2/4\sigma^2}}{(2\pi\sigma^2)^{1/4}}, \quad \sigma > 0, \quad p > 0.$$

Solution: We use Poisson's relation (4.18) to find the Fourier series

$$u(x,0) = \frac{(8\pi\sigma^2)^{1/4}}{p} \sum_{k=-\infty}^{\infty} e^{-4\pi^2\sigma^2 k^2/p^2} e^{2\pi i kx/p},$$

and then use (113) to write

$$u(x,t) = \frac{(8\pi\sigma^2)^{1/4}}{p} \sum_{k=-\infty}^{\infty} e^{-4\pi^2\sigma^2 k^2/p^2} e^{-i\pi\lambda t k^2/p^2} e^{2\pi i kx/p}. \tag{115}$$

Selected t-slices for the intensity $|u(x,t)|^2$ are shown in Fig. 9.16.   ∎

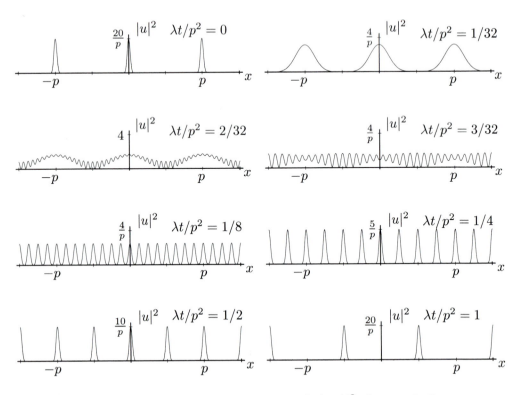

Figure 9.16. Selected t-slices for the intensity $|u(x,t)|^2$ of a p-periodic array of diffracting gaussian beams corresponding to (115) with $\sigma = .02p$.

The reduced wave function $u(x, t)$ for a laser beam must vanish when it meets an ideal silver mirror. (The electric field of the electromagnetic wave is zero at the surface of a perfect conductor.) We can use odd symmetry to eliminate one or two such boundary conditions.

Example: A coherent laser beam illumines a slit of width $2a$ centered at the point $x = x_0$ between ideal silver mirrors at $x = 0$, $x = L$. (Here $0 < x_0 - a < x_0 + a < L$.) Solve $u_t = (i\lambda/4\pi)u_{xx}$ with the boundary conditions $u(0, t) = 0$, $u(L, t) = 0$, $t \geq 0$ and the initial source

$$u(x, 0) = f(x) = \Pi\left(\frac{x - x_0}{2a}\right), \qquad 0 \leq x \leq L$$

to determine the intensity $|u(x, t)|^2$ of the diffraction pattern on a screen that is at distance t from the source.

Solution: We extend the initial source from $[0, L]$ to $\mathbb{R}$ by writing

$$f(x) = \sum_{m=-\infty}^{\infty}\left\{\Pi\left(\frac{x - x_0 - m\,2L}{2a}\right) - \Pi\left(\frac{x + x_0 - m\,2L}{2a}\right)\right\}.$$

By construction this $2L$-periodic function has odd symmetry with respect to $x = 0$, $x = L$, so f satisfies the given boundary conditions. We use Poisson's relation (4.18) to find the Fourier series

$$u(x, 0) = \frac{-2ia}{L}\sum_{k=-\infty}^{\infty}\sin\left(\frac{k\pi x_0}{L}\right)\mathrm{sinc}\left(\frac{ka}{L}\right)e^{2\pi ikx/2L},$$

and then use (113) to write

$$f(x) = \frac{4a}{L}\sum_{k=1}^{\infty}\mathrm{sinc}\left(\frac{ka}{L}\right)\sin\left(\frac{k\pi x_0}{L}\right)\sin\left(\frac{k\pi x}{L}\right)e^{-i\pi\lambda tk^2/4L^2}. \tag{116}$$

Some unusual piecewise constant t-slices of the corresponding intensity $|u(x, t)|^2$ are shown in Fig. 9.17. Exercise 9.46 will help you understand this phenomenon. ∎

There is a variation of the Fraunhofer approximation (109) that is useful for analyzing the diffraction pattern that results when we window a p-periodic source. Let g be a piecewise smooth p-periodic function with the Fourier series

$$g(x) = \sum_{k=-\infty}^{\infty}G[k]e^{2\pi ikx/p}. \tag{117}$$

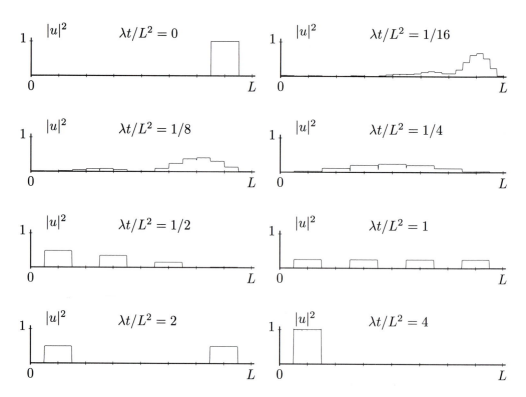

Figure 9.17. Selected t-slices for the intensity $|u(x,t)|^2$ obtained from 4000 terms of (116) when $x_0 = 7L/8$ and $a = L/16$.

The function g might correspond to some p-periodic pattern (*i.e.*, a univariate *crystal*) on a 35-mm slide. We will choose some small $\sigma > 0$ and illumine the slide with a wide coherent gaussian laser beam to produce the localized source function

$$u(x, 0) = f(x) := (8\pi\sigma^2)^{1/4} e^{-4\pi^2\sigma^2 x^2} \cdot g(x). \tag{118}$$

The function f has the Fourier transform

$$F(s) = \frac{e^{-s^2/4\sigma^2}}{(2\pi\sigma^2)^{1/4}} * \sum_{k=-\infty}^{\infty} G[k]\delta\left(s - \frac{k}{p}\right) = \sum_{k=-\infty}^{\infty} G[k]\frac{e^{-(s-k/p)^2/4\sigma^2}}{(2\pi\sigma^2)^{1/4}}. \tag{119}$$

We will assume that the laser beam illumines many periods of g, *i.e.*, that $\sigma p \ll 1$. This hypothesis makes it impossible for the gaussians from (119) to have appreciable overlap, allowing us to write

$$|F(s)|^2 \approx \sum_{k=-\infty}^{\infty} |G[k]|^2 \frac{e^{-(s-k/p)^2/2\sigma^2}}{\sqrt{2\pi}\,\sigma}. \tag{120}$$

We will also assume that $\sigma^2 \lambda t \gg 1$. This hypothesis allows us to use the Fraunhofer approximation (109) with (120) to see that

$$|u(x,t)|^2 \approx \frac{1}{\lambda t} \sum_{k=-\infty}^{\infty} |G[k]|^2 \frac{e^{-(x/\lambda t - k/p)^2/2\sigma^2}}{\sqrt{2\pi}\,\sigma}.$$

For simplicity, we will replace each normal density with a Dirac δ, *i.e.*,

$$\frac{1}{\lambda t} \frac{e^{-(x/\lambda t - k/p)^2/2\sigma^2}}{\sqrt{2\pi}\,\sigma} \approx \frac{1}{\lambda t}\delta\left(\frac{x}{\lambda t} - \frac{k}{p}\right) = \delta\left(x - \frac{k\lambda t}{p}\right),$$

[*cf.* Ex. 7.60(a)] and write

$$|u(x,t)|^2 \approx \sum_{k=-\infty}^{\infty} |G[k]|^2\delta\left(x - \frac{k\lambda t}{p}\right). \tag{121}$$

In this way we see that the far-field diffraction pattern for (118) is a train of spots. We can determine p from the spot-to-spot separation $\lambda t/p$, and we can determine the modulus (but not the phase) of the Fourier coefficient $G[k]$ from the intensity of the corresponding spot, $k = 0, \pm 1, \pm 2, \ldots$. It is surprising how much we can infer about the *crystal g* from the *diffraction pattern* (121)!

9.5 Fast Computation of Frames for Movies

We have included a few graphs of t-slices of $u(x,t)$ to help you follow the evolution of selected solutions of the wave equation, the diffusion equation, and the diffraction equation (*cf.* Figs. 9.2, 9.8, and 9.14), but these static illustrations are not nearly so memorable as computer-generated movies that show hundreds of such t-slices in quick succession. You can produce such a movie by using a corresponding formula for $u(x,t)$ (as given in the preceding sections of this chapter) together with the animation feature of a good mathematical software package. You will quickly discover that it takes a prodigious amount of computer time to generate a movie when we simply evaluate $u(x,t)$ at each point (x,t) of a suitably chosen grid. We will show you a much faster way to do the numerical analysis when the function u is p-periodic. (You can simulate the early evolution of any localized u on $\mathbb{R}$ by showing the movie for some corresponding periodization, *cf.* Fig. 1.22.)

Let $u(x,t)$ be a p-periodic generalized function of x for each $t \geq 0$. The coefficients of the corresponding Fourier series

$$u(x,t) = \sum_{k=-\infty}^{\infty} C_k(t)e^{2\pi ikx/p} \tag{122}$$

are then functions of t. In particular, the solutions (34), (75), and (113) of the wave, diffusion, and diffraction equations have this form with

$$C_k(t) = F[k]\cos(2\pi kt/T) + t\,G[k]\mathrm{sinc}(2kt/T), \quad T := \frac{p}{c}, \tag{123}$$

$$C_k(t) = F[k]e^{-k^2 t/T}, \quad T := \frac{p^2}{4\pi^2 a^2}, \tag{124}$$

$$C_k(t) = F[k]e^{-2\pi i k^2 t/T}, \quad T := \frac{2p^2}{\lambda}, \tag{125}$$

respectively. [The parameter T serves as a period in (123) and (125).]

We will choose a suitably large discretization index $K = 1, 2, \ldots$, set $N := 2K$, define

$$x_n := \frac{np}{N}, \quad n = 0, 1, \ldots, N - 1,$$

and approximate $u(x_n, t)$ with the truncated series

$$\begin{aligned}
u_n(t) :&= \sum_{k=-K+1}^{K} C_k(t)e^{2\pi i k x_n/p} \\
&= \sum_{k=-(N/2)+1}^{N/2} C_k(t)e^{2\pi i k n/N}, \quad n = 0, 1, \ldots, N - 1.
\end{aligned} \tag{126}$$

You will immediately see that the coefficient vector

$$[C_0(t), C_1(t), \ldots, C_{N/2}(t), C_{-N/2+1}(t), \ldots, C_{-1}(t)]^T \tag{127}$$

is the DFT of the vector

$$[u_0(t), u_1(t), u_2(t), \ldots, u_{N-1}(t)]^T. \tag{128}$$

You can generate N approximate samples for a t-slice of u by evaluating the components of the vector (127) [e.g., by using the formulas (123), (124), or (125)] and then using the FFT to find the N components of the inverse DFT (128), cf. Ex. 5.11. When you are working with a resolution of $N = 512$ samples per t-slice this procedure will be about 50 times faster than direct evaluation. You can use your PC to produce movies that simulate a vibrating string, heat flow in a rod, a diffracting laser beam, and various other natural phenomena, cf. Exs. 9.18, 9.19, 9.20, and 9.47.

References

M. Born and E. Wolf, *Principles of Optics, 6th ed.*, Pergamon Press, Oxford, 1993.
The physics of diffraction is in Chapter 8 of this treatise.

J.W. Brown and R.V. Churchill, *Fourier Series and Boundary Value Problems, 5th ed.*, McGraw-Hill, New York, 1993.
Traditional treatments of the diffusion and wave equations can be found in this highly evolved elementary text.

H.S. Carslaw and J.C. Jaeger, *Conduction of Heat in Solids*, Clarendon Press, Oxford, 1959.
The definitive classical treatise on the diffusion equation.

R. Haberman, *Elementary Applied Partial Differential Equations, 3rd ed.*, Prentice Hall, Upper Saddle River, NJ, 1998.
A well-written introduction to PDEs for scientists and engineers.

I.G. Main, *Vibrations and Waves in Physics, 3rd ed.*, Cambridge University Press, New York, 1993.
An exceptionally well motivated elementary introduction to wave motion.

M.A. Morrison, *Understanding Quantum Physics*, Prentice Hall, Englewood Cliffs, NJ, 1990.
The mathematics of the diffraction equation is interpreted within the context of quantum mechanics in Chapter 4 of this undergraduate physics text.

B.E.A. Saleh and M.C. Teich, *Fundamentals of Photonics*, John Wiley & Sons, New York, 1991.
Gaussian beams are analyzed in Chapter 3 of this graduate-level engineering text. A nice exposition of the use of Fourier analysis in optics is in Chapter 4.

R.A.R. Tricker, *Bores, Breakers, Waves and Wakes*, American Elsevier, New York, 1965.
An exceptionally well written simplified introduction to water waves (with striking illustrations!).

J.S. Walker, *Fast Fourier Transforms, 2nd ed.*, CRC Press, Boca Raton, FL, 1996.
Chapter 4 shows how the FFT is used to evaluate the filtered Fourier series solutions of the wave equation, the diffusion equation, and the diffraction equation.

Exercise Set 9

▶ **EXERCISE 9.1:**   Let $u(x,t)$ be a twice continuously differentiable ordinary function that satisfies the wave equation $u_{tt} = c^2 \cdot u_{xx}$. In this exercise you will use a classical argument of d'Alembert to show that u is a sum of traveling waves, $i.e.$,

$$u(x,t) = w_+(x+ct) + w_-(x-ct)$$

for suitably chosen functions w_+, w_-. (A more general modern argument is required for Ex. 9.5.)

(a) Let $\xi := x + ct$, $\eta := x - ct$. Verify that

$$\frac{\partial}{\partial \eta}\frac{\partial}{\partial \xi}\, u\left(\frac{\xi+\eta}{2}, \frac{\xi-\eta}{2c}\right) = 0.$$

(b) Use (a) to show in turn that

$$\frac{\partial}{\partial \xi}\, u\left(\frac{\xi+\eta}{2}, \frac{\xi-\eta}{2c}\right) = w'_+(\xi), \qquad u\left(\frac{\xi+\eta}{2}, \frac{\xi-\eta}{2c}\right) = w_+(\xi) + w_-(\eta).$$

Hint. When $(\partial/\partial\eta)w(\xi, \eta) = 0$, you can conclude that $w(\xi, \eta)$ is a function of ξ (that is not necessarily a constant).

▶ **EXERCISE 9.2:**   Verify the expression (22) for the convolution product $(1/2c)\Pi(x/2ct) * \Pi(x)$.

▶ **EXERCISE 9.3:**   Let $u(x,t)$ satisfy the wave equation $u_{tt} = c^2 u_{xx}$ for $-\infty < x < \infty$, $t \geq 0$. Sketch slices of u at suitably chosen times $0 = t_0 < t_1 < t_2 < \cdots$ to show the motion when:

(a) $u(x,0) = \Lambda(x)$, $u_t(x,0) = 0$;   (b) $u(x,0) = 0$, $u_t(x,0) = \Lambda(x)$;

(c) $u(x,0) = \text{sgn}(x)$, $u_t(x,0) = 0$;   (d) $u(x,0) = 0$, $u_t(x,0) = \text{sgn}(x)$.

▶ **EXERCISE 9.4:**   The bivariate *wave polynomials* $u^{[n]}(x,t)$, $v^{[n]}(x,t)$ are defined to be the solutions of

$$u_{tt}^{[n]} = u_{xx}^{[n]}, \quad u^{[n]}(x,0) = x^n, \quad u_t^{[n]}(x,0) = 0, \quad n = 0,1,2,\ldots$$

$$v_{tt}^{[n]} = v_{xx}^{[n]}, \quad v^{[n]}(x,0) = 0, \quad v_t^{[n]}(x,0) = x^n, \quad n = 0,1,2,\ldots\ .$$

(a) Derive a simple formula for $u^{[n]}(x,t)$ and thereby obtain

$$1, \quad x, \quad x^2 + t^2, \quad x^3 + 3xt^2, \quad x^4 + 6x^2t^2 + t^4$$

when $n = 0,1,2,3,4$.

Hint. You can specialize the d'Alembert formula (15) or you can simplify the synthesis equation that corresponds to (14).

(b) Derive a simple formula for $v^{[n]}(x, t)$ and thereby obtain

$$t, \quad xt, \quad x^2 t + t^3/3, \quad x^3 t + xt^3, \quad x^4 t + 2x^2 t^3 + t^5/5$$

when $n = 0, 1, 2, 3, 4$.

(c) How must we modify $u^{[n]}$, $v^{[n]}$ if we want to work with the wave equation $u_{tt} = c^2 u_{xx}$ instead of $u_{tt} = u_{xx}$?

Note. You can use linear combinations of the dilated functions $u^{[n]}$, $v^{[n]}$, $n = 0, 1, 2, \ldots$ to solve the wave equation $u_{tt} = c^2 u_{xx}$ when $u(x, 0)$ and $u_t(x, 0)$ are arbitrary polynomials.

▶ **EXERCISE 9.5:**   Let u be a generalized solution of the wave equation $u_{tt} = c^2 u_{xx}$ for $-\infty < x < \infty$, $t \geq 0$. Show that $u(x, t) = w_+(x + ct) + w_-(x - ct)$ for suitably chosen generalized functions w_+, w_-, i.e., show that u is a sum of traveling waves.

Hint. Begin by replacing g by $\delta' * g^{(-1)}$ in (15) where $g^{(-1)}$ is an antiderivative of g.

▶ **EXERCISE 9.6:**   Let $u(x, t)$ be the solution of the wave equation $u_{tt} = c^2 u_{xx}$ for $-\infty < x < \infty$, $t \geq 0$ with $u(x, 0) = f(x)$, $u_t(x, 0) = g(x)$. What can you infer about $u(x, t)$ at time $t > 0$ if you know that:

(a) $f(x) = f(-x)$, $g(x) = g(-x)$?

(b) $f(x) = -f(-x)$, $g(x) = -g(-x)$?

(c) $f(x_0 + x) = f(x_0 - x)$,
     $g(x_0 + x) = g(x_0 - x)$?

(d) $f(x_0 + x) = -f(x_0 - x)$,
     $g(x_0 + x) = -g(x_0 - x)$?

(e) $f(x + p) = f(x)$, $g(x + p) = g(x)$?

(f) $g(x) = \pm c f'(x)$?

▶ **EXERCISE 9.7:**   Let $u(x, t; f, g)$ denote the solution of the wave equation $u_{tt} = c^2 u_{xx}$, for $-\infty < x < \infty$, $-\infty < t < \infty$ that satisfies the initial conditions $u(x, 0) = f(x)$, $u_t(x, 0) = g(x)$.

(a) Show that $u(x, -t; f, g) = u(x, t; f, -g)$.

(b) What conditions, if any, must be imposed on f, g to ensure that the movie showing $u(x, t; f, g)$ from $t = 0$ to $t = +\infty$ will be identical to the time-reversed movie showing $u(x, t; f, g)$ from $t = 0$ to $t = -\infty$?

▶ **EXERCISE 9.8:**   Let r be the wave kernel (16).

(a) Show that r has the generalized partial derivative (17) by evaluating the generalized limit $\lim_{h \to 0} [r(x, t + h) - r(x, t)]/h$.

    Hint. Sketch the graph of $[r(x, t + h) - r(x, t)]/h$.

(b) Show that r has the generalized partial derivative (17) by finding the Fourier transform $R(s, t)$, computing the partial derivative $R_t(s, t)$, and finding the inverse Fourier transform of $R_t(s, t)$.

(c) Show that r satisfies the wave equation $r_{tt} = c^2 r_{xx}$.

▶ **EXERCISE 9.9:**   Let r be the wave kernel (17) and let $t_1 > 0$, $t_2 > 0$.

(a) Show that $r(x, t_1 + t_2) = r(x, t_1) * r_t(x, t_2) + r_t(x, t_1) * r(x, t_2)$.

   Hint. Use (17) and (18) to evaluate and sketch $r(x, t_1 + t_2)$, $r(x, t_1) * r_t(x, t_2)$, $r_t(x, t_1) * r(x, t_2)$, or work with the corresponding identity for the Fourier transforms.

(b) Show that $r_t(x, t_1 + t_2) = r_t(x, t_1) * r_t(x, t_2) + r_{tt}(x, t_1) * r(x, t_2)$.

   Hint. Suitably differentiate the identity from (a).

(c) Let f, g be generalized functions. We can use (18) with $t = t_1$ to produce

$$u(x, t_1) = r_t(x, t_1) * f(x) + r(x, t_1) * g(x),$$

differentiate to obtain

$$u_t(x, t_1) = r_{tt}(x, t_1) * f(x) + r_t(x, t_1) * g(x),$$

and then use (18) with $t = t_2$ to obtain

$$u(x, t_1 + t_2) = r_t(x, t_2) * u(x, t_1) + r(x, t_2) * u_t(x, t_1).$$

We can also use (18) with $t = t_1 + t_2$ to produce

$$u(x, t_1 + t_2) = r_t(x, t_1 + t_2) * f(x) + r(x, t_1 + t_2) * g(x).$$

Show that these two processes give the same function $u(x, t_1 + t_2)$.

▶ **EXERCISE 9.10:**   The localized stimulation of a taut string produces a disturbance that can be detected at some distant point of observation. Let u be a solution of the wave equation $u_{tt} = c^2 u_{xx}$ with an initial position $u(x, 0) = f(x)$ and an initial velocity $u_t(x, 0) = g(x)$ that vanish outside some finite interval $a \le x \le b$. Let x_0 be our point of observation, and let

$$t_{min} := \max\left\{0, \left|\frac{x_0}{c} - \frac{a+b}{2c}\right| - \frac{b-a}{2c}\right\}, \qquad t_{max} := \left|\frac{x_0}{c} - \frac{a+b}{2c}\right| + \frac{b-a}{2c}.$$

(a) Show that we cannot detect the disturbance before the time t_{min}, i.e., $u(x_0, t) = 0$ if $0 \le t < t_{min}$.

(b) Show that the string is quiescent after the time t_{max}, i.e., $u(x_0, t) = A$ if $t > t_{max}$, where A is a constant that depends on g.

▶ **EXERCISE 9.11:**   Suppose you are holding one end of a semi-infinite motionless taut string and that you wish to create a traveling triangle wave (having the shape of Λ) that moves off to infinity. How should you choose the forcing function d (*i.e.*, how should you move the free end of the string) so that the solution of

$$u_{tt}(x, t) = c^2 u_{xx}(x, t) \quad \text{for } 0 < x < \infty, \ t \ge 0$$
$$u(0, t) = d(t), \qquad\qquad t > 0$$
$$u(x, 0) = u_t(x, 0) = 0, \qquad 0 \le x < \infty$$

has the desired form?

Hint. Begin with a traveling triangle on a taut string that stretches from $-\infty$ to $+\infty$.

▶ **EXERCISE 9.12:**   A taut string with ends at $x = 0$, $x = L$ is governed by the wave equation $u_{tt} = c^2 u_{xx}$. The end at $x = 0$ is fixed so that $u(0, t) = 0$, $t \geq 0$. You hold the end at $x = L$ in your hand and move it so that

$$u(L, t) = d(t), \quad t \geq 0$$

where d is some control function. In this exercise you will learn how to choose d (in terms of the initial position and velocity) so that you can shake the string to rest in some finite time.

(a) Find the displacements $u(0, t)$, $u(L, t)$, $t \geq 0$ for a taut string that satisfies the wave equation $u_{tt} = c^2 u_{xx}$ for $-\infty < x < \infty$, $t \geq 0$ when

$$u(x, 0) = \Lambda\left(\frac{x - L/2}{L/2}\right) - \Lambda\left(\frac{x + L/2}{L/2}\right), \qquad u_t(x, 0) = 0.$$

Hint. Use the d'Alembert formula (15) and the odd symmetry of $u(x, 0)$.

(b) Let f be a continuous function that vanishes at $x = 0$, $x = L$, and let

$$u(x, 0) = f(x), \ u_t(x, 0) = 0 \quad \text{for} \quad 0 \leq x \leq L.$$

Find a continuous control function d that will make

$$u(x, 2L/c) = 0, \ u_t(x, 2L/c) = 0 \quad \text{for} \quad 0 \leq x \leq L.$$

Hint. Suitably generalize the result of (a).

Note. If $f(L) \neq 0$, it takes a bit more time to shake the string to rest with a continuous control function d.

(c) Let g be a bounded piecewise continuous function, and let

$$u(x, 0) = 0, \ u_t(x, 0) = g(x) \quad \text{for} \quad 0 \leq x \leq L.$$

Find a continuous control function d that will make

$$u(x, 2L/c) = 0, \ u_t(x, 2L/c) = 0 \quad \text{for} \quad 0 \leq x \leq L.$$

Hint. You may wish to use the result of Ex. 9.10.

▶ **EXERCISE 9.13:**   A vibrating string with fixed ends at $x = 0$, $x = L$ has the initial position and velocity $u(x, 0) = \sin^2(\pi x/L)$, $u_t(x, 0) = 0$, $0 \leq x \leq L$.

(a) Show that the displacement function (35) has the coefficients

$$a_k = \begin{cases} -\dfrac{8}{\pi k(k^2 - 4)} & \text{if } k = 1, 3, 5, \ldots \\ 0 & \text{if } k = 2, 4, 6, \ldots, \end{cases} \qquad b_k = 0, \quad k = 1, 2, \ldots .$$

(b) Let $T := 2L/c$ be the period of (35). Show that there are 6 times $0 < t_1 < t_2 < \cdots < t_6 < T$ when $u(x, t)$ has a substantial flat spot. Express these times in terms of T and sketch the corresponding time slices.

Hint. Odd extend $u(x, 0)$ from $[0, L]$ to $[-L, L]$, and then $2L$-periodically extend u to $\mathbb{R}$. You can then use the d'Alembert formula.

Note. This vibrational mode makes a good movie!

▶ **EXERCISE 9.14:**   A taut string has its ends fixed at $x = 0, L$. When the string vibrates, the resulting audio wave is a mixture of pure sinusoids having the frequencies $f_1, f_2, \ldots$ as given by (40). Let $k = 2, 3, \ldots$ be chosen, let $\kappa = 1, 2, \ldots, k - 1$, and let $x_0 := \kappa L/k$ [so that x_0 is a node of the kth vibrational mode (39)]. Show that if we cause the string to vibrate by striking, plucking, or bowing it at the point x_0, then the frequencies $f_k, f_{2k}, f_{3k}, \ldots$ will be missing from the corresponding audio wave.

Note. This observation is due to Helmholtz.

▶ **EXERCISE 9.15:**   Let u be an ordinary solution of the wave equation $u_{tt} = c^2 u_{xx}$ for $-\infty < x < \infty$, $t \geq 0$.

(a) Assume that $u(x, 0)$ and $u_t(x, 0)$ are bounded with

$$|u(x, 0)| \leq M_0, \quad |u_t(x, 0)| \leq M_1 \quad \text{for} \quad -\infty < x < \infty.$$

Show that

$$|u(x, t)| \leq M_0 + M_1 t \quad \text{for} \quad -\infty < x < \infty, \quad t \geq 0.$$

(b) Let $u(x, 0)$ be bounded as in (a) and assume that

$$\left| \int_a^b u_t(x, 0) dx \right| \leq I_1 \quad \text{whenever} \quad -\infty \leq a < b \leq \infty.$$

Show that

$$|u(x, t)| \leq M_0 + \frac{I_1}{2c} \quad \text{for} \quad -\infty < x < \infty, \quad t > 0.$$

(c) Let $u(x, 0)$, $u_t(x, 0)$ be bounded as in (a), let $u(x, 0)$ and $u_t(x, 0)$ be p-periodic, and assume that

$$\int_0^p u_t(x, 0) dx = 0.$$

Show that

$$|u(x, t)| \leq M_0 + \frac{M_1 p}{2c} b\left(\frac{2ct}{p}\right) \quad \text{for} \quad -\infty < x < \infty, \quad t > 0$$

where

$$b(\tau) := \sum_{m=-\infty}^{\infty} \Lambda(\tau - 1 - 2m).$$

▶ **EXERCISE 9.16:** Let $u(x,t)$ satisfy the wave equation $u_{tt} = c^2 u_{xx}$ for $0 < x < 1$, $t \geq 0$, with the boundary conditions $u(0,t) = u(1,t) = 0$ for $t \geq 0$, and the initial conditions $u(x,0) = f(x)$, $u_t(x,0) = 0$. Sketch the *orbit*

$$u(x,t) + iu_t(x,t), \quad 0 \leq x \leq 1$$

(in the complex plane) at suitably chosen times $0 = t_0 < t_1 < t_2 < \cdots < t_m = T$, $T := 1/c$ to show the motion when:

(a) $f(x) = \sin(\pi x)$;   (b) $f(x) = \sin(2\pi x)$;   (c) $f(x) = \Lambda(2x - 1)$.

Hint. You may wish to prepare slices of $u(x,t)$, $u_t(x,t)$ analogous to those from Fig. 9.6 and use them to produce the orbit for (c).

Note. You can use your computer to generate movies that show the time evolution of such orbits. The one that corresponds to $f(x) = \sin^2(\pi x)$ is most unusual!

▶ **EXERCISE 9.17:** This exercise will help you formulate a *vectorized* procedure for generating time slices for a movie of a vibrating string. We will use the binary operators $+$, $-$, $\circ$ for the *componentwise* sum, difference, product of two vectors, we will apply cos, sinc *componentwise*, and we will use $\mathcal{F}$ for the DFT operator, *e.g.*,

$$\begin{bmatrix} 1 \\ 2 \end{bmatrix} + \begin{bmatrix} 3 \\ 4 \end{bmatrix} = \begin{bmatrix} 4 \\ 6 \end{bmatrix}, \qquad \begin{bmatrix} 1 \\ 2 \end{bmatrix} \circ \begin{bmatrix} 3 \\ 4 \end{bmatrix} = \begin{bmatrix} 3 \\ 8 \end{bmatrix},$$

$$\cos \begin{bmatrix} 1 \\ 2 \end{bmatrix} = \begin{bmatrix} \cos(1) \\ \cos(2) \end{bmatrix}.$$

(a) Let $\mathbf{C}(t)$ be the vector (127) with the components (123), and let $\mathbf{u}(t)$ be the vector (128) with the components (126). Convince yourself that we can write

$$\mathcal{F}\mathbf{u}(t) = \mathbf{a} \circ \cos(2\pi t\mathbf{f}) + t\,\mathbf{b} \circ \mathrm{sinc}(2t\mathbf{f})$$

where $\mathbf{f} = T^{-1}[0, 1, 2, \ldots, K-1, K, K-1, \ldots, 1]^T$, and $\mathbf{a}$, $\mathbf{b}$ are certain N component vectors.

Note. You can use the subsequent analysis to model the discrete string of Ex. 9.18 and the stiff string of Ex. 9.19 if you suitably modify the components of $\mathbf{f}$.

(b) Show that $\mathbf{a} = \mathcal{F}\mathbf{u}(0)$, $\mathbf{b} = \mathcal{F}\dot{\mathbf{u}}(0)$. If you are given K samples of the position and velocity of the string from $[0, L]$, you can odd extend them to produce the N component vectors $\mathbf{u}(0)$, $\dot{\mathbf{u}}(0)$ and then use the FFT to obtain the coefficient vectors $\mathbf{a}$, $\mathbf{b}$.

(c) Given $t \geq 0$ we compute in turn

$$\mathbf{a} \circ \cos(2\pi t\mathbf{f}) + t\,\mathbf{b} \circ \mathrm{sinc}(2t\mathbf{f}) \quad \text{and} \quad \mathbf{u}(t) = \mathcal{F}^{-1}\{\mathbf{a} \circ \cos(2\pi t\mathbf{f}) + t\,\mathbf{b} \circ \mathrm{sinc}(2t\mathbf{f})\}.$$

Estimate the number of operations that we expend to produce the time slice $\mathbf{u}(t)$ (assuming that we use the FFT to apply $\mathcal{F}^{-1}$).

Note. You can simulate the motion of the string by computing time slices at $t = mT/M$, $m = 0, 1, \ldots, M-1$ and showing them periodically!

(d) You can generate time slices for the orbit $\mathbf{u}(t) + i\,\dot{\mathbf{u}}(t)$, cf. Ex. 9.16. Show that

$$\mathbf{u}(t) + i\,\dot{\mathbf{u}}(t) = \mathcal{F}^{-1}\{(\mathbf{a} + i\,\mathbf{b})\cos(2\pi t\mathbf{f}) + t(\mathbf{b} - i\,4\pi^2\mathbf{f}\circ\mathbf{f}\circ\mathbf{a})\mathrm{sinc}(2t\mathbf{f})\}.$$

Hint. You can use the expression from (a) to find $\mathcal{F}\,\dot{\mathbf{u}}(t)$.

▶ **EXERCISE 9.18:**   A massless thread of length L with endpoints fixed at $x = 0$, $x = L$ is stretched with tension τ. Tiny beads, each with mass m, are attached at the points $x = d, 2d, \dots, (M-1)d$ where $d := L/M$, cf. Fig. 9.18. In this exercise you will analyze the plane vibration of the system using $u[n, t]$ for the displacement of the nth bead at time $t \geq 0$, $n = 1, 2, \dots, M - 1$.

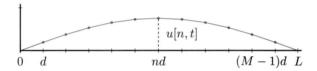

$0 \quad d \qquad\qquad nd \qquad\qquad (M-1)d \quad L$

Figure 9.18. The thread and bead model for a vibrating string that is used for Ex. 9.18.

(a) Suitably modify (10) and thereby show that the time evolution of the system is governed by the coupled differential equations

$$u_{tt}[n, t] = \left(\frac{\tau}{md}\right)\{u[n-1, t] - 2u[n, t] + u[n+1, t]\}, \quad n = 1, 2, \dots, M-1,$$

provided that we introduce the boundary conditions $u[0, t] = 0$, $u[M, t] = 0$ for the ends of the thread at $x = 0$, $x = Md$. You must now find the solution that has the initial position and velocity $u[n, 0] = f[n]$, $u_t[n, 0] = g[n]$, $n = 0, 1, \dots, M$ when the discrete functions f, g are given.

(b) Odd extend u, f, g to obtain functions on $\mathbb{P}_{2M}$ and then show that the DFT of u satisfies the uncoupled differential equations

$$U_{tt}[k, t] = \left(\frac{\tau}{md}\right)\{e^{-2\pi ik/2M} - 2 + e^{2\pi ik/2M}\}U[k, t] = -\omega[k]^2 U[k, t],$$

$$U[k, 0] = F[k], \quad U_t[k, 0] = G[k]$$

where

$$\omega[k] := 2\left(\frac{\tau}{md}\right)^{1/2}\left|\sin\left(\frac{k\pi}{2M}\right)\right|, \quad k = 0, 1, \dots, 2M - 1.$$

(c) Use (b) to find $U[k, t]$, $k = 0, 1, 2, \ldots, 2M - 1$, and thereby show that

$$u[n, t] = \sum_{k=1}^{M-1} \left\{ 2i\, F[k] \cos(\omega[k]t) + 2i\, G[k] \frac{\sin(\omega[k]t)}{\omega[k]} \right\} \sin\left(\frac{\pi k n}{M}\right)$$

where

$$2i\, F[k] = \frac{2}{M} \sum_{n=0}^{M-1} f[n] \sin\left(\frac{\pi k n}{M}\right), \quad 2i\, G[k] = \frac{2}{M} \sum_{n=0}^{M-1} g[n] \sin\left(\frac{\pi k n}{M}\right).$$

You may wish to compare these identities to (35) and (38).

(d) Find all separable solutions $u[n, t] = v[n] \cdot w(t)$ of the coupled differential equations from (a).

(e) Verify that

$$w_k[n, t] := \exp\left\{ \frac{2\pi i k n}{2M} - i\omega[k]t \right\}$$

is a solution of the coupled differential equations [from (a)] that travels with the velocity

$$c_k := \frac{Md}{\pi k} \omega[k], \quad k = 1, 2, \ldots, M - 1.$$

Since $c_1 > c_2 > \cdots$ we cannot synthesize a traveling wave from two or more such solutions, *cf.* (23). There is no d'Alembert formula within this discrete setting!

▶ **EXERCISE 9.19:** The motion of a vibrating string that is *not* perfectly flexible can be modeled by using the partial differential equation

$$u_{tt}(x, t) = a\, u_{xx}(x, t) - b\, u_{xxxx}(x, t)$$

where a, b are positive constants.

(a) Show that this PDE has the particular solutions $e^{2\pi i s[x - c(s)t]}$, $e^{2\pi i s[x + c(s)t]}$ where $c(s) := \sqrt{a + 4\pi^2 b s^2}$ is the corresponding phase velocity, *cf.* Ex. 9.40.

(b) Show that the solution of the PDE with the initial position and velocity $u(x, 0) = f(x)$, $u_t(x, 0) = g(x)$ is given by

$$u(x, t) = \int_{-\infty}^{\infty} \{F(s) \cos[2\pi\, s\, c(s)t] + G(s)\, t \operatorname{sinc}[2s\, c(s)t]\} e^{2\pi i s x}\, ds.$$

(c) Show that an odd p-periodic solution of the PDE can be represented by a Fourier series

$$u(x, t) = \sum_{k=1}^{\infty} \left\{ a_k \cos\left[2\pi \frac{k}{p} c\left(\frac{k}{p}\right) t\right] + b_k \sin\left[2\pi \frac{k}{p} c\left(\frac{k}{p}\right) t\right] \right\} \sin\left(\frac{2\pi k x}{p}\right).$$

(d) A perfectly flexible string with ends fixed at $x = 0$, $x = L$ has the frequencies (40). What are the corresponding frequencies for the stiff string?

▶ **EXERCISE 9.20:**   A stiff string (governed by the PDE of Ex. 9.19) has fixed ends at $x = 0$ and $x = L$. We can make the string vibrate by bowing it, by plucking it, or by striking it at some point x_0, $0 < x_0 < L$. Find the solutions $u^b(x, t)$, $u^p(x, t)$, and $u^s(x, t)$ that correspond to the initial conditions (41), (42), and (43).

Hint. Make suitable use of (44)–(46) and the analysis from Ex. 9.19.

▶ **EXERCISE 9.21:**   A number of elementary PDEs have traveling wave solutions

$$e^{2\pi i[sx - \nu(s)t]}$$

that can be used to synthesize more general solutions, *cf.* (5) and (7). When the PDE has the form $\mathcal{P}(\mathbf{D}_t, \mathbf{D}_x)u = 0$ where $\mathcal{P}$ is a bivariate polynomial in the operators

$$\mathbf{D}_t = \frac{\partial}{\partial t}, \quad \mathbf{D}_x = \frac{\partial}{\partial x},$$

we can use the characteristic equation $\mathcal{P}(-2\pi i\nu, 2\pi i s) = 0$ to find the function $\nu(s)$. Find $\nu(s)$ for each of the following, assuming $a > 0$ and $b \geq 0$.

(a) $u_t = a\, u_x$ (transport equation)

(b) $u_{tt} = a\, u_{xx} - b\, u$ (Klein-Gordon equation)

(c) $u_{tt} = -a\, u_{xxxx}$ (vibrating bar equation, *cf.* Ex. 9.19)

▶ **EXERCISE 9.22:**   Let v be the solution of the diffusion equation $v_t = a^2 v_{xx}$ for $-\infty < x < \infty$, $t \geq 0$ with $v(x, 0) = \mathrm{sgn}(x)$.

(a) Set $f(x) = \mathrm{sgn}\, x$ in (55) and thereby show that

$$v(x, t) = \mathrm{erf}\left(\frac{x}{\sqrt{4a^2 t}}\right) \quad \text{where} \quad \mathrm{erf}(x) := \frac{2}{\sqrt{\pi}} \int_0^x e^{-\xi^2}\, d\xi.$$

Most mathematical software packages include code for the *error function* erf and a calculator approximation is given in Appendix 6.

(b) Let $x_1 < x_2 < \cdots < x_n$ and let $c_0, c_1, \ldots, c_n$ be real. Find the solution of the diffusion equation $u_t(x, t) = a^2 u_{xx}(x, t)$ for $-\infty < x < \infty$, $t \geq 0$ with

$$u(x, 0) = \begin{cases} c_0 & \text{if} & x < x_1 \\ c_1 & \text{if} & x_1 < x < x_2 \\ \vdots & & \\ c_{n-1} & \text{if} & x_{n-1} < x < x_n \\ c_n & \text{if} & x_n < x. \end{cases}$$

Hint. Observe that

$$u(x, 0) = \frac{1}{2}(c_0 + c_n) + \frac{1}{2}\sum_{m=1}^{n}(c_m - c_{m-1})\mathrm{sgn}(x - x_m)$$

and make suitable use of the function v from (a).

Note. The diffusion kernel (53) is given by $k = (1/2)v_x$.

▶ **EXERCISE 9.23:** Let u, v be obtained by solving

$$v_t = v_{xx} \quad \text{for } -\infty < x < \infty, \ t \geq 0 \text{ with } v(x, 0) = f(x)$$
$$u_t = a^2 u_{xx} \text{ for } -\infty < x < \infty, \ t \geq 0 \text{ with } u(x, 0) = f\left(\frac{x - x_0}{h}\right).$$

Here f is a generalized function, $-\infty < x_0 < \infty$ and $h > 0$. Express u in terms of v.

▶ **EXERCISE 9.24:** Let u be the solution of the diffusion equation $u_t = a^2 u_{xx}$ for $-\infty < x < \infty$, $t \geq 0$ with

$$u(x, 0) = \sum_{n=-\infty}^{\infty} c_n \Lambda\left(\frac{x - nh}{h}\right).$$

Here $h > 0$ and the coefficients c_n, $n = 0, \pm 1, \pm 2, \ldots$ are absolutely summable so that $u(x, 0)$ is continuous, piecewise linear, and absolutely integrable. Express u in terms of translates of dilates of the basis function

$$v(x, t) := \int_{x-1}^{x+1} \Lambda(x - \xi) \frac{e^{-\xi^2/4t}}{\sqrt{4\pi t}} d\xi$$

that satisfies $v_t = v_{xx}$ for $-\infty < x < \infty$, $t \geq 0$ with $v(x, 0) = \Lambda(x)$. (You can express v in terms of the exponential function and the error function of Ex. 9.22.)

Hint. Use the analysis of Ex. 9.23.

▶ **EXERCISE 9.25:** Let f_1, f_2 be piecewise smooth and absolutely integrable functions on $\mathbb{R}$ and assume that

$$|f_1(x) - f_2(x)| \leq C \text{ when } -\infty < x < \infty$$

for some choice of $C > 0$. Let u_1, u_2 be solutions of the diffusion equation (2) with the initial temperatures $u_1(x, 0) = f_1(x)$, $u_2(x, 0) = f_2(x)$. Show that

$$|u_1(x, t) - u_2(x, t)| \leq C \text{ when } -\infty < x < \infty, \ t \geq 0.$$

Note. Solutions of the diffraction equation do not have this continuity property. You can use the analysis from Ex. 9.42 to construct a counterexample.

▶ **EXERCISE 9.26:**   The bivariate *diffusion polynomial* $w^{[n]}(x,t)$ is defined to be the solution of

$$w_t^{[n]} = w_{xx}^{[n]}, \quad w^{[n]}(x,0) = x^n, \quad n = 0, 1, 2, \ldots .$$

(a) Use (52) to show that

$$w^{[n]}(x,t) = \frac{1}{(-2\pi i)^n} \int_{-\infty}^{\infty} e^{-4\pi^2 s^2 t + 2\pi i s x} \delta^{(n)}(s) ds$$

and thereby obtain the formula

$$w^{[n]}(x,t) = \frac{\partial^n}{\partial \alpha^n} \{e^{\alpha x + \alpha^2 t}\} \Big|_{\alpha=0} .$$

(b) Derive the explicit representation

$$w^{[n]}(x,t) = n! \sum_{k=0}^{\lfloor n/2 \rfloor} \frac{t^k}{k!} \frac{x^{n-2k}}{(n-2k)!},$$

and thereby obtain

$$1, \; x, \; x^2 + 2t, \; x^3 + 6xt, \; x^4 + 12x^2t + 12t^2$$

when $n = 0, 1, 2, 3, 4$.

Hint. Multiply the power series for $e^{\alpha x}$ and $e^{\alpha^2 t}$.

(c) How must we modify $w^{[n]}$ if we want to work with the diffusion equation $u_t = a^2 u_{xx}$ instead of $u_t = u_{xx}$?

Note. You can use linear combinations of the dilated functions $w^{[n]}$, $n = 0, 1, 2, \ldots$ to solve the diffusion equation $u_t = a^2 u_{xx}$ when $u(x,0)$ is an arbitrary polynomial.

▶ **EXERCISE 9.27:**   Let $u(x,t)$ be the solution of the diffusion equation $u_t = a^2 u_{xx}$ for $-\infty < x < \infty$, $t \geq 0$ with $u(x,0) = f(x)$. What can you infer about $u(x,t)$ at time $t > 0$ if you know that:

(a) $f(x) = f(-x)$?   (b) $f(x) = -f(-x)$?   (c) $f(x_0+x) = f(x_0-x)$?

(d) $f(x_0+x) = -f(x_0-x)$?   (e) $f(x+p) = f(x)$?   (f) $f(x+p/2) = -f(x)$?

▶ **EXERCISE 9.28:**   When we solve the diffusion equation $u_t = a^2 u_{xx}$, $-\infty < x < \infty$, $t \geq 0$ with the initial temperature

$$f(x) = \Pi(x), \; \Lambda(x), \; e^{-\pi x^2}, \; \delta(x)$$

we observe that u is well approximated by

$$u(x,t) \approx \frac{e^{-x^2/4a^2 t}}{\sqrt{4\pi a^2 t}}$$

when t is large. Sharpen this observation by showing that

$$\lim_{t \to \infty} \max_{-\infty < x < \infty} \left| \sqrt{2a^2 t}\, u(\sqrt{2a^2 t}\, x, t) - \frac{e^{-x^2/2}}{\sqrt{2\pi}} \right| = 0$$

when f has a bounded, continuous Fourier transform with $F(0) = 1$.

Hint. Use (52) with the synthesis equation to write

$$\left| \sqrt{2a^2 t}\, u(\sqrt{2a^2 t}\, x, t) - \frac{e^{-x^2/2}}{\sqrt{2\pi}} \right| = \left| \int_{-\infty}^{\infty} \left\{ F\left(\frac{r}{\sqrt{2a^2 t}} \right) - 1 \right\} e^{-2\pi^2 r^2 + 2\pi i r x}\, dr \right|$$

$$\leq \max_{|r| \leq L} \left| F\left(\frac{r}{\sqrt{2a^2 t}} \right) - 1 \right| \frac{1}{\sqrt{2\pi}} + \left\{ 1 + \max_{-\infty < s < \infty} |F(s)| \right\} \int_{|r| \geq L} e^{-2\pi^2 r^2}\, dr.$$

▶ **EXERCISE 9.29:**   A rod of length $L = 2$ has the initial temperature

$$u(x, 0) = \Pi(x - 1), \quad 0 \leq x \leq 2.$$

Find the Fourier series for the temperature $u(x, t)$, $0 \leq x \leq 2$, $t \geq 0$ when u satisfies the boundary conditions:

(a) $u(0, t) = 0$, $u(2, t) = 0$, $t \geq 0$;   (b) $u(0, t) = 0$, $u_x(2, t) = 0$, $t \geq 0$;

(c) $u_x(0, t) = 0$, $u_x(2, t) = 0$, $t \geq 0$.

Hint. Use symmetry to eliminate the boundary conditions, *e.g.*, when solving (b) create an 8-periodic function of x on $\mathbb{R}$ using

$$\begin{aligned} u(4 - x, t) &:= u(x, t), & 0 \leq x \leq 2 \\ u(-x, t) &:= -u(x, t), & 0 \leq x \leq 4 \\ u(x + 8n, t) &:= u(x, t), & -4 \leq x \leq 4,\ n = \pm 1, \pm 2, \ldots. \end{aligned}$$

▶ **EXERCISE 9.30:**   In this exercise you will find the solution of the diffusion equation $u_t = a^2 u_{xx}$ for $0 < x < 1$, $t \geq 0$ that has the boundary values

$$u(0, t) = 0, \quad u(1, t) = 1 + t, \quad t \geq 0$$

and the initial values

$$u(x, 0) = x, \quad 0 \leq x \leq 1.$$

(a) Let $w^{[n]}$, $n = 0, 1, \ldots$ be the bivariate diffusion polynomials of Ex. 9.26, and let

$$v(x, t) := u(x, t) - \frac{1}{6a^2} \{ (6a^2 - 1) w^{[1]}(x, a^2 t) + w^{[3]}(x, a^2 t) \}.$$

Verify that

$$\begin{aligned} v_t(x, t) &= a^2 v_{xx}(x, t), & 0 \leq x \leq 1,\ t \geq 0, \\ v(0, t) &= v(1, t) = 0, & t \geq 0,\ \text{and} \\ v(x, 0) &= (x - x^3)/6a^2, & 0 \leq x \leq 1. \end{aligned}$$

(b) Find the Fourier coefficients $c_1, c_2, \ldots$ for

$$v(x, t) = \sum_{k=1}^{\infty} c_k e^{-\pi^2 a^2 k^2 t} \sin(\pi k x),$$

e.g., by using Kronecker's rule.

Note. You can use this procedure to solve the diffusion equation when the boundary values are polynomials in t.

▶ **EXERCISE 9.31:**   Let $f(x) := e^{-2\pi(1+i)|x|}$. Show that

$$f''(x) = 4\pi^2 (1 + i)^2 f(x) - 4\pi(1 + i)\delta(x)$$

and thereby obtain the Fourier transform (68) used in the derivation of (70).

▶ **EXERCISE 9.32:**   Assume that the temperature of the soil in a certain neighborhood can be well approximated by a translated sinusoid (63) that has a maximum of $26°\mathrm{C}$ on July 1 and a minimum of $-10°\mathrm{C}$ on January 1 of each year.

(a) Use (70) with $L = 13.8$ m to determine how deep you should bury the water pipes to keep them from freezing.

Hint. Water freezes at $0°\mathrm{C}$.

(b) If the pipes are buried at this depth, when are they most likely to freeze?

Hint. There is a phase shift in (70).

▶ **EXERCISE 9.33:**   Find the steady-state solution of the diffusion equation $u_t = a^2 u_{xx}$ on the half line $x \geq 0$ when the temperature at the end is a T-periodic square wave with

$$u(0, t) = \begin{cases} 1 & \text{if } 0 < t < T/2 \\ -1 & \text{if } T/2 < t < T. \end{cases}$$

Hint. Use (69) (or a direct computation) to see that

$$e^{-2\pi\sqrt{|k|}\, x/L} e^{2\pi i(k\, t/T - \mathrm{sgn}\, k\, \sqrt{|k|}\, x/L)}, \quad k = 0, \pm 1, \pm 2, \ldots$$

is a solution of the diffusion equation when T and L are related by (67), and then use a series of such terms to synthesize $u(x, t)$.

▶ **EXERCISE 9.34:**   Let k be the diffusion kernel (53) and let γ be the diffraction kernel (92).

(a) Show that k, γ have the weak limits $\lim_{t \to 0+} k(x, t) = \delta(x)$, $\lim_{t \to 0} \gamma(x, t) = \delta(x)$.

(b) Show that

$$k(x, t_1) * k(x, t_2) = k(x, t_1 + t_2), \quad t_1 \geq 0, \ t_2 \geq 0,$$
$$\gamma(x, t_1) * \gamma(x, t_2) = \gamma(x, t_1 + t_2), \quad -\infty < t_1 < \infty, \ -\infty < t_2 < \infty.$$

(c) Why are the identities of (a), (b) important?

Hint. Begin with (54) and (93).

▶ **EXERCISE 9.35:** Let f be a generalized function, let k be the diffusion kernel (53), and let $u(x,t)$ be given by (54). In this exercise you will establish the weak limits

$$\lim_{h \to 0} \frac{u(x, t+h) - u(x,t)}{h} = a^2 u_{xx}(x,t), \quad t \geq 0, \quad \text{and} \quad \lim_{t \to 0+} u(x,t) = f(x),$$

i.e., you will prove that u satisfies (2) with the initial condition (51).

(a) Let f have the Fourier transform $F = G^{(n)}$ where G is CSG and $n = 0, 1, 2, \ldots$. Explain why it is sufficient to show that

$$\lim_{h \to 0} \frac{1}{h} \int_{-\infty}^{\infty} G(s) \left\{ \left[e^{-4\pi^2 a^2 s^2 h} - 1 + 4\pi^2 a^2 s^2 h \right] e^{-4\pi^2 a^2 s^2 t} \phi(s) \right\}^{(n)} ds = 0, t \geq 0,$$

$$\lim_{h \to 0+} \int_{-\infty}^{\infty} G(s) \left\{ \left[e^{-4\pi^2 a^2 s^2 h} - 1 \right] \phi(s) \right\}^{(n)} ds = 0$$

whenever ϕ is a Schwartz function.

(b) Let m, m_1, m_2 be nonnegative integers, let $t \geq 0$ be fixed, and let $|h| \leq 1$ with $t + h \geq 0$. Show that there are polynomials $\mathcal{P}_{m_1,m_2}(s), \mathcal{Q}_m(s)$ such that

$$|\{e^{-4\pi^2 a^2 s^2 t}\}^{(m_1)} \{e^{-4\pi^2 a^2 s^2 h} - 1 + 4\pi^2 a^2 s^2 h\}^{(m_2)}| \leq h^2 \mathcal{P}_{m_1,m_2}(s)$$

(c) Use (b) with the Leibnitz rule (2.29) to establish the limits of (a).

Note. Similar arguments can be used to show that (18) satisfies (1), (12), (13) and that (93) satisfies (3), (90).

▶ **EXERCISE 9.36:** Let (86) be written in the form $U(s,z) = \mathcal{A}(z)e^{-i\pi\lambda z s^2} U(s,0)$. This exercise will help you understand why we set $\mathcal{A}(z) = 1$ to obtain (87).

(a) Show that if $\mathcal{A}(z)$ is differentiable, then the consistency relation $\mathcal{A}(z_1 + z_2) = \mathcal{A}(z_1)\mathcal{A}(z_2)$ (from the text) implies that $\mathcal{A}(z) = e^{i\alpha z}$, $\alpha := \mathcal{A}'(0)$.

(b) Show that if we want the energy integral

$$\int_{-\infty}^{\infty} |u(x,z)|^2 dx$$

to be independent of z [when $u(x,0)$ is piecewise smooth and square integrable], then we must have $|\mathcal{A}(z)| = 1$, *i.e.*, α must be real. We removed such a factor $e^{2\pi i z/\lambda}$ when we passed from $\mathcal{U}(x,z,t)$ to the reduced wave function $u(x,z)$, so we choose $\mathcal{A}(z) = 1$.

Note. It is not so easy to find the phase shift $e^{-i\pi/4}$ from (87) if we do not make use of the Fourier transform (7.98), *cf.* Born and Wolf, pp. 370–375.

▶ **EXERCISE 9.37:**   Let $v(x, t)$ be the solution of the diffraction equation $v_t = (i\lambda/4\pi)v_{xx}$ for $-\infty < x < \infty$, $t \geq 0$ with $v(x, 0) = \text{sgn}(x)$.

(a) Set $f(x) = \text{sgn}(x)$ in (94) and thereby show that

$$v(x, t) = (1 - i)\mathcal{F}\left(\frac{x}{\sqrt{\lambda t/2}}\right), \quad t > 0$$

where $\mathcal{F}$ is the Fresnel function (95).

Note. The diffraction kernel (92) is given by $\gamma = (1/2)v_x$.

(b) Use (a) to find the diffraction pattern of a straight edge, *i.e.*, find the intensity $|u(x, t)|^2$ that results when we solve the diffraction equation with the initial wave function $u(x, 0) = h(x)$. Here h is the Heaviside function.

▶ **EXERCISE 9.38:**   Let $u(x, t)$ satisfy the diffraction equation $u_t = (i\lambda/4\pi)u_{xx}$ for $-\infty < x < \infty$ and $t \geq 0$ with the initial wave function

$$u(x, 0) = \frac{1}{h}\Pi\left(\frac{x - a}{h}\right) + \frac{1}{h}\Pi\left(\frac{x + a}{h}\right)$$

where $0 < h < 2a$. (This is a more accurate mathematical model for Young's double slit experiment than the one given in the text.)

(a) Express $u(x, t)$ in terms of the Fresnel function (95).

Hint. Use translates of dilates of (96) or use the analysis from Ex. 9.37(a).

(b) Find the Fraunhofer approximation for $u(x, t)$ when $t \gg a^2/\lambda$.

(c) Sketch the intensity $|u(x, t)|^2$ as a function of x when $t \gg a^2/\lambda$ is fixed, showing the effects of the parameters a, h.

▶ **EXERCISE 9.39:**   Let $f(x) = e^{-\pi(\alpha + i\beta)^2 x^2}$ where α, β are real and $\alpha > |\beta|$.

(a) Show that the Fourier transform of f has the form $F(s) = F(0)e^{-\pi s^2/(\alpha + i\beta)^2}$.

Hint. $f'(x) = i(\alpha + i\beta)^2 2\pi i f(x)$.

(b) Show that $F(0)^2 = 1/(\alpha + i\beta)^2$.

Hint. Use polar coordinates to integrate $f(x) \cdot f(y)$.

(c) Show that Re $F(0) > 0$ and thereby obtain the identity

$$F(s) = \frac{e^{-\pi s^2/(\alpha + i\beta)^2}}{\alpha + i\beta}$$

that was used to derive (98).

Note. This formula for F can also be used when $\alpha = |\beta| \neq 0$.

▶ **EXERCISE 9.40:**   This exercise will help you determine the *phase velocity* of a sinusoidal solution

$$w(x, t, s) := e^{2\pi i[sx - \nu(s)t]}, \quad \nu(s) := \lambda s^2/2$$

of the diffraction equation and the *group velocity* of a wave packet

$$u(x, t) = \int_{-\infty}^{\infty} A(s)e^{2\pi i[sx - \nu(s)t]} ds$$

corresponding to an amplitude function A that is localized near $s = s_0$.

(a) Show that $w(x, t, s_0)$ travels with the phase velocity

$$c(s_0) := \frac{\nu(s_0)}{s_0} = \frac{\lambda s_0}{2}, \quad s_0 \neq 0.$$

The high-frequency (large $|s|$) sinusoids travel faster than the low-frequency (small $|s|$) sinusoids.

(b) When A is suitably localized near $s = s_0$ we can use the linear approximation

$$\nu(s) \approx \nu(s_0) + (s - s_0)\nu'(s_0)$$

to write

$$u(x, t) \approx \int_{-\infty}^{\infty} A(s)e^{2\pi i\{sx - [\nu(s_0) + (s - s_0)\nu'(s_0)]t\}} ds.$$

Show that this implies

$$|u(x, t)| \approx |u(x - \nu'(s_0)t, 0)|$$

and thereby show that $|u|$ travels with the group velocity

$$c_{group}(s_0) := \nu'(s_0) = \lambda s_0$$

when t is small.

Note. You will observe this phenomenon in (105).

(c) Let $u_1(x, t), \ldots, u_4(x, t)$ be the solutions of the diffraction equation with

$$u_1(x, 0) = \Pi(x), \ u_2(x, 0) = \text{sinc}(x), \ u_3(x, 0) = e^{-|x|}, \ u_4(x, 0) = e^{-\pi x^2}.$$

The functions $|u_1(x, t)|^2, \ldots, |u_4(x, t)|$ will all spread as t increases. Which one will disperse the fastest? Which one will disperse the slowest?

(d) When we use $A(s) = \delta(s - s_0)$ we find $u(x, t) = w(x, t, s_0)$. Show that the analysis of (b) can be reconciled with the analysis of (a) in this case.

▶ **EXERCISE 9.41:**   Let $u(x, t)$ be a solution of the diffraction equation $u_t = (i\lambda/4\pi)u_{xx}$ for $-\infty < x < \infty$, $t \geq 0$. Does $|u(x, t)|^2$ evolve into $|U(x/\lambda t, 0)|^2/\lambda t$ as $t \to +\infty$ (*i.e.*, is the Fraunhofer approximation valid) when:

(a) $u(x, 0) = \delta(x - a) + \delta(x + a), \quad a > 0$?   (b) $u(x, 0) = \cos(2\pi x/p), \quad p > 0$?

(c) $u(x, 0) = e^{-\pi x^2/\alpha^2} \cos(2\pi x/p), \quad \alpha > 0, \ p > 0$?

▶ **EXERCISE 9.42:**   Let $N = 1, 2, \ldots$. Show how to construct a solution $u(x, t)$ of the diffraction equation $u_t = (i\lambda/4\pi)u_{xx}$ for $-\infty < x < \infty$, $t \geq 0$ such that

$$\max_{-\infty < x < \infty} |u(x, 0)| \leq 1, \quad \text{and} \quad \max_{-\infty < x < \infty} |u(x, 1)| \geq N.$$

Hint. Use (104) with x replaced by $x + \beta\lambda$ to construct a gaussian beam that travels (with spreading) from $x = -\beta\lambda$ to $x = 0$ as t increases from $t = 0$ to $t = 1$. You can combine such gaussian beams having $\beta = M\sqrt{2/\lambda}\, n$, $n = 1, 2, \ldots$ where M is a large positive integer.

Note. In view of this analysis, you cannot formulate a bound for a solution of the diffraction equation that is comparable to the bound of Ex. 9.15 for the wave equation or the bound of (58)–(59) for the diffusion equation.

▶ **EXERCISE 9.43:**   Let $u(x, t)$ be the solution of the diffraction equation $u_t = (i\lambda/4\pi)u_{xx}$ for $-\infty < x < \infty$, $t \geq 0$ with $u(x, 0) = f(x)$. What can you infer about $|u(x, t)|^2$ at time $t > 0$ if you know that:

(a) $f(x) = f(-x)$?   (b) $f(x) = -f(-x)$?   (c) $f(x_0 + x) = f(x_0 - x)$?

(d) $f(x_0 + x) = -f(x_0 - x)$?   (e) $f(x + p) = f(x)$?   (f) $f(x + p/2) = -f(x)$?

▶ **EXERCISE 9.44:**   Let $u(x, t; f)$ denote the solution of the diffraction equation $u_t = (i\lambda/4\pi)u_{xx}$ for $-\infty < x < \infty$, $-\infty < t < \infty$ with the initial source $u(x, 0) = f(x)$.

(a) Show that $u(x, -t; f) = \overline{u(x, t; \overline{f})}$.

(b) What conditions, if any, must be imposed on f to ensure that the movie showing $|u(x, t; f)|^2$ from $t = 0$ to $t = +\infty$ will be identical to the time-reversed movie showing $|u(x, t; f)|^2$ from $t = 0$ to $t = -\infty$?

▶ **EXERCISE 9.45:**   Let $u(x, t)$ satisfy the diffraction equation $u_t = (i\lambda/4\pi)u_{xx}$ for $0 < x < L$, $t \geq 0$ with the boundary conditions $u(0, t) = 0$, $u(L, t) = 0$ for $t \geq 0$. Sketch t-slices of $|u(x, t)|^2$ to show how the intensity evolves when

(a) $u(x, 0) = (2/L)^{1/2} \sin(\pi x/L)$;   (b) $u(x, 0) = (2/L)^{1/2} \sin(2\pi x/L)$;

(c) $u(x, 0) = (1/L)^{1/2}\{\sin(\pi x/L) + \sin(2\pi x/L)\}$.

▶ **EXERCISE 9.46:**   Let $u(x, t)$ be given by the Fourier series (113) and let $T := 2p^2/\lambda$ (so that u is T-periodic in t as well as p-periodic in x). Show that:

(a) $u(x, t + T/2) = u(x + p/2, t)$;

(b) $u(x, t + T/4) = \frac{1}{2}\{(1 - i)u(x, t) + (1 + i)u(x + p/2, t)\}$;

(c) $u(x, t + T/8) = \frac{1}{4}\{(\sqrt{2} - \sqrt{2}i)u(x, t) + 2u(x + p/4, t) + (-\sqrt{2} + \sqrt{2}i)u(x + p/2, t)$
$+ 2u(x + 3p/4, t)\}$.

Hint. First observe that

$$\sum_{n=0}^{3} c_n u(x + np/4, t) = u(x, t + T/8)$$

when

$$\sum_{n=0}^{3} c_n e^{2\pi i n k/4} = e^{-2\pi i k^2/8}, \quad k = 0, \pm 1, \pm 2, \dots .$$

Note. This analysis will help you understand the piecewise constant structure of the graphs of $|u(x, T/2^n)|^2$, $n = 1, 2, 3, \dots$ shown in Fig. 9.17.

▶ **EXERCISE 9.47:**   When a small-amplitude sinusoidal wave travels on water that is much deeper than the wavelength λ, the velocity is given by the formula

$$c = \sqrt{\frac{g\lambda}{2\pi}}$$

where g is the gravitational constant, *cf.* I.G. Main, pp. 236–257.

(a) Calculate the velocity for such water waves with $\lambda = 1$ cm, 1 m, 100 m, and 10 km.

Note. When an earthquake occurs in midocean, the long low-frequency waves are the first to arrive at a distant detector!

(b) Derive the expressions

$$c(s) = \sqrt{\frac{g}{2\pi s}}, \quad c_{group}(s) = \frac{1}{2}\sqrt{\frac{g}{2\pi s}}, \quad s > 0$$

for the phase velocity and the group velocity, *cf.* Ex. 9.40.

(c) Explain why we might use the formula

$$u(x, t) = \int_{-\infty}^{\infty} \{F(s) \cos[2\pi s\, c(s)t] + G(s)\, t \operatorname{sinc}[2s\, c(s)t]\} e^{2\pi i s x}\, dx$$

to simulate the water level at time t and coordinate x in a long deep canal when

$$u(x, 0) = f(x) \quad \text{and} \quad u_t(x, 0) = g(x).$$

Note. The choices $f(x) = \Pi(x)$, $g(x) = 0$ and $f(x) = 0$, $g(x) = \Pi(x)$ make interesting movies!

▶ **EXERCISE 9.48:**   In this exercise you will use Fourier analysis to find a solution of Laplace's equation $u_{xx}(x, y) + u_{yy}(x, y) = 0$ for $-\infty < x < \infty$, $y \geq 0$, that takes specified values $u(x, 0+) = f(x)$ on the boundary line $y = 0$.

(a) Let $g(x, y)$ be a generalized function of x for each $y \geq 0$ and assume that

$$g_{xx}(x, y) + g_{yy}(x, y) = 0 \quad \text{for} \quad -\infty < x < \infty, \ y \geq 0, \quad g(x, 0+) = \delta(x).$$

Show that the Fourier transform of g (with respect to x) satisfies

$$G_{yy}(s, y) - 4\pi^2 s^2 G(s, y) = 0, \quad G(s, 0) = 1.$$

(b) Within the context of (a), show that

$$g(x,y) = \frac{1}{\pi} \frac{y}{x^2 + y^2}.$$

You may assume that $g(x,y)$ and $G(s,y)$ are slowly growing as $x \to \pm\infty$, $s \to \pm\infty$, and $y \to +\infty$.

(c) Using (b), show that

$$u(x,y) = \frac{1}{\pi} \int_{-\infty}^{\infty} \frac{y\, f(\xi)}{(x - \xi)^2 + y^2}\, d\xi, \quad -\infty < x < \infty, k \ \ y > 0$$

satisfies Laplace's equation with the boundary values $u(x, 0+) = f(x)$, $-\infty < x < \infty$, when f is a suitably regular function of x.

Chapter 10

Wavelets

10.1 The Haar Wavelets

Introduction

We use dilates of the complex exponential *wave*

$$w(x) := e^{2\pi i x}, \quad -\infty < x < \infty$$

when we write the familiar Fourier synthesis equation

$$f(x) = \int_{-\infty}^{\infty} F(s)w(sx)dx$$

for a suitably regular function f on $\mathbb{R}$. (The function F is the Fourier transform of f.) For example, the identity

$$\frac{e^{-x^2/2\sigma^2}}{\sqrt{2\pi}\,\sigma} = \int_{-\infty}^{\infty} e^{-2\pi^2\sigma^2 s^2} e^{2\pi i s x} ds = \int_{-\infty}^{\infty} e^{-2\pi^2\sigma^2 s^2} \cos(2\pi s x) ds$$

shows how to synthesize a normal density by combining *waves* of constant amplitude,

$$e^{-2\pi^2\sigma^2 s^2} \cos(2\pi s x),$$

which stretch from $x = -\infty$ to $x = +\infty$. When $0 < \sigma \ll 1$, this density is an approximate delta with 99.7% of its integral in the tiny interval $-3\sigma \leq x \leq 3\sigma$. Almost perfect destructive interference must occur at every point $|x| > 3\sigma$. Such a synthesis has mathematical validity (a proof is given in Section 1.5), but it is physically unrealistic. In practice, we cannot produce the audio signal for a 1-ms "click" by having tubas, trombones, ... , piccolos play sinusoidal tones of constant amplitude for all eternity!

In this chapter you will learn to synthesize a function f with localized, oscillatory basis functions called *wavelets*. We use the prototype *Haar wavelet*

$$\psi(x) := \begin{cases} 1 & \text{if } 0 \le x < \frac{1}{2} \\ -1 & \text{if } \frac{1}{2} \le x < 1 \\ 0 & \text{otherwise} \end{cases} \tag{1}$$

to introduce the fundamental concepts from an exciting new branch of analysis created by mathematiciains, electrical engineers, physicists, ... during the past two decades. More sophisticated wavelets for audio signal processing, image compression, etc. are described in the following sections.

You will observe that the wavelet (1) *oscillates*. The positive and negative parts of ψ have the same size in the sense that

$$\int_{-\infty}^{\infty} \psi(x)dx = 0. \tag{2}$$

You will also notice that the wavelet (1) has the width or *scale* $a = 1$. This being the case, the *dyadic* (power of 2) dilates

$$\dots, \ \psi(2^{-2}x), \ \psi(2^{-1}x), \ \psi(x), \ \psi(2x), \ \psi(2^2x), \ \dots$$

have the widths

$$\dots, \ 2^2, \ 2^1, \ 1, \ 2^{-1}, \ 2^{-2}, \ \dots$$

as shown in Fig. 10.1. Since the dilate $\psi(2^m x)$ has width 2^{-m}, its translates

$$\psi(2^m x - k) = \psi(2^m[x - k\,2^{-m}]), \quad k = 0, \pm1, \pm2, \dots \tag{3}$$

nicely cover the whole x-axis. We use the complete collection of all such dyadic dilates of integer translates of ψ to write the *synthesis equation*

$$f(x) = \sum_{m=-\infty}^{\infty} \sum_{k=-\infty}^{\infty} F[m, k]\psi(2^m x - k) \tag{4}$$

for a suitably regular function f on $\mathbb{R}$. With somewhat picturesque language we refer to ψ as the *mother wavelet* for this expansion. The coefficient function F is a real, complex-valued function on $\mathbb{Z}^2$ when f is a real, complex-valued function on $\mathbb{R}$, respectively. We say that F is the *discrete wavelet transform* of f.

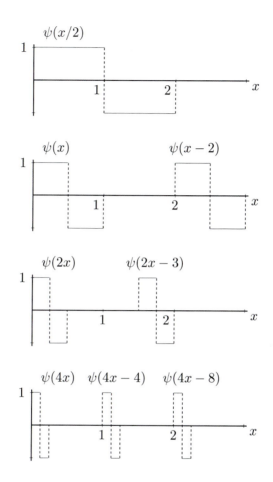

Figure 10.1. Dyadic dilates of integer translates of the Haar wavelet (1).

Interpretation of $F[m,k]$

It is natural to associate some meaning with the coefficient $F[m, k]$ that specifies the amount of the wavelet $\psi(2^m x - k)$ needed for the synthesis (4). Since this function vanishes except when

$$0 \le 2^m x - k < 1, \quad i.e., \quad k\,2^{-m} \le x < (k+1)2^{-m},$$

we see that $F[m, k]$ gives us information about the behavior of f near the *point* $x = k\,2^{-m}$ at the *scale* 2^{-m}. For example, the coefficients $F[-10, k]$, $k = 0, \pm 1, \pm 2, \ldots$ correspond to variations of f that take place over intervals of length $2^{10} = 1024$ while the coefficients $F[10, k]$, $k = 0, \pm 1, \pm 2, \ldots$ reveal fluctuations of f over intervals of length $2^{-10} = .000976\ldots\,$.

   This observation will help you understand why (4) gives us an exceptionally efficient scheme for representing a support-limited signal f. If f does not fluctuate wildly near $x = x_0$, neither the large- nor small-scale basis functions from (4) will make a significant contribution to the sum $f(x_0)$. (Basis functions having widths of 1 km or 1 mm are of little use when f is a half circle with a 1-m radius!) Moreover, for each fixed choice of the scale index m, there are only finitely many coefficients $F[m, k]$ that contribute to the value of f at $x = x_0$. Thus $f(x_0)$ can be well approximated by using a relatively small number of terms from (4).

   When we work with waves, the spatial frequency (*e.g.*, as measured in cycles per meter) is the reciprocal of the wavelength, a natural measure of scale. Since the wavelet $\psi(2^m x - k)$ has the scale 2^{-m}, we might reasonably expect the modulus of its Fourier transform to exhibit peaks near one or both of the points $s = \pm 2^m$. The Haar wavelet (1) has the Fourier transform

$$\Psi(s) = i\, e^{-i\pi s} \sin(\pi s/2)\, \mathrm{sinc}(s/2),$$

cf. Ex. 10.1, and

$$\psi(2^m x - k) \quad \text{has the FT} \quad e^{-2\pi i k\, 2^{-m} s}\, 2^{-m}\Psi(2^{-m}s).$$

We plot $|\Psi(2^{-m}s)|^2$ to see how "energy" is distributed across the spectrum of the wavelet $\psi(2^m x - k)$. This "energy" is more or less concentrated in bands of width 2^m centered near $s = \pm 2^m$, as shown in Fig. 10.2.

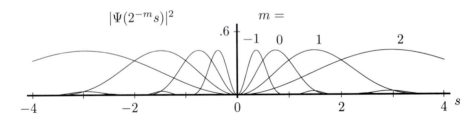

Figure 10.2. The functions $|\Psi(2^{-m}s)|^2$, $m = -1, 0, 1, 2$ that correspond to the Haar wavelet (1).

   You may find it helpful to interpret (4) within a musical context where we compose f from "notes" that correspond to the wavelets $\psi(2^m x - k)$. We are not allowed to use all of the keys from the piano keyboard (as specified in Appendix 7); we must produce our composition from the notes ..., $C2$, $C3$, $C4$, $C5$, ... with the (approximate) frequencies F $= 2^m$, $m = \ldots$, 2^6, 2^7, 2^8, 2^9, ..., *cf.* Section 11.1. Furthermore, we are not allowed to vary the duration of these tones; we must always play ..., $C2$, $C3$, $C4$, $C5$, ... as ..., whole, half, quarter, eighth, ... notes. Indeed, we always play from the fixed musical score shown in Fig. 10.3, but

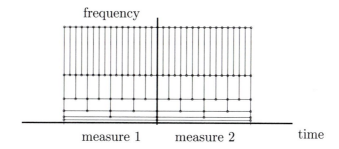

Figure 10.3. Two measures from a musical score for discrete wavelet synthesis and a corresponding equal-area tiling of the time-frequency plane using the points $(k\,2^m, 2^m)$.

we do not always play the same thing. We are free to choose the coefficient $F[m, k]$ that specifies how loudly (and with what phase, up-down or down-up) we sound the "note" $\psi(2^m x - k)$. As astonishing as it may seem, this gives us enough flexibility to play anything we please! We will now give a proof of this remarkable assertion.

Arbitrarily good approximation

Let f be a continuous, absolutely integrable, real-valued function on $\mathbb{R}$, and assume that $f(x) \to 0$ as $x \to \pm\infty$. We will show how to construct an arbitrarily good discrete wavelet approximation for f. The argument, which is adopted from the multiresolution analysis of S. Mallat and Y. Meyer, makes use of the *scaling function* (or *father wavelet*)

$$\phi(x) := \begin{cases} 1 & \text{if } 0 \leq x < 1 \\ 0 & \text{otherwise} \end{cases} \tag{5}$$

that goes with the Haar wavelet (1). Later on we will carry out an analogous construction with other choices of ψ and ϕ.

We begin by using ϕ to produce a good piecewise constant approximation for f. The function $\phi(2^m x - k)$ vanishes except when $k\, 2^{-m} \le x < (k+1)2^{-m}$, so

$$\alpha_m[k] := 2^m \int_{-\infty}^{\infty} f(x)\phi(2^m x - k)dx = 2^m \int_{k\, 2^{-m}}^{(k+1)2^{-m}} f(x)dx \tag{6}$$

is the average value of f on the interval $[k\, 2^{-m}, (k+1)2^{-m}]$. This average lies between the minimum and maximum of f on the interval, and it follows that

$$
\begin{aligned}
|f(x) - \alpha_m[k]| &\le \max\{|f(x) - f(x')| : k\, 2^{-m} \le x' \le (k+1)2^{-m}\} \\
&\le \max\{|f(x'') - f(x')| : x', x'' \in \mathbb{R},\ |x' - x''| \le 2^{-m}\} \tag{7}\\
&=: \omega_m \quad \text{when} \quad k\, 2^{-m} \le x \le (k+1)2^{-m}.
\end{aligned}
$$

The step function approximation or *frame*

$$f_m(x) := \sum_{k=-\infty}^{\infty} \alpha_m[k]\phi(2^m x - k) \tag{8}$$

takes the constant value $\alpha_m[k]$ on the interval $k\, 2^{-m} \le x < (k+1)2^{-m}$, so we can use (7) to write

$$|f(x) - f_m(x)| \le \omega_m, \quad -\infty < x < \infty. \tag{9}$$

Since f is a continuous function that vanishes at $x = \pm\infty$, it follows that $\omega_m \to 0$ as $m \to +\infty$. In this way we see that it is possible to produce an arbitrarily good uniform approximation for f on $\mathbb{R}$ by using (6) and (8) with some sufficiently large scaling index m.

We are now ready to introduce the simple (but exceptionally powerful) idea that underlies wavelet analysis: the "wide" dilates $\phi(x/2), \psi(x/2)$ can be written as linear combinations of the "narrow" translates $\phi(x), \phi(x-1)$, *i.e.*,

$$
\begin{aligned}
\phi(x/2) &= \phi(x) + \phi(x-1), \\
\psi(x/2) &= \phi(x) - \phi(x-1).
\end{aligned}
\tag{10}
$$

By inspection, we invert the system (10) and write

$$
\begin{aligned}
\phi(x) &= \frac{1}{2}\phi(x/2) + \frac{1}{2}\psi(x/2), \\
\phi(x-1) &= \frac{1}{2}\phi(x/2) - \frac{1}{2}\psi(x/2).
\end{aligned}
\tag{11}
$$

For our present purpose we combine these equations to obtain the identity

$$\alpha_0\phi(x) + \alpha_1\phi(x-1) = \frac{\alpha_0 + \alpha_1}{2}\phi(x/2) + \frac{\alpha_0 - \alpha_1}{2}\psi(x/2),$$

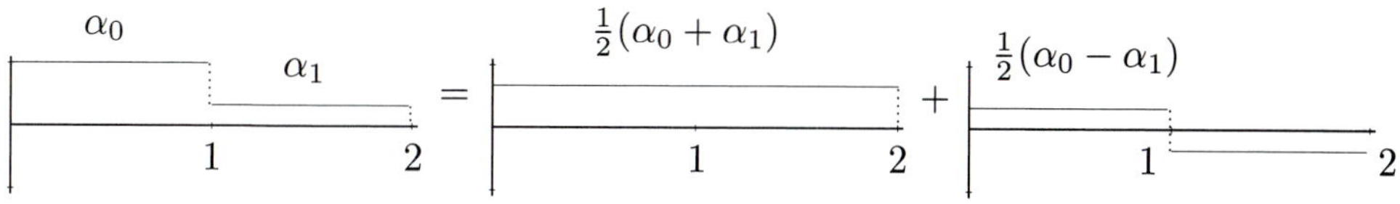

Figure 10.4. A pictorial proof of the identity (12) when $m = k = 0$.

cf. Fig. 10.4, or more generally,

$$\alpha_0 \phi(2^m x - 2k) + \alpha_1 \phi(2^m x - 2k - 1)$$
$$= \frac{\alpha_0 + \alpha_1}{2} \phi(2^{m-1} x - k) + \frac{\alpha_0 - \alpha_1}{2} \psi(2^{m-1} x - k) \qquad (12)$$

where α_0, α_1 are arbitrary scalars. In this way we derive a *splitting*

$$f_m(x) = \sum_{k=-\infty}^{\infty} \{\alpha_m[2k]\phi(2^m x - 2k) + \alpha_m[2k+1]\phi(2^m x - 2k - 1)\}$$

$$= \sum_{k=-\infty}^{\infty} \frac{1}{2}\{\alpha_m[2k] + \alpha_m[2k+1]\}\phi(2^{m-1} x - k) \qquad (13)$$

$$+ \sum_{k=-\infty}^{\infty} \frac{1}{2}\{\alpha_m[2k] - \alpha_m[2k+1]\}\psi(2^{m-1} x - k)$$

$$=: f_{m-1}(x) + d_{m-1}(x)$$

of the frame f_m. Here

$$f_{m-1}(x) := \sum_{k=-\infty}^{\infty} \alpha_{m-1}[k]\phi(2^{m-1} x - k)$$

with

$$\alpha_{m-1}[k] := \frac{1}{2}\{\alpha_m[2k] + \alpha_m[2k+1]\}$$

$$= 2^{m-1} \int_{2k2^{-m}}^{(2k+1)2^{-m}} f(x)dx + 2^{m-1} \int_{(2k+1)2^{-m}}^{(2k+2)2^{-m}} f(x)dx$$

$$= 2^{m-1} \int_{k\,2^{1-m}}^{(k+1)2^{1-m}} f(x)dx$$

$$= 2^{m-1} \int_{-\infty}^{\infty} f(x)\phi(2^{m-1} x - k)dx$$

in keeping with (6), and analogously,

$$d_{m-1}(x) := \sum_{k=-\infty}^{\infty} \beta_{m-1}[k]\psi(2^{m-1}x - k) \tag{14}$$

with

$$\beta_{m-1}[k] := \frac{1}{2}\{\alpha_m[2k] - \alpha_m[2k+1]\}$$
$$= 2^{m-1}\int_{-\infty}^{\infty} f(x)\psi(2^{m-1}x - k)dx. \tag{15}$$

Figure 10.5 illustrates this splitting process. (The graphs are filled to make it easier for you to see the steps.) An approximating frame f_m from the left side of one row is the sum of the frame f_{m-1} (left) and the *detail* d_{m-1} (right) from the following row. If you observe the way a pair of adjacent "narrow" ribbons from the graph of f_m is replaced by a correspond "wide" ribbon in the graph of f_{m-1} and a corresponding "wide" Haar wavelet in the graph of d_{m-1}, you will understand how we can produce these functions by using (12).

The splitting (13) leads at once to the synthesis equation (4). Indeed, given integers $I \leq J$ we write

$$f_{J+1} = f_J + d_J = f_{J-1} + d_{J-1} + d_J = \cdots = f_I + d_I + d_{I+1} + \cdots + d_J, \tag{16}$$

and thereby see that

$$\left| f - \sum_{m=I}^{J} d_m \right| = |f - f_{J+1} + f_I| \leq |f - f_{J+1}| + |f_I|.$$

In conjunction with (14), (9) and the inequality

$$|f_I(x)| \leq \max_k |\alpha_I[k]| \leq 2^I \int_{-\infty}^{\infty} |f(x)|dx$$

that follows from (8) and (6), this allows us to write

$$\left| f(x) - \sum_{m=I}^{J} \sum_{k=-\infty}^{\infty} \beta_m[k]\psi(2^m x - k) \right| \leq \omega_{J+1} + 2^I \int_{-\infty}^{\infty} |f(\xi)|d\xi, \quad -\infty < x < \infty.$$

Since $\omega_{J+1} \to 0$ as $J \to +\infty$ and $2^I \to 0$ as $I \to -\infty$, this shows that the synthesis equation (4) holds at each point x (with the convergence being uniform) provided that we identify $F[m, k]$ with $\beta_m[k]$, *i.e.*, provided that we use the *analysis equation*

$$F[m, k] := 2^m \int_{-\infty}^{\infty} f(x)\psi(2^m x - k)dx, \quad m, k = 0, \pm 1, \pm 2, \ldots \tag{17}$$

to obtain F from f. We really *can* play anything we like from the musical score of Fig. 10.3 (if we use sufficiently many measures, *cf.* Ex. 10.2)!

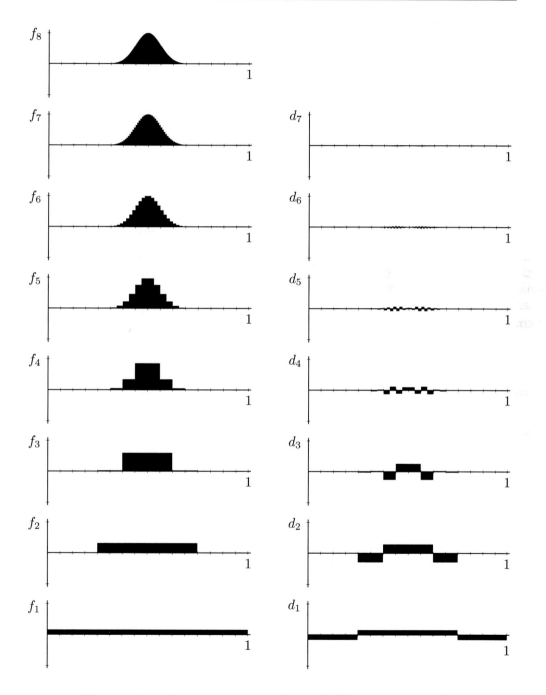

Figure 10.5. Piecewise constant frames (left) and corresponding details (right) for the Haar wavelet approximation of a normal density.

Wavelets are used to analyze audio signals, EKGs, images, ... that do not have the perfect smoothness of the normal density. Figure 10.6 shows frames and details for a function f that represents the silhouette of a city skyline. You will again observe the local use of (12) as we move from top to bottom, constructing each pair f_{m-1}, d_{m-1} from the previous f_m. You will also notice a new phenomenon. At every scale there are large wavelet coefficients that correspond to the points where f has jump discontinuities (*cf.* Ex. 10.3). We can use these large coefficients to locate the edges of the buildings!

You can develop some feeling for the efficiency of the wavelet synthesis (4) by studying Figs. 10.5 and 10.6. Each detail d_m shown in these illustrations is an inner sum from (4). As your eye focuses on the ribbon to the right of the point $x = k\,2^{-m}$ from the graph of $d_m(x)$, you see the size and sign of the summand $F[m,k]\psi(2^m x - k)$ from (4)! When f is localized, only finitely many terms from an inner sum of (4) make a significant contribution to f. When f is smooth and slowly varying on some interval $[a,b]$, only finitely many terms from the outer sum of (4) make a significant contribution to f on any $[c,d] \subset (a,b)$.

From your study of linear algebra, you will recognize that the integral representation (17) for the coefficients $F[m,k]$ follows from the *orthogonality relations*

$$\int_{-\infty}^{\infty} \psi(2^m x - k)\psi(2^{m'} x - k')dx = \begin{cases} 2^{-m} & \text{if } m' = m \text{ and } k' = k \\ 0 & \text{otherwise} \end{cases} \quad (18)$$

for the Haar wavelets that appear in (4). It is a simple matter to verify (18). Indeed, the wavelet $\psi(2^m x)$ has width 2^{-m}, so its 2^{-m} translates (3) have nonoverlapping supports, *cf.* Fig. 10.1. In this way we see that (18) holds when $m' = m$. On the other hand, when the wavelets from the integrand of (18) have different scales, the "wide" wavelet takes the same constant value (0, 1, or -1) at each point of the interval where the "narrow" wavelet is nonzero, again *cf.* Fig. 10.1. Since the integral of the "narrow" wavelet is zero, (18) also holds when $m' \neq m$.

Successive approximation

The above analysis facilitates a brief discussion of an exceptionally important feature of the wavelet representation (4). Suppose that we want to transmit a function f over a digital communication channel. We will assume that the approximation

$$f \approx f_{J+1} = f_I + d_I + d_{I+1} + \cdots + d_J$$

from (8) and (16) is sufficiently accurate for our purposes. When f is suitably localized, only finitely many of the coefficients $\alpha_I[k]$ for the frame f_I and only finitely many of the coefficients $\beta_m[k]$ for the detail d_m, $m = I, I+1, \ldots, J$ will make a significant contribution to the approximation. You can see this as you examine the frames and details shown in Figs. 10.5 and 10.6.

Within this context we suitably quantize and transmit in turn the nonzero coefficients for $f_I, d_I, d_{I+1}, \ldots, d_J$. At the other end of the communication channel,

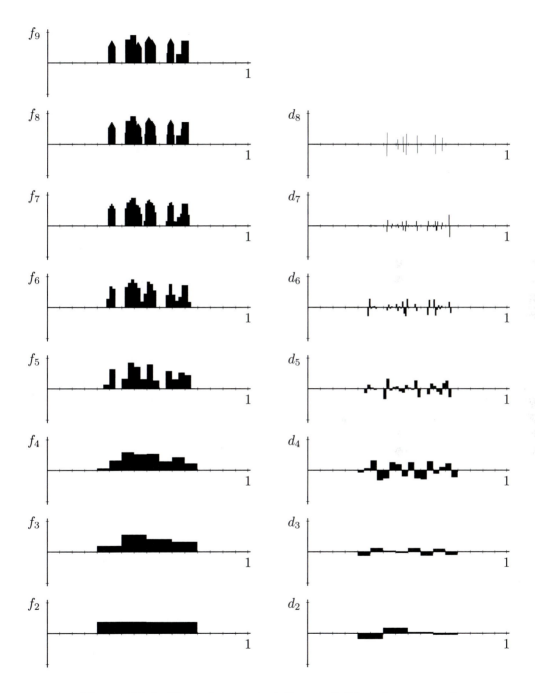

Figure 10.6. Piecewise constant frames (left) and corresponding details (right) for the Haar wavelet approximation of a city skyline.

the coarse approximation f_I is constructed just as soon as the required coefficients for this frame arrive. The refinements

$$f_{I+1} = f_I + d_I, \quad f_{I+2} = f_{I+1} + d_{I+1}, \ldots, f_{J+1} = f_J + d_J$$

(as shown from bottom to top on the left side of Figs. 10.5 and 10.6) are generated one by one as the coefficients for the details $d_I, d_{I+1}, \ldots, d_J$ are received. You may well recall viewing such successive approximations while downloading a large image file from a site on the Internet.

Coded approximation

Figure 10.5 shows selected frames and details for the normal density

$$f(x) := \frac{e^{-(x-\mu)^2/2\sigma^2}}{\sqrt{2\pi}\,\sigma} \quad \text{with} \quad \mu = \frac{1}{2} \quad \text{and} \quad \sigma = \frac{1}{16}\ .$$

When x lies outside the interval $[0,1]$ this density is relatively small, with

$$\frac{f(x)}{f(1/2)} \le e^{-32} = 1.26\ldots \cdot 10^{-14} \quad \text{for} \quad |x - 1/2| \ge 1/2.$$

To this level of precision $f_m, d_m, m = 0,1,2,\ldots$ all vanish outside $[0,1]$, so we can truncate the series (8), (14) and write

$$f_m(x) = \sum_{k=0}^{2^m-1} \alpha_m[k]\phi(2^m x - k),$$

$$d_m(x) = \sum_{k=0}^{2^m-1} \beta_m[k]\psi(2^m x - k). \tag{19}$$

We facilitate numerical analysis by introducing the coefficient vectors

$$\boldsymbol{\alpha_m} := (\alpha_m[0], \alpha_m[1], \ldots, \alpha_m[2^m - 1])^T,$$
$$\boldsymbol{\beta_m} := (\beta_m[0], \beta_m[1], \ldots, \beta_m[2^m - 1])^T \tag{20}$$

to *encode* the functions (19). Such vectors for the frames and details of Fig. 10.5 are displayed in Fig. 10.7. Of course, if we are given the vectors $\boldsymbol{\alpha_m}$, $\boldsymbol{\beta_m}$, we can use (19) to construct the functions f_m, d_m. Conversely, if we are given the step functions f_m, d_m, we can produce the vectors (20) by using the right hand limits

$$\alpha_m[k] = f_m(k\,2^{-m}+), \qquad \beta_m[k] = d_m(k\,2^{-m}+)$$

$\boldsymbol{\alpha_5} = (0, \ldots, 0, 1, 3, 15, 53, 142, 294, 480, 612, 612, 480, 294, 142, 53, 15, 3, 1, 0, \ldots, 0)$

$\boldsymbol{\alpha_4} = (0, 0, 0, 0,\ \ 2,\ \ \ \ 34,\ \ \ 218,\ \ \ 546,\ \ \ 546,\ \ \ 218,\ \ \ \ 34,\ \ \ 2, 0, 0, 0, 0)$
$\boldsymbol{\beta_4} = (0, 0, 0, 0, -1,\ \ -19,\ \ -76,\ \ -66,\ \ \ \ 66,\ \ \ \ 76,\ \ \ \ 19,\ \ \ 1, 0, 0, 0, 0)$

$\boldsymbol{\alpha_3} = (\ \ \ \ 0,\ \ \ \ 0,\ \ \ 18,\ 382,\ 382,\ \ \ 18,\ \ \ \ 0,\ \ \ \ 0)$
$\boldsymbol{\beta_3} = (\ \ \ \ 0,\ \ \ \ 0, -16, -164,\ 164,\ \ \ 16,\ \ \ \ 0,\ \ \ \ 0)$

$\boldsymbol{\alpha_2} = (\ \ \ \ 0,\ \ 200,\ 200,\ \ \ \ 0)$
$\boldsymbol{\beta_2} = (\ \ \ \ 0, -182,\ 182,\ \ \ \ 0)$

$\boldsymbol{\alpha_1} = (\ \ 100,\ \ \ 100)$
$\boldsymbol{\beta_1} = (-100,\ \ \ 100)$

$\boldsymbol{\alpha_0} = (\ \ 100)$
$\boldsymbol{\beta_0} = (\ \ \ \ 0)$

Figure 10.7. Vectors of coefficients $\boldsymbol{\alpha_5}, \ldots, \boldsymbol{\alpha_1}$ and $\boldsymbol{\beta_4}, \ldots, \boldsymbol{\beta_1}$ that encode the frames $f_5, \ldots, f_1$ and details $d_4, \ldots, d_1$ shown in Fig. 10.5. Each coefficient has been scaled by 100 and rounded to an integer.

or by using the integrals

$$\alpha_m[k] = 2^m \int_{-\infty}^{\infty} f_m(x)\phi(2^m x - k)dx = \int_0^1 f_m(2^{-m}k + 2^{-m}u)\phi(u)du,$$

$$\beta_m[k] = 2^m \int_{-\infty}^{\infty} d_m(x)\psi(2^m x - k)dx = \int_0^1 d_m(2^{-m}k + 2^{-m}u)\psi(u)du. \tag{21}$$

Such expressions follow from the fact that

$$f_m(x) = \alpha_m[k]\phi(2^m x - k) \text{ when } k\,2^{-m} \le x < (k+1)2^{-m},$$

$$d_m(x) = \beta_m[k]\psi(2^m x - k) \text{ when } k\,2^{-m} \le x < (k+1)2^{-m}.$$

We will use the splitting $f_m = f_{m-1} + d_{m-1}$ with (19) to establish identities that link certain components of the coefficient vectors $\boldsymbol{\alpha_m}, \boldsymbol{\alpha_{m-1}}, \boldsymbol{\beta_{m-1}}$. Indeed, by restricting the splitting to the interval $2k\,2^{-m} \le x < (2k+2)2^{-m}$ we see that

$$
\begin{aligned}
\alpha_m[2k]\phi(2^m x - 2k) &+ \alpha_m[2k+1]\phi(2^m x - 2k - 1) \\
&= \alpha_{m-1}[k]\phi(2^{m-1}x - k) + \beta_{m-1}[k]\psi(2^{m-1}x - k).
\end{aligned}
\tag{22}
$$

In conjunction with (12) (*cf.* Fig. 10.4) this gives us *Mallat's relations*

$$
\begin{aligned}
\alpha_m[2k] &= \alpha_{m-1}[k] + \beta_{m-1}[k], \\
\alpha_m[2k+1] &= \alpha_{m-1}[k] - \beta_{m-1}[k]
\end{aligned}
\tag{23}
$$

and

$$
\begin{aligned}
\alpha_{m-1}[k] &= \tfrac{1}{2}\{\alpha_m[2k] + \alpha_m[2k+1]\}, \\
\beta_{m-1}[k] &= \tfrac{1}{2}\{\alpha_m[2k] - \alpha_m[2k+1]\}
\end{aligned}
\tag{24}
$$

for these coefficients. Make sure that you can see the sums and differences (23) as you synthesize $\boldsymbol{\alpha_m}$ from $\boldsymbol{\alpha_{m-1}}$ and $\boldsymbol{\beta_{m-1}}$ in Fig. 10.7. For example, the $k = 4, 5$ components of $\boldsymbol{\alpha_3}$ are obtained from the $k = 2$ components of $\boldsymbol{\alpha_2}, \boldsymbol{\beta_2}$ by writing $382 = 200 + 182$, $18 = 200 - 182$. Likewise, make sure that you see how (24) produces $\boldsymbol{\alpha_{m-1}}$ and $\boldsymbol{\beta_{m-1}}$ from $\boldsymbol{\alpha_m}$. For example, the $k = 4$ components of $\boldsymbol{\alpha_3}, \boldsymbol{\beta_3}$ are obtained from the $k = 8, 9$ components of $\boldsymbol{\alpha_4}$ by writing $382 = (546 + 218)/2$, $164 = (546 - 218)/2$.

Figure 10.8 will help you visualize how we do analysis and synthesis with the coefficient vectors from Fig. 10.7. We use (24) at each node of Mallat's *herringbone algorithm* to produce the $16, 8, 4, 2, 1, 1$ component vectors $\boldsymbol{\beta_4}, \boldsymbol{\beta_3}, \boldsymbol{\beta_2}, \boldsymbol{\beta_1}, \boldsymbol{\beta_0}, \boldsymbol{\alpha_0}$ from the 32-component vector $\boldsymbol{\alpha_5}$. We use (23) at each node of Mallat's *reverse herringbone algorithm* when we reconstruct $\boldsymbol{\alpha_5}$ from $\boldsymbol{\alpha_0}, \boldsymbol{\beta_0}, \boldsymbol{\beta_1}, \boldsymbol{\beta_2}, \boldsymbol{\beta_3}, \boldsymbol{\beta_4}$. Equivalently, the herringbone algorithm enables us to split the initial frame f_5 into its constituent parts, the details d_4, d_3, d_2, d_1, d_0, and the frame f_0. The reverse herringbone algorithm allows us to reconstruct the frame f_5 from $f_0, d_0, d_1, d_2, d_3, d_4$.

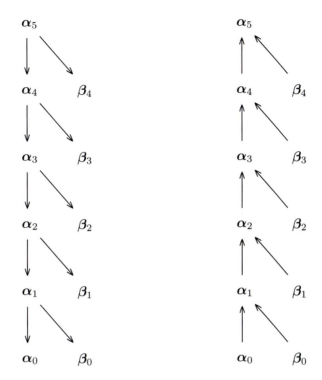

Figure 10.8. Mallat's herringbone algorithm (left) is used with (24) to split a frame into its constituent details. The reverse herringbone algorithm (right) is used with (23) when we assemble a frame from its constituent details.

Suppose now that we are given an arbitrary vector $\boldsymbol{\alpha_m}$ with $N := 2^m$ components. We concatenate the corresponding $\boldsymbol{\alpha_0}, \boldsymbol{\beta_0}, \boldsymbol{\beta_1}, \ldots, \boldsymbol{\beta_{m-1}}$ to form a vector

$$\boldsymbol{\alpha_m^w} := \boldsymbol{\alpha_0} \ \& \ \boldsymbol{\beta_0} \ \& \ \boldsymbol{\beta_1} \ \& \ \cdots \ \& \ \boldsymbol{\beta_{m-1}}$$

$$:= (\alpha_0[0], \beta_0[0], \beta_1[0], \beta_1[1], \ldots, \beta_{m-1}[2^{m-1} - 1])^T$$

(25)

having $1 + 1 + 2 + 4 + \cdots + 2^{m-1} = 2^m$ components, *cf.* Exs. 6.7 and 6.9. We say that $\boldsymbol{\alpha_m^w}$ is the *discrete wavelet transform* (DWT) of $\boldsymbol{\alpha_m}$. The tag, ^{w}, is analogous to the tag, $^\wedge$, for the DFT.

As we use (24) to generate $\boldsymbol{\alpha_{m-1}}, \boldsymbol{\beta_{m-1}}$ from $\boldsymbol{\alpha_m}$, we perform one addition and one multiplication (by $1/2$) per component. Since each of these half-sized vectors has 2^{m-1} components the cost is 2^m operations. This being the case, when we use the herringbone algorithm

$$
\begin{aligned}
\boldsymbol{\alpha_m} &\curvearrowright \boldsymbol{\alpha_{m-1}} \ \& \ \boldsymbol{\beta_{m-1}} \\
&\curvearrowright \boldsymbol{\alpha_{m-2}} \ \& \ \boldsymbol{\beta_{m-2}} \ \& \ \boldsymbol{\beta_{m-1}} \\
&\vdots \\
&\curvearrowright \boldsymbol{\alpha_1} \ \& \ \boldsymbol{\beta_1} \ \& \ \boldsymbol{\beta_2} \ \& \ \cdots \ \& \ \boldsymbol{\beta_{m-1}} \\
&\curvearrowright \boldsymbol{\alpha_0} \ \& \ \boldsymbol{\beta_0} \ \& \ \boldsymbol{\beta_1} \ \& \ \cdots \ \& \ \boldsymbol{\beta_{m-1}} =: \boldsymbol{\alpha_m^w}
\end{aligned}
\tag{26}
$$

to compute the DWT we expend $2^m + 2^{m-1} + \cdots + 2 = 2 \cdot 2^m - 2 \approx 2N$ operations. Each mapping from the chain (26) is represented by a sparse matrix (with one or two nonzero entries per row). The product of these sparse matrices gives the dense matrix for the DWT, *cf.* Ex. 10.4.

Analogously, as we use (23) to generate $\boldsymbol{\alpha_1}$ from $\boldsymbol{\alpha_0}$ and $\boldsymbol{\beta_0}$, $\boldsymbol{\alpha_2}$ from $\boldsymbol{\alpha_1}$ and $\boldsymbol{\beta_1}, \ldots$ we perform one addition per component. This being the case, when we use the reverse herringbone algorithm

$$
\begin{aligned}
\boldsymbol{\alpha_m^w} &= \boldsymbol{\alpha_0} \ \& \ \boldsymbol{\beta_0} \ \& \ \boldsymbol{\beta_1} \ \& \ \cdots \ \& \ \boldsymbol{\beta_{m-1}} \\
&\curvearrowright \boldsymbol{\alpha_1} \ \& \ \boldsymbol{\beta_1} \ \& \ \boldsymbol{\beta_2} \ \& \ \cdots \ \& \ \boldsymbol{\beta_{m-1}} \\
&\vdots \\
&\curvearrowright \boldsymbol{\alpha_{m-1}} \ \& \ \boldsymbol{\beta_{m-1}} \\
&\curvearrowright \boldsymbol{\alpha_m}
\end{aligned}
\tag{27}
$$

to invert the DWT we expend $2 + 4 + \cdots + 2^m \approx 2N$ additions. The matrices that correspond to the mappings (27) give a sparse factorization for the inverse of the DWT.

It is very easy and surprisingly inexpensive to do analysis and synthesis with Haar wavelets when we use (26), (27) with Mallat's relations (24), (23) that follow from the dilation equations (10). Once you see how these thoroughly modern constructs fit together within this elementary context you will be well prepared to study generalizations that lead to exceptionally powerful new tools for signal processing. [Haar did not forsee such developments in 1910 when he introduced the synthesis equation (4) and used (18) to obtain the coefficients (17).]

10.2 Support-Limited Wavelets

The dilation equation

Let $c_0, c_1, \ldots, c_M$ be real numbers with

$$\sum_{m=0}^{M} c_m = 2. \tag{28}$$

We will construct a support-limited generalized scaling function ϕ by solving the *dilation equation*

$$\phi(x) = \sum_{m=0}^{M} c_m \phi(2x - m) \tag{29}$$

subject to the normalization condition

$$\Phi(0) = 1. \tag{30}$$

We use this function ϕ to produce a corresponding wavelet

$$\psi(x) := \sum_{m=0}^{M} (-1)^m c_{M-m} \phi(2x - m). \tag{31}$$

When we choose the coefficients

$$M = 1, \quad c_0 = 1, \quad c_1 = 1 \tag{32}$$

the equations (29), (31) give the relations (10) that link the scaling function (5) and the Haar wavelet (1) from the preceding section. The generalization makes it possible for us to find other functions ϕ, ψ that work together in much the same way. (You will understand why we reverse the order of the coefficients and alternate the signs later on when you work Ex. 10.44.)

We will solve (29) by iteration. With this in mind we define

$$\phi_0(x) := \delta(x),$$
$$\phi_n(x) := \sum_{m=0}^{M} c_m \phi_{n-1}(2x - m), \quad n = 1, 2, \ldots, \tag{33}$$

observing that each of these generalized functions vanishes outside the interval $[0, M]$. The Fourier transforms

$$\Phi_0(s) := 1,$$
$$\Phi_n(s) := \frac{1}{2} \sum_{m=0}^{M} c_m e^{-2\pi i m s / 2} \Phi_{n-1}\left(\frac{s}{2}\right), \quad n = 1, 2, \ldots, \tag{34}$$

are all M-bandlimited. We use the recursion (34) with the trigonometric polynomial

$$T(s) := \frac{1}{2} \sum_{m=0}^{M} c_m e^{-2\pi ims} \tag{35}$$

to see that

$$\Phi_1(s) = T\left(\frac{s}{2}\right), \quad \Phi_2(s) = T\left(\frac{s}{2}\right) T\left(\frac{s}{4}\right), \quad \Phi_3(s) = T\left(\frac{s}{2}\right) T\left(\frac{s}{4}\right) T\left(\frac{s}{8}\right), \quad \ldots \tag{36}$$

or equivalently,

$$\phi_1(x) = 2\tau(2x), \quad \phi_2(x) = 2\tau(2x) * 4\tau(4x), \quad \phi_3(x) = 2\tau(2x) * 4\tau(4x) * 8\tau(8x), \quad \ldots \tag{37}$$

where

$$\tau(x) := \frac{1}{2} \sum_{m=0}^{M} c_m \delta(x - m) \tag{38}$$

is the inverse Fourier transform of (35), *cf.* Ex. 10.7.

Example: Let $M = 0$ and $c_0 = 2$. Find the corresponding sequences (37), (36).

Solution: When $M = 0$, $c_0 = 2$ we use (38) and (35) to find

$$\tau(x) = \delta(x), \quad T(s) = 1.$$

Since $2^\nu \delta(2^\nu x) = \delta(x)$, $\nu = 1, 2, \ldots$, the convolution products of (37) and the products of (36) give

$$\phi_n(x) = \delta(x), \quad \Phi_n(s) = 1, \quad n = 1, 2, \ldots.$$

The limit $\phi(x) = \delta(x)$ is the solution of the dilation equation

$$\phi(x) = 2\phi(2x)$$

with $\Phi(0) = 1$.   ∎

Figure 10.9 shows the sequences (37), (36) that correspond to the coefficients (32) (with deltas represented by vertical line segments). The functions $\phi_1, \phi_2, \phi_3, \ldots$ converge weakly to the scaling function (5) for the Haar wavelet. Figure 10.10 shows the sequences (37), (36) that correspond to the coefficients

$$M = 3, \quad c_0 = \frac{1 + \sqrt{3}}{4}, \quad c_1 = \frac{3 + \sqrt{3}}{4}, \quad c_2 = \frac{3 - \sqrt{3}}{4}, \quad c_3 = \frac{1 - \sqrt{3}}{4}. \tag{39}$$

In this case $\phi_1, \phi_2, \phi_3, \ldots$ converge weakly to the scaling function for a wavelet that was designed by Ingrid Daubechies in 1988.

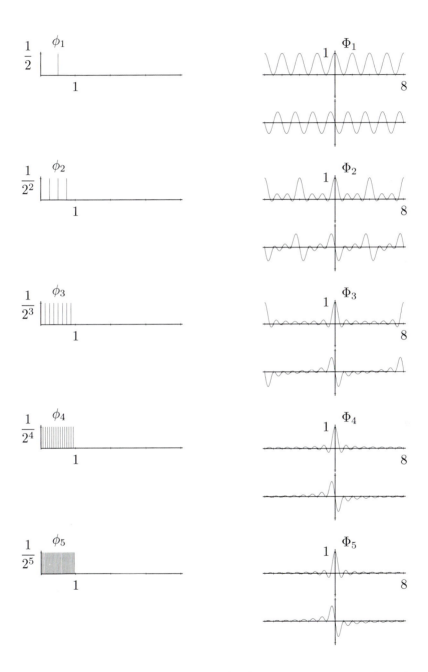

Figure 10.9. The functions ϕ_n (left) and Re Φ_n, Im Φ_n (right), $n = 1, 2, 3, 4, 5$, from the sequences (37), (36) that correspond to the Haar coefficients (32).

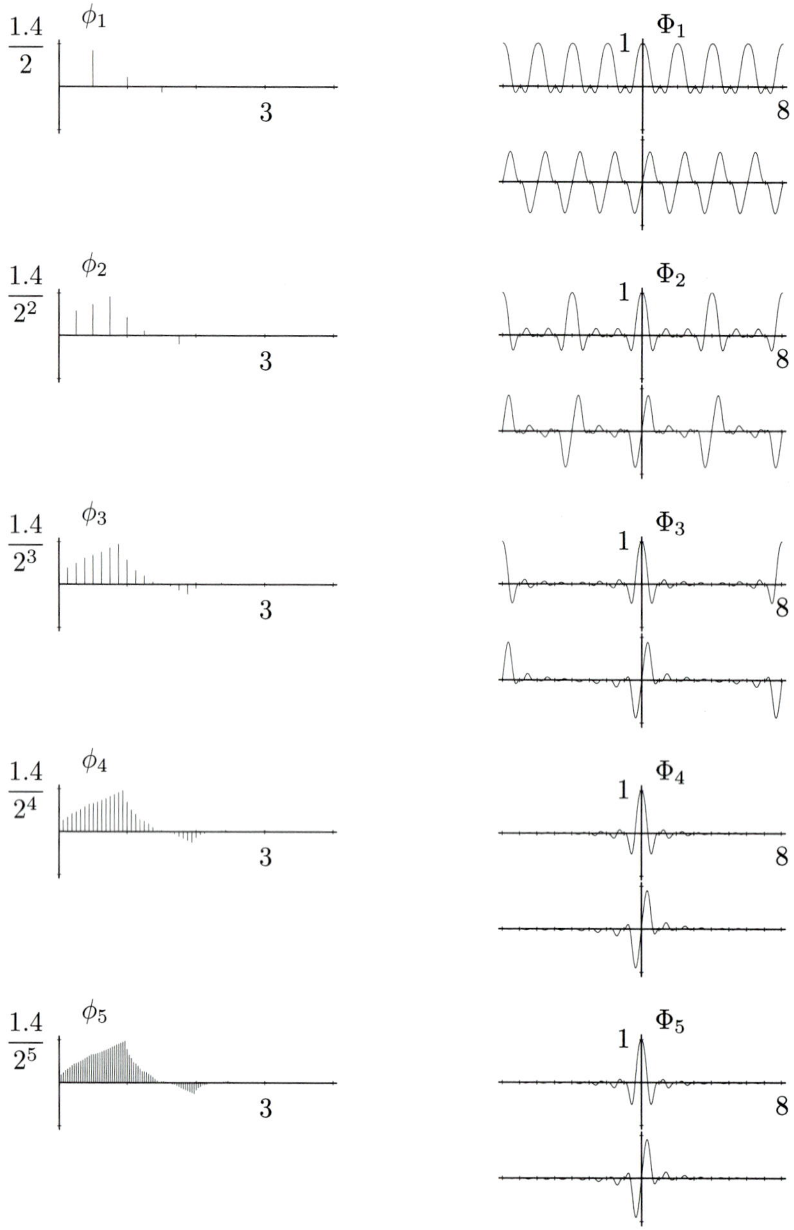

Figure 10.10. The functions ϕ_n (left) and Re Φ_n, Im Φ_n (right), $n = 1, 2, 3, 4, 5$, from the sequences (37), (36) that correspond to the Daubechies coefficients (39).

Example: Show that the sequences (37), (36) have the weak limits

$$\phi(x) = \Pi\left(x - \tfrac{1}{2}\right), \quad \Phi(s) = e^{-\pi i s}\operatorname{sinc}(s)$$

when we use the coefficients (32).

Solution: We use the trigonometric polynomial

$$T(s) = \tfrac{1}{2}\{1 + 1\cdot e^{-2\pi i s}\} = e^{-\pi i s}\cos(\pi s)$$

with (36) to write

$$\Phi_n(s) = T\left(\frac{s}{2}\right) T\left(\frac{s}{4}\right)\cdots T\left(\frac{s}{2^n}\right)$$
$$= e^{-\pi i s(1-2^{-n})}\cos\left(\frac{\pi s}{2}\right)\cos\left(\frac{\pi s}{4}\right)\cdots\cos\left(\frac{\pi s}{2^n}\right),$$

and thereby infer that

$$|\Phi_n(s)| \le 1 \text{ for } -\infty < s < \infty \text{ and } n = 1, 2, \ldots .$$

The cosine product telescopes when we use the identity

$$\cos(u) = \frac{\sin(2u)}{2\sin(u)}$$

to write

$$\Phi_n(s) = e^{-\pi i s(1-2^{-n})}\frac{\sin(\pi s)}{2\sin(\pi s/2)}\frac{\sin(\pi s/2)}{2\sin(\pi s/4)}\cdots\frac{\sin(\pi s/2^{n-1})}{2\sin(\pi s/2^n)}$$
$$= e^{-\pi i s}\operatorname{sinc}s\,\frac{e^{\pi i s/2^n}}{\operatorname{sinc}(s/2^n)}\,. \tag{40}$$

This enables us to see that $\Phi_1(s), \Phi_2(s), \Phi_3(s), \ldots$ converges uniformly to the function $e^{-\pi i s}\operatorname{sinc}(s)$ on every finite interval. Since these functions are uniformly bounded, the sequence has the weak limit

$$\lim_{n\to\infty} \Phi_n(s) = e^{-\pi i s}\operatorname{sinc}(s) =: \Phi(s).$$

Weak limits are preserved by the Fourier transform (and its inverse), so we also have

$$\lim_{n\to\infty} \phi_n(x) = \Pi\left(x - \tfrac{1}{2}\right) = \phi(x).$$

[An alternative proof within a probabilistic context follows (12.71).] Knowing ϕ, we use (31) with the coefficients (32) to produce the Haar wavelet (1). ∎

We will present an analogous argument to establish the weak convergence of (36), (37) when $c_0, c_1, \ldots, c_M$ satisfy (28). Let $A \geq 1$, $B > 0$ be chosen so that

$$|T(s)| \leq A, \quad 0 \leq s \leq 2\pi, \tag{41}$$
$$|T'(s)| \leq B, \quad 0 \leq s \leq 2\pi. \tag{42}$$

We use (28) and (42) with the identity

$$T(s) = T(0) + \int_0^s T'(\sigma)d\sigma$$

to see that

$$|T(s)| \leq 1 + B|s|, \quad -\infty < s < \infty. \tag{43}$$

We then majorize the products (36) by writing

$$|\Phi_n(s)| = \prod_{\nu=1}^n \left| T\left(\frac{s}{2^\nu}\right) \right| \leq \prod_{\nu=1}^n \left\{ 1 + \frac{B|s|}{2^\nu} \right\} \leq \prod_{\nu=1}^n e^{B|s|/2^\nu} \leq e^{B|s|}, \tag{44}$$
$$-\infty < s < \infty, \quad n = 1, 2, \ldots .$$

An argument of Daubechies allows us to replace the exponential envelope from (44) with one that is slowly growing. We use (44) to see that

$$|\Phi_n(s)| \leq e^B, \quad n = 1, 2, \ldots \text{ when } |s| \leq 1.$$

If $|s| > 1$, we will choose the positive integer $N = N(s)$ so that

$$2^{N-1} < |s| \leq 2^N$$

and consider in turn the cases where $n \leq N$, $n > N$ as we write

$$|\Phi_n(s)| = \prod_{\nu=1}^n \left| T\left(\frac{s}{2^\nu}\right) \right| \leq A^N \leq A^{1+\log_2 |s|} = A|s|^{\log_2 A}, \quad |s| > 1, \ n \leq N,$$

$$|\Phi_n(s)| = \left| \prod_{\nu=1}^N T\left(\frac{s}{2^\nu}\right) \right| \left| \prod_{\nu=1}^{n-N} T\left(\frac{s}{2^{N+\nu}}\right) \right| \leq A^N \left| \Phi_{n-N}\left(\frac{s}{2^N}\right) \right| \leq A|s|^{\log_2 A} e^B,$$

$$|s| > 1, \quad n > N.$$

We combine these three bounds for $\Phi_n(s)$ to obtain the envelope

$$|\Phi_n(s)| \leq C\{1 + |s|^{\log_2 A}\}, \quad -\infty < s < \infty, \quad n = 1, 2, \ldots \tag{45}$$

where $C := Ae^B$.

We now have the tools we need to show that $\Phi_1, \Phi_2, \Phi_3, \ldots$ converges both pointwise and weakly. Given $b > 0$ and positive integers $m < n$, we use (36), (44), (45) to write

$$
\begin{aligned}
|\Phi_n(s) - \Phi_m(s)| &= \left| \prod_{\nu=1}^{m} T\left(\frac{s}{2^\nu}\right) \right| \cdot \left| \prod_{\nu=1}^{n-m} T\left(\frac{s}{2^{m+\nu}}\right) - 1 \right| \\
&= |\Phi_m(s)| \, |\Phi_{n-m}(s/2^m) - 1| \\
&\leq C\{1 + |s|^{\log_2 A}\}\{e^{B|s|/2^m} - 1\} \\
&\leq C\{1 + b^{\log_2 A}\}\{e^{Bb/2^m} - 1\} \text{ when } |s| \leq b.
\end{aligned}
\tag{46}
$$

Since $e^{Bb/2^m} \to 1$ as $m \to \infty$, we can use this bound to infer that $\Phi_1(s), \Phi_2(s), \Phi_3(s), \ldots$ is a Cauchy sequence that converges uniformly on $[-b, b]$ to a continuous function Φ represented by the infinite product

$$
\Phi(s) := \prod_{\nu=1}^{\infty} T\left(\frac{s}{2^\nu}\right), \quad -\infty < s < \infty.
\tag{47}
$$

By using (45) we see that the limit function is slowly growing with

$$
|\Phi(s)| \leq C\{1 + |s|^{\log_2 A}\}, \quad -\infty < s < \infty.
\tag{48}
$$

It takes a little extra effort to establish the weak convergence. Let χ be a Schwartz function. (The symbol ϕ is reserved for the scaling function when we work with wavelets!) We use (46), (45), (48) to see that

$$
\begin{aligned}
\left| \int_{-\infty}^{\infty} \Phi(s)\chi(s)ds - \int_{-\infty}^{\infty} \Phi_n(s)\chi(s)ds \right| & \\
&\leq \int_{|s|\leq b} |\Phi(s) - \Phi_n(s)| \, |\chi(s)|ds + \int_{|s|\geq b} |\Phi(s) - \Phi_n(s)| \, |\chi(s)|ds \\
&\leq \max_{-\infty < s < \infty} |\chi(s)| \cdot C\{1 + b^{\log_2 A}\}\{e^{Bb/2^n} - 1\}2b \\
&\quad + 2C \int_{|s|\geq b} \{1 + |s|^{\log_2 A}\}|\chi(s)|ds.
\end{aligned}
$$

Now let $\epsilon > 0$ be given. Since $\{1 + |s|^{\log_2 A}\}|\chi(s)|$ rapidly approaches zero as $s \to \pm\infty$, the integral term will be less than $\epsilon/2$ when b is sufficiently large. For any such b, the first term will also be less than $\epsilon/2$ for all sufficiently large n. In this way we see that

$$
\lim_{n\to\infty} \Phi_n\{\chi\} = \Phi\{\chi\} \text{ when } \chi \in \mathbb{S},
\tag{49}
$$

and thereby prove that (47) is the weak limit of the sequence (36).

Three important properties of the generalized function

$$\phi := (\Phi)^{\wedge\vee} = \left(\lim_{n\to\infty} \Phi_n\right)^{\wedge\vee} \tag{50}$$

follow at once from (49). First, weak limits are preserved by reflection and Fourier transformation, so ϕ is the weak limit of the sequence $\phi_1, \phi_2, \phi_3, \ldots$ from (37), *i.e.*,

$$\lim_{n\to\infty} \phi_n\{\chi\} = \phi\{\chi\} \text{ when } \chi \in \mathbb{S}. \tag{51}$$

Second, weak limits are also preserved by dilation, translation, scaling, and addition, so we can use (50) and (33) to see that $\phi = \lim \phi_n$ satisfies the dilation equation (29). Third, each of the functions (37) vanishes outside the interval $[0, M]$, so the same must be true of the weak limit ϕ, *i.e.*,

$$\phi(x) = 0 \text{ for } x < 0 \text{ and for } x > M. \tag{52}$$

You may wish to show that (50) is the only support-limited solution of (29)–(30), *cf.* Ex. 10.9.

Smoothness constraints

The Fourier transform (47) of the solution of the dilation equation (29)–(30) is bandlimited, so Φ and each of its derivatives is continuous and slowly growing, *cf.* (7.130). In contrast, a ϕ produced from (28), (35), (47), and (50) is almost never an ordinary function much less one that is continuous or continuously differentiable. If we want ϕ to be a smooth ordinary function, we must choose coefficients $c_0, c_1, \ldots, c_M$ that make the infinite product (47) go to zero rapidly at $\pm\infty$, *cf.* Exs. 3.41, 3.42, and 12.14. We will now show how this is done.

The trigonometric polynomial (35) is 1-periodic with $T(0) = 1$, so we can use (47) to see that

$$\Phi(q\,2^m) = \prod_{\nu=1}^{m} T\left(\frac{q\,2^m}{2^\nu}\right) \prod_{\nu=1}^{\infty} T\left(\frac{q\,2^m}{2^{m+\nu}}\right) = T(0)^m\,\Phi(q) = \Phi(q) \tag{53}$$

when $q = \pm 1, \pm 2, \ldots$ and $m = 1, 2, \ldots$. From this identity we see that a necessary condition for $\Phi(s)$ to vanish in the limit as $s \to \pm\infty$ is that $\Phi(q) = 0$ for each $q = \pm 1, \pm 2, \ldots$. Now any nonzero integer q has the factorization $q = (2r + 1)2^m$ for some $r = 0, \pm 1, \pm 2, \ldots$ and $m = 0, 1, 2, \ldots$. We use this representation with (53), (47), and the 1-periodicity of T to see that

$$\Phi(q) = \Phi(2r + 1)$$
$$= T\left(\frac{2r+1}{2}\right) T\left(\frac{2r+1}{4}\right) T\left(\frac{2r+1}{8}\right) \cdots$$
$$= T\left(\frac{1}{2}\right) T\left(\frac{2r+1}{4}\right) T\left(\frac{2r+1}{8}\right) \cdots .$$

Every such product includes the factor $T(\frac{1}{2})$, so we can force Φ to vanish at every nonzero integer q by choosing $c_0, c_1, \ldots, c_M$ to make

$$T\left(\tfrac{1}{2}\right) = \frac{1}{2} \sum_{m=0}^{M} (-1)^m c_m = 0. \tag{54}$$

The constraint (54) is a necessary (but not sufficient) condition for ϕ to be piecewise smooth.

The trigonometric polynomial (35) vanishes at $s = \frac{1}{2}$ if and only if T has $1 + e^{-2\pi i s}$ as a factor, *i.e.*,

$$T(s) = \tfrac{1}{2}(1 + e^{-2\pi i s})T_1(s) \tag{55}$$

where

$$T_1(s) = \frac{1}{2} \sum_{m=0}^{M-1} c_{1m} e^{-2\pi i m s}$$

is a trigonometric polynomial of degree $M - 1$, *cf.* Ex. 10.14. This factorization of T produces a corresponding factorization of Φ. Indeed, we can use (47), (55), and (40) to write

$$\Phi(s) = \prod_{\nu=1}^{\infty} \left\{ \tfrac{1}{2}(1 + e^{-2\pi i s/2^\nu}) \cdot T_1\left(\frac{s}{2^\nu}\right) \right\}$$

$$\stackrel{?}{=} \left\{ \prod_{\nu=1}^{\infty} \tfrac{1}{2}(1 + e^{-2\pi i s/2^\nu}) \right\} \cdot \left\{ \prod_{\nu=1}^{\infty} T_1\left(\frac{s}{2^\nu}\right) \right\} \tag{56}$$

$$= e^{-\pi i s} \mathrm{sinc}(s) \prod_{\nu=1}^{\infty} T_1\left(\frac{s}{2^\nu}\right).$$

From (55) we see that $T_1(0) = 1$ when $T(0) = 1$, so our previous analysis shows that this infinite product of T_1's converges weakly. The remaining details associated with the factorization $\stackrel{?}{=}$ are left for Ex. 10.15.

The sinc factor on the right side of (56) approaches zero like $1/s$ when $s \to \pm\infty$. The bandlimited product of T_1's is slowly growing, and we can use (45) to write

$$\left| \prod_{\nu=1}^{\infty} T_1\left(\frac{s}{2^\nu}\right) \right| \leq C_1\{1 + |s|^{p_1}\}, \quad -\infty < s < \infty$$

where C_1 is a constant and

$$p_1 := \log_2\{ \max_{0 \leq s \leq 2\pi} |T_1(s)| \}.$$

If $p_1 < 1$, we can use this growth estimate with (56) to show that $\Phi(s) \to 0$ at least as fast as $|s|^{p_1-1}$ when $s \to \pm\infty$.

We will generalize (56). Suppose that

$$T\left(\tfrac{1}{2}\right) = T'\left(\tfrac{1}{2}\right) = \cdots = T^{(K-1)}\left(\tfrac{1}{2}\right) = 0 \tag{57}$$

for some $K = 1, 2, \ldots, M$, i.e., that the coefficients $c_0, c_1, \ldots, c_M$ satisfy the *moment constraints*

$$\sum_{m=0}^{M} (-1)^m m^k c_m = 0 \text{ for } k = 0, 1, \ldots, K-1. \tag{58}$$

The equivalent conditions (57), (58) hold if and only if T has the factorization

$$T(s) = \left\{ \tfrac{1}{2}(1 + e^{-2\pi i s}) \right\}^K T_K(s) \tag{59}$$

where

$$T_K(s) = \frac{1}{2} \sum_{m=0}^{M-K} c_{Km} e^{-2\pi i m s}$$

is a trigonometric polynomial of degree $M - K$ with $T_K(0) = 1$, *cf.* Ex. 10.14. In such a situation we can use (56) to see that

$$\Phi(s) = \{e^{-\pi i s} \text{sinc}(s)\} \prod_{\nu=1}^{\infty} T_1\left(\frac{s}{2^\nu}\right) = \{e^{-\pi i s} \text{sinc}(s)\}^2 \prod_{\nu=1}^{\infty} T_2\left(\frac{s}{2^\nu}\right)$$

$$\cdots = \{e^{-\pi i s} \text{sinc}(s)\}^K \prod_{\nu=1}^{\infty} T_K\left(\frac{s}{2^\nu}\right). \tag{60}$$

If the exponent

$$p_K := \log_2\{ \max_{0 \le s \le 2\pi} |T_K(s)| \} \tag{61}$$

is less than K, we can use (60) and the bound (45) (applied to T_K) to see that $\Phi(s) \to 0$ at least as fast as $|s|^{p_K - K}$ when $s \to \pm\infty$.

Example: Let ϕ be the scaling function that corresponds to the Daubechies coefficients (39), *cf.* Fig. 10.10. Show that ϕ is continuous.

Solution: With a bit of algebra we verify that

$$T(s) := \frac{1}{8}\{(1 + \sqrt{3}) + (3 + \sqrt{3})e^{-2\pi i s} + (3 - \sqrt{3})e^{-4\pi i s} + (1 - \sqrt{3})e^{-6\pi i s}\}$$

$$= \{\tfrac{1}{2}(1 + e^{-2\pi i s})\}^2 \cdot \tfrac{1}{2}\{(1 + \sqrt{3}) + (1 - \sqrt{3})e^{-2\pi i s}\}.$$

The factor

$$T_2(s) := \tfrac{1}{2}\{(1 + \sqrt{3}) + (1 - \sqrt{3})e^{-2\pi i s}\}$$

takes its maximum modulus $\sqrt{3}$ at $s = \frac{1}{2}$, so (61) gives the exponent $p_2 = \log_2 \sqrt{3} = .79248\ldots$, and $\Phi(s)$ decays at least as fast as $|s|^{p_2 - 2} = |s|^{-1.207\cdots}$ when $s \to \pm\infty$. In this way we see that the continuous function Φ is absolutely integrable, and thereby infer that ϕ is continuous, *cf.* Ex. 1.38. ∎

Example: Let $M = 1, 2, \ldots$ and let

$$c_m = \frac{2}{2^M} \binom{M}{m}, \quad m = 0, 1, \ldots, M. \tag{62}$$

Find the corresponding scaling function ϕ.

Solution: The coefficients (62) have been chosen so that

$$T(s) = \frac{1}{2^M} \sum_{m=0}^{M} \binom{M}{m} e^{-2\pi i m s} = \left\{ \tfrac{1}{2}(1 + e^{-2\pi i s}) \right\}^M.$$

We use (60) to see that

$$\Phi(s) = \left\{ e^{-\pi i s} \operatorname{sinc}(s) \right\}^M = e^{-2\pi i (M/2) s} \operatorname{sinc}^M(s)$$

and then use the Fourier transform calculus to write

$$\phi(x) = B_{M-1}(x - M/2). \tag{63}$$

The *B*-spline is defined by

$$B_0 := \Pi, \quad B_1 := \Pi * \Pi, \quad B_2 := \Pi * \Pi * \Pi, \quad \ldots,$$

cf. Ex. 2.7. The function

$$\phi^{(M-1)} = \Pi' * \Pi' * \cdots * \Pi' * \Pi \quad \text{(with } M \text{ factors)}$$

is piecewise constant and $\phi^{(k)}$ is continuous when $k = 0, 1, \ldots, M - 2$. ∎

We introduced the constraints (58) in an attempt to generate a smooth scaling function ϕ and a smooth wavelet ψ. We will now show that these constraints produce a wavelet ψ that has K vanishing moments. With this in mind, we define

$$R(s) := \frac{1}{2} \sum_{m=0}^{M} (-1)^m c_{M-m} e^{-2\pi i m s} \tag{64}$$

and use (35) to write

$$R(s) = \frac{1}{2} \sum_{\mu=0}^{M} (-1)^{M-\mu} c_\mu e^{-2\pi i (M-\mu) s} = (-1)^M e^{-2\pi i M s} T\left(\tfrac{1}{2} - s\right). \tag{65}$$

In this way we see that the K-fold zero of T at $s = \frac{1}{2}$ produces a K-fold zero of R at $s = 0$, *i.e.*, (58) implies that

$$R(0) = R'(0) = \cdots = R^{(K-1)}(0) = 0. \tag{66}$$

We Fourier transform (31) and use (64) to write

$$\Psi(s) = \frac{1}{2} \sum_{m=0}^{M-1} (-1)^m c_{M-m} e^{-2\pi i m s/2} \Phi\left(\frac{s}{2}\right) = R\left(\frac{s}{2}\right) \Phi\left(\frac{s}{2}\right). \tag{67}$$

Since Φ is bandlimited, the functions $\Phi, \Phi', \Phi'', \ldots$ are continuous, and we can use (66)–(67) to see that

$$\Psi(0) = \Psi'(0) = \cdots = \Psi^{(K-1)}(0) = 0 \tag{68}$$

when $c_0, c_1, \ldots, c_M$ satisfy (58). Of course, this K-fold zero of Ψ at $s = 0$ is equivalent to the moment conditions

$$\int_{-\infty}^{\infty} x^k \psi(x) dx = 0, \quad k = 0, 1, \ldots, K-1 \tag{69}$$

when these integrals are well defined, *e.g.*, as is the case when ϕ and therefore ψ is piecewise smooth. From (69) we see that every dilate of every translate of ψ is orthogonal to a given polynomial of degree $K-1$ or less, so all of the corresponding detail functions must vanish. [We cannot use the Haar wavelet (1) to reduce the error when we approximate a constant function!] Indeed, we will now show that any such polynomial can be synthesized from dyadic dilates of integer translates of the scaling function ϕ.

Order of approximation

Let the support-limited scaling function ϕ be obtained from (29)–(30) using coefficients $c_0, c_1, \ldots, c_M$ that satisfy both (28) and (58). From (60) we see that the Fourier transform has a factorization

$$\Phi(s) = \{\text{sinc}(s)\}^K G(s)$$

where G and each of its derivatives is continuous and slowly growing on $\mathbb{R}$. Since the sinc function vanishes at every nonzero integer, we can use the Leibnitz formula (2.29) to see that

$$\Phi(n) = \Phi'(n) = \cdots = \Phi^{(K-1)}(n) = 0 \quad \text{for each} \quad n = \pm 1, \pm 2, \ldots. \tag{70}$$

We will use this zero structure to develop a fundamental identity that makes it possible for us to analyze the quality of a wavelet approximation.

Let $\mathcal{P}$ be a polynomial of degree $K-1$ or less. Since $\mathcal{P}$ is bandlimited and ϕ is support-limited, the generalized function

$$f := \phi * (\mathcal{P} \cdot \text{III})$$

is well defined, and we can use properties of the weak limit to write

$$f(x) = \phi(x) * \sum_{k=-\infty}^{\infty} \mathcal{P}(x)\delta(x-k) = \phi(x) * \sum_{k=-\infty}^{\infty} \mathcal{P}(k)\delta(x-k) = \sum_{k=-\infty}^{\infty} \mathcal{P}(k)\phi(x-k).$$

From (7.82) and (70) we see that

$$\Phi(s)\delta^{(\nu)}(s-n) = 0 \quad \text{when} \quad \nu = 0,1,\ldots,K-1 \quad \text{and} \quad n = \pm 1, \pm 2, \ldots,$$

and thereby infer that

$$\Phi(s)\mathcal{P}^\wedge(s-n) = 0 \quad \text{when} \quad n = \pm 1, \pm 2, \ldots .$$

We use this relation with the definition of f to write

$$F(s) = \Phi(s) \cdot \left\{ \mathcal{P}^\wedge(s) * \sum_{n=-\infty}^{\infty} \delta(s-n) \right\} = \Phi(s) \cdot \sum_{n=-\infty}^{\infty} \mathcal{P}^\wedge(s-n)$$

$$= \sum_{n=-\infty}^{\infty} \Phi(s) \cdot \mathcal{P}^\wedge(s-n) = \Phi(s) \cdot \mathcal{P}^\wedge(s)$$

or equivalently,

$$f = \phi * \mathcal{P}.$$

We equate these expressions for f to obtain an identity

$$\sum_{k=-\infty}^{\infty} \mathcal{P}(k)\phi(x-k) = (\phi * \mathcal{P})(x) \quad \text{when} \quad \mathcal{P}^{(K)} \equiv 0 \qquad (71)$$

that uses samples of $\mathcal{P}$ to synthesize $\phi * \mathcal{P}$.

Now since $\Phi(0) \neq 0$, the convolution product $\mathcal{Q} = \phi * \mathcal{P}$ is a polynomial that has the same degree as $\mathcal{P}$, and we can solve this linear equation to find $\mathcal{P}$ when $\mathcal{Q}$ is given, cf. Ex. 10.17. This being the case, (71) shows that we can *exactly* represent any polynomial of degree $K-1$ or less with a linear combination of the integer translates of ϕ.

Example: Let the support-limited scaling function ϕ be constructed from coefficients $c_0, c_1, \ldots, c_M$ that satisfy (28), (58), and let

$$\mu_k := \frac{\Phi^{(k)}(0)}{(-2\pi i)^k} = \int_{-\infty}^{\infty} x^k \phi(x)dx, \quad k = 0, 1, 2, \ldots . \qquad (72)$$

[The derivatives of the bandlimited function Φ are always defined. We can use the integrals when ϕ is piecewise smooth, *cf.* (12.38) and (12.40).] Show that

$$\sum_{k=-\infty}^{\infty} \phi(x-k) = 1 \qquad \text{if } K \geq 1, \tag{73}$$

$$\sum_{k=-\infty}^{\infty} \{k+\mu_1\}\phi(x-k) = x \qquad \text{if } K \geq 2, \tag{74}$$

$$\sum_{k=-\infty}^{\infty} \{k^2 + 2\mu_1 k + [2\mu_1^2 - \mu_2]\}\phi(x-k) = x^2 \qquad \text{if } K \geq 3. \tag{75}$$

Solution: In view of (71) we must show that

$$\phi(x) * 1 = 1,$$
$$\phi(x) * \{x + \mu_1\} = x,$$
$$\phi(x) * \{x^2 + 2\mu_1 x + [2\mu_1^2 - \mu_2]\} = x^2$$

when $K \geq 1, 2, 3$, respectively. We work with the corresponding Fourier transforms, using (7.82), (30), (72) as we write

$$\Phi(s) \cdot \delta(s) = \Phi(0)\delta(s) = \delta(s),$$

$$\Phi(s) \cdot \left\{ \frac{\delta'(s)}{-2\pi i} + \mu_1 \delta(s) \right\} = -\frac{1}{2\pi i}\{\Phi(0)\delta'(s) - \Phi'(0)\delta(s)\} + \mu_1 \Phi(0)\delta(s) = \frac{-\delta'(s)}{2\pi i},$$

$$\Phi(s) \cdot \left\{ \frac{\delta''(s)}{(-2\pi i)^2} + 2\mu_1 \frac{\delta'(s)}{(-2\pi i)} + [2\mu_1^2 - \mu_2]\delta(s) \right\}$$

$$= \frac{1}{(-2\pi i)^2}\{\Phi(0)\delta''(s) - 2\Phi'(0)\delta'(s) + \Phi''(0)\delta(s)\}$$

$$- \frac{2\mu_1}{2\pi i}\{\Phi(0)\delta'(s) - \Phi'(0)\delta(s)\} + [2\mu_1^2 - \mu_2]\Phi(0)\delta(s) = \frac{\delta''(s)}{(-2\pi i)^2}.$$

An alternative analysis is given in Ex. 10.18, and Ex. 10.19 shows how to express $\mu_1, \mu_2, \ldots$ in terms of the coefficients $c_0, c_1, \ldots, c_M$. ∎

A polynomial f of degree $K-1$ or less can be synthesized from a linear combination of the translates $\phi(x-k)$, $k = 0, \pm 1, \pm 2, \ldots$. Since the scaling function ϕ vanishes when $x < 0$ or when $x > M$, at most M such translates contribute to the sum on any interval $(k, k+1)$ with integer endpoints. In particular, when $K \geq 1$ we can write

$$\phi(x-k) + \phi(x-k+1) + \cdots + \phi(x-k+M-1) = 1 \quad \text{for } k < x < k+1$$

even in cases where ϕ is not an ordinary function! We will now use these observations to study the error that results when the function f is piecewise polynomial or piecewise smooth.

Suppose that we wish to approximate a piecewise linear function f with a linear combination of the translates $\phi(x - k)$, $k = 0, \pm 1, \pm 2, \ldots$ of the scaling function that corresponds to the Daubechies coefficients (39), *cf.* Fig. 10.10. In this case $K = 2$, $M = 3$, and we can use (73), (74) to write

$$Ax + B = [A(k + \mu_1) + B]\phi(x - k) + [A(k - 1 + \mu_1) + B]\phi(x - k + 1)$$
$$+ [A(k - 2 + \mu_1) + B]\phi(x - k + 2) \quad \text{for} \quad k \le x \le k + 1 \tag{76}$$

where

$$\mu_1 := \int_0^3 x\,\phi(x)dx = \frac{3 - \sqrt{3}}{2} = .63397\ldots,$$

cf. Ex. 10.20. Figure 10.11 shows an approximation

$$f_0(x) = \sum_{k=-2}^{31} \alpha_0[k]\phi(x - k)$$

to a discontinuous broken line on $[0, 32]$. The function f has jumps at the knots $x = 4, 28$, and the derivative f' has jumps at the knots $x = 12, 20$. You will observe that f_0 coincides with f on the intervals $[0, 2]$, $[6, 10]$, $[14, 18]$, $[22, 26]$, and $[30, 32]$. The line segments (76) that lie to the left and to the right of a knot at $x = \kappa$ require different coefficients for the basis functions $\phi(x - k + 1)$, $\phi(x - k + 2)$ that straddle the knot, so we cannot expect f_0 to match f on intervals $|x - 4| < 2$, $|x - 12| < 2$, $|x - 20| < 2$, $|x - 28| < 2$ centered at the four knots.

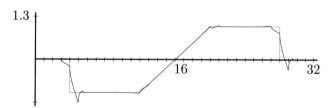

Figure 10.11. Approximation of a discontinuous broken line (with knots at $x = 4, 12, 20, 28$) using integer translates of the Daubechies scaling function ϕ from Fig. 10.10.

It is easy to generalize this observation. Let f be a piecewise polynomial function with $f^{(K)}(x) = 0$ except at certain knots $x = \kappa_1, \kappa_2, \ldots$. We will approximate f with the frame

$$f_0(x) = \sum_{k=-\infty}^{\infty} \alpha_0[k]\phi(x - k)$$

using a scaling function ϕ obtained from coefficients satisfying (28) and (58). In view of (71), we can choose coefficients $\alpha_0[k]$ to make f_0 coincide with f except on certain intervals that contain the knots. If the knot κ_ν is an integer, we must exclude the interval $|x - \kappa_\nu| < M - 1$ (as done in the preceding paragraph). If the knot κ_ν is not an integer, we must expand this error interval, but it is never necessary to exclude an interval larger than $|x - \kappa_\nu| < M$.

We can reduce the size of the error intervals by using a more accurate frame

$$f_m(x) = \sum_{k=-\infty}^{\infty} \alpha_m[k]\phi(2^m x - k) \tag{77}$$

with basis functions of width $M/2^m$, $m = 1, 2, \ldots$. It is now possible to choose the coefficients $\alpha_m[k]$ to make f_m coincide with f except on the intervals $|x - \kappa_\nu| < M/2^m$, $\nu = 1, 2, \ldots$ centered at the knots. The size of the error $f(x) - f_m(x)$ on the interval $|x - \kappa_\nu| < M/2^m$ depends on ϕ and the jumps in $f, f', \ldots$ at $x = \kappa_\nu$. (This is the wavelet analog of the Gibbs phenomenon for Fourier series!)

The accuracy index K determines how well a wavelet frame (77) can approximate a smooth function f. Indeed, if $f, f', \ldots, f^{(K-1)}$ are continuous on some interval $|x - x_0| < r$, we can use Taylor's formula

$$f(x) = f(x_0) + (x - x_0)\frac{f'(x_0)}{1!} + \cdots + \frac{(x - x_0)^{K-1}}{(K-1)!}f^{(K-1)}(x_0)$$
$$+ \frac{1}{(K-1)!}\int_{x_0}^{x} f^{(K)}(u)(x - u)^{K-1}du$$

to analyze the error, *cf.* Ex. 2.28. It is possible to choose coefficients $\alpha_m[k]$ to make (77) coincide with the Taylor polynomial in some neighborhood of $x = x_0$ provided that $M\,2^{-m} < r$. The remainder term from Taylor's formula gives the error in this approximation, so we have

$$|f(x) - f_m(x)| \le \frac{1}{(K-1)!}\int_{x_0}^{x} |f^{(K)}(u)(x - u)^{K-1}|du$$
$$\le C_K 2^{-mK} \quad \text{when} \ \ |x - x_0| < 2^{-m}.$$

Here C_K is a bound for $f^{(K)}(x)/K!$. The local error decays at least as fast as 2^{-mK}, $m = 1, 2, \ldots$ at points where f is sufficiently smooth.

Of course, we must expect a slower rate of decay when we relax the smoothness hypothesis. If $f, f', \ldots, f^{(L-1)}$ are continuous but $f^{(L)}$ has an isolated jump discontinuity at $x = x_0$, then we can produce a wavelet approximation (77) with a local error bounded by $C_L 2^{-mL}$ on some interval surrounding the jump when m is sufficiently large and $L = 1, 2, \ldots, K - 1$.

Orthogonality constraints

Throughout this subsection we will assume that the ϕ, ψ of (29)–(31) are piecewise continuous ordinary functions that vanish outside the interval $[0, M]$. We introduce this hypothesis to ensure that certain *inner product* integrals are well defined. We will also assume that the corresponding coefficients $c_0, c_1, \ldots, c_M$ satisfy (54). [Later on we will see that this additional condition follows from the normalization constraint (28) and certain orthogonality constraints (87) that are needed for the analysis.] These hypotheses are satisfied by the Daubechies scaling functions we will introduce presently, *cf.* Ex. 10.28.

We want to produce a scaling function ϕ that is orthogonal to its integer translates in the sense that

$$\int_{-\infty}^{\infty} \phi(x)\phi(x-k)dx = \begin{cases} 1 & \text{if } k = 0 \\ 0 & \text{if } k = \pm 1, \pm 2, \ldots \, . \end{cases} \tag{78}$$

We will deduce corresponding (necessary) conditions for the coefficients $c_0, c_1, \ldots, c_M$ and then obtain the complete set of *orthogonality relations*

$$\int_{-\infty}^{\infty} \phi(2^m x - k)\phi(2^m x - k')dx = \begin{cases} 2^{-m} & \text{if } k' = k \\ 0 & \text{otherwise,} \end{cases} \tag{79}$$

$$\int_{-\infty}^{\infty} \phi(2^m x - k)\psi(2^{m'} x - k')dx = 0 \qquad \text{if } m' \geq m, \tag{80}$$

$$\int_{-\infty}^{\infty} \psi(2^m x - k)\psi(2^{m'} x - k')dx = \begin{cases} 2^{-m} & \text{if } m' = m \text{ and } k' = k \\ 0 & \text{otherwise} \end{cases} \tag{81}$$

which make it possible for us to generalize the analysis from Section 10.1.

We use the 1-periodic auxiliary function

$$\mathcal{A}(s) := \sum_{m=-\infty}^{\infty} |\Phi(s+m)|^2, \quad -\infty < s < \infty, \tag{82}$$

to establish a connection between the orthogonality relations (78) and the coefficients $c_0, c_1, \ldots, c_M$. [In view of our hypothesis, both $\Phi(s)$ and $s\Phi(s)$ are bounded on $\mathbb{R}$, so this series converges uniformly on every finite interval.] We obtain the initial value

$$\mathcal{A}(0) = 1 \tag{83}$$

by setting $s = 0$ in (82) and using (30), (70). By using Parseval's identity we see that $\mathcal{A}$ has the Fourier coefficients

$$\int_0^1 \mathcal{A}(s)e^{-2\pi iks}ds = \int_0^1 \sum_{m=-\infty}^{\infty} |\Phi(s+m)|^2 e^{-2\pi iks}ds$$

$$= \sum_{m=-\infty}^{\infty} \int_0^1 |\Phi(s+m)|^2 e^{-2\pi ik(s+m)}ds$$

$$= \int_{-\infty}^{\infty} \overline{\Phi(s)}\,\Phi(s)e^{-2\pi iks}ds$$

$$= \int_{-\infty}^{\infty} \phi(x)\phi(x-k)dx, \quad k = 0, \pm 1, \pm 2, \ldots,$$

and since ϕ vanishes outside the interval $[0, M]$ we can write

$$\mathcal{A}(s) = \sum_{k=-(M-1)}^{M-1} \left\{ \int_0^M \phi(x)\phi(x-k)dx \right\} e^{2\pi iks}.$$

In this way we prove that ϕ satisfies (78) if and only if $\mathcal{A}(s) \equiv 1$.

The trigonometric polynomials $T(s)$ and $\mathcal{A}(s)$ are inextricably linked. Using the identity $\Phi(2s) = T(s)\Phi(s)$ from (47) with the 1-periodicity of T, we find

$$\mathcal{A}(2s) = \sum_{m=-\infty}^{\infty} |\Phi(2s+m)|^2$$

$$= \sum_{m=-\infty}^{\infty} \left| T\left(s+\frac{m}{2}\right) \Phi\left(s+\frac{m}{2}\right) \right|^2$$

$$= \sum_{\mu=-\infty}^{\infty} \left\{ \left| T\left(s+\frac{2\mu}{2}\right) \Phi\left(s+\frac{2\mu}{2}\right) \right|^2 \right. \tag{84}$$

$$\left. + \left| T\left(s+\frac{2\mu+1}{2}\right) \Phi\left(s+\frac{2\mu+1}{2}\right) \right|^2 \right\}$$

$$= |T(s)|^2 \mathcal{A}(s) + \left| T\left(s+\tfrac{1}{2}\right) \right|^2 \mathcal{A}\left(s+\tfrac{1}{2}\right).$$

In particular, if ϕ satisfies the orthogonality relations (78), then $\mathcal{A}(s) \equiv 1$ and

$$|T(s)|^2 + \left| T\left(s+\tfrac{1}{2}\right) \right|^2 = 1, \quad -\infty < s < \infty. \tag{85}$$

Example: Find T and $\mathcal{A}$ when we choose the coefficients

$$M = 3, \quad c_0 = 1, \quad c_1 = 0, \quad c_2 = 0, \quad c_3 = 1. \tag{86}$$

Solution: We have

$$T(s) := \frac{1}{2} \sum_{m=0}^{M} c_m e^{-2\pi i m s} = \frac{1}{2}\{1 + e^{-6\pi i s}\} = e^{-3\pi i s} \cos(3\pi s),$$

and by using (40) (with s replaced by $3s$ on the right side) we see that

$$\Phi(s) = e^{-3\pi i s} \operatorname{sinc}(3s).$$

Knowing Φ, we find in turn

$$\phi(x) = \frac{1}{3}\Pi\left(\frac{x - 3/2}{3}\right),$$

$$\int_0^3 \phi(x)\phi(x-k)dx = \frac{1}{9}\begin{cases} 3 & \text{if } k = 0 \\ 2 & \text{if } k = \pm 1 \\ 1 & \text{if } k = \pm 2 \\ 0 & \text{if } k = \pm 3, \pm 4, \dots, \end{cases}$$

$$\mathcal{A}(s) = \frac{1}{9}\{e^{-4\pi i s} + 2e^{-2\pi i s} + 3 + 2e^{2\pi i s} + e^{4\pi i s}\}$$

$$= \frac{1}{9}\{3 + 4\cos(2\pi s) + 2\cos(4\pi s)\}.$$

You will observe that

$$|T(s)|^2 + \left|T\left(s + \tfrac{1}{2}\right)\right|^2 = \cos^2(3\pi s) + \sin^2(3\pi s) = 1,$$

but $\mathcal{A}(s) \not\equiv 1$ in this case. ∎

We will use (35) to express (85) in terms of the coefficients $c_0, c_1, \dots, c_M$. We set $c_m := 0$ when $m < 0$ or $m > M$ so that we can use $\pm\infty$ for the limits of summation. We display the summation index, but condense the notation by omitting the limits. For our present purpose, we write

$$|T(s)|^2 + \left|T\left(s + \tfrac{1}{2}\right)\right|^2 = \frac{1}{4}\sum_{m=0}^{M}\sum_{\mu=0}^{M}\{c_m e^{2\pi i m s} \cdot c_\mu e^{-2\pi i \mu s}$$

$$+ c_m e^{2\pi i m(s+1/2)} \cdot c_\mu e^{-2\pi i \mu(s+1/2)}\}$$

$$= \frac{1}{4}\sum_{m}\sum_{\mu} c_m c_\mu [1 + (-1)^{m-\mu}] e^{2\pi i (m-\mu)s}$$

$$= \frac{1}{4}\sum_{m}\sum_{\kappa} c_m c_{m-\kappa} [1 + (-1)^{\kappa}] e^{2\pi i \kappa s}$$

$$= \frac{1}{2}\sum_{k}\left\{\sum_{m} c_m c_{m-2k}\right\} e^{4\pi i k s}$$

by using the transformations $\kappa := m - \mu$, $\kappa := 2k$ and exchanging the summation order. In this way we see that (85) is equivalent to the *orthogonality constraints*

$$\sum_m c_m c_{m-2k} = \begin{cases} 2 & \text{if } k = 0 \\ 0 & \text{if } k = \pm 1, \pm 2, \dots \end{cases} \tag{87}$$

We can now summarize the main result of this analysis. If we want ϕ to satisfy (78), then we must choose coefficients $c_0, c_1, \dots, c_M$ that satisfy (87)!

Unfortunately, this necessary condition is not sufficient. The coefficients (86) satisfy the constraints (87), but the corresponding scaling function is not orthogonal to its translates by $\pm 1, \pm 2$. If you examine the above argument you will quickly identify the problem: $\mathcal{A}(s) \equiv 1$ implies but is not implied by (85). In practice, we almost always have

$$T(s) \neq 0 \quad \text{for} \quad -1/2 < s < 1/2, \tag{88}$$

and we can use this hypothesis with (85) to prove that $\mathcal{A}(s) \equiv 1$, *cf.* Ex. 10.24. If we choose coefficients $c_0, c_1, \dots, c_M$ that satisfy (28), (54), (87), if the corresponding T satisfies (88), and if the corresponding ϕ is piecewise continuous, then ϕ will be orthogonal to its translates by $\pm 1, \pm 2, \dots$.

The $k = 0$ constraint from (87) serves to normalize $c_0, c_1, \dots, c_M$. We will verify that this normalization is compatible with that of (28).

Example: Let $c_0, c_1, \dots, c_M$ satisfy the orthogonality constraints

$$\sum_m c_m c_{m-2k} = 0 \quad \text{for} \quad k = \pm 1, \pm 2, \dots .$$

Show that any two of the conditions

$$\sum_m c_m = \pm 2, \quad \sum_m (-1)^m c_m = 0, \quad \sum_m c_m^2 = 2$$

[from (28), (54), and (87)] implies the third.

Solution:

$$\left\{ \sum_m c_m \right\}^2 + \left\{ \sum_m (-1)^m c_m \right\}^2 = \sum_m \{ c_m c_m + (-1)^m c_m (-1)^m c_m \}$$

$$+ 2 \sum_m \{ c_m c_{m+1} + (-1)^m c_m (-1)^{m+1} c_{m+1} \}$$

$$+ 2 \sum_m \{ c_m c_{m+2} + (-1)^m c_m (-1)^{m+2} c_{m+2} \}$$

$$+ \cdots$$

$$= 2 \sum_m c_m^2 + 4 \sum_m c_m c_{m+2} + 4 \sum_m c_m c_{m+4} + \cdots = 2 \sum_m c_m^2. \quad \blacksquare$$

You will notice that when M is even, (87) includes the constraint

$$c_M c_0 = c_M c_{M-2(M/2)} = 0.$$

Since we routinely use c_0 for the first *nonzero* coefficient, we will always take $M = 1, 3, 5, \ldots$ when we choose coefficients that satisfy (87).

We will show that the orthogonality constraints

$$\sum_m c_m \cdot (-1)^m c_{M-m-2k} = 0 \quad \text{if } k = 0, \pm 1, \pm 2, \ldots \tag{89}$$

are satisfied automatically.

Example: Let $M = 1, 3, 5, \ldots$. Show that any $c_0, c_1, \ldots, c_M$ and the corresponding *alternating flip* $c_M, -c_{M-1}, c_{M-2}, \ldots, -c_0$ satisfy (89).

Solution: Using the transformation $\mu := M - m - 2k$ we write

$$\sum_m (-1)^m c_m c_{M-m-2k} = \sum_\mu (-1)^{M-\mu-2k} c_{M-\mu-2k} c_\mu = -\sum_\mu (-1)^\mu c_\mu c_{M-\mu-2k}. \quad \blacksquare$$

Given (78), (87), and (89) it is a simple matter to establish (79)–(81). In so doing, it is always sufficient to consider the cases with $m = 0$ and $k = 0$. For example, by setting $u := 2^m x - k$ and writing

$$\int_{-\infty}^{\infty} \phi(2^m x - k)\phi(2^m x - k')dx = 2^{-m} \int_{-\infty}^{\infty} \phi(u)\phi[u - (k' - k)]du$$

we see that (79) follows from (78).

We show that ϕ is orthogonal to ψ and its translates by using (29), (31), (79), and (89) as we verify that

$$\int_{-\infty}^{\infty} \phi(x)\psi(x - k)dx = \int_{-\infty}^{\infty} \sum_m c_m \phi(2x - m) \sum_\mu (-1)^\mu c_{M-\mu} \phi(2x - 2k - \mu)dx$$

$$= \sum_m \sum_\mu c_m (-1)^\mu c_{M-\mu} \int_{-\infty}^{\infty} \phi(2x - m)\phi(2x - 2k - \mu)dx$$

$$= \sum_m \sum_\mu c_m (-1)^\mu c_{M-\mu} \begin{cases} 1/2 & \text{if } m = \mu + 2k \\ 0 & \text{otherwise} \end{cases}$$

$$= \frac{1}{2} \sum_m c_m (-1)^m c_{M-m+2k} = 0.$$

Using (29), we then write in turn

$$\int_{-\infty}^{\infty} \phi(x)\psi(2x-k)dx = \sum_m c_m \int_{-\infty}^{\infty} \phi(2x-m)\psi(2x-k)dx = 0,$$

$$\int_{-\infty}^{\infty} \phi(x)\psi(4x-k)dx = \sum_m c_m \int_{-\infty}^{\infty} \phi(2x-m)\psi(4x-k)dx = 0,$$

$$\vdots$$

and thereby obtain (80).

Analogously, we use (31), (79), and (87) to see that

$$\int_{-\infty}^{\infty} \psi(x)\psi(x-k)dx = \sum_m \sum_\mu (-1)^m c_{M-m}(-1)^\mu c_{M-\mu} \int_{-\infty}^{\infty} \phi(2x-m)\phi(2x-2k-\mu)dx$$

$$= \sum_m \sum_\mu (-1)^{m+\mu} c_{M-m}c_{M-\mu} \begin{cases} 1/2 & \text{if } m = \mu + 2k \\ 0 & \text{otherwise} \end{cases}$$

$$= \frac{1}{2}\sum_m c_{M-m}c_{M-m+2k}$$

$$= \begin{cases} 1 & \text{if } k = 0 \\ 0 & \text{if } k = \pm 1, \pm 2, \dots \end{cases}.$$

We then use (31) and (80) to write

$$\int_{-\infty}^{\infty} \psi(x)\psi(2^{m'}x-k)dx = \sum_m (-1)^m c_{M-m} \int_{-\infty}^{\infty} \phi(2x-m)\psi(2^{m'}x-k)dx$$

$$= 0 \quad \text{when} \quad m' = 1, 2, \dots$$

and thereby complete the proof of (81).

The dilation identities (10) and (11) that link the Haar wavelet (1) and its scaling function (5) give rise to Mallat's clever herringbone algorithm as illustrated in Fig. 10.7 and Fig. 10.8. You will recognize (29), (31) as natural generalizations of (10). We will now identify the corresponding generalizations of (11).

Example: Let $c_0, c_1, \dots, c_M$ satisfy (28) and the orthogonality constraints (87), and let ϕ, ψ be the support-limited scaling function and wavelet from (29)–(31). Show that

$$\phi(2x) = \frac{1}{2}\sum_{k=0}^{\lfloor M/2\rfloor} \{c_{2k}\phi(x+k) + c_{M-2k}\psi(x+k)\}, \tag{90}$$

$$\phi(2x-1) = \frac{1}{2}\sum_{k=0}^{\lfloor M/2\rfloor} \{c_{2k+1}\phi(x+k) - c_{M-2k-1}\psi(x+k)\}. \tag{91}$$

Solution: We use (29), (31) and the transformation $m = 2k + \mu$ to write

$$\sum_k \{c_{2k}\phi(x+k) + c_{M-2k}\psi(x+k)\}$$

$$= \sum_k \sum_m \{c_{2k}c_m + (-1)^m c_{M-2k}c_{M-m}\}\phi(2x + 2k - m)$$

$$= \sum_\mu \sum_k \{c_{2k}c_{2k+\mu} + (-1)^\mu c_{M-2k-\mu}c_{M-2k}\}\phi(2x - \mu).$$

If μ is even, then $M - 2k - \mu$ is odd (since M is odd) and we can use (87) to write

$$\sum_k \{c_{2k}c_{2k+\mu} + c_{M-2k-\mu}c_{M-2k}\} = \sum_k \{c_{2k}c_{2k+\mu} + c_{2k+1}c_{2k+1+\mu}\}$$

$$= \begin{cases} 2 & \text{if } \mu = 0 \\ 0 & \text{if } \mu = \pm 2, \pm 4, \dots \end{cases}.$$

If μ is odd, then $M - 2k - \mu$ is even and

$$\sum_k \{c_{2k}c_{2k+\mu} - c_{M-2k-\mu}c_{M-2k}\} = \sum_k \{c_{2k}c_{2k+\mu} - c_{2k}c_{2k+\mu}\}$$

$$= 0, \quad \mu = \pm 1, \pm 3, \dots .$$

In this way we verify (90). A similar argument gives (91). ∎

Daubechies' wavelets

For each $M = 1, 3, 5, \dots$ there are coefficients $c_0, c_1, \dots, c_M$ that are normalized by (28) [so we can solve the dilation equation (29)], that satisfy the moment constraints (58) with $K := (M + 1)/2$ [so we can exactly represent any polynomial of degree $K - 1$ or less using $\phi(x - k)$, $k = 0, \pm 1, \pm 2, \dots$], and that satisfy the orthogonality constraints (87) [a necessary condition for the functions $\phi(x - k)$, $k = 0, \pm 1, \pm 2, \dots$ to be pairwise orthogonal]. It is enough to use $k = 1, 2, \dots, K - 1$ with (87) (since $c_m = 0$ when $m < 0$ or $m > M$ and since the $k = 0$ normalization is implied by the other constraints). Thus we must find $c_0, c_1, \dots, c_M$ by solving a system of $1 + K + (K - 1) = M + 1$ equations. We will show how this can be done.

Example: Use (28), (58) to find c_0, c_1 when $K = 1$, $M = 1$.

Solution: We solve the linear equations

$$c_0 + c_1 = 2$$
$$c_0 - c_1 = 0$$

associated with the normalization and moment constraints to obtain the Haar coefficients (32). There are no additional orthogonality constraints in this case. ∎

Example: Use (28), (58), and (87) to find c_0, c_1, c_2, c_3 when $K = 2$, $M = 3$.

Solution: We must solve the system of equations

$$
\begin{aligned}
c_0 + \quad c_1 + \quad c_2 + \quad c_3 &= 2 \\
c_0 - \quad c_1 + \quad c_2 - \quad c_3 &= 0 \\
0\,c_0 - \quad 1\,c_1 + \quad 2\,c_2 - \quad 3\,c_3 &= 0 \\
c_0 c_2 + c_1 c_3 \qquad\qquad &= 0.
\end{aligned}
$$

From (59) we see that the three linear equations are satisfied when

$$
\tfrac{1}{2}\{c_0 + c_1 e^{-2\pi i s} + c_2 e^{-4\pi i s} + c_3 e^{-6\pi i s}\} = \left\{\tfrac{1}{2}(1 + e^{-2\pi i s})\right\}^2 \left\{\tfrac{1}{2}(\gamma_0 + \gamma_1 e^{-2\pi i s})\right\},
$$

i.e., when

$$
c_0 = \frac{\gamma_0}{4}, \quad c_1 = \frac{2\gamma_0 + \gamma_1}{4}, \quad c_2 = \frac{\gamma_0 + 2\gamma_1}{4}, \quad c_3 = \frac{\gamma_1}{4} \tag{92}
$$

for certain real numbers γ_0, γ_1 with $\gamma_0 + \gamma_1 = 2$ (to make (92) satisfy (28).) We use the nonlinear orthogonality constraint with (92) and this sum to write

$$
0 = \gamma_0(\gamma_0 + 2\gamma_1) + (2\gamma_0 + \gamma_1)\gamma_1 = \gamma_0(2 + \gamma_1) + (\gamma_0 + 2)\gamma_1 = 4 + 2\gamma_0\gamma_1,
$$

and thereby see that $\gamma_0\gamma_1 = -2$. From the quadratic polynomial

$$
(\gamma - \gamma_0)(\gamma - \gamma_1) = \gamma^2 - (\gamma_0 + \gamma_1)\gamma + \gamma_0\gamma_1 = \gamma^2 - 2\gamma - 2
$$

we find $\gamma_0 = 1 + \sqrt{3}$, $\gamma_1 = 1 - \sqrt{3}$, and by using these roots with (92) we obtain the Daubechies coefficients (39). You may wish to analyze the effect of choosing $\gamma_0 = 1 - \sqrt{3}$, $\gamma_1 = 1 + \sqrt{3}$ by solving Ex. 10.12. ∎

A more sophisticated analysis is needed when $M = 2K - 1$ with $K = 3, 4, \ldots$. Using (59) (as in the preceding example) we see that the coefficients $c_0, c_1, \ldots, c_M$ satisfy the linear constraints (28), (58) if and only if the trigonometric polynomial (35) has the factorization

$$
T(s) = \left\{\tfrac{1}{2}(1 + e^{-2\pi i s})\right\}^K \Gamma(s) \tag{93}
$$

where

$$
\Gamma(s) = \tfrac{1}{2}\{\gamma_0 + \gamma_1 e^{-2\pi i s} + \gamma_2 e^{-4\pi i s} + \cdots + \gamma_{K-1} e^{-2\pi i (K-1)s}\}, \tag{94}
$$

and the K real coefficients $\gamma_0, \gamma_1, \ldots, \gamma_{K-1}$ have the sum 2 so that $T(0) = \Gamma(0) = 1$. We define

$$
Q(s) := |T(s)^2| = T(s)T(-s) = \{\cos(\pi s)\}^{2K}\Gamma(s)\Gamma(-s), \tag{95}
$$

and observe that Q is a real, nonnegative, even, 1-periodic trigonometric polynomial of degree $M = 2K - 1$ with

$$Q^{(k)}\left(\tfrac{1}{2}\right) = 0 \quad \text{for} \quad k = 0, 1, \ldots, 2K - 1. \tag{96}$$

We want the coefficients of T to satisfy the orthogonality constraints (87), so we can use the equivalent identity (85) to write

$$Q(s) = 1 - Q\left(s + \tfrac{1}{2}\right). \tag{97}$$

In conjunction with (96) this enables us to see that

$$Q(0) = 1, \quad Q^{(k)}(0) = 0 \quad \text{for} \quad k = 1, 2, \ldots, 2K - 1. \tag{98}$$

The conditions (96), (98) uniquely determine the 1-periodic trigonometric polynomial Q. Indeed, since Q' is odd, since Q' has degree $2K - 1$, and since Q' has zeros of multiplicity at least $2K - 1$ at $s = 0$ and $s = \tfrac{1}{2}$, it follows that

$$Q'(s) = -I^{-1}\{\sin(2\pi s)\}^{2K-1}$$

for some constant $I \neq 0$. Since $Q(0) = 1$ and $Q(1/2) = 0$, we must have

$$Q(s) = 1 - I^{-1} \int_0^s \{\sin(2\pi u)\}^{2K-1} du \tag{99}$$

where

$$I = \int_0^{1/2} \{\sin(2\pi u)\}^{2K-1} du = \frac{1}{\pi}\frac{2}{3}\frac{4}{5}\frac{6}{7} \cdots \frac{2K-2}{2K-1}, \quad K = 2, 3, \ldots .$$

(We use the Wallis formula from calculus to evaluate this integral.) Figure 10.12 shows that the Q we obtain when $K = 3$ and $M = 5$. You can see the symmetry (97)!

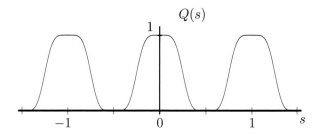

Figure 10.12. The maxflat trigonometric polynomial (93) that satisfies (96), (98) when $K = 3$, $M = 5$.

For the subsequent analysis we need an alternative formula. We will show that

$$Q(s) = \{\cos^2(\pi s)\}^K \sum_{k=0}^{K-1} \binom{2K-1}{k} \{\cos^2(\pi s)\}^{K-1-k} \{\sin^2(\pi s)\}^k. \tag{100}$$

By using the familiar identities

$$\cos^2(\pi s) = \{1 + \cos(2\pi s)\}/2, \quad \sin^2(\pi s) = \{1 - \cos(2\pi s)\}/2$$

it is easy to see that (100) is a real, nonnegative, even, 1-periodic trigonometric polynomial of degree $M = 2K - 1$ that satisfies (96). Moreover, since $\sin^2(\pi s)$ and $\cos^2(\pi s)$ are interchanged when we replace s by $s + \frac{1}{2}$, we also have

$$Q(s) + Q\left(s + \tfrac{1}{2}\right) = \{\cos^2(\pi s)\}^K \sum_{k=0}^{K-1} \binom{2K-1}{k} \{\cos^2(\pi s)\}^{K-1-k} \{\sin^2(\pi s)\}^k$$

$$+ \{\sin^2(\pi s)\}^K \sum_{k=0}^{K-1} \binom{2K-1}{k} \{\sin^2(\pi s)\}^{K-1-k} \{\cos^2(\pi s)\}^k$$

$$= \sum_{k=0}^{2K-1} \binom{2K-1}{k} \{\cos^2(\pi s)\}^{2K-1-k} \{\sin^2(\pi s)\}^k$$

$$= \{\cos^2(\pi s) + \sin^2(\pi s)\}^{2K-1} = 1.$$

In this way we verify that (100) satisfies (97) and thereby establish the equivalence of (99), (100).

We will now remove the factor $\{\cos^2(\pi s)\}^K$ from both (95) and (100). The remaining task is to find coefficients $\gamma_0, \gamma_1, \ldots, \gamma_{K-1}$ for (94) so that the reduced trigonometric polynomial

$$Q_0(s) := \sum_{k=0}^{K-1} \binom{2K-1}{k} \{\cos^2(\pi s)\}^{K-1-k} \{\sin^2(\pi s)\}^k \tag{101}$$

has the *spectral factorization*

$$Q_0(s) = \Gamma(s)\Gamma(-s). \tag{102}$$

To simplify the exposition, we will consider the case where $K = 3$, $M = 5$, and

$$Q_0(s) = \sum_{k=0}^{2} \binom{5}{k} \{\cos^2(\pi s)\}^{2-k} \{\sin^2(\pi s)\}^k \tag{103}$$

$$= \cos^4(\pi s) + 5\cos^2(\pi s)\sin^2(\pi s) + 10\sin^4(\pi s).$$

The analysis extends at once to cases where $K = 4, 5, \ldots$.

The function (103) is even, strictly positive, and 1-periodic. By using Euler's formula and a bit of algebra we verify that

$$Q_0(s) = \frac{1}{8}\{38 - 18[e^{2\pi i s} + e^{-2\pi i s}] + 3[e^{4\pi i s} + e^{-4\pi i s}]\}.$$

We replace the complex exponential $e^{2\pi i s}$ by z to obtain the auxiliary function

$$\mathcal{P}_0(z) = \frac{1}{8}\{38 - 18[z + z^{-1}] + 3[z^2 + z^{-2}]\}$$

with $\mathcal{P}_0(1) = Q_0(0) = 1$. Since Q_0 is strictly positive, $\mathcal{P}_0$ has no root z with $|z| = 1$. From symmetry we see that if z is a root of $\mathcal{P}_0$, then so is z^{-1}. It follows that $\mathcal{P}_0$ has the factorization

$$\mathcal{P}_0(z) = \frac{z^{-2}(z - \varsigma_1)(z - \varsigma_2)(z - \varsigma_1^{-1})(z - \varsigma_2^{-1})}{(1 - \varsigma_1)(1 - \varsigma_2)(1 - \varsigma_1^{-1})(1 - \varsigma_2^{-1})}$$

$$= \left\{\frac{(1 - \varsigma_1 z^{-1})(1 - \varsigma_2 z^{-1})}{(1 - \varsigma_1)(1 - \varsigma_2)}\right\}\left\{\frac{(1 - \varsigma_1 z)(1 - \varsigma_2 z)}{(1 - \varsigma_1)(1 - \varsigma_2)}\right\} \tag{104}$$

where ς_1, ς_2 are the complex roots of the polynomial $z^2\mathcal{P}_0(z)$ that lie *inside* the unit circle $|z| = 1$. In view of (104) the trigonometric polynomial

$$\Gamma(s) := \frac{(1 - \varsigma_1 e^{-2\pi i s})(1 - \varsigma_2 e^{-2\pi i s})}{(1 - \varsigma_1)(1 - \varsigma_2)}$$

gives us a spectral factorization (102).

We can use Newton's method to compute the roots

$$\zeta_1 = .28725\ldots + .15289\ldots i, \quad \zeta_2 = .28725\ldots - .15289\ldots i$$

of $\mathcal{P}_0$, and then expand the factored form of Γ to obtain the explicit display

$$\Gamma(s) = \frac{1}{2}\{3.76373\ldots - 2.16227\ldots e^{-2\pi is} + .39853\ldots e^{-4\pi is}\}$$

of the coefficients $\gamma_0, \gamma_1, \gamma_2$ from (94). Finally, we convolve the string $\gamma_0, \gamma_1, \gamma_2$ with the weighted binomial coefficients $1/8$, $3/8$, $3/8$, $1/8$ [*i.e.*, we use (93)] to produce the Daubechies coefficients

$$\begin{aligned}
c_0 &= .47046\ldots, & c_1 &= 1.14111\ldots, & c_2 &= .65036\ldots, \\
c_3 &= -.19093\ldots, & c_4 &= -.12083\ldots, & c_5 &= .04981\ldots
\end{aligned} \tag{105}$$

that satisfy (28), (58) (with $K = 3$) and (87).

It is easy to understand the above numerical analysis, but it is somewhat cumbersome to produce a corresponding computer code that will work with an arbitrary K. We will describe a much more efficient algorithm that uses the FFT. From (102), (104) and the Maclaurin series

$$-\log(1 - w) = w + \frac{w^2}{2} + \frac{w^3}{3} + \cdots, \quad |w| < 1$$

[with a complex argument, *cf.* (4.25), (4.26) and Ex. 4.7], we obtain the Fourier series

$$\log Q_0(s) = \log\{\Gamma(s)\Gamma(-s)\}$$

$$= -\log\{(1 - \zeta_1)(1 - \zeta_2)\} - \sum_{k=1}^{\infty} \frac{\zeta_1^k + \zeta_2^k}{k} e^{-2\pi iks}$$

$$- \log\{(1 - \zeta_1)(1 - \zeta_2)\} - \sum_{k=1}^{\infty} \frac{\zeta_1^k + \zeta_2^k}{k} e^{+2\pi iks}.$$

The Fourier coefficients for $\log\Gamma(s)$, $\log\Gamma(-s)$ appear in the left, right half of the spectrum. These series converge rapidly, so the approximation

$$\log Q_0\left(\frac{n}{N}\right) \approx -\log\{(1 - \zeta_1)(1 - \zeta_2)\} - \sum_{k=1}^{\lfloor (N-1)/2 \rfloor} \left(\frac{\zeta_1^k + \zeta_2^k}{k}\right)' e^{-2\pi ikn/N}$$

$$- \log\{(1 - \zeta_1)(1 - \zeta_2)\} - \sum_{k=1}^{\lfloor (N-1)/2 \rfloor} \left(\frac{\zeta_1^k + \zeta_2^k}{k}\right) e^{2\pi ikn/N},$$

$$n = 0, 1, \ldots, N - 1$$

will be good to working precision when N is sufficiently large. (Since $|\zeta_1| = |\zeta_2| = .325\ldots$, you can verify that the modulus of the error does not exceed 10^{-16} when $N \geq 64$.)

Figure 10.13 will help you understand how this observation makes it possible for us to compute $\gamma_0, \gamma_1, \gamma_2$. We use (103) to create a real, even vector with components

$$\log Q_0 \left(\frac{n}{N}\right), \quad n = 0, 1, \ldots, N-1,$$

and then use the FFT to compute the corresponding real, even discrete Fourier transform. We discard the left half of this DFT (*i.e.*, we divide the index 0 component by 2 and replace components having the indices $\lfloor N/2 \rfloor, \ldots, N-1$ by 0), thereby producing a very accurate approximation for the discrete Fourier transform of the vector with components

$$\log \Gamma \left(\frac{-n}{N}\right), \quad n = 0, 1, \ldots, N-1.$$

We use the FFT to compute the inverse transform and apply the exponential function componentwise to obtain a very accurate approximation for the vector with components

$$\Gamma \left(\frac{-n}{N}\right) = \frac{1}{2} \sum_{m=0}^{2} \gamma_m e^{2\pi imn/N}, \qquad n = 0, 1, \ldots, N-1.$$

A final application of the FFT gives $\gamma_0/2$, $\gamma_1/2$, $\gamma_2/2$. Knowing $\gamma_0, \gamma_1, \gamma_2$, we compute the corresponding coefficients $c_0, c_1, \ldots, c_5$ as described above. An alternative scheme that avoids the final convolution is described in Ex. 10.27. It is easy to write a computer code to carry out this calculation using an arbitrary $K = 1, 2, \ldots$. You may need to experiment a bit to determine a suitable N. The choice $N \approx 20K$ is a good place to start.

You now know how to compute coefficients $c_0, c_1, \ldots, c_M$ that satisfy (28), (58) [with $K = (M+1)/2$], and (87) when $M = 1, 3, 5, \ldots$ is given. Knowing these coefficients you can use (37) as illustrated in Fig. 10.10 (or the more efficient algorithms from Ex. 10.40) to find the corresponding scaling function ϕ and then use (31) to generate the wavelet ψ that goes with this ϕ. Figure 10.14 shows the Daubechies functions ϕ, ψ for $M = 1, 3, 5, 7$. These functions are continuous when $M = 3, 5, 7, \ldots$, *cf.* Ex. 10.28.

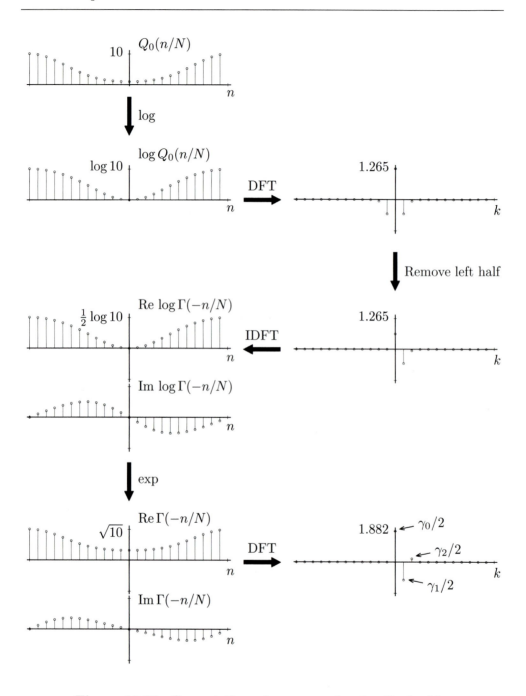

Figure 10.13. Computation of $\gamma_0, \gamma_1, \gamma_2$ for the Daubechies wavelet with $M = 5$. The trigonometric polynomial Q_0 is given by (103) and (for illustrative purposes) we use $N = 24$ samples.

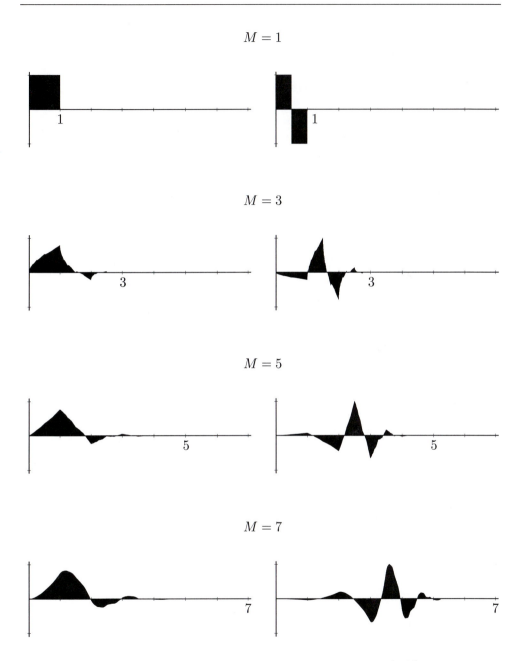

Figure 10.14. The Daubechies scaling function ϕ (left) and wavelet ψ (right) for $M = 1, 3, 5, 7$.

10.3 Analysis and Synthesis with Daubechies' Wavelets

Coefficients for frames and details

We have expended a great deal of effort constructing coefficients $c_0, c_1, \ldots, c_M$ that give rise to support-limited functions ϕ, ψ satisfying the two scale relations (29), (31), (90), (91), the moment conditions (69), and the orthogonality relations (79)–(81). Such functions can be used in much the same way as the Haar wavelet (1) and its scaling function (5) from Section 10.1. We will now fill in some of the details associated with this generalization.

Given a suitably regular function f on $\mathbb{R}$, we define corresponding *frames*

$$f_m(x) := \sum_k \alpha_m[k]\phi(2^m x - k), \quad m = 0, \pm 1, \pm 2, \ldots, \tag{106}$$

and *details*

$$d_m(x) := \sum_k \beta_m[k]\psi(2^m x - k), \quad m = 0, \pm 1, \pm 2, \ldots, \tag{107}$$

using the coefficients

$$\alpha_m[k] := \frac{\int_{-\infty}^{\infty} f(x)\phi(2^m x - k)dx}{\int_{-\infty}^{\infty} \phi(2^m x - k)^2 dx} = \int_0^M f(2^{-m}k + 2^{-m}u)\phi(u)du, \tag{108}$$

$$\beta_m[k] := \frac{\int_{-\infty}^{\infty} f(x)\psi(2^m x - k)dx}{\int_{-\infty}^{\infty} \psi(2^m x - k)^2 dx} = \int_0^M f(2^{-m}k + 2^{-m}u)\psi(u)du. \tag{109}$$

We can use (106)–(109) with any piecewise continuous function f on $\mathbb{R}$. There may be infinitely many nonzero terms in the series (106)–(107), but only finitely many contribute to the values taken by the sums on any bounded interval, *cf.* Ex. 10.29. In practice, f is almost always support-limited, so only finitely many terms of these series are nonzero.

Figure 10.15 shows certain frames (106) and details (107) that correspond to a normal density f and the jagged functions ϕ, ψ constructed from the Daubechies coefficients (39). Powers of 4 have been applied to $d_4, d_5, \ldots, d_8$ so we can show all of these graphs with a common scale. You should compare these $M = 3$ approximations with the corresponding $M = 1$ approximations from Fig. 10.5.

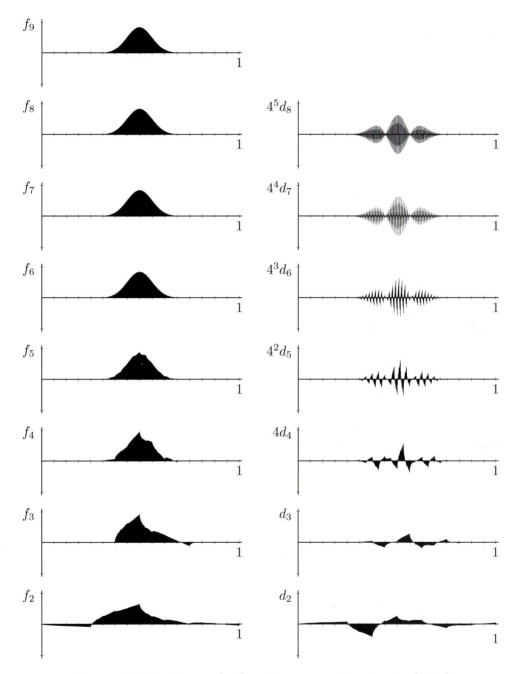

Figure 10.15. Frames (left) and corresponding details (right) for a Daubechies wavelet approximation of a normal density using the coefficients (39). Observe that $d_4, d_5, \ldots, d_8$ have been scaled by $4, 4^2, \ldots, 4^5$.

There are three things you should observe about f_m and d_m. First, the coefficients (108) and (109) have been chosen to make f_m and d_m the *orthogonal projections* of f onto the linear spaces spanned by the pairwise orthogonal functions $\phi(2^m x - k)$, $k = 0, \pm 1, \pm 2, \ldots$ and $\psi(2^m x - k)$, $k = 0, \pm 1, \pm 2, \ldots$, respectively. Second, f_m is *orthogonal* to d_m in the sense that

$$\int_{-\infty}^{\infty} f_m(x) d_m(x) dx = 0, \quad m = 0, \pm 1, \pm 2, \ldots \tag{110}$$

(provided that this integral is well defined). You can verify this by using the orthogonality relations (80) with the representations (106), (107). Third,

$$f_{m+1} = f_m + d_m, \quad m = 0, \pm 1, \pm 2, \ldots . \tag{111}$$

This important identity makes it possible for us to perform a *multiresolution analysis* using ψ and ϕ in place of the prototypes (1) and (5). You can verify (111) by using the two scale identities (29), (31), (90), and (91), cf. Ex. 10.30.

The coefficients (108)–(109) are determined by the values f takes on the interval $[2^{-m}k, 2^{-m}(k + M)]$. When m is large, this interval is small and simple estimates produce good approximations for the integrals. Indeed, if f is continuous in some neighborhood of $x = 2^{-m}k$, then we can use (108) with (30) to see that

$$\alpha_m[k] = \int_{-\infty}^{\infty} f(2^{-m}k + 2^{-m}u)\phi(u)du \approx f(2^{-m}k)\int_0^M \phi(u)du = f(2^{-m}k) \tag{112}$$

when m is sufficiently large. Analogously, if $f^{(K)}$ is continuous in a neighborhood of $x = 2^{-m}k$, then we can use (109) with Taylor's formula and (69) to see that

$$\begin{aligned}
\beta_m[k] &= \int_{-\infty}^{\infty} \left\{ \sum_{n=0}^{K-1} f^{(n)}(2^{-m}k)\frac{(2^{-m}u)^n}{n!} \right. \\
&\quad + \left. \frac{1}{(K-1)!}\int_0^{2^{-m}u} f^{(K)}(2^{-m}k + v)(2^{-m}u - v)^{K-1}dv \right\} \psi(u)du \\
&\approx \frac{f^{(K)}(2^{-m}k)}{(K-1)!}\int_0^M \psi(u)\int_0^{2^{-m}u}(2^{-m}u - v)^{K-1}dv\,du \\
&= 2^{-mK}\frac{f^{(K)}(2^{-m}k)}{K!}\int_0^M u^K\psi(u)du
\end{aligned} \tag{113}$$

when m is sufficiently large. Figures 10.5 and 10.15 use $K = 1$ and $K = 2$, respectively. You can see the frame coefficient approximation (112) on the left sides of these illustrations, and you can observe the 2^{-mK} decay of the detail coefficients [as predicted by (113)] on the right.

It is easy to bound the error in the frame approximation $f \approx f_m$. We will assume that f is a continuous, support-limited function on $\mathbb{R}$. This guarantees that the *modulus of continuity*

$$\omega(h) := \max\{|f(x) - f(x')| : x, x' \in \mathbb{R}, \ |x - x'| \le h\}, \quad h \ge 0, \qquad (114)$$

is well defined and that $\omega(h) \to 0$ as $h \to 0+$. When $0 \le x \le 2^{-m}$, we use (106), (73), (30), (108), in turn to write

$$|f(x) - f_m(x)| = \left| f(x) - \sum_k \alpha_m[k] \phi(2^m x - k) \right|$$

$$= \left| \sum_k \{f(x) - \alpha_m[k]\} \phi(2^m x - k) \right|$$

$$\le \max |\phi| \cdot \sum_{k=-M+1}^{0} \left| \int_{u=0}^{M} [f(x) - f(2^{-m}k + 2^{-m}u)] \phi(u) du \right|$$

$$\le \max |\phi| \cdot M \cdot \omega(2^{-m}M) \cdot \int_0^M |\phi(u)| du.$$

The argument immediately extends to the case where x lies in the interval $[k\, 2^{-m}, (k+1)2^{-m}]$, $k = 0, \pm 1, \pm 2, \ldots$, and in this way we prove that

$$|f(x) - f_m(x)| \le M \cdot \max |\phi| \cdot \int_0^M |\phi(u)| du \cdot \omega(2^{-m}M), \quad -\infty < x < \infty, \quad (115)$$

cf. (9). You can use (115) to show that it is possible to replace the Haar wavelet (1) by the Daubechies wavelet ψ for some $M = 3, 5, 7, \ldots$ when we use the synthesis equation (4) and the analysis equation (17), *cf.* Ex. 10.31.

If f has some additional smoothness, *e.g.*, if $f^{(p)}$ is continuous and bounded for some $p = 1, 2, \ldots, K$, then we can sharpen the bound (115). From (107) and a slight refinement of (113) we see that

$$|d_m(x)| = \left| \sum_k \beta_m[k] \psi(2^m x - k) \right| \le M \cdot \max |\psi| \cdot \max_k |\beta_m[k]| \le C_p\, 2^{-mp} \quad (116)$$

where

$$C_p := M \cdot \max |\psi| \cdot \int_0^M u^p |\psi(u)| du \cdot \frac{\max |f^{(p)}|}{p!}.$$

We use (111) and (116) with $n = m + 1, m + 2, \ldots$ to write

$$|f(x) - f_m(x)| = |f(x) - f_n(x) + d_m(x) + d_{m+1}(x) + \cdots + d_{n-1}(x)|$$

$$\le |f(x) - f_n(x)| + C_p\{2^{-mp} + 2^{-(m+1)p} + 2^{-(m+2)p} + \cdots\}$$

$$\le |f(x) - f_n(x)| + 2C_p 2^{-mp}.$$

Using (115) we see that $|f(x) - f_n(x)| \to 0$ as $n \to \infty$, so

$$|f(x) - f_m(x)| \leq 2M \cdot \max |\psi| \cdot \int_0^M u^p |\psi(u)| du \cdot \frac{\max |f^{(p)}|}{p!} \cdot 2^{-mp},$$

$$-\infty < x < \infty . \quad (117)$$

The error in the frame approximation (106) will decay at least as fast as 2^{-mK} when we use the smoothness index $K = 1, 2, \ldots$ to construct ϕ, ψ provided that f is sufficiently smooth!

The operators $\mathbf{L_-}$, $\mathbf{H_-}$, $\mathbf{L_+}$, $\mathbf{H_+}$

In practice, we never use the *functions* ϕ, ψ and we never use the *integrals* (108)–(109) for computation. Serious numerical analysis is done with the *coefficients* $\alpha_m[k]$, $\beta_m[k]$ that encode the frames and details (106)–(107). The algorithms are based on a suitable generalization of Mallat's relations (23), (24) for the Haar wavelets. We will now derive these extensions using four linear operators constructed from the coefficients $c_0, c_1, \ldots, c_M$.

We begin by expressing $\alpha_{m-1}[k]$ and $\beta_{m-1}[k]$ in terms of the coefficients $\alpha_m[n]$, $n = 0, \pm 1, \pm 2, \ldots$, *cf.* (24). Using (108), (29) and the transformations $v := 2u - \mu$, $n := 2k + \mu$ we find

$$\alpha_{m-1}[k] := \int_{-\infty}^{\infty} f(2^{-m+1}k + 2^{-m+1}u)\phi(u)du$$

$$= \sum_{\mu} c_\mu \int_{-\infty}^{\infty} f(2^{-m} \cdot 2k + 2^{-m} \cdot 2u)\phi(2u - \mu)du$$

$$= \frac{1}{2} \sum_{\mu} c_\mu \int_{-\infty}^{\infty} f[2^{-m} \cdot (2k + \mu) + 2^{-m}v]\phi(v)dv$$

$$= \frac{1}{2} \sum_{\mu} c_\mu \alpha_m[2k + \mu]$$

$$= \frac{1}{2} \sum_{n} c_{n-2k} \alpha_m[n].$$

Analogously, by using (109) and (31) we find

$$\beta_{m-1}[k] := \int_{-\infty}^{\infty} f(2^{-m+1}k + 2^{-m+1}u)\psi(u)du$$

$$= \sum_{\mu} (-1)^\mu c_{M-\mu} \int_{-\infty}^{\infty} f(2^{-m} \cdot 2k + 2^{-m} \cdot 2u)\phi(2u - \mu)du$$

$$= \frac{1}{2} \sum_{n} (-1)^n c_{M-n+2k} \alpha_m[n].$$

We introduce the linear operators

$$(\mathbf{L}_-\alpha)[k] := \frac{1}{2}\sum_n c_{n-2k}\alpha[n], \qquad\qquad k = 0, \pm 1, \pm 2, \dots, \qquad (118)$$

$$(\mathbf{H}_-\alpha)[k] := \frac{1}{2}\sum_n (-1)^n c_{M-n+2k}\alpha[n], \quad k = 0, \pm 1, \pm 2, \dots \qquad (119)$$

on the linear space of sequences. (Here α, $\mathbf{L}_-\alpha$, and $\mathbf{H}_-\alpha$ are functions on $\mathbb{Z}$.) We then condense the above notation and write

$$\alpha_{m-1} = \mathbf{L}_-\alpha_m, \qquad m = 0, \pm 1, \pm 2, \dots, \qquad (120)$$

$$\beta_{m-1} = \mathbf{H}_-\alpha_m, \qquad m = 0, \pm 1, \pm 2, \dots. \qquad (121)$$

In this way we obtain the desired generalization of Mallat's relations (24).

We can also express $\alpha_m[k]$ in terms of the coefficients $\alpha_{m-1}[n], \beta_{m-1}[n]$, $n = 0, \pm 1, \pm 2, \dots$, cf. (23). Indeed, by using the defining identity (108) with (111), the series (106), (107), the two scale equations (29), (31), and the orthogonality relations (78) we see that

$$\alpha_m[k] = \int_{-\infty}^{\infty} f_m(2^{-m}k + 2^{-m}u)\phi(u)du$$

$$= \int_{-\infty}^{\infty} \{f_{m-1}(2^{-m}k + 2^{-m}u) + d_{m-1}(2^{-m}k + 2^{-m}u)\}\phi(u)du$$

$$= \int_{-\infty}^{\infty} \Big\{\sum_n \alpha_{m-1}[n]\phi(2^{-1}k + 2^{-1}u - n)$$

$$+ \sum_n \beta_{m-1}[n]\psi(2^{-1}k + 2^{-1}u - n)\Big\}\phi(u)du$$

$$= \sum_n \sum_\mu \{c_\mu\alpha_{m-1}[n] + (-1)^\mu c_{M-\mu}\beta_{m-1}[n]\}\int_{-\infty}^{\infty}\phi(u + k - 2n - \mu)\phi(u)du$$

$$= \sum_n \{c_{k-2n}\alpha_{m-1}[n] + (-1)^k c_{M-k+2n}\beta_{m-1}[n]\}.$$

We introduce the linear operators

$$(\mathbf{L}_+\alpha)[k] := \sum_n c_{k-2n}\alpha[n], \qquad\qquad k = 0, \pm 1, \pm 2, \dots, \qquad (122)$$

$$(\mathbf{H}_+\beta)[k] := \sum_n (-1)^k c_{M-k+2n}\beta[n], \quad k = 0, \pm 1, \pm 2, \dots \qquad (123)$$

on the linear space of sequences. (Here α, $\mathbf{L}_+\alpha$, β, and $\mathbf{H}_+\beta$ are functions on $\mathbb{Z}$.) In this way we obtain the desired generalization

$$\alpha_m = \mathbf{L}_+\alpha_{m-1} + \mathbf{H}_+\beta_{m-1} \tag{124}$$

of Mallat's relations (23).

We use the operators (118)–(119), (122)–(123) when we move down, up the herringbones on the left, right of Fig. 10.16. Each of the coefficients $c_0, c_1, \dots, c_M$ appears once in each of the sums (118), (119), so we expend $M + 1$ operations per component as we generate $\alpha_{m-1}, \beta_{m-1}$ from α_m. Exactly half of these coefficients appear in each of the sums (122), (123), so we also expend $M + 1$ operations per component as we generate α_m from α_{m-1} and β_{m-1}.

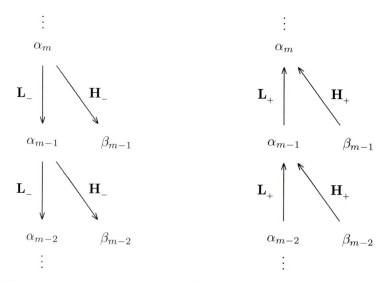

Figure 10.16. The operators (118)–(119) are used with Mallat's herringbone (left) when we split a frame into its constituent details. The operators (122)–(123) are used with the reverse herringbone (right) to reassemble a frame from its details.

The operators (118), (119), (122), and (123) that we construct from the Daubechies coefficients $c_0, c_1, \dots, c_M$ satisfy the identities

$$\mathbf{L}_+\mathbf{L}_- + \mathbf{H}_+\mathbf{H}_- = \mathbf{I}, \tag{125}$$

$$\mathbf{L}_-\mathbf{L}_+ = \mathbf{I}, \qquad \mathbf{H}_-\mathbf{L}_+ = \mathbf{0}, \tag{126}$$

$$\mathbf{L}_-\mathbf{H}_+ = \mathbf{0}, \qquad \mathbf{H}_-\mathbf{H}_+ = \mathbf{I}. \tag{127}$$

We will give simple arguments based on (120), (121), and (124). Straightforward (but somewhat tedious) manipulations of the sums can be used to show that

(125)–(127) hold whenever $c_0, c_1, \ldots, c_M$ satisfy the orthogonality constraints (87), *cf.* Ex. 10.33. The piecewise smoothness of ϕ [that was used to prove (120), (121), (124)] is not required.

Example: Use (120), (121), and (124) to establish (125).

Solution: Let α_m be an arbitrary function on $\mathbb{Z}$. [The series (106) gives the corresponding piecewise smooth function on $\mathbb{R}$.] We use (120), (121) to define

$$\alpha_{m-1} := \mathbf{L}_-\alpha_m, \qquad \beta_{m-1} := \mathbf{H}_-\alpha_m,$$

and then use (124) to recover α_m by writing

$$\begin{aligned}
\alpha_m &= \mathbf{L}_+\alpha_{m-1} + \mathbf{H}_+\beta_{m-1} \\
&= \mathbf{L}_+(\mathbf{L}_-\alpha_m) + \mathbf{H}_+(\mathbf{H}_-\alpha_m) \\
&= (\mathbf{L}_+\mathbf{L}_- + \mathbf{H}_+\mathbf{H}_-)\alpha_m.
\end{aligned}$$
∎

Example: Use (120), (121), and (124) to establish (126), (127).

Solution: Let α_{m-1} be an arbitrary function on $\mathbb{Z}$, and let β_{m-1} be the zero function on $\mathbb{Z}$. We use (124) to define

$$\alpha_m := \mathbf{L}_+\alpha_{m-1} + \mathbf{H}_+0,$$

and then use (120), (121) to recover

$$\alpha_{m-1} = \mathbf{L}_-\alpha_m \; = \mathbf{L}_-\mathbf{L}_+\alpha_{m-1},$$
$$0 =: \beta_{m-1} = \mathbf{H}_-\alpha_m \; = \mathbf{H}_-\mathbf{L}_+\alpha_{m-1}.$$

In this way we prove (126). A similar argument (with an arbitrary β_{m-1} and with $\alpha_{m-1} := 0$) gives (127). ∎

We use the operators $\mathbf{L}_+$, $\mathbf{H}_+$ when we want to replace the basis functions $\phi(2^m x - k)$, $\psi(2^m x - k)$, $k = 0, \pm1, \pm2, \ldots$ from the series (106)–(107) with the "narrower" basis functions $\phi(2^{m+p}x - k)$, $k = 0, \pm1, \pm2, \ldots$ associated with a more accurate frame f_{m+p}, $p = 1, 2, \ldots$.

Example: Let α, β be functions on $\mathbb{Z}$ and let $m = 0, \pm1, \pm2, \ldots$. Show that

$$\sum_k \alpha[k]\phi(2^m x - k) = \sum_k (\mathbf{L}_+^p\alpha)[k]\phi(2^{m+p}x - k), \quad p = 1, 2, \ldots, \tag{128}$$

$$\sum_k \beta[k]\psi(2^m x - k) = \sum_k (\mathbf{L}_+^{p-1}\mathbf{H}_+\beta)[k]\phi(2^{m+p}x - k), \quad p = 1, 2, \ldots . \tag{129}$$

Solution: Let $\alpha_m := \alpha$, $\beta_m := 0$ so that (124) gives $\alpha_{m+1} = \mathbf{L}_+\alpha$. We use these coefficients with the series (106)–(107) for the corresponding functions from (111) and thereby write

$$\sum_k (\mathbf{L}_+\alpha)[k]\phi(2^{m+1}x - k) = \sum_k \alpha[k]\phi(2^m x - k).$$

In this way we can see that (128) holds when $p = 1$. It follows that

$$\sum_k (\mathbf{L}_+^p\alpha)[k]\phi(2^{m+p}x - k) = \sum_k (\mathbf{L}_+^{p-1}\alpha)[k]\phi(2^{m+p-1}x - k), \quad p = 1, 2, \dots,$$

so (128) holds for every $p = 1, 2, \dots$.

An analogous argument with $\alpha_m := 0$, $\beta_m := \beta$, $\alpha_{m+1} = \mathbf{H}_+\beta$ shows that (129) holds when $p = 1$. We then use (128) to see that (129) also holds when $p = 2, 3, \dots$. ∎

We use the operators $\mathbf{L}_-, \mathbf{H}_-$ when we want to replace the basis functions $\phi(2^m x - k)$, $k = 0, \pm 1, \pm 2, \dots$ from (106) with the "wider" functions $\psi(2^{m-1}x - k)$, $\psi(2^{m-2}x - k), \dots, \psi(2^{m-p}x - k)$, $\phi(2^{m-p}x - k)$, $k = 0, \pm 1, \pm 2, \dots, p = 1, 2, \dots$.

Example: Let α be a function on $\mathbb{Z}$, and let $p = 1, 2, \dots$. Show that

$$\sum_k \alpha[k]\phi(2^m x - k) = \sum_k (\mathbf{L}_-^p\alpha)[k]\phi(2^{m-p}x - k)$$

$$+ \sum_{n=1}^{p}\sum_k (\mathbf{H}_-\mathbf{L}_-^{n-1}\alpha)[k]\psi(2^{m-n}x - k).$$

(130)

Solution: We set $\alpha_m := \alpha$ and use (120)–(121) to write

$$\begin{aligned}
\alpha_{m-1} &= \mathbf{L}_-\alpha, & \beta_{m-1} &= \mathbf{H}_-\alpha, \\
\alpha_{m-2} &= \mathbf{L}_-^2\alpha, & \beta_{m-2} &= \mathbf{H}_-\mathbf{L}_-\alpha, \\
\alpha_{m-3} &= \mathbf{L}_-^3\alpha, & \beta_{m-3} &= \mathbf{H}_-\mathbf{L}_-^2\alpha, \\
&\ \ \vdots & & \\
\alpha_{m-p} &= \mathbf{L}_-^p\alpha, & \beta_{m-p} &= \mathbf{H}_-\mathbf{L}_-^{p-1}\alpha,
\end{aligned}$$

cf. Fig. 10.16. We use these expressions as we replace each function from

$$f_m = f_{m-p} + d_{m-p} + d_{m-p+1} + \cdots + d_{m-1}$$

by the corresponding series (106) or (107), and thereby obtain (130). ∎

Samples for frames and details

We often have occasion to produce samples of a frame or detail associated with the wavelet analysis of some function f, *e.g.*, as shown in Fig. 10.15. As a rule, we cannot use the defining relations (106)–(109) for this purpose since we do not have a simple formula for evaluating ϕ and ψ. [The Haar prototypes (5), (1) are notable exceptions!] Instead, we use the operators $\mathbf{L}_-$, $\mathbf{H}_-$, $\mathbf{L}_+$, $\mathbf{H}_+$ with Mallat's herringbones from Fig. 10.16 to generate good approximate samples. We will illustrate the process using the gaussian

$$f(x) := e^{-128(x-1/2)^2} \tag{131}$$

and the Daubechies coefficients (39) with $M = 3$. Graphs of the corresponding ϕ, ψ are shown in Fig. 10.14.

We begin by computing the samples $f(2^{-8}k)$, $k = 0, \pm 1, \pm 2, \ldots$. The top central illustration from Fig. 10.17 shows the points

$$(2^{-8}k, f(2^{-8}k)), \quad k = 0, \pm 1, \pm 2, \ldots .$$

The graph of f is well represented by the broken line through these points. (The modulus of the error is bounded by .0005.) The approximation

$$\alpha_8[k] \approx f(2^{-8}k), \quad k = 0, \pm 1, \pm 2, \ldots$$

from (112) is somewhat less accurate, but you may not notice the difference on the scale used for the illustration. (The modulus of this error is bounded by .024, *cf.* Ex. 10.35.)

From (120) we see that

$$\alpha_7 = \mathbf{L}_- \alpha_8, \ \alpha_6 = \mathbf{L}_- \alpha_7, \ \ldots$$

so we can generate approximate coefficients for the frames $f_7, f_6, \ldots$ by applying $\mathbf{L}_-, \mathbf{L}_-^2, \ldots$ to the approximate coefficients α_8 for the frame f_8. Plots of the corresponding points

$$(2^{-8+n}k, (\mathbf{L}_-^n \alpha_8)[k]), \quad k = 0, \pm 1, \pm 2, \ldots$$

are shown in the center (and lower left) of Fig. 10.17 for each $n = 0, 1, 2, 3, 4$. There are approximately 64, 32, 16, 8, 4 significant ordinates when $n = 0, 1, 2, 3, 4$, respectively. We *halve* the number of significant ordinates each time we apply $\mathbf{L}_-$.

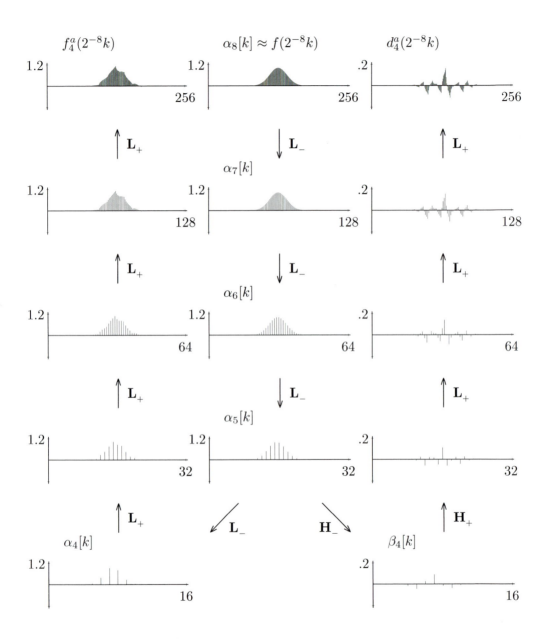

Figure 10.17. Construction of approximate samples f_4^a, d_4^a for the frame f_4 and the detail d_4 when f is a gaussian and ϕ, ψ correspond to the Daubechies coefficients (39).

The operator $\mathbf{L}_+$ has the opposite effect. From (128) we see that

$$f_4(x) = \sum_k \alpha_4[k]\phi(2^4 x - k) = \sum_k (\mathbf{L}_+\alpha_4)[k]\phi(2^5 x - k)$$

$$= \sum_k (\mathbf{L}_+^2 \alpha_4)[k]\phi(2^6 x - k) = \cdots .$$

Plots of the corresponding points

$$\left(2^{-4-n}k, (\mathbf{L}_+^n \alpha_4)[k]\right), \quad k = 0, \pm 1, \pm 2, \ldots$$

are shown on the left side of Fig. 10.17 for each $n = 0, 1, 2, 3, 4$. There are approximately 4, 8, 16, 32, 64 significant ordinates when $n = 0, 1, 2, 3, 4$, respectively. We *double* the number of significant ordinates each time we apply $\mathbf{L}_+$. From (112) (with f replaced by f_4) we see that

$$f_4(2^{-4-n}k) \approx (\mathbf{L}_+^n \alpha_4)[k]$$

when n is large. The upper left illustration from Fig. 10.17 gives a reasonable approximation for the frame f_4. (The modulus of the error is bounded by .13.)

We use a slight modification of this process when we compute samples for a detail. From (121) we have

$$\beta_4 = \mathbf{H}_-\alpha_5,$$

and the points

$$\left(2^{-4}k, \beta_4[k]\right) = \left(2^{-4}k, (\mathbf{H}_-\mathbf{L}_-^3 \alpha_8)[k]\right), \quad k = 0, \pm 1, \pm 2, \ldots$$

are plotted in the lower right illustration from Fig. 10.17. You will observe that $\mathbf{H}_-$, like $\mathbf{L}_-$, halves the number of significant ordinates. Using (129) we see that

$$d_4(x) = \sum_k \beta_4[k]\psi(2^4 x - k) = \sum_k (\mathbf{H}_+\beta_4)[k]\phi(2^5 x - k)$$

$$= \sum_k (\mathbf{L}_+\mathbf{H}_+\beta_4)[k]\phi(2^6 x - k) = \cdots .$$

Plots of the corresponding points

$$\left(2^{-5-n}k, (\mathbf{L}_+^n\mathbf{H}_+\beta_4)[k]\right), \quad k = 0, \pm 1, \pm 2, \ldots$$

are shown on the right side of Fig. 10.17 for each $n = 0, 1, 2, 3$. You will observe that $\mathbf{H}_+$, like $\mathbf{L}_+$, doubles the number of significant ordinates. In this case we have

$$d_4(2^{-5-n}k) \approx \mathbf{L}_+^n\mathbf{H}_+\beta_4[k]$$

when n is large. The upper right illustration from Fig. 10.17 gives a reasonable approximation to the detail d_4. (The modulus of the error is bounded by .10.)

We summarize the above discussion with the following generalization. We can generate approximate samples

$$
\begin{aligned}
f_m(2^{-m-p}k) &\approx \mathbf{L}_+^p\mathbf{L}_-^q f(2^{-m-q}k), \\
d_m(2^{-m-p}k) &\approx \mathbf{L}_+^{p-1}\mathbf{H}_+\mathbf{H}_-\mathbf{L}_-^{q-1} f(2^{-m-q}k)
\end{aligned}
\tag{132}
$$

for the frame f_m and detail d_m that correspond to a given function f (and the coefficients $c_0, c_1, \ldots, c_M$) by applying the operators

$$
\mathbf{L}_+^p\mathbf{L}_-^q, \quad \mathbf{L}_+^{p-1}\mathbf{H}_+\mathbf{H}_-\mathbf{L}_-^{q-1}
$$

to the sample sequence

$$
f(2^{-m-q}k), \quad k = 0, \pm1, \pm2, \ldots .
$$

The precision index $p = 1, 2, \ldots$ allows us to specify the density of the ordinates we compute for f_m, d_m. The precision index $q = 1, 2, \ldots$ specifies the density of the initial samples of f and controls the overall accuracy of the process. [The errors in the approximations (132) go to zero like 2^{-q} when f is a smooth function.]

The operators $\mathbf{P}_-, \mathbf{P}_+$

The approximation (112) introduces errors at the beginning and at the ends of the chains of mappings shown in Fig. 10.17. We start with samples of f (instead of the coefficients α_8) and we end up with the coefficients $\mathbf{L}_+^4\alpha_4$ (instead of samples of f_4) or with the coefficients $\mathbf{L}_+^3\mathbf{H}_+\beta_4$ (instead of samples of d_4). We will now show how to reduce the size of the first error and eliminate the second by using a *preprocessing operator* $\mathbf{P}_-$ and a *postprocessing operator* $\mathbf{P}_+$.

We work with Daubechies coefficients $c_0, c_1, \ldots, c_M$ for some $M = 3, 5, \ldots$. The corresponding scaling function ϕ is continuous, and $\phi(x) = 0$ when $x \leq 0$ or $x \geq M$. We need the samples $\phi(n)$, $n = 0, \pm1, \pm2, \ldots$ for our analysis. We will determine $\phi(1), \phi(2)$ when $M = 3$. Exercise 10.38 shows how to find numerical values for $\phi(1), \phi(2), \ldots, \phi(M-1)$ when $M = 5, 7, \ldots$.

Example: Find the ordinates $\phi(1), \phi(2)$ for the scaling function that corresponds to the Daubechies coefficients (39) with $M = 3$.

Solution: Using the dilation equation (29) we write

$$
\begin{aligned}
\phi(1) &= c_0\,\phi(2) + c_1\,\phi(1) + c_2\,\phi(0) + c_3\,\phi(-1) = c_1\,\phi(1) + c_0\,\phi(2) \\
\phi(2) &= c_0\,\phi(4) + c_1\,\phi(3) + c_2\,\phi(2) + c_3\,\phi(1) \ \ = c_3\,\phi(1) + c_2\,\phi(2)
\end{aligned}
$$

or equivalently,

$$\begin{bmatrix} \phi(1) \\ \phi(2) \end{bmatrix} = \frac{1}{4} \begin{bmatrix} 3 + \sqrt{3} & 1 + \sqrt{3} \\ 1 - \sqrt{3} & 3 - \sqrt{3} \end{bmatrix} \begin{bmatrix} \phi(1) \\ \phi(2) \end{bmatrix}.$$

We solve this homogeneous system of linear equations subject to the constraint $\phi(1) + \phi(2) = 1$ from (73) to obtain

$$\phi(1) = \frac{1 + \sqrt{3}}{2}, \quad \phi(2) = \frac{1 - \sqrt{3}}{2} \quad \text{when} \quad M = 3. \tag{133}$$

∎

From (106) we obtain the identity

$$f_m(2^{-m}k) = \sum_\ell \alpha_m[\ell]\phi(2^m \cdot 2^{-m}k - \ell)$$
$$= \phi(1)\alpha_m[k-1] + \phi(2)\alpha_m[k-2] + \cdots + \phi(M-1)\alpha_m[k - M + 1]$$

that links the coefficients $\alpha_m[k]$ and the samples $f_m(2^{-m}k)$. This relation is independent of m, so if we define the operator $\mathbf{P}_+$ by writing

$$(\mathbf{P}_+\alpha)[k] := \phi(1)\alpha[k-1] + \phi(2)\alpha[k-2] + \cdots + \phi(M-1)\alpha[k - M + 1] \tag{134}$$

(when α is a function on $\mathbb{Z}$), then

$$f_m(2^{-m}k) = (\mathbf{P}_+\alpha_m)[k], \quad k = 0, \pm 1, \pm 2, \ldots \tag{135}$$

for each index m. The operator $\mathbf{P}_+$ maps α_m to the 2^{-m}-samples of f_m (with no error).

We would like to construct an inverse $\mathbf{P}_-$ for (134) that maps the 2^{-m}-samples of f_m to the coefficients α_m (with no error), so that

$$\alpha_m[k] = \mathbf{P}_- f_m(2^{-m}k), \quad k = 0, \pm 1, \pm 2, \ldots \tag{136}$$

for each index m. We will derive an explicit representation for $\mathbf{P}_-$ when $M = 3$. Exercise 10.39 shows how we can apply $\mathbf{P}_-$ numerically when $M = 5, 7, \ldots$.

Example: Find an explicit representation for the inverse of the operator $\mathbf{P}_+$ given by (134) when the ordinates (135) correspond to the Daubechies coefficients (39) with $M = 3$.

Solution: Using (134) and (133) we see that

$$(\mathbf{P}_+\alpha)[k] = \frac{1 + \sqrt{3}}{2}\alpha[k-1] + \frac{1 - \sqrt{3}}{2}\alpha[k-2], \quad k = 0, \pm 1, \pm 2, \ldots \tag{137}$$

when $M = 3$. We will set $\gamma := \mathbf{P}_+\alpha$ and solve this equation for $\alpha = \mathbf{P}_-\gamma$.

When α, γ are suitably regular, their 1-periodic Fourier transforms satisfy

$$\sum_{k=-\infty}^{\infty} \gamma[k]e^{-2\pi iks} = \tfrac{1}{2}\{(\sqrt{3}+1)e^{-2\pi is} - (\sqrt{3}-1)e^{-4\pi is}\}\sum_{k=-\infty}^{\infty} \alpha[k]e^{-2\pi iks}$$

so that

$$\sum_{k=-\infty}^{\infty} \alpha[k]e^{-2\pi iks} = \frac{2e^{2\pi is}}{\sqrt{3}+1}\left\{1 - \left(\frac{\sqrt{3}-1}{\sqrt{3}+1}\right)e^{-2\pi is}\right\}^{-1} \sum_{\kappa=-\infty}^{\infty} \gamma[\kappa]e^{-2\pi i\kappa s}$$

$$= \frac{2}{\sqrt{3}+1}\sum_{n=0}^{\infty}\left(\frac{\sqrt{3}-1}{\sqrt{3}+1}\right)^{n} e^{-2\pi i(n-1)s} \sum_{\kappa=-\infty}^{\infty} \gamma[\kappa]e^{-2\pi i\kappa s}$$

$$= \frac{2}{\sqrt{3}+1}\sum_{k=-\infty}^{\infty}\left\{\sum_{n=0}^{\infty}\left(\frac{\sqrt{3}-1}{\sqrt{3}+1}\right)^{n} \gamma[k+1-n]\right\}e^{-2\pi iks}.$$

In this way we obtain the formula

$$(\mathbf{P}_-\gamma)[k] := \frac{2}{\sqrt{3}+1}\sum_{n=0}^{\infty}\left(\frac{\sqrt{3}-1}{\sqrt{3}+1}\right)^{n} \gamma[k+1-n], \quad k = 0, \pm 1, \pm 2, \ldots \qquad (138)$$

that gives α in terms of γ when $M = 3$. [You can use (137), (138) to verify that

$$\mathbf{P}_+\mathbf{P}_- = \mathbf{P}_-\mathbf{P}_+ = \mathbf{I}$$

when these operators are applied to slowly growing functions on $\mathbb{Z}$.]   ∎

We derive an *exact* representation

$$f_m(2^{-m-p}k) = \mathbf{P}_+\mathbf{L}_+^p\alpha_m[k] = \mathbf{P}_+\mathbf{L}_+^p\mathbf{L}_-^q\alpha_{m+q}[k] = \mathbf{P}_+\mathbf{L}_+^p\mathbf{L}_-^q\mathbf{P}_-f_{m+q}(2^{-m-q})$$

for samples of the frame f_m by using the identities (135), (128), (120), (136) for $\mathbf{P}_+, \mathbf{L}_+, \mathbf{L}_-, \mathbf{P}_-$. The corresponding exact representation

$$d_m(2^{-m-p}k) = \mathbf{P}_+\mathbf{L}_+^{p-1}\mathbf{H}_+\mathbf{H}_-\mathbf{L}_-^{q-1}\mathbf{P}_-f_{m+q}(2^{-m-q})$$

for samples of the detail d_m also uses the identities (129), (121) for $\mathbf{H}_+, \mathbf{H}_-$. We replace f_{m+q} by the known function f on the right side of these expressions to obtain approximations

$$f_m(2^{-m-p}k) \approx \mathbf{P}_+\mathbf{L}_+^p\mathbf{L}_-^q\mathbf{P}_-f(2^{-m-q}k), \quad k = 0, \pm 1, \pm 2, \ldots ,$$

$$d_m(2^{-m-p}k) \approx \mathbf{P}_+\mathbf{L}_+^{p-1}\mathbf{H}_+\mathbf{H}_-\mathbf{L}_-^{q-1}\mathbf{P}_-f(2^{-m-q}k), \quad k = 0, \pm 1, \pm 2, \ldots , \qquad (139)$$

that are much more accurate than those from (132). When f is sufficiently smooth, the errors in (132) go to zero like 2^{-q} while the errors in (139) go to zero like 2^{-Kq}, *cf.* Ex. 10.34. The improvement can be quite dramatic. By using $\mathbf{P}_-, \mathbf{P}_+$ we can reduce the maximum error modulus for the approximate components of $f_4(2^{-8}k)$, $\alpha_8[k]$, $d_4(2^{-8}k)$ at the top of Fig. 10.17 from .13, .024, .090 to the unobservable .0011, .00092, .00041, respectively.

10.4 Filter Banks

Introduction

The flow diagram of Fig. 10.18 shows how we use the operators $\mathbf{L}_-, \mathbf{H}_-$ and $\mathbf{L}_+, \mathbf{H}_+$ from (118)–(119) and (122)–(123) to do analysis and synthesis. We split a frame coefficient sequence α_m into certain constituent parts, *i.e.*, the frame coefficient sequence α_{m-1} and the detail coefficient sequence β_{m-1} from the next coarser level. We then reassemble α_m from these constituent parts. (In practice, we do not tear a sequence apart and immediately reassemble it. That is why we include $\cdots$ in the diagram!) We combine analysis forks and synthesis forks to form the herringbone structures shown in Fig. 10.16. We use these herringbone algorithms to prepare data for coarse-to-fine transmission, to compress data, to filter data, etc. Such applications are possible because the synthesis fork is a *perfect inverse* for the analysis fork.

Frame coefficients

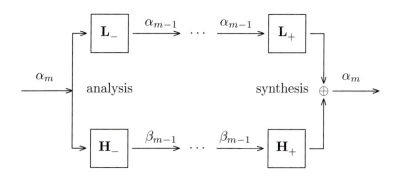

Detail coefficients

Figure 10.18. Flow diagram for analysis and synthesis using the operators (118)–(119) and (122)–(123).

Factorization of $\mathbf{L_-}, \mathbf{H_-}, \mathbf{L_+}, \mathbf{H_+}$

We will factor the operators $\mathbf{L_-}, \mathbf{H_-}, \mathbf{L_+}, \mathbf{H_+}$. Each block from Fig. 10.18 is replaced by two blocks that correspond to the factors. Figure 10.19 shows the resulting two-stage *filter bank*. Such filter banks (originally created by electrical engineers to compress speech signals for digital communication systems) facilitate the design of hardware for wavelet-based signal processing. Mathematical structures from wavelet analysis can be interpreted within a filter bank context and *vice versa*. Indeed, Daubechies produced the coefficients (39) by designing a corresponding filter bank!

Low frequency channel

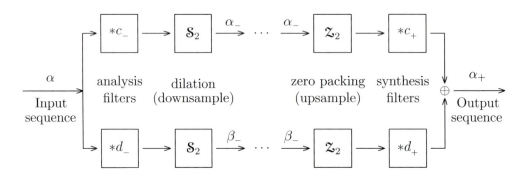

High frequency channel

Figure 10.19. Flow diagram for analysis and synthesis with a two-channel filter bank that uses convolution, dilation (delete each odd index component), and zero packing (insert a zero after each component).

We will use the discrete *dilation* or *downsampling* operator

$$(\mathbf{S}_2\alpha)[k] := \alpha[2k], \quad k = 0, \pm 1, \pm 2, \ldots, \tag{140}$$

and the *zero packing* or *upsampling* operator

$$(\mathbf{Z}_2\alpha)[k] := \begin{cases} \alpha[k/2] & \text{if } k = 0, \pm 2, \pm 4, \ldots \\ 0 & \text{if } k = \pm 1, \pm 3, \ldots \end{cases} \tag{141}$$

to produce these factorizations. The operator $\mathbf{S}_2$ compresses the data stream by deleting the components with odd indices. The operator $\mathbf{Z}_2$ expands the data stream by inserting a zero after each component. You will immediately observe that

$$(\mathbf{S}_2\mathbf{Z}_2\alpha)[k] = \alpha[k], \quad k = 0, \pm 1, \pm 2, \ldots \quad \text{and}$$

$$(\mathbf{Z}_2\mathbf{S}_2\alpha)[k] = \begin{cases} \alpha[k] & \text{if } k = 0, \pm 2, \pm 4, \ldots \\ 0 & \text{if } k = \pm 1, \pm 3, \ldots \end{cases} \tag{142}$$

In the first case $\mathbf{S}_2$ removes zeros introduced by the initial application of $\mathbf{Z}_2$; in the second, $\mathbf{Z}_2$ inserts zeros to replace the odd components deleted by the initial application of $\mathbf{S}_2$.

The operators $\mathbf{L}_-, \mathbf{H}_-$ are best described in terms of convolution and down-sampling. Using (118) and the identity

$$(c^\vee * \alpha)[2k] = \sum_n c^\vee_{2k-n}\alpha[n] = \sum_n c_{n-2k}\alpha[n],$$

we see that

$$\mathbf{L}_-\alpha = \tfrac{1}{2}\mathbf{S}_2(c^\vee * \alpha). \tag{143}$$

Here $^\vee$ is the reflection tag from (5.35), *i.e.*,

$$b^\vee_n := b_{-n}, \quad n = 0, \pm 1, \pm 2, \ldots$$

when b is a sequence. Analogously, using (119) we see that

$$\mathbf{H}_-\alpha = \tfrac{1}{2}\mathbf{S}_2(d^\vee * \alpha) \tag{144}$$

where

$$d_n := (-1)^n c_{M-n}, \quad n = 0, \pm 1, \pm 2, \ldots . \tag{145}$$

The operators $\mathbf{L}_+, \mathbf{H}_+$ are best described in terms of upsampling and convolution. Using (122) and the identity

$$\{c * (\mathbf{Z}_2\alpha)\}[k] = \sum_m c_{k-m}(\mathbf{Z}_2\alpha)[m] = \sum_n c_{k-2n}\alpha[n],$$

we see that

$$\mathbf{L}_+\alpha = c * (\mathbf{Z}_2\alpha). \tag{146}$$

Analogously, using (123) and (145) we see that

$$\mathbf{H}_+\alpha = d * (\mathbf{Z}_2\alpha). \tag{147}$$

We produce the desired factorizations of $\mathbf{L}_-, \mathbf{H}_-, \mathbf{L}_+, \mathbf{H}_+$ by using the operator

$$(\mathbf{C}_b\alpha)[k] := \sum_n b[k-n]\alpha[n]$$

to display the convolution products from (143), (144), (146), (147). (The notation $*b$ is used in Fig. 10.19.) Indeed, we set

$$c_-[n] := \frac{1}{\sqrt{2}}c_{-n}, \qquad c_+[n] := \frac{1}{\sqrt{2}}c_n,$$

$$d_-[n] := \frac{(-1)^n}{\sqrt{2}}c_{M+n}, \quad d_+[n] := \frac{(-1)^n}{\sqrt{2}}c_{M-n}, \quad n = 0, \pm 1, \pm 2, \dots, \tag{148}$$

and write

$$\mathbf{L}_- = \frac{1}{\sqrt{2}}\,\mathcal{S}_2\mathcal{C}_{c_-}, \qquad \mathbf{L}_+ = \sqrt{2}\,\mathcal{C}_{c_+}\mathcal{Z}_2, \tag{149}$$

$$\mathbf{H}_- = \frac{1}{\sqrt{2}}\,\mathcal{S}_2\mathcal{C}_{d_-}, \qquad \mathbf{H}_+ = \sqrt{2}\,\mathcal{C}_{d_+}\mathcal{Z}_2. \tag{150}$$

There is a troublesome $\frac{1}{2}$ that appears in (143), (144) but not in (146), (147). We have chosen to associate a factor $1/\sqrt{2}$ with each of the sequences c_-, d_-, c_+, d_+ in (148). The resulting normalization

$$\sum_n c_\pm[n] = \sqrt{2} \tag{151}$$

is commonly used with filter banks. (This is different from the normalization (28) we have been using for $c_0, c_1, \dots, c_M$.) We routinely drop the $\sqrt{2}$ factors from (149), (150) when we display the corresponding filter bank in Fig. 10.19.

You should understand how this filter bank generalizes the data flow for frame and detail coefficients as shown in Fig. 10.18. We are free to choose "arbitrary" sequences c_-, c_+, d_-, d_+ when we design a filter bank. The data flow does not correspond to the wavelet analysis of (120), (121), and (124) unless we use (148) to produce c_-, c_+, d_-, d_+ from suitably chosen coefficients $c_0, c_1, \dots, c_M$.

Fourier analysis of a filter bank

We will analyze the filter bank from Fig. 10.19. You may find it helpful to interpret the analysis within a context where the input sequence is obtained by sampling a bandlimited audio signal f, e.g., as shown in Fig. 8.1. The sample $\alpha[n] = f(n\mathrm{T})$ is taken at time $t = n\mathrm{T}$ where $\mathrm{T} > 0$ is the sampling interval. When f is σ-bandlimited and $2\sigma\mathrm{T} < 1$, the spectrum of the original signal can be found by restricting the 1-periodic Fourier transform

$$A(s) := \sum_n \alpha[n]e^{-2\pi ins} \tag{152}$$

to the interval $|s| < \frac{1}{2}$. The edges $s = \pm\frac{1}{2}$ correspond to the Nyquist frequency $F = 1/2T$. [To see this, replace s by s/T in (8.9).]

There are four *filters* specified by the sequences c_-, d_-, c_+, d_+. We will assume that these sequences are real and that they have finitely many nonzero components. The Fourier transform converts convolution into multiplication, so the filters $c_\pm, d_\pm$ act by introducing the factors

$$C_\pm(s) := \sum_n c_\pm[n]e^{-2\pi ins}, \qquad D_\pm(s) := \sum_n d_\pm[n]e^{-2\pi ins} \tag{153}$$

in the Fourier transform domain.

Figure 10.20 shows graphs of the trigonometric polynomials constructed by using (153) and (148) with the Daubechies coefficients (39). By design,

$$|C_\pm(s)|^2 \approx \begin{cases} 2 & \text{if} \quad |s| < \frac{1}{4} \\ 0 & \text{if } \frac{1}{4} < |s| < \frac{1}{2}, \end{cases} \qquad |D_\pm(s)|^2 \approx \begin{cases} 0 & \text{if} \quad |s| < \frac{1}{4} \\ 2 & \text{if } \frac{1}{4} < |s| < \frac{1}{2}, \end{cases}$$

but these max flat approximations are admittedly crude. When we multiply $A(s)$ by $C_\pm(s)$, $D_\pm(s)$ we suppress (but do not eliminate) the high- and low-frequency portions of the spectrum. We say that $c_\pm, d_\pm$ produce low- and high-pass filters. The phases of these trigonometric polynomials are also important, as we shall see presently.

Figure 10.21 shows the Fourier transform of the filter bank from Fig. 10.19. You will recognize the trigonometric polynomials $C_\pm, D_\pm$ that come from the four filters. The *grouping* operator

$$\tfrac{1}{2}(\mathcal{G}A)(s) := \tfrac{1}{2}\{A(\tfrac{1}{2}s) + A(\tfrac{1}{2}s + \tfrac{1}{2})\} \tag{154}$$

and the *dilation* operator

$$(\mathcal{S}_2 A)(s) := A(2s) \tag{155}$$

for functions on $\mathbb{T}_1$ are the Fourier transforms of the downsampling and upsampling operators (140), (141) for functions on $\mathbb{Z}$, *cf.* (4.30) and (4.29). You will observe that

$$(\tfrac{1}{2}\mathcal{G}_2\mathcal{S}_2 A)(s) = A(s)$$

when A is 1-periodic but

$$(\mathcal{S}_2\tfrac{1}{2}\mathcal{G}_2 A)(s) = \tfrac{1}{2}\{A(s) + A(s + \tfrac{1}{2})\}, \tag{156}$$

cf. (142). (We introduce aliasing when we replace the odd samples of α with zeros!)

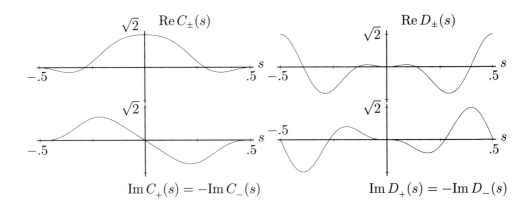

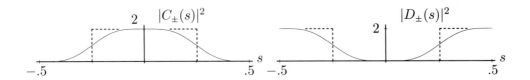

Figure 10.20. Graphs of the 1-periodic hermitian trigonometric polynomials $C_\pm(s)$, $D_\pm(s)$ constructed from the Daubechies coefficients (39) by using (148) and (153).

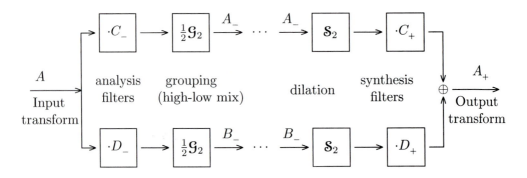

Figure 10.21. Fourier transform of the two-channel filter bank from Fig. 10.19.

Figure 10.22 shows what happens when we process an obviously contrived spectrum $A(s)$ with a filter bank that uses the trigonometric polynomials from Fig. 10.20. The initial analysis filters split $A(s)$ into low- and high-frequency components $C_-(s)A(s)$, $D_-(s)A(s)$. You can see the low-frequency "half circle" in the first and the (phase distorted) high-frequency "L's" in the second. The filters are not perfect, however, so you can also see a small admixture of high- and low-frequency components in these spectra.

Aliasing is introduced when we halve the sampling rate. In accordance with (154) we average the high and low portions of the filtered spectra to obtain the low- and high-frequency outputs

$$A_-(s) := \tfrac{1}{2}\{C_-(\tfrac{s}{2})A(\tfrac{s}{2}) + C_-(\tfrac{s}{2} + \tfrac{1}{2})A(\tfrac{s}{2} + \tfrac{1}{2})\},$$
$$B_-(s) := \tfrac{1}{2}\{D_-(\tfrac{s}{2})A(\tfrac{s}{2}) + D_-(\tfrac{s}{2} + \tfrac{1}{2})A(\tfrac{s}{2} + \tfrac{1}{2})\} \tag{157}$$

of the analysis fork of the filter bank. We have more or less separated the low-frequency "half circle" from the high-frequency "L's," but both of these spectra seem to be hopelessly contaminated with errors.

In accordance with (155), we dilate the inputs $A_-(s)$, $B_-(s)$ to the synthesis fork to form $A_-(2s)$, $B_-(2s)$. We then apply the synthesis filter factors $C_+(s)$, $D_+(s)$ and add the products to obtain the output

$$A_+(s) = \tfrac{1}{2}\{C_-(s)C_+(s) + D_-(s)D_+(s)\}A(s)$$
$$+ \tfrac{1}{2}\{C_-(s + \tfrac{1}{2})C_+(s) + D_-(s + \tfrac{1}{2})D_+(s)\}A(s + \tfrac{1}{2}). \tag{158}$$

Figure 10.22 shows perfect reconstruction at this point, *i.e.*, $A_+ = A$. This will be the case for an arbitrary initial spectrum when

$$C_-(s + \tfrac{1}{2})C_+(s) + D_-(s + \tfrac{1}{2})D_+(s) = 0 \tag{159}$$

[*i.e.*, the aliasing term $A(s + \tfrac{1}{2})$ does not appear in (158)] and

$$C_-(s)C_+(s) + D_-(s)D_+(s) = 2 \tag{160}$$

[*i.e.*, the spectrum $A(s)$ is not distorted]. The trigonometric polynomials from Fig. 10.20 satisfy (159)–(160), so the output is the same as the input in Fig. 10.22.

Perfect reconstruction filter banks

A *perfect reconstruction* filter bank has four real sequences c_-, c_+, d_-, d_+ (each with finitely many nonzero components) that satisfy the antialiasing constraint (159) and the antidistortion constraint (160). (A relaxed definition that allows translation of the output is given in Ex. 10.41.) There is a vast literature dealing with the design of such filter banks. We will give a few examples to introduce these ideas.

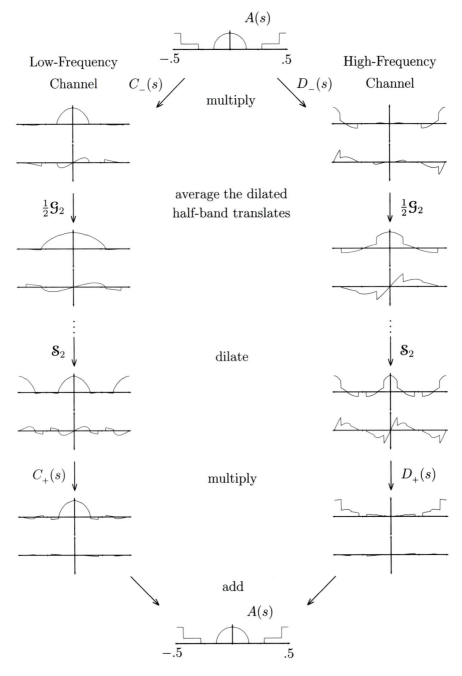

Figure 10.22. Flow of a contrived spectrum $A(s)$ through the filter bank of Fig. 10.21 when $C_\pm, D_\pm$ are the trigonometric polynomials shown in Fig. 10.20.

Example: Let m_-, m_+, n_-, n_+ be integers and let

$$c_-[n] := \delta[n - m_-], \qquad d_-[n] := \delta[n - n_-],$$
$$c_+[n] := \delta[n - m_+], \qquad d_+[n] := \delta[n - n_+]. \tag{161}$$

Show that these *translation filters* satisfy the constraints (159)–(160) for a perfect reconstruction filter bank if and only if

$$m_- + m_+ = 0, \quad n_- + n_+ = 0, \quad \text{and} \quad m_- + n_- \text{ is odd.} \tag{162}$$

Solution: These translation filters have the Fourier transforms

$$C_-(s) = e^{-2\pi i m_- s}, \; C_+(s) = e^{-2\pi i m_+ s}, \; D_-(s) = e^{-2\pi i n_- s}, \; D_+(s) = e^{-2\pi i n_+ s}.$$

We observe that

$$C_-\left(s + \tfrac{1}{2}\right) = e^{-2\pi i m_-(s+1/2)} = (-1)^{m_-} e^{-2\pi i m_- s},$$
$$D_-\left(s + \tfrac{1}{2}\right) = e^{-2\pi i n_-(s+1/2)} = (-1)^{n_-} e^{-2\pi i n_- s},$$

and use these expressions to write the antialiasing, antidistortion constraints (159), (160) in the form

$$(-1)^{m_-} e^{-2\pi i (m_-+m_+)s} + (-1)^{n_-} e^{-2\pi i (n_-+n_+)s} = 0,$$
$$e^{-2\pi i (m_-+m_+)s} + e^{-2\pi i (n_-+n_+)s} = 2.$$

The second constraint holds (for all s) if and only if each exponential is identically 1, *i.e.*, $m_- + m_+ = 0$ and $n_- + n_+ = 0$. The first holds when, in addition, precisely one of the integers m_-, n_- is odd. (The integers m_-, n_- must have opposite parity to ensure that we retain the odd indexed components of α in one branch of the filter bank and the even indexed components in the other. The indices m_-, m_+ and n_-, n_+ must each have sum zero to ensure that the initial $\alpha[n]$ ends up in the nth component of the output when n is even and when n is odd.) ∎

We motivated the introduction of a filter bank by factoring the operators $\mathbf{L}_-, \mathbf{L}_+, \mathbf{H}_-, \mathbf{H}_+$ we use to manipulate the frame and detail coefficients (108)–(109). We are no longer constrained by this interpretation. The coefficients $c_0, c_1, \ldots, c_M$ from (148) need not correspond to piecewise smooth functions ϕ, ψ that we can use for the integrals (79)–(81) and (108)–(109). What matters now is perfect reconstruction, and the following example shows that this filter bank property is a logical consequence of the orthogonality constraints (87).

Example: Let real coefficients $c_0, c_1, \ldots, c_M$ (with $M = 1, 3, 5, \ldots$) be selected subject to the orthogonality constraints (87). Show that the corresponding filters c_-, c_+, d_-, d_+ from (148) satisfy the constraints (159)–(160) for perfect reconstruction.

Solution: We express C_-, C_+, D_-, D_+ in terms of the trigonometric polynomial (35) by using (148) to see that

$$C_+(s) = \frac{1}{\sqrt{2}} \sum_n c_n e^{-2\pi i n s} = \sqrt{2}\, T(s),$$

$$D_+(s) = \frac{1}{\sqrt{2}} \sum_n (-1)^n c_{M-n} e^{-2\pi i n s} = \frac{1}{\sqrt{2}} \sum_m (-1)^{M-m} c_m e^{-2\pi i(M-m)s}$$

$$= -\frac{e^{-2\pi i M s}}{\sqrt{2}} \sum_m c_m e^{2\pi i m(s-1/2)} = -\sqrt{2}\, e^{-2\pi i M s} T(-s + \tfrac{1}{2}),$$

and analogously

$$C_-(s) = \sqrt{2}\, T(-s), \qquad D_-(s) = -\sqrt{2} e^{2\pi i M s} T(s + \tfrac{1}{2}).$$

Since M is odd and T is 1-periodic, we can use these expressions to write

$$C_-(s + \tfrac{1}{2}) C_+(s) + D_-(s + \tfrac{1}{2}) D_+(s) = 2\, T(-s + \tfrac{1}{2}) T(s)[1 + (-1)^M] = 0.$$

Since $c_0, c_1, \dots, c_M$ are real and (87) is equivalent to (85), we can also write

$$C_-(s) C_+(s) + D_-(s) D_+(s) = 2\{T(-s)T(s) + T(s + \tfrac{1}{2})T(-s - \tfrac{1}{2})\} = 2.$$

(You may wish to compare this analysis with that of Ex. 10.32.) ∎

There are many filter banks that do not have the coefficient symmetry of (148). We will consider the less restrictive case where

$$d_-[n] = \sigma(-1)^n c_+[n+N], \qquad d_+[n] = \sigma(-1)^n c_-[n-N]. \tag{163}$$

Here $\sigma = \pm 1$ and N is an odd integer. The filters d_-, d_+ are obtained by translating $c_+, c_-, -N, +N$ steps, respectively, and then changing the signs of the components that have even indices (if $\sigma = -1$) or odd indices (if $\sigma = +1$), *cf.* Fig. 10.23. Using the translation and modulation rules for functions on $\mathbb{Z}$, we find

$$D_-(s) = \sigma(-1)^N e^{2\pi i N s} C_+(s + \tfrac{1}{2}), \qquad D_+(s) = \sigma(-1)^N e^{-2\pi i N s} C_-(s + \tfrac{1}{2}). \tag{164}$$

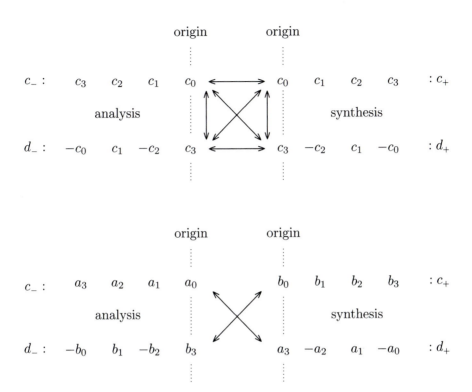

Figure 10.23. The coefficient symmetry (148) (top) and (163) (bottom). The diagonal mappings use translation (illustrated here with $N = 3$) and sign oscillation. The horizontal, vertical mappings from the more symmetric upper diagram use reflection, reflection with sign oscillation, respectively.

Since C_+ is 1-periodic, $\sigma^2 = 1$, and N is odd, we can use these expressions to write

$$C_-(s + \tfrac{1}{2})C_+(s) + D_-(s + \tfrac{1}{2})D_+(s) = C_-(s + \tfrac{1}{2})C_+(s)\{1 + (-1)^N\} = 0,$$
$$C_-(s)C_+(s) + D_-(s)D_+(s) = C_-(s)C_+(s) + C_-(s + \tfrac{1}{2})C_+(s + \tfrac{1}{2}).$$

In this way we see that filters with the symmetry (163) automatically satisfy the antialiasing constraint (159). They also satisfy the antidistortion constraint (160) when the product

$$Q(s) := \tfrac{1}{2}C_-(s)C_+(s) \tag{165}$$

and its half-period translate have the sum

$$Q(s) + Q(s + \tfrac{1}{2}) = 1. \tag{166}$$

You can verify that (166) is equivalent to the *biorthogonality constraint*

$$\sum_m c_+[m]c_-[2k-m] = \begin{cases} 1 & \text{if } k = 0 \\ 0 & \text{if } k = \pm 1, \pm 2, \ldots \end{cases} \tag{167}$$

on the filters c_-, c_+, *cf.* Ex. 10.43. [This generalizes (87).]

In view of this analysis, we can use the following procedure to find real co-efficients for a perfect reconstruction filter bank. We first choose some 1-periodic hermitian trigonometric polynomial Q that satisfies (166) and that has the hermi-tian factors C_-, C_+ of (165). The coefficients (*i.e.*, inverse Fourier transforms) of C_-, C_+ give us the filters c_-, c_+. We then use (163) to obtain the filters d_-, d_+. You may wish to show that every perfect reconstruction filter bank can be found in this way, *cf.* Ex. 10.44.

Example: A 1-periodic hermitian trigonometric polynomial

$$Q(s) = \cos^4(\pi s)[\cos^2(\pi s) + 3\sin^2(\pi s)]$$

that satisfies (166) is obtained by setting $K = 2$ in (100). Factor Q and thereby find filters c_-, c_+ for a perfect reconstruction filter bank. [You can then use (163) to find corresponding d_-, d_+.]

Solution: We use Euler's identities and a bit of algebra (*cf.* Ex. 10.45) to write

$$
\begin{aligned}
2Q(s) &= \frac{1}{16}[e^{i\pi s} + e^{-i\pi s}]^4[-e^{2\pi is} + 4 - e^{-2\pi is}] \\
&= \frac{1}{32}[e^{i\pi s} + e^{-i\pi s}]^4 \cdot [(1 - \sqrt{3})e^{2\pi is} + (1 + \sqrt{3})] \\
&\quad\quad \cdot [(1 - \sqrt{3})e^{-2\pi is} + (1 + \sqrt{3})] \\
&= 2e^{-6\pi is} \cdot \left(\frac{e^{2\pi is} + 1}{2}\right)^4 \left(\frac{e^{2\pi is} - \zeta}{1 - \zeta}\right)\left(\frac{e^{2\pi is} - \zeta^{-1}}{1 - \zeta^{-1}}\right)
\end{aligned}
\tag{168}
$$

where

$$\zeta := \frac{\sqrt{3} - 1}{\sqrt{3} + 1} = .26794\ldots, \quad \zeta^{-1} = \frac{\sqrt{3} + 1}{\sqrt{3} - 1} = 3.73205\ldots.$$

There are twenty distinct ways we can form an ordered pair of products from the factors

$$e^{2\pi is} + 1, \ e^{2\pi is} + 1, \ e^{2\pi is} + 1, \ e^{2\pi is} + 1, \ e^{2\pi is} - \zeta, e^{2\pi is} - \zeta^{-1}$$

of (168). The coefficient strings for ten such pairs of trigonometric polynomials are as follows.

$$1, 4, 6, 4, 1 \qquad\qquad \text{and} \quad -1, 4, 1$$
$$1, 3, 3, 1 \qquad\qquad \text{and} \quad -1, 3, 3, -1$$
$$1, 2, 1 \qquad\qquad\qquad \text{and} \quad -1, 2, 6, 2, -1$$
$$1, 1 \qquad\qquad\qquad\quad \text{and} \quad -1, 1, 8, 8, 1, -1$$
$$1 \qquad\qquad\qquad\qquad \text{and} \quad -1, 0, 9, 16, 9, 0, -1$$
$$-\zeta, 1 \qquad\qquad\qquad \text{and} \quad -\zeta^{-1}, 1-4\zeta^{-1}, 4-6\zeta^{-1}, 6-4\zeta^{-1}, 4-\zeta^{-1}, 1$$
$$-\zeta, 1-\zeta, 1 \qquad\qquad \text{and} \quad -\zeta^{-1}, 1-3\zeta^{-1}, 3-3\zeta^{-1}, 3-\zeta^{-1}, 1$$
$$-\zeta, 1-2\zeta, 2-\zeta, 1 \qquad \text{and} \quad -\zeta^{-1}, 1-2\zeta^{-1}, 2-\zeta^{-1}, 1$$
$$-\zeta, 1-3\zeta, 3-3\zeta, 3-\zeta, 1 \quad \text{and} \quad -\zeta^{-1}, 1-\zeta^{-1}, 1$$
$$-\zeta, 1-4\zeta, 4-6\zeta, 6-4\zeta, 4-\zeta, 1 \quad \text{and} \quad -\zeta^{-1}, 1$$

(You can see the binomial coefficients that come from 1, 2, 3, 4 powers of the factor $e^{2\pi i s} + 1$.) We reverse the order of these pairs to obtain the other ten.

We must account for the remaining factor

$$\frac{e^{-6\pi i s}}{8(1-\zeta)(1-\zeta^{-1})} = -\frac{e^{-6\pi i s}}{16}$$

from (168). We do this by partitioning the $-1/16$ and three left shifts (corresponding to $e^{-2\pi i 3s}$) between a pair of coefficient strings from the above list. For example, we can use the second pair to produce

$$c_-[-3] = 1, \quad c_-[-2] = 3, \quad c_-[-1] = 3, \quad c_-[0] = 1,$$
$$c_+[0] = \frac{1}{16}, \quad c_+[1] = \frac{-3}{16}, \quad c_+[2] = \frac{-3}{16}, \quad c_+[3] = \frac{1}{16},$$

and we can use the eighth pair to produce

$$c_-[-3] = \frac{-\zeta}{\sqrt{8}(1-\zeta)}, \qquad\qquad c_+[0] = \frac{-\zeta^{-1}}{\sqrt{8}(1-\zeta^{-1})},$$

$$c_-[-2] = \frac{1-2\zeta}{\sqrt{8}(1-\zeta)}, \qquad\qquad c_+[1] = \frac{1-2\zeta^{-1}}{\sqrt{8}(1-\zeta^{-1})},$$

$$c_-[-1] = \frac{2-\zeta}{\sqrt{8}(1-\zeta)}, \qquad\qquad c_+[2] = \frac{2-\zeta^{-1}}{\sqrt{8}(1-\zeta^{-1})},$$

$$c_-[0] = \frac{1}{\sqrt{8}(1-\zeta)}, \qquad\qquad c_+[3] = \frac{1}{\sqrt{8}(1-\zeta^{-1})},$$

or equivalently,

$$c_-[-3] = \frac{1-\sqrt{3}}{4\sqrt{2}}, \quad c_-[-2] = \frac{3-\sqrt{3}}{4\sqrt{2}}, \quad c_-[-1] = \frac{3+\sqrt{3}}{4\sqrt{2}}, \quad c_-[0] = \frac{1+\sqrt{3}}{4\sqrt{2}},$$
$$c_+[0] = \frac{1+\sqrt{3}}{4\sqrt{2}}, \quad c_+[1] = \frac{3+\sqrt{3}}{4\sqrt{2}}, \quad c_+[2] = \frac{3-\sqrt{3}}{4\sqrt{2}}, \quad c_+[3] = \frac{1-\sqrt{3}}{4\sqrt{2}}$$

(169)

as expected from (39) and (148). (All unspecified components are set to 0.) You should have no trouble finding c_-, c_+ pairs for the other 18 factorizations. ∎

We can use this procedure to design various filter banks. In practice, we prefer to use filters with a "small" number of nonzero coefficients, but we also want filters with the "smoothness" that comes from sufficiently many factors of $e^{2\pi i s} + 1$. We can experiment with several choices for c_-, c_+, d_-, d_+ to see how they perform for a particular application. For example, the FBI has a massive collection of digitized fingerprint images scanned at 500 pixels/inch. A filter bank is used to compress these files so that they can be stored and transmitted more efficiently. The filters c_-, c_+, d_-, d_+ that work best for this purpose were obtained from a factorization of (100) (with $K = 4$) as described in Ex. 10.46.

Compression and reconstruction

Using a digitizing tablet and a computer, I generated points

$$g[n] = x[n] + i\,y[n], n = 0, 1, \dots, 512,$$

along the handwritten "Gauss" from a signature of the great mathematician. The corresponding broken line image is shown at the top of Fig. 10.24 and Fig. 10.25. (Germans often write ß in place of ss.) We use the analysis fork of a filter bank to compress a signal so that it can be stored or transmitted more efficiently. We then process the compressed file with the synthesis fork to obtain a good (but not perfect) reconstruction. We will use the univariate Gauss signature to illustrate these ideas.

The operators $\mathbf{L}_-$, $\mathbf{H}_-$, $\mathbf{L}_+$, $\mathbf{H}_+$ are applied to functions on $\mathbb{Z}$, so before we begin the analysis we must suitably extend g. There are several ways to do this. The zero extension

$$g[n] = 0 \quad \text{if} \quad n < 0 \text{ or } n > 512$$

that we have used elsewhere does not work well in this context. The jumps in g at $n = 0$, $n = 512$ introduce undesirable ripples in the reconstructed images, *cf.* Fig. 10.11. We could eliminate such ripples by using the extension

$$g[n] = \begin{cases} g[0] & \text{if } n < 0 \\ g[512] & \text{if } n > 512, \end{cases}$$

but this would force us to make special provision for storing components of $\mathbf{L}_-g$, $\mathbf{H}_-g$, $\mathbf{L}_-\mathbf{L}_-g$, $\mathbf{H}_-\mathbf{L}_-g$, ... that lie beyond the edges of the original image.

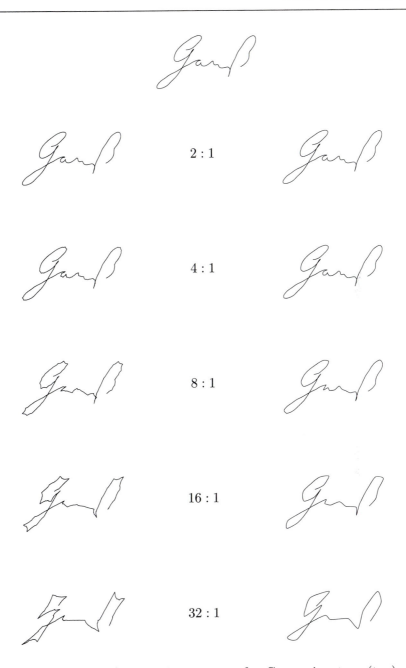

2 : 1

4 : 1

8 : 1

16 : 1

32 : 1

Figure 10.24. A 512-point trace g of a Gauss signature (top) and the reconstruction $\mathbf{L}_+^p \mathbf{L}_-^p g$ corresponding to the $2^p : 1$ compression $\mathbf{L}_-^p g$, $p = 1, 2, 3, 4, 5$, where the filter bank uses the Daubechies coefficients from (39), (148) (left) and the 9/7 FBI coefficients from Ex. 10.46 (right).

Figure 10.25. A 512-point trace g of a Gauss signature (top) and reconstructions from the $1/4$, $1/\sqrt{32}$, $1/8$, $1/\sqrt{128}$, $1/16$ most significant components of (170) when the filter bank uses the Daubechies coefficients from (39), (148) (left) and the 9/7 FBI coefficients from Ex. 10.46 (right).

We will use symmetry to avoid such boundary conditions. We define

$$g[-n] := g[n], \quad n = 1, 2, \ldots, 511$$

and then 1024-periodically extend g to all of $\mathbb{Z}$. When we apply $\mathbf{L}_-$, $\mathbf{H}_-$ to a function on $\mathbb{P}_{2N}$ we obtain functions on $\mathbb{P}_N$. When we apply $\mathbf{L}_+$, $\mathbf{H}_+$ to a function on $\mathbb{P}_N$ we obtain functions on $\mathbb{P}_{2N}$. This being the case, we will always store the components from one period and forget about any edge effects. (If each filter preserves even symmetry, we need only store the components from a half period!)

In particular, the periodic functions

$$g, \mathbf{L}_- g, \mathbf{H}_- g, \mathbf{L}_-^2 g, \mathbf{H}_- \mathbf{L}_- g, \ldots, \mathbf{L}_-^5 g, \mathbf{H}_- \mathbf{L}_-^4 g$$

from Mallat's herringbone (*cf.* Fig. 10.16) are represented by vectors

$$\boldsymbol{\alpha}_{10}, \boldsymbol{\alpha}_9, \boldsymbol{\beta}_9, \boldsymbol{\alpha}_8, \boldsymbol{\beta}_8, \ldots, \boldsymbol{\alpha}_5, \boldsymbol{\beta}_5$$

that have

$$1024, 512, 512, 256, 256, \ldots, 32, 32$$

components, respectively. We form 1024 component vectors

$$
\begin{aligned}
\boldsymbol{\alpha}_{10} \quad &\curvearrowright \quad \boldsymbol{\alpha}_9 \,\&\, \boldsymbol{\beta}_9 \\
&\curvearrowright \quad \boldsymbol{\alpha}_8 \,\&\, \boldsymbol{\beta}_8 \,\&\, \boldsymbol{\beta}_9 \\
&\vdots \\
&\curvearrowright \quad \boldsymbol{\alpha}_5 \,\&\, \boldsymbol{\beta}_5 \,\&\, \boldsymbol{\beta}_6 \,\&\, \boldsymbol{\beta}_7 \,\&\, \boldsymbol{\beta}_8 \,\&\, \boldsymbol{\beta}_9
\end{aligned}
$$

cf. (26). These maps are all reversible since the operators $\mathbf{L}_-$, $\mathbf{H}_-$, $\mathbf{L}_+$, $\mathbf{H}_+$ come from a perfect reconstruction filter bank. We expend approximately μN operations as we implement

$$\boldsymbol{\alpha}_{m+1} \curvearrowright \boldsymbol{\beta}_m \& \boldsymbol{\alpha}_m \quad \text{or} \quad \boldsymbol{\beta}_m \& \boldsymbol{\alpha}_m \curvearrowright \boldsymbol{\alpha}_{m+1}$$

when the filters c_-, c_+ together have μ nonzero coefficients and $\boldsymbol{\alpha}_m, \boldsymbol{\beta}_m$ each have N components, *cf.* Ex. 10.47.

We compress the Gauss signature file by discarding coefficients from some of the high-frequency details. The approximation

$$g \approx \mathbf{L}_+ \mathbf{L}_- g$$

results when we replace the components of $\boldsymbol{\beta}_9$ with zeros. Since $\boldsymbol{\alpha}_9$ has only half as many coefficients as $\boldsymbol{\alpha}_{10}$, we achieve a $2:1$ compression ratio. The approximation

$$g \approx \mathbf{L}_+^2 \mathbf{L}_-^2 g$$

results when we replace the components of $\boldsymbol{\beta}_9$ and $\boldsymbol{\beta}_8$ with zeros. Since $\boldsymbol{\alpha}_8$ has only a fourth as many components as $\boldsymbol{\alpha}_{10}$, we achieve a $4:1$ compression ratio. Analogously,

$$g \approx \mathbf{L}_+^p \mathbf{L}_-^p g$$

gives a $2^p:1$ compression ratio $p = 1, 2, 3, 4, 5$. Figure 10.24 shows such broken-line approximations for the Gauss signature. The Daubechies filters from (39) and (148) were used to produce the images on the left. The scaling function for this c_+ (as shown in Fig. 10.14) has jagged corners so the same is true of the reconstructed signatures. The images on the right side of Fig. 10.24 were produced with the 9/7 FBI filters from Ex. 10.46. The smooth $4:1$ and $8:1$ approximations are clearly superior to the corresponding ones that come from the four tap filters.

There are natural limits to the compression ratios we can achieve in this way. Our initial file seems to oversample the Gauss signature, so almost any reasonable scheme gives us a $2:1$ compression with no loss of quality. At the other extreme, neither of the $32:1$ compressions at the bottom of Fig. 10.24 are legible. (They do a surprisingly good job, however, considering they come from only 17 "points"!)

We will now describe a way to generate improved approximations with compression ratios in the intermediate range from $4:1$ to $16:1$. We first use the analysis operators $\mathbf{L}_-$, $\mathbf{H}_-$ to find the 1024 components of the vector

$$\boldsymbol{\alpha}_5 \,\&\, \boldsymbol{\beta}_5 \,\&\, \boldsymbol{\beta}_6 \,\&\, \boldsymbol{\beta}_7 \,\&\, \boldsymbol{\beta}_8 \,\&\, \boldsymbol{\beta}_9 \tag{170}$$

as described above. We produced the $4:1$ compressions of Fig. 10.24 by replacing the components of $\boldsymbol{\beta}_9$, $\boldsymbol{\beta}_8$ with zeros. A few of the discarded coefficients correspond to significant localized details of the signature while some of the retained coefficients make no essential contribution to the final image. This time we will locate the largest (by modulus) 256 components and replace all of the others with zeros. Analogously, we will produce $\sqrt{32}:1$, $8:1$, $\sqrt{128}:1$, $16:1$ compressions by locating the largest 181, 128, 91, 64 components and replacing the others with zeros. After this is done, we use the synthesis operators $\mathbf{L}_+$, $\mathbf{H}_+$ to generate the corresponding broken-line approximations shown in Fig. 10.25. Most of these reconstructed images are legible, and the $8:1$ reconstruction from the 9/7 FBI filters is as good as the original!

Audio files and two-dimensional image files (that we process row by row and column by column) can be compressed and reconstructed in a similar fashion. You can find additional details in the references for this chapter. At some point, however, you will want to use your computer to experiment with these ideas. The subject is still relatively new, and you may very well make a discovery that does not yet appear in the scientific literature!

References

I. Daubechies, *Ten Lectures on Wavelets*, SIAM, Philadelphia, 1992.

> A definitive graduate level introduction to the mathematical theory of wavelets.

B.B. Hubbard, *The World According to Wavelets, 2nd ed.*, A.K. Peters, Wellesley, MA, 1998.

> A charming nontechnical introduction to the ideas of wavelet analysis and the people who created them.

Y. Meyer, *Wavelets: Algorithms and Applications*, SIAM, Philadelphia, 1993.

> An interdisciplinary introduction to the use of wavelets for signal and image processing.

M. Misiti, Y. Misiti, G. Oppenheim, and J.-M. Poggi, *Wavelet Toolbox for Use with MATLAB*, The Math Works, Inc., Natick, MA, 1996.

> An illustrated introduction to sophisticated software for signal and image processing with wavelets.

E.J. Stollnitz, T.D. Derose, and D.H. Salesin, *Wavelets for Computer Graphics*, Morgan Kaufmann Publishers, Inc., San Francisco, 1996.

> An introduction to the use of wavelets for representing images, curves, and surfaces.

G. Strang and T. Nuygen, *Wavelets and Filter Banks*, Wellesley-Cambridge Press, Wellesley, MA, 1996.

> A widely used textbook written for engineers and scientists.

Wavelet Digest, http://www.wavelet.org/wavelet

> This Internet site will provide you with a listing of books, software, bibliographies, current research papers, on-line tutorials, etc., that deal with wavelets.

Exercise Set 10

▶ **EXERCISE 10.1:**   Let ψ be the Haar wavelet (1).

(a) Show that ψ has the Fourier transform

$$\Psi(s) = i\, e^{-i\pi s}\, \sin\left(\frac{\pi s}{2}\right) \operatorname{sinc}\left(\frac{s}{2}\right).$$

(b) Show that

$$\int_0^\infty |\Psi(s)|^2 ds = \frac{1}{2}.$$

Hint. Use Plancherel's identity (1.15).

(c) A numerical integration can be used to verify that

$$\int_0^2 |\Psi(s)|^2 ds = .4278\dots .$$

What does this imply about the functions shown in Fig. 10.2?

▶ **EXERCISE 10.2:**   Two Fourier analysis students are trying to make sense of the musical score from Fig. 10.3.

"I am an accomplished pianist," says the first, "and I know from experience that you cannot play anything either of us would call music by pressing only the keys $C1, C2, \dots, C8$ from this unimaginative, repetitive score."

"A mathematical proof compels belief!" says the second. "How can you argue with the picture proof from Fig. 10.5 that even shows us the amplitude and phase of the notes?"

"I argue because the conclusion is ridiculous!" responds the pianist. "There is something about the proof that makes no sense within the normal context of music.

Resolve this conflict.

Hint. Musicians are accustomed to beat rates that range from 40/min to 200/min. What (quarter note) beat rate corresponds to the score of Fig. 10.3? Assume that the quarter note $C4$ is represented by a suitably dilated Haar wavelet (1) with a support of $1/261.6$ Hz.

▶ **EXERCISE 10.3:** Use the analysis equation (17) to find the coefficients $F[m, k]$ for the Haar wavelet synthesis (4) of the function

$$f(x) = \begin{cases} 1 & \text{if } 0 < x < \frac{1}{3} \\ 0 & \text{otherwise.} \end{cases}$$

Hint. The nonzero coefficient is $2^m/3$ when $m = 0, -1, -2, \ldots$ and $1/3$ when $m = 1, 2, \ldots$. You may find it helpful to observe that

$$\frac{1}{3} = (.010101\ldots)_2$$

and that the critical wavelet is centered at

$$x = (.010)_2, \ (.011)_2, \ (.01010)_2, \ (.01011)_2, \ \ldots$$

when $m = 1, 2, 3, 4, \ldots$.

▶ **EXERCISE 10.4:** Let the real column vectors $\boldsymbol{\alpha}_m, \boldsymbol{\beta}_m, \ m = 0, 1, 2, \ldots$, from (20) have components linked by (24).

(a) Write down the 8×8 matrices $M_{3,1}, M_{3,2}, M_{3,3}$ for the linear mappings

$$\boldsymbol{\alpha}_3 \overset{M_{3,1}}{\rightsquigarrow} \boldsymbol{\alpha}_2 \,\&\, \boldsymbol{\beta}_2 \overset{M_{3,2}}{\rightsquigarrow} \boldsymbol{\alpha}_1 \,\&\, \boldsymbol{\beta}_1 \,\&\, \boldsymbol{\beta}_2 \overset{M_{3,3}}{\rightsquigarrow} \boldsymbol{\alpha}_0 \,\&\, \boldsymbol{\beta}_0 \,\&\, \boldsymbol{\beta}_1 \,\&\, \boldsymbol{\beta}_2$$

of (26).

Hint. Use 0,1, and $h := \frac{1}{2}$ for the elements.

(b) Find the 8×8 product matrix

$$M_3 := M_{3,3} \, M_{3,2} \, M_{3,1}$$

for the Haar wavelet transform

$$\boldsymbol{\alpha}_3 \overset{M_3}{\rightsquigarrow} \boldsymbol{\alpha}_3^w.$$

Hint. You should obtain $(h^3, h^3, h^2, 0, h, 0, 0, 0)^T$ for the first column.

(d) Write down the 16×16 matrix M_4 for the Haar wavelet transform

$$\boldsymbol{\alpha}_4 \overset{M_4}{\rightsquigarrow} \boldsymbol{\alpha}_4^w.$$

Note. The matrix M_n has $(n + 1)2^n$ nonzero elements, $n = 1, 2, \ldots$, so a direct application of this matrix requires approximately $N \log_2 N$ operations ($N := 2^n$), about the same as an N-point FFT. Mallat's fast herringbone algorithm does the same task with about N operations!

▶ **EXERCISE 10.5:**   We usually use the data structures (25)–(27) when we work with Mallat's relations (23), (24). If we have been given a long train of samples (*e.g.*, from an audio signal) we may prefer to perform the computation in place. For example, when $N = 8$ we can write

$$\alpha_3 = (\alpha_3[0], \alpha_3[1], \alpha_3[2], \alpha_3[3], \alpha_3[4], \alpha_3[5], \alpha_3[6], \alpha_3[7])$$
$$\curvearrowright (\alpha_2[0], \beta_2[0], \alpha_2[1], \beta_2[1], \alpha_2[2], \beta_2[2], \alpha_2[3], \beta_2[3])$$
$$\curvearrowright (\alpha_1[0], \beta_2[0], \beta_1[0], \beta_2[1], \alpha_1[1], \beta_2[2], \beta_1[1], \beta_2[3])$$
$$\curvearrowright (\alpha_0[0], \beta_2[0], \beta_1[0], \beta_2[1], \beta_0[0], \beta_2[2], \beta_1[1], \beta_2[3]).$$

At each step $\alpha_m[2k], \alpha_m[2k+1]$ are overwritten by $\alpha_{m-1}[k], \beta_{m-1}[k]$, respectively.

(a) Let α have $N = N_0 2^m$ components for some $m = 1, 2, \dots$ and $N_0 = 1, 2, \dots$. Formulate an in place algorithm that overwrites α with the partial discrete Haar wavelet transform obtained from m steps of the above scheme.

  Note. Algorithm 6.5 performs an in place calculation of the discrete Fourier transform.

(b) Formulate an in-place inverse for the algorithm of (a).

▶ **EXERCISE 10.6:**   In this exercise you will study ordinary solutions of the dilation equation

$$\phi(x) = 2\phi(2x), \quad -\infty < x < \infty.$$

(The generalized solution $\phi = \delta$ was found in the text.)

(a) Let ϕ be an ordinary solution of the above dilation equation, let $a > 0$, and assume that you know

$$\phi_a(x) := \begin{cases} \phi(x) & \text{if } a \le x < 2a \\ 0 & \text{otherwise.} \end{cases}$$

  Show how to find ϕ at each point of $(0, \infty)$.

  Hint. Sum suitably scaled dilates of ϕ_a.

(b) Find or construct several essentially different ordinary solutions of the above dilation equation.

(c) Show that $\phi \equiv 0$ is the only continuous solution of the above dilation equation.

(d) Show that no support-limited ordinary solution of the above dilation equation satisfies (30).

▶ **EXERCISE 10.7:**   In this exercise you will derive (37) by using a dilation rule for the convolution product.

(a) Let $\phi_0, \phi_1, \phi_2, \dots$ and τ be given by (33) and (38). Show that

$$\phi_n(x) = 2\tau(2x) * 2\phi_{n-1}(2x), \quad n = 1, 2, \dots .$$

(b) Show that if $g := f_1 * f_2 * \cdots * f_m$, then

$$2g(2x) = 2f_1(2x) * 2f_2(2x) * \cdots * 2f_m(2x).$$

(c) Use (a) and (b) to derive (37).

   Note. In the text we obtain (37) from (36).

▶ **EXERCISE 10.8:**   Let ϕ satisfy the dilation equation (29) and let τ be given by (38).

(a) Show that
$$\phi(x) = 2\tau(2x) * 2\phi(2x).$$

(b) More generally, show that
$$a\phi(ax) = 2a\tau(2ax) * 2a\phi(2ax) \quad \text{when} \quad a > 0.$$

(c) Use (b) to derive the identity
$$\phi(x) = \phi_n(x) * 2^n \phi(2^n x), \quad n = 0, 1, 2, \ldots,$$

   where $\phi_0, \phi_1, \phi_2, \ldots$ are given by (37).

▶ **EXERCISE 10.9:**   Let $c_0, c_1, \ldots, c_M$ satisfy (28), and let ϕ be a support-limited generalized solution of the dilation equation (29) with $\Phi(0) = 0$. Show that $\phi = 0$. [In this way you prove that there is at most one support-limited solution of (29)–(30).]

Hint. Use the factorization
$$\Phi(s) = \Phi_n(s) \cdot \Phi\left(\frac{s}{2^n}\right), \quad n = 0, 1, 2, \ldots$$

from Ex. 10.8 with the uniform bound (45) to show that Φ vanishes on every interval $[-b, b]$.

▶ **EXERCISE 10.10:**   Let $c_0, c_1, \ldots, c_M$ satisfy (28) and let ϕ be the support-limited generalized solution of (29), (30). Let $p = 1, 2, \ldots$ be chosen, and let
$$\theta_0(x) := \delta^{(p)}(x)$$
$$\theta_n(x) := \sum_{m=0}^{M} 2^p c_m \theta_{n-1}(2x - m), \quad n = 1, 2, \ldots,$$

cf. (33). Show that $\phi^{(p)}$ is the weak limit of $\theta_0, \theta_1, \theta_2, \ldots$.

▶ **EXERCISE 10.11:**   Let $M \geq 1$, let $c_0, c_1, \ldots, c_M$ satisfy (28), and assume that $c_0 \neq 0$, $c_M \neq 0$. Show that the dilation equation (29) has no solution of the form
$$\phi(x) = \sum_{n=1}^{N} a_n \delta(x - x_n)$$

with $0 \leq x_1 < x_2 < \cdots < x_N$ and $a_n \neq 0$ for each n.

▶ **EXERCISE 10.12:**   Let $c_0, c_1, \ldots, c_M$ satisfy (28) (with $c_m = 0$ when $m < 0$ or $m > M$) and let ϕ, ψ be the corresponding support-limited generalized functions from (29)–(31). Express in terms of ϕ, ψ the solution of (29)–(31) that we obtain when we replace c_m by:

(a) the reflection $c_m^R = c_{-m}$, $m = 0, \pm 1, \pm 2, \ldots$;

(b) the translate $c_m^T = c_{m-m_0}$, $m = 0, \pm 1, \pm 2, \ldots$ where m_0 is an integer;

(c) the reflected translate $c_m^{RT} = c_{M-m}$, $m = 0, \pm 1, \pm 2, \ldots$.

▶ **EXERCISE 10.13:**   Let $c_0, c_1, \ldots, c_M$ satisfy (28) and assume that the corresponding support-limited generalized solution of (29)–(30) has the symmetry $\phi(x) = \phi(M - x)$, $-\infty < x < \infty$. Show that $c_m = c_{M-m}$, $m = 0, 1, \ldots, M$.

Note. Such symmetry is incompatible with the orthogonality constraints (87) when $M > 1$. (Additional details can be found in Chapter 8 of Daubechies' monograph.)

▶ **EXERCISE 10.14:**   Let $\mathcal{P}$ be a polynomial and let $K = 1, 2, \ldots$.

(a) Show that $(1 + z)^K$ evenly divides $\mathcal{P}(z)$ if and only if

$$\mathcal{P}(-1) = \mathcal{P}'(-1) = \cdots = \mathcal{P}^{(K-1)}(-1) = 0.$$

Hint. Taylor's formula.

(b) Show that $(1 + e^{-2\pi is})^K$ evenly divides $T(s) := \mathcal{P}(e^{-2\pi is})$ if and only if

$$T\left(\tfrac{1}{2}\right) = T'\left(\tfrac{1}{2}\right) = \cdots = T^{(K-1)}\left(\tfrac{1}{2}\right) = 0.$$

Note. We use this result to produce the factorizations (55) and (59).

▶ **EXERCISE 10.15:**   Let $c_0, c_1, \ldots, c_M$ satisfy (28) and assume that (35) has the factorization (55). For $n = 1, 2, \ldots$ let

$$\Phi_n(s) := \prod_{\nu=1}^{n} T\left(\frac{s}{2^\nu}\right), \quad F_n(s) := \prod_{\nu=1}^{n} T_1\left(\frac{s}{2^\nu}\right), \quad P_n(s) := \prod_{\nu=1}^{n} \frac{1}{2}(1 + e^{-2\pi is/2^\nu}).$$

Establish the identity

$$\lim \Phi_n = \lim P_n \cdot \lim F_n$$

from (56) when these limits are taken pointwise and when they are taken weakly.

Hint. You know from the text that $\Phi_1, \Phi_2, \ldots$ converges pointwise as well as weakly to a bandlimited function Φ.

▶ **EXERCISE 10.16:** In this exercise you will show that "graphical convergence" of $\phi_1, \phi_2, \ldots$ that you observe in Figs. 10.9 and 10.10 cannot occur unless the corresponding coefficients $c_0, c_1, \ldots, c_M$ satisfy the regularity condition $T\left(\frac{1}{2}\right) = 0$.

(a) Let $N = 1, 2, 3, \ldots$, let $a_0, a_1, \ldots, a_N$ be real, let

$$f^{[e]}(x) := \sum_{m=even} a_m \delta\left(x - \frac{m}{N}\right), \quad f^{[o]}(x) := \sum_{m=odd} a_m \delta\left(x - \frac{m}{N}\right), \quad f(x) := f^{[e]}(x) + f^{[0]}(x),$$

and let $F^{[e]}, F^{[o]}, F$ be the corresponding Fourier transforms. Show that

$$F^{[e]}(s) = \frac{1}{2}\left\{F(s) + F\left(s + \frac{N}{2}\right)\right\}, \quad F^{[o]}(s) = \frac{1}{2}\left\{F(s) - F\left(s + \frac{N}{2}\right)\right\}.$$

(b) Let $c_0, c_1, \ldots, c_M$ satisfy (28) and let $\phi_1, \phi_2, \ldots$ be given by (37)–(38). Let $\phi_n^{[e]}$ be obtained by removing from ϕ_n the δ's at $x = (2m+1)/2^n$, $m = 0, 1, \ldots$. Show that $\phi_n^{[e]}$ has the Fourier transform

$$\Phi_n^{[e]}(s) = \Phi_{n-1}(s) \cdot \frac{1}{2}\left\{T\left(\frac{s}{2^n}\right) + T\left(\frac{s}{2^n} + \frac{1}{2}\right)\right\}.$$

Hint. Use (a), (36), and the 1-periodicity of (35).

(c) After examining Figs. 10.9 and 10.10 you might reasonably conjecture that $\phi_1^{[e]}, \phi_2^{[e]}, \ldots$ has the weak limit $\frac{1}{2}\phi$. Show that this is the case only if $T\left(\frac{1}{2}\right) = 0$.

Hint. Use (b) to show that

$$\tfrac{1}{2}\Phi(s) = \Phi(s) \cdot \tfrac{1}{2}\left\{T(0) + T\left(\tfrac{1}{2}\right)\right\}$$

when $\lim \phi_n^{[e]} = \frac{1}{2}\phi$.

Note. When $M = 2$, $c_0 = 1$, $c_1 = 0$, $c_2 = 1$ we have $T(\frac{1}{2}) = 1$. You may wish to find the weak limits of the corresponding sequences $\phi_1^{[e]}, \phi_2^{[e]}, \ldots$ and $\phi_1^{[o]}, \phi_2^{[o]}, \ldots$.

(d) Let $M = 1$ and let $c_0 + c_1 = 2$. What additional constraints must you impose on c_0, c_1 to ensure that
$$\lim \phi_n^{[e]} = \lim \phi_n^{[o]}?$$

▶ **EXERCISE 10.17:** Let ϕ be a support-limited generalized function with $\phi^\wedge(0) = 1$ [*e.g.*, as is the case for any scaling function obtained from (28)–(30)]. In this exercise you will study the convolution equation $\phi * \mathcal{P} = \mathcal{Q}$ associated with (71).

(a) Show that if $\mathcal{P}$ is a polynomial of degree n, then so is $\phi * \mathcal{P}$.

Hint. Use the differentiation rule $(\phi * \mathcal{P})^{(k)} = \phi * \mathcal{P}^{(k)}$, $k = 0, 1, \ldots$ from Ex. 7.19.

(b) Let $\mathcal{Q}$ be a polynomial of degree n. Show that there is a unique polynomial $\mathcal{P}$ (of degree n) such that $\phi * \mathcal{P} = \mathcal{Q}$.

▶ **EXERCISE 10.18:**   Derive the formula

$$\sum_{k=-\infty}^{\infty} \{k^3 + 3\mu_1 k^2 + [6\mu_1^2 - 3\mu_2]k + [6\mu_1^3 - 6\mu_1\mu_2 + \mu_3]\} = x^3 \quad \text{if } K \geq 4$$

that follows (73), (74), and (75).

Hint. In view of (71) you can find a_0, a_1, a_2 to make

$$\int_{-\infty}^{\infty} \{(x-u)^3 + a_2(x-u)^2 + a_1(x-u) + a_0\}\phi(u)du = x^3$$

or you can simplify

$$\mathcal{P}^\wedge(s) = \frac{1}{\Phi(s)} \cdot \frac{\delta'''(s)}{(-2\pi i)^3}$$

by using (7.82) and (72).

▶ **EXERCISE 10.19:**   Let $c_0, c_1, \ldots, c_M$ satisfy (28) and let

$$M_k := \sum_{m=0}^{M} c_m m^k, \quad k = 0, 1, 2, \ldots .$$

Let ϕ be the support-limited scaling function of (29)–(30) and let $\mu_0, \mu_1, \mu_2, \ldots$ be the moments of (72).

(a) Show that

$$\mu_k = \frac{1}{2^{k+1}} \sum_{\ell=0}^{k} \binom{k}{\ell} \mu_\ell M_{k-\ell}, \quad k = 0, 1, \ldots .$$

   Hint. Use the chain rule and the Leibnitz rule (2.29) for the kth derivative of

$$\Phi(2s) = \Phi(s) \cdot T(s).$$

(b) Using (a), derive the recursion

$$\mu_0 = 1, \qquad \mu_k = \frac{1}{2(2^k - 1)} \sum_{\ell=0}^{k-1} \binom{k}{\ell} \mu_\ell M_{k-\ell}, \quad k = 1, 2, \ldots$$

   for expressing the moments (72) in terms of $c_0, c_1, \ldots, c_M$.

▶ **EXERCISE 10.20:**   Let μ_1, μ_2 be moments (72) of the scaling function ϕ for the Daubechies coefficients (39). Show that

$$\mu_1 = \frac{3 - \sqrt{3}}{2}, \quad \mu_2 = \frac{6 - 3\sqrt{3}}{2}.$$

Hint. First use the analysis from Ex. 10.19 to derive the identities

$$\mu_1 = \frac{1}{2}M_1, \quad \mu_2 = \frac{1}{6}[M_2 + M_1^2].$$

▶ **EXERCISE 10.21:** Let μ_1, μ_2 be the moments (72) for a piecewise continuous support-limited scaling function ϕ constructed from coefficients $c_0, c_1, \ldots, c_M$ that satisfy (28), (58) with $K \geq 3$, and (87). Show that $\mu_2 = \mu_1^2$.

Hint. Multiply (75) by $\phi(x)$ and integrate.

▶ **EXERCISE 10.22:** Let the scaling function ϕ be constructed from (28)–(30), and assume that ϕ is continuous. Show that

$$\sum_{k=-\infty}^{\infty} \phi\left(\frac{k}{2^p}\right) = 2^p \quad \text{for each} \quad p = 0, 1, 2, \ldots .$$

Hint. Use (73) with a suitable induction.

▶ **EXERCISE 10.23:** Let ϕ be the support-limited scaling function of (28)–(30) corresponding to the Daubechies coefficients (39).

(a) Use (72)–(74) to verify that

$$\phi(x) + \qquad \phi(x+1) + \qquad \phi(x+2) = 1$$
$$\mu_1 \phi(x) + (\mu_1 - 1)\phi(x+1) + (\mu_1 - 2)\phi(x+2) = x$$

when $0 \leq x \leq 1$.

(b) Express $\phi(x+1)$ and $\phi(x+2)$ in terms of $\phi(x)$ when $0 \leq x \leq 1$ and thereby show that ϕ is determined by its values in $[0, 1]$.

Hint. The moment μ_1 is given in Ex. 10.20.

Note. More generally, the Daubechies scaling function ϕ with $M = 2K - 1$ is determined by its values on $[0, K - 1]$, $K = 2, 3, \ldots .$

▶ **EXERCISE 10.24:** Let the trigonometric polynomial (35) be chosen so that $T(0) = 1$, $T(-s) = \overline{T(s)}$, $|T(s)|^2 + |T(s + \frac{1}{2})|^2 = 1$, and $T(s) \neq 0$ for $0 \leq s \leq \frac{1}{2}$. Assume further that the series (82) converges uniformly on every finite interval to a continuous function $\mathcal{A}$. Show that $\mathcal{A}(s) \equiv 1$ [and thereby establish the orthogonality relations (78)].

Hint. Suppose that $\mathcal{A}$ takes its maximum or minimum value at some point s_0 with $0 < s_0 < 1$. Use (84) and (85) to show that

$$\min\left\{\mathcal{A}\left(\frac{s_0}{2}\right), \mathcal{A}\left(\frac{s_0}{2} + \frac{1}{2}\right)\right\} \leq \mathcal{A}(s_0) \leq \max\left\{\mathcal{A}\left(\frac{s_0}{2}\right), \mathcal{A}\left(\frac{s_0}{2} + \frac{1}{2}\right)\right\}$$

and thereby infer that $\mathcal{A}(s_0) = \mathcal{A}(s_0/2) = \mathcal{A}(s_0/4) = \cdots .$

▶ **EXERCISE 10.25:**   Let $c_0, c_1, \ldots, c_M$ satisfy the orthogonality constraints (87) with $c_0 \neq 0$, $c_M \neq 0$ and $M = 1, 3, 5, \ldots$. Show that

$$\sum_n \{c_{k-2n}c_{m-2n} + (-1)^{m-k}c_{M-m-2n}c_{M-k-2n}\} = \begin{cases} 2 & \text{if } k = m \\ 0 & \text{otherwise.} \end{cases}$$

Hint. Observe that you can write the sum in the form

$$\sum_{\ell=even} c_{\ell+k}c_{\ell+m} + (-1)^{m-k}\sum_{\ell=odd} c_{\ell-m}c_{\ell-k},$$

and then consider separately the cases where k is even, odd and m is even, odd.

▶ **EXERCISE 10.26:**   Let $c_0, c_1, \ldots, c_M$ satisfy (28) and the orthogonality constraints (87). Let

$$\theta_0(x) := \Pi(x), \qquad \theta_n(x) := \sum_{m=0}^{M} c_m \theta_{n-1}(2x - m), \quad n = 1, 2, \ldots,$$

and assume that $\theta_0, \theta_1, \theta_2, \ldots$ converges uniformly on $\mathbb{R}$ to a piecewise continuous function θ.

(a) Show that $\theta_1, \theta_2, \ldots$ and θ vanish outside the interval $[0, M]$.

(b) Show that θ satisfies the dilation equation (29)–(30).

(c) Show that

$$\int_{-\infty}^{\infty} \theta_n(x - k)\theta_n(x - \ell)dx = \begin{cases} 1 & \text{if } k = \ell \\ 0 & \text{if } k \neq \ell, \ n = 1, 2, \ldots \end{cases}$$

and

$$\int_{-\infty}^{\infty} \theta(x - k)\theta(x - \ell)dx = \begin{cases} 1 & \text{if } k = \ell \\ 0 & \text{if } k \neq \ell. \end{cases}$$

(d) What goes wrong when we try to use this *cascade algorithm* with the coefficients (86)?

▶ **EXERCISE 10.27:**   Two Fourier analysis students are trying to modify the algorithm of Fig. 10.13 so that it will generate the Daubechies coefficients $c_0, c_1, \ldots, c_M$ without using the discrete convolution product

$$(c_0, c_1, \ldots, c_M) = 2^{-K}(1, 1, 0, \ldots, 0) * \cdots * (1, 1, 0, \ldots, 0) * (\gamma_0, \gamma_1, \ldots, \gamma_{K-1}, 0, \ldots, 0).$$

"The algorithm produces the coefficients of Γ from samples of $Q_0(s) = \Gamma(s) \cdot \Gamma(-s)$," says the first student, "so if we begin with samples of $Q(s) = T(s) \cdot T(-s)$ from (100) we will end up with the desired coefficients of T."

"Your program will abort" responds the second. "The best way to avoid the convolution is to apply the factor $\cos^K(n\pi/N)$ to the sample $\Gamma(-n/N)$ from the penultimate step of the algorithm."

(a) What is wrong with the first suggestion?

(b) Show that the second suggestion will work (when N is sufficiently large).

▶ **EXERCISE 10.28:**   In this exercise you will show that the functions ϕ, ψ from Fig. 10.14 are continuous when $M = 2K - 1$ with $K = 2, 3, 4, \ldots$.

(a) Show that the trigonometric polynomial (101) has the maximum

$$Q_0 \left(\frac{1}{2} \right) = \binom{2K - 1}{K - 1}.$$

(b) Show that

$$\binom{2K - 1}{K - 1} \leq 3 \cdot 4^{K-2} \quad \text{when} \quad K = 2, 3, 4, \ldots \ .$$

(c) Use (a)–(b) to establish the bound

$$|\Gamma(s)| \leq \sqrt{3} \, 2^{K-2}$$

for the (hermitian) trigonometric polynomial Γ from the spectral factorization (102).

(d) Use (61) (with $T_K = \Gamma$) and (c) to show that

$$p_K := \log_2 \max_{0 \leq s \leq 2\pi} |\Gamma(s)| \leq K - 1.2.$$

(e) Use (45) with (d) to bound the infinite product from (60) and thereby show that $|\Phi(s)|$ and $|s^{1.2} \cdot \Phi(s)|$ are bounded on $\mathbb{R}$ when $K \geq 2$.

(f) Use (e) with Ex. 1.38 to show that ϕ is continuous.

(g) Show that ψ is continuous.

Note. You can sharpen the analysis slightly by using

$$K - \log_2 \binom{2K - 1}{K - 1}$$

in place of 1.2 when $K = 3, 4, \ldots$.

▶ **EXERCISE 10.29:**   Let $\phi(x)$ be a piecewise continuous function that vanishes when $x \leq 0$ or $x \geq M$.

(a) Show that at most $M - 2 + \lfloor b \rfloor - \lceil a \rceil + \lceil \lceil a \rceil - a \rceil + \lceil b - \lfloor b \rfloor \rceil$ terms contribute to the sum

$$f_0(x) = \sum_{k=-\infty}^{\infty} \alpha_0[k] \phi(x - k)$$

on the finite interval $a \leq x \leq b$. Here $\lfloor \ \rfloor$, $\lceil \ \rceil$ are the floor, ceiling functions, *cf.* Ex. 4.29.

(b) Find a similar estimate for the number of terms that contribute to the sum

$$f_m(x) = \sum_{k=-\infty}^{\infty} \alpha_m[k] \phi(2^m x - k)$$

on $[a, b]$.

▶ **EXERCISE 10.30:**   Let ϕ, ψ be piecewise continuous support-limited functions that satisfy the orthogonality relations (79)–(81) and the two scale relations (29), (31),(90), (91).   Let f be a piecewise continuous function on $\mathbb{R}$, and let f_m, d_m be given by (106)–(109). Establish the multiresolution identity (111) by working out the details for the following argument.

(a) Use (90), (91) to establish the identity

$$\phi(2^{m+1}x - k) = \frac{1}{2}\sum_n \{c_{k-2n}\phi(2^m x - n) + (-1)^k c_{M-k+2n}\psi(2^m x - n)\}$$

when $m = 0, \pm 1, \pm 2, \ldots$ and $k = 0, \pm 1, \pm 2, \ldots$.

Hint. Consider separately the cases where k is even and where k is odd.

(b) Use (a) with (108)–(109) to show that

$$\alpha_{m+1}[k] = \sum_n \{c_{k-2n}\alpha_m[n] + (-1)^k c_{M-k+2n}\beta_m[n]\}.$$

(c) Use (106)–(107) with (29), (31), and (b) to show that

$$f_m(x) + d_m(x) = f_{m+1}(x), \quad -\infty < x < \infty, \quad m = 0, \pm 1, \pm 2, \ldots .$$

▶ **EXERCISE 10.31:**   Let f be a continuous, absolutely integrable function on $\mathbb{R}$, and assume that $f(x) \to 0$ as $x \to \pm\infty$. Let ϕ, ψ be the Daubechies scaling function, wavelet for some $M = 3, 5, \ldots$ (as described in Section 10.2). Let the wavelet transform F be defined by (17) [using Daubechies' ψ in place of the Haar wavelet (1)]. In this exercise you will establish the synthesis equation (4) [using Daubechies' ψ in place of the Haar wavelet (1)].

(a) Let f_m be defined by (106) and (108). Show that

$$|f_m(x)| \leq M 2^m \cdot \{\max |\phi|\}^2 \cdot \int_{-\infty}^{\infty} |f(x)|dx.$$

(b) Let d_m be defined by (107) and (109). Show that

$$f(x) = \lim_{\substack{I \to -\infty \\ J \to +\infty}} \sum_{m=I}^{J} d_m(x) =: \sum_{m=-\infty}^{\infty} \sum_{k=-\infty}^{\infty} \beta_m[k]\psi(2^m x - k)$$

and thereby establish (4) (with $F[m, k] := \beta_m[k]$).

Hint. Use (115) and (a) with the inequality

$$\left| f(x) - \sum_{m=I}^{J} d_m(x) \right| \leq |f(x) - f_{J+1}(x)| + |f_I(x)|.$$

▶ **EXERCISE 10.32:**   Let $c_0, c_1, \dots, c_M$ satisfy the orthogonality constraints (87) with $c_0 \neq 0$, $c_M \neq 0$ and $M = 1, 3, 5, \dots$. Give a direct argument to show that the operators $\mathbf{L}_-$, $\mathbf{H}_-$ $\mathbf{L}_+$, $\mathbf{H}_+$ from (118), (119), (122), (123) satisfy the identities:

(a) (125);   (b) (126);   (c) (127).

Hint. You may wish to use (89) and the identity of Ex. 10.25.

▶ **EXERCISE 10.33:**   Use (125)–(127) to verify the following identities for the operators (118)–(119), (122)–(123) constructed from coefficients $c_0, c_1, \dots, c_M$ that satisfy (87), *cf.* Ex. 10.32.

(a) $(\mathbf{L}_+\mathbf{L}_-)(\mathbf{L}_+\mathbf{L}_-) = \mathbf{L}_+\mathbf{L}_-$ (c) $(\mathbf{L}_+\mathbf{L}_-)(\mathbf{H}_+\mathbf{H}_-) = (\mathbf{H}_+\mathbf{H}_-)$

(b) $(\mathbf{H}_+\mathbf{H}_-)(\mathbf{H}_+\mathbf{H}_-) = \mathbf{H}_+\mathbf{H}_-$ (d) $\mathbf{L}_+\mathbf{L}_+\mathbf{L}_-\mathbf{L}_- + \mathbf{L}_+\mathbf{H}_+\mathbf{H}_-\mathbf{L}_- + \mathbf{H}_+\mathbf{H}_- = \mathbf{I}$

▶ **EXERCISE 10.34:**   Let f be a suitably smooth function on $\mathbb{R}$, and let f_m be the frame (106) having coefficients (108) computed with the Daubechies scaling function ϕ obtained from the coefficients (39). In this exercise you will analyze the errors associated with the approximations

$$\alpha_m[k] \approx f(2^{-m}k), \quad \alpha_m[k] \approx f_m(2^{-m}k), \quad \text{and} \quad f(2^{-m}k) \approx f_m(2^{-m}k).$$

(a) Use (108) (with Taylor's formula) to show that

$$\alpha_m[k] - f(2^{-m}k) = \mu_1 f'(2^{-m}k) \cdot 2^{-m} + \mu_2 \frac{f''(2^{-m}k)}{2!} \cdot 2^{-2m} + \cdots$$

$$\approx \frac{3 - \sqrt{3}}{2} f'(2^{-m}k) \cdot 2^{-m} \quad \text{when } m \text{ is large}.$$

Hint. The moments μ_1, μ_2 are evaluated in Ex. 10.20.

(b) Show that

$$\alpha_m[k] - f_m(2^{-m}k) = \{\phi(1) + 2\phi(2)\} f'(2^{-m}k) \cdot 2^{-m}$$

$$+ \frac{1}{2}\{(2\mu_1 - 1)\phi(1) + (4\mu_1 - 4)\phi(2)\} f''(2^{-m}k) \cdot 2^{-2m} + \cdots$$

$$\approx \frac{3 - \sqrt{3}}{2} f'(2^{-m}k) \cdot 2^{-m} \quad \text{when } m \text{ is large}.$$

Hint. Begin by deriving the identity

$$\alpha_m[k] - f_m(2^{-m}k) = \phi(1)\{\alpha_m[k] - \alpha_m[k-1]\} + \phi(2)\{\alpha_m[k] - \alpha_m[k-2]\}$$

and then use (108) with Taylor's formula. The samples of ϕ are given in (133).

(c) Use (a) and (b) to show that

$$f(2^{-m}k) - f_m(2^{-m}k) = \tfrac{1}{2}\{(2\mu_1 - 1)\phi(1) + (4\mu_1 - 4)\phi(2)$$
$$- \mu_2\}f''(2^{-m}k) \cdot 2^{-2m} + \cdots$$
$$\approx \tfrac{1}{4}f''(2^{-m}k) \cdot 2^{-2m} \quad \text{when } m \text{ is large.}$$

Note. If f is a suitably smooth function and ϕ is a piecewise continuous scaling function constructed from coefficients $c_0, c_1, \ldots, c_M$ that satisfy (28) and (58), then we can use a similar analysis to show that

$$\alpha_m[k] - f(2^{-m}k) \approx C_1 f'(2^{-m}k) \cdot 2^{-m}$$
$$\alpha_m[k] - f_m(2^{-m}k) \approx C_2 f'(2^{-m}k) \cdot 2^{-m}$$
$$f(2^{-m}k) - f_m(2^{-m}k) \approx C_3 f^{(K)}(2^{-m}k) \cdot 2^{-mK}$$

when m is sufficiently large. Here C_1, C_2, C_3 are constants that depend on ϕ (but not on m, k, or f).

▶ **EXERCISE 10.35:**   Let f_m, α_m be constructed from $f(x) := e^{-128x^2}$ (a translate of the function f from Fig. 10.17) as described in Ex. 10.34.

(a) Show that $\max|f'(x)| = 16e^{-1/2}$, $\max|f''(x)| = 256$.

(b) Use (a) with the analysis from Ex. 10.34 to show that

$$\max|\alpha_m[k] - f(2^{-m}k)| \approx 6.15 \cdot 2^{-m}$$
$$\max|\alpha_m[k] - f_m(2^{-m}k)| \approx 6.15 \cdot 2^{-m}$$
$$\max|f(2^{-m}k) - f_m(2^{-m}k)| \approx 64 \cdot 4^{-m}$$

when m is large.

Note. These approximations are quite good even when $m = 8$.

▶ **EXERCISE 10.36:**   Let f be a continuous support-limited function on $\mathbb{R}$ with the modulus of continuity (114). Let ϕ be a piecewise continuous support-limited scaling function constructed from (29)–(30) using coefficients $c_0, c_1, \ldots, c_M$ that satisfy (28) and (58) with $K \geq 1$. Show that

$$\left| f(x) - \sum_k f(2^{-m}k)\phi(2^m x - k) \right| \leq M \cdot \max|\phi| \cdot \omega(2^{-m}M), \quad -\infty < x < \infty.$$

Hint. Begin by studying the proof of (115).

▶ **EXERCISE 10.37:**   Let ϕ be a piecewise continuous scaling function that vanishes outside of the interval $[0, M]$, $M = 1, 2, 3, \ldots$, and assume that ϕ satisfies the orthogonality relations (79) as well as (73) and the constraint

$$\int_0^M \phi(x)dx = 1$$

from (30). Let f_m be the approximation of (106) and (108) for the Heaviside function

$$f(x) := \begin{cases} 1 & \text{if } x > 0 \\ 0 & \text{if } x < 0. \end{cases}$$

(a) Show that

$$f(x) - f_0(x) = \begin{cases} I_1\phi(x + 1) + I_2\phi(x + 2) + \cdots + I_{M-1}\phi(x + M - 1) \\ \qquad\qquad\qquad \text{when } x > 0 \\ (I_1 - 1)\phi(x + 1) + (I_2 - 1)\phi(x + 2) + \cdots + (I_{M-1} - 1)\phi(x + M - 1) \\ \qquad\qquad\qquad \text{when } x < 0. \end{cases}$$

Here

$$I_n := \int_0^n \phi(x)dx, \quad n = 1, 2, \ldots$$

(with $I_n = 1$ when $n \geq M$).

(b) Find a corresponding expression for $f(x) - f_m(x)$, $m = 0, \pm 1, \pm 2, \ldots$.

Note. You can use this result to study the Gibbs ripple illustrated in Fig. 10.11.

▶ **EXERCISE 10.38:**   Let $M = 2K - 1$ for some $K = 2, 3, 4, \ldots$, let $c_0, c_1, \ldots, c_M$ be obtained from the spectral factorization of (100), and let ϕ be the corresponding support-limited scaling function from (29)–(30). This exercise will show you how to compute the samples $\phi(n)$, $n = 1, 2, \ldots, M - 1$ that are used to define the operator (134) and its inverse.

(a) Let $\alpha[n] := \phi(n)$, $n = 0, \pm 1, \pm 2, \ldots$ (so that α is a function on $\mathbb{Z}$). Show that

$$\alpha = \mathcal{S}_2(c * \alpha)$$

where $\mathcal{S}_2$ is the dilation operator (140).

Hint. Use the dilation equation (29) with x replaced by n.

(b) Show that when $M = 7$,

$$\begin{bmatrix} \phi[1] \\ \phi[2] \\ \phi[3] \\ \phi[4] \\ \phi[5] \\ \phi[6] \end{bmatrix} = \begin{bmatrix} c_1 & c_0 & 0 & 0 & 0 & 0 \\ c_3 & c_2 & c_1 & c_0 & 0 & 0 \\ c_5 & c_4 & c_3 & c_2 & c_1 & c_0 \\ c_7 & c_6 & c_5 & c_4 & c_3 & c_2 \\ 0 & 0 & c_7 & c_6 & c_5 & c_4 \\ 0 & 0 & 0 & 0 & c_7 & c_6 \end{bmatrix} \begin{bmatrix} \phi[1] \\ \phi[2] \\ \phi[3] \\ \phi[4] \\ \phi[5] \\ \phi[6] \end{bmatrix}.$$

Analogously, when $M = 3, 5, 7, \ldots$ we can find $c_0, c_1, \ldots, c_M$ numerically, form the matrix $[c_{2i-j}]_{i,j=1}^{M-1}$, and obtain $\phi(1), \phi(2), \ldots, \phi(M-1)$ by suitably normalizing the components of an eigenvector that corresponds to the eigenvalue $\lambda = 1$.

(c) Use the identity of (a) with (143) to show that

$$\alpha^\vee = 2\mathbf{L}_-\alpha^\vee$$

where α is the sample function from (a), $^\vee$ is the reflection tag (5.35), and $\mathbf{L}_-$ is the operator (118).

Note. If you work with wavelets you will have code that allows you to apply the operator $\mathbf{L}_-$ to a function on $\mathbb{P}_{2M}$. Here is an easy way to compute $\phi(1), \ldots, \phi(M-1)$. Set

$$\boldsymbol{\alpha}_0 := \frac{1}{M-1}(0, \ 1, \ 1, \ \ldots, \ 1) \text{ on } \mathbb{P}_M$$

and compute

$$\boldsymbol{\alpha}_{n+1} := \left\{2\mathbf{L}_-(\mathcal{P}_2\boldsymbol{\alpha}_n)^\vee\right\}^\vee, \quad n = 0, 1, 2, \ldots .$$

Here

$$\mathcal{P}_2(x_0, x_1, \ldots, x_{M-1}) := (x_0, x_1, \ldots, x_{M-1}, 0, 0, \ldots, 0)$$

is the end padding operator of Ex. 5.35 (that doubles the number of components). The sequence $\boldsymbol{\alpha}_0, \boldsymbol{\alpha}_1, \ldots$ rapidly converges to

$$\boldsymbol{\alpha} := (0, \phi[1], \phi[2], \ldots, \phi[M-1])$$

when M is of modest size, *e.g.*, $M \leq 21$.

▶ **EXERCISE 10.39:**   The operator (134) has the representation

$$\mathbf{P}_+\alpha = p * \alpha$$

where

$$p[n] := \begin{cases} \phi(n) & \text{if } n = 1, 2, \ldots, M-1 \\ 0 & \text{otherwise.} \end{cases}$$

When ϕ is one of the Daubechies scaling functions from Section 10.2 with $M = 3, 5, \ldots, 21$, you can (numerically) verify that

$$P(s) := \sum_{n=1}^{M-1} \phi[n]e^{-2\pi ins}$$

has the factorization

$$P(s) = e^{-2\pi is} \prod_{n=1}^{M-2} \left(\frac{1 - \zeta_n e^{-2\pi is}}{1 - \zeta_n}\right)$$

with $|\zeta_n| < 1$ for each n. Within this context you will learn to solve $\mathbf{P}_+\alpha = \gamma$ to find $\alpha = \mathbf{P}_-\gamma$.

(a) Let α, γ be bounded functions on $\mathbb{R}$ and let $|\zeta| < 1$. Show that

$$\gamma[k] = \alpha[k] - \zeta\,\alpha[k-1], \quad k = 0, \pm 1, \pm 2, \ldots$$

if and only if

$$\alpha[k] = \gamma[k] + \zeta\,\gamma[k-1] + \zeta^2\gamma[k-2] + \cdots, \quad k = 0, \pm 1, \pm 2, \ldots\ .$$

(b) Let γ be a bounded sequence on $\mathbb{Z}$ and let p, P be as described above. Show that there is a unique bounded solution α to the equation $p * \alpha = \gamma$.

Hint. Write $p = C \cdot (\ldots, 0, 1, \ldots) * (\ldots, 1, -\zeta_1, \ldots) * \cdots * (\ldots, 1, -\zeta_{M-1}, \ldots)$ where $C \neq 0$ is a constant, and then use (a).

(c) Let γ be a suitably localized function on $\mathbb{Z}$. Explain why the same is true of the solution α of $p * \alpha = \gamma$.

(d) Let γ be a function on $\mathbb{Z}$ that is suitably localized in $[0, N]$ for some sufficiently large N. Explain how you would use the FFT to compute a good approximate solution of $p * \alpha = \gamma$ and thereby obtain $\alpha = \mathbf{P}_-\gamma$.

Note. Exercise 10.38 shows how to compute the samples $\phi(1), \phi(2), \ldots, \phi(M-1)$ that you need for this purpose.

▶ **EXERCISE 10.40:**   This exercise will show you two ways to compute samples $\phi(n/2^m)$, $\psi(n/2^m)$ for the Daubechies scaling functions and wavelets shown in Fig. 10.14. Both schemes make use of the samples $\phi(n)$, $n = 0, \pm 1, \pm 2, \ldots$ from Ex. 10.38.

(a) Knowing $\phi(n)$, $n = 0, \pm 1, \pm 2, \ldots$, explain how to compute in turn the samples $\phi(n/2^m)$, $n = 0, \pm 1, \pm 2, \ldots$ for each $m = 1, 2, \ldots\ .$

Hint. Use the dilation equation (29).

(b) Knowing the samples $\phi(n/2^{m-1})$, $n = 0, \pm 1, \pm 2, \ldots$, explain how to compute the samples $\psi(n/2^m)$, $n = 0, \pm 1, \pm 2, \ldots\ .$

(c) For an alternative calculation, show that

$$\phi(n/2^m) = (\mathbf{P}_+\mathbf{L}_+^m\delta)[n], \qquad n = 0, \pm 1, \pm 2, \ldots, \quad m = 1, 2, \ldots$$
$$\psi(n/2^m) = (\mathbf{P}_+\mathbf{L}_+^{m-1}\mathbf{H}_+\delta)[n], \quad n = 0, \pm 1, \pm 2, \ldots, \quad m = 1, 2, \ldots,$$

where $\mathbf{L}_+$, $\mathbf{H}_+$, $\mathbf{P}_+$ are the operators (122), (123), and (134).

Hint. Use (128), (129), and (135).

▶ **EXERCISE 10.41:**   Suppose that the output of the filter bank from Fig. 10.19 is a translate of the input, *i.e.*, $\alpha_+[n] = \alpha[n - n_0]$, $n = 0, \pm 1, \pm 2, \ldots$ for some integer n_0. (In practice, the output is always delayed, *i.e.*, $n_0 > 0$.)

(a) Suitably generalize the constraints (159)–(160) (that were derived for the case where $n_0 = 0$).

(b) Express the constraints from (a) in terms of convolution products of the filter coefficient sequences c_-, d_-, c_+, d_+.

Hint. Let $c_-^a[n] := (-1)^n c_-[n]$, $d_-^a[n] := (-1)^n d_-[n]$.

▶ **EXERCISE 10.42:**   Let c_-, c_+, d_-, d_+ be real sequences (each with finitely many nonzero components) and assume that the Fourier transforms (153) satisfy the constraints (159), (160) for a perfect reconstruction filter bank. Let m_-, m_+, n_-, n_+ be integers. Show that the translated sequences

$$c_-^t[n] := c_-[n - m_-], \qquad c_+^t[n] := c_+[n - m_+],$$
$$d_-^t[n] := d_-[n - n_-], \qquad d_+^t[n] := d_+[n - n_+]$$

produce a perfect reconstruction filter bank if and only if $m_- + n_-$ is even and $m_- + m_+ = n_- + n_+ = 0$.

Hint. When c_-^t, c_+^t, d_-^t, d_+^t correspond to a perfect reconstruction filter bank you can use the antialiasing constraint (159) to write

$$\left[\begin{array}{cc} 1 & 1 \\ (-1)^{m_-}e^{-2\pi i(m_-+m_+)s} & (-1)^{n_-}e^{-2\pi i(n_-+n_+)s} \end{array} \right] \left[\begin{array}{c} C_-\left(s + \tfrac{1}{2}\right)C_+(s) \\ D_-\left(s + \tfrac{1}{2}\right)D_+(s) \end{array} \right] = \left[\begin{array}{c} 0 \\ 0 \end{array} \right],$$

and thereby infer that $(-1)^{m_-} = (-1)^{n_-}$ and that $m_- + m_+ = n_- + n_+$.

▶ **EXERCISE 10.43:**   Let c_-, c_+ be real sequences (each with finitely many nonzero components), let C_-, C_+ be the Fourier transforms (153), and let Q be given by (165). Show that the biorthogonality constraint (167) is equivalent to (166).

▶ **EXERCISE 10.44:**   Let c_-, c_+, d_-, d_+ be real sequences (each with finitely many nonzero components) and assume that the Fourier transforms (153) satisfy the constraints (159), (160) for a perfect reconstruction filter bank. In this exercise you will show that d_-, d_+ and c_-, c_+ must be linked by a suitably scaled variation of the alternating flip relations (163). For this purpose we will set $z := e^{2\pi i s}$ and write

$$C_-(s) = \frac{\mathcal{C}_-(z)}{z^{m_-}}, \quad C_+(s) = \frac{\mathcal{C}_+(z)}{z^{m_+}}, \quad D_-(s) = \frac{\mathcal{D}_-(z)}{z^{n_-}}, \quad D_+(s) = \frac{\mathcal{D}_+(z)}{z^{n_+}}$$

where m_-, m_+, n_-, n_+ are integers and $\mathcal{C}_-$, $\mathcal{C}_+$, $\mathcal{D}_-$, $\mathcal{D}_+$ are polynomials that do not vanish at $z = 0$.

(a) Use (159) to show that $m_- + m_+ = n_- + n_+$ and that

$$(-1)^{m_-}\mathcal{C}_-(-z)\mathcal{C}_+(z) + (-1)^{n_-}\mathcal{D}_-(-z)\mathcal{D}_+(z) = 0.$$

(b) Use (160) (and the identity $m_- + m_+ = n_- + n_+$) to show that

$$\mathcal{C}_-(z)\mathcal{C}_+(z) + \mathcal{D}_-(z)\mathcal{D}_+(z) = 2z^{m_- + m_+}.$$

(c) Show that

$$\mathcal{C}_-(-z) = \quad (-1)^{m_-}\mathcal{D}_+(z) \cdot \mathcal{P}(z)$$
$$\mathcal{D}_-(-z) = -(-1)^{n_-}\mathcal{C}_+(z) \cdot \mathcal{P}(z)$$

for some polynomial $\mathcal{P}$.

Hint. From (b) you know that $\mathcal{C}_+$ and $\mathcal{D}_+$ have no common factors.

(d) Show that $\mathcal{P}(z) = -az^p$ for some constant $a \neq 0$ and some integer p.

Hint. Use (b) with the factorizations from (c).

(e) Show that

$$\frac{\mathcal{C}_-(z)}{z^{m_-}} \cdot \frac{\mathcal{C}_+(z)}{z^{m_+}} - (-1)^{p+m_+-n_-} \frac{\mathcal{C}_-(-z)}{(-z)^{m_-}} \cdot \frac{\mathcal{C}_+(-z)}{(-z)^{m_+}} = 2$$

and thereby infer that $N := p + m_+ - n_-$ is odd.

(f) Show that

$$D_-(s) = -aC_+\left(s + \tfrac{1}{2}\right) e^{2\pi i N s}$$
$$D_+(s) = -a^{-1}C_-\left(s + \tfrac{1}{2}\right) e^{-2\pi i N s}.$$

(g) Show that

$$d_-[n] = a(-1)^n c_+[n + N]$$
$$d_+[n] = a^{-1}(-1)^n c_-[n - N].$$

▶ **EXERCISE 10.45:** In this exercise you will use three methods to find the trigono-metric polynomial Q that corresponds to the Daubechies scaling function with $M = 3$, *cf.* Fig. 10.14.

(a) Use (35) with the coefficients (39) to verify that

$$Q(s) := |T(s)|^2 = \frac{1}{16}\{8 + 9\cos(2\pi s) - \cos(6\pi s)\}.$$

(b) Derive the above formula for Q by using (99) with $K = 2$.

(c) Derive the above formula for Q by using (100) with $K = 2$.

▶ **EXERCISE 10.46:** In this exercise you will find the coefficients for the 9/7 FBI filters that are used to compress digitized fingerprint files.

(a) Let $K = 4$. Show that the trigonometric polynomial

$$Q(s) = \cos^8(\pi s)\{\cos^6(\pi s) + 7\cos^4(\pi s)\sin^2(\pi s) + 21\cos^2(\pi s)\sin^4(\pi s) + 35\sin^6(\pi s)\}$$

from (100) can be written in the form

$$Q(s) = \frac{z^{-7}}{2^{12}} \cdot (z + 1)^8 \cdot \{-5 + 40z - 131z^2 + 208z^3 - 131z^4 + 40z^5 - 5z^6\}$$

where $z := e^{2\pi i s}$.

(b) Verify that the 6th-degree polynomial from (a) has the roots

$$\zeta = .32887\ldots, \quad \eta = .28409\ldots + .24322\ldots i$$

and thereby produce the factorization (165) with

$$C_-(s) = \sqrt{2}\, z^{-4} \left(\frac{z+1}{2}\right)^4 \frac{(z-\eta)(z-\overline{\eta})(z-\eta^{-1})(z-\overline{\eta}^{-1})}{(1-\eta)(1-\overline{\eta})(1-\eta^{-1})(1-\overline{\eta}^{-1})}$$

$$C_+(s) = \sqrt{2}\, z^{-3} \left(\frac{z+1}{2}\right)^4 \frac{(z-\zeta)(z-\zeta^{-1})}{(1-\zeta)(1-\zeta^{-1})}.$$

(c) Use a computer to obtain the coefficients

$$
\begin{aligned}
c_-[0] &= &.85269867\ldots, \qquad c_+[0] &= &.78848561\ldots, \\
c_-[\pm 1] &= &.37740285\ldots, \qquad c_+[\pm 1] &= &.41809227\ldots, \\
c_-[\pm 2] &= &-.11062440\ldots, \qquad c_+[\pm 2] &= &-.04068941\ldots, \\
c_-[\pm 3] &= &-.02384946\ldots, \qquad c_+[\pm 3] &= &-.06453888\ldots, \\
c_-[\pm 4] &= &.03782845\ldots,
\end{aligned}
$$

of the polynomials C_-, C_+ from (b).

(d) Find corresponding coefficients d_-, d_+ for a perfect reconstruction filter bank.

▶ **EXERCISE 10.47:**   Let c_-, d_- and c_+, d_+ be coefficients for a perfect reconstruction filter bank, and let $\mathbf{L}_-, \mathbf{H}_-$ and $\mathbf{L}_+, \mathbf{H}_+$ be the corresponding analysis and synthesis operators from (149)–(150). Let μ be the number of nonzero coefficients from c_- and d_- (as well as from c_+ and d_+, cf. Ex. 10.44). Let α be a function on $\mathbb{P}_N$ and assume that $N = N_0 \cdot 2^m$ for some positive integers N_0, m.

(a) Show that we expend approximately μN operations when we split α into its constituent parts

$$\mathbf{H}_-\alpha, \ \mathbf{H}_-\mathbf{L}_-\alpha, \ \mathbf{H}_-\mathbf{L}_-^2\alpha, \ \ldots, \ \mathbf{H}_-\mathbf{L}_-^{m-1}\alpha, \ \mathbf{L}_-^m\alpha.$$

(b) Show that we expend approximately μN operations when we reassemble α from the constituent parts of (a).

Hint. Suitably generalize the analysis associated with (26) and (27).

Chapter 11

Musical Tones

11.1 Basic Concepts

Introduction

In 1799 Ludwig Van Beethoven composed his Sonata No. 8 in C minor *Pathétique* and published the corresponding 18-page music score, *cf.* Fig. 11.1. A few years ago John O'Connor used a Hamburg Steinway piano to create an audio signal from Beethoven's score, and TELARC® produced a digital recording of the concert. I purchased a compact disk with this recording, and my home stereo system converts the 13 million samples into a faithful representation of O'Connor's rendition of Beethoven's work.

Figure 11.1. A portion of the *Adagio* from the 18-page score for Beethoven's Sonata No. 8 in C minor.

As I listen to the sonata, I hear a collage of *musical tones* that correspond to the individual chords of the composition. When O'Connor strikes the keys for a chord, the piano produces an audio signal $w(t)$. From your study of sampling in Chapter 8, you know that we can reconstruct a musical tone from its T-samples (provided that T satisfies the Nyquist condition). We can synthesize the audio signal for the sonata by adding the w's for the individual musical tones, and we can obtain T-samples for the sonata by adding corresponding T-samples of these w's.

In this chapter we will explore a natural extension of these ideas. We will *mathematically* define some function $w(t)$, use a computer to prepare a file of T-samples, and produce the corresponding audio signal by sending these samples to the computer's sound card. For example, if we play the samples

$$w_n := \sin\left\{2\pi \cdot 440 \text{ Hz} \cdot n\frac{1 \text{ sec}}{8000}\right\}, \quad n = 0, 1, \dots, 8000$$

at a rate of 8000 samples/sec we produce a 440-Hz sine tone that lasts for 1 sec. (If your sound card expects an input of 1-byte integers, you can replace each w_n by $w_n' := \lfloor 128(w_n + 1)\rfloor$.)

As you study this material you will find it very beneficial to create and play such sound files. You might think that the musical tone from the above samples $w_0, w_1, \dots, w_{8000}$ is comparable to that produced by striking the $A4$ key from a Hamburg-Steinway piano since both correspond to the same 440-Hz pitch, *cf.* Appendix 7. The pitch *is* the same, but your ear can easily distinguish between the simplistic computer-generated sinusoid and the rich auditory sensation produced by the time-varying overtones in the sound that comes from a concert piano!

Perception of pitch and loudness

The human auditory system will respond to a sinusoidal pressure wave

$$w(t) := A\sin(2\pi f t) \tag{1}$$

when the amplitude $A > 0$ and the frequency $f > 0$ are suitably chosen. We will briefly describe how these parameters affect our perception of the pitch and loudness of such a tone. You can find additional details in the psychoacoustics chapter of *The Computer Music Tutorial* cited in the references.

The *pitch* of (1) is determined by the frequency parameter f. As a rule, you can hear the tone when

$$20 \text{ Hz} \leq f \leq 20000 \text{ Hz}.$$

(Of course, there are individual differences: The lower limit varies from 20 Hz to 60 Hz and the upper limit varies from 8000 Hz to 25000 Hz.) Within this range, the ear perceives frequency on a logarithmic scale, *i.e.*, the *musical interval* between tones having the frequencies f_1, f_2 is determined by the *ratio* f_2/f_1. For example, when you press in turn the piano keys for the frequencies

$$\ldots, \ 110 \text{ Hz}, \ 220 \text{ Hz}, \ 440 \text{ Hz}, \ 880 \text{ Hz}, \ \ldots$$

(*cf.* Appendix 7), you will find that successive tones seem to be evenly spaced in pitch. You can use any consecutive tones from this sequence to sing the first two notes of Dorothy's "*Somewhere, over the rainbow, ...* " from the *Wizard of Oz.*

The physical intensity of the pressure wave (1) is proportional to the *square* of the amplitude A (and independent of the frequency). When $f \approx 3000$ Hz, your ear will respond to intensities that range from 10^{-12} watt/m^2 at the threshhold of hearing to 1 watt/m^2 near the threshold of pain. (Comparable light intensities can be obtained by viewing a 100-watt light bulb from distances of 1000 km and 1 m!) When f is fixed, the ear perceives *loudness* on a logarithmic scale, and we use

$$10 \log_{10} \left(\frac{I_2}{I_1} \right) = 10 \log_{10} \left(\frac{A_2^2}{A_1^2} \right) = 20 \log_{10} \left(\frac{A_2}{A_1} \right)$$

to compare the loudness of tones having the intensities I_1, I_2 and corresponding amplitudes A_1, A_2. We attach the dimensionless *decibel* (dB) unit to such expressions. For example, when

$$w_1(t) := A \sin(2\pi ft) \quad \text{and} \quad w_2(t) := 2A \sin(2\pi ft)$$

we say that the tone w_2 is

$$20 \log_{10} \left(\frac{2A}{A} \right) = 6.02 \text{ dB}$$

louder than the tone w_1. As a rule of thumb (based on countless experiments with human subjects) the loudness of the tone (1) seems to double whenever the intensity increases by 10 dB, *i.e.*, when the amplitude A increases by the factor $\sqrt{10} = 3.16 \ldots$.

The Fletcher-Munsen contours from Fig. 11.2 show how the perception of loudness varies with f as well as A, *cf.* Fig. 4 from H. Fletcher and W.A. Munsen, Loudness: Its definition, measurement, and calculation, *J. Acous. Soc. Am.* **5**(1933), 82–108. The bottom curve specifies the amplitude $A_0(f)$ for a just audible tone (1) at the frequency f. [We measure $A_0(f)$ in dB relative to the minimum amplitude that occurs near $f = 3000$ Hz between $C7$ and $C8$.] For example, as f increases 3 octaves from 130.8 Hz at $C3$ to 1046 Hz at $C6$, $A_0(f)$ decreases by approximately 30 dB. Thus the amplitude for a just audible $C3$ tone is approximately $10^{30/20} \approx 31.6$ times larger than that for a just audible $C6$ tone. When we jump from one of these contours to the one that lies directly above, the loudness more or less doubles. You can use such contours to determine the amplitude for a tone (1) that has a desired loudness level.

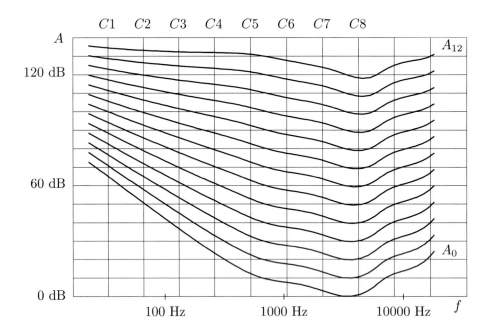

Figure 11.2. Relative intensity (in dB) as a function of frequency (in octaves) for tones (1) that seem to be equally loud. The frequency for $C4$ (middle C) is 261.6 Hz.

Ohm's law

When you listen to a sinusoidal tone

$$w(t) := A\sin(2\pi ft + \phi)$$

you perceive a pitch that depends on the frequency f and a loudness that depends on both f and the amplitude $A > 0$. The phase parameter ϕ has no effect on what you hear, *cf.* Ex. 11.4. The same is (more or less) true when you listen to sums of such sinusoids. Your eye can easily see the difference between the graphs of

$$w_1(t) := \sin(2\pi\,400t) + \sin(2\pi\,500t) + \sin(2\pi\,600t)$$

$$w_2(t) := \sin(2\pi\,400t) + \sin(2\pi\,500t) - \sin(2\pi\,600t),$$

(2)

cf. Fig. 11.3, but you cannot hear any difference between the corresponding audio signals, *i.e.*, your ear is *phase deaf*. This phenomenon, known as *Ohm's law* of acoustics, allows us to simplify the mathematical expressions we use to synthesize a musical tone.

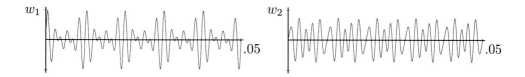

Figure 11.3. The audio signals w_1 and w_2 from (2).

For example, suppose that we want to create a sound file for some p-periodic audio signal w. We can use Fourier's representation to write

$$w(t) = A_0 + \sum_{k=1}^{\infty} A_k \sin(2\pi k f t + \phi_k)$$

where $f := 1/p$, where $A_0, A_1, \ldots$ are nonnegative, and where $\phi_1, \phi_2, \ldots$ are real. We only hear the terms with $20 \text{ Hz} \leq kf \leq 20000 \text{ Hz}$ and the ear is phase deaf, so we lose no generality by taking

$$w(t) = \sum_{20/f \leq k \leq 20000/f} A_k \sin(2\pi k f t). \tag{3}$$

(In particular, when $f > 10000$ Hz the tone we hear is a pure sinusoid!) We can use Goertzel's algorithm, *cf.* Ex. 6.1, to compute the T-samples of such a sin sum.

Scales

At some time you have probably learned to sing the familiar diatonic scale

| *do* | *re* | *mi* | *fa* | *sol* | *la* | *ti* | *do* |

to consecutive tones

| *C* | *D* | *E* | *F* | *G* | *A* | *B* | *C* |

from a piano keyboard. Choir directors often specify a *musical interval* by using the tones from this familiar sequence, *e.g.*, the *fourth*, *fifth*, and *eighth* (or *octave*), is the musical interval between the initial *do* and the following *fa*, *sol*, and *do*, respectively. You use these musical intervals when you sing

> *Here* ↑fourth *comes the bride* ... ,
> *Twinkle,* ↑fifth *twinkle, little star* ... ,
> *Some-* ↑eighth *where, over the rainbow* ... ,

(whether or not you begin at the pitch associated with some *C*).

The octave, which corresponds to a 2/1 frequency ratio, is the fundamental musical interval in every culture. Most people instinctively identify tones that differ by an octave. You may have observed that when men and women sing together in "unison," they usually sing an octave apart. You may have also noticed that you tend to shift up or shift down by one octave when you are singing a melody that becomes too low or too high for your voice.

Musicians from different cultures subdivide the octave by adding 5 or 7 or 11 or more intermediate tones to form a *scale*. (Functioning flutes from the time of the Egyptian pharaohs allow us to hear such scales that are more than 4000 years old.) The theoretical development of the familiar diatonic scale of Western music dates from the time of Pythagoras. You may be surprised to learn that *music* was one of the four major divisions of Greek mathematics! (The others were *arithmetic*, *geometry*, and *astronomy*. Add *rhetoric*, *logic*, and *grammar* and you have the original seven liberal arts.)

A simple instrument provided the link between mathematics (as you know it) and music. A *monochord* consists of a string stretched between two fixed supports on top of a resonating box. The tension is adjusted so that the frequency of vibration is in the audible range. By placing a movable bridge between the two fixed supports you can shorten the effective length of the vibrating string without changing the tension so that you can produce different tones, *cf.* Ex. 11.6. From your study of the vibrating string in Chapter 9, you know that the frequency and the length are inversely proportional, *cf.* (9.40). The Pythagoreans of ancient Greece did not know this, however, for they had no way to determine the actual frequency of an audible tone. They could *hear* the musical interval associated with measurable lengths L_1, L_2, however, and relate this musical interval to the ratio L_1/L_2.

For example, suppose that the initial tone is produced by an unstopped monochord having the length $L_1 = L$. If we stop the string at its midpoint, the new length is $L_2 = (1/2)L$, the frequency ratio is $f_2/f_1 = L_1/L_2 = 2/1$, and the musical interval is an octave. If you place the bridge 1/3 or 1/4 of the distance from one end, the longer segment has the length $L_2 = (2/3)L$ or $L_2 = (3/4)L$, the frequency ratio is 3/2 or 4/3, and the pitch rises by a fifth or a fourth. [After observing that $(2/1) = (3/2) \cdot (4/3)$ you should have no trouble understanding what a musician means by saying that an eighth is a fifth more than a fourth!]

After discovering the link between the ratios 4/3, 3/2, 2/1 and certain musical intervals (that they called the *diatesseron*, the *diapente*, and the *diapason*) the Pythagoreans quickly developed the eight-tone *Pythagorean scale* with the following relative frequencies, *cf.* Ex. 11.7.

do	*re*	*mi*	*fa*	*sol*	*la*	*ti*	*do*
1	$\frac{9}{8}$	$\frac{81}{64}$	$\frac{4}{3}$	$\frac{3}{2}$	$\frac{27}{16}$	$\frac{243}{128}$	2
	T	T	s	T	T	T	s

There are five steps T with a frequency ratio $9/8 = 1.125$ and two smaller steps s with a frequency ratio $256/243 = 1.053\ldots$. This TTsTTTs division of the octave corresponds to the major scale of Western music. (You can hear a close approximation by playing $C - C$ on the white keys of a modern piano.) Cyclic permutations of the TTsTTTs steps gave the Pythagoreans six other scales for music composition, *e.g.*, the sequence TsTTsTT (produced by playing $A - A$ on the white keys of a piano) corresponds to the minor scale of Western music.

Some five centuries later Claudius Ptolemy (who wrote *Concerning Harmonics* as well as *The Almagest*) adjusted three of the Pythagorean tones to produce the following *just scale* that includes the ratios $5/4$ and $5/3$ for a *perfect third* and a *perfect sixth*.

do		re		mi		fa		sol		la		ti		do
1		$\frac{9}{8}$		$\frac{5}{4}$		$\frac{4}{3}$		$\frac{3}{2}$		$\frac{5}{3}$		$\frac{15}{8}$		$\frac{2}{1}$
	T		T$'$		s'		T		T$'$		T		s'	

This scale has 3, 2, 2 steps T, T$'$ s' with corresponding frequency ratios $9/8 = 1.125$, $10/9 = 1.100$, $16/15 = 1.066\ldots$.

The rapid evolution of keyboard instruments and keyboard music during the 17th century led to the development of the *equal-tempered scale*

do		re		mi		fa		sol		la		ti		do
1		ρ^2		ρ^4		ρ^5		ρ^7		ρ^9		ρ^{11}		2
	T$''$		T$''$		s''		T$''$		T$''$		T$''$		s''	

which uses the frequency ratios

$$\rho := 2^{1/12} = 1.0594\ldots, \quad \rho^2 = 1.1224\ldots$$

for the steps s'' and T$''$. You hear these tones as you play $C - C$ on the white notes of a properly tuned piano. The black notes correspond to the missing frequency ratios $\rho, \rho^3, \rho^6, \rho^8$, and ρ^{10}. By using the black keys as well as the white ones, you can begin at any key and play the equal-tempered scale! In exchange for this flexibility, you must accept the approximations

$$\rho^4 = 1.2599\ldots, \quad \rho^5 = 1.3348\ldots, \quad \rho^7 = 1.4983\ldots, \quad \rho^9 = 1.6817\ldots$$

for the frequency ratios

$$\frac{5}{4} = 1.2500\ldots, \quad \frac{4}{3} = 1.3333\ldots, \quad \frac{3}{2} = 1.5000\ldots, \quad \frac{5}{3} = 1.6666\ldots$$

of the perfect third, fourth, fifth, and sixth.

In 1939 the International Standards Association (which is responsible for the S.I. definition of the meter, kilogram, second, ...) agreed to a worldwide standard of 440 Hz for *Concert A*, the note *A4* from the piano keyboard. Thus the 88 keys of a properly tuned piano have the frequencies $440\rho^n$ Hz, $n = -48, -47, \ldots, 39$, as given in Appendix 7.

Musical notation

The system that we use for writing music (as illustrated in Fig. 11.1) reached its present form approximately four centuries ago. The notation allows us to specify the pitch, the loudness, the duration, and the moment in time for each tone in a composition. We will briefly explain how this is done so that you can use a music score to prepare a sound file to play on your computer!

A musician measures time with a rather flexible unit known as a *beat*. (When you sing

Twink–le, twink–le, lit–tle star ⋯

each syllable gets one beat until you reach *star*, which gets two.) For simplicity, we will limit our discussion to cases where the beat is assigned to a quarter note as described below. Italian words are used to specify an approximate beat rate, *e.g.*,

Larghissimo, Adagio, Andante, Moderato, Allegro, Presto

(very slow)   (very fast)

are associated with rates that run from 40 to 200 beats per minute. (An "equation" can be used for more precision, *e.g.*, ♩ $= 90$ specifies 90 beats/minute.)

A *note* pictogram

whole note	half note	quarter note	eighth note	sixteenth note
4 beats	2 beats	1 beat	$\frac{1}{2}$ beat	$\frac{1}{4}$ beat

is used to specify the duration of a tone and a corresponding *rest* pictogram

whole rest	half rest	quarter rest	eighth rest	sixteenth rest
4 beats	2 beats	1 beat	$\frac{1}{2}$ beat	$\frac{1}{4}$ beat

is used to specify the duration of a period of silence. When we append a dot to such a pictogram, the duration is increased by 50%, *e.g.*, ♩. gets 3 beats.

The pitch of a tone is specified by the vertical position of the corresponding note on a *staff* of five horizontal lines as shown in Fig. 11.4. You will observe that *middle C* (*i.e.*, C4 with a frequency of 261.6 Hz) occurs on the first ledger line that lies below, above a staff that bears the *treble clef, bass clef* symbol, respectively. We prefix a *sharp* sign, ♯, or a *flat* sign, ♭, to a note to indicate that the pitch must be raised, lowered by one semitone. A collection of sharps or flats placed immediately after the clef symbol is known as a *key signature*. All of the corresponding notes (including those that differ by one, two, ... octaves) are to be raised or lowered by a semitone. A *natural* sign, ♮, is used to circumvent this convention.

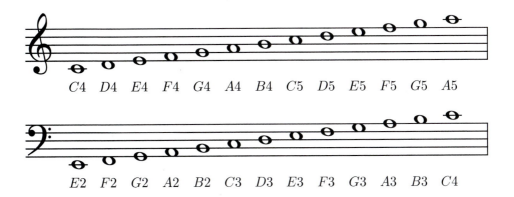

Figure 11.4. The pitch of notes having various vertical positions on a five-line staff bearing the treble clef (top) and the bass clef (bottom). Corresponding piano keys and numerical values for the frequencies are given in Appendix 7.

We write a sequence of notes and rests from left to right along a staff (continuing the sequence on subsequent staffs in the same way that you read the lines of this text). The staff provides a Cartesian coordinate system with frequency increasing (exponentially) along the vertical axis and with time increasing (more or less linearly) along the horizontal axis.

We specify the loudness of a section of music by using the symbols

$$ppp \quad pp \quad p \quad mp \quad mf \quad f \quad ff \quad fff$$

very soft   very loud

that serve as abbreviations for the Italian *pianississimo*, ... , *fortississimo*. Such prescriptions (like those used for the beat rate) are very elastic and subject to artistic interpretation.

Example: Find the frequency for each note in the score from Fig. 11.5.

Figure 11.5. A simple music score.

Solution: We use Fig. 11.4 to identify the notes

$$C4, \quad C5, \quad B4, \quad G4, \quad A4, \quad B4, \quad C5,$$

and then use Appendix 7 to obtain the corresponding frequencies

$$261.6, \quad 523.2, \quad 493.9, \quad 392.0, \quad 440.0, \quad 493.9, \quad 523.2 \, \text{Hz.}$$

11.2 Spectrograms

Introduction

In this section you will learn how to produce a *spectrogram* from T-samples $\ldots, w_0, w_1, w_2, \ldots$ of a real audio signal $w(t)$. A spectrogram is a computer-generated picture that shows the frequencies present in an audio signal at each moment of time, and in this sense it is much like a music score. Figure 11.6 shows a seven-note music score and spectrograms made from three corresponding audio signals. Computer-generated samples of a gaussian damped sinusoid were used to generate each tone in the first sound file, and you can easily identify each note from the music score with a corresponding spot in the spectrogram. Tones produced by real musical instruments are not so simple, as you will discover when you examine the spectrograms prepared from digital recordings of a piano and of a singing soprano. You should expect to find integral multiples of the fundamental frequency in the spectrum of a tone produced by a vibrating piano string, and you can clearly see them in the second spectrogram from Fig. 11.6. As you study the third spectrogram, you will identify the overtones from the soprano's voice and observe how they changed with time as she adjusted her mouth and throat to sing various vowels and semivowels.

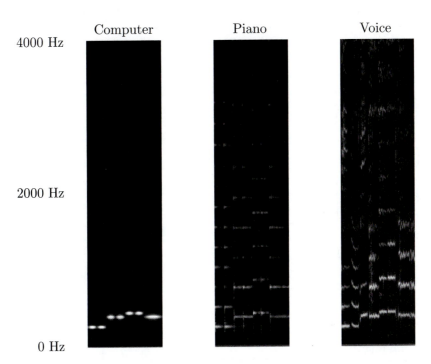

Figure 11.6. A music score and spectrograms from three corresponding sound files. In each case, time increases from 0 to 6 seconds along the horizontal axis, and frequency ranges from 0 to 4000 Hz along the vertical axis. (The notes $C4$, $G4$, $A4$ have the frequencies 261.6, 392.0, 440 Hz, respectively.)

The computation

We will now describe the numerical process that we use to produce a spectrogram. We begin by selecting a *resolution index* $N = 2, 3, \ldots$ and defining corresponding limits of summation

$$N_- := \left\lfloor \frac{-N + 1}{2} \right\rfloor, \quad N_+ := \left\lfloor \frac{N}{2} \right\rfloor.$$

We will use the N samples w_{m+n}, $n = N_-, \ldots, N_+$ (with indices centered at m) to estimate the frequency content of $w(t)$ at time $m\tau$, *cf.* Fig. 11.7.

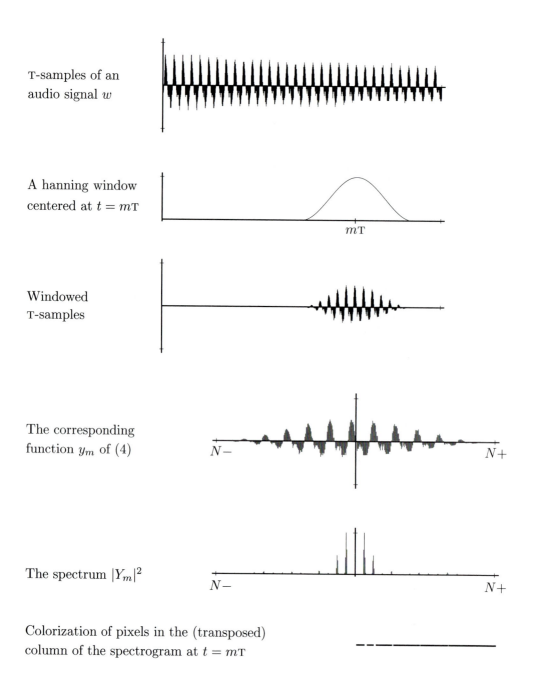

T-samples of an
audio signal w

A hanning window
centered at $t = m\mathrm{T}$

$m\mathrm{T}$

Windowed
T-samples

The corresponding
function y_m of (4)

$N-$ $N+$

The spectrum $|Y_m|^2$

$N-$ $N+$

Colorization of pixels in the (transposed)
column of the spectrogram at $t = m\mathrm{T}$

Figure 11.7. Colorization of the pixels for a spectrogram.

We weight these samples with a suitable *window* function to form

$$y_m[n] := \frac{1}{2}\left\{1 + \cos\left(\frac{2\pi n}{N}\right)\right\} \cdot w_{m+n}, \quad n = N_-, \ldots, N_+, \tag{4}$$

and then use the FFT to obtain the discrete Fourier transform

$$Y_m[k] = \frac{1}{N} \sum_{n=N_-}^{N_+} y_m[n]e^{-2\pi ikn/N}. \tag{5}$$

(The window suppresses ringing in the discrete transform, *cf.* Ex. 4.29.) Since y_m is real, Y_m is hermitian, and we take

$$p_{m,k} := |Y_m[k]|^2 = |Y_m^-[-k]|^2, \quad k = 1, 2, \ldots, N_+ \tag{6}$$

as a measure of the energy in the audio signal $w(t)$ at time $t = m\text{T}$ and frequency $f = k/N\text{T}$.

The sampling interval T determines the highest (Nyquist) frequency, $1/2\text{T}$, that we can hope to detect by processing the samples $\ldots, w_0, w_1, w_2, \ldots$. The frequency separation, $1/N\text{T}$, is determined jointly by T and the choice of N, *cf.* Ex. 1.17. (Recall that $k = 1$ corresponds to one oscillation during a time interval of length $N\text{T}$.) The spectrograms from Fig. 11.6 were made with $\text{T} = 1/8000$ sec and $N = 640$. Thus the Nyquist frequency is 4000 Hz and the frequency separation is 12.5 Hz. (This is a bit less than one semitone at the 261.6-Hz frequency for the lowest note, $C4$, from the score.)

We use the nonnegative numbers $p_{m,1}, p_{m,2}, \cdots, p_{m,N_+}$ to color the mth column of pixels for the spectrogram of w. If we want to produce a high-quality gray-tone picture, we must assign a color index

$$\text{pure black: } 0, 1, 2, \ldots, 255: \text{ pure white}$$

to each pixel, *cf.* Fig. 11.7. A colorization based on a straightforward quantization of $p_{m,k}$ produces a spectrogram with color intensities that correspond to energy. The spectrograms for this text were all produced by quantizing $\log p_{m,k}$ since the resulting color intensities correspond more closely to the response of the human ear.

Slowly varying frequencies

When the phase function θ from a mathematically defined audio signal

$$w(t) := \sin\{\theta(t)\} \tag{7}$$

has a smooth and slowly varying derivative, we can use Taylor's formula to write

$$w(t + \tau) = \sin\left\{\theta(t) + \tau\,\theta'(t) + \left(\frac{\tau^2}{2}\right)\theta''(t) + \cdots\right\}$$

$$\approx \sin\left\{2\pi\left[\frac{\theta'(t)}{2\pi}\right]\tau + \theta(t)\right\},$$

(when τ is small) and thereby see that w has the local frequency

$$f(t) := \frac{\theta'(t)}{2\pi} \tag{8}$$

at time t. For example, the audio signals

$$
\begin{aligned}
w_1(t) &:= \sin\{2\pi \cdot 500t\}, \\
w_2(t) &:= \sin\{2\pi \cdot [500t + (50/\pi)\sin(2\pi t)]\}, \\
w_3(t) &:= \sin\{2\pi \cdot [500t + (125/2)t^2]\}
\end{aligned}
\tag{9}
$$

for a sinusoid, a vibrato, and a chirp have the slowly varying frequencies

$$
\begin{aligned}
f_1(t) &= 500, \\
f_2(t) &= 500 + 100\cos(2\pi t), \\
f_3(t) &= 500 + 125t.
\end{aligned}
\tag{10}
$$

Spectrograms from corresponding computer-generated sound files with $0 \le t \le 4$ sec are shown in Fig. 11.8. You can see the graphs of the local frequency functions (10)!

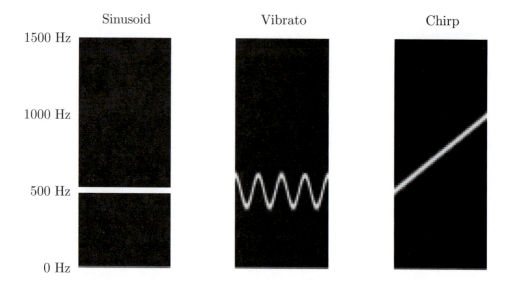

Figure 11.8. Spectrograms from computer-generated sound files for the audio signals (9) using $0 \le t \le 4$ sec, $\mathrm{T} = 1/8000$ sec, $N = 640$. (These spectrograms were cropped at 1500 Hz.)

11.3 Additive Synthesis of Tones

Introduction

You can easily tell the difference between tones produced by a flute, a xylophone, an oboe, a piano, and a trumpet, even when they all have the same nominal pitch, loudness, and duration. Musicians use the term *timbre* to denote the attributes of a sound that facilitate such classifications. During the mid-19th century, Hermann Helmholtz devised a way to identify the harmonics in sustained musical tones by using cleverly designed glass resonators. In this way, he showed that the timbre of the periodic tone (3) is determined by the relative sizes of the (nonnegative) amplitudes $A_1, A_2, \ldots$.

The tones we use for music do not always have perfect harmonics, however, and they correspond to notes that are localized in time. For these reasons we replace (3) with a finite sum of the form

$$w(t) = \sum_{k=1}^{K} A_k(t) \sin(2\pi f_k t) \tag{11}$$

where the amplitude envelopes $A_1(t), A_2(t), \ldots$ are suitably chosen functions of time and $f_1, f_2, \ldots$ are corresponding frequencies. If you record samples of a tone produced by a flute, a xylophone, $\ldots$, you can use the spectrogram analysis from the preceding section to obtain approximations for $A_1(t), A_2(t), \ldots$ and $f_1, f_2, \ldots$. [Although many natural tones are well modeled by (11), you should not expect to find an exact fit to the recorded data.] Once you know the approximate amplitude envelopes and frequencies, you can compute samples of (11) to create a sound file for a tone with a timbre that is quite similar to the one from your recording.

Of course, there is no reason why you must be restricted to envelopes and frequencies that have been obtained in this way. You can arbitrarily choose $A_1(t), A_2(t), \ldots$ and $f_1, f_2, \ldots$, compute samples of (11), and use your sound card to play the corresponding tone. With a bit of experimentation you can create tones that have musically interesting (but unfamiliar!) timbres.

Amplitude envelopes

Figure 11.9 shows four amplitude envelopes and the audio waves that result when they are used to modulate a 320-Hz sinusoid. The envelope A_1 has an impulsive attack (that may cause the sound card to emit an annoying little click) followed by an exponentially decaying release. The corresponding tone has a timbre much like the tone produced by a xylophone. The gaussian envelope A_2 gives a tone with a timbre similar to that produced by a flute. This envelope was used to create the "notes" you see in the first spectrogram of Fig. 11.6. The trapezoid envelope A_3 allows us to hear the sustained central portion of the corresponding tone without

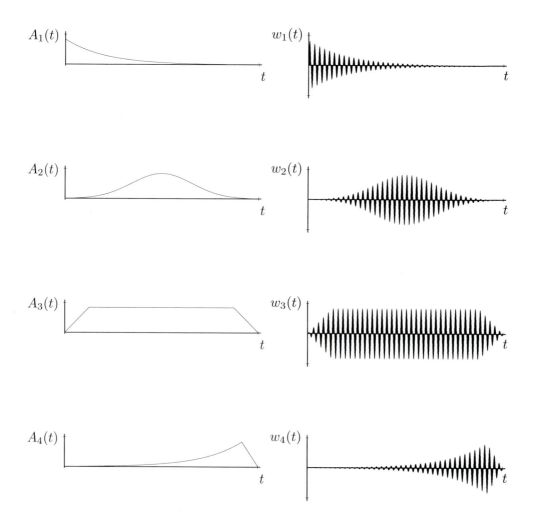

Figure 11.9. Amplitude functions and the corresponding audio
waves for tones that have a frequency of 320 Hz and a duration
of approximately 1/8 sec.

modulation. Short linear attack and release segments smoothly turn the sound on
and off. The envelope A_4 has a long exponentially growing attack followed by a
short linear release. The timbre of this tone is pleasing but unfamiliar (unless you
listen to xylophone recordings played backwards!).

Synthesis of a bell tone

When a large brass bell is struck by its clapper, the impulse excites a number of vibrational modes. The energy in each mode decays more or less independently due to sound radiation and internal losses, *i.e.*, each mode has an exponential envelope as shown in Fig. 11.9. The audio signal is well modeled by a wave

$$w(t) = \sum_{k=1}^{K} A_k e^{-t/\tau_k} \sin(2\pi f_k t) \tag{12}$$

where f_k, τ_k, A_k, $k = 1, 2, \ldots, K$ are suitably chosen parameters. During the manufacturing process the bell is tuned (by using a large bell lathe to remove bits of metal from the inner surface of the casting) so that the five lowest frequencies have the approximate ratios $1 : 2 : 2.4 : 3 : 4$. The lowest of these frequencies is that of the residual hum; the nominal pitch is two octaves higher. A tone with the timbre of a large bell can be produced from samples of (12) with the parameters shown in Fig. 11.10, *cf.* T.D. Rossing, The acoustics of bells, *Amer. Sci.* **72**(1982), 440–447. A spectrogram allows you to visualize the evolution of the corresponding audio wave.

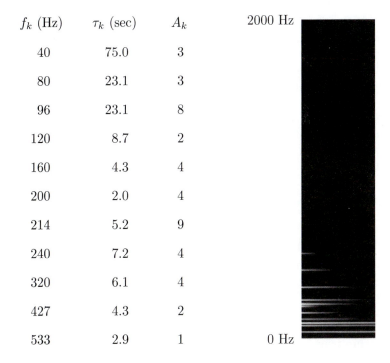

f_k (Hz)	τ_k (sec)	A_k
40	75.0	3
80	23.1	3
96	23.1	8
120	8.7	2
160	4.3	4
200	2.0	4
214	5.2	9
240	7.2	4
320	6.1	4
427	4.3	2
533	2.9	1

Figure 11.10. Parameters for the bell tone (12) and a corresponding spectrogram using $0 \le t \le 30$ sec, $\mathrm{T} = 1/4000$ sec, $N = 1000$.

Synthesis of string tones

Let f be the fundamental frequency of a vibrating string with fixed ends at $x = 0$, $x = L$, and let $0 < \alpha < 1$. If we strike, pluck, bow the string at the excitation point $x_0 := \alpha L$, the displacement function has the form

$$u(x, t) = A \sum_{k=1}^{\infty} \frac{\sin(\pi k \alpha)}{k^\ell} \sin\left(\frac{\pi k x}{L}\right) \sin(2\pi k f t + \phi_\ell),$$

where A, ϕ_ℓ are constants and $\ell = 1, 2, 3$, respectively, *cf.* (9.44)–(9.46). You can produce various string tones from computer-generated samples of

$$w(t) = A(t) \sum_{k=1}^{K} \frac{\sin(\pi k \alpha)}{k^\ell} \sin(2\pi k f t). \tag{13}$$

The timbre depends on the choice of α, ℓ, K and the amplitude envelope $A(t)$. Try using a trapezoidal envelope with $\alpha = 1/7$ and $\ell = 1, 2, 3$ to approximate the timbre of a piano string, a guitar string, and a bowed violin string, respectively. Exercise 11.12 will show you an efficient way to compute the samples in the case where $K = \infty$.

You will quickly discover that tones generated from samples of (13) do not have the richness of those produced by a real stringed instrument. When a string vibrates, it exerts forces [proportional to $u_x(0+, t)$, $u_x(L-, t)$] on the end supports. These forces excite vibrational modes of sounding boards, air cavities, etc. that then radiate acoustic energy to the surrounding air. The wave (13) models the sound emitted by the string (a very poor radiator!) but neglects the sound that comes from the other parts of the instrument. More sophisticated wave forms can be developed from mathematical models that use the principles of musical acoustics, *e.g*, as given in Benade's text.

Synthesis of a brass tone

Shortly after the discovery of the FFT, several mathematically inclined musicians used windowed DFTs to analyze the time evolution of various musical tones. They found that brass tones are well approximated by a sum (11) with harmonic frequencies $f, 2f, 3f, \ldots$. Figure 11.11 shows the (smoothed) amplitude envelopes from one such study, *cf.* J.A. Moorer, J. Grey, and J. Strawn, Lexicon of analyzed tones, Part 3: Trumpet, *Comp. Music J.* **2**(1978), No. 2, 23–31. The tone is colored by the delayed attack and early decay of the higher harmonics (from the portion of the spectrum where the ear is most sensitive). You can use these envelope functions (or simple broken-line approximations) with (11) to generate samples for a tone having a timbre much like that of a trumpet. Goertzel's algorithm, *cf.* Ex. 6.1, can be used to expedite the computation of the samples.

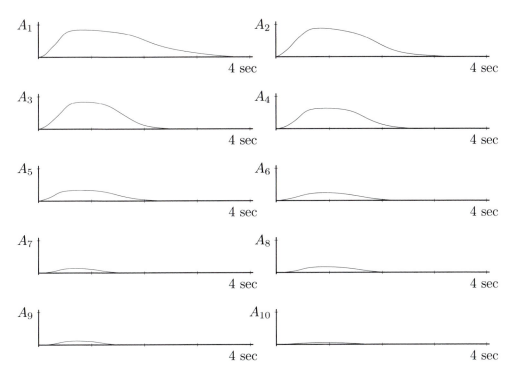

Figure 11.11. Amplitude envelopes for synthesizing a trumpet tone from (11) (when using the harmonic frequencies $f_k = k \cdot 311$ Hz, $k = 1, \ldots, 10$).

11.4 FM Synthesis of Tones

Introduction

In the early 1970s the musician John Chowning was experimenting with audio signals of the form

$$w(t) := \sin\{2\pi f_c t + \mu \sin(2\pi f_m t)\}. \tag{14}$$

The *carrier frequency*, f_c, the *modulation frequency*, f_m, and the *modulation index*, μ, are real parameters. When f_c is in the audible range and μf_m is "small," w is locally sinusoidal with a time-varying frequency

$$f(t) = f_c + \mu f_m \cos(2\pi f_m t)$$

that oscillates between $f_c - \mu f_m$ and $f_c + \mu f_m$, *e.g.*, as illustrated in the middle spectrogram from Fig. 11.8. Chowning discovered that (14) has a radically different sound when μf_m is "large." His analysis of this phenomenon led to a U.S. patent for the technological innovation that underlies Yamaha's DX7 digital synthesizer.

The spectral decomposition

The *Bessel function* $J_k(\mu)$ of the first kind with order $k = 0, \pm 1, \pm 2, \ldots$ and real argument μ can be defined by writing

$$e^{i\mu \sin(x)} =: \sum_{k=-\infty}^{\infty} J_k(\mu) e^{ikx}, \quad -\infty < x < \infty, \tag{15}$$

i.e., $J_k(\mu)$ is the kth Fourier coefficient of the 2π periodic function of x on the left side of (15). These functions are real valued and have the symmetries

$$J_k(-\mu) = (-1)^k J_k(\mu) = J_{-k}(\mu), \quad -\infty < \mu < \infty, \quad k = 0, \pm 1, \pm 2, \ldots, \tag{16}$$

cf. Ex. 4.19. Graphs of J_0, J_1, J_2, J_3 are shown in Fig. 11.12.

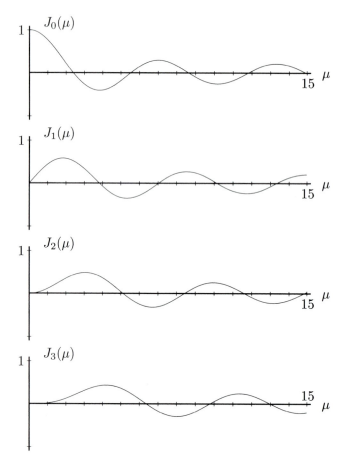

Figure 11.12. The Bessel functions J_0, J_1, J_2, J_3 on $[0, 15]$.

We use (15) to write

$$e^{2\pi i f_c t} e^{i\mu \sin(2\pi f_m t)} = \sum_{k=-\infty}^{\infty} J_k(\mu) e^{2\pi i f_c t} e^{2\pi i k f_m t},$$

and from the imaginary part of this identity we obtain the spectral decomposition

$$\sin\{2\pi f_c t + \mu \sin(2\pi f_m t)\} = \sum_{k=-\infty}^{\infty} J_k(\mu) \sin\{2\pi(f_c + k\, f_m)t\}. \qquad (17)$$

In this way we see that the audio signal (14) can be synthesized from pure sinusoids having the frequencies $f_c + k\, f_m$, $k = 0, \pm 1, \pm 2, \ldots$.

The modulation parameter μ controls the *number* of significant terms that are present in the sums (15) and (17). From graphs such as those shown in Fig. 11.13, Chowning observed that

$$J_k(\mu) \approx 0 \quad \text{when} \quad |k| > |\mu| + 1 \qquad (18)$$

so that

$$\sin\{2\pi f_c t + \mu \sin(2\pi f_m t)\} \approx \sum_{|k| \le |\mu|+1} J_k(\mu) \sin\{2\pi(f_c + k\, f_m)t\}. \qquad (19)$$

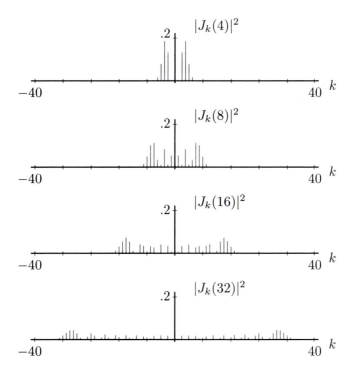

Figure 11.13. Plots of $|J_k(\mu)|^2$, $k = 0, \pm 1, \ldots, \pm 40$ when $\mu = 4, 8, 16, 32$.

By applying Plancherel's identity (1.17) to (15), we see that

$$\sum_{k=-\infty}^{\infty} |J_k(\mu)|^2 = 1, \quad -\infty < \mu < \infty,$$

and with a bit of numerical analysis we can verify that 96% of this sum comes from the terms with $|k| \le |\mu| + 1$ when $|\mu| \le 100$, *cf.* Ex. 11.15. This allows us to assess the quality of the Chowning approximations (18) and (19).

Example: Use Chowning's rule to determine the (significant) frequencies present in the audio signal $w(t) := \sin\{2\pi \cdot 200t + .5 \cdot \sin(2\pi \cdot 10t)\}$.

Solution: When $\mu = .5$ the sum from (19) has terms with $k = 0, \pm 1$, so we obtain the frequencies

$$f = 200 + k \cdot 10 = 190,\ 200,\ 210 \text{ Hz}.$$

(The three sinusoids combine to produce a vibrato that oscillates between 195 Hz and 205 Hz with a frequency of 10 Hz!)   ∎

Example: Use Chowning's rule to determine the (significant) frequencies present in the audio signal $w(t) := \sin\{2\pi \cdot 200t + 2 \cdot \sin(2\pi \cdot 400t)\}$.

Solution: When $\mu = 2$ the sum from (19) has terms with $k = 0, \pm 1, \pm 2, \pm 3$, so we obtain the frequencies

$$f = 200 + k \cdot 400 = -1000,\ -600,\ -200,\ 200,\ 600,\ 1000,\ 1400.$$

By using (16) we see that

$$J_{-k}(\mu)\sin(-2\pi f t) = (-1)^{k+1} J_k(\mu)\sin(2\pi f t),$$

so we can combine the terms from (19) with $f = \pm 200$, with $f = \pm 400$, and with $f = \pm 600$. In this way we see that the audio signal has the (significant) frequencies

$$f = 200,\ 600,\ 1000,\ 1400 \text{ Hz}.$$

(Such $1:3:5:7$ harmonics are characteristic of tones produced by woodwinds!) ∎

Dynamic spectral enrichment

A variety of interesting musical tones can be generated from T-samples of an audio signal

$$w(t) := A(t) \cdot \sin\{2\pi f_c t + \mu(t) \cdot \sin(2\pi f_m t)\} \tag{20}$$

where the amplitude envelope A and the modulation index μ are now suitably regular functions of time, *cf.* (14). The attack and release of the tone are determined by A as described in the preceding section. The remarkable new feature is the use of the slowly varying function $\mu(t)$ to control the *number* of frequencies

$$f_c, \ f_c \pm f_m, \ f_c \pm 2f_m, \dots$$

that sound at time t.

Example: Describe the tone associated with

$$w(t) := \sin\{2\pi \cdot 100t + .3125t \cdot \sin(2\pi \cdot 100t)\}, \quad 0 \le t \le 32 \text{ sec}. \tag{21}$$

Solution: The modulation index $\mu(t) := .3125t$ increases from 0 to 10 as t increases from 0 to 32, so the tone begins as a pure 100-Hz sinusoid and evolves into an electronic twang as the frequencies 200 Hz, 300 Hz, ... are added, *cf.* Fig. 11.14. Chowning's rule predicts that a new frequency will arise whenever μ increases by 1, *i.e.*, every 3.2 sec, and this is confirmed by the spectrogram. Since $\mu(32) + 1 = 11$, Chowning's rule predicts that the 12 frequencies 100 Hz, 200 Hz, ... , 1200 Hz will sound at time $t = 32$ sec. [The spectrogram was produced by using a sensitive log quantization that shows 3 additional frequencies from the 4% of the energy neglected by the approximation (19).]   ∎

2000 Hz

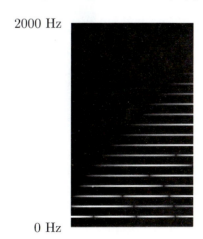

0 Hz

Figure 11.14. A spectrogram for the tone (21) using $0 \le t \le 32$ sec, T $= 1/4000$ sec, $N = 640$.

After a good deal of experimentation, Chowning and his coworkers learned to choose functions A and μ to produce tones having the timbre (but not always the spectrum) of those from familiar musical instruments. For example, a surprisingly good bell tone can be produced from samples of

$$w(t) := e^{-\alpha t} \cdot \sin\{2\pi f_c t + b e^{-\beta t} \cdot \sin(2\pi f_m t)\} \tag{22}$$

using the parameters given with Fig. 11.15. The spectrogram allows you to see how the initial clang (from a cluttered spectrum) evolves into an exponentially damped sinusoid. This bell tone is more lively and realistic than the one produced from (12) (using the parameters from Fig. 11.10) even though it takes less effort to compute samples of (22) than it does to compute samples of (12), *cf.* Ex. 11.13.

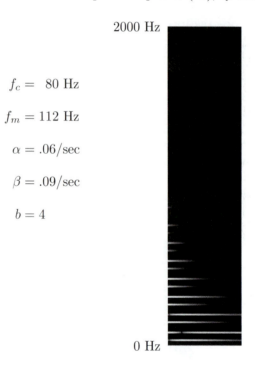

$$f_c = 80 \text{ Hz}$$

$$f_m = 112 \text{ Hz}$$

$$\alpha = .06/\text{sec}$$

$$\beta = .09/\text{sec}$$

$$b = 4$$

Figure 11.15. Parameters for the Chowning bell tone (22) and a corresponding spectrogram using $0 \le t \le 30$ sec, $\mathrm{T} = 1/4000$ sec, $N = 1000$.

Once you understand these basic ideas, you can use FM synthesis to design unusual new musical tones. For example, if you use a positive α and a negative β in (22), the spectral bandwidth will grow with time but all of the components will be exponentially damped as shown in Fig. 11.16.

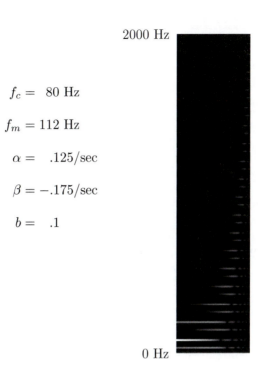

2000 Hz

$f_c = 80$ Hz

$f_m = 112$ Hz

$\alpha = .125/\text{sec}$

$\beta = -.175/\text{sec}$

$b = .1$

0 Hz

Figure 11.16. Parameters for an ethereal musical tone (22) and a corresponding spectrogram using $0 \le t \le 30$ sec, $\text{T} = 1/4000$ sec, $N = 1000$.

You can have a lot of fun producing tones from T-samples of (20), but you should always be aware of the constraints imposed by the Nyquist condition (8.6). A local application of Chowning's rule (19) shows that (20) can contain the frequencies

$$f_c, \; f_c \pm f_m, \; \ldots, \; f_c \pm (\lfloor |\mu(t)| \rfloor + 1) f_m$$

at time t, so you must choose T so that

$$2\text{T} \left\{ f_c + (\lfloor |\mu(t)| \rfloor + 1) f_m \right\} < 1 \tag{23}$$

to avoid the unexpected effects of aliasing.

11.5 Synthesis of Tones from Noise

Introduction

You can produce a hissing sound by sending a stream of *random numbers* $r_1, r_2, \ldots, r_N$ to the sound card of your computer. The audio signal is given by the cardinal series

$$r(t) := \sum_{n=1}^{N} r_n \operatorname{sinc}\left(\frac{t - n\mathrm{T}}{\mathrm{T}}\right) \tag{24}$$

where $\mathrm{T} > 0$ is the sampling interval.

A random number generator is used to produce $r_1, r_2, \ldots, r_N$ as follows. We carefully choose three positive integers M, a, and c. (A table of suitable values, *e.g.*, $M = 714025$, $a = 4096$, $b = 150889$, can be found in W.H. Press *et al.*, *Numerical Recipes*, Cambridge University Press, New York, 1986, pp. 191–199.) Given an initial *seed* $k_0 = 0, 1, \ldots, M - 1$ we generate the integers

$$k_n := ak_{n-1} + c \pmod{M}, \quad n = 1, 2, \ldots, N. \tag{25}$$

When $N \ll M$ these integers *seem* to be independently and randomly chosen from $\{0, 1, \ldots, M - 1\}$ [*cf.* (4.35)] even though they are completely determined by (25)! Likewise, the scaled translates

$$r_n := 2k_n/M - 1, \qquad n = 1, 2, \ldots, N$$

seem to be independently and randomly chosen from $[-1, 1]$ with a common uniform density. (You will find a thorough study of such random sequences in D. Knuth, *The Art of Computer Programming, Vol. 2, Seminumerical Algorithms*, Addison-Wesley, Reading, MA, 1969, pp. 1–160.)

We will analyze the spectral characteristics of (24) and show how to produce interesting musical tones by processing such stochastic signals.

White noise

The audio signal (24) has the Fourier transform

$$R(s) = \left\{ \sum_{n=1}^{N} r_n e^{-2\pi i n \mathrm{T} s} \right\} \mathrm{T}\Pi(\mathrm{T}s). \tag{26}$$

We will use

$$\int_{-\infty}^{\infty} |r(t)|^2 dt = \int_{-\infty}^{\infty} |R(s)|^2 ds$$

as a measure of the total energy in the signal (24) so that

$$|R(s)|^2 = \left| \sum_{n=1}^{N} r_n e^{-2\pi i n \text{T}s} \right|^2 \text{T}^2 \Pi(\text{T}s) \tag{27}$$

is a density that shows how energy is distributed across the spectrum. Figure 11.17 shows a representative signal $r(t)$ and the corresponding spectral density $|R(s)|^2$.

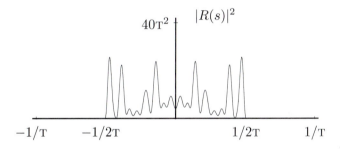

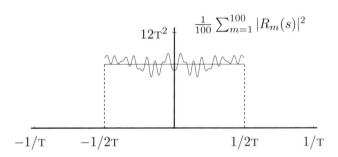

Figure 11.17. An audio signal (24) constructed with $N = 25$ random numbers from $[-1, 1]$ and the corresponding spectral density (27). An average of 100 such spectral density curves (bottom) is well approximated by the expected spectral density (30).

Suppose now that the $r_1, r_2, \ldots, r_N$ from (27) are independent random variables. The *expected spectral density*

$$P_r(s) := \langle |R(s)|^2 \rangle = \left\langle \left| \sum_{n=1}^{N} r_n e^{-2\pi i n \mathrm{T}s} \right|^2 \mathrm{T}^2 \Pi(\mathrm{T}s) \right\rangle \tag{28}$$

[as defined in (12.33)] then provides us with a representative function that depends on the probability densities associated with $r_1, r_2, \ldots, r_N$ (but not on the particular samples that we have selected). We will further assume that r_n has the mean $\mu = 0$ and the common variance σ^2, $0 < \sigma < \infty$, for each $n = 1, 2, \ldots, N$, i.e.,

$$\langle r_n \rangle = 0, \quad n = 1, 2, \ldots, N, \text{ and}$$
$$\langle r_m r_n \rangle = \begin{cases} \sigma^2 & \text{if } n = m = 1, 2, \ldots, N \\ 0 & \text{otherwise.} \end{cases} \tag{29}$$

After writing

$$\left| \sum_{n=1}^{N} r_n e^{-2\pi i n \mathrm{T}s} \right|^2 = \left\{ \sum_{m=1}^{N} r_m e^{2\pi i m \mathrm{T}s} \right\} \left\{ \sum_{n=1}^{N} r_n e^{-2\pi i n \mathrm{T}s} \right\}$$
$$= \sum_{m=1}^{N} \sum_{n=1}^{N} r_m r_n e^{2\pi i (m-n)\mathrm{T}s}$$

we can use (29) to see that

$$P_r(s) = N\sigma^2 \mathrm{T}^2 \Pi(\mathrm{T}s). \tag{30}$$

In particular, every frequency in the band $[-1/2\mathrm{T},\ 1/2\mathrm{T}]$ is equally likely to be found in the signal r, and the energy integral

$$\int_{-\infty}^{\infty} P_N(s)\,ds = N\sigma^2 \mathrm{T}$$

increases by $\sigma^2 \mathrm{T}$ each time we add another term to (24) or (28). You can approximate the function (30) by averaging a large number of independently chosen functions $|R(s)|^2$, *cf.* Fig. 11.17 (provided that the higher moments of the r_n's are finite).

A signal (24) that has the expected power spectrum (30) is said to be *white noise*. When $\mathrm{T} \approx 10^{-4}$ sec, such a signal has a hissing sound like a radio tuned to static. When $\mathrm{T} \approx 10^{-2}$ sec, the sound rumbles like thunder. The first graph from Fig. 11.17 illustrates the structure of such white noise (but we must use a much larger value of N to produce representative signals).

You can use an amplitude envelope to change the timbre of white noise. For example, when we apply an exponentially growing amplitude envelope to the signal (24), the sound changes from a hiss to a whoosh (like that emitted by a rocket at a fireworks display!). Corresponding spectrograms are shown in Fig. 11.18.

4000 Hz   4000 Hz

0 Hz   0 Hz

Figure 11.18. Spectrograms for white noise (left) and for white noise with an exponentially growing amplitude envelope (right) using $0 \le t \le 16$ sec, $\text{T} = 1/8000$ sec, $N = 640$.

Filtered noise

Let $r_1, r_2, \ldots, r_N$ be random numbers as in the preceding section, and let $r_n = 0$ when $n < 1$ or $n > N$ (to simplify the limits of summation in the subsequent discussion). We convolve this sequence with certain real *filter coefficients* $\ldots, c_{-1}, c_0, c_1, \ldots$ to obtain a sequence

$$w_n := \sum_{m=-\infty}^{\infty} c_m r_{n-m}, \quad n = 0, \pm 1, \pm 2, \ldots \tag{31}$$

which has some additional regularity. The corresponding audio signal

$$w(t) := \sum_{n=-\infty}^{\infty} w_n \operatorname{sinc}\left(\frac{t - n\text{T}}{\text{T}}\right) \tag{32}$$

can be produced by sending the stream of w_n's to a sound card. In practice, only finitely many of the c_m's are nonzero, so the sums (31) and (32) are finite. An analysis of (32) will show how we can choose the c_n's to create interesting new musical tones.

By using the convolution rule

$$\sum_{n=-\infty}^{\infty} w_n e^{-2\pi i n \text{T}s} = \sum_{m=-\infty}^{\infty} c_m e^{-2\pi i m \text{T}s} \cdot \sum_{n=-\infty}^{\infty} r_n e^{-2\pi i n \text{T}s}$$

we see that (32) has the Fourier transform

$$W(s) = \left\{ \sum_{n=-\infty}^{\infty} w_n e^{-2\pi i n \text{T}s} \right\} \text{T}\Pi(\text{T}s) = C(s) \cdot R(s) \tag{33}$$

where R is given by (26) and

$$C(s) := \sum_{m=-\infty}^{\infty} c_m e^{-2\pi i m \mathrm{T} s}. \tag{34}$$

The spectral density

$$|W(s)|^2 = |C(s) \cdot R(s)|^2 = |C(s)|^2 \cdot \left| \sum_{n=1}^{N} r_n e^{-2\pi i n \mathrm{T} s} \right|^2 \mathrm{T}^2 \Pi(\mathrm{T} s)$$

depends on $r_1, r_2, \dots, r_N$, but within the context that led to (30) we see that w has the expected spectral density

$$P_w(s) = |C(s)|^2 \cdot N\sigma^2 \mathrm{T}^2 \Pi(\mathrm{T} s). \tag{35}$$

We can vary the timbre by using C to shape the spectral density and by using an amplitude envelope to control the attack and release of the tone!

Example: Let $a > 0$ be fixed. Find the function (34) that results when we filter a random sequence using the gaussian coefficients

$$c_m := \frac{1}{a} e^{-\pi m^2/a^2}, \quad m = 0, \pm 1, \pm 2, \dots . \tag{36}$$

Solution: We use Poisson's relation (4.18) with $p = 1/\mathrm{T}$ as we write

$$C(s) := \sum_{m=-\infty}^{\infty} \frac{\mathrm{T}}{a\mathrm{T}} e^{-\pi(m\mathrm{T})^2/(a\mathrm{T})^2} e^{-2\pi i m \mathrm{T} s} = \sum_{m=-\infty}^{\infty} e^{-\pi a^2 \mathrm{T}^2 (s - m/\mathrm{T})^2}.$$

In particular,

$$C(s) \approx e^{-\pi a^2 \mathrm{T}^2 s^2} \quad \text{when } |s| \leq 1/2\mathrm{T} \text{ and } a \gg 1. \tag{37}$$

∎

You can use the coefficients (36) to remove most of the high-frequency components from (24), leaving a spectrum centered at the origin. You can then use simple modulation to translate this spectrum, *e.g.*, the tone (32) produced from

$$w_n := \left\{ \sum_{m=-\infty}^{\infty} c_m r_{n-m} \right\} \cos(2\pi f n \mathrm{T}), \quad n = 0, \pm 1, \pm 2, \dots \tag{38}$$

will have a spectrum be centered at the frequencies $\pm f$, $0 < f < 1/2\mathrm{T}$. Figure 11.19 shows spectrograms for two tones produced in this way together with the corresponding expected spectral density functions. The first gives a steady (mid- to low-frequency) roar and the second gives a warble that fluctuates about a frequency of 1000 Hz. You should compare these spectrograms to those given in Fig. 11.18.

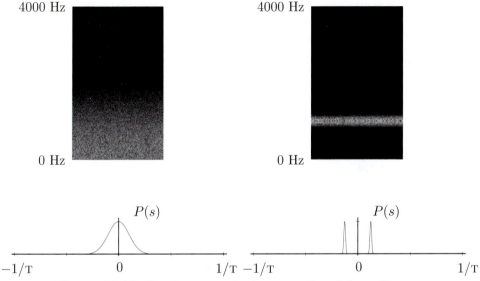

Figure 11.19. Spectrograms for tones produced from the samples (32) and (38) with the gaussian filter coefficients (36) using $a = 4$ (left) and $a = 50$, $f = 1000$ (right). In both cases $0 \leq t \leq 16$ sec, $\text{T} = 1/8000$ sec, and $N = 640$.

11.6 *Music with Mathematical Structure*

Introduction

You can use your computer to prepare sound files for a variety of musical tones and combine them to produce some aesthetic effect, *i.e.*, you can create *music*! At the most elementary level this involves forming a tone sequence for a simple melody, *cf.* Fig. 11.6, using the amplitude envelope to adjust the timbre. Musicians have various rules for forming chords, for constructing chord sequences, and for transforming chord sequences, and you may wish to use your computer to study such basic principles of harmony, *cf.* Robert W. Ottman, *Elementary Harmony, Theory and Practice, 5th ed.*, Prentice Hall, Englewood Cliffs, NJ, 1998. The computer allows you to go far beyond such classical notions, however. You can imagine some auditory phenomenon and generate samples for the corresponding sound file. We will show that it is possible to use a bit of mathematics as you work with these ideas.

Transformation of frequency functions

We consider a simple music theme consisting of a succession of notes played one by one. We construct a corresponding step function by defining $f(t)$ to be the frequency of the note that is played at time t. (We do not bother to define f at points that separate consecutive notes.) Such a *frequency function* for the *Twinkle, twinkle, little star* theme of Fig. 11.6 is shown at the top of Fig. 11.20.

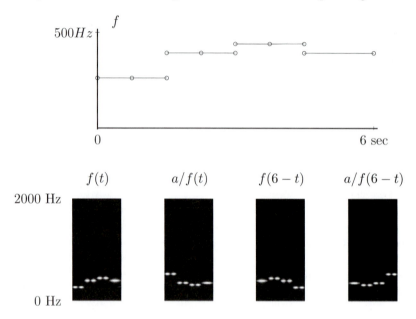

Figure 11.20. A frequency function $f(t)$, $0 \le t \le 6$ sec, for the score from Fig. 11.6 and spectrograms that correspond to the theme, $f(t)$, the inversion $a/f(t)$, the retrogression, $f(6-t)$, and the inverted retrogression $a/f(6-t)$. The constant a is the produce of the frequencies 261.6 Hz, 523.2 Hz for $C4$, $C5$ and $\text{T} = 1/8000$ sec, $N = 640$.

We can *mathematically transform* a frequency function f, *e.g.*, form

$$cf(t), \ f(ct), \ \frac{a}{f(t)}, \ f(b-t), \ \frac{a}{f(b-t)} \tag{39}$$

and generate a sound file for the corresponding transformed theme. Musicians routinely use the transformations (39) that preserve or invert the frequency ratio for two consecutive notes. When the ratio is preserved the musical interval is the same before and after the transformation. When the ratio is inverted, the size of the interval is preserved but the direction of movement is reversed, *e.g.*, when the ratio changes from $3/2$ to $2/3$ we go up, down by a fifth before, after the transformation.

The frequency scaling $f(t) \to cf(t)$, $c > 0$, preserves all of the musical intervals, so the transformed theme is identical to the original except for being higher, lower in pitch when $c > 1$, $c < 1$. Musicians use the term *transposition* to refer to this process. If each note of the theme has a frequency $440\rho^n$, $n = 0, \pm 1, \pm 2, \ldots$ from the equal-tempered scale and if $c = \rho^m$ for some $m = 0, \pm 1, \pm 2, \ldots$, then each note of the transposition will have a frequency from the equal-tempered scale. Of course, this is the reason for the development and widespread use of this scale.

The time dilation $f(t) \to f(ct)$, $c > 0$, uses exactly the same musical intervals, but the transformed theme is played at a faster, slower rate than the original when $c > 1$, $c < 1$. Musicians use the terms *diminution*, *augmentation* for such transformations.

When we subject a theme to the mapping $f(t) \to a/f(t)$, $a > 0$, we produce what musicians call an *inversion*. The inversion of a familiar tune is melodic (since we use the same musical intervals) but quite unlike the original (since we reverse the direction of movement at each step). If each note from the theme has a frequency from the equal-tempered scale and $a = 440^2 \rho^m$ for some $m = 0, \pm 1, \pm 2, \ldots$, then each note of the inversion will have a frequency from the equal-tempered scale. J.S. Bach had an uncanny ability to combine themes with their inversions in his keyboard compositions, *cf.* Ex. 11.20.

We obtain the *retrogression* of a theme corresponding to the frequency function $f(t)$, $0 \le t \le b$, by using the transformation $f(t) \to f(b-t)$. A sound file for the retrogression can be obtained from a sound file for the theme by reversing the order of the samples. The mapping $f(t) \to a/f(b-t)$ produces the inversion of the retrogression, *cf.* Fig. 11.20. Of course, when the theme and its inversion have frequencies from the equal-tempered scale, the same is true of the retrogression and inverted retrogression. You will find systematic use of inversion, retrogression, and inverted retrogression in the classical music of Bach, Haydn, Mozart, and Beethoven as well as in the contemporary music of Arnold Schoenberg.

You can experiment with transformations other than those given in (39), but it is not so easy to produce aesthetically pleasing effects. For example, the mapping $f(t) \to f(t) + d$, $d > 0$, raises the pitch of each note, but in most cases the transformed theme will contain unfamiliar intervals that are offensive to the musically trained ear.

Risset's endless glissando

We will describe an intriguing auditory anomaly that was created by Jean Claude Risset for his computer-generated composition *Mutations I*. Let

$$y(t) := \sin(2\pi f_0 \cdot e^{\alpha t}/\alpha)$$

where $f_0 := 784$ Hz (the frequency of $G5$) and $\alpha := \log(1.5)/8$ sec. By design, y has the local frequency

$$f(t) = f_0\, e^{\alpha t}$$

which increases by a factor of 1.5 (*i.e.*, one musical fifth) during any 8-sec interval. We apply a gaussian amplitude envelope to y and obtain

$$g(t) = e^{-\beta t^2} \sin(2\pi f_0 \cdot e^{\alpha t} / \alpha). \tag{40}$$

We use $\beta := 4/32^2$ to effectively localize the tone in the interval $-32 \text{ sec} \le t \le 32$ sec. The frequency of g increases from 155 Hz (inaudibly soft) to 784 Hz (loud) to 3964 Hz (inaudibly soft) during this time, as shown in the left spectrogram of Fig. 11.21.

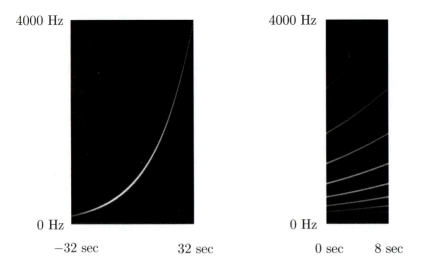

Figure 11.21. Spectrograms for the function g of (40) and the 8-periodic function w of (41) using $\text{T} = 1/8000$ sec, $N = 640$.

We now periodize g by defining

$$w(t) := \sum_{m=-\infty}^{\infty} g(t + 8\,\text{m}). \tag{41}$$

As we listen to w we hear a chord formed from the 8-translates of g. No translate of g is audible for more than 64 sec, so the chord contains at most 8 notes at any time t. By design, these notes are separated by 1,2, ... musical fifths, so the chord is harmonious. And by design, the local frequency of each note is strictly increasing so we hear an endless glissando (even though we know that a strictly increasing function cannot be periodic!).

The spectrogram on the right side of Fig. 11.21 will help you make sense of this auditory anomaly. The 8 curves come from the translates of g that sound during the interval $0 \leq t \leq 8$. These curves join to form a smooth spiral when we wrap the spectrogram around a right circular cylinder with circumference 8. (The same spiral is obtained when we wrap the left spectrogram around such a cylinder.) Risset's glissando is analogous to the visual illusion of a spinning barber pole!

References

A.H. Benade, *Fundamentals of Musical Acoustics, 2nd ed.*, Oxford University Press, New York, 1976; reprinted by Dover Publications, New York, 1990.

An exceptionally well written elementary introduction to the physics of music.

C. Dodge and T.A. Jerse, *Computer Music: Synthesis, Composition, and Performance*, Macmillan, New York, 1985.

One of the first textbooks to present the fundamentals of computer music for musicians.

H.L.F. Helmholtz, *On the Sensations of Tone* (English translation by A.J. Ellis), Longmans, London, 1885; reprinted by Dover Publications, New York, 1954.

The classic 19th-century treatise on the physics and the psychological perception of musical tones.

J.R. Pierce, *The Science of Musical Sound*, W.H. Freeman, New York, 1983.

A well-illustrated elementary exposition of musical tones.

C. Roads and J.Strawn, eds., *Foundations of Computer Music*, MIT Press, Cambridge, MA, 1985.

You will find John Chowning's original paper on FM synthesis reprinted in this collection.

C. Roads *et al.*, *The Computer Music Tutorial*, MIT Press, Cambridge, MA, 1996.

A comprehensive elementary introduction to computer music (with a 92-page bibliography!).

Exercise Set 11

▶ **EXERCISE 11.1:**   An 8-bit binary number $(b_8 b_7 \cdots b_1)_2$ can represent each integer $0, 1, \ldots, 255$. After shifting the origin to the center of this range we obtain the relative amplitudes $\pm.5, \pm 1.5, \ldots, \pm 127.5$ for driving a speaker attached to an 8-bit sound card.

(a) Explain why the dynamic range (between the softest and the loudest sounds) for an 8-bit digital recording is approximately 48 dB.

(b) How many bits per sample must we use if we want to reproduce the full 140-dB dynamic range of human hearing as shown in Fig. 11.2?

▶ **EXERCISE 11.2:**   Two Fourier analysis students are trying to determine the information content of a musical signal.

"You have to use 40,000 samples/sec if you want to reproduce all the frequencies the ear can hear, and there is no way to encode the incredible dynamic range of an auditory signal with less than 3 bytes/sample," says the first. "A good recording has to contain 120,000 bytes/sec."

"Nonsense!" responds the second. "There are only 88 piano keys, so with 2 bytes I can easily specify which note to play and how loud to play it. Nobody can play more than 25 notes/sec, so I can record anything that anybody can play on a piano by using no more than 50 bytes/sec."

Are these arguments valid? Why or why not?

Hint. Compare the music produced by a player piano to that you might hear at a Van Cliburn concert.

Note. One byte of information allows us to specify one item from 256 possible choices, two bytes of information allow us to specify one item from 256^2 possible choices, etc.

▶ **EXERCISE 11.3:**   Two Fourier analysis students are trying to distinguish seismic signals that come from earthquakes from those that are generated by an underground nuclear explosion. In both cases the energy is carried by wave packets having frequencies in the range .01 to 10 Hz.

"The seismograms look so much alike that I don't see how we can tell one from the other," says the first.

"Maybe we can hear the difference," says the second. "We could sample the seismic signals at one rate and then play them at another to make them audible."

(a) Determine suitable sampling and playback rates for this proposal.

(b) Do you think that it will work? Why or why not?

▶ **EXERCISE 11.4:**   The piano notes $C4$ (middle C), $E4$, $G4$ have the fundamental frequencies $f_C = 261.6$ Hz, $f_E = 329.6$ Hz, $f_G = 392.0$ Hz. We press the corresponding keys at times τ_C, τ_E, τ_G to produce

$$w(t) = \sin\{2\pi f_C(t - \tau_C)\} + \sin\{2\pi f_E(t - \tau_E)\} + \sin\{2\pi f_G(t - \tau_G)\}$$

when $t > \max\{\tau_C, \tau_E, \tau_G\}$. (Our model neglects amplitude variation, overtones, etc.)

(a) How small must τ_C, τ_E, τ_G be to ensure that the phase variations $|2\pi f_C \tau_C|$, $|2\pi f_E \tau_E|$, $|2\pi f_G \tau_G|$ do not exceed .1 radian?

(b) Use (a) and your auditory experiences with such chords to argue that, "The human ear is phase deaf."

▶ **EXERCISE 11.5:**   Show that the Fourier coefficients of

$$w_1(t) := \frac{1}{2} - \lfloor ft \rfloor, \qquad w_2(t) := -\frac{1}{\pi}\log|2\sin(\pi ft)|$$

have the same modulii $|W_1[k]| = |W_2[k]|$, $k = 0, \pm 1, \pm 2, \ldots$ (so that these audio waves sound the same!).

Hint. cf. Ex. 4.7.

▶ **EXERCISE 11.6:**   Attach small binder clips to the ends of a 300-mm ruler and stretch a rubber band (having a width of approximately 1 mm and an unstretched circumference of approximately 150 mm) around the ruler and clips lengthwise. The clips position the rubber band *string* away from the ruler so that it can vibrate. When you hold the ruler against the top of a desk or table (to serve as a resonator) you should get an audible tone when you pluck the string. Place a third binder clip between the ruler and the string to serve as a bridge. You can use the scale on the ruler to position the bridge precisely so that you can produce a tone with a predictable pitch. You now have a functioning monochord!

(a) Convince yourself that you can produce the musical intervals of an octave (*e.g.*, using bridge positions of 200 mm and 100 mm) and of a fifth (*e.g.*, using bridge positions of 200 mm and 133 mm).

(b) Calculate positions for the bridge that will enable you to play *do, re, ... , do* using the scale of Pythagoras.

(c) Calculate positions for the bridge that will enable you to play *Twinkle, twinkle, little star* using the scale of Pythagoras.

▶ **EXERCISE 11.7:**   In this exercise you will explore several features of the scale of Pythagoras.

(a) When we pass from the frequency f to the frequency $(3/2)f$, $(2/3)f$ we say that we go up, down a musical fifth. When we pass from the frequency f to the frequency $(2/1)f$, $(1/2)f$ we say that we go up, down a musical octave. Show that we can pass from *do* to *re, mi, ... , do* by going up or down by fifths and octaves.

Hint. To reach *re* you go up two fifths and down one octave: $F^2 O^{-1}$.

(b) How many fifths, fourths (with ratios 3/2, 4/3) can you play with the notes *do, re, ... , do* from the scale of Pythagoras?

(c) Find the first seven terms of the following sequence. Begin at *fa* and generate tones that are one, two, ... fifths higher. Then lower each such tone by some multiple of the octave to produce a result in the *do, re, ... , do* range. Any two consecutive tones from this sequence will harmonize.

(d) Adjacent notes of the Pythagorean scale differ by a tone (ratio 9/8) or a hemitone (ratio 256/243). Show that two hemitones give a tone to the same accuracy that 12 fifths $\approx$ 7 octaves. [The sequence from (c) is 12-periodic to the same accuracy.]

Note. The musical interval associated with $3^{12}/2^{19} = 1.0136\ldots$ is known as the *comma of Pythagoras*. It amounts to about a quarter of a hemitone.

▶ **EXERCISE 11.8:**   Let $w(t) := \sin\{2\pi(100t + 3900t^2)\}$, $0 \le t \le 1$ sec, and for $N = 1, 2 \ldots$ let

$$w_N(t) := \sum_{n=0}^{N} w\left(\frac{n}{N}\right) \cdot \mathrm{sinc}(Nt - n)$$

be the cardinal series approximation that uses N samples/sec.

(a) Use (8) to find the local frequency function for w.

(b) Sketch spectrograms to show what you would hear when you listen to w_{8000}, w_{12000}, and w_{16000}.

Hint. There are aliasing effects.

▶ **EXERCISE 11.9:**   Create an audio wave $w(t)$, $0 \le t \le 40$ sec for a sound having a local frequency $f(t)$ that rises at a constant rate of $1/4$ octave per second from an initial value of $f(0) = 20$ Hz.

Hint. Use (8) with the differential equation $[\log_2 f(t)]' = \frac{1}{4}$.

▶ **EXERCISE 11.10:**   Doppler's formula $f = f_0/(1 + v/v_s)$ gives the frequency we hear when a sound source with frequency f_0 moves away from us with velocity v. Here v_s is the velocity of sound. This gives the expression

$$f(t) = \frac{f_0}{1 + \gamma\cos(2\pi f_b t)}$$

for the frequency we hear at time t when a steady hum with frequency f_0 is emitted from the tip of a helicopter blade that rotates with velocity $v_b = \gamma v_s$ and frequency f_b. Show that this leads to the audio wave

$$w(t) := \sin\left\{\frac{f_0}{f_b}\frac{2}{\sqrt{1-\gamma^2}}\arctan\left[\left(\frac{1-\gamma}{1+\gamma}\right)^{1/2}\tan(\pi f_b t)\right]\right\}$$

for the sound of a chopper.

Note. Try $f_0 = 23.5$ Hz, $f_b = 8$ Hz, $\gamma = .956$.

▶ **EXERCISE 11.11:**   In this exercise you will analyze an auditory anomaly associated with the *Weierstrass function*

$$w(t) := \sum_{k=0}^{\infty} \alpha^k \sin(2\pi\beta^k f_0 t).$$

We choose $0 < \alpha < 1$ (to make the sum converge rapidly), $0 < f_0 < 20$ Hz (to make the first term inaudible), and $\beta := 2^{13/12} = 2.1189\ldots$ (so that successive terms differ in pitch by an octave plus a semitone).

(a) Show that $w(2t) = \alpha^{-1} w(2^{-1/12}t) - \alpha^{-1} \sin(2\pi f_0 2^{-1/12}t)$.

  Note. The second term on the right is inaudible.

(b) When we record speech, music, ... at 40,000 samples/sec and then play the recording at 80,000 samples/sec, we hear the speech, music, ... transposed up in pitch by one octave. What happens when we try this with w?

▶ **EXERCISE 11.12:**   You can use additive synthesis with (13) to produce a variety of tones having string like timbres. This exercise will show you an efficient way to compute the samples.

(a) Let $0 < \alpha < 1$ and let $\mathrm{box}_\alpha(x)$, $\mathrm{tri}_\alpha(x)$ be obtained by 1-periodically extending the functions shown in Fig. 11.22 from $[-1/2, 1/2]$ to all of $\mathbb{R}$.

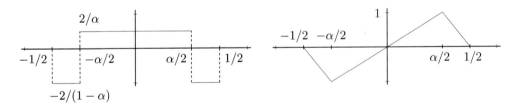

Fig. 11.22. Functions that we periodize to form box_α and tri_α.

Show that

$$\mathrm{box}_\alpha(x) = \frac{2}{\alpha(1-\alpha)} \left\{ \left\lfloor x + \frac{\alpha}{2} \right\rfloor - \left(x + \frac{\alpha}{2} \right) + 1 + \alpha \right\rfloor - \alpha \right\},$$

$$\mathrm{tri}_\alpha(x) = \min \left\{ \frac{2}{\alpha} \left(x + \frac{\alpha}{2} - \left\lfloor x + \frac{\alpha}{2} \right\rfloor \right), \frac{2}{1-\alpha} \left(-x - \frac{\alpha}{2} - \left\lfloor -x - \frac{\alpha}{2} \right\rfloor \right) \right\} - 1$$

where $\lfloor \ \rfloor$ is the floor function.

Hint. Verify that these expressions give 1-periodic functions that take the correct values when $-1/2 < x < 1/2$.

(b) Verify that box_α, tri_α have the Fourier series

$$\mathrm{box}_\alpha(x) = \frac{4}{\pi\alpha(1-\alpha)}\left\{ \frac{\sin(\pi\alpha)\cos(2\pi x)}{1} + \frac{\sin(2\pi\alpha)\cos(4\pi x)}{2} + \frac{\sin(3\pi\alpha)\cos(6\pi x)}{3} + \cdots \right\},$$

$$\mathrm{tri}_\alpha(x) = \frac{2}{\pi^2\alpha(1-\alpha)}\left\{ \frac{\sin(\pi\alpha)\sin(2\pi x)}{1^2} + \frac{\sin(2\pi\alpha)\sin(4\pi x)}{2^2} + \frac{\sin(3\pi\alpha)\sin(6\pi x)}{3^2} + \cdots \right\}.$$

Hint. Observe that

$$\mathrm{tri}'_\alpha(x) = \mathrm{box}_\alpha(x), \qquad \mathrm{box}'_\alpha(x) = \mathrm{III}\left(x + \frac{\alpha}{2}\right) - \mathrm{III}\left(x - \frac{\alpha}{2}\right).$$

Note. When you use the expressions from (a) to generate samples of $\mathrm{box}_\alpha(ft)$ or $\mathrm{tri}_\alpha(ft)$, you get all of the terms from the $\ell = 1$ or $\ell = 2$ series from (13) without evaluating a single sin or cos!

▶ **EXERCISE 11.13:**   Estimate the number of real operations we must do to generate

$$30 \text{ sec} \cdot 44,100 \, \frac{\text{samples}}{\text{sec}} = 1,323,000 \text{ samples}$$

for a CD quality digital sound file of a bell tone when we use:

(a) (12) with the parameters of Fig. 11.10 (additive synthesis);

(b) (22) with the parameters of Fig. 11.15 (FM synthesis).

Hint. An operation is the work we do as we evaluate $S := S + a \cdot b$. Assume that each evaluation of sin or exp requires 8 operations.

▶ **EXERCISE 11.14:**   You can use (16) to simplify (17) when the frequencies f_c, f_m are commensurate.

(a) Verify that

$$\sin\{2\pi ft + \mu\sin(2\pi ft)\}$$
$$= \{J_0(\alpha) - J_2(\alpha)\}\sin(2\pi ft) + \{J_1(\alpha) + J_3(\alpha)\}\sin(4\pi ft)$$
$$+ \{J_2(\alpha) - J_4(\alpha)\}\sin(6\pi ft) + \cdots .$$

(b) Find the analogous expansion for $\sin\{2\pi ft + \mu\sin(4\pi ft)\}$.

▶ **EXERCISE 11.15:**   In this exercise you will quantify Chowning's rule (19) by developing a numerical bound for the fraction

$$S_m(\mu) := \sum_{|k|\geq m} |J_k(\mu)|^2, \quad \mu \geq 0, \ m = 0, 1, \ldots$$

of the "energy" in the FM kernel

$$f(x) := e^{i\mu\sin(2\pi x)} = \sum_{k=-\infty}^{\infty} J_m(\mu)e^{2\pi ikx}$$

that lies outside the frequency band $-m < k < m$.

(a) Let N be a positive integer, and let

$$g_\mu[n] := e^{i\mu \sin(2\pi n/N)}, \quad n = 0, \pm 1, \pm 2, \ldots$$

be obtained by sampling f. Show that g has the discrete Fourier transform

$$G_\mu[k] = \sum_{\nu=-\infty}^{\infty} J_{k+\nu N}(\mu).$$

(b) Derive the bound

$$|J_k(\mu)| \leq \frac{\mu^2 + \mu}{k^2}, \quad \mu \geq 0, \ k = \pm 1, \pm 2, \ldots .$$

Hint. Begin by writing

$$|(2\pi i k)^2 J_k(\mu)| = \left| \int_0^1 f''(x) e^{-2\pi i k x} \, dx \right| \leq \int_0^1 |f''(x)| \, dx.$$

Note. An analogous argument can be used to show that

$$|J_k(\mu)| \leq \frac{\mu^3 + 3\mu^2 + \mu}{|k|^3}, \quad |J_k(\mu)| \leq \frac{\mu^4 + 6\mu^3 + 4\mu^2 + \mu}{k^4}, \ldots .$$

(c) Let $N = 2M$. Use (a) and (b) to show that

$$|J_k(\mu) - G_\mu[k]| \leq \frac{2(\mu^2 + \mu)}{M^2},$$

when $k = 0, \pm 1, \pm 2, \ldots, \pm M$ and $\mu \geq 0$.

Note. In view of (c) we can compute

$$S_m(\mu) \approx \sum_{m \leq |k| < M} |G_\mu[k]|^2$$

to any desired precision by choosing a sufficiently large $N = 2M$. The table

m	$\displaystyle\max_{0 \leq \mu \leq 5} S_m(\mu)$	$\displaystyle\max_{0 \leq \mu \leq 10} S_m(\mu)$	$\displaystyle\max_{0 \leq \mu \leq 100} S_m(\mu)$
$\lfloor \mu \rfloor + 2$	$4.08 \cdot 10^{-2}$	$4.10 \cdot 10^{-2}$	$4.10 \cdot 10^{-2}$
$\lfloor \mu \rfloor + 3$	$6.44 \cdot 10^{-3}$	$1.00 \cdot 10^{-2}$	$1.60 \cdot 10^{-2}$
$\lfloor \mu \rfloor + 4$	$7.43 \cdot 10^{-4}$	$2.01 \cdot 10^{-3}$	$8.33 \cdot 10^{-3}$

shows that Chowning's rule accounts for 96% of the "energy" of the FM kernel (when $|\mu| \leq 100$). [The approximation $J_k(\mu) \approx G_\mu[k]$ for $|k| < N/2$ is much more accurate than the bound (c) would suggest. Indeed, the above table can be computed with $N = 256$!]

▶ **EXERCISE 11.16:**   Derive the following generalizations of (17) that use 2 modulating frequencies f_1, f_2.

(a)
$$\sin\{2\pi f_c t + \mu_1 \sin(2\pi f_1 t) + \mu_2 \sin(2\pi f_2 t)\}$$
$$= \sum_{k_1=-\infty}^{\infty} \sum_{k_2=-\infty}^{\infty} J_{k_1}(\mu_1) J_{k_2}(\mu_2) \sin\{2\pi (f_c + k_1 f_1 + k_2 f_2)t\}$$

(b)
$$\sin\{2\pi f_c t + \mu_1 \sin[2\pi f_1 t + \mu_2 \sin(2\pi f_2 t)]\}$$
$$= \sum_{k_1=-\infty}^{\infty} \sum_{k_2=-\infty}^{\infty} J_{k_1}(\mu_1) J_{k_2}(k_1 \mu_2) \sin\{2\pi (f_c + k_1 f_1 + k_2 f_2)t\}$$

You should then be able to write down corresponding formulas for *parallel* FM synthesis and *cascade* FM synthesis that use $3, 4, \ldots$ modulating frequencies. These audio signals can have very wide bandwidths (that depend on $\mu_1, \mu_2, \ldots$), and we must use some care when we prepare digital sound files. If the coefficient of $\sin\{2\pi (f_c + k_1 f_1 + k_2 f_2 + \cdots)\}$ is nonnegligible, then we must choose a sampling interval T that satisfies the Nyquist condition

$$2(f_c + k_1 f_1 + k_2 f_2 + \cdots)T < 1.$$

▶ **EXERCISE 11.17:**   Sketch a spectrogram for each of the following, with time on the horizontal axis ($0 \le t \le 4$ sec) and with frequency on the vertical axis ($0 \le f \le 4000$ Hz).

(a) $w(t) = \sin(2\pi \cdot 400t) + \sin(2\pi \cdot 500t) + \sin(2\pi \cdot 600t)$

   Hint. The spectrogram for this major chord has 3 horizontal lines!

(b) $w(t) = \sin(2\pi \cdot 1000t) \cdot \sin(2\pi \cdot 200t)$

(c) $w(t) = \sin(2\pi \cdot 1000e^{t/2})$

(d) $w(t) = \sin\{2\pi \cdot 2000t + 500 \cdot \sin(2\pi t)\}$

(e) $w(t) = \sin\{2\pi \cdot 2000t + 100t \cdot \sin(2\pi t)\}$

(f) $w(t) = \sin\{2\pi \cdot 2000t + 4 \cdot \sin(2\pi \cdot 200t)\}$

   Hint. Use Chowning's rule (19).

(g) $w(t) = \sin\{2\pi \cdot 200t + 5 \cdot \sin(2\pi t/2) \cdot \sin(2\pi \cdot 400t)\}$

▶ **EXERCISE 11.18:**   Let $r_1, r_2, \ldots, r_N$ be independently chosen random numbers with mean $\mu = 0$ and variance σ^2, $0 < \sigma < \infty$, as in the text. Find the expected spectral density for (32) when the w_n's are given by:

(a) r_1, $r_1 + r_2$, $r_2 + r_3$, $\ldots, r_{N-1} + r_N$, r_N;

(b) r_1, $2r_1 + r_2$, $r_1 + 2r_2 + r_3$, $\ldots, r_{N-2} + 2r_{N-1} + r_N$, $r_{N-1} + 2r_N$, r_N;

(c) r_1, 0, r_2, 0, r_3, 0, $\ldots, r_{N-1}$, 0, r_N;

(d) r_1, r_1, r_2, r_2, r_3, r_3, $\ldots, r_{N-1}, r_{N-1}$, r_N, r_N.

▶ **EXERCISE 11.19:**   Let $M = 1, 2, \ldots$. The weighted binomial coefficients

$$c_m := \frac{1}{2^M} \binom{M}{m}, \qquad m = 0, \pm 1, \pm 2, \ldots$$

can be used in place of the coefficients (36) to remove high frequencies from the expected power spectrum (35) of (32).

(a) Show that
$$C(s) = \{e^{-\pi i s \mathrm{T}} \cos(\pi s \mathrm{T})\}^M.$$

(b) Sketch $|C(s)|^2$, $-1/2\mathrm{T} \le s \le 1/2\mathrm{T}$ when $M = 1$ and when $M \gg 1$.

▶ **EXERCISE 11.20:**   The theme from Goldberg's *Aria Grand* uses the half-second quarter notes

$$A^{\#}5 \quad A5 \quad G5 \quad F5 \quad D5 \quad D^{\#}5 \quad F5 \quad A^{\#}4$$

that are

$$13 \quad 12 \quad 10 \quad 8 \quad 5 \quad 6 \quad 8 \quad 1$$

semitones (ratio $2^{1/12}$) above the $f_A := 440$ Hz frequency of A4. Find the notes for the transformation that results when we replace the frequency function $f(t)$, $0 \le t \le 4$ sec by:

(a) $2f(t)$;   (b) $(2/3)f(t)$;   (c) $2f_A^2/f(t)$;   (d) $f(4-t)$;   (e) $2f_A^2/f(4-t)$.

You can then play the theme and these transformations on a piano!

Note. Johann Sebastian Bach used time translates of this theme and its transpositions (a), (b), its inversion (c), its retrogression (d), and its inverted retrogression (e) to construct his *Diverse Kanons*, cf. Don Dorsey, *Bach Busters*, TELARC®, 1995.

▶ **EXERCISE 11.21:**   Let $\mathfrak{T}$ be a mapping that has the frequencies $440\rho^n$, $n = 0, \pm 1, \pm 2, \ldots$ from the equal tempered keyboard for its domain and range.

(a) Show that $\mathfrak{T}(f) = 440\rho^{\kappa[\log_\rho(f/440)]}$ where κ is a function that maps $\mathbb{Z}$ into $\mathbb{Z}$.

(b) Find the κ functions that correspond to transposition and to inversion.

▶ **EXERCISE 11.22:**   You must hear a number of cycles of a sinusoidal audio signal before you can recognize its pitch. This exercise will help you quantify this observation. For simplicity, we will work with a complex wave function

$$w(t) := \frac{e^{-t^2/4\sigma^2}}{\sqrt[4]{2\pi\sigma^2}} e^{2\pi i f t}$$

where $0 < \sigma < \infty$ and $f > 0$.

(a) Using Table 2 of Appendix 6, convince yourself that 98% of the integral of $|w(t)|^2$ comes from the interval

$$|t| \leq 2.326\sigma$$

and that 98% of the integral of $|W(s)|^2$ comes from the interval

$$|s - f| \leq \frac{2.326}{4\pi\sigma} .$$

We will say that the tone sounds on the first interval and that the spectrum is contained in the second.

(b) Show that if we want the spectrum to be contained in the interval

$$2^{-1/12} f \leq s \leq 2^{1/12} f$$

(*i.e.*, within one semitone of f), then we must choose σ and f so that

$$\sigma f \geq \frac{2.326}{4\pi(1 - 2^{-1/12})} = 3.29\ldots$$

and the length of the interval during which the tone sounds is at least

$$2 \cdot 2.326\sigma \geq \frac{15.3\ldots}{f} ,$$

i.e., we must hear approximately 15 cycles of the tone. [This drops to 8 cycles if we replace 98% by 90% in (a) and to 4 cycles if we also relax the semitone to a tone.]

(c) Can a tuba play C2 (65.4 Hz) as a hemidemisemiquaver (64th note) during a march with a tempo of 120 quarter-note beats per minute?

▶ **EXERCISE 11.23:**   Almost all of the sound for the gaussian damped sinusoid

$$\gamma_{\tau,f}(t) := e^{-2\pi^2 t^2/\tau^2} \sin(2\pi f t), \quad -\infty < t < \infty$$

with nominal frequency $f > 0$ is emitted during the interval $-\tau/2 < t < \tau/2$, $\tau > 0$. Can you distinguish between computer-generated graphs of $|W_1(s)|^2$ and $|W_2(s)|^2$ when:

(a) $w_1(t) = \gamma_{1,100}(t)$, $w_2(t) = \gamma_{1,200}(t)$?

(b) $w_1(t) = \gamma_{1,100}(t)$, $w_2(t) = \gamma_{2,100}(t)$?

(c) $w_1(t) = \gamma_{1,100}(t)$, $w_2(t) = \gamma_{1,100}(t-2)$?

(d) $w_1(t) = \gamma_{1,100}(t) + \gamma_{1,200}(t)$, $w_2(t) = \gamma_{1,100}(t) + \gamma_{1,200}(t-2)$?

Chapter 12

Probability

12.1 Probability Density Functions on $\mathbb{R}$

Introduction

A liter bottle filled with air (at room temperature and pressure) holds approximately $2.7 \cdot 10^{22}$ molecules of N_2, O_2, CO_2, ... that fly about at high-speed colliding with one another and with the walls of the bottle. We cannot hope to keep track of every particle in such a large ensemble, but in the mid-19th century, Maxwell showed that we can deduce many statistical properties about such a system. For example, he found the *probability density*

$$f(v) := \frac{4}{\sqrt{\pi}} \frac{v^2}{\alpha^3} e^{-v^2/\alpha^2}, \quad v \geq 0 \tag{1}$$

for the speed V of a given gas molecule, *cf.* Fig. 12.1. (The parameter

$$\alpha = \sqrt{2\kappa T/m}$$

depends on the mass m of the molecule, the absolute temperature T and *Boltzmann's constant* $\kappa = 1.3806\ldots \cdot 10^{-23}$ joule/K.) We use the integral

$$\int_{v_1}^{v_2} f(v)dv$$

to find the probability that a molecule will have a speed in the interval $v_1 < V < v_2$.

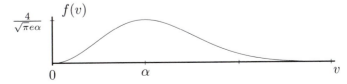

Figure 12.1. The Maxwell density (1) with parameter $\alpha > 0$.

Example: Use Maxwell's density to find the fraction of N_2 molecules (with $m = 4.65 \cdot 10^{-26}$ kg) in a warm room (with $T = 300$ K) that have speeds less than $v_0 = 1000$ km/hr $= 278$ m/sec.

Solution: We use (1) with the parameter

$$\alpha = \left\{ \frac{2 \cdot 1.3806 \cdot 10^{-23} \text{ joule/K} \cdot 300 \text{ K}}{4.65 \cdot 10^{-26} \text{ kg}} \right\}^{1/2} = 422 \text{ m/sec}$$

and the Maclaurin series for the exponential to compute

$$\int_0^{v_0} f(v)dv = \frac{4}{\sqrt{\pi}} \int_0^{v_0/\alpha} u^2 e^{-u^2} du$$

$$= \frac{4}{\sqrt{\pi}} \left\{ \frac{u^3}{3} - \frac{u^5}{5 \cdot 1!} + \frac{u^7}{7 \cdot 2!} - \frac{u^9}{9 \cdot 3!} + \cdots \right\} \Bigg|_0^{.658...} \approx .166.$$

Approximately one-sixth of the N_2 molecules have speeds less than 1000 km/hr! ∎

We can use the density function (1) to compute the average (or *expected*) value of certain functions of the molecular speed V.

Example: Use Maxwell's density to find the average speed and the average kinetic energy for an N_2 molecule when $T = 300$ K.

Solution: The *average speed* is given by the integral

$$\langle V \rangle := \int_0^\infty v f(v)dv = \frac{4\alpha}{\sqrt{\pi}} \int_0^\infty u^3 e^{-u^2} du = \frac{2}{\sqrt{\pi}}\alpha,$$

and the *average kinetic energy* is given by

$$\left\langle \frac{1}{2}mV^2 \right\rangle := \int_0^\infty \frac{1}{2}mv^2 f(v)dv = \frac{2}{\sqrt{\pi}} m\alpha^2 \int_0^\infty u^4 e^{-u^2} du = \frac{3}{4}m\alpha^2.$$

When $T = 300$ K we find $\alpha = 422$ m/sec (as in the above example), so

$$\langle V \rangle = 476 \text{ m/sec}, \qquad \left\langle \frac{1}{2}mV^2 \right\rangle = 6.21 \cdot 10^{-21} \text{ joule.} \qquad \blacksquare$$

Random processes occur in most areas of science and engineering, and you will find it advantageous to learn certain basic skills for manipulating probability densities and evaluating (or closely approximating) related integrals. This chapter will show you how Fourier analysis can be used to expedite such calculations. After learning a few basic concepts you will routinely write down integrals that solve problems analogous to those given above. You might even enjoy deducing the form of Maxwell's density (1) from considerations of symmetry.

Generalized probability densities

We say that an ordinary function f is a *probability density* on $\mathbb{R}$ provided that

$$f(x) \geq 0 \qquad \text{for } -\infty < x < \infty, \text{ and}$$

$$\int_{-\infty}^{\infty} f(x)dx = 1.$$

We will immediately extend this definition so that we can assign positive atoms of probability to isolated points of the real line. We say that a generalized function f is a *probability density* on $\mathbb{R}$ provided that

$$\int_{-\infty}^{\infty} f(x)\phi(x)dx \geq 0 \text{ when } \phi \in \mathbb{S} \text{ is nonnegative, and} \tag{2}$$

$$\lim_{n \to \infty} \int_{-\infty}^{\infty} f(x)e^{-\pi x^2/n^2} dx = 1. \tag{3}$$

For example,

$$f(x) := \frac{1}{a}\Pi\left(\frac{x-\mu}{a}\right), \quad -\infty < \mu < \infty, \ a > 0, \qquad \text{(uniform density)}$$

$$f(x) := \frac{e^{-(x-\mu)^2/2\sigma^2}}{\sqrt{2\pi}\,\sigma}, \quad -\infty < \mu < \infty, \ \sigma > 0, \qquad \text{(normal density)}$$

$$f(x) := \frac{e^{-|x|/\alpha}}{2\alpha}, \ \alpha > 0, \qquad \text{(Laplace density)}$$

$$f(x) := \frac{1}{6}\sum_{k=1}^{6} \delta(x-k), \qquad \text{(die-toss density)}$$

$$f(x) := \frac{1}{2^n}\sum_{k=0}^{n} \binom{n}{k}\delta(x-k), \ n = 0,1,2,\dots, \qquad \text{(binomial density)}$$

$$f(x) := e^{-\alpha}\sum_{k=0}^{\infty} \frac{\alpha^k}{k!}\delta(x-k), \ \alpha > 0, \qquad \text{(Poisson density)}$$

are all probability densities on $\mathbb{R}$.

An antiderivative

$$g = f^{-1}$$

of a probability density f can be represented by an ordinary function that has finite limits at $\pm\infty$, *cf.* Ex. 12.12. When we select the constant of integration so that

$$\lim_{x \to -\infty} g(x) = 0, \qquad \lim_{x \to +\infty} g(x) = 1$$

and assign function values so that

$$g(x+) = g(x) \text{ for } -\infty < x < \infty$$

we will say that g is the *distribution function* for f. We can always recover a probability density f from its distribution function g by writing

$$f = g'.$$

Of course, when f is an ordinary probability density we have

$$g(x) = \int_{-\infty}^{x} f(\xi)d\xi.$$

You will observe such relations as you examine the f, g pairs shown in Fig. 12.2 or when you use the table for the distribution function of the standard normal density that is given in Appendix 6.

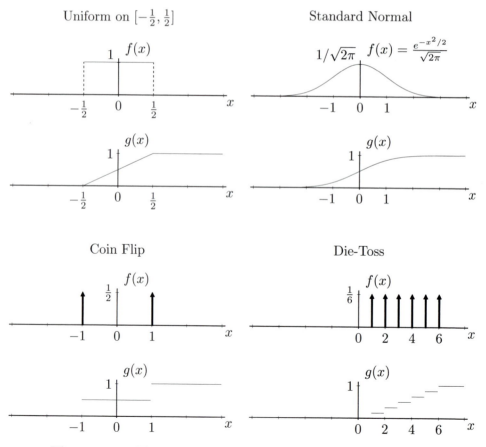

Figure 12.2. The uniform, standard normal, coin flip, and die-toss densities together with the corresponding distribution functions.

The *characteristic function* of a probability density f is defined to be the Fourier transform

$$F := f^\wedge.$$

(Most authors use the $+i$ instead of the $-i$ Fourier transform for the characteristic function, and many use a different 2π convention, *cf.* Ex. 1.4. You must exercise some caution when you compare formulas for characteristic functions in this text with those you find in a probability text or a statistics text!) You can use the Fourier transform calculus from Chapters 3 and 7 to find the characteristic functions for common probability densities. We will presently show that F is always a bounded, continuous ordinary function on $\mathbb{R}$ as illustrated in Fig. 12.3. We can always recover a probability density f from its characteristic function F by writing

$$f = F^{\wedge\vee}.$$

Our primary purpose for introducing F is to facilitate the calculation of convolution products (e.g., as done in Section 3.3), but we will also show that characteristic functions can be used to find moments and to construct new probability densities.

12.2 Some Mathematical Tools

Introduction

We will now present several specialized limits and inequalities that are needed for our study of characteristic functions, expectations, and the central limit theorem. You may wish to skim over the material in this section during your first reading. After you see how we use these tools you will have a bit more motivation to sort through the details.

Gaussian mollification and tapering

For each $n = 1, 2, \ldots$ we define the gaussian density

$$\gamma_n(x) := n\, e^{-\pi(nx)^2}, \tag{4}$$

noting that

$$\Gamma_n(s) = e^{-\pi(s/n)^2} \tag{5}$$

[as used in (3)] is the Fourier transform. It is fairly easy to show that

$$\lim_{n\to\infty} \gamma_n = \delta \quad \text{and} \quad \lim_{n\to\infty} \Gamma_n = 1,$$

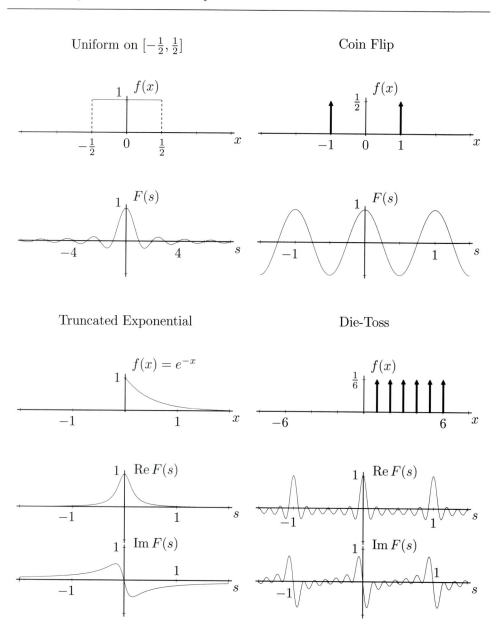

Figure 12.3. The uniform, coin flip, truncated exponential, and die-toss densities together with the corresponding characteristic functions.

cf. Fig. 12.4. These are special cases ($f = \delta$, $F = 1$) of the weak limits

$$\lim_{n \to \infty} \gamma_n * f = f \tag{6}$$

$$\lim_{n \to \infty} \Gamma_n \cdot F = F \tag{7}$$

that hold for every choice of the generalized functions f, F (not just when f is a probability density and F is a characteristic function).

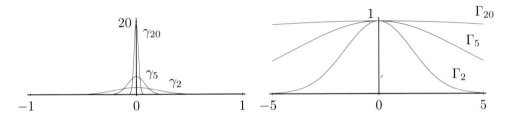

Figure 12.4. The gaussians γ_n, Γ_n from (4), (5) with $n = 2, 5, 20$.

Example: Establish the weak limits (6) and (7).

Solution: Weak limits are preserved by the Fourier transform so it is sufficient to prove (7). Let $F = G^{(m)}$ where G is CSG and m is a nonnegative integer. Given $\phi \in \mathbb{S}$ we must verify that

$$\lim_{n \to \infty} \int_{-\infty}^{\infty} G(x)[\Gamma_n(x)\phi(x)]^{(m)} dx = \int_{-\infty}^{\infty} G(x)\phi^{(m)}(x) dx.$$

By using the Leibnitz rule (2.29) we see that it is sufficient to show that

$$\lim_{n \to \infty} \int_{-\infty}^{\infty} G(x)\Gamma_n^{(k)}(x)\phi^{(m-k)}(x) dx = 0 \text{ when } k = 1, 2, \ldots, m \text{ and} \tag{8}$$

$$\lim_{n \to \infty} \int_{-\infty}^{\infty} G(x)[1 - \Gamma_n(x)]\phi^{(m)}(x) dx = 0. \tag{9}$$

From the Rodrigues formula (3.28) for the Hermite polynomials we see that

$$\Gamma_n^{(k)}(x) = \frac{d^k}{dx^k} e^{-(\sqrt{\pi}x/n)^2} = \left(\frac{\sqrt{\pi}}{n}\right)^k \cdot (-1)^k H_k\left(\frac{\sqrt{\pi}x}{n}\right) e^{-(\sqrt{\pi}x/n)^2}.$$

It follows that

$$|\Gamma_n^{(k)}(x)| \le \frac{C_k}{n^k}, \qquad n = 1, 2, \ldots$$

for some constant C_k. Using this bound we write

$$\left| \int_{-\infty}^{\infty} G(x) \Gamma_n^{(k)}(x) \phi^{(m-k)}(x) dx \right| \le \frac{C_k}{n^k} \int_{-\infty}^{\infty} |G(x) \phi^{(m-k)}(x)| dx$$

and thereby obtain (8).

Given $L > 0$ we majorize the integral from (9) by writing

$$\left| \int_{-\infty}^{\infty} G(x)[1 - \Gamma_n(x)] \phi^{(m)}(x) dx \right| \le \int_{|x| \le L} |G(x)[1 - \Gamma_n(x)] \phi^{(m)}(x)| dx$$

$$+ \int_{|x| \ge L} |G(x)[1 - \Gamma_n(x)] \phi^{(m)}(x)| dx \qquad (10)$$

$$\le \max_{-\infty < x < \infty} |G(x) \phi^{(m)}(x)| \cdot 2L[1 - \Gamma_n(L)] + \int_{|x| \ge L} |G(x) \phi^{(m)}(x)| dx.$$

Let $\epsilon > 0$ be selected. Since $G\phi^{(m)}$ rapidly approaches 0 at $\pm\infty$, we can choose $L > 0$ so that

$$\int_{|x| \ge L} |G(x) \phi^{(m)}(x)| dx < \frac{\epsilon}{2}.$$

With L so chosen, we will then have

$$\max_{-\infty < x < \infty} |G(x) \phi^{(m)}(x)| \cdot 2L[1 - \Gamma_n(L)] < \frac{\epsilon}{2}$$

for all sufficiently large n. In conjunction with (10), this proves (9). ∎

Fundamental inequalities

Let f be a probability density and let ϕ_1, ϕ_2 be real Schwartz functions. We obtain the *monotonicity relation*

$$\int_{-\infty}^{\infty} f(x) \phi_1(x) dx \le \int_{-\infty}^{\infty} f(x) \phi_2(x) dx \quad \text{when } \phi_1 \le \phi_2 \qquad (11)$$

by using (2) (with $\phi := \phi_2 - \phi_1$).

Suppose that ϕ is a real Schwartz function with $b \le \phi(x) \le B$, $-\infty < x < \infty$ for suitably chosen constants b, B. We use (11) to see that

$$b \int_{-\infty}^{\infty} f(x) \Gamma_n(x) dx \le \int_{-\infty}^{\infty} f(x) \Gamma_n(x) \phi(x) dx \le B \int_{-\infty}^{\infty} f(x) \Gamma_n(x) dx, \quad n = 1, 2, \dots,$$

and then use (3), (7) to write

$$b \le \int_{-\infty}^{\infty} f(x) \phi(x) dx \le B. \qquad (12)$$

This allows us to deduce that a probability density is real, *i.e.*,

$$\overline{f} = f, \tag{13}$$

and to establish the *max bound*

$$\left| \int_{-\infty}^{\infty} f(x)\phi(x)dx \right| \leq \max_{-\infty < x < \infty} |\phi(x)|. \tag{14}$$

Let $\phi_1, \phi_2 \in \mathbb{S}$ be complex valued. We choose $\beta \in \mathbb{C}$ with $|\beta| = 1$ so that

$$\beta \int_{-\infty}^{\infty} f(x)\phi_1(x)\phi_2(x)dx = \left| \int_{-\infty}^{\infty} f(x)\phi_1(x)\phi_2(x)dx \right|.$$

For each real value of t

$$|t\,\phi_1(x) + \overline{\beta\,\phi_2(x)}|^2 = t^2\phi_1(x)\overline{\phi_1(x)} + t\{\beta\,\phi_1(x)\phi_2(x) + \overline{\beta\,\phi_1(x)\phi_2(x)}\} + \phi_2(x)\overline{\phi_2(x)}$$

is a nonnegative Schwartz function, so we can use (2) [and (13)] to see that

$$t^2 \int_{-\infty}^{\infty} f(x)|\phi_1(x)|^2 dx + 2t \left| \int_{-\infty}^{\infty} f(x)\phi_1(x)\phi_2(x)dx \right| + \int_{-\infty}^{\infty} f(x)|\phi_2(x)|^2 dx \geq 0.$$

A nonnegative quadratic polynomial must have a nonpositive discriminant, and within the present context this observation gives us the *Cauchy-Schwartz inequality*

$$\left| \int_{-\infty}^{\infty} f(x)\phi_1(x)\phi_2(x)dx \right|^2 \leq \int_{-\infty}^{\infty} f(x)|\phi_1(x)|^2 dx \cdot \int_{-\infty}^{\infty} f(x)|\phi_2(x)|^2 dx. \tag{15}$$

Example: Let f be a probability density and let $\phi \in \mathbb{S}$. Show that

$$\left| \int_{-\infty}^{\infty} f(x)\phi(x)dx \right|^2 \leq \int_{-\infty}^{\infty} f(x)|\phi(x)|^2 dx. \tag{16}$$

Solution: We use (15) (with $\phi_1 = \Gamma_n$, $\phi_2 = \phi$) and (11) to write

$$\left| \int_{-\infty}^{\infty} f(x)\Gamma_n(x)\phi(x)dx \right|^2 \leq \int_{-\infty}^{\infty} f(x)\Gamma_n^2(x)dx \cdot \int_{-\infty}^{\infty} f(x)|\phi(x)|^2 dx$$

$$\leq \int_{-\infty}^{\infty} f(x)\Gamma_n(x)dx \cdot \int_{-\infty}^{\infty} f(x)|\phi(x)|^2 dx.$$

In conjunction with the limits (3), (7) this gives (16).   ∎

Example: Let f be a probability density. Show that the max bound (14) holds even when $\phi \in \mathbb{S}$ is complex valued.

Solution: We use (16) with the real version of (14) to write

$$\left| \int_{-\infty}^{\infty} f(x)\phi(x) \right|^2 \leq \int_{-\infty}^{\infty} f(x)|\phi(x)|^2 dx \leq \max_{-\infty < x < \infty} |\phi(x)|^2 = \{ \max_{-\infty < x < \infty} |\phi(x)| \}^2$$

and thereby prove (14) when $\phi \in \mathbb{S}$ is complex. ∎

Example: Let f be a probability density and let $\phi \in \mathbb{S}$ with $\phi \geq 0$. Show that

$$\left| \int_{-\infty}^{\infty} f(x)\phi(x)e^{-2\pi isx} dx \right| \leq \int_{-\infty}^{\infty} f(x)\phi(x)dx, \quad -\infty < s < \infty. \tag{17}$$

Solution: When $f\phi = 0$, (17) holds trivially. In all other cases $f\phi$ is a scaled probability density, so we can use (7) with the complex version of (14) to write

$$\left| \int_{-\infty}^{\infty} f(x)\phi(x)e^{-2\pi isx} dx \right| = \lim_{n \to \infty} \left| \int_{-\infty}^{\infty} f(x)\phi(x)\Gamma_n(x)e^{-2\pi isx} dx \right|$$

$$\leq \lim_{n \to \infty} \int_{-\infty}^{\infty} f(x)\phi(x)dx \cdot \max_{-\infty < x < \infty} |\Gamma_n(x)e^{-2\pi isx}|$$

$$= \int_{-\infty}^{\infty} f(x)\phi(x)dx. \qquad ∎$$

12.3 The Characteristic Function

F is bounded and continuous

Let f be a probability density and let $F := f^\wedge$ be the corresponding characteristic function. For $n = 1, 2, \ldots$ and $-\infty < s < \infty$ we define

$$F_n(s) := \int_{-\infty}^{\infty} f(x)\Gamma_n(x)e^{-2\pi isx} dx. \tag{18}$$

We use the max bound to verify that F_n is bounded and continuous by writing

$$|F_n(s)| \leq \max_{-\infty < x < \infty} |\Gamma_n(x)e^{-2\pi isx}| = 1$$

and

$$|F_n(s+h) - F_n(s)| = \left| \int_{-\infty}^{\infty} f(x)\Gamma_n(x)\{e^{-2\pi i(s+h)x} - e^{-2\pi isx}\}dx \right|$$

$$\leq \max_{-\infty < x < \infty} \Gamma_n(x)|e^{-2\pi ihx} - 1| \to 0 \quad \text{as} \quad h \to 0.$$

We will show that $F_1, F_2, \ldots$ converges to F and thereby show that the characteristic function is bounded and continuous.

As a first step we use (17) and (3) to write

$$|F_m(s) - F_n(s)| = \left| \int_{-\infty}^{\infty} f(x)\{\Gamma_m(x) - \Gamma_n(x)\}e^{-2\pi i s x} dx \right|$$

$$\leq \int_{-\infty}^{\infty} f(x)|\Gamma_m(x) - \Gamma_n(x)| dx$$

$$= \left| \int_{-\infty}^{\infty} f(x)\Gamma_m(x) dx - \int_{-\infty}^{\infty} f(x)\Gamma_n(x) dx \right|$$

$$\to 0 \quad \text{as} \quad m, n \to \infty.$$

This proves that $F_1(s), F_2(s), \ldots$ is a Cauchy sequence that has a *pointwise* limit

$$F_\infty(s) := \lim_{n \to \infty} F_n(s), \qquad -\infty < s < \infty,$$

with $F_\infty(s)| \leq 1$ for each s and with

$$|F_\infty(s) - F_n(s)| \leq 1 - \int_{-\infty}^{\infty} f(x)\Gamma_n(x) dx, \quad -\infty < s < \infty. \tag{19}$$

Using (19) and (3) we see that F_∞ is the *uniform* limit of the continuous functions $F_1, F_2, \ldots$, so F_∞ must be continuous. Knowing that $F_1, F_2, \ldots$ and F_∞ are all bounded and continuous we again use (19) and (3) to see that

$$\left| \int_{-\infty}^{\infty} \{F_\infty(s) - F_n(s)\}\phi(s) ds \right| \leq \left\{ 1 - \int_{-\infty}^{\infty} f(x)\Gamma_n(x) dx \right\} \cdot \int_{-\infty}^{\infty} |\phi(x)| dx$$

$$\to 0 \quad \text{as} \quad n \to \infty$$

when $\phi \in \mathbb{S}$. In this way we establish the *weak* limit

$$\lim_{n \to \infty} F_n = F_\infty. \tag{20}$$

We will now identify F_∞ with the characteristic function F. The convolution product $\gamma_m * f$ is a bounded continuous probability density, *cf.* Ex. 12.10, so $(\gamma_m * f) \cdot \Gamma_n$ has the ordinary Fourier transform

$$F_{mn}(s) := \int_{-\infty}^{\infty} (\gamma_m * f)(x)\Gamma_n(x)e^{-2\pi i s x} dx. \tag{21}$$

We will establish the weak limit

$$\lim_{m \to \infty} F_{mn} = F_n, \tag{22}$$

so we can use (20), (22), (21), (6), (7) in turn to write

$$F_\infty = \lim_{n\to\infty} F_n \overset{?}{=} \lim_{n\to\infty}\lim_{m\to\infty} F_{mn} = \lim_{n\to\infty}\lim_{m\to\infty}\{(\gamma_m * f)\cdot\Gamma_n\}^\wedge = \lim_{n\to\infty}\{f\cdot\Gamma_n\}^\wedge =: F$$

and thereby obtain the representation

$$F(s) = \lim_{n\to\infty}\int_{-\infty}^{\infty} f(x)\Gamma_n(x)e^{-2\pi isx}dx, \qquad -\infty < s < \infty \tag{23}$$

for the characteristic function F. This identification of F with F_∞ allows us to see that F is continuous and bounded with

$$|F(s)| \le 1, \qquad -\infty < s < \infty. \tag{24}$$

Example: Show that $F_{1n}, F_{2n}, \ldots$ converges weakly to F_n [when these functions are constructed from the probability density f by using (21) and (18)].

Solution: We use (14) with the Schwartz function $\Gamma_{ns}(x) := \Gamma_n(x)e^{-2\pi isx}$ to write

$$|F_{mn}(s) - F_n(s)| = \left|\int_{-\infty}^{\infty}\{(\gamma_m * f)(x) - f(x)\}\Gamma_{ns}(x)dx\right|$$

$$= \left|\int_{-\infty}^{\infty} f(x)\{(\gamma_m * \Gamma_{ns})(x) - \Gamma_{ns}(x)\}dx\right|$$

$$\le \max_{-\infty < x < \infty}|(\gamma_m * \Gamma_{ns})(x) - \Gamma_{ns}(x)|.$$

Given $\epsilon > 0$ we will choose $M > 0$ to ensure that

$$\max_{|x|\ge M}|(\gamma_m * \Gamma_{ns})(x) - \Gamma_{ns}(x)| \le \max_{|x|\ge M}\{(\gamma_m * \Gamma_n)(x) + \Gamma_n(x)\} < \epsilon$$

for all $m = 1, 2, \ldots$ and $-\infty < s < \infty$. If s is constrained to lie in some finite interval $-L \le s \le L$, we will also have

$$\max_{|x|\le M, |s|\le L}|(\gamma_m * \Gamma_{ns})(x) - \Gamma_{ns}(x)| < \epsilon$$

for all sufficiently large m. In this way we see that

$$|F_{mn}(s) - F_n(s)| < \epsilon \text{ when } -L \le s \le L$$

and m is sufficiently large, *i.e.*, $F_{1n}, F_{2n}, \ldots$ converges uniformly to F_n on every finite interval $[-L, L]$.

We will use this uniform convergence to establish the weak limit (22). Let $\phi \in \mathbb{S}$ be selected. Since $|F_{mn}(s)| \leq 1$, $|F_n(s)| \leq 1$ we can write

$$\left| \int_{-\infty}^{\infty} F_{mn}(s)\phi(s)ds - \int_{-\infty}^{\infty} F_n(s)\phi(s)ds \right| \leq \int_{-\infty}^{\infty} |F_{mn}(s) - F_n(s)| \cdot |\phi(s)|ds$$

$$\leq \max_{-\infty < s < \infty} |\phi(s)| \cdot \int_{|s| \leq L} |F_{mn}(s) - F_n(s)|ds + 2 \int_{|s| \geq L} |\phi(s)|ds.$$

Given $\epsilon > 0$, we can choose $L > 0$ to make

$$2 \int_{|s| \geq L} |\phi(s)|ds < \epsilon/2.$$

Since $F_{1n}, F_{2n}, \ldots$ converges uniformly to F_n on $[-L, L]$ we will also have

$$\max_{-\infty < s < \infty} |\phi(s)| \cdot \int_{|s| \leq L} |F_{mn}(s) - F_n(s)|ds < \epsilon/2$$

when m is sufficiently large. It follows that

$$\left| \int_{-\infty}^{\infty} F_{mn}(s)\phi(s)ds - \int_{-\infty}^{\infty} F_n(s)\phi(s)ds \right| < \epsilon$$

when m is sufficiently large. Since $\epsilon > 0$ is arbitrary, the proof of (22) is complete. ∎

Bochner's characterization of F

We now know that the characteristic function (23) of a probability density f is continuous and bounded with

$$F(0) = 1. \tag{25}$$

A characteristic function is also *nonnegative semidefinite* in the sense that

$$\int_{-\infty}^{\infty} F(s)(\psi * \psi^{\dagger})(s)ds \geq 0 \text{ for all } \psi \in \mathbb{S} \tag{26}$$

[as we see by using the Parseval relation

$$\int_{-\infty}^{\infty} F(s)(\psi * \psi^{\dagger})(s)ds = \int_{-\infty}^{\infty} f(x)|\psi^{\wedge}(x)|^2 dx$$

with (2)]. Bochner discovered that any F having these four properties must be a characteristic function. We will give a proof and then use this characterization to show that the set of generalized probability densities is closed under convolution.

Let F be a continuous and bounded function on $\mathbb{R}$ that satisfies (25) and (26). We will prove that $f := F^{\wedge\vee}$ is a probability density. Let $\phi \in \mathbb{S}$ be selected with $\phi \geq 0$. For each $m, n = 1, 2, \ldots$ we define ψ_{mn} so that

$$\psi_{mn}^{\wedge} = \{\Gamma_n \cdot [\phi + 1/m]\}^{1/2}. \tag{27}$$

We verify that $\psi_{mn} \in \mathbb{S}$, and then show that f is nonnegative by using (7), (27), the Parseval relation, and (26) in turn to write

$$\int_{-\infty}^{\infty} f(x)\phi(x)dx = \lim_{n\to\infty} \lim_{m\to\infty} \int_{-\infty}^{\infty} f(x)\Gamma_n(x)[\phi(x) + 1/m]dx$$

$$= \lim_{n\to\infty} \lim_{m\to\infty} \int_{-\infty}^{\infty} f(x)|\psi_{mn}^{\wedge}(x)|^2 dx$$

$$= \lim_{n\to\infty} \lim_{m\to\infty} \int_{-\infty}^{\infty} F(s)(\psi_{mn} * \psi_{mn}^{\dagger})(s)ds \geq 0.$$

Since F is continuous and bounded, we can use the Parseval relation, the fact that γ_n is an approximate δ, and the hypothesis (25) to see that

$$\lim_{n\to\infty} \int_{-\infty}^{\infty} f(x)\Gamma_n(x)dx = \lim_{n\to\infty} \int_{-\infty}^{\infty} F(s)\gamma_n(s)ds = F(0) = 1.$$

In this way we verify that f has the properties (2), (3) of a probability density.

Products of characteristic functions

Let f_1, f_2 be probability densities and let F_1, F_2 be the corresponding characteristic functions. Since F_1, F_2 are continuous and bounded, the same is true of

$$F(s) := F_1(s) \cdot F_2(s), \qquad -\infty < s < \infty, \tag{28}$$

so this pointwise product is a well-defined generalized function represented by the fundamental functional

$$F\{\phi\} := \int_{-\infty}^{\infty} F_1(s)F_2(s)\phi(s)ds, \qquad \phi \in \mathbb{S}.$$

[We do not assume that $F_1 \cdot \phi \in \mathbb{S}$ whenever $\phi \in \mathbb{S}$, so the discussion following (7.75) does not apply within the present context.] The convolution rule

$$(f_1 * f_2)^{\wedge} = F_1 \cdot F_2$$

and the corresponding Parseval relation will be valid within this context if we define the convolution product

$$f = f_1 * f_2 \tag{29}$$

of the probability densities f_1, f_2 by writing

$$f_1 * f_2 := (F_1 \cdot F_2)^{\wedge\vee}. \tag{30}$$

We will show that (29) is a probability density by verifying that (28) satisfies the Bochner conditions for a characteristic function.

The function (28) is continuous and bounded with $F(0) = 1$ (since these properties are possessed by the characteristic functions F_1, F_2). We must show that F also satisfies (26). Let $\psi \in \mathbb{S}$ be selected and let $n = 1, 2, \dots$. The convolution product of a generalized probability density and a probability density from $\mathbb{S}$ is a smooth bounded probability density (*cf.* Ex. 12.10), so

$$\{f_1 * \gamma_n\} \cdot \{f_2^\vee * |\psi^\wedge|^2\}$$

is a smooth, nonnegative, integrable function on $\mathbb{R}$. This being the case, we can use (7) with Parseval's relation (and the convolution rule) to write

$$\int_{-\infty}^{\infty} F(s)(\psi * \psi^\dagger)(s)ds = \lim_{n \to \infty} \int_{-\infty}^{\infty} \{F_1(s)\Gamma_n(s)\} \cdot \{F_2(s)(\psi * \psi^\dagger)(s)\}ds$$

$$= \lim_{n \to \infty} \int_{-\infty}^{\infty} (f_1 * \gamma_n)(x) \cdot (f_2^\vee * |\psi^\wedge|^2)(x)dx \geq 0.$$

In this way we see in turn that (26) holds, that $F = F_1 \cdot F_2$ is a characteristic function, and that $f = f_1 * f_2$ is a probability density.

More generally, if $f_1, f_2, f_3, f_4, \dots$ are probability densities with the characteristic functions $F_1, F_2, F_3, F_4, \dots$, then

$$F_1 \cdot F_2, \quad F_1 \cdot F_2 \cdot F_3, \quad F_1 \cdot F_2 \cdot F_3 \cdot F_4, \dots$$

are characteristic functions and the corresponding

$$f_1 * f_2, \quad f_1 * f_2 * f_3, \quad f_1 * f_2 * f_3 * f_4, \dots$$

are probability density functions. Such products and convolution products are always commutative and associative, *cf.* Ex. 2.36.

Periodic characteristic functions

Let $p > 0$. From your work with p-periodic generalized functions in Section 7.7, you know that when the probability density f is a weighted sum of $\delta(x - k/p)$, $k = 0, \pm 1, \pm 2, \dots$, then the characteristic function F must be p-periodic. More generally, if $0 \leq x_0 < 1/p$ and f is a weighted sum of $\delta(x - x_0 - k/p)$, $k = 0, \pm 1, \pm 2, \dots$, then $e^{2\pi i s x_0} F(s)$ must be p-periodic. We will show that this is the case precisely when

$$|F(p)| = F(0) = 1 \tag{31}$$

and

$$|F(s)| < 1 \quad \text{for} \quad 0 < s < p, \tag{32}$$

as illustrated by the characteristic functions on the right side of Fig. 12.3.

Indeed, let F be a characteristic function and assume that (31) holds for some $p > 0$. To simplify the analysis we will further assume that $F(p) = 1$. [If this is not the case, we apply the following argument to the translate $f(x + x_0)$ where $0 \leq x_0 < 1/p$ is chosen to make $e^{2\pi i p x_0} F(p) = 1$.] From (13), it follows that $F(-p) = \overline{F(p)} = 1$, so we can use the pointwise limit (23) to see that

$$\lim_{n \to \infty} \int_{-\infty}^{\infty} f(x)\Gamma_n(x)[1 - \cos(2\pi p x)]dx = F(0) - F(p)/2 - F(-p)/2 = 0.$$

Since the integrand is nonnegative, we can use this limit with (7) and the max bound to write

$$\left| \int_{-\infty}^{\infty} f(x)[1 - \cos(2\pi p x)]\phi(x)dx \right|$$

$$= \left| \lim_{n \to \infty} \int_{-\infty}^{\infty} f(x)\Gamma_n(x)[1 - \cos(2\pi p x)]\phi(x)dx \right|$$

$$\leq \lim_{n \to \infty} \max_{-\infty < x < \infty} |\phi(x)| \cdot \int_{-\infty}^{\infty} f(x)\Gamma_n(x)[1 - \cos(2\pi p x)]dx$$

$$= 0 \quad \text{when} \quad \phi \in \mathbb{S},$$

i.e.,

$$f(x)[1 - \cos(2\pi p x)] = 0.$$

We use the modulation rule to see that

$$F(s + p) - 2F(s) + F(s - p) = 0, \qquad -\infty < s < \infty,$$

and (since F is bounded) thereby deduce that F is p-periodic, *cf.* Ex. 7.78.

In view of this analysis, we see that a characteristic function F can be classified as follows. Either

$$|F(s)| < 1 \quad \text{for} \quad 0 < s < \infty$$

(in which case $|F|$ is aperiodic) or there is some $p > 0$ such that

$$|F(s)| < 1 \text{ for } 0 < s < p \text{ and } |F(p)| = 1$$

[in which case $e^{2\pi i s x_0} F(s)$ is p-periodic for some choice of $0 \leq x_0 < 1/p$], or

$$|F(s)| = 1 \quad \text{for all} \quad 0 < s < \infty$$

[in which case $F(s) = e^{2\pi i s x_0}$ for some choice of $-\infty < x_0 < \infty$].

12.4 Random Variables

Probability integrals

Let f be a probability density on $\mathbb{R}$. In practice we use f to compute *probabilities* and to compute certain averages, *i.e.*, *statistics*, as illustrated by the examples in Section 12.1. We will now introduce some notation to facilitate such calculations.

When f is an ordinary function, the integral

$$\int_a^b f(x)dx$$

gives us the portion of the total probability mass that f assigns to the interval (a, b) as illustrated in Fig. 12.5. We say that

$$P\{a < X < b\} = \int_a^b f(x)dx$$

is the probability of finding a *random variable* X in the interval (a, b) when X has the density f. In cases where f is a generalized function, we must introduce notation that allows us to specify whether we should include atoms of probability, if any, at the endpoints of the interval, *e.g.*,

$$P\{X \le b\} = \int_{-\infty}^{b+} f(x)dx, \ \ P\{a < X < b\} = \int_{a+}^{b-} f(x)dx, \ \ P\{X = a\} = \int_{a-}^{a+} f(x)dx.$$

Figure 12.5. The probability $P\{a < X < b\}$ is obtained by integrating the probability density f over (a, b) when f is an ordinary function (left) or by summing the atoms of probability within (a, b) when f is a train of weighted deltas (right).

We use the *mesa functions* from Fig. 12.6 to give a precise meaning to such improper integrals by writing

$$\int_{-\infty}^{b+} f(x)dx := \lim_{\epsilon \to 0+} \lim_{a \to -\infty} \int_{-\infty}^{\infty} f(x)m_{-+}(x; a, b, \epsilon)dx,$$

$$\int_{a+}^{b-} f(x)dx := \lim_{\epsilon \to 0+} \int_{-\infty}^{\infty} f(x)m_{+-}(x; a, b, \epsilon)dx,$$

$$\int_{a-}^{a+} f(x)dx := \lim_{\epsilon \to 0+} \int_{-\infty}^{\infty} f(x)m_{-+}(x; a, a, \epsilon)dx.$$

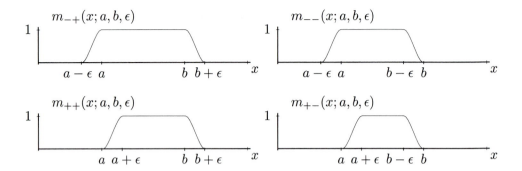

Figure 12.6. Mesa functions used to define integrals of a generalized probability density f.

[These mesa functions are defined by suitably modifying the parameters that appear in (7.15).] Such notation is seldom required for practical purposes, however, and we will routinely work with improper integrals of f in the remainder of this chapter.

Example: Find $P\{X < 3\}$ when X is a random variable with the die-toss density

$$f(x) := \frac{1}{6} \sum_{k=1}^{6} \delta(x - k).$$

Solution: The density f is shown in Fig. 12.2. It has atoms of probability, each with mass $1/6$, at the points $x = 1, 2, \ldots, 6$. Only those at $x = 1, 2$ lie to the left of 3, so

$$P\{X < 3\} = \frac{1}{6} + \frac{1}{6} = \frac{1}{3}.$$

Equivalently,

$$P\{X < 3\} = g(3-) = \frac{1}{3}$$

where g is the distribution function for the die-toss density as shown in Fig. 12.2. Of course, we can also write

$$P\{X < 3\} := \int_{-\infty}^{3-} f(x)dx := \lim_{\epsilon \to 0+} \lim_{a \to -\infty} \int_{-\infty}^{\infty} f(x)m_{--}(x; a, 3, \epsilon)dx$$

$$= \lim_{\epsilon \to 0+} \lim_{a \to -\infty} \frac{1}{6} \sum_{k=1}^{6} m_{--}(k; a, 3, \epsilon) = \frac{1}{6} + \frac{1}{6} = \frac{1}{3}. \qquad \blacksquare$$

Example: Let X be a random variable with the standard normal density

$$f(x) = \frac{e^{-x^2/2}}{\sqrt{2\pi}}.$$

Find $P\{-1 < X < 1\}$, $P\{-2 < X < 2\}$, and $P\{-3 < X < 3\}$.

Solution: The distribution function for the standard normal density is usually called Φ. Since $\Phi' = f$ we can use the fundamental theorem of calculus to write

$$P\{-1 < X < 1\} = \int_{-1}^{1} f(x)dx = \Phi(1) - \Phi(-1).$$

Appendix 6 has a table for Φ, and you can evaluate Φ on many scientific calculators. (If yours does not include this function, you may wish to write a program to approximate Φ by using the formula given in Appendix 6.) Using the table we find

$$P\{-1 < X < 1\} = .8413 - .1587 = .6826,$$

and analogously,

$$P\{-2 < X < 2\} = .9772 - .0228 = .9544$$
$$P\{-3 < X < 3\} = .9987 - .0013 = .9974.$$

The standard normal density (as shown in Fig. 12.2 and Appendix 6) has 68%, 95%, 99.7% of its probability mass within 1, 2, 3 units of the origin! ∎

Expectation integrals

Let f be a probability density on $\mathbb{R}$, and let X be a corresponding random variable. We will frequently have occasion to compute an average value for some suitably regular function $q(X)$. For this purpose we use the *expectation integral*

$$\langle q(X) \rangle := \int_{-\infty}^{\infty} q(x)f(x)dx. \tag{33}$$

[You may recall that we computed the average speed $\langle V \rangle$ and the average kinetic energy $\langle (1/2)mV^2 \rangle$ for a molecule of N_2 by using (33) with Maxwell's density (1).] We limit our consideration to cases where the product $q \cdot f$ is a well-defined generalized function that is subject to a max bound

$$\left| \int_{-\infty}^{\infty} q(x)f(x)\phi(x)dx \right| \leq M \cdot \max_{-\infty < x < \infty} |\phi(x)|, \quad \phi \in \mathbb{S}. \tag{34}$$

Here $0 \leq M < \infty$ is a constant that depends on f and q but not on ϕ. [When $q \cdot f$ is an ordinary function, we take

$$M = \int_{-\infty}^{\infty} |q(x)| f(x)dx$$

provided that this integral is finite.] We can then use mollification, mesa functions, and suitable limits to interpret (33), *e.g.*, as done in Ex. 12.13.

We easily verify that the expectation integral is linear by observing that it is both homogeneous and additive

$$\langle \alpha q(X) \rangle = \alpha \langle q(X) \rangle, \quad \alpha \in \mathbb{C},$$
$$\langle q_1(X) + q_2(X) \rangle = \langle q_1(X) \rangle + \langle q_2(X) \rangle$$

(when all of these expressions are well defined). Since $f \geq 0$, we also have

$$\langle q(X) \rangle \geq 0 \quad \text{when} \quad q \geq 0$$

and

$$|\langle q_1(X) \cdot q_2(X) \rangle|^2 \leq \langle q_1(X)^2 \rangle \cdot \langle q_2(X)^2 \rangle. \tag{35}$$

We prove the Cauchy-Schwartz inequality (35) by suitably modifying the argument used to establish (15).

When we are given a probability density f we will routinely compute two fundamental statistics, the *mean*

$$\mu := \langle X \rangle \tag{36}$$

and the *variance*

$$\sigma^2 := \langle (X - \mu)^2 \rangle = \langle X^2 \rangle - 2\mu\langle X \rangle + \mu^2 \langle 1 \rangle = \langle X^2 \rangle - \langle X \rangle^2 \tag{37}$$

(provided that $\langle X \rangle$, $\langle X^2 \rangle$ are well defined). The nonnegative square root of the variance is called the *standard deviation*, $\sigma \geq 0$. The mean locates the "center" (more precisely the *centroid*) of the density f and the standard deviation serves as a measure of its effective "width."

Example: Find the mean and standard deviation for the die-toss density.

Solution: We suitably extend the sifting property of δ to write

$$\langle q(X) \rangle = \frac{1}{6} \int_{-\infty}^{\infty} \{\delta(x-1) + \delta(x-2) + \cdots + \delta(x-6)\} q(x) dx$$
$$= \frac{1}{6} \{q(1) + q(2) + \cdots + q(6)\}$$

when q is continuous at the points $x = 1, 2, \ldots, 6$. [Since each atom of probability has the same weight, $\langle q(X) \rangle$ is the ordinary arithmetic average of $q(1), q(2), \ldots, q(6)$ when we work with the die-toss density!] We compute

$$\langle X \rangle = \frac{1}{6} \{1 + 2 + \cdots + 6\} = \frac{7}{2}, \quad \langle X^2 \rangle = \frac{1}{6} \{1^2 + 2^2 + \cdots + 6^2\} = \frac{91}{6},$$

and then use (36), (37) to write

$$\mu = \langle X \rangle = 3.5, \quad \sigma = \{\langle X^2 \rangle - \langle X \rangle^2\}^{1/2} = \left\{\frac{91}{6} - \frac{49}{4}\right\}^{1/2} = 1.7078 \ldots \quad \blacksquare$$

Example: Find the mean and standard deviation for the truncated exponential density $f(x) := h(x)e^{-x}$.

Solution: Using the formula

$$\int_0^\infty x^n e^{-x} dx = n!, \qquad n = 0, 1, 2, \ldots$$

from calculus, we compute

$$\langle X \rangle = \int_0^\infty x\, e^{-x} dx = 1, \qquad \langle X^2 \rangle = \int_0^\infty x^2 e^{-x} dx = 2,$$

and then use (36), (37) to write

$$\mu = \langle X \rangle = 1, \qquad \sigma = \{\langle X^2 \rangle - \langle X \rangle^2\}^{1/2} = 1. \qquad \blacksquare$$

Let $n = 0, 1, \ldots$. We define the nth *moment*

$$m^{(n)} := \langle X^n \rangle = \int_{-\infty}^\infty x^n f(x) dx \tag{38}$$

of the probability density f provided that

$$\left| \int_{-\infty}^\infty x^n f(x) \phi(x) dx \right| \leq M_n \cdot \max_{-\infty < x < \infty} |\phi(x)|, \quad \phi \in \mathbb{S} \tag{39}$$

for some constant $0 \leq M_n < \infty$ *(e.g., we take*

$$M_n = \int_{-\infty}^\infty |x^n| f(x) dx$$

in cases where f is an ordinary function and the integral is finite). When $x^n f(x)$ is subject to the max bound (39), the characteristic function F has n uniformly continuous derivatives and we can use the power scaling rule to see that

$$m^{(n)} = \frac{F^{(n)}(0)}{(-2\pi i)^n}, \tag{40}$$

cf. Ex. 12.14. Sometimes this formula gives us the moments $m^{(0)}, m^{(1)}, \ldots$ with minimal effort.

Example: Find the moments for the Dirac density $f(x) := \delta(x)$.

Solution: Since $F(s) = 1$, we can use (40) to see that

$$m^{(0)} = 1, \qquad m^{(n)} = 0 \quad \text{for} \quad n = 1, 2, \ldots . \qquad \blacksquare$$

Example: Find the moments (38) for the standard normal density

$$f(x) := \frac{e^{-x^2/2}}{\sqrt{2\pi}}.$$

Solution: Using (3.14) and the power series for the exponential we see that

$$F(s) = e^{-2\pi^2 s^2} = 1 - \frac{2\pi^2 s^2}{1!} + \frac{(2\pi^2 s^2)^2}{2!} - \frac{(2\pi^2 s^2)^3}{3!} + \cdots$$

$$= 1 + \frac{(-2\pi i s)^2}{2 \cdot 1!} + \frac{(-2\pi i s)^4}{2^2 \cdot 2!} + \frac{(-2\pi i s)^6}{2^3 \cdot 3!} + \cdots.$$

We equate this series to the one from Maclaurin's formula

$$F(s) = F(0) + \frac{F'(0)s}{1!} + \frac{F''(0)s^2}{2!} + \frac{F'''(0)s^3}{3!} + \cdots$$

and thereby obtain

$$m^{(n)} = \frac{F^{(n)}(0)}{(-2\pi i)^n} = \begin{cases} 0 & \text{if } n = 1, 3, 5, \ldots \\ \dfrac{n!}{2^{n/2}(n/2)!} & \text{if } n = 0, 2, 4, \ldots . \end{cases}$$

In particular, the standard normal density has the mean and variance

$$\mu = \langle X \rangle = m^{(1)} = 0, \qquad \sigma^2 = \langle X^2 \rangle - \langle X \rangle^2 = m^{(2)} = 1. \qquad \blacksquare$$

Example: Find the moments for the Cauchy density $f(x) := 1/[\pi(1 + x^2)]$.

Solution: We have $m^{(0)} = 1$, but $m^{(1)}, m^{(2)}, \ldots$ are not defined! The absolute nth moment

$$\int_{-\infty}^{\infty} |x^n| f(x) dx = \int_{-\infty}^{\infty} \frac{|x^n|}{\pi(1 + x^2)} dx$$

is infinite for each $n = 1, 2, \ldots$, and the characteristic function $F(s) = e^{-2\pi|s|}$ is not differentiable at $s = 0$. $\blacksquare$

Functions of a random variable

Let X be a random variable with the probability density f_X. We often have occasion to form a new random variable

$$Y := r(X) \tag{41}$$

by using a suitably regular real-valued function r on $\mathbb{R}$. In such a situation we can often find the corresponding probability density f_Y as follows. We compute

$$\langle e^{-2\pi i s r(X)} \rangle := \int_{-\infty}^{\infty} e^{-2\pi i s r(x)} f_X(x) dx \tag{42}$$

and thereby obtain the characteristic function

$$F_Y(s) = \langle e^{-2\pi i s Y} \rangle = \langle e^{-2\pi i s r(X)} \rangle \tag{43}$$

for Y. The inverse Fourier transform

$$f_Y = F_Y^{\wedge\vee} \tag{44}$$

then gives us the desired density. Analogously, we compute

$$\langle r(X) \rangle := \int_{-\infty}^{\infty} r(x) f_X(x) dx, \qquad \langle r(X)^2 \rangle := \int_{-\infty}^{\infty} r(x)^2 f_X(x) dx$$

(when they exist), and thereby obtain the mean and variance

$$\mu_Y := \langle Y \rangle = \langle r(X) \rangle, \tag{45}$$
$$\sigma_Y^2 := \langle Y^2 \rangle - \langle Y \rangle^2 = \langle r(X)^2 \rangle - \langle r(X) \rangle^2 \tag{46}$$

for Y. We will give three examples to illustrate these ideas.

Example: Let X be a random variable with the mean μ, the variance σ^2, and the probability density f_X. Find the mean, variance, and probability density for the random variable

$$Y := aX + b. \tag{47}$$

Here a, b are real parameters with $a \neq 0$.

Solution: We use the linearity of the expectation integral to write

$$\langle aX + b \rangle = a\langle X \rangle + b,$$
$$\langle (aX + b)^2 \rangle = a^2 \langle X^2 \rangle + 2ab\langle X \rangle + b^2,$$

and then use (45), (46) with (36), (37) to obtain

$$\mu_Y = a\mu + b, \tag{48}$$
$$\sigma_Y^2 = a^2 \sigma^2. \tag{49}$$

We use (43) to find the characteristic function

$$F_Y(s) = \langle e^{-2\pi i s(aX+b)} \rangle = e^{-2\pi i s b} \langle e^{-2\pi i (as)X} \rangle = e^{-2\pi i s b} F_X(as).$$

We then use the dilation and translation rules to see that F_Y is the Fourier transform of the probability density

$$f_Y(y) = \frac{1}{|a|} f_X \left(\frac{y-b}{a} \right). \tag{50}$$

∎

Example: Let X be a random variable with the uniform density $f_X(x) := (1/\pi)\Pi(x/\pi)$. Find the mean, variance, and probability density for $Y := \sin(X)$.

Solution: We easily find

$$\mu_Y = \langle \sin(X) \rangle = \frac{1}{\pi} \int_{-\pi/2}^{\pi/2} \sin x \, dx = 0, \quad \sigma_Y^2 = \langle \sin^2(X) \rangle = \frac{1}{\pi} \int_{-\pi/2}^{\pi/2} \sin^2 x \, dx = \frac{1}{2}.$$

For the characteristic function F_Y we write

$$F_Y(s) = \langle e^{-2\pi i s \sin(X)} \rangle = \frac{1}{\pi} \int_{-\pi/2}^{\pi/2} e^{-2\pi i s \sin(x)} dx = \frac{1}{\pi} \int_{-1}^{1} \frac{e^{-2\pi i s y}}{\sqrt{1-y^2}} dy \tag{51}$$

(using the transformation $y = \sin x$ at the last step). The second integral from (51) is found in the analysis equation for

$$f_Y(y) = \begin{cases} \dfrac{1}{\pi\sqrt{1-y^2}} & \text{if } -1 < y < 1 \\ 0 & \text{otherwise.} \end{cases} \tag{52}$$

[You can obtain the explicit formula

$$F_Y(s) = J_0(2\pi s) \tag{53}$$

for the characteristic function F_Y by using the generating function

$$e^{i\alpha \sin x} = \sum_{k=-\infty}^{\infty} J_k(\alpha) e^{ikx}$$

for the Bessel functions (*cf.* Ex. 4.19) to evaluate the first integral from (51).] ∎

Example: Let X be a random variable with the Cauchy density $f_X(x) := 1/[\pi(1+x^2)]$. Find the probability density for the reciprocal $Y := 1/X$.

Solution: We use (42)–(43) with a careful change of variables to write

$$
\begin{aligned}
F_Y(s) &= \langle e^{-2\pi i s/X} \rangle \\
&= \int_{-\infty}^{0-} \frac{e^{-2\pi i s/x}}{\pi(1+x^2)}\,dx + \int_{0+}^{+\infty} \frac{e^{-2\pi i s/x}}{\pi(1+x^2)}\,dx \\
&= -\int_{0-}^{-\infty} \frac{e^{-2\pi i s y}}{\pi(1+y^{-2})} \cdot \frac{dy}{y^2} - \int_{+\infty}^{0+} \frac{e^{-2\pi i s y}}{\pi(1+y^{-2})} \cdot \frac{dy}{y^2} \\
&= \int_{-\infty}^{\infty} \frac{e^{-2\pi i s y}}{\pi(1+y^2)}\,dy.
\end{aligned}
\tag{54}
$$

In this way we see that Y has the same Cauchy density $f_Y = f_X$ as X. ∎

The uncertainty relation

It is impossible to synthesize a highly localized function f on $\mathbb{R}$ without a substantial contribution from the high-frequency components. When the "width" of f is small, the "width" of F must be large. We will quantify this observation within a context where $|f|^2$ (not f!) serves as a probability density. For example, when f is the reduced wave function for a diffracting laser beam (with finite energy) we can normalize $|f|^2$ to obtain a probability density that shows where the energy is concentrated. The t-slices from Fig. 9.14 show the evolution of such a density from the Π of a slit to the dilated sinc^2 of the far-field diffraction pattern.

Let f be a continuous piecewise smooth function on $\mathbb{R}$ with a continuous piecewise smooth Fourier transform F. We will assume that f is square integrable with

$$
\int_{-\infty}^{\infty} |f(x)|^2 dx = 1.
$$

In view of Plancherel's identity (1.15) we must also have

$$
\int_{-\infty}^{\infty} |F(s)|^2 ds = 1.
$$

We will assume that the means

$$
\mu_f := \int_{-\infty}^{\infty} x\,|f(x)|^2 dx, \qquad \mu_F := \int_{-\infty}^{\infty} s\,|F(x)|^2 ds
\tag{55}
$$

and the variances

$$
\sigma_f^2 := \int_{-\infty}^{\infty} (x-\mu_f)^2 |f(x)|^2 ds, \qquad \sigma_F^2 := \int_{-\infty}^{\infty} (s-\mu_F)^2 |F(s)|^2 ds
\tag{56}
$$

of the probability densities $|f|^2$, $|F|^2$ are finite, *cf.* Fig. 12.7. We will show that the standard deviations σ_f, σ_F of such densities satisfy the *uncertainty relation*

$$\sigma_f \sigma_F \geq \frac{1}{4\pi}. \tag{57}$$

(Within an appropriate context, this inequality is equivalent to the Heisenberg uncertainty principle of quantum mechanics!)

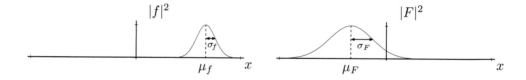

Figure 12.7. The standard deviations σ_f, σ_F of the probability densities $|f|^2, |F|^2$ must satisfy the uncertainty relation (57).

We first establish (57) for the case where $\mu_f = \mu_F = 0$. By using Plancherel's identity, the Cauchy-Schwartz inequality

$$\left| \int_{-\infty}^{\infty} f(x)g(x)dx \right|^2 \leq \int_{-\infty}^{\infty} |f(x)|^2 dx \cdot \int_{-\infty}^{\infty} |g(x)|^2 dx \tag{58}$$

for square integrable functions, and an integration by parts we write

$$\sigma_f^2 \sigma_F^2 = \int_{-\infty}^{\infty} x^2 |f(x)|^2 dx \cdot \int_{-\infty}^{\infty} s^2 |F(s)|^2 ds$$

$$= \frac{1}{4\pi^2} \int_{-\infty}^{\infty} |x\, f(x)|^2 dx \cdot \int_{-\infty}^{\infty} |2\pi i s\, F(s)|^2 ds$$

$$= \frac{1}{4\pi^2} \int_{-\infty}^{\infty} |x\, f(x)|^2 dx \cdot \int_{-\infty}^{\infty} |f'(x)|^2 dx$$

$$\geq \frac{1}{4\pi^2} \left| \int_{-\infty}^{\infty} x\, f(x)\overline{f'(x)}dx \right|^2$$

$$\geq \frac{1}{4\pi^2} \left| \mathrm{Re}\left\{ \int_{-\infty}^{\infty} x\, f(x)\overline{f'(x)}dx \right\} \right|^2$$

$$= \frac{1}{16\pi^2} \left| \int_{-\infty}^{\infty} x\, \frac{d}{dx}|f(x)|^2 dx \right|^2$$

$$= \frac{1}{16\pi^2} \left| \int_{-\infty}^{\infty} |f(x)|^2 dx \right|^2$$

$$= \frac{1}{16\pi^2} \,.$$

For the general case where $\mu_f \neq 0$ or $\mu_F \neq 0$, we define the modulated translate

$$g(x) := e^{-2\pi\mu_F x} f(x + \mu_f)$$

of f, and verify in turn that

$$\int_{-\infty}^{\infty} |g(x)|^2 dx = \int_{-\infty}^{\infty} |f(x + \mu_f)|^2 dx = 1,$$

$$\mu_g := \int_{-\infty}^{\infty} x \, |g(x)|^2 dx = \int_{-\infty}^{\infty} (x + \mu_f - \mu_f)|f(x + \mu_f)|^2 dx = 0,$$

$$\sigma_g^2 := \int_{-\infty}^{\infty} x^2 |g(x)|^2 dx = \int_{-\infty}^{\infty} (x + \mu_f - \mu_f)^2 |f(x + \mu_f)|^2 dx = \sigma_f^2.$$

We use the modulation and translation rules to write

$$G(s) = e^{2\pi i \mu_f (s + \mu_F)} F(s + \mu_F),$$

and verify in turn that

$$\int_{-\infty}^{\infty} |G(s)|^2 ds = \int_{-\infty}^{\infty} |F(s + \mu_F)|^2 ds = 1,$$

$$\mu_G := \int_{-\infty}^{\infty} s \, |G(s)|^2 ds = \int_{-\infty}^{\infty} (s + \mu_F - \mu_F)|F(s + \mu_F)|^2 ds = 0,$$

$$\sigma_G^2 := \int_{-\infty}^{\infty} s^2 |G(s)|^2 ds = \int_{-\infty}^{\infty} (s + \mu_F - \mu_F)^2 |F(s + \mu_F)|^2 ds = \sigma_F^2.$$

We use these identities with the above zero means version of (57) to write

$$\sigma_f \sigma_F = \sigma_g \sigma_G \geq \frac{1}{4\pi}$$

and thereby show that (57) holds when $\mu_f \neq 0$ or $\mu_F \neq 0$. You may wish to verify that equality occurs if and only if

$$f(x) = A \, e^{-(x-\mu)^2/4\sigma^2} \tag{59}$$

with $|A| = (2\pi\sigma^2)^{-1/4}$, $\sigma > 0$, and $-\infty < \mu < \infty$, cf. Ex. 12.27 and (9.97).

12.5 The Central Limit Theorem

Sums of independent random variables

Let $f(x_1, x_2)$ be a suitably regular nonnegative function on $\mathbb{R}^2$ with

$$\int_{-\infty}^{\infty} \int_{-\infty}^{\infty} f(x_1, x_2) dx_1 dx_2 = 1.$$

We use this bivariate density to describe *two* random variables X_1, X_2. We say that

$$P\{(X_1, X_2) \in \mathcal{D}\} = \iint_{\mathcal{D}} f(x_1, x_2) dx_1 dx_2$$

is the probability of finding (X_1, X_2) in some suitably regular domain $\mathcal{D} \subseteq \mathbb{R}^2$ when X_1, X_2 have the *joint probability density f*. Since

$$P\{a < X_1 < b\} = \int_{x_1=a}^{b} \int_{x_2=-\infty}^{\infty} f(x_1, x_2) dx_2 dx_1,$$

$$P\{a < X_2 < b\} = \int_{x_2=a}^{b} \int_{x_1=-\infty}^{\infty} f(x_1, x_2) dx_1 dx_2,$$

we immediately see that X_1, X_2 have the univariate probability densities

$$f_{X_1}(x_1) = \int_{x_2=-\infty}^{\infty} f(x_1, x_2) dx_2, \quad f_{X_2}(x_2) = \int_{x_1=-\infty}^{\infty} f(x_1, x_2) dx_1.$$

We will consider only bivariate probability densities that can be written as a product

$$f(x_1, x_2) = f_{X_1}(x_1) f_{X_2}(x_2)$$

of the corresponding univariate probability densities. When this is the case we say that the random variables X_1, X_2 are *independent*. Within this context we have

$$\begin{aligned}
\langle X_1 \cdot X_2 \rangle &= \int_{-\infty}^{\infty} \int_{-\infty}^{\infty} x_1 x_2 \, f_{X_1}(x_1) f_{X_2}(x_2) dx_1 dx_2 \\
&= \int_{-\infty}^{\infty} x_1 f_{X_1}(x_1) dx_1 \cdot \int_{-\infty}^{\infty} x_2 f_{X_2}(x_2) dx_2 \\
&= \langle X_1 \rangle \cdot \langle X_2 \rangle,
\end{aligned} \tag{60}$$

or more generally,

$$\langle q_1(X_1) \cdot q_2(X_2) \rangle = \langle q_1(X_1) \rangle \cdot \langle q_2(X_2) \rangle$$

when q_1, q_2 are suitably regular functions on $\mathbb{R}$ and these expectation integrals are well defined.

We often have occasion to form the sum of independent random variables X_1, X_2. When the corresponding probability densities f_{X_1}, f_{X_2} are suitably regular (*e.g.*, bounded and piecewise continuous), we can write

$$P\{a < X_1 + X_2 < b\} = \iint\limits_{a < x_1 + x_2 < b} f_{X_1}(x_1) f_{X_2}(x_2) dx_1 dx_2$$

$$\overset{?}{=} \int_{x=a}^{b} \int_{u=-\infty}^{\infty} f_{X_1}(u) f_{X_2}(x - u) du\, dx,$$

cf. Fig. 12.8. In this way we see that $X_1 + X_2$ has the probability density

$$f_{X_1 + X_2} = f_{X_1} * f_{X_2}. \tag{61}$$

When f_{X_1}, f_{X_2} are arbitrary generalized probability densities we work with the corresponding characteristic functions and write

$$\begin{aligned}
F_{X_1 + X_2}(s) &= \langle e^{-2\pi i s(X_1 + X_2)} \rangle \\
&= \langle e^{-2\pi i s X_1} \rangle \cdot \langle e^{-2\pi i s X_2} \rangle \\
&= F_{X_1}(s) \cdot F_{X_2}(s).
\end{aligned} \tag{62}$$

Since we have shown that a product of characteristic functions is also a characteristic function, we can again use (61) to obtain the probability density for $X_1 + X_2$ provided that we use (30) to define the convolution product.

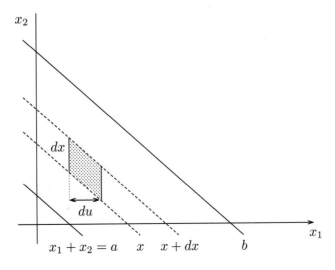

Figure 12.8. The differential area $du\, dx$ associated with the transformation $x_1 = u$, $x_2 = x - u$ that we use to derive (61).

Example: Let X_1, X_2 be independent random variables with the means μ_{X_1}, μ_{X_2} and the variances $\sigma^2_{X_1}, \sigma^2_{X_2}$. Show that $X_1 + X_2$ has the mean

$$\mu_{X_1 + X_2} = \mu_{X_1} + \mu_{X_2} \tag{63}$$

and the variance

$$\sigma^2_{X_1 + X_2} = \sigma^2_{X_1} + \sigma^2_{X_2}. \tag{64}$$

Solution: We can use expectation integrals with (60) to obtain (63), (64) by writing

$$\langle X_1 + X_2 \rangle = \langle X_1 \rangle + \langle X_2 \rangle,$$

$$\langle (X_1 + X_2)^2 \rangle - \langle X_1 + X_2 \rangle^2 = \langle X_1^2 \rangle + 2\langle X_1 X_2 \rangle + \langle X_2^2 \rangle$$
$$- \langle X_1 \rangle^2 - 2\langle X_1 \rangle \langle X_2 \rangle - \langle X_2 \rangle^2$$
$$= \langle X_1^2 \rangle - \langle X_1 \rangle^2 + \langle X_2^2 \rangle - \langle X_2 \rangle^2.$$

We can also use (62) with (40) and the Leibnitz formula (2.29) to see that $X_1 + X_2$ has the nth moment

$$m^{(n)}_{X_1 + X_2} = \frac{F^{(n)}_{X_1 + X_2}(0)}{(-2\pi i)^n} = \sum_{\nu=0}^{n} \binom{n}{\nu} \frac{F^{(\nu)}_{X_1}(0)}{(-2\pi i)^\nu} \frac{F^{(n-\nu)}_{X_2}(0)}{(-2\pi i)^{n-\nu}}$$

$$= \sum_{\nu=0}^{n} \binom{n}{\nu} m^{(\nu)}_{X_1} m^{(n-\nu)}_{X_2} \tag{65}$$

when all of these moments are defined. In particular,

$$m^{(1)}_{X_1 + X_2} = m^{(1)}_{X_1} + m^{(1)}_{X_2}, \qquad m^{(2)}_{X_1 + X_2} = m^{(2)}_{X_1} + 2m^{(1)}_{X_1} m^{(1)}_{X_2} + m^{(2)}_{X_2},$$

and we obtain (63), (64) by using these moment relations with the identities

$$\mu_X = m^{(1)}_X, \qquad \sigma^2_X = m^{(2)}_X - [m^{(1)}_X]^2. \qquad \blacksquare$$

More generally, we will say that the random variables $X_1, X_2, \ldots, X_n$ are independent when their joint probability density on $\mathbb{R}^n$ is the product

$$f(x_1, x_2, \ldots, x_n) = f_{X_1}(x_1) f_{X_2}(x_2) \cdots f_{X_n}(x_n)$$

of the corresponding univariate probability densities $f_{X_1}, f_{X_2}, \ldots, f_{X_n}$. The sum $X_1 + X_2 + \cdots + X_n$ will then have the probability density

$$f_{X_1 + X_2 + \cdots + X_n} = f_{X_1} * f_{X_2} * \cdots * f_{X_n}, \tag{66}$$

the characteristic function

$$F_{X_1+X_2+\cdots+X_n} = F_{X_1} \cdot F_{X_2} \cdot \cdots \cdot F_{X_n}, \tag{67}$$

the mean

$$\mu_{X_1+X_2+\cdots+X_n} = \mu_{X_1} + \mu_{X_2} + \cdots + \mu_{X_n}, \tag{68}$$

(when the means $\mu_{X_1}, \mu_{X_2}, \ldots, \mu_{X_n}$ are all defined), and the variance

$$\sigma^2_{X_1+X_2+\cdots+X_n} = \sigma^2_{X_1} + \sigma^2_{X_2} + \cdots + \sigma^2_{X_n} \tag{69}$$

(when the variances $\sigma^2_{X_1}, \sigma^2_{X_2}, \ldots, \sigma^2_{X_n}$ are all defined).

Example: Let each of the independent random variables $X_1, X_2, \ldots, X_n$ have the Bernoulli density $f(x) := q\,\delta(x) + p\,\delta(x-1)$ where $p \geq 0$, $q \geq 0$, and $p + q = 1$. Find the probability density, the mean, and the variance for the sum $X_1 + X_2 + \cdots + X_n$.

Solution: The probability density f has the Fourier transform $F(s) = q + p\,e^{-2\pi is}$, so we can use (67) to obtain the characteristic function

$$F_{X_1+X_2+\cdots+X_n}(s) = (q + p\,e^{-2\pi is})^n = \sum_{k=0}^{n} \binom{n}{k} p^k q^{n-k} e^{-2\pi iks}.$$

The inverse Fourier transform gives the binomial density

$$f_{X_1+X_2+\cdots+X_n}(x) = \sum_{k=0}^{n} \binom{n}{k} p^k q^{n-k} \delta(x-k).$$

The random variables $X_1, X_2, \ldots, X_n$ have the common mean

$$\mu = q \cdot 0 + p \cdot 1 = p$$

and the common variance

$$\sigma^2 = \{q \cdot 0^2 + p \cdot 1^2\} - \{q \cdot 0 + p \cdot 1\}^2 = pq,$$

so we can use (68), (69) to write

$$\mu_{X_1+X_2+\cdots+X_n} = np, \qquad \sigma^2_{X_1+X_2+\cdots+X_n} = npq. \qquad \blacksquare$$

Example: Let each of the independent random variables $X_1, X_2, \ldots, X_n$ have the standard normal density

$$f(x) := \frac{e^{-x^2/2}}{\sqrt{2\pi}}$$

(with mean $\mu = 0$ and variance $\sigma^2 = 1$). Show that

$$X := \frac{1}{\sqrt{n}}(X_1 + X_2 + \cdots + X_n)$$

also has the standard normal density f.

Solution: Since $X_1, X_2, \ldots, X_n$ are independent and since f has the Fourier transform $F(s) = e^{-2\pi^2 s^2}$ we can write

$$\begin{aligned}
F_X(s) &= \langle e^{-2\pi i s(X_1 + X_2 + \cdots + X_n)/\sqrt{n}} \rangle \\
&= \langle e^{-2\pi i s X_1/\sqrt{n}} \rangle \langle e^{-2\pi i s X_2/\sqrt{n}} \rangle \cdots \langle e^{-2\pi i s X_n/\sqrt{n}} \rangle \\
&= F(s/\sqrt{n})^n \\
&= F(s).
\end{aligned}$$

∎

Example: Let each of the independent random variables $X_1, X_2, \ldots, X_n$ have the Cauchy density

$$f(x) := \frac{1}{\pi(1 + x^2)}.$$

Show that

$$X := \frac{1}{n}(X_1 + X_2 + \cdots + X_n)$$

also has the Cauchy density f.

Solution: Since $X_1, X_2, \ldots, X_n$ are independent and since $F(s) = e^{-2\pi|s|}$ we can write

$$\begin{aligned}
F_X(s) &= \langle e^{-2\pi i s(X_1 + X_2 + \cdots + X_n)/n} \rangle \\
&= \langle e^{-2\pi i s X_1/n} \rangle \langle e^{-2\pi i s X_2/n} \rangle \cdots \langle e^{-2\pi i s X_n/n} \rangle \\
&= F(s/n)^n \\
&= F(s).
\end{aligned}$$

∎

Example: Let each of the independent random variables $B_1, B_2, B_3, \ldots$ have the Bernoulli density

$$b(x) := \frac{1}{2}\delta(x) + \frac{1}{2}\delta(x-1),$$

and for $n = 1, 2, \ldots$ let

$$S_n := \frac{1}{2}B_1 + \frac{1}{4}B_2 + \cdots + \frac{1}{2^n}B_n.$$

(We obtain $S_n = (.B_1 B_2 \ldots B_n)_2$ by randomly choosing the bits $B_1, B_2, \ldots, B_n$ from $\{0, 1\}$.) Find the probability density functions $f_{S_1}, f_{S_2}, f_{S_3}, \ldots$, and show that this sequence has the weak limit

$$f(x) := \begin{cases} 1 & \text{if } 0 < x < 1 \\ 0 & \text{otherwise.} \end{cases}$$

Solution: We define $X_n := 2^{-n}B_n$ and use (50) with the dilation identity (7.65) for δ to see that

$$f_{X_n}(x) = 2^n b(2^n x) = \frac{1}{2}\delta(x) + \frac{1}{2}\delta\left(x - \frac{1}{2^n}\right). \tag{70}$$

Since S_n is the sum of the independent random variables $X_1, X_2, \ldots, X_n$ we can use (66) and the sifting property of δ to write

$$f_{S_1}(x) = f_{X_1}(x)$$
$$= \frac{1}{2}\left\{\delta(x) + \delta\left(x - \frac{1}{2}\right)\right\},$$
$$f_{S_2}(x) = (f_{X_1} * f_{X_2})(x) = f_{S_1}(x) * \frac{1}{2}\left\{\delta(x) + \delta\left(x - \frac{1}{4}\right)\right\}$$
$$= \frac{1}{2}\left\{f_{S_1}(x) + f_{S_1}\left(x - \frac{1}{4}\right)\right\}$$
$$= \frac{1}{4}\left\{\delta(x) + \delta\left(x - \frac{1}{4}\right) + \delta\left(x - \frac{2}{4}\right) + \delta\left(x - \frac{3}{4}\right)\right\}, \tag{71}$$

$$\vdots$$

$$f_{S_n}(x) = (f_{X_1} * f_{X_2} * \cdots * f_{X_n})(x)$$
$$= \frac{1}{2}\left\{f_{S_{n-1}}(x) + f_{S_{n-1}}\left(x - \frac{1}{2^n}\right)\right\}$$
$$= \frac{1}{2^n}\left\{\delta(x) + \delta\left(x - \frac{1}{2^n}\right) + \delta\left(x - \frac{2}{2^n}\right) + \cdots + \delta\left(x - \frac{2^n - 1}{2^n}\right)\right\}.$$

The probability mass for S_n is evenly distributed over the 2^n points $x = k/2^n$, $k = 0, 1, \ldots, 2^n - 1$, so we might reasonably expect f_{S_n} to have the weak limit f, cf. Figs. 12.9 and 10.9. We prove this by using (71) and the definition of the Riemann integral to write

$$\lim_{n\to\infty} \int_{-\infty}^{\infty} f_{S_n}(x)\phi(x)dx = \lim_{n\to\infty} \frac{1}{2^n} \sum_{k=0}^{2^n-1} \phi\left(\frac{k}{2^n}\right) = \int_0^1 \phi(x)dx$$

$$= \int_{-\infty}^{\infty} f(x)\phi(x)dx \text{ when } \phi \in \mathbb{S}.$$

[We obtain *Vieta's formula*

$$\cos\left(\frac{2\pi s}{2}\right) \cdot \cos\left(\frac{2\pi s}{2^2}\right) \cdot \cos\left(\frac{2\pi s}{2^3}\right) \cdot \cdots = \frac{\sin 2\pi s}{2\pi s} \tag{72}$$

from the corresponding

$$\lim_{n\to\infty} F_{X_1} \cdot F_{X_2} \cdots \cdot F_{X_n} = F.] \tag{73}$$

∎

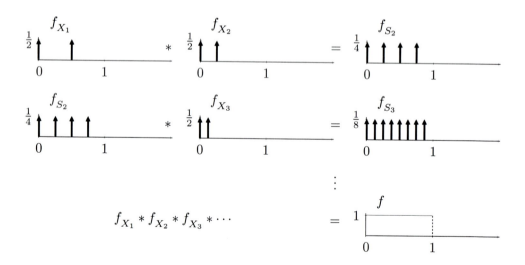

Figure 12.9. The convolution products $f_{S_2} = f_{X_1} * f_{X_2}$, $f_{S_3} = f_{X_1} * f_{X_2} * f_{X_3}$ and the weak limit of $f_{X_1} * f_{X_2} * f_{X_3} * \cdots$ when $f_{X_1}, f_{X_2}, f_{X_3}, \ldots$ are given by (70).

The ubiquitous bell curve

Let f be the common probability density for the independent random variables $X_1, X_2, X_3, \dots$. We will assume that f has a well-defined mean μ and variance σ^2 with $0 < \sigma^2 < \infty$. From (66), (68), and (69) we know that the sums

$$S_1 := X_1, \quad S_2 := X_1 + X_2, \quad S_3 := X_1 + X_2 + X_3, \dots \tag{74}$$

have the probability density functions

$$f_1 := f, \quad f_2 := f * f, \quad f_3 := f * f * f, \dots \tag{75}$$

with the means

$$\mu_1 = \mu, \quad \mu_2 = 2\mu, \quad \mu_3 = 3\mu, \dots \tag{76}$$

and the variances

$$\sigma_1^2 = \sigma^2, \quad \sigma_2^2 = 2\sigma^2, \quad \sigma_3^2 = 3\sigma^2, \dots . \tag{77}$$

Example: Find simple expressions for f_n, μ_n, σ_n [as given by (75)–(77)] when we begin with the truncated exponential density $f(x) := h(x)e^{-x}$.

Solution: The probability density f_n has the characteristic function

$$F_n(s) = F(s)^n = (1 + 2\pi i s)^{-n} = \frac{F^{(n-1)}(s)}{(n-1)!(-2\pi i)^{n-1}},$$

so we can use the power scaling rule to see that

$$f_n(x) = \frac{1}{(n-1)!} h(x) x^{n-1} e^{-x}, \tag{78}$$

(*cf.* Exs. 2.4 and 3.10). We have shown that the truncated exponential density f has the mean $\mu = 1$ and the variance $\sigma^2 = 1$, so we use (76)–(77) to see that f_n has the mean $\mu_n = n$ and the standard deviation $\sigma_n = \sqrt{n}$. Plots of $f_1, f_2, f_4, f_8, f_{16}, f_{32}$ are shown in Fig. 12.10. When n is "large," f_n is well approximated by a normal density (a gaussian bell curve!) with center $x = n$ and "width" $\sqrt{n}$. ∎

It is a remarkable fact that the probability density f_n from the sequence (75) always assumes the shape of a gaussian bell curve when n is "large". This observation (first made by Gauss, Laplace, ... at the beginning of the 19th century) has many practical implications, and we will give a precise analysis of this important phenomenon. With this in mind we introduce the random variable

$$U_n := \frac{S_n - n\mu}{\sqrt{n}\,\sigma} \tag{79}$$

and use (47)–(49) with (76)–(77) to see that the corresponding probability density

$$\beta_n(x) = \sqrt{n}\,\sigma\, f_n(\sqrt{n}\,\sigma\, x + n\mu), \quad n = 1, 2, \dots \tag{80}$$

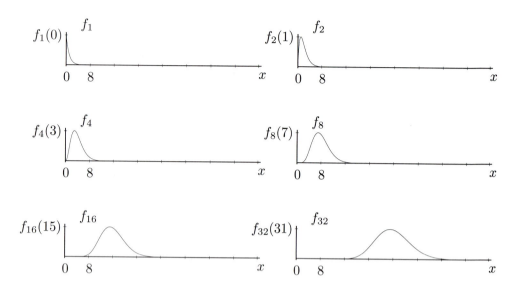

Figure 12.10. The probability densities $f_1, f_2, f_4, f_8, f_{16}, f_{32}$ from (78).

has the mean 0 and the variance 1. We have chosen the translation parameter $n\mu$ to "center" β_n at the origin. (This eliminates the drifting you see in Fig. 12.10.) We have chosen the dilation parameter $\sqrt{n}\,\sigma$ to normalize the "width" of β_n. (This eliminates the spreading that you see in Fig. 12.10.) Figure 12.11 shows the probability densities β_4 that correspond to four choices for f. In the two cases where f is piecewise smooth, β_4 is well approximated by the standard normal density

$$\beta(x) := \frac{e^{-x^2/2}}{\sqrt{2\pi}}$$

(with mean 0 and variance 1). In the two cases where f is discrete the same is true of β_4, and the amplitudes of the δ's are well approximated by some constant multiple of uniformly spaced samples of β. We will show that the sequence $\beta_1, \beta_2, \beta_3, \ldots$ converges weakly to β, i.e.,

$$\lim_{n\to\infty} \int_{-\infty}^{\infty} \beta_n(x)\phi(x)dx = \int_{-\infty}^{\infty} \beta(x)\phi(x)dx \text{ for all } \phi \in \mathbb{S}. \tag{81}$$

Using (75) and (80) we see that f_n, β_n have the characteristic functions

$$F_n(s) = F(s)^n$$

$$B_n(s) = e^{2\pi i n\mu s/\sqrt{n}\,\sigma} F_n\left(\frac{s}{\sqrt{n}\,\sigma}\right) = \left\{e^{2\pi i \mu s/\sqrt{n}\,\sigma} F\left(\frac{s}{\sqrt{n}\,\sigma}\right)\right\}^n,$$

$$n = 1, 2, \ldots . \tag{82}$$

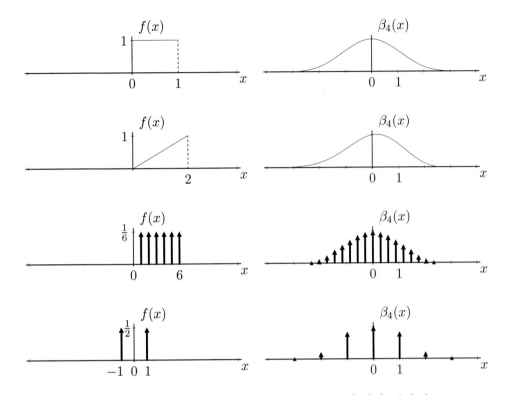

Figure 12.11. The probability density β_4 from (80) (right) that corresponds to f (left) when f is the uniform density on $[0, 1]$, the ramp density on $[0, 2]$, the die-toss density, and the coin flip density.

We will establish (81) by showing that $B_1, B_2, B_3, \ldots$ converge weakly to the characteristic function

$$B(s) = e^{-2\pi^2 s^2}$$

of the standard normal density. Since the expression within the braces of (82) is the $1/\sqrt{n}$ dilated characteristic function of the probability density $\sigma f(\sigma x + \mu)$ (which has mean 0 and variance 1), we lose no generality if we assume that $\mu = 0$, $\sigma = 1$ and we will do so in the remainder of this proof.

Let $\phi \in \mathbb{S}$ and let $\epsilon > 0$ be given. From Section 12.3 we know that the characteristic functions B_n, B are continuous and bounded (with maximum modulus 1), so we can write

$$\left| \int_{-\infty}^{\infty} B_n(s)\phi(s)\,ds - \int_{-\infty}^{\infty} B(s)\phi(s)\,ds \right| \leq \int_{-\infty}^{\infty} |B_n(s) - B(s)| \cdot |\phi(s)|\,ds$$

$$\leq \max_{-\infty < s < \infty} |\phi(s)| \cdot \int_{-L}^{L} |B_n(s) - B(s)|\,ds + 2 \int_{|s|>L} |\phi(s)|\,ds$$

(83)

when $L > 0$. We will choose L so that

$$2 \int_{|s|>L} |\phi(s)| ds < \frac{\epsilon}{2}. \tag{84}$$

Using the inequality

$$|a^n - b^n| = |a - b| \cdot |a^{n-1} + a^{n-2}b + \cdots + b^{n-1}| \le n|a - b| \text{ when } |a| \le 1, |b| \le 1$$

we write

$$|B_n(s) - B(s)| = \left| F\left(\frac{s}{\sqrt{n}}\right)^n - B\left(\frac{s}{\sqrt{n}}\right)^n \right| \le n \left| F\left(\frac{s}{\sqrt{n}}\right) - B\left(\frac{s}{\sqrt{n}}\right) \right| \tag{85}$$

$$\le n \left| F\left(\frac{s}{\sqrt{n}}\right) - 1 + \frac{2\pi^2 s^2}{n} \right| + n \left| B\left(\frac{s}{\sqrt{n}}\right) - 1 + \frac{2\pi^2 s^2}{n} \right|.$$

Since the mean and variance of f are well defined, the derivatives F', F'' of the characteristic function F are defined and continuous on $\mathbb{R}$, *cf.* Ex. 12.14. We use

$$F(0) = 1, \quad F'(0) = -2\pi i\mu = 0, \quad F''(0) = (-2\pi i)^2(\sigma^2 + \mu^2) = -4\pi^2$$

from (25), (40) with Taylor's formula (from Ex. 2.28) to write

$$F(s) = F(0) + sF'(0) + \frac{s^2}{2}F''(0) + \int_0^s [F''(u) - F''(0)](s - u)du$$

$$= 1 - 2\pi^2 s^2 + \int_0^s [F''(u) - F''(0)](s - u)du.$$

It follows that

$$\left| F\left(\frac{s}{\sqrt{n}}\right) - 1 + \frac{2\pi^2 s^2}{n} \right| \le \frac{L^2}{2n} \cdot \max_{|u| \le L/\sqrt{n}} |F''(u) - F''(0)| \text{ when } |s| \le L, \tag{86}$$

and analogously,

$$\left| B\left(\frac{s}{\sqrt{n}}\right) - 1 + \frac{2\pi^2 s^2}{n} \right| \le \frac{L^2}{2n} \cdot \max_{|u| \le L/\sqrt{n}} |B''(u) - B''(0)| \text{ when } |s| \le L. \tag{87}$$

The functions F'', B'' are continuous, so we can use (85)–(87) to see that

$$\max_{-\infty < s < \infty} |\phi(s)| \cdot \int_{-L}^L |B_n(s) - B(s)| ds < \frac{\epsilon}{2}$$

for all sufficiently large n. In conjunction with (83) and (84) this shows that

$$\left| \int_{-\infty}^{\infty} B_n(s)\phi(s)ds - \int_{-\infty}^{\infty} B(s)\phi(s)ds \right| < \epsilon$$

for all sufficiently large n. Since $\epsilon > 0$ is arbitrary, we can write

$$\lim_{n\to\infty} \int_{-\infty}^{\infty} B_n(s)\phi(s)ds = \int_{-\infty}^{\infty} B(s)\phi(s)ds, \quad \phi \in \mathbb{S},$$

and since the Fourier transform preserves weak limits, the proof of (81) is complete!

The hypothesis $0 < \sigma^2 < \infty$ is essential for this argument. The function f_n from (75) does *not* assume the shape of a gaussian bell when we use the Dirac density $f(x) = \delta(x)$ (with $\sigma = 0$) or the Cauchy density $f(x) = 1/[\pi(1 + x^2)]$ (with $\sigma = +\infty$).

The law of errors

We will now interpret (81) within a probabilistic context. Since β_n is the probability density for the random variable U_n of (79), we can use the mesa functions of Fig. 12.6 to write

$$\int_{-\infty}^{\infty} \beta_n(x)m_{+-}(x; a, b, \epsilon)dx \le P\{a < U_n < b\} \le \int_{-\infty}^{\infty} \beta_n(x)m_{-+}(x; a, b, \epsilon)dx$$

when $-\infty < a < b < \infty$ and $\epsilon > 0$ is sufficiently small. Using (81) we see that

$$\lim_{\epsilon\to 0+} \lim_{n\to\infty} \int_{-\infty}^{\infty} \beta_n(x)m_{+-}(x; a, b, \epsilon)dx = \int_a^b \beta(x)dx.$$

The same is true when we replace m_{+-} by m_{-+}, so we can use the squeeze theorem for limits to see that

$$\lim_{n\to\infty} P\left\{ a < \frac{X_1 + X_2 + \cdots + X_n - n\mu}{\sqrt{n}\,\sigma} < b \right\} = \int_a^b \frac{e^{-x^2/2}}{\sqrt{2\pi}}dx, \tag{88}$$
$$-\infty \le a < b \le \infty$$

(when $X_1, X_2, X_3, \ldots$ are independent identically distributed random variables with mean μ and variance $0 < \sigma^2 < \infty$). This result is known as the *central limit theorem* (or as the *law of errors*). It explains why the standard normal density is the most important probability density that you will encounter in the study of statistics!

In practice we routinely use the approximation

$$P\left\{ a < \frac{X_1 + X_2 + \cdots + X_n - n\mu}{\sqrt{n}\,\sigma} < b \right\} \approx \int_a^b \frac{e^{-x^2/2}}{\sqrt{2\pi}}dx, \tag{89}$$

which corresponds to (88). (An analysis of the error can be found in Feller, Vol. 2, pp. 505–509.) Equivalent rearrangements of (89), *e.g.*,

$$P\{a < X_1 + X_2 + \cdots + X_n < b\} \approx \int_{(a-n\mu)/\sqrt{n}\,\sigma}^{(b-n\mu)/\sqrt{n}\,\sigma} \frac{e^{-x^2/2}}{\sqrt{2\pi}} dx \qquad (90)$$

often facilitate the analysis for a particular application.

Example: What is the probability of finding at most 950 zeros in a string of 10,000 randomly chosen digits? Assume that the digits are selected independently and that a given digit is equally likely to be $0, 1, \ldots, 9$.

Solution: Let $X_1, X_2, X_3, \ldots$ be independent random variables with each having the Bernoulli density

$$f(x) = \frac{9}{10}\delta(x) + \frac{1}{10}\delta(x-1).$$

These random variables serve as counters with $X_k = 1, 0$ when the kth digit is, is not a zero. The density f has the moments

$$\langle X \rangle = \frac{9}{10} \cdot 0 + \frac{1}{10} \cdot 1 = \frac{1}{10}, \quad \langle X^2 \rangle = \frac{9}{10} \cdot 0^2 + \frac{1}{10} \cdot 1^2 = \frac{1}{10}$$

that we use with (36), (37) to find the mean $\mu = .1$ and standard deviation $\sigma = .3$. The sum $X_1 + X_2 + \cdots + X_{10000}$ gives the number of zeros that appear in the string of 10,000 digits, so we compute

$$\frac{950 - n\mu}{\sqrt{n}\,\sigma} = \frac{950 - 10000 \cdot .1}{100 \cdot .3} = -\frac{5}{3}$$

and use (90) to write

$$P\{X_1 + X_2 + \cdots + X_{10000} < 950\} \approx \int_{-\infty}^{-5/3} \frac{e^{-x^2/2}}{\sqrt{2\pi}} = .0478. \qquad \blacksquare$$

Example: Assume that the time T it takes a checkout clerk to process a customer is a random variable having the truncated exponential density

$$f(t) := \frac{1}{\alpha}\begin{cases} e^{-t/\alpha} & \text{if } t > 0 \\ 0 & \text{if } t < 0 \end{cases}$$

with $\alpha = 2$ min. What is the probability that the clerk will service at least 70 customers during a 120-minute shift?

Solution: Using (36), (37) we find the mean $\mu = \alpha$ and the standard deviation $\sigma = \alpha$ from the moments

$$\langle T \rangle = \int_0^\infty t\,f(t)dt = \alpha \int_0^\infty u\,e^{-u}du = \alpha,$$

$$\langle T^2 \rangle = \int_0^\infty t^2 f(t)dt = \alpha^2 \int_0^\infty u^2 e^{-u}du = 2\alpha^2.$$

Let the independent random variables $T_1, T_2, \ldots, T_{70}$ be the times for processing the 1st,2nd, ... ,70th customers. Assuming that the queue is never empty, we compute

$$\frac{120\ \text{min} - 70\alpha}{\sqrt{70}\ \alpha} = -1.195 \ldots$$

and use (90) to estimate the desired probability

$$P\{T_1 + T_2 + \cdots + T_{70} < 120\} \approx \int_{-\infty}^{-1.195} \frac{e^{-x^2/2}}{\sqrt{2\pi}}dx = .116. \qquad \blacksquare$$

Example: A marker is initially at the origin of the x-axis. Each time a fair coin is tossed the marker is moved 1 unit to the right, left when the coin comes up heads, tails, respectively. What is the probability of finding the marker at some coordinate $|x| \leq 5$ after 100 tosses?

Solution: Let $X_1, X_2, X_3, \ldots$ be independent random variables with each having the coin flip density

$$f(x) := \frac{1}{2}\delta(x+1) + \frac{1}{2}\delta(x-1).$$

The sum $S_{100} := X_1 + X_2 + \cdots + X_{100}$ gives the position of the marker after $n = 100$ tosses. The coin flip density has the mean $\mu = 0$ and the standard deviation $\sigma = 1$, so we compute

$$\frac{\pm 5 - n\mu}{\sqrt{n}\ \sigma} = \pm .5$$

and use (90) to write

$$P\{|X_1 + X_2 + \cdots + X_{100}| \leq 5\} \approx \int_{-.5}^{.5} \frac{e^{-x^2/2}}{\sqrt{2\pi}} = .3829 \ldots.$$

You can use (66) to show that S_{100} has the binomial density

$$f_{100}(x) = \frac{1}{2^{100}} \sum_{k=-50}^{50} \binom{100}{50+k} \delta(x - 2k),$$

and thereby obtain the exact probability

$$P\{|S_{100}| \le 5\} = \frac{1}{2^{100}} \sum_{k=-2}^{2} \binom{100}{50+k} = .3827\ldots .$$

The estimate based on (90) is remarkably accurate! ∎

We often have occasion to estimate some parameter associated with a probability density function f by observing corresponding independent random variables $X_1, X_2, \ldots, X_n$. For example, we might reasonably use

$$M_n := \frac{1}{n}\{X_1 + X_2 + \cdots + X_n\} \tag{91}$$

to estimate the mean μ and use

$$V_n := \frac{1}{n-1}\{(X_1 - M_n)^2 + (X_2 - M_n)^2 + \cdots + (X_n - M_n)^2\} \tag{92}$$

to represent the variance σ^2, *cf.* Ex. 12.20. The following examples show how the central limit theorem can help us assess the quality of such estimates.

Example: Let the independent random variables $X_1, X_2, X_3, \ldots$ have a common probability density function f with mean μ and variance $0 < \sigma^2 < \infty$, and let $\epsilon > 0$. Prove the *law of large numbers*

$$\lim_{n \to \infty} P\left\{\left|\frac{X_1 + X_2 + \cdots + X_n}{n} - \mu\right| < \epsilon\right\} = 1, \tag{93}$$

and thereby show that we can confidently use the arithmetic mean (91) to estimate μ when n is "large."

Solution: Let $L > 0$. When $n > (L\sigma/\epsilon)^2$ we have

$$1 \ge P\left\{\left|\frac{X_1 + X_2 + \cdots + X_n}{n} - \mu\right| < \epsilon\right\}$$

$$= P\left\{\left|\frac{X_1 + X_2 + \cdots + X_n - n\mu}{\sqrt{n}\,\sigma}\right| < \frac{\sqrt{n}\,\epsilon}{\sigma}\right\}$$

$$\ge P\left\{\left|\frac{X_1 + X_2 + \cdots + X_n - n\mu}{\sqrt{n}\,\sigma}\right| < L\right\}.$$

We obtain (93) by using this inequality together with

$$\lim_{L \to +\infty} \lim_{n \to \infty} P\left\{\left|\frac{X_1 + X_2 + \cdots + X_n - n\mu}{\sqrt{n}\,\sigma}\right| < L\right\} = \lim_{L \to +\infty} \int_{-L}^{L} \frac{e^{-x^2/2}}{\sqrt{2\pi}} = 1$$

and a suitable modification of the squeeze theorem for limits. ∎

Example: A *thick coin* has the shape of a right circular cylinder with diameter $d > 0$ and height $h > 0$. When we drop the coin on the floor, it will land on one of its flat faces with probability p and it will land on its edge (the curved surface) with probability $q := 1 - p$. (The unknown parameter p increases from 0 to 1 as the ratio d/h increases from 0 to ∞.) We can estimate p by dropping the coin n times, observing the number of times m that it lands on a flat face, and taking $p \approx m/n$. How large should we choose n if we want this estimate of p to be within $\pm 10\%$ of its true value with a probability of 95%?

Solution: Let $X_1, X_2, X_3, \ldots$ be independent random variables with each having the Bernoulli density $f(x) := q \, \delta(x) + p \, \delta(x - 1)$. The sum $X_1 + X_2 + \cdots + X_n$ corresponds to the number of times in n tosses that the coin lands on one of its faces. The density f has the mean $\mu = p$ and the variance $\sigma^2 = pq$, so we can use the central limit theorem to write

$$P\left\{\left|\frac{X_1 + X_2 + \cdots + X_n}{n} - p\right| < .1\,p\right\}$$

$$= P\left\{\left|\frac{X_1 + X_2 + \cdots + X_n - n\mu}{\sqrt{n}\,\sigma}\right| \leq \frac{.1\,p\sqrt{n}}{\sqrt{pq}}\right\} \approx \int_{-.1\sqrt{np/q}}^{.1\sqrt{np/q}} \frac{e^{-x^2/2}}{\sqrt{2\pi}}\,dx.$$

From Table 2 of Appendix 6 we see that $\Phi(1.960) = .975$, so that

$$\int_{-1.960}^{1.960} \frac{e^{-x^2/2}}{\sqrt{2\pi}}\,dx = .950.$$

To the accuracy of the above approximation we will have

$$P\left\{\left|\frac{X_1 + X_2 + \cdots + X_n}{n} - p\right| < .1p\right\} \geq .95$$

when

$$.1\sqrt{np/q} \geq 1.960,$$

i.e.

$$n \geq 384(1 - p)/p.$$

Assuming that $p \geq 1/2$, we can accomplish our purpose using $n = 384$ tosses.

You might try to find p for the plastic spool from a role of Scotch^{TM} tape. (I used $n = 400$ tosses to obtain the estimate $p = .80$ for such a spool with $d = 36$ mm and $h = 19$ mm!) ∎

References

W. Feller, *An Introduction to Probability Theory and Its Applications, Vol. 2*, John Wiley & Sons, New York, 1966.

> A classic graduate-level text on probability. Chapter 15 contains a thorough discussion of characteristic functions and a proof of the central limit theorem.

R.R. Goldberg, *Fourier Transforms*, Cambridge University Press, Cambridge, 1965.

> The traditional proof of Bochner's theorem can be found in Chapter 5.

T.W. Körner, *Fourier Analysis*, Cambridge University Press, Cambridge, 1988.

> An elegant elementary proof of the central limit theorem (for bounded continuous densities) is given on pp. 347–361.

E. Lukacs, *Characteristic Functions, 2nd ed.*, Griffin, London, 1970.

> A detailed study of characteristic functions for their intrinsic mathematical interest.

A. Papoulis, *Probability, Random Variables, and Stochastic Processes*, McGraw-Hill, New York, 1965.

> An exceptionally well written introduction to probability theory for scientists and engineers.

Exercise Set 12

▶ **EXERCISE 12.1:** The faces of an equilateral tetrahedron are marked with the spots •, ••, •••, ••••. When we roll the tetrahedron we count all of the spots that are visible on the three upper faces.

(a) Sketch the generalized probability density f on $\mathbb{R}$ for rolling X spots.

(b) Sketch the corresponding distribution function g.

(c) Find a simple formula for the characteristic function F.

(d) Find the mean μ and the variance σ^2 for f.

(e) Sketch the probability density $f * f$ for rolling S_2 spots with two such tetrahedra.

(f) Find $p\{S_3 \leq 20\}$ when S_3 is the number of spots that you roll with three such tetrahedra.

▶ **EXERCISE 12.2:**   Let X_1, X_2 be independent random variables with each having the truncated exponential density $f(x) := \alpha^{-1}e^{-x/\alpha}\cdot h(x)$, $\alpha > 0$. Show that $Y := X_1 - X_2$ has the Laplace density

$$f_Y(y) = \frac{1}{2\alpha}\, e^{-|y|/\alpha} :$$

(a) by computing the convolution product $f(x) * f(-x)$;

(b) by inverting the characteristic function $F(s) \cdot F(-s)$.

▶ **EXERCISE 12.3:**   The Poisson density with parameter $\alpha > 0$ is defined by

$$f(x) := e^{-\alpha}\sum_{k=0}^{\infty}\frac{\alpha^k}{k!}\delta(x-k).$$

(a) Use the Maclaurin series for the exponential to find the characteristic function

$$F(s) = \exp\{\alpha(e^{-2\pi i s}-1)\}.$$

(b) Find the moments $m^{(1)}$, $m^{(2)}$ by evaluating the series

$$m^{(1)} = e^{-\alpha}\sum_{k=0}^{\infty}\frac{k\alpha^k}{k!}, \quad m^{(2)} = e^{-\alpha}\sum_{k=0}^{\infty}\frac{k^2\alpha^k}{k!}$$

from (38) and by using the characteristic function of (a) with (40).

(c) Show that f has the mean $\mu = \alpha$ and the variance $\sigma^2 = \alpha$.

(d) Show that if X_1, X_2 are independent random variables that have Poisson densities with the parameters α_1, α_2, then $X_1 + X_2$ has a Poisson density with the parameter $\alpha_1 + \alpha_2$.

▶ **EXERCISE 12.4:**   The gamma density with parameters $\nu > 0$, $\alpha > 0$ is defined by

$$f(x) := \begin{cases} \dfrac{x^{\nu-1}e^{-x/\alpha}}{\alpha^\nu\Gamma(\nu)} & \text{if } x > 0 \\ 0 & \text{if } x < 0 \end{cases} \quad \text{where} \quad \Gamma(\nu) := \int_0^\infty u^{\nu-1}e^{-u}du, \quad \nu > -\frac{1}{2}$$

is the *gamma function*. [You may choose to remember that $\Gamma(n+1) = n!$, $n = 0, 1, 2, \dots$ and $\Gamma\left(\frac{1}{2}\right) = \sqrt{\pi}$.]

(a) Show that f has the characteristic function $F(s) = 1/(1+2\pi i\alpha s)^\nu$.

    Hint. Suitably adapt the argument from Ex. 3.33.

(b) Show that f has the moments $m^{(n)} = [\nu(\nu+1)\cdots(\nu+n-1)]\alpha^n$, $n = 0, 1, 2, \dots$.

    Hint. You can use (38) with the recursion $\Gamma(u+1) = u\Gamma(u)\dots$ or you can use (40).

(c) Show that f has the mean $\mu = \alpha\nu$ and the variance $\sigma^2 = \alpha^2\nu$.

(d) Show that if X_1, X_2 are independent random variables that have gamma densities with parameters ν_1, α and ν_2, α, then $X_1 + X_2$ has a gamma density with the parameters $\nu_1 + \nu_2, \alpha$.

▶ **EXERCISE 12.5:**   Let $T_1, T_2, T_3, \ldots$ be the times between successive events that occur at random but that occur at some constant average rate

$$\lambda \approx \frac{n}{T_1 + T_2 + \cdots + T_n} \quad \text{(when } n \text{ is large)}.$$

For example, the events might be clicks of a Geiger counter that register the disintegration of atomic nuclei, earthquakes along some geologic fault line, or cars passing some fixed point along a busy expressway. In this exercise you will study such a *Poisson process*.

(a) Let $f(t)$ be the probability density for the random time T between successive events. The probability that an event does, does not occur in a small interval of length Δt will then be approximately $\lambda \Delta t$, $1 - \lambda \Delta t$, respectively (provided that we can neglect the possibility of having 2 or more events). Explain why

$$f(t)\Delta t \approx (1 - \lambda \Delta t)^{N-1} \cdot \lambda \Delta t \quad \text{when } \Delta t = t/N$$

and $N = 1, 2, \ldots$ is large, and then use a suitable limit to infer that

$$f(t) := \lambda\, e^{-\lambda t} h(t)$$

where h is the Heaviside function.

(b) Let $T_1, T_2, T_3, \ldots$ each have the truncated exponential density f with parameter $\lambda > 0$ as derived in (a). Show that $S_k := T_1 + T_2 + \cdots + T_k$ has the probability density

$$f_k(t) := \frac{\lambda(\lambda t)^{k-1}}{(k-1)!} e^{-\lambda t} h(t), \qquad k = 1, 2, \ldots \,.$$

(Selected graphs for the case where $\lambda = 1$ are shown in Fig. 12.10.)

(c) Show that the probability $p_k(t)$ for exactly k events in some interval of length $t > 0$ is given by the Poisson density

$$p_k(t) = \frac{(\lambda t)^k}{k!} e^{-\lambda t}, \qquad k = 0, 1, \ldots$$

with parameter $\alpha = \lambda t$, cf. Ex. 12.3.

Hint. $p_k(t) = P\{S_k \le t\} - P\{S_{k+1} \le t\}$.

▶ **EXERCISE 12.6:**   A random variable $X(t)$ has the Poisson density

$$p(x) = \sum_{k=0}^{\infty} \frac{(\lambda t)^k}{k!} e^{-\lambda t} \delta(x - k).$$

Within the context of Ex. 12.5, $X(t)$ is the number of events that occur in some interval of length $t \ge 0$. Use such a density with a suitable choice of λ to answer each of the following.

(a) Earthquakes occur along the New Madrid fault at an average rate of $1/200$ years. What is the probability of having no earthquake during the next 25 years?

(b) A 1000-page book has 100 typographical errors. What is the probability of having 2 or more errors on a single page?

(c) A Fourier analysis class has 30 students. What is the probability that 2 or more of them have birthdays on the day the final exam is scheduled?

▶ **EXERCISE 12.7:**   Let f be a probability density on $\mathbb{R}$, let $\gamma \in \mathbb{S}$ with $0 \le \gamma(x) \le 1$ for $-\infty < x < \infty$, and let $\phi \in \mathbb{S}$. Show that:

(a) $\left| \displaystyle\int_{-\infty}^{\infty} f(x)\gamma(x)\phi(x)dx \right| \le \max_{-\infty < x < \infty} |\phi(x)| \cdot \int_{-\infty}^{\infty} f(x)\gamma(x)dx;$

(b) $\left| \displaystyle\int_{-\infty}^{\infty} (f * \gamma)(x)\phi(x)dx \right| \le \max_{-\infty < x < \infty} |\phi(x)| \cdot \int_{-\infty}^{\infty} \gamma(x)dx;$

(c) $\left| \displaystyle\int_{-\infty}^{\infty} f(x)[1 - \gamma(x)]\phi(x)dx \right| \le \max_{-\infty < x < \infty} |\phi(x)| \cdot \left\{ 1 - \int_{-\infty}^{\infty} f(x)\gamma(x)dx \right\}.$

▶ **EXERCISE 12.8:**   Let f be a probability density on $\mathbb{R}$. Show that the sequence of translates $f(x)$, $f(x - 1)$, $f(x - 2)$, $f(x - 3)$, $\ldots$ converges weakly to 0.

Hint. Use the max bounds of (14) and Ex. 12.7(c) with

$$\int_{-\infty}^{\infty} f(x - k)\phi(x)dx = \int_{-\infty}^{\infty} f(x)[1 - \Gamma_n(x)]\phi(x + k)dx + \int_{-\infty}^{\infty} f(x)\Gamma_n(x)\phi(x + k)dx.$$

▶ **EXERCISE 12.9:**   Let f be a probability density on $\mathbb{R}$ and let

$$M_k(x) := m_{-+}(x; -k, k, \epsilon), \quad k = 1, 2, \ldots$$

where m_{-+} is the mesa function of Fig. 12.6 and $0 < \epsilon \le 1$ is fixed.

(a) Show that the sequence $M_1 f, M_2 f, M_3 f, \ldots$ converges weakly to f.

 Hint. Use the max bound (14).

 Note.   A somewhat more sophisticated argument can be used to show that $M_1 f, M_2 f, M_3 f, \ldots$ converges weakly to f when f is a generalized function that is not a probability density.

(b) Show that $\displaystyle\lim_{k \to \infty} \int_{-\infty}^{\infty} M_k(x)f(x)dx = 1$.

 Hint. Observe that

$$\int_{-\infty}^{\infty} M_k(x)f(x)\Gamma_n(x)dx \le \int_{-\infty}^{\infty} M_k(x)f(x)dx \le 1$$

 and then use the result from (a).

▶ **EXERCISE 12.10:**   Let the generalized function f and the Schwartz function γ be probability densities on $\mathbb{R}$, and let

$$f_\gamma(x) := \int_{-\infty}^{\infty} f(u)\gamma(x-u)du, \quad -\infty < x < \infty,$$

i.e., the ordinate $f_\gamma(x)$ obtained by applying the functional f to the x-translate of the Schwartz function $\gamma^\vee$.

(a) Use the max bound (14) to show that f_γ is bounded and continuous.

(b) Show that $f_\gamma = \gamma * f$, *i.e.*, show that

$$\int_{-\infty}^{\infty} f_\gamma(x)\phi(x)dx = \int_{-\infty}^{\infty} f(x)(\phi * \gamma^\vee)(x)dx, \quad \phi \in \mathbb{S}.$$

Hint. Let $f = g^{(m)}$ where g is continuous and $|g(x)| \leq B(1+x^2)^q$ for some $B > 0$ and $q = 0, 1, \ldots$. The function $g(u)\gamma^{(m)}(x-u)\phi(x)$ is then continuous and absolutely integrable on $\mathbb{R}^2$.

(c) Show that f_γ is a probability density.

(d) Show that $f_\gamma', f_\gamma'', \ldots$ are defined, bounded, and continuous on $\mathbb{R}$.

Hint. You may wish to use the results of Exs. 7.18 and 7.19.

(e) Is $f_\gamma \in \mathbb{S}$?

▶ **EXERCISE 12.11:**   A generalized function f is said to be max-bounded if there is some constant M such that

$$|f\{\phi\}| \leq M \cdot \max_{-\infty < x < \infty} |\phi| \quad \text{when} \quad \phi \in \mathbb{S}.$$

[The inequality (14) shows that every probability density is max bounded with $M = 1$.]

(a) Let f_0, f_1, f_2, f_3 be probability densities on $\mathbb{R}$, let $\alpha_0, \alpha_1, \alpha_2, \alpha_3$ be nonnegative constants and let $f(x) = \alpha_0 f_0(x) + \alpha_1 i\, f_1(x) + \alpha_2 i^2 f_2(x) + \alpha_3 i^3 f_3(x)$. Show that f is max-bounded.

Note. F. Riesz has shown that every max-bounded generalized function has this form.

(b) Give an example of a generalized function $f \geq 0$ that is not max bounded.

▶ **EXERCISE 12.12:**   Let f be a probability density on $\mathbb{R}$, and let the mesa function

$$h_{kn}(x) := m_{-+}(x; 0, n, 1/k), \quad n = 1, 2, \ldots, \quad k = 1, 2, \ldots$$

with flat width n and with edge width $\epsilon = 1/k$ serve as an approximation to the Heaviside step. (Figure 12.6 shows how m_{-+} is constructed.) In this exercise you will show that we can construct the distribution function g for f by writing

$$g_{kn}(x) := \int_{-\infty}^{\infty} h_{kn}(x-\xi)f(\xi)d\xi, \quad g_k(x) := \lim_{n \to \infty} g_{kn}(x), \quad g(x) := \lim_{k \to \infty} g_k(x).$$

(a) Sketch the $g_k(x)$ that corresponds to the coin flip density of Fig. 12.2. Be sure to show the effect of the edge width parameter $\epsilon = 1/k$.

(b) Let k be fixed. Show that $g_{k1}(x), g_{k2}(x), \ldots$ converges pointwise to a nondecreasing function g_k with $g_k(-\infty) = 0$ and $g_k(+\infty) = 1$.

Hint. Use the bound from Ex. 12.9(b).

(c) Let $x \in \mathbb{R}$, let $\alpha > 0$, and let $h_k := \lim_{n \to \infty} h_{kn}$. Show that

$$g_k(x) - g_k(x - \alpha) \leq \max_{-1/k \leq \xi \leq \alpha} |h_k(\xi) - h_k(\xi - \alpha)|$$

and thereby show that g_k is continuous.

Hint. Use the max bound (14).

(d) Show that $g_{k1}, g_{k2}, \ldots$ converges weakly to g_k.

Hint. $h_k(x) = h_{kn}(x) + h_k(x - n - 1/k)$.

(e) Show that $g_1, g_2, \ldots$ converges pointwise to a nondecreasing function g with $g(-\infty) = 0$, $g(+\infty) = 1$.

Hint. You may wish to verify that $g_k(x - 1/k) \leq g(x) \leq g_k(x)$.

(f) Show that $g_1, g_2, \ldots$ converges weakly to g.

Hint. Write

$$|g(x) - g_k(x)| \leq \left| g_k(x) - g_k\left(x - \frac{1}{k}\right) \right| \leq \left| f(x) * \left\{ h_k(x) - h_k\left(x - \frac{1}{k}\right) \right\} \right|$$

and use the bound from Ex. 12.7(b).

(g) Show that $f = g'$.

Hint. Observe that

$$\left| \int_{-\infty}^{\infty} f(x)\phi(x)dx - \int_{-\infty}^{\infty} g'_{kn}(x)\phi(x)dx \right| = \left| \int_{-\infty}^{\infty} f(x)\{\phi(x) + (h^{\vee}_{kn} * \phi')(x)\}dx \right|.$$

▶ **EXERCISE 12.13:**   Let f be a probability density on $\mathbb{R}$, let $n = 1, 2, \ldots$ and assume that $x^n f(x)$ satisfies the max bound (39). Show that the nth moment $m^{(n)}$ is well defined by writing

$$m^{(n)} := \lim_{a \to -\infty} \lim_{b \to +\infty} \int_{-\infty}^{\infty} x^n f(x) m_{-+}(x; a, b, 1)dx,$$

and thereby infer that

$$|m^{(n)}| \leq M_n.$$

Here m_{-+} is the mesa function of Fig. 12.6.

Hint. The integral is a monotonic function of a and of b when $a < 0$ and $b > 0$.

▶ **EXERCISE 12.14:**   Let f be a probability density on $\mathbb{R}$ and assume that $x\,f(x)$ is max bounded [*i.e.*, (39) holds with $n = 1$] so that f has a well-defined first moment, *cf.* Ex. 12.13. In this exercise you will show that the characteristic function $F = f^\wedge$ is continuously differentiable. An analogous argument shows that $F^{(n)}$ is continuous when $x^n f(x)$ is max bounded, $n = 2, 3, \ldots$.

(a) When we approximate e^{it} by a Taylor polynomial, the remainder is

$$R_n(t) := e^{it} - \{1 + it + (it)^2/2! + \cdots + (it)^n/n!\}, \quad n = 0, 1, 2, \ldots .$$

Show that $|R_n(t)| \le |t|^{n+1}/(n+1)!$, $-\infty < t < \infty$.

Hint. $R_0(t) = i \int_0^t e^{iu}\,du$, $R_{n+1}(t) = i \int_0^t R_n(u)\,du$.

(b) Let F_n be defined by (18). Show that F_n has the derivative

$$F_n'(s) = \int_{-\infty}^{\infty} -2\pi i x f(x)\Gamma_n(x)e^{-2\pi i s x}\,dx$$

by verifying that

$$\lim_{h \to 0} \int_{-\infty}^{\infty} f(x)\left\{\frac{e^{-2\pi i x h} - 1 + 2\pi i x h}{h}\right\}\Gamma_n(x)e^{-2\pi i s x}\,dx = 0.$$

Hint. Use the inequality $|R_1(t)| \le |t|^2/2$ from (a) and the max bound (14).

(c) Show that F_n' is continuous.

Hint. Use the hypothesis that $x\,f(x)$ is max bounded.

(d) Let m_{--}, m_{-+}, m_{++} be the mesa functions of Fig. 12.6, and let

$$h_L(x) = \lim_{a \to -\infty} m_{--}(x; a, 0, 1),$$
$$h_0(x) = m_{-+}(x; 0, 0, 1),$$
$$h_R(x) = \lim_{b \to +\infty} m_{++}(x; 0, b, 1)$$

so that $h_L(x) + h_0(x) + h_R(x) = 1$. Show that

$$\int_{-\infty}^{\infty} x\,f(x)\Gamma_n(x)h_L(x)\,dx, \quad \int_{-\infty}^{\infty} x\,f(x)\Gamma_n(x)h_R(x)\,dx, \quad n = 1, 2, \ldots$$

are bounded nonincreasing, nondecreasing sequences that have limits μ_L, μ_R, respectively.

(e) Show that the sequence $F_1', F_2', F_3', \ldots$ converges uniformly on $\mathbb{R}$.

Hint. When $n > m$ you can use (17) to write

$$|F_n'(s) - F_m'(s)| \leq \int_{-\infty}^{\infty} -2\pi x \, f(x)[\Gamma_n(x) - \Gamma_m(x)]h_L(x)dx$$

$$+ \left| \int_{-\infty}^{\infty} 2\pi x \, f(x)[\Gamma_n(x) - \Gamma_m(x)]h_0(x)dx \right|$$

$$+ \int_{-\infty}^{\infty} 2\pi x \, f(x)[\Gamma_n(x) - \Gamma_m(x)]h_R(x)dx.$$

Use (d) to bound the first and third terms and use (14) to bound the second term.

(f) Show that F' is continuous.

Hint. Use (e) and the weak limit (23).

Note. You may wish to compare this analysis with that given in Ex. 3.41.

▶ **EXERCISE 12.15:** Let F be the characteristic function of a probability density f on $\mathbb{R}$, and let $a > 0$, $-\infty < x_0 < \infty$, $n = 2, 3, \ldots$. Show that the following are also characteristic functions, and express the corresponding probability densities in terms of f.

(a) $e^{-2\pi i x_0 s} F(s)$ (b) $F(as)$ (c) $\overline{F}(s)$

(d) $F(s)^n$ (e) $|F(s)|^2$ (f) $F(s)/(2 - F(s))$

▶ **EXERCISE 12.16:** Which of the following generalized functions, if any, satisfy Bochner's nonnegativity condition (26)?

(a) $F(s) := e^{-s^2}$ (b) $F(s) := (-1)^n e^{-s^2} H_{2n}(s)$, cf. (3.28)

(c) $F(s) := -\delta''(s)$ (d) $F(s) := -1/s^2$

(e) $F(s) := \text{III}(s)$ (f) $F(s) := \sum_{k=-\infty}^{\infty} \text{sinc}(k/5)\delta(s - k/5)$

▶ **EXERCISE 12.17:** Find the smallest $p > 0$, if any, such that F is p-periodic and the smallest $q > 0$, if any, such that $|F(q)| = 1$) [cf. (31)–(32)] when F is the characteristic function of the probability density:

(a) $f(x) := \delta(x - 1)$;

(b) $f(x) := a_0\delta(x) + a_1\delta(x - 1)$, $a_0 > 0$, $a_1 > 0$, $a_0 + a_1 = 1$;

(c) $f(x) := a_1\delta(x - 1) + a_2\delta(x - \sqrt{2})$, $a_1 > 0$, $a_2 > 0$, $a_1 + a_2 = 1$;

(d) $f(x) := a_0\delta(x) + a_1\delta(x - 1) + a_2\delta(x - \sqrt{2})$, $a_0 > 0$, $a_1 > 0$, $a_2 > 0$, $a_0 + a_1 + a_2 = 1$.

▶ **EXERCISE 12.18:**   Let f be a probability density on $\mathbb{P}_N$, *i.e.*, $f[n] \geq 0$ for $n = 0, 1, \ldots, N - 1$ and $f[0] + f[1] + \cdots + f[N - 1] = 1$. In this exercise you will develop properties of the characteristic function $F := f^\wedge$ on $\mathbb{P}_N$.

(a) Show that $NF[0] = 1$ and $|NF[k]| \leq 1$ for $k = 1, 2, \ldots, N - 1$.

(b) Show that $F[-k] = \overline{F[k]}$, $k = 0, 1, \ldots, N - 1$.

(c) Let q be the smallest integer $1, 2, \ldots, N$ such that $|NF[q]| = 1$, and let n_0 be the smallest index $0, 1, \ldots, N - 1$ with $f[n_0] > 0$. Show that

$$f[n + n_0]e^{-2\pi iqn/N} = f[n + n_0] \text{ for each } n = 0, 1, \ldots, N - 1$$

and thereby deduce that $F[k]e^{2\pi ikn_0/N}$ is q-periodic, that $q|N$, and that

$$f[n] = \sum_{\ell=0}^{q-1} \alpha[\ell] \, \delta\left[n - n_0 - \ell\frac{N}{q}\right]$$

for some choice of the nonnegative numbers $\alpha[0], \alpha[1], \ldots, \alpha[q - 1]$ with sum 1.

Hint. Observe that

$$|N\,F[q]| = \left|\sum_{n=0}^{N-1} f[n + n_0]e^{-2\pi iq(n+n_0)/N}\right| = \left|\sum_{n=0}^{N-1} f[n + n_0]e^{-2\pi iqn/N}\right| \leq 1$$

with $f[n_0] > 0$.

(d) Show that

$$\sum_{k=0}^{N-1} F[k](\psi * \psi^\dagger)[k] \geq 0$$

when ψ is any function on $\mathbb{P}_N$, *cf.* (26).

Note. A function F on $\mathbb{P}_N$, is a characteristic function if and only if $NF[0] = 1$ and F is nonnegative in this sense.

▶ **EXERCISE 12.19:**   Find all of the moments $m^{(n)}$, $n = 0, 1, 2, \ldots$ for the uniform density $f(x) = \Pi(x)$:

(a) by evaluating the integrals (38);

(b) by using the formula (40) in conjunction with the Maclaurin series

$$\frac{\sin u}{u} = 1 - \frac{u^2}{3!} + \frac{u^4}{5!} - \frac{u^6}{7!} + \cdots .$$

▶ **EXERCISE 12.20:** Let $X_1, X_2, X_3, \ldots$ be independent random variables with each having the probability density f that has the mean μ and the variance $0 < \sigma^2 < \infty$. Show that the sample mean (91) and the sample variance (92) have the expectations

$$\langle M_n \rangle = \mu, \quad \langle V_n \rangle = \sigma^2.$$

Hint. First derive the algebraic identity

$$V_n = \frac{1}{n(n-1)} \left\{ (n-1) \sum_{k=1}^{n} X_k^2 - \sum_{\substack{k,\ell=1 \\ k \neq \ell}}^{n} X_k X_\ell \right\}.$$

▶ **EXERCISE 12.21:** A certain car runs a 600-km race at the constant velocity V. The velocity is a random variable with the probability density

$$f_V(v) = \begin{cases} 1/100 & \text{if } 200 \text{ km/hr} < v < 300 \text{ km/hr} \\ 0 & \text{otherwise,} \end{cases}$$

i.e., all velocities in the interval from 200 km/hr to 300 km/hr are equally likely.

(a) Find the probability density $f_T(t)$ for the time $T := 600/V$ it will take the car to finish the race.

   Hint. The random variable T does *not* have a uniform density.

(b) Find $\langle V \rangle, \langle T \rangle$ (and observe that $\langle T \rangle \neq 600/\langle V \rangle$).

▶ **EXERCISE 12.22:** Let X be a random variable with the density f_X and let $Y := X^2$. The generalized derivative of the distribution function

$$P\{Y \leq y\} = P\{-\sqrt{y} \leq X \leq \sqrt{y}\} = \int_{-\sqrt{y}-}^{\sqrt{y}+} f_X(x) dx, \quad y \geq 0$$

gives the formula

$$f_Y(y) = \frac{d}{dy} \left\{ \int_{-\sqrt{y}-}^{\sqrt{y}+} f_X(x) dx \right\}$$

for the density f_Y. Use this approach to find f_Y when:

(a) $f_X(x) = e^{-x^2/2}/\sqrt{2\pi}$, *cf.* Ex. 12.25;

(b) $f_X(x) = \frac{1}{4}\delta(x+1) + \frac{1}{2}\delta(x) + \frac{1}{4}\delta(x-1)$.

   Hint. First sketch the "integral" as a function of y.

▶ **EXERCISE 12.23:**   Let $X_1, X_2, \ldots, X_n$ be independent random variables with the probability densities $f_1, f_2, \ldots, f_n$ and the distribution functions $g_1, g_2, \ldots, g_n$.

(a) Show that $X_{\max} := \max\{X_1, X_2, \ldots, X_n\}$ has the probability density

$$f_{\max} = \{g_1 \cdot g_2 \cdot \;\cdots\; \cdot g_n\}'.$$

Hint. $P\{X_{\max} \le x\} = P\{X_1 \le x, \ X_2 \le x, \ \ldots, \ X_n \le x\}.$

(b) Show that $X_{\min} := \min\{X_1, X_2, \ldots, X_n\}$ has the probability density

$$f_{\min} = -\{(1 - g_1) \cdot (1 - g_2) \cdot \;\cdots\; \cdot (1 - g_n)\}'.$$

Hint. $P\{X_{\min} \le x\} = 1 - P\{X_1 > x, \ X_2 > x, \ \ldots, \ X_n > x\}.$

(c) Find a simple expression for $f_{\max}$ and $f_{\min}$ when $X_1, X_2, \ldots, X_n$ each have the die-toss density.

(d) Find a simple expression for $f_{\max}$ and $f_{\min}$ when $X_1, X_2, \ldots, X_n$ each have the standard normal density function.

▶ **EXERCISE 12.24:**   Three Fourier analysis students are discussing the probability densities from Ex. 12.23 for the case where $n = 2$.

"I know that $X_1 + X_2$ has the probability density $f_1 * f_2$," says the first.

"Yes, and since $X_{\min} + X_{\max} = X_1 + X_2$ we must have $f_{\min} * f_{\max} = f_1 * f_2$," adds the second.

"That's what I thought until I computed the two convolution products and got different results in the case where X_1, X_2 have the coin flip density," said the third.

(a) Verify that the third student is correct.

(b) What is wrong with the argument of the second student?

Appendices

Appendix 0: The Impact of Fourier Analysis

Is it possible to prove that the definitions, equations, patterns, theorems, ...
that comprise the body of mathematics known as Fourier analysis have *value* as
well as *validity*? Consider the following.

- The function concept; the Riemann and Lebesgue integrals; the concepts of
  pointwise, uniform, and mean square convergence; Cantor's theory of sets; the
  Schwartz theory of distributions; ... were all invented in an attempt to answer
  the question: When is Fourier's representation valid?, *cf.*

  > E.B. VanVleck, The influence of Fourier's series on the development of
  > mathematics, *Science* **39**(1914), 113–124;

  > W.A. Coppel, J.B. Fourier — On the occasion of his two hundredth birth-
  > day, *Amer. Math. Monthly* **76**(1969), 468–483;

  > S. Bochner, Fourier series came first, *Amer. Math. Monthly* **86**(1979), 197–
  > 199;

  > E.A. González-Velasco, Connections in mathematical analysis: The case
  > of Fourier series, *Amer. Math. Monthly* **99**(1992), 427–441.

- The most frequently cited mathematics paper ever written,

  > J.W. Cooley and J.W. Tukey, *Math. Comp.* **19**(1965), 297–301,

  describes a clever scheme for doing Fourier analysis on a computer. The FFT,
  which Gilbert Strang calls the most important algorithm of the 20th century,
  initiated a revolution in scientific computation. The bibliography in

  > E.O. Brigham, *The Fast Fourier Transform and Its Applications*, Prentice
  > Hall, Englewood Cliffs, NJ, 1988

  will point you to some of the applications.

- The first U.S. patent for a mathematical algorithm was assigned to Stanford
  University for R. Bracewell's FHT, a variation of the FFT. After reading

  > E.N. Zalta, Are algorithms patentable? *Notices AMS* **35**(1988), 796–799,

  see if you can figure out why knowledgeable individuals would invest thousands
  of dollars seeking patent protection for this particular algorithm.

- Approximately 3/4 of the Nobel prizes in physics were awarded for work done using the tools and concepts of Fourier analysis. Examine the abstracts from

  F.N. Magill, ed., *The Nobel Prize Winners — Physics, Vols. 1–3*, Salem Press, Englewood Cliffs, NJ, 1989

  with a knowledgeable physicist and see how close you come to this estimate!

- Herbert Hauptman (a mathematician) and Jerome Karle shared the 1985 Nobel prize in chemistry) for showing how to use Fourier analysis to determine the structure of large molecules from X-ray diffraction data, *cf*.

  W.A. Hendrickson, The 1985 Nobel Prize in Chemistry, *Science* **231**(1986), 362–364;

  Mathematics: The unifying thread in science, *Notices AMS* **33**(1986), 716–733.

- Francis Crick, James Watson, and Maurice Wilkins won the 1962 Nobel prize in medicine and physiology for discovering the molecular structure of DNA. Fourier analysis of X-ray diffraction data played an essential role in this work, *cf*.

  F. Crick, *What Mad Pursuit*, Basic Books, Inc., New York, 1988, pp. 39–61.

- The sophisticated instruments of modern science now produce *signals* instead of *numbers* as the basic data for scientific research. It is impossible to describe the function of an FT-NMR spectrum analyzer, an X-ray diffraction machine, a seismic recorder, ... without using the vocabulary and concepts of Fourier analysis, *cf*.

  A.G. Marshall and F.R. Verdun, *Fourier Transforms in NMR, Optical, and Mass Spectroscopy*, Elsevier, New York, 1990

  to develop some appreciation for what is involved in learning to use an FT-NMR spectrum analyzer.

- The power and flexibility of Fourier analysis have facilitated an incredibly diverse range of applications to modern mathematics, science, and engineering. It is easy to verify this assertion: Subject search your local university library using the key word "Fourier."

- The amazing technology and consumer products associated with digital signal processing (compact disk players, high-definition TV, digital phones, ...) rest on the mathematical base of Fourier analysis. You can confirm this by examining the *sampling* chapter from an introductory text such as

  A.V. Oppenheim, A.S. Willsky, and I.T. Young, *Signals and Systems*, Prentice Hall, Englewood Cliffs, NJ, 1983.

- There is an extraordinary high rate of return on your investment of time in learning Fourier analysis. You will *experience* the Fourier advantage as you "speed learn" the applications chapters of this text and when you take subsequent courses in PDEs, Quantum Mechanics, Signals and Systems, Modern Optics, ... !

Appendix 1: Functions and Their Fourier Transforms

REGULAR FUNCTIONS ON $\mathbb{R}$

$f(x)$, $-\infty < x < \infty$:	$F(s)$, $-\infty < s < \infty$:
$\Pi(x) := \begin{cases} 1 & \text{if } |x| < 1/2 \\ 0 & \text{if } |x| > 1/2 \end{cases}$	$\text{sinc}(s) := \dfrac{\sin(\pi s)}{\pi s}$
$\text{sinc}(x)$	$\Pi(s)$
$\Lambda(x) := \begin{cases} 1 - |x| & \text{if } |x| \le 1 \\ 0 & \text{otherwise} \end{cases}$	$\text{sinc}^2(s)$
$\text{sinc}^2 x$	$\Lambda(s)$
$e^{-x}h(x)$ Here $h(x) := \begin{cases} 1 & \text{if } x > 0 \\ 0 & \text{if } x < 0. \end{cases}$	$\dfrac{1}{1 + 2\pi i s}$
$x^n e^{-\alpha x} h(x)$, $\alpha > 0$ and $n = 0, 1, 2, \ldots$.	$\dfrac{n!}{(\alpha + 2\pi i s)^{n+1}}$
$\dfrac{2}{1 + 4\pi^2 x^2}$	$e^{-|s|}$
$e^{-|x|}$	$\dfrac{2}{1 + 4\pi^2 s^2}$
$e^{-(1+i\beta)|x|}$, $-\infty < \beta < \infty$	$\dfrac{2(1 + i\beta)}{(1 + i\beta)^2 + 4\pi^2 s^2}$
$\dfrac{1}{(x + \alpha - i\beta)^{n+1}}$, $-\infty < \alpha < \infty$, $\beta > 0$, and $n = 0, 1, 2, \ldots$	$\dfrac{2\pi i(-2\pi i s)^n}{n!} e^{2\pi i(\alpha - i\beta)s} h(-s)$
$\dfrac{1}{(x + \alpha + i\beta)^{n+1}}$, $-\infty < \alpha < \infty$, $\beta > 0$, and $n = 0, 1, 2, \ldots$	$\dfrac{(-2\pi i)(-2\pi i s)^n}{n!} e^{2\pi i(\alpha + i\beta)s} h(s)$
$e^{-\pi x^2}$	$e^{-\pi s^2}$
$H_n(\sqrt{2\pi}\, x)e^{-\pi x^2}$, $n = 0, 1, 2, \ldots$ Here $H_n(u) := e^{u^2}(-D)^n e^{-u^2}$ is the nth Hermite polynomial.	$(-i)^n H_n(\sqrt{2\pi}\, s)e^{-\pi s^2}$

REGULAR FUNCTIONS ON $\mathbb{R}$ – Continued

$\mathbf{f(x)}$, $-\infty < \mathbf{x} < \infty$:	$\mathbf{F(s)}$, $-\infty < \mathbf{s} < \infty$:

$b_\alpha(x) := \begin{cases} 1 & \text{if } 0 < x < \alpha \\ 0 & \text{if } x < 0 \text{ or } x > \alpha, \ \alpha > 0 \end{cases}$

$\alpha \operatorname{sinc}(\alpha s) e^{-\pi i \alpha s}$

$x^n b_\alpha(x), \ n = 0, 1, \ldots$

$n! \, \alpha^{n+1} \dfrac{R_n(2\pi i \alpha s) e^{-2\pi i \alpha s}}{(2\pi i \alpha s)^{n+1}}$

Here $R_n(z) := e^z - \Big\{ 1 + z$
$\qquad\qquad + \dfrac{z^2}{2!} + \ldots + \dfrac{z^n}{n!} \Big\}.$

$t_\alpha(x) := \begin{cases} \alpha - |x - \alpha| & \text{if } 0 < x < 2\alpha \\ 0 & \text{otherwise}, \ \alpha > 0 \end{cases}$

$\alpha^2 \operatorname{sinc}^2(\alpha s) e^{-2\pi i \alpha s}$

$\dfrac{\Pi(x/2)}{\pi \sqrt{1 - x^2}}$

$J_0(2\pi s)$

Here $J_0(u) := \displaystyle\int_0^1 e^{iu \sin 2\pi t} \, dt$
is the regular zero-order
Bessel function.

$2\sqrt{1 - x^2}$

$\dfrac{J_1(2\pi s)}{s}$

Here $J_1(u) := \displaystyle\int_0^1 e^{iu \sin 2\pi t - 2\pi i t} \, dt$
is the regular first-order
Bessel function.

$\operatorname{sech} \pi x$

$\operatorname{sech} \pi s$

$\operatorname{sech}^2 \pi x$

$2s \operatorname{cosech} \pi s$

$\arctan\left(\dfrac{1}{2x^2}\right)$

$\pi e^{-|\pi s|} \operatorname{sinc} s$

$\arctan \alpha x - \arctan \beta x, \ \alpha > \beta > 0$

$\dfrac{-i}{2s}\{ e^{-2\pi |s|/\alpha} - e^{-2\pi |s|/\beta} \}$

$\dfrac{\alpha^\nu x^{\nu - 1} e^{-\alpha x}}{\Gamma(\nu)}, \ \alpha > 0, \ \nu > 0$

$\left(\dfrac{\alpha}{\alpha + 2\pi i s}\right)^\nu$

Here $\Gamma(z) := \displaystyle\int_0^\infty u^{z-1} e^{-u} \, du$ is Euler's
gamma function.

$e^x e^{-e^x}$

$\Gamma(1 + 2\pi i s)$

GENERALIZED FUNCTIONS ON $\mathbb{R}$

$f(x), \; -\infty < x < \infty:$	$F(s), \; -\infty < s < \infty:$
$\delta(x)$	1
$\delta^{(n)}(x), \; n = 0, 1, \ldots$	$(2\pi i s)^n$
1	$\delta(s)$
$x^n, \; n = 0, 1, \ldots$	$(-2\pi i)^{-n} \delta^{(n)}(s)$
$e^{2\pi i s_0 x}, \; -\infty < s_0 < \infty$	$\delta(s - s_0)$
$\cos(2\pi s_0 x), \; -\infty < s_0 < \infty$	$\dfrac{1}{2}\delta(s + s_0) + \dfrac{1}{2}\delta(s - s_0)$
$\sin(2\pi s_0 x), \; -\infty < s_0 < \infty$	$\dfrac{i}{2}\delta(s + s_0) - \dfrac{i}{2}\delta(s - s_0)$
$\text{III}(x) = \displaystyle\sum_{m=-\infty}^{\infty} \delta(x - m)$	$\text{III}(s)$
$\dfrac{1}{x} := \{x \log|x| - x\}''$	$-\pi i \, \text{sgn}\, s$
$\dfrac{1}{(x + \alpha)^{n+1}}, \; -\infty < \alpha < \infty, \; n = 0, 1, 2, \ldots$	$\dfrac{-\pi i (-2\pi i s)^n e^{2\pi i \alpha s}\, \text{sgn}\, s}{n!}$
$\text{sgn}\, x$	$\dfrac{1}{\pi i s}$
$h(x) := \dfrac{1}{2}(1 + \text{sgn}\, x)$	$\dfrac{1}{2}\delta(s) + \dfrac{1}{2\pi i s}$
$x^n h(x), \; n = 0, 1, \ldots$	$\dfrac{\delta^{(n)}(s)}{2(-2\pi i)^n} + \dfrac{n!}{(2\pi i s)^{n+1}}$
$|x|^{-1/2}$	$|s|^{-1/2}$
$|x|^\alpha, \; \alpha > -1$	$-2\dfrac{\sin(\alpha \pi/2)\alpha!}{(2\pi|s|)^{\alpha+1}}$
$|x|^\alpha \text{sgn}\, x, \; \alpha > -1$	$-2i\dfrac{\cos(\alpha \pi/2)\alpha!\, \text{sgn}\, s}{(2\pi|s|)^{\alpha+1}}$
$\tanh \pi x$	$-i \, \text{cosech}(\pi s)$

GENERALIZED FUNCTIONS ON $\mathbb{R}$ – Continued

| $f(x)$, $-\infty < x < \infty$: | $F(s)$, $-\infty < s < \infty$: |

$$\mathcal{G}(x) := \int_0^x \frac{\sin \pi u}{\pi u}\, du \qquad\qquad \frac{\Pi(s)}{2\pi i s}$$

$$\operatorname{sgn} x \operatorname{sinc} x \qquad\qquad -\frac{i}{\pi} \log \left| \frac{s+1/2}{s-1/2} \right|$$

$$e^{i\pi x^2} \qquad\qquad \frac{1+i}{\sqrt{2}} e^{-i\pi s^2}$$

$$\cos \pi x^2 \qquad\qquad \cos \pi \left(s^2 - \frac{1}{4} \right)$$

$$\sin \pi x^2 \qquad\qquad -\sin \pi \left(s^2 - \frac{1}{4} \right)$$

$$\mathcal{E}(x) := \int_0^x e^{i\pi u^2}\, du \qquad\qquad \frac{1+i}{\sqrt{2}} \frac{e^{-i\pi s^2}}{2\pi i s}$$

$$e^{-\pi(\alpha+i\beta)^2 x^2}, \quad \alpha \ge |\beta| \qquad\qquad \frac{1}{\alpha+i\beta} e^{-\pi s^2/(\alpha+i\beta)^2}$$
$$\text{and } \alpha + i\beta \neq 0$$

REGULAR FUNCTIONS ON $\mathbb{T}_p$

| $f(x)$, $0 < x < p$: | $F[k]$, $k = 0, \pm 1, \pm 2, \dots$: |

$$e^{2\pi i k_0 x/p}, \ k_0 = 0, \pm 1, \pm 2, \dots \qquad \delta[k - k_0] := \begin{cases} 1 & \text{if } k = k_0 \\ 0 & \text{otherwise} \end{cases}$$

$$\cos(2\pi k_0 x/p), \ k_0 = 0, \pm 1, \pm 2, \dots \qquad \frac{1}{2}\delta[k + k_0] + \frac{1}{2}\delta[k - k_0]$$

$$\sin(2\pi k_0 x/p), \ k_0 = 0, \pm 1, \pm 2, \dots \qquad \frac{i}{2}\delta[k + k_0] - \frac{i}{2}\delta[k - k_0]$$

$$b_\alpha(x) := \begin{cases} 1 & \text{if } 0 < x < \alpha \le p \\ 0 & \text{if } \alpha < x < p \end{cases} \qquad \left(\frac{\alpha}{p}\right) \operatorname{sinc}\left(\frac{k\alpha}{p}\right) e^{-\pi i k\alpha/p}$$

$$t_\alpha(x) := \begin{cases} \alpha - |x - \alpha| \\ \quad \text{if } 0 < x < 2\alpha \le p \\ 0 \quad \text{if } 2\alpha < x < p \end{cases} \qquad \left(\frac{\alpha^2}{p}\right) \operatorname{sinc}^2\left(\frac{k\alpha}{p}\right) e^{-2\pi i k\alpha/p}$$

REGULAR FUNCTIONS ON $\mathbb{T}_p$ – Continued

f(x), 0 < x < p: | **F[k], k = 0, ±1, ±2, ... :**

$x^n b_\alpha(x), \ n = 0, 1, 2, \ldots$

$$\begin{cases} \dfrac{\alpha^{n+1} e^{-2\pi i k \alpha/p}}{p\,(n+1)} \\ \qquad \text{if } k = 0 \\[2mm] \dfrac{n!\,\alpha^{n+1} e^{-2\pi i k \alpha/p}\, R_n(2\pi i k \alpha/p)}{(2\pi i k \alpha/p)^{n+1}} \\ \qquad \text{otherwise} \end{cases}$$

Here $R_n(z) := e^z - \left\{ 1 + z + \dfrac{z^2}{2!} + \ldots + \dfrac{z^n}{n!} \right\}.$

$\left| \sin \dfrac{\pi x}{p} \right|$

$$\begin{cases} \dfrac{2}{\pi(1 - k^2)} & \text{if } k = 0, \pm 2, \pm 4, \ldots \\ 0 & \text{if } k = \pm 1, \pm 3, \pm 5, \ldots \end{cases}$$

$-\log \left| 2 \sin \left(\dfrac{\pi x}{p} \right) \right|$

$$\begin{cases} 0 & \text{if } k = 0 \\ \dfrac{1}{2|k|} & \text{otherwise} \end{cases}$$

$e^{i\alpha \sin(2\pi x/p)}, \ -\infty < \alpha < \infty$

$J_k(\alpha)$
Here J_k is the regular kth-order Bessel function.

Broken line connecting the points:
$\left(\dfrac{np}{N}, \ e^{2\pi i n/N} \right), \ n = 0, 1, \ldots, N$

$$\begin{cases} \dfrac{\text{sinc}^2(1/N)}{k^2} & \text{if } k \equiv 1 \ (\text{mod } N) \\ 0 & \text{otherwise} \end{cases}$$

$\dfrac{1 - r^2}{1 - 2r \cos(2\pi x/p) + r^2}, \ 0 < |r| < 1$

(Poisson kernel)

$r^{|k|}$

$(2m + 1) \dfrac{\text{sinc}\{(2m+1)x/p\}}{\text{sinc}(x/p)}, \ m = 0, 1, 2, \ldots$

(Dirichlet kernel)

$$\begin{cases} 1 & \text{if } k = 0, \pm 1, \ldots, \pm m \\ 0 & \text{otherwise} \end{cases}$$

$(2m + 1) \dfrac{\text{sinc}^2\{(2m+1)x/p\}}{\text{sinc}^2(x/p)}, \ m = 0, 1, 2, \ldots$

(Fejér kernel)

$$\begin{cases} 1 - |k|/(2m+1) \\ \qquad \text{if } k = 0, \pm 1, \ldots, \pm 2m \\ 0 \qquad \text{otherwise} \end{cases}$$

REGULAR FUNCTIONS ON $\mathbb{T}_p$ – Continued

f(x), $0 < x < p$:	**F[k], $k = 0, \pm 1, \pm 2, \ldots$:**

$$\frac{4^m \cos^{2m}\left(\dfrac{\pi x}{p}\right)}{\dbinom{2m}{m}}, \quad m = 0, 1, 2, \ldots$$

(de la Vallée-Poussin power kernel)

$$\frac{\dbinom{2m}{m-k}}{\dbinom{2m}{m}}$$

$$\frac{1}{1 + \epsilon \cos(2\pi x/p)}, \quad 0 < \epsilon < 1$$

$$\frac{1}{\sqrt{1-\epsilon^2}} \left\{\frac{-\epsilon}{1 + \sqrt{1-\epsilon^2}}\right\}^{|k|}$$

$$\omega_0\left(\frac{x}{p}\right) := \frac{1}{2} - \frac{x}{p}$$

$$\begin{cases} 0 & \text{if } k = 0 \\ (2\pi i k)^{-1} & \text{otherwise} \end{cases}$$

$$\omega_1\left(\frac{x}{p}\right) := -\frac{1}{12} + \frac{1}{2}\left(\frac{x}{p}\right) - \frac{1}{2}\left(\frac{x}{p}\right)^2$$

$$\begin{cases} 0 & \text{if } k = 0 \\ (2\pi i k)^{-2} & \text{otherwise} \end{cases}$$

$$\omega_2\left(\frac{x}{p}\right) := -\frac{1}{12}\left(\frac{x}{p}\right) + \frac{1}{4}\left(\frac{x}{p}\right)^2 - \frac{1}{6}\left(\frac{x}{p}\right)^3$$

$$\begin{cases} 0 & \text{if } k = 0 \\ (2\pi i k)^{-3} & \text{otherwise} \end{cases}$$

$$\omega_3\left(\frac{x}{p}\right) := \frac{1}{720} - \frac{1}{24}\left(\frac{x}{p}\right)^2 + \frac{1}{12}\left(\frac{x}{p}\right)^3 - \frac{1}{24}\left(\frac{x}{p}\right)^4$$

$$\begin{cases} 0 & \text{if } k = 0 \\ (2\pi i k)^{-4} & \text{otherwise} \end{cases}$$

$$\vdots$$

$$\omega_{n-1}\left(\frac{x}{p}\right) := \frac{-1}{n!} B_n\left(\frac{x}{p}\right), \quad n = 1, 2, \ldots$$

$$\begin{cases} 0 & \text{if } k = 0 \\ (2\pi i k)^{-n} & \text{otherwise} \end{cases}$$

Here B_n is the nth Bernoulli polynomial.

REGULAR FUNCTIONS ON $\mathbb{T}_p$ – Continued

$f(x), \ -p/2 < x < p/2$:	$F[k], \ k = 0, \pm 1, \pm 2, \ldots$:

$\text{sgn } x$
$$\begin{cases} 0 & \text{if } k = 0, \pm 2, \pm 4, \ldots \\ -2i/(\pi k) & \text{if } k = \pm 1, \pm 3, \pm 5, \ldots \end{cases}$$

x
$$p \begin{cases} 0 & \text{if } k = 0 \\ i\,(-1)^k/(2\pi k) & \text{if } k = \pm 1, \pm 2, \ldots \end{cases}$$

$|x|$
$$p \begin{cases} 1/4 & \text{if } k = 0 \\ 2/(\pi^2 k^2) & \text{if } k = \pm 2, \pm 4, \pm 6, \ldots \\ 0 & \text{if } k = \pm 1, \pm 3, \pm 5, \ldots \end{cases}$$

x^2
$$p^2 \begin{cases} 1/12 & \text{if } k = 0 \\ (-1)^k/(\pi^2 k^2) & \text{if } k = \pm 1, \pm 2, \ldots \end{cases}$$

$e^{\alpha x}, \ -\infty < \alpha < \infty$
$$\frac{(-1)^k \ \sinh(\alpha p/2)}{(\alpha p/2) - i\pi k}$$

$\Pi\left(\dfrac{x}{\alpha}\right), \ 0 < \alpha \le p$
$$\frac{\alpha}{p} \operatorname{sinc}\left(\frac{\alpha k}{p}\right)$$

$\Lambda\left(\dfrac{x}{\alpha}\right), \ 0 < \alpha \le p/2$
$$\frac{\alpha}{p} \operatorname{sinc}^2\left(\frac{\alpha k}{p}\right)$$

$\Pi\left(\dfrac{x - p/4}{\alpha}\right) - \Pi\left(\dfrac{x + p/4}{\alpha}\right), \ 0 < \alpha \le p/2$
$$\frac{-2i\alpha}{p} \sin\left(\frac{k\pi}{2}\right) \operatorname{sinc}\left(\frac{\alpha k}{p}\right)$$

$\Lambda\left(\dfrac{x - p/4}{\alpha}\right) - \Lambda\left(\dfrac{x + p/4}{\alpha}\right), \ 0 < \alpha \le p/4$
$$\frac{-2i\alpha}{p} \sin\left(\frac{k\pi}{2}\right) \operatorname{sinc}^2\left(\frac{\alpha k}{p}\right)$$

Broken line connecting the points:
$\left(-\dfrac{p}{2}, 0\right), \ \left(-\dfrac{\alpha p}{2}, -1\right), \ \left(\dfrac{\alpha p}{2}, 1\right), \ \left(\dfrac{p}{2}, 0\right),$
$$\frac{2}{\pi^2 \alpha(1 - \alpha)} \frac{\sin(\alpha k \pi)}{k^2}, \ 0 < \alpha < 1$$

REGULAR FUNCTIONS ON $\mathbb{Z}$

f[n], n $= 0, \pm 1, \pm 2, \dots$: | **F(s), $0 < s < p$:**

$$\delta[n] := \begin{cases} 1 & \text{if } n = 0 \\ 0 & \text{otherwise} \end{cases}$$

$$\frac{1}{p}$$

$$\Pi\left(\frac{n}{2m+1}\right), \quad m = 0, 1, \dots$$

$$\frac{(2m+1)\operatorname{sinc}\{(2m+1)s/p\}}{p\operatorname{sinc}(s/p)}$$

$$\Lambda\left(\frac{n}{2m+1}\right), \quad m = 0, 1, \dots$$

$$\frac{(2m+1)^2 \operatorname{sinc}^2\{(2m+1)s/p\}}{p\operatorname{sinc}^2(s/p)}$$

$$r^{|n|}, \quad |r| < 1$$

$$\frac{1 - r^2}{p\{1 - 2r\cos(2\pi s/p) + r^2\}}$$

$$r^n h(n) = \begin{cases} r^n, & n = 0, 1, \dots \\ 0, & n = -1, -2, \dots, |r| < 1 \end{cases}$$

$$\frac{1 - re^{2\pi i s/p}}{p\{1 - 2r\cos(2\pi s/p) + r^2\}}$$

$$\binom{m}{n}, \quad m = 1, 2, \dots$$

$$\frac{2^m}{p}\cos^m(\pi s/p)\, e^{-\pi i m s/p}$$

$$J_n(\alpha) := \int_0^1 e^{i\alpha \sin 2\pi t - 2\pi i n t}\, dt$$

Here J_n is the regular nth-order Bessel function.

$$\frac{1}{p} e^{-i\alpha \sin(2\pi s/p)}$$

$$b_m[n] := \begin{cases} 1 & \text{if } n = 0, 1, \dots, m-1 \\ 0 & \text{otherwise}, m = 1, 2, \dots \end{cases}$$

$$\frac{m\operatorname{sinc}(ms/p)}{p\operatorname{sinc}(s/p)} e^{-\pi i(m-1)s/p}$$

$$t_m[n] := \begin{cases} m - |n - m + 1| \\ \quad \text{if } n = 0, 1, \dots, 2m-2 \\ 0 \quad \text{otherwise}, m = 1, 2, \dots \end{cases}$$

$$\frac{m^2 \operatorname{sinc}^2(ms/p)}{p\operatorname{sinc}(s/p)} e^{-2\pi i(m-1)s/p}$$

$$n\, b_m[n],$$

$$\frac{e^{-\pi i(m-1)s/p}}{2p\sin^2(\pi s/p)}\left\{(m-1)\sin\left(\frac{\pi m s}{p}\right)\sin\left(\frac{\pi s}{p}\right)\right.$$
$$+ im\sin\left(\frac{\pi s}{p}\right)\cos\left(\frac{\pi m s}{p}\right)$$
$$\left. - i\sin\left(\frac{\pi m s}{p}\right)\cos\left(\frac{\pi s}{p}\right)\right\}$$

REGULAR FUNCTIONS ON $\mathbb{P}_N$

$\mathbf{f[n]}, \ \mathbf{n} = 0, 1, \ldots, N-1:$	$\mathbf{F[k]}, \ \mathbf{k} = 0, 1, \ldots, N-1:$

$\delta[n] := \begin{cases} 1 & \text{if } n = 0 \\ 0 & \text{otherwise} \end{cases}$

$\dfrac{1}{N}$

1

$\delta[k]$

$c_m[n] := \begin{cases} 1 & \text{if } m|n \\ 0 & \text{otherwise} \end{cases}$

$\dfrac{c_{m'}[k]}{m}$

Here m, m' are positive integers and $m\,m' = N$.

$\delta[n - n_0], \ n_0 = 0, \pm 1, \pm 2, \ldots$

$\dfrac{e^{-2\pi i k n_0/N}}{N}$

$e^{2\pi i k_0 n/N}, \ k_0 = 0, \pm 1, \pm 2, \ldots$

$\delta[k - k_0]$

$e^{2\pi i \alpha n/N}, \ 0 \le \alpha \le N$

$e^{-\pi i (N-1)(k-\alpha)/N} \dfrac{\operatorname{sinc}\{k - \alpha\}}{\operatorname{sinc}\{(k - \alpha)/N\}}$

n

$\dfrac{1}{2} \begin{cases} N - 1 & \text{if } k = 0 \\ -1 + i\,\cot(\pi k/N), \\ \qquad \text{if } k = 1, 2, \ldots, N-1 \end{cases}$

$r^n, \ r \ne 1$

$\dfrac{1 - r^N}{1 - r\,e^{-2\pi i k/N}}$

$b_m[n] := \begin{cases} 1 & \text{if } n = 0, 1, \ldots, m-1 \\ 0 & \text{if } n = m, m+1, \ldots, N-1 \end{cases}$

$\dfrac{m \operatorname{sinc}(mk/N)}{N \operatorname{sinc}(k/N)} e^{-\pi i k(m-1)/N}$

Here $m = 1, 2, \ldots, N$ is the width of the box.

$\Pi\left(\dfrac{n}{2m + 1}\right) := b_{2m+1}[n + m]$

$\dfrac{(2m + 1) \operatorname{sinc}\{(2m + 1)k/N\}}{N \operatorname{sinc}(k/N)}$

Here m is a positive integer with $2m + 1 \le N$.

$\operatorname{sgn}[n] := \begin{cases} 0 & \text{if } n = 0 \\ 1 & \text{if } n = 1, 2, \ldots, M \\ -1 & \text{if } n = M+1, \ldots, 2M-1 \end{cases}$

$\begin{cases} i\,\tan(k\pi/2N) \\ \qquad \text{if } k = 0, 2, \ldots, 2M \\ -i\,\cot(k\pi/2N) \\ \qquad \text{if } k = 1, 3, \ldots, 2M-1 \end{cases}$

Here $N = 2M + 1$ is odd.

REGULAR FUNCTIONS ON $\mathbb{P}_N$ – Continued

f[n], n = 0, 1, ..., N − 1:   **F[k], k = 0, 1, ..., N − 1:**

$$\text{sgn}[n] := \begin{cases} 0 & \text{if } n = 0, M \\ 1 & \text{if } n = 1, 2, \ldots, M-1 \\ -1 & \text{if } n = M+1, \ldots, 2M-1 \end{cases}$$

Here $N = 2M$ is even.

$$\begin{cases} -(2i/N)\cot(k\pi/N) \\ \qquad \text{if } k = 1, 3, \ldots, 2M-1 \\ 0 \qquad \text{if } k = 0, 2, \ldots, 2M-2 \end{cases}$$

$$t_m[n] := \begin{cases} m - |n - m + 1| \\ \qquad \text{if } n = 0, 1, \ldots, 2m-2 \\ 0 \qquad \text{if } n = 2m-1, \ldots, N-1 \end{cases}$$

Here m is a positive integer with
$2m - 1 \le N$.

$$\frac{m^2}{N}\left\{\frac{\text{sinc}(mk/N)}{\text{sinc}(k/N)}\right\}^2 e^{-2\pi i k(m-1)/N}$$

$$\Lambda(n/m) := \begin{cases} 1 - (n/m) \\ \qquad \text{if } n = 0, 1, \ldots, m-1 \\ 0 \qquad \text{if } n = m, m+1, \ldots, N-m \\ 1 - (N-n)/m \\ \qquad \text{if } n = N-m+1, \ldots, N-1 \end{cases}$$

$$\frac{m}{N}\left\{\frac{\text{sinc}(mk/N)}{\text{sinc}(k/N)}\right\}^2$$

$$h_m[n] := \sum_{\mu=-\infty}^{\infty} H_m\left\{\sqrt{\frac{2\pi}{N}}(n - \mu N)\right\} e^{-\pi(n-\mu N)^2/N},$$

$$\frac{(-i)^m}{\sqrt{N}} h_m[k]$$

Here $H_m(u) := e^{u^2}(-D)^m(e^{-u^2})$ is the
mth Hermite polynomial.

$$e^{2\pi i n^2/N}$$

$$\frac{(1+i)\{1 + (-1)^k i^{-N}\}}{2\sqrt{N}} e^{-\pi i k^2/(2N)}$$

$$\left(\frac{n}{P}\right)$$

Here $N = P$ is an odd prime, and
the Legendre symbol $\left(\frac{n}{P}\right)$
takes the values $0, 1, -1$
when $n^{(P-1)/2} \equiv 0, 1, -1 \pmod{P}$.

$$\frac{\overline{\sigma}_P}{\sqrt{P}}\left(\frac{k}{P}\right)$$

Here $\sigma_P := \begin{cases} 1 & \text{if } P \equiv 1 \pmod 4 \\ i & \text{if } P \equiv 3 \pmod 4 \end{cases}$

$$e^{2\pi i m n^2/P}$$

Here $N = P$ is an odd prime
and $m = 1, 2, \ldots, P-1$.

$$\left(\frac{m}{P}\right)\frac{\sigma_P}{\sqrt{P}}e^{2\pi i m' k^2/P}$$

Here $m' = 1, 2, \ldots, P-1$ is chosen
so that $4m\, m' \equiv -1 \pmod{P}$.

Appendix 2: The Fourier Transform Calculus

BASIC RELATIONS: REGULAR FUNCIONS ON $\mathbb{R}$

Name	Relation
Synthesis	$f(x) = \int_{s=-\infty}^{\infty} F(s)e^{2\pi isx}ds$
Analysis	$F(s) := \int_{x=-\infty}^{\infty} f(x)e^{-2\pi isx}dx$
Parseval	$\int_{x=-\infty}^{\infty} f(x)\overline{g(x)}dx = \int_{s=-\infty}^{\infty} F(s)\overline{G(s)}ds$
Convolution	$(f * g)(x) := \int_{u=-\infty}^{\infty} f(u)g(x-u)du$

FOURIER TRANSFORM RULES: REGULAR FUNCTIONS ON $\mathbb{R}$

Name	Function	Transform
Linearity	$c_1 f_1(x) + c_2 f_2(x)$	$c_1 F_1(s) + c_2 F_2(s)$
Reflection	$f(-x)$	$F(-s)$
Conjugation	$\overline{f(x)}$	$\overline{F(-s)}$
Translation	$f(x - x_0)$	$e^{-2\pi isx_0} F(s)$
Modulation	$e^{2\pi is_0 x} \cdot f(x)$	$F(s - s_0)$
Convolution	$(f_1 * f_2)(x)$	$F_1(s) \cdot F_2(s)$
Multiplication	$f_1(x) \cdot f_2(x)$	$(F_1 * F_2)(s)$
Inversion	$F(x)$	$f(-s)$
Derivative	$f'(x)$	$2\pi is \cdot F(s)$
Power scaling	$x \cdot f(x)$	$(-1/2\pi i)\, F'(s)$
Dilation	$f(ax)$	$|a|^{-1} F(s/a)$

$a \neq 0$ is real

BASIC RELATIONS: REGULAR FUNCTIONS ON $\mathbb{T}_p$

Name	Relation

Synthesis

$$f(x) = \sum_{k=-\infty}^{\infty} F[k]e^{2\pi ikx/p}$$

Analysis

$$F[k] := \frac{1}{p}\int_{x=0}^{p} f(x)e^{-2\pi ikx/p}dx$$

Parseval

$$\int_{x=0}^{p} f(x)\overline{g(x)}dx = p\sum_{k=-\infty}^{\infty} F[k]\overline{G[k]}$$

Convolution

$$(f * g)(x) := \int_{u=0}^{p} f(u)g(x-u)du$$

FOURIER TRANSFORM RULES: REGULAR FUNCTIONS ON $\mathbb{T}_p$

Name	Function	Transform
Linearity	$c_1 f_1(x) + c_2 f_2(x)$	$c_1 F_1[k] + c_2 F_2[k]$
Reflection	$f(-x)$	$F[-k]$
Conjugation	$\overline{f(x)}$	$\overline{F[-k]}$
Translation	$f(x-x_0)$	$e^{-2\pi ikx_0/p}\,F[k]$
Modulation	$e^{2\pi ik_0 x/p}\cdot f(x)$	$F[k-k_0]$
Convolution	$(f_1 * f_2)(x)$	$p\,F_1[k]\cdot F_2[k]$
Multiplication	$f_1(x)\cdot f_2(x)$	$(F_1 * F_2)[k]$
Inversion f on $\mathbb{Z}$	$F(x)$	$(1/p)\,f[-k]$
Derivative	$f'(x)$	$(2\pi ik/p)\cdot F[k]$
Dilation	$f(mx),\ m = 1,2,\ldots$	$\begin{cases} F[k/m] & \text{if } m\|k \\ 0 & \text{otherwise} \end{cases}$
Grouping	$\displaystyle\sum_{\ell=0}^{m-1} f(x/m - \ell p/m),$ $m = 1,2,\ldots$	$m\,F[k\,m]$
Summation f on $\mathbb{R}$	$\displaystyle\sum_{m=-\infty}^{\infty} f(x-mp)$	$(1/p)\,F(k/p)$

BASIC RELATIONS: REGULAR FUNCTIONS ON $\mathbb{Z}$

Name	Relation
Synthesis	$f[n] = \displaystyle\int_{s=0}^{p} F(s)e^{2\pi isn/p}ds$
Analysis	$F(s) := \dfrac{1}{p} \displaystyle\sum_{n=-\infty}^{\infty} f[n]e^{-2\pi isn/p}$
Parseval	$\displaystyle\sum_{n=-\infty}^{\infty} f[n]\overline{g[n]} = p \int_{0}^{p} F(s)\overline{G(s)}ds$
Convolution	$(f * g)[n] := \displaystyle\sum_{m=-\infty}^{\infty} f[m]g[n-m]$

FOURIER TRANSFORM RULES: REGULAR FUNCTIONS ON $\mathbb{Z}$

Name	Function	Transform
Linearity	$c_1 f_1[n] + c_2 f_2[n]$	$c_1 F_1(s) + c_2 F_2(s)$
Reflection	$f[-n]$	$F(-s)$
Conjugation	$\overline{f[n]}$	$\overline{F(-s)}$
Translation	$f[n - n_0]$	$e^{-2\pi isn_0/p} \cdot F(s)$
Modulation	$e^{2\pi is_0 n/p} \cdot f[n]$	$F(s - s_0)$
Convolution	$(f_1 * f_2)[n]$	$p\,F_1(s) \cdot F_2(s)$
Multiplication	$f_1[n] \cdot f_2[n]$	$(F_1 * F_2)(s)$
Inversion f on $\mathbb{T}_p$	$F[n]$	$(1/p)\,f(-s)$
Power scaling	$n \cdot f[n]$	$(-p/2\pi i)\,F'(s)$
Dilation	$f[m\,n],\ m = 1, 2, \ldots$	$\dfrac{1}{m} \displaystyle\sum_{\ell=0}^{m-1} F(s/m - \ell p/m)$
Zero packing	$\begin{cases} f[n/m] & \text{if } m \mid n \\ 0 & \text{otherwise,} \end{cases}$ $m = 1, 2, \ldots$	$F(ms)$
Sampling f on $\mathbb{R}$	$f(n/p)$	$\displaystyle\sum_{m=-\infty}^{\infty} F(s - mp)$

BASIC RELATIONS: FUNCTIONS ON $\mathbb{P}_N$

Name	Relation
Synthesis	$f[n] = \displaystyle\sum_{k=0}^{N-1} F[k] e^{2\pi i k n/N}$
Analysis	$F[k] := \dfrac{1}{N} \displaystyle\sum_{n=0}^{N-1} f[n] e^{-2\pi i k n/N}$
Parseval	$\displaystyle\sum_{n=0}^{N-1} f[n]\overline{g[n]} = N \sum_{k=0}^{N-1} F[k]\overline{G[k]}$
Convolution	$(f * g)[n] := \displaystyle\sum_{m=0}^{N-1} f[m] g[n-m]$

FOURIER TRANSFORM RULES: FUNCTIONS ON $\mathbb{P}_N$

Name	Function	Transform
Linearity	$c_1 f_1[n] + c_2 f_2[n]$	$c_1 F_1[k] + c_2 F_2[k]$
Reflection	$f[-n]$	$F[-k]$
Conjugation	$\overline{f[n]}$	$\overline{F[-k]}$
Translation	$f[n - n_0]$	$e^{-2\pi i k n_0/N} \cdot F[k]$
Modulation	$e^{2\pi i k_0 n/N} \cdot f[n]$	$F[k - k_0]$
Convolution	$(f_1 * f_2)[n]$	$N\, F_1[k] \cdot F_2[k]$
Multiplication	$f_1[n] \cdot f_2[n]$	$(F_1 * F_2)[k]$
Inversion	$F[n]$	$(1/N)\, f[-k]$
Repeat f on $\mathbb{P}_{N/m}$	$f[n]$	$\begin{cases} F[k/m] & \text{if } m\vert k \\ 0 & \text{otherwise} \end{cases}$
Zero packing f on $\mathbb{P}_{N/m}$	$\begin{cases} f[n/m] & \text{if } m\vert n \\ 0 & \text{otherwise} \end{cases}$	$(1/m)\, F[k]$

FOURIER TRANSFORM RULES: FUNCTIONS ON $\mathbb{P}_N$ – Continued

Name	Function	Transform
Decimation f on $\mathbb{P}_{m \cdot N}$	$f[mn]$	$\displaystyle\sum_{\ell=0}^{m-1} F[k - \ell N]$
Summation f on $\mathbb{P}_{m \cdot N}$	$\displaystyle\sum_{\ell=0}^{N-1} f[n - \ell N]$	$m\, F[mk]$
Dilation f on $\mathbb{P}_N$	$f[mn]$, Here m, N are relatively prime.	$F[m'k]$, Here $mm' \equiv 1 \pmod{N}$
Dilation f on $\mathbb{P}_N$	$f[mn]$ Here $m \mid N$.	$\begin{cases} \displaystyle\sum_{\ell=0}^{m-1} F[k/m - \ell N/m] \\ \quad \text{if } m \mid k \\ 0 \quad \text{otherwise} \end{cases}$
Sampling f on $\mathbb{T}_p$	$f(np/N)$	$\displaystyle\sum_{m=-\infty}^{\infty} F[k - mN]$
Summation f on $\mathbb{Z}$	$\displaystyle\sum_{m=-\infty}^{\infty} f[n - mN]$	$(p/N)\, F(kp/N)$
Sample-sum f on $\mathbb{Z}$	$\displaystyle\sum_{m=-\infty}^{\infty} f(\alpha\,[n - mN])$ $\alpha := a/\sqrt{N},\ a \neq 0$	$\displaystyle\sum_{m=-\infty}^{\infty} |\beta|\, F(\beta[k - mN])$ $\beta := a^{-1}/\sqrt{N}$

Appendix 3: Operators and Their Fourier Transforms

OPERATORS USED TO DESCRIBE COMMON SYMMETRIES

Name	$\mathcal{A}$	$\mathcal{A}\,f$	$\mathcal{A}^\wedge$
Reflection	**R**	$f^\vee$	**R**
Bar	$\mathcal{B}$	f^-	$\mathcal{D}$
Dagger	$\mathcal{D}$	$f^\dagger$	$\mathcal{B}$
Even projection	$\mathbf{P}_e$	$\frac{1}{2}(f + f^\vee)$	$\mathbf{P}_e$
Odd projection	$\mathbf{P}_o$	$\frac{1}{2}(f - f^\vee)$	$\mathbf{P}_o$
Real projection	$\mathbf{P}_r$	$\frac{1}{2}(f + f^-)$	$\mathbf{P}_h$
Imaginary projection	$\mathbf{P}_i$	$\frac{1}{2}(f - f^-)$	$\mathbf{P}_a$
Hermitian projection	$\mathbf{P}_h$	$\frac{1}{2}(f + f^\dagger)$	$\mathbf{P}_r$
Antihermitian projection	$\mathbf{P}_a$	$\frac{1}{2}(f - f^\dagger)$	$\mathbf{P}_i$

Note. $f^\vee(x) := f(-x)$, $f^-(x) := \overline{f(x)}$ when f is defined on $\mathbb{R}$, $\mathbb{T}_p$, $f^\vee[n] := f[-n]$, $f^-[n] := \overline{f[n]}$ when f is defined on $\mathbb{Z}$, $\mathbb{P}_N$, and $f^\dagger := f^{\vee-}$.

OPERATORS THAT ARE POLYNOMIALS IN $\mathcal{F}$

Name	$\mathcal{A}$	$\mathcal{A}f$	$\mathcal{A}^{\wedge}$
Identity	$\mathbf{I}$	f	$\mathbf{I}$
Reflection	$\mathbf{R}$	$f^{\vee}$	$\mathbf{R}$
Fourier transform	$\mathcal{F}$	$f^{\wedge}$	$\mathcal{F}$
Normalized exponential	$\mathbf{E}_-$	$\beta^{1/2} f^{\wedge}$	$\mathbf{E}_-$
Normalized exponential	$\mathbf{E}_+$	$\beta^{1/2} f^{\wedge\vee}$	$\mathbf{E}_+$
Normalized cosine	$\mathbf{C}$	$\mathbf{P}_e \mathbf{E}_- f$	$\mathbf{C}$
Normalized sin	$\mathbf{S}$	$i\mathbf{P}_o \mathbf{E}_- f$	$\mathbf{S}$
Normalized Hartley	$\mathbf{H}_+$	$(\mathbf{C} + \mathbf{S})f$	$\mathbf{H}_+$
Normalized Hartley	$\mathbf{H}_-$	$(\mathbf{C} - \mathbf{S})f$	$\mathbf{H}_-$
Fundamental projection	$\mathbf{Q}_0$	$\frac{1}{2}(\mathbf{P}_e + \mathbf{C})f$	$\mathbf{Q}_0$
Fundamental projection	$\mathbf{Q}_1$	$\frac{1}{2}(\mathbf{P}_o + \mathbf{S})f$	$\mathbf{Q}_1$
Fundamental projection	$\mathbf{Q}_2$	$\frac{1}{2}(\mathbf{P}_e - \mathbf{C})f$	$\mathbf{Q}_2$
Fundamental projection	$\mathbf{Q}_3$	$\frac{1}{2}(\mathbf{P}_o - \mathbf{S})f$	$\mathbf{Q}_3$

Note. We use $\beta := 1, p, p, N$ when f is defined on $\mathbb{R}, \mathbb{T}_p, \mathbb{Z}, \mathbb{P}_N$, respectively.

OTHER OPERATORS WHERE f AND $\mathcal{A}$ f
HAVE THE SAME DOMAIN

Name	$\mathcal{A}$	$\mathcal{A}$ f	Domain f, $\mathcal{A}$ f	$\mathcal{A}^\wedge$
Translation	$\mathcal{T}_a$	$f(x+a)$	$\mathbb{R}, \mathbb{T}_p$	$\mathcal{E}_a$
Translation	$\mathcal{T}_m$	$f[n+m]$	$\mathbb{Z}, \mathbb{P}_N$	$\mathcal{E}_m$
Exponential modulation	$\mathcal{E}_a$	$e^{2\pi i a x} \cdot f(x)$	$\mathbb{R}$	$\mathcal{T}_{-a}$
Exponential modulation	$\mathcal{E}_m$	$e^{2\pi i m x/p} \cdot f(x)$	$\mathbb{T}_p$	$\mathcal{T}_{-m}$
Exponential modulation	$\mathcal{E}_a$	$e^{2\pi i a n/p} \cdot f[n]$	$\mathbb{Z}$	$\mathcal{T}_{-a}$
Exponential modulation	$\mathcal{E}_m$	$e^{2\pi i m n/N} \cdot f[n]$	$\mathbb{P}_N$	$\mathcal{T}_{-m}$
Convolution	$\mathcal{C}_g$	$g * f$	$\mathbb{R}, \mathbb{T}_p, \mathbb{Z}, \mathbb{P}_N$	$\beta\,\mathcal{M}_{g^\wedge}$
Multiplication	$\mathcal{M}_g$	$g \cdot f$	$\mathbb{R}, \mathbb{T}_p, \mathbb{Z}, \mathbb{P}_N$	$\mathcal{C}_{g^\wedge}$
Derivative	$\mathbf{D}$	f'	$\mathbb{R}, \mathbb{T}_p$	$\mathcal{P}$
Power scaling	$\mathcal{P}$	$2\pi i x \cdot f(x)$	$\mathbb{R}$	$-\mathbf{D}$
Power scaling	$\mathcal{P}$	$(2\pi i n/p) \cdot f[n]$	$\mathbb{Z}$	$-\mathbf{D}$
Grouping	$\mathcal{G}_m,\ m \neq 0$	$\displaystyle\sum_{\ell=0}^{|m|-1} f\left(\frac{x}{m} + \frac{\ell p}{m}\right)$	$\mathbb{T}_p$	$\|m\|\mathcal{S}_m$
Zero packing	$\mathcal{Z}_m,\ m \neq 0$	$\begin{cases} f[n/m] & \text{if } m\|n \\ 0 & \text{otherwise} \end{cases}$	$\mathbb{Z}$	$\mathcal{S}_m$
Dilation	$\mathcal{S}_a,\ a \neq 0$	$f(ax)$	$\mathbb{R}$	$\|a\|^{-1}\mathcal{S}_{1/a}$
Dilation	$\mathcal{S}_m,\ m \neq 0$	$f(mx)$	$\mathbb{T}_p$	$\mathcal{Z}_m$
Dilation	$\mathcal{S}_m,\ m \neq 0$	$f[mn]$	$\mathbb{Z}$	$\|m\|^{-1}\mathcal{G}_m$
Dilation	$\mathcal{S}_m,\ m\,m' \equiv 1 \pmod{N}$	$f[mn]$	$\mathbb{P}_N$	$\mathcal{S}_{m'}$
Dilation	$\mathcal{S}_m,\ m\|N$	$f[mn]$	$\mathbb{P}_N$	$\mathcal{Z}_m\,\Sigma_{N/m}$

Note. Each subscript a is real and each subscript m is an integer.

OTHER OPERATORS WHERE f AND $\mathcal{A}$ f HAVE DIFFERENT DOMAINS

Name	$\mathcal{A}$	$\mathcal{A}$ f	f	$\mathcal{A}$ f	$\mathcal{A}^{\wedge}$
Decimation	Ξ_m	$f[mn]$	$\mathbb{P}_{N\cdot m}$	$\mathbb{P}_N$	Σ_N
Summation	Σ_N	$\displaystyle\sum_{\mu=0}^{m-1} f[n - \mu N]$	$\mathbb{P}_{N\cdot m}$	$\mathbb{P}_N$	$m\,\Xi_m$
Repeat	$\mathcal{R}_m$	$f[n]$	$\mathbb{P}_{N/m}$	$\mathbb{P}_N$	$\mathcal{Z}_m$
Zero packing	$\mathcal{Z}_m$	$\begin{cases} f[n/m] & \text{if } m\mid n \\ 0 & \text{otherwise} \end{cases}$	$\mathbb{P}_{N/m}$	$\mathbb{P}_N$	$(1/m)\,\mathcal{R}_m$
Sampling	$\Xi_{1/p}$	$f(n/p)$	$\mathbb{R}$	$\mathbb{Z}$	Σ_p
Summation	Σ_p	$\displaystyle\sum_{m=-\infty}^{\infty} f(x - mp)$	$\mathbb{R}$	$\mathbb{T}_p$	$(1/p)\,\Xi_{1/p}$
Sampling	$\Xi_{p/N}$	$f(np/N)$	$\mathbb{T}_p$	$\mathbb{P}_N$	Σ_N
Summation	Σ_N	$\displaystyle\sum_{m=-\infty}^{\infty} f[n - mN]$	$\mathbb{Z}$	$\mathbb{P}_N$	$(p/N)\,\Xi_{p/N}$
Sample-sum	$\mathcal{X}_a$	$\displaystyle\sum_{m=-\infty}^{\infty} f\left(\frac{a}{\sqrt{N}}[n - mN]\right)$	$\mathbb{R}$	$\mathbb{P}_N$	$(a^2 N)^{-1/2}\mathcal{X}_{1/a}$

Note. Each subscript m is a positive integer, $p > 0$, $n = 1, 2, \ldots$, and $a < 0$ or $a > 0$.

Appendix 4: The Whittaker-Robinson Flow Chart for Harmonic Analysis

$\theta =$ 0° $u_0 =$............ =............ $\times 10$............

30° $u_1 =$............ =............

60° $u_2 =$............ =............

90° $u_3 =$............ =............

120° $u_4 =$............ =............

150° $u_5 =$............ =............

180° $u_6 =$............ =............

210° $u_7 =$............ =............

240° $u_8 =$............ =............

270° $u_9 =$............ =............

300° $u_{10} =$............ =............

330° $u_{11} =$............ =............

1st-order sums and differences

(u_0 to u_6)   ....

(u_{11} to u_7)   ....

Sums   (v_0 to v_6)   ....

Diffs.   (w_1 to w_5)   ....

2nd-order sums and differences

$(v_0$ to $v_3)$ ....

$(v_6$ to $v_4)$ ....

Sums $(p_0$ to $p_3)$ ....

Diffs. $(q_0$ to $q_2)$ ....

$(w_1$ to $w_3)$ ....

$(w_5$ to $w_4)$ ....

Sums $(r_1$ to $r_3)$ ....

Diffs. $(s_1$ to $s_2)$ ....

Scaling by 1/2 and $\sqrt{3}/2$

	p_1 = ...	p_2 = ...	q_2 = ...	r_1 = ...
(0.500 × line above)	h_1 = ...	h_2 = ...	l_2 = ...	m_1 = ...

	q_1 = ...	r_2 = ...	s_1 = ...	s_2 = ...
(0.866 × line above[†])	l_1 = ...	m_2 = ...	n_1 = ...	n_2 = ...

[†] For mental computation use $.866\,x \approx x - (1/10)x - (1/30)x$.

3rd-order sums and differences

p_0 =............ p_1 =............ q_0 =............ l_1 =............
p_2 =............ p_3 =............ l_2 =............

Sum of 1st column   ....
Sum of 2nd column   ....

Sums   .... $= 12a_0$ .... $= 6a_1$
Diffs.   .... $= 12a_6$ .... $= 6a_5$

p_0 =............ h_1 =............ p_0 =............ p_3 =............
p_3 =............ h_2 =............ h_1 =............ h_2 =............

Sum of 1st column   ....
Sum of 2nd column   ....
Sums
Diffs.   .... $= 6a_4$ .... $= 6a_2$

m_1 =............ m_2 =............ q_0 =............
r_3 =............ q_2 =............

Sum of 1st column   ....
Sum of 2nd column   ....

Sums   .... $= 6b_1$
Diffs.   .... $= 6b_5$ .... $= 6a_3$

n_1 =............ r_1 =............
n_2 =............ r_3 =............

Sums   .... $= 6b_2$
Diffs.   .... $= 6b_4$ .... $= 6b_3$

Final scaling

$a_0 =$ .... $a_1 =$ .... $a_2 =$ .... $a_3 =$ .... $a_4 =$ .... $a_5 =$ .... $a_6 =$ ....

$b_1 =$ .... $b_2 =$ .... $b_3 =$ .... $b_4 =$ .... $b_5 =$ ....

Result

$$u = a_0 + a_1 \cos \theta + a_2 \cos 2\theta + a_3 \cos 3\theta + a_4 \cos 4\theta + a_5 \cos 5\theta + a_6 \cos 6\theta + b_1 \sin \theta$$
$$+ b_2 \sin 2\theta + b_3 \sin 3\theta + b_4 \sin 4\theta + b_5 \sin 5\theta$$

Checks

$$u_0 = a_0 + a_1 + a_2 + a_3 + a_4 + a_5 + a_6$$
$$w_1 = b_1 + 2b_3 + b_5 + 1.732(b_2 + b_4)$$

E. Whittaker and G. Robinson, *The Calculus of Observations, 4th ed.*, Blackie & Son Ltd., London, 1944, insert at p. 270.

Appendix 5: FORTRAN *Code for a* Radix 2 FFT

```fortran
      subroutine fft(fr,fi,mp,isw)
c
c This FORTRAN subroutine computes the discrete Fourier transform
c
c         ft(k+1)=scale*sum f(j+1)*exp(isn*2*pi*i*j*k/n), k=0,1,...,n-1
c
c of the complex array
c
c         f(j+1)=fr(j+1)+i*fi(j+1), j=0,1,...,n-1, where
c
c         n     =2**mp
c         i     =sqrt(-1)
c         isn   =+1 if isw.gt.0
c               =-1 if isw.lt.0
c         scale=1          if isw=-1 or +1
c               =1/n        if isw=-2 or +2
c               =1/sqrt(n) if isw=-3 or +3.
c
c Computations are done in place, and at the time of return the real
c arrays fr,fi have been overwritten with the real and imaginary parts
c of the desired Fourier transform.
c
c The code is based on the presentation of the FFT given in the
c text 'A First Course in Fourier Analysis' by David W. Kammler.
c
      implicit real*8 (a-h,o-z)
      parameter(zero=0.d0,one=1.d0,two=2.d0)
c
c To change the code to single precision, delete the above implicit
c real*8 statement, delete the d0's in the above parameter statement
c that defines zero,one,two, and replace dsqrt by sqrt in the equations
c defining rrootn, rroot2, and in the first equation of the do 6 loop.
c
      parameter(maxmp=10,maxds=257,maxdir=32)
      dimension fr(*),fi(*),s(maxds),ir(maxdir)
c
c To accomodate a larger vector f, insert a larger value of maxmp and
c corresponding values of maxds=2**(maxmp-2)+1 and
c maxdir=2**[(maxmp+1)/2)] in the above parameter statement.
c
      data lastmp,maxs/0,0/
      isws=isw**2
      if((mp.lt.0).or.(mp.gt.maxmp).or.(isws.eq.0).or.(isws.gt.9)) then
```

```
          write(*,2)
    2     format(' Improper argument in FFT Subroutine')
          stop
          endif
       if(mp.eq.0) return
       if(mp.eq.lastmp) then
             if(mp.eq.1) go to 20
             if(mp.gt.1) go to 14
          endif
c
c At the time of the first call of the subroutine with a given value of
c mp.gt.0, initialize various constants and arrays that depend on mp
c but not fr,fi.  On subsequent calls with the same mp, bypass this
c initialization process.
c
       lastmp=mp
       n      =2**mp
       temp   =n
       rn     =one/temp
       rrootn=dsqrt(rn)
       nh     =n/2
       if(mp.eq.1) go to 20
       if(mp.eq.2) go to 8
       if(mp.lt.maxs) then
             nsn=2**(maxs-mp)
          else
             nsn=1
          endif
c
c When mp.le.maxs, the spacing parameter nsn is used to retreive sine
c values from a previously computed table.
c
       if(mp.le.maxs) go to 8
c
c When n=8,16,32,...  and mp.gt.maxs, precompute
c s(j+1)=sin((pi/2)*(j/nq)), j=0,1,2,...,nq, for use in
c the subsequent calculations.
c
       nq     =n/4
       ne     =n/8
       rroot2 =one/dsqrt(two)
       maxs   =mp
       s(1)   =zero
c                =sin(0*pi/4)
       s(ne+1)=rroot2
c                =sin(1*pi/4))
```

```
         s(nq+1)=one
c                =sin(2*pi/4)
         if(mp.eq.3)go to 8
            h=rroot2
c             =.5*sec(pi/4)
c
c Pass from the course grid to a finer one by using the trig identity:
c
c                 sin(a)=(.5*sec(b))*(sin(a-b)+sin(a+b)).
c
c (This clever idea is due to O.Buneman, cf.  SIAM J. SCI. STAT.
c COMPUT. 7 (1986), pp.  624-638.)
c
         k=ne
         do 6 i=4,mp
            h =one/dsqrt(two+one/h)
c               =half secant of half the previous angle
            kt2=k
            k =k/2
c               =n/2**i
            do 4 j=k,nq,kt2
    4          s(j+1)=h*(s(j-k+1)+s(j+k+1))
    6       continue
    8 continue
c
c Prepare a short table of bit reversed integers to use in the
c subsequent bit reversal permutation.
c
         muplus=(mp+1)/2
         m     =1
         ir(1) =0
         do 12 nu=1,muplus
            do 10 k=1,m
               it        =2*ir(k)
               ir(k)   =it
   10          ir(k+m)=it+1
   12       m=m+m
c
c If mp is odd, then m=m/2.
c
         itemp=2*(mp/2)
         if(itemp.ne.mp) then
               m=m/2
            endif
c
c The parameters n,nh,nq,ne,m,lastmp,maxs,nsn,rn,rrootn,rroot2 and the
```

```
c arrays ir(*), s(*) are now suitably initialized.
c
   14 continue
c Apply the bit reversal permutation to the array f using the
c Bracewell-Buneman scheme.
c
      do 18 iq=1,m-1
         npr   =iq-m
         irpp =ir(iq+1)*m
         do 16 ip=0,ir(iq+1)-1
            npr      =npr+m
            npr1     =npr+1
            irp1     =irpp+ir(ip+1)+1
            tempr    =fr(npr1)
            tempi    =fi(npr1)
            fr(npr1) =fr(irp1)
            fi(npr1) =fi(irp1)
            fr(irp1) =tempr
   16       fi(irp1) =tempi
   18 continue
   20 continue
c
c Apply the mp Q-matrices.
c
c Carry out stage 1 of the FFT using blocks of size 2x2 and apply
c the desired scale factor.
c
      if(isws.eq.1) then
            scale=one
         elseif(isws.eq.4) then
            scale=rn
         else
            scale=rrootn
         endif
      do 22 k=0,nh-1
         k1    =2*k+1
         k2    =k1+1
         tempr =(fr(k1)-fr(k2))*scale
         tempi =(fi(k1)-fi(k2))*scale
         fr(k1)=(fr(k1)+fr(k2))*scale
         fi(k1)=(fi(k1)+fi(k2))*scale
         fr(k2)=tempr
   22    fi(k2)=tempi
      if(mp.eq.1) return
c
c Carry out stages 2,3,...,mp of the FFT using blocks
```

```
c of size mxm = 4x4,8x8,...,nxn.
c
      mcap=1
      kcap=n/4
      do 32 mu=2,mp
c
c At this point mcap=2**(mu-2) and kcap=2**(mp-mu).
c
c Deal first with the quadruplet of components where sin=0 or cos=0.
c
      do 24 k=0,kcap-1
         k0=k*4*mcap+1
         k1=k0+mcap
         k2=k0+2*mcap
         k3=k0+3*mcap
         tempr =fr(k0)-fr(k2)
         tempi =fi(k0)-fi(k2)
         fr(k0)=fr(k0)+fr(k2)
         fi(k0)=fi(k0)+fi(k2)
         fr(k2)=tempr
         fi(k2)=tempi
         if(isw.lt.0) then
                fr(k3)=-fr(k3)
                fi(k3)=-fi(k3)
            endif
         temp1 =fr(k1)+fi(k3)
         temp2 =fi(k1)-fr(k3)
         fr(k1)=fr(k1)-fi(k3)
         fi(k1)=fi(k1)+fr(k3)
         fr(k3)=temp1
         fi(k3)=temp2
   24 continue
      if(mcap.eq.1) go to 30
c
c Now deal with the remaining mcap-1 quadruplets of components where sin,
c cos are both nonzero.
c
```

```
      do 28 lamda=1,mcap-1
        indx =nsn*lamda*kcap+1
        sn   =s(indx)
c            =sin((2*pi)*(lamda/mcap))
        indx =nq-indx+2
        cs   =s(indx)
c            =cos((2*pi)*(lamda/mcap))
        if(isw.lt.0) then
            sn=-sn
        endif
        do 26 k=0,kcap-1
          k4m=k*4*mcap+1
          k0 =k4m+lamda
          k1 =k4m+2*mcap-lamda
          k2 =k4m+2*mcap+lamda
          k3 =k4m+4*mcap-lamda
          r1    =cs*fr(k2)-sn*fi(k2)
          r2    =cs*fi(k2)+sn*fr(k2)
          temp1 =fr(k0)-r1
          temp2 =fi(k0)-r2
          fr(k0)=fr(k0)+r1
          fi(k0)=fi(k0)+r2
          fr(k2)=temp1
          fi(k2)=temp2
          r1    =cs*fr(k3)+sn*fi(k3)
          r2    =cs*fi(k3)-sn*fr(k3)
          temp1 =fr(k1)+r1
          temp2 =fi(k1)+r2
          fr(k1)=fr(k1)-r1
          fi(k1)=fi(k1)-r2
          fr(k3)=temp1
26        fi(k3)=temp2
28    continue
30    mcap=mcap*2
      kcap=kcap/2
32 continue
   return
   end
```

Appendix 6: The Standard Normal Probability Distribution

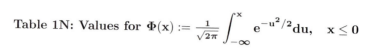

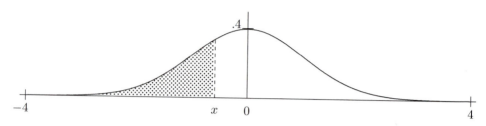

x	9	8	7	6	5	4	3	2	1	0
-3.9	.0000	.0000	.0000	.0000	.0000	.0000	.0000	.0000	.0000	.0000
-3.8	.0001	.0001	.0001	.0001	.0001	.0001	.0001	.0001	.0001	.0001
-3.7	.0001	.0001	.0001	.0001	.0001	.0001	.0001	.0001	.0001	.0001
-3.6	.0001	.0001	.0001	.0001	.0001	.0001	.0001	.0001	.0002	.0002
-3.5	.0002	.0002	.0002	.0002	.0002	.0002	.0002	.0002	.0002	.0002
-3.4	.0002	.0003	.0003	.0003	.0003	.0003	.0003	.0003	.0003	.0003
-3.3	.0003	.0004	.0004	.0004	.0004	.0004	.0004	.0005	.0005	.0005
-3.2	.0005	.0005	.0005	.0006	.0006	.0006	.0006	.0006	.0007	.0007
-3.1	.0007	.0007	.0008	.0008	.0008	.0008	.0009	.0009	.0009	.0010
-3.0	.0010	.0010	.0011	.0011	.0011	.0012	.0012	.0013	.0013	.0013
-2.9	.0014	.0014	.0015	.0015	.0016	.0016	.0017	.0018	.0018	.0019
-2.8	.0019	.0020	.0021	.0021	.0022	.0023	.0023	.0024	.0025	.0026
-2.7	.0026	.0027	.0028	.0029	.0030	.0031	.0032	.0033	.0034	.0035
-2.6	.0036	.0037	.0038	.0039	.0040	.0041	.0043	.0044	.0045	.0047
-2.5	.0048	.0049	.0051	.0052	.0054	.0055	.0057	.0059	.0060	.0062
-2.4	.0064	.0066	.0068	.0069	.0071	.0073	.0075	.0078	.0080	.0082
-2.3	.0084	.0087	.0089	.0091	.0094	.0096	.0099	.0102	.0104	.0107
-2.2	.0110	.0113	.0116	.0119	.0122	.0125	.0129	.0132	.0136	.0139
-2.1	.0143	.0146	.0150	.0154	.0158	.0162	.0166	.0170	.0174	.0179
-2.0	.0183	.0188	.0192	.0197	.0202	.0207	.0212	.0217	.0222	.0228
-1.9	.0233	.0239	.0244	.0250	.0256	.0262	.0268	.0274	.0281	.0287
-1.8	.0294	.0301	.0307	.0314	.0322	.0329	.0336	.0344	.0351	.0359
-1.7	.0367	.0375	.0384	.0392	.0401	.0409	.0418	.0427	.0436	.0446
-1.6	.0455	.0465	.0475	.0485	.0495	.0505	.0516	.0526	.0537	.0548
-1.5	.0559	.0571	.0582	.0594	.0606	.0618	.0630	.0643	.0655	.0668
-1.4	.0681	.0694	.0708	.0721	.0735	.0749	.0764	.0778	.0793	.0808
-1.3	.0823	.0838	.0853	.0869	.0885	.0901	.0918	.0934	.0951	.0968
-1.2	.0985	.1003	.1020	.1038	.1056	.1075	.1093	.1112	.1131	.1151
-1.1	.1170	.1190	.1210	.1230	.1251	.1271	.1292	.1314	.1335	.1357
-1.0	.1379	.1401	.1423	.1446	.1469	.1492	.1515	.1539	.1562	.1587
-.9	.1611	.1635	.1660	.1685	.1711	.1736	.1762	.1788	.1814	.1841
-.8	.1867	.1894	.1922	.1949	.1977	.2005	.2033	.2061	.2090	.2119
-.7	.2148	.2177	.2206	.2236	.2266	.2296	.2327	.2358	.2389	.2420
-.6	.2451	.2483	.2514	.2546	.2578	.2611	.2643	.2676	.2709	.2743
-.5	.2776	.2810	.2843	.2877	.2912	.2946	.2981	.3015	.3050	.3085
-.4	.3121	.3156	.3192	.3228	.3264	.3300	.3336	.3372	.3409	.3446
-.3	.3483	.3520	.3557	.3594	.3632	.3669	.3707	.3745	.3783	.3821
-.2	.3859	.3897	.3936	.3974	.4013	.4052	.4090	.4129	.4168	.4207
-.1	.4247	.4286	.4325	.4364	.4404	.4443	.4483	.4522	.4562	.4602
-.0	.4641	.4681	.4721	.4761	.4801	.4840	.4880	.4920	.4960	.5000

Table 1P: Values for $\Phi(x) := \frac{1}{\sqrt{2\pi}} \int_{-\infty}^{x} e^{-u^2/2} du, \quad x \geq 0$

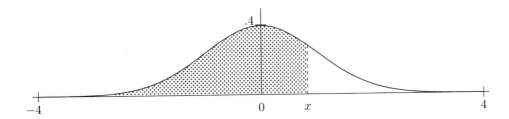

x	0	1	2	3	4	5	6	7	8	9
.0	.5000	.5040	.5080	.5120	.5160	.5199	.5239	.5279	.5319	.5359
.1	.5398	.5438	.5478	.5517	.5557	.5596	.5636	.5675	.5714	.5753
.2	.5793	.5832	.5871	.5910	.5948	.5987	.6026	.6064	.6103	.6141
.3	.6179	.6217	.6255	.6293	.6331	.6368	.6406	.6443	.6480	.6517
.4	.6554	.6591	.6628	.6664	.6700	.6736	.6772	.6808	.6844	.6879
.5	.6915	.6950	.6985	.7019	.7054	.7088	.7123	.7157	.7190	.7224
.6	.7257	.7291	.7324	.7357	.7389	.7422	.7454	.7486	.7517	.7549
.7	.7580	.7611	.7642	.7673	.7704	.7734	.7764	.7794	.7823	.7852
.8	.7881	.7910	.7939	.7967	.7995	.8023	.8051	.8078	.8106	.8133
.9	.8159	.8186	.8212	.8238	.8264	.8289	.8315	.8340	.8365	.8389
1.0	.8413	.8438	.8461	.8485	.8508	.8531	.8554	.8577	.8599	.8621
1.1	.8643	.8665	.8686	.8708	.8729	.8749	.8770	.8790	.8810	.8830
1.2	.8849	.8869	.8888	.8907	.8925	.8944	.8962	.8980	.8997	.9015
1.3	.9032	.9049	.9066	.9082	.9099	.9115	.9131	.9147	.9162	.9177
1.4	.9192	.9207	.9222	.9236	.9251	.9265	.9279	.9292	.9306	.9319
1.5	.9332	.9345	.9357	.9370	.9382	.9394	.9406	.9418	.9429	.9441
1.6	.9452	.9463	.9474	.9484	.9495	.9505	.9515	.9525	.9535	.9545
1.7	.9554	.9564	.9573	.9582	.9591	.9599	.9608	.9616	.9625	.9633
1.8	.9641	.9649	.9656	.9664	.9671	.9678	.9686	.9693	.9699	.9706
1.9	.9713	.9719	.9726	.9732	.9738	.9744	.9750	.9756	.9761	.9767
2.0	.9772	.9778	.9783	.9788	.9793	.9798	.9803	.9808	.9812	.9817
2.1	.9821	.9826	.9830	.9834	.9838	.9842	.9846	.9850	.9854	.9857
2.2	.9861	.9864	.9868	.9871	.9875	.9878	.9881	.9884	.9887	.9890
2.3	.9893	.9896	.9898	.9901	.9904	.9906	.9909	.9911	.9913	.9916
2.4	.9918	.9920	.9922	.9925	.9927	.9929	.9931	.9932	.9934	.9936
2.5	.9938	.9940	.9941	.9943	.9945	.9946	.9948	.9949	.9951	.9952
2.6	.9953	.9955	.9956	.9957	.9959	.9960	.9961	.9962	.9963	.9964
2.7	.9965	.9966	.9967	.9968	.9969	.9970	.9971	.9972	.9973	.9974
2.8	.9974	.9975	.9976	.9977	.9977	.9978	.9979	.9979	.9980	.9981
2.9	.9981	.9982	.9982	.9983	.9984	.9984	.9985	.9985	.9986	.9986
3.0	.9987	.9987	.9987	.9988	.9988	.9989	.9989	.9989	.9990	.9990
3.1	.9990	.9991	.9991	.9991	.9992	.9992	.9992	.9992	.9993	.9993
3.2	.9993	.9993	.9994	.9994	.9994	.9994	.9994	.9995	.9995	.9995
3.3	.9995	.9995	.9995	.9996	.9996	.9996	.9996	.9996	.9996	.9997
3.4	.9997	.9997	.9997	.9997	.9997	.9997	.9997	.9997	.9997	.9998
3.5	.9998	.9998	.9998	.9998	.9998	.9998	.9998	.9998	.9998	.9998
3.6	.9998	.9998	.9999	.9999	.9999	.9999	.9999	.9999	.9999	.9999
3.7	.9999	.9999	.9999	.9999	.9999	.9999	.9999	.9999	.9999	.9999
3.8	.9999	.9999	.9999	.9999	.9999	.9999	.9999	.9999	.9999	.9999
3.9	1.0000	1.0000	1.0000	1.0000	1.0000	1.0000	1.0000	1.0000	1.0000	1.0000

Table 2: Critical values for $\Phi(x) := \frac{1}{\sqrt{2\pi}} \int_{-\infty}^{x} e^{-u^2/2} du$

x	$\Phi(x)$	x	$\Phi(x)$	x	$\Phi(x)$	x	$\Phi(x)$
-3.719	.0001	-0.706	.24	0.025	.51	0.772	.78
-3.090	.001	-0.674	.25	0.050	.52	0.806	.79
-2.576	.005	-0.643	.26	0.075	.53	0.842	.80
-2.326	.01	-0.613	.27	0.100	.54	0.878	.81
-2.054	.02	-0.583	.28	0.126	.55	0.915	.82
-1.960	.025	-0.553	.29	0.151	.56	0.954	.83
-1.881	.03	-0.524	.30	0.176	.57	0.994	.84
-1.751	.04	-0.496	.31	0.202	.58	1.036	.85
-1.645	.05	-0.468	.32	0.228	.59	1.080	.86
-1.555	.06	-0.440	.33	0.253	.60	1.126	.87
-1.476	.07	-0.412	.34	0.279	.61	1.175	.88
-1.405	.08	-0.385	.35	0.305	.62	1.227	.89
-1.341	.09	-0.358	.36	0.332	.63	1.282	.90
-1.282	.10	-0.332	.37	0.358	.64	1.341	.91
-1.227	.11	-0.305	.38	0.385	.65	1.405	.92
-1.175	.12	-0.279	.39	0.412	.66	1.476	.93
-1.126	.13	-0.253	.40	0.440	.67	1.555	.94
-1.080	.14	-0.228	.41	0.468	.68	1.645	.95
-1.036	.15	-0.202	.42	0.496	.69	1.751	.96
-0.994	.16	-0.176	.43	0.524	.70	1.881	.97
-0.954	.17	-0.151	.44	0.553	.71	1.960	.975
-0.915	.18	-0.126	.45	0.583	.72	2.054	.98
-0.878	.19	-0.100	.46	0.613	.73	2.326	.99
-0.842	.20	-0.075	.47	0.643	.74	2.576	.995
-0.806	.21	-0.050	.48	0.674	.75	3.090	.999
-0.772	.22	-0.025	.49	0.706	.76	3.290	.9995
-0.739	.23	0.000	.50	0.739	.77	3.719	.9999

Approximation for $\Phi(\mathbf{x}) := \frac{1}{\sqrt{2\pi}} \int_{-\infty}^{\mathbf{x}} e^{-u^2/2} d\mathbf{u}$

$$\Phi(x) = \frac{1}{2}\left\{1 + \mathrm{erf}\left(\frac{x}{\sqrt{2}}\right)\right\}$$

$$\mathrm{erf}(x) := \frac{2}{\sqrt{\pi}} \int_0^x e^{-u^2} du$$

$$\approx \mathrm{sgn}(x) \cdot \left\{1 - \frac{P_0 + P_1|x| + P_2|x|^2}{Q_0 + Q_1|x| + Q_2|x|^2 + |x|^3} \cdot e^{-x^2}\right\}$$

(to 6 decimals when $-10 \le x \le 10$)

$P_0 = 3.53221659121$	$Q_0 = 3.53221617055$
$P_1 = 2.153997718487$	$Q_1 = 6.139719523819$
$P_2 = 0.57404837033548$	$Q_2 = 3.9690911552337$

J.F. Hart, *et al.*, *Computer Approximations*, John Wiley & Sons, New York, 1968, pp. 136–140, 290.

Appendix 7: Frequencies of the Piano Keyboard

The standard piano keyboard has 88 keys with A4 tuned to 440 Hz and with the uniform semitone ratio $2^{1/12} = 1.05946\ldots$.   The 7+ octave range (27.5 to 4186 Hz) includes the 5+ octave range (60 to 2000 Hz) of the human voice, and is included in the 10- octave range (20 to 20000 Hz) of the human ear.

A0	27.50		A$^\#$0	29.14
B0	30.87			
C1	32.70		C$^\#$1	34.85
D1	36.71		D$^\#$1	38.89
E1	41.20			
F1	43.65		F$^\#$1	46.25
G1	49.00		G$^\#$1	51.91
A1	55.00		A$^\#$1	58.27
B1	61.74			
C2	65.41		C$^\#$2	69.30
D2	73.42		D$^\#$2	77.78
E2	82.41			
F2	87.31		F$^\#$2	92.50
G2	98.00		G$^\#$2	103.8
A2	110.0		A$^\#$2	116.5
B2	123.5			
C3	130.8		C$^\#$3	138.6
D3	146.8		D$^\#$3	155.6
E3	164.8			
F3	174.6		F$^\#$3	185.0
G3	196.0		G$^\#$3	207.6
A3	220.0		A$^\#$3	233.1
B3	246.9			
C4	261.6		C$^\#$4	277.2
D4	293.7		D$^\#$4	311.1
E4	329.6			
F4	349.2		F$^\#$4	370.0
G4	392.0		G$^\#$4	415.3
A4	440.0		A$^\#$4	466.2
B4	493.9			
C5	523.2		C$^\#$5	554.4
D5	587.3		D$^\#$5	622.2
E5	659.3			
F5	698.5		F$^\#$5	740.0
G5	784.0		G$^\#$5	830.6
A5	880.0		A$^\#$5	932.3
B5	987.8			
C6	1046.		C$^\#$6	1109.
D6	1175.		D$^\#$6	1245.
E6	1319.			
F6	1397.		F$^\#$6	1480.
G6	1568.		G$^\#$6	1661.
A6	1760.		A$^\#$6	1865.
B6	1976.			
C7	2093.		C$^\#$7	2217.
D7	2349.		D$^\#$7	2489.
E7	2637.			
F7	2794.		F$^\#$7	2960.
G7	3136.		G$^\#$7	3322.
A7	3520.		A$^\#$7	3729.
B7	3951.			
C8	4186.			

Index

Abel-Dirichlet argument, 43
absolutely integrable, 48–50, 57, 81–82, 90, 122, 126–127, 164, 169–171, 179
absolutely summable, 38, 40–41, 127, 210
additive synthesis, *ix*
  of solution to PDE, 15, 524
  of tone, 707–711, 730–732
alias, 7, 486, 495, 506, 659, 717, 730
Almagast, 12, 699
almost bandlimited, 505
almost periodic, 88
alternating Ωip, 609, 629, 665, 690
analysis equation
  for almost periodic f, 88
  for Dirac δ, 429
  of Fourier, 3, 6–7, 10
  for periodic generalized f, 446
  using sin, cos, 12–13, 66–68, 247
  for wavelets, 600, 640
antiderivative
  of generalized function, 390
  of probability density, 379
antiderivative rule, 164
arginine spectrum, 22, 165
arrow notation for δ, 381
audio signal, *xiii*, 483–484, 509, 593, 602, 672, 693–694
autocorrelation, 109, 162, 238
automobile suspension, 418
averaging operator, 72

B-spline, 117, 164, 171, 193, 462, 794
Bach, J.S., 725, 735
backward difference, 230 (*see also* forward difference*)
bandlimited function, 426
  almost, 505
  bound for derivatives, 517
  convolution of, 511
  discrete, 512, 515
  for ear, 485, 728
  for eye, 510
  limit of sequence, 511
  Maclaurin series for, 486, 517
  multiplication by, 477, 487
  recovery from samples, 489, 491, 497, 511
  recovery from filtered samples, 499, 503
  series for derivative, 515
  structure of, 427, 485
bar operator, 251
  tag, 252
Beethoven sonata, 693
bell tone, 709, 716, 732
Benford's density for first digits, 798
Bernoulli discretized string, 580

Bernoulli functions, 41–42, 177, 182
  for creating comb, 432
  discrete, 230
  and Eagle's method, 184, 479
  for Euler-Maclaurin formula, 212, 225
  for evaluating sums, 221
  properties, 43–44, 182–183, 212, 225
  for synthesis of Fourier series, 184
Bernstein's bound, 516
Bessel functions, 224, 712–714, 760
Bessel's inequality, 26, 77
biorthoganality constraint, 666, 690
bit reversal permutation, 303, 354
  algorithms, 305–307, 310
  via even-odd sorts, 303, 314–315, 351
  via Kronecker products, 342, 365
  operation count, 354
  via perfect shuffles, 314–315, 342
Bluestein's chirp FFT, 358
Bochner, S., 59, A1
  analysis of Gibbs phenomenon, 80
  characterization of densities, 749
Bohr, H., 88
Born and Wolf, 553, 573, 587
boundary conditions for PDE, 535, 545, 550–552, 569, 585
box function, 130
Bracewell, R., 115
  bit reversal, 307–310
  FHT, 323–326, A1
bunched samples, 519
Buneman's tricks, 308, 319–320

$\mathbb{C}$ **for complex numbers**, 41, 165
cardinal series, 491, 513
  basis functions, 141, 160, 493, 514–515
  convergence, 492–493
  generalization, 497, 515
Carlson's theorem, 57
carrier frequency, 711
cascade algorithm, 682
Cauchy-Schwartz inequality, 491, 507, 745, 756, 762
ceiling function $\lceil\ \rceil$, 229, 683
central limit theorem, 771–775
  for probability densities on $\mathbb{P}_N$ 796
  for probability densities on $\mathbb{T}_p$, 797
centroid, 74, 756
characteristic functions, 741
  Bochner's characterization, 749
  boundedness, 748
  for computing moments, 757
  continuity, 748
  convention for 2π, 741
  from expectation integrals, 759
  periodic, 752
  for probability densities on $\mathbb{P}_N$ 788

characteristic functions (*cont.*)
  for probability densities on $\mathbb{T}_p$, 797
  products of, 750
  smoothness, 786
  for sums of random variables, 765, 767
chirp signal, 109, 487, 706, 797
  for Bluestein's FFT, 358
Chowning, J., *xiii*, 711, 716
Chowning's rule, 713–717, 732
circulant matrix, 122
closure
  of $\mathbb{G}$, 390, 396, 459
  of $\mathbb{S}$, 375–376, 454
  of probability densities, 787
  of symmetries, 277
comb function on $\mathbb{P}_N$ 197
comb function III on $\mathbb{R}$, 383, 393, 432, 437, 448
commensurate, 88
commuting diagram, 37, 205, 256
commuting operators, 122
complex exponentials
  on $\mathbb{R}$, $\mathbb{T}p$, $\mathbb{Z}$, $\mathbb{P}_N$ 2, 4, 6, 9
  as eigenfunctions of LTI systems, 18
compression of signal
  via filter bank, 668–670
  via interpolation, 71, 234
concatenation of vectors, 351, 352, 607–608, 671, 675
Concerning Harmonics, 699
conjugation, 62
  operator, 251
  tag, 292
conjugation rule, 136, 177, 199, 413
conservation of energy, 24, 531, 544, 565
convention for 2π, 63, 741
convolution, 89–90
  algebraic properties, 103, 120, 122, 126–127, 233
  of bandlimited functions, 511
  and correlation, 91
  differentiation, 106, 121, 126, 401
  via FFT, 113, 357
  via FHT, 265
  as filtering, 497
  of generalized functions, 398, 400, 461
  via generating function, 101
  identity, 104, 118–120, 168, 196, 400, 408 (*see also* Dirac delta*)
  indirect, 103, 149, 158
  and LTI systems, 105, 282, 470, 481
  of ordinary functions, 89–90
  of periodic generalized functions, 447
  of probability densities, 110, 123, 150, 751
  via sliding strip, 94
  as smearing, 107

convolution (*cont.*)
and smoothness, 107, 123, 126, 128, 401
for solving ODEs, 145, 470, 481
for solving PDEs, 528, 534, 542, 548, 559, 567
square root, 105, 232
support, 121
translation, 120, 162
convolution rule, 103, 143, 170, 177, 199, 257, 286, 413
for Hartley transforms, 265, 286
for Hilbert transforms, 268, 289
Cooley-Tukey, 295, 362
correlation, 90, 109, 162
cos operator, 246, 287
cos transform, 63, 66, 247, 287
cosine signal, 511
CSG function, 376

dagger operator, 251
tag, 252
d'Alembert, 523
d'Alembert formula, 528, 574, 575
Daubechies, I., *xiii*, 610, 614, 631
Daubechies wavelets, 631, 639
decay of Fourier transform, 153, 170, 193, 376, 786
decibel, 695
decimation (downsampling) rule, 188, 204, 258
deconvolution, 108, 679
definition chasing, 391, 471
delta function on $\mathbb{P}_N$ 196
density (*see* probability density)
derivative operator, 121, 251, 285
tag, 251
derivative rule
for convolution, 106, 401
for Fourier transform, 141, 169, 413
for generalized functions, 377, 405, 413
for Hartley transform, 265, 285
for Hilbert transform, 269
of Leibnitz, 107
detail for wavelet approximation, 600, 640
computation of, 606, 608, 649, 652
DFT, 11, 291
eigenvalues, 279
fast algorithms (*see* FFT)
via geometric progression, 198, 227
via Horner's algorithm, 293
via paper strips, 294, 349
via Poisson's relations, 209
for problems of harmonic analysis, 348
of real vector, 276, 352
via rules, 199–208
via summation, 196
table, A11
DHT, 249
fast algorithms (*see* FHT)
dice, 110, 123, 793
loaded, 796
tetrahedral, 780
differential equation, 163, 281, 408
via convolution, 470
for finding Fourier transform, 132, 164, 420–425

differential equation (*cont.*)
homogeneous, 408, 416, 460
inhomogeneous, 146, 367, 408, 410–412, 417, 419, 461, 470, 477
differentiation of
generalized function, 377, 405
piecewise smooth function, 406–408, 525
diffraction, 553
from edge, 588
of gaussian beam, *ii*, 561–564
from periodic source, 568–571
from slits, 560–561, 588
diffraction equation, 524
with boundary conditions, 569
conservation of energy, 565
derivation, 553–558, 587
Fourier synthesis, 524, 567, 571–572
Fraunhofer approximation, 565, 569
initial condition, 557, 559, 566, 587
kernel, 466, 559, 566, 586
no extreme value principle, 590
symmetry, 590
diffusion equation, 15, 524
with boundary conditions, 545, 550, 551, 552, 585
conservation of energy, 544
derivation, 540
extreme value principle, 545
forced, 546, 585–586
Fourier synthesis, 524, 549, 571–572
initial condition, 541, 548, 587
kernel, 542, 548, 586
polynomial solutions, 544, 584
smoothness of solutions, 542, 584
symmetry, 584
digitized sound file, *xiii*, 483–484, 508–509, 593, 602, 672, 693–694
dilate of wavelet, 594
dilation equation, 163, 281, 598, 609
and multiresolution analysis, 599, 645–646, 650
solution via infinite product, 615, 677
symmetry, 678
uniqueness of solution, 616, 677
dilation operator, 258, 283, 656, 659
dilation rule, 138, 140, 187, 205–207, 236, 257, 413
dimensional analysis, 69
dipoles, 429
Dirac delta, 371, 380, 392–393
approximations, 28, 77–78, 168, 474
dilation property, 392
discrete, 196
and Eagle's method, 421–425, 479
as identity for convolution product, 400
not an ordinary function, 168, 453
sifting property, 392
as solution of dilation equation, 610
Dirichlet conditions, 57, 77
Dirichlet kernel, 66, 187, 474
discrete function (*see also* function on $\mathbb{Z}$)
Fourier transform (*see* DFT)
wavelet transform, 594
discretization (sampling), 32, 483–484
dispersion
of heat, 543
of light waves, 566
of water waves, 591

distribution (tempered), *x*, 368, 451
distribution function, 740
via limits, 784
for max, min of random variables, 790
for standard normal density, 755, A32
dual rule, 260, 280
duality for Fourier analysis and synthesis, 8, 66
DuBois-Reymond, 57
DWT, 607, 675
via herringbone, 608, 646
operation count, 608, 692
using operators, 646
in place, 676
dyadic dilate, 594

Eagle's method, 184, 422, 479
eavesdrop, 512, 520
echo location, 109
eigenvalues, 279–280, 282
of $\mathcal{F}$, 151, 167, 212
of Kronecker product, 363
of LTI system, 18, 282
end padding operator, 234, 284, 688
equidistribution of arithmetic sequence, 194
error function, 582, A35
errors
for computation of ω^k, 357
for computing sum with round off, 793
for fast arithmetic, 124
for FFT, 301
for frame, detail coefficients, 649–655, 685–686
for least squares, 75–76, 84
for sampling theorems, 492, 497, 499, 507
Euclidean algorithm for gcd, 227
even
function, 62, 64, 247
generalized function, 397, 455
projection, 245
expectation integral, 720, 738, 755–761, 764
for independent random variables, 764
for spectral density, 720
Euler
gamma function, 164
identity for sin, cos, 1, 67
-Maclaurin sum formula, 213
exponent notation, 313
via Kronecker products, 339
exponential operators, 245
extreme value principle, 545

factorial powers, 440
factorization
of convolution operator, 257
of DFT matrix, 314, 329, 338
of DHT matrix, 324, 359
of $\mathbf{L}_\pm$, $\mathbf{H}_\pm$ from filter bank, 656
of Q for filter bank, 666
fast arithmetic, 113–114, 124
fast convolution, 113, 357
FBI filters, 672, 691

Fejer kernel, 78, 474
Fejer example of divergent Fourier
    series, 57, 85
Fermat theorem, 236
FFT
    Bluestein's scheme, 358
    Cooley-Tukey, 295, 349
    decimation in frequency, 299, 318,
        338
    decimation in time, 296, 316, 322,
        331
    via DFT rules, 296, 299, 353
    FORTRAN code, A26
    for frames of movies, 572
    Gauss discovery, 70, 295, 360
    impact, 294–295, A1
    via Kronecker product, 344, 365–366
    via Mason Ωow diagram, 355
    via matrix factorization, 310, 329,
        356
    operations, 295, 323, 332, 353, 358
    in place, 311
    with precomputed sines, 318, 322
    via recursive algorithm, 301, 350,
        351
    via segments, 352
    for spectral factorization, 637
    for spectrogram, 705
    Stockham's autosort, 344, 366
    three loop algorithm, 313, 318, 322
    two loop algorithm, 365
    via zipper identity, 312, 314, 328
FHT
    advantages, 325, 327
    patent, 325, A1
    three loop algorithm, 326
    via zipper identity, 324, 356, 359
filter
    via convolution, 497
    FBI, 672, 691
    for filter bank, 659
    high-, low-pass, 281, 511, 659
    for sampling, 497, 519
    for shaping noise, 721
    translation, 663
filter bank, 655
    Fourier analysis, 658
    perfect reconstruction, 661, 689
    using up, down sampling, 656
Fletcher-Munsen contours, 695–696
Ωoor function ⌊ ⌋, 124, 229, 383, 467,
    683, 703, 717, 731
FM synthesis of tone, 711–717
    parallel and cascade, 734
forward difference, 118, 440, 480
Fourier
    analysis and synthesis, 3–11, 15, 73,
        86
        validity, 37–58
    big pixel image, 510
    and dimensional analysis, 69
    and heat conduction, 15, 72, 541
    impact of work, A1–A2
    quote, *iii*
    sketch, *xv*, 134
    spoken word, 483, 509
    FOURIER, *ii*, *xiv*, *xv* (check author's
        web site for additional details)
Fourier coefficient, 5, 10, 441, 446
    rate of decay, 193, 441

Fourier-Poisson cube, 31, 36–37, 205
Fourier series, 5, 173
    via Bernoulli functions, 184, 218, 437
    convergence, 39–48, 75, 77
    via differentiation, 422, 479
    for generalized functions, 440–441
    via integration, 174
    via Laurent series, 185
    via Poisson's formula, 179, 478
    via Riemann sum, 79
    via rules, 176–179, 191–192
    to solve PDEs, 532, 535, 538, 549,
        567
    uniqueness, 30, 78
    weak convergence, 58, 433, 441
Fourier transform ($\mathbb{R}$, $\mathbb{T}_p$, $\mathbb{Z}$, $\mathbb{P}_N$), 3, 6,
    7, 11
    rules, A13–A17
    tables, A1–A12
Fourier transform ($\mathbb{R}$)
    calculus, 129, 114, 146
    decay at infinity, 153, 170
    via differentiation, 132, 421–425
    of generalized function, 413
    via integration, 129–131
    of periodic functions, 444
    of probability density, 746–752
    via rules, 134–147, 413
    smoothness, 48, 81, 153, 169, 786
    table, A1–A6
Fourier transform operator, 240, 243
    tag, 251
Fourier transform of operator, 255, 259
fractional derivatives, 154
fragmentation of Π, 495, 518
frame for wavelet approximation, 598,
    640
    computation, 606, 608, 649, 652
    illustrations, 601, 603, 650, 669
Fraunhofer approximation, 565, 569,
    588–589
frequency
    carrier-modulation, 711
    of concert A, 700
    function for tune, 723
    local, 706, 730
    for piano keyboard, A36
    and pitch, 694
    via spectrogram, 705
    units for, 73
    of vibrating string, 536
    via wavelet coefficients, 596
Fresnel
    approximation, 555
    convolution equation, 558
    function, 165, 420
        discrete, 214
    integrals, 165, 215, 420, 466
FT-NMR spectrum, 22, 165, A2
function
    on $\mathbb{R}$, $\mathbb{T}_p$, $\mathbb{Z}$, $\mathbb{P}_N$ 3–5, 8
    bandlimited, 426
    CSG, 376
    entire, 517
    frequency, 724
    generalized, 58, 368, 378–379, 524
    locally integrable, 389, 467
    of operator, 244, 272, 277–278, 281,
        283
    probability density, 739

function (*cont.*)
    of random variable, 758–760, 789
    Schwartz, 372–374
    slowly growing, 376
    support-limited, 426
functional, 369
    continuous, 451
    fundamental, 370, 377–378
    integral notation, 372, 378
    linear, 390, 451
FWT
    via coefficients, 606
    via matrix factorization, 675
    via operators, 645–646
    in place, 676

$\mathbb{G}$ **for generalized functions**, 390
Gauss
    asteroid orbit, 14, 70
    discovery of FFT, 71, 295, 360
    interpolation, 8, 14, 360
    law of errors, 771, 794
    Poisson sum formula, 179
    signature, 669–670
    sums, 215
gaussian function, 132, 741
    for mollification and tapering, 743
gaussian laser beam, 561
    interfering, 564
    pointing, 562
    spreading, 562
gcd, 206, 227
generalized function, 367–372, 378
    bandlimited, 426
    closure, 390, 459
    "continuity", 431
    convolution, 398–399
    via CSG functions, 378
    division, 402–405, 459
    Fourier transform rules, 413
    as functional, 369–372, 376, 382, 384,
        388, 451
    integral notation, 371, 378, 390–391
    via limit of Schwartz functions, 450,
        482
    limits, 427–439
    via locally integrable functions, 389,
        467
    multiplication, 398–399, 402
    as ordinary function, 58, 376, 379
    partial derivatives, 438–439, 525
    periodic, 440–448
    as probability densities, 739
    as scaling function, 616, 623
    as solution to PDE, 524–525, 528,
        541, 559, 587
    special structure, 408–410, 427, 441,
        459–464
    table of Fourier transforms, A3–A5
    transformations, 389–405, 458
    "values", 378, 382
generating function for
    Bernoulli polynomials, 225
    Bessel functions, 224
    function on $\mathbb{Z}$, 101
    Hermite polynomials, 166
geometric progression, 25, 65–66, 228
GFT, 450
Gibbs phenomenon, 44, 47–48, 80, 84
    for wavelets, 624, 687
glissando of Risset, 725

Goertzel algorithm, 347
grouping operator, 659
grouping rule, 187

Haar wavelet, 594
  analysis, 600, 606
  Fourier transform, 596
  scaling function, 597
  synthesis, 594, 606
hanning window, 229
Hartley transform, 248, 249
  advantages, 248, 255, 263, 279
  via Fourier transform, 249, 251
  via rules, 263, 265–266, 285
  tag, 251
heat Ωow, *xiii*, 15, 72, 540–553, 582–587
Heaviside function, 116, 131, 424
Helmholtz, H., 707
Hermite functions, 151, 166
  discrete, 211
Hermite polynomials, 151, 160
hermitian conjugation, 62
  operator, 251
  tag, 252
hermitian conjugation rule, 136
Hilbert transform, 266, 471
  via analytic function, 270
  via Kramers-Kronig relations, 269
  via rules, 267, 288
  for sampling theorem, 519
  tag, 267
Hipparchus-Ptolemy model, 12, 70
Horner's algorithm
  for DFT, 293
  for other tasks, 345–347
Huygens synthesis of waves, 555

i for $\sqrt{-1}$, 2
impulse response, 282, 481
  automobile suspension, 418
  mass on spring, 367-368, 408, 417, 453
  for ODE, 419, 470
  for PDE (*see* kernel)
independent random variables, 764
infinite product, 615
infinite series, 1, 5
  of bandlimited functions, 489, 491, 497, 503, 505
  of generalized functions, 431, 435
  of sinusoids, 5, 440
  of solutions for PDE, 16, 532, 549, 567
  weak convergence of, 431, 440
initial "conditions" for PDEs, 587
integral notation, 371, 378
interpolation
  of bandlimited function, 485, 489, 491, 497, 503, 505
  using FFT, 234, 284
  of Pallas orbit, 14, 70, 360
  by piecewise linear function, 145, 234, 485
  by trigonometric polynomials, 70, 284, 360, 514
inverse power function, 387, 395
inversion rule, 141, 174, 199, 240, 413
involution, 251, 279
isoperimetric inequality, 226, 235

jumps in $f, f', \ldots$, 46, 55, 123, 184
  and Eagle's method, 184, 479
  and generalized derivatives, 406
  and Gibbs phenomenon, 47, 84, 624
  removing, 46, 55

Kasner's problem, 125
kernel
  de la Vallée-Poussin, 27
  diffraction, 466, 559, 566, 586
  diffusion, 542, 548, 586
  Dirichlet, 66, 187, 474
  Fejer, 78, 474
  Poisson, 65
  wave, 528, 532, 575, 576
Kramers-Kronig relations, 269, 289–290
Kronecker product, 338
  algebraic properties, 339, 363
  eigenvalues, 363
  rearrangement, 341
Kronecker rule, 174

Laplace's equation, 525, 591
Laplace function, 135
laser beam, 553
Laurent series
  for ellipse, 222
  for Fourier series, 185, 218
law of large numbers, 778
least squares approximation
  and Fourier synthesis, 75–76, 84
  and sampling theory, 492, 497, 522
Legendre function, 237
Leibnitz notation, 372
Leibnitz rule for differentiation, 107
Lighthill, M.J., 451–452
likelihood function, 794
lolipop plots, 6
Lorenzian, 22
LTI system, 16, 72, 282, 470

Maclaurin series, 1
  for bandlimited function, 486, 517
  and weak convergence, 438
Mallat's herringbone algorithm, 606–607, 645–646 (*see also* FWT)
Mars orbit, 13, 70
Mason Ωow diagram for FFT, 355
max bound
  for generalized function, 784
  for probability density, 745, 783
max Ωat trigonometric polynomial, 633
mean μ, 756, 759
  for sum of random variables, 766–767
Mersenne's formula for frequency, 536
mesa function, 373, 454, 754 (*see also* tapered box)
Michelson and Stratton harmonic analyzer, 87
midpoint regularization, 45
mirror
  for boundary condition, 569
  for Fourier Transform rules, 134
  for reverse carry algorithm, 305
mnemonic, 137, 177, 200
modulation
  frequency, 711
  index, 711, 715

modulation rule, 137, 177, 199, 260, 413
modulus of coninuity, 598, 643
moments
  for probability density and smoothness, 757, 785
  for sum of random variables, 766
  for wavelet, 618, 621, 680–681
monochord, 698, 729
monotonicity relation, 744
Monticello, 107
mother wavelet, 594
movies
  computing frames with FFT, 571
  for diffracting laser beam, 572, 590
  for heat Ωow, 572
  for vibrating string, 572, 575, 577, 579–581
  for water waves, 591
multiplication using FFT, 113
multiplication of generalized functions, 398
multiplication rule, 144, 170, 177, 199, 413
multiresolution analysis, 597, 642, 684
music
  beat, 700
  interval, 694, 697–699
  loudness, 701
  pictogram, 700
  pitch, 694, A36
  samples, 693, 707–708, 715, 732
  scales, 697–700
  score, 693, 701–703
    for wavelets, 597, 600, 674
  spectrogram, 703, 724
  timbre, 707
  transformation, 725, 735
musical tone, 694
  via additive synthesis, 73, 707–711
  for bell, 709, 716
  for brass, 710
  via computer, 694, 700, 715, 718
  via FM synthesis, 711–717, 734
  information content, 728
  loudness, 695, 728
  from monochord, 729
  from noise, 718–723
  for string, 710, 731

N$_2$ molecule, 738
Newton's "method", 1
Nobel prize, 22, A2
noise
  filtered, 721
  sound of, 720, 723
  white, 718
normal density, 139, 150, 739
  distribution function, 740, A32
Nyquist condition, 486, 705, 717

odd
  function, 62, 64, 247
  generalized function, 397, 455
  projection, 245
Ohm's law of acoustics, 696, 729

orbit
  for cardioid, rose, . . ., 225
  for Mars, 13, 70
  for Pallas, 14, 70
  symmetry of, 225
  for vibrating string, 579
order of approximation, 620, 642–644
operation, 292
operation count
  for additive, FM synthesis of tone,
    732
  for bit reversal permutation, 354
  for FFT, 299, 323, 327, 358
  for FHT, 325
  for FWT, 608, 646, 692
  for naive DFT, 292, 294
operators, 16, 239, A18–A21
  blanket hypotheses, 241–242
  from complex conjugation, 251–253
  factorization of, 257
  for filter bank, 656–660
  Fourier transform of, 255–256, A18–
    A21
  Hartley transform of, 263
  LTI, 16, 72, 282, 470
  for Mallat's herringbone, 645
  from powers of $\mathcal{F}$, 243
  for pre-, post-processing, 652
  projection, 245, 253–254, 277
  symmetry preserving, 254, 282
orthogonal projection, 642
orthogonality relations
  via centroid, 74
  for complex exponentials, 24–25
  for Hermite functions, 166
  for sin, cos, 75
  for sinc functions, 160
  for wavelets, 602, 625, 642

$\mathbb{P}_N$ **for polygon**, 8
Paley-Wiener theorem, 517
Pallas, 14, 70, 361
Papoulis sampling theorem, 503–504,
  519
parallel operation for FFT, 343
Parseval identities, 23–24, 73–74
  for evaluating integrals, 83, 149
  for evaluating sums, 190, 221, 226
  for generalized functions, 391, 466
  link to convolution, 170
  validity, 24, 82 (*see also* Plancherel)
partial derivative of generalized function, 438
partial fractions, 143, 416
partition of unity, 171, 446, 622
PDE, *xiii*, 523, 587
periodic function, 4, 8, 33, 440
  for ear, 697
periodization, 32, 535, 550, 569, 671
phase deaf, 697, 729
pi, computation of, 113–114
piano
  equitempered scale, 699
  for harmonic synthesis, 73
  keyboard frequencies, A36
piecewise
  constant, 81
  continuous, 26, 76–77, 81, 121–123,
    126–128

piecewise (*cont.*)
  polynomial, 117, 145, 170, 380, 382–
    383, 463, 623
  smooth, 39, 42, 45, 55, 57, 83–85,
    123, 406, 491, 505, 623
pitch perception, 694
Plancherel identities, 24
  via autocorrelation, 162
  for evaluating integrals, 148
  for evaluating sums, 190, 221, 224,
    226
  validity, 30, 76, 82–83
Poisson probability density, 781–782
Poisson process, 782
Poisson relations, 33–36
  for evaluating sums, 149
  for finding Fourier series, 179, 262,
    478, 488
  for unifying Fourier analysis, 36–37
Poisson sum formula, 39, 50, 393, 488
polarization identity, 24, 74
polygon function, 191
power functions, 329, 395, 401, 456
  truncated, 382
power scaling rule, 142, 413
primitive root, 238
probability density function, 739 (*see
  also* random variable)
  Benford, 798
  Bernoulli, 767, 769, 776, 779
  binomial, 239
  bivariate, 764
  Cauchy, 758, 760, 775
  from characteristic function, 741
  chi squared, 791
  closure, 787
  coin Ωip, 740, 742, 777
  convolution of, 765
  die-toss, 739, 742, 754, 756, 773
  Dirac, 757, 775
  from distribution function, 740
  gamma, 781
  Laplace, 739, 781
  Maxwell, 737, 791
  Poisson, 739, 781
  standard normal, 139, 150, 739–740,
    754, 758, 768
  for sum of random numbers, 765–766,
    768
  truncated exponential, 742, 757, 771,
    777
  uniform, 739, 742
projection operators, 245, 253–254, 277–
  278, 281, 642
Ptolemy, C., 12, 699
pulse amplitude, 511
Pythagoras and music, 698, 729–730

quadratic residue, 237
quantization of samples, 484
quantum mechanics
  Schrödinger equation, 558
  uncertainty relation, 762
  wave packet for free particle, 563

$\mathbb{R}$ **for real numbers**, 3
ramp function, 227, 381
random number generator, 718

random variables, 753
  characteristic function for, 759, 767
  generation of, 795
  independent, 764
  joint density, 764
  max, min of, 790
  via probability density, 253
  sum of independent, 764
random walk, 777
rational function, 143, 159, 416–419,
  465
real world sampling theorem, 507
reciprocity relations, 69
recursion (*see* recursion)
recursive algorthm for FFT, 350–351
reduced wave function, 556
reΩection of light at mirror, 569
reΩection operator, 240, 243
  tag, 251
reΩection rule, 135, 177, 199, 413
regular tails, 48, 53, 55, 82, 85
relatively prime, 206, 236
repeat rule, 202, 232, 257, 261
response of LTI system, 282, 419, 470
Riemann sum, 39, 44
  for Fourier coefficient, 79
  for Fourier transform, 79
  and Gibbs phenomenon, 44
Riemann-Lebesgue lemma, 81
Risset's glissando, 725
rules
  for derivatives of generalized func-
    tions, 405
  for Fourier transforms of functions
    on $\mathbb{P}_N$ 199–212, A16–A17
    on $\mathbb{R}$, 132–147, A13
    on $\mathbb{T}_p$, $\mathbb{Z}$, 176–182, 187–190, A14–
      A15
  for Fourier transforms of generalized
    functions, 413
  for manipulation of generalized func-
    tions, 389

$\mathbb{S}$ **for Schwartz functions**, 372
same sign shift, 137 (*see also*
  mnemonic)
sample-sum rule, 210, 212, 284
samples for Daubechies wavelet, 652,
  687, 689
sampling, 32, 483
  rate for audio, 484
  for wavelet analysis, 649, 668, 685
sampling function (*see* comb function)
sampling rule, 210, 214, 262
sampling theorem
  for almost bandlimited functions, 505
  when F is piecewise smooth, 491,
    497
  using filters, 498, 501, 503
  using fragments of Π, 495
  for generalized functions, 487
scale for music, 697–700, 729
scale for wavelet approximation, 595–
  596
scaling function for wavelets, 597, 609
Schoenberg, I., 19, 72, 125
Schrödinger's equation, *xiii*, 558
Schwartz, L., *xii*, 368, 451

Schwartz functions, 372–374, 482
  closure of, 375
semitone, 699
Shah function (*see* comb function)
Shannon, C., *xiii*, 484, 491
Shannon's sampling theorem, 491
shift rule (*see* translation rule)
shuffle permutation, 312, 332
  action, 315, 332, 341–342
  operator identities, 314, 333, 342, 364
  products for FFT, 314, 365
signum function, 228, 424
  for Hilbert transform, 267
sin operator, 63, 67, 246–247
sin transform, 63, 67, 247, 287
sinc function, 130
  properties, 141, 160, 493, 514–515
singularity function
  on $\mathbb{R}$, 51–52, 55, 82–83
  on $\mathbb{T}_p$, 41–42, 46
slowly growing function, 376
slowly growing sequence, 444
smoothness of
  B-spline, 117, 193
  convolution product, 107, 123, 126, 128, 401
  Fourier transform, 48, 81, 153, 169, 454
  solution of diffusion equation, 542
  solution of dilation equation, 616–619
soil temperature, 547, 586
sparce matrix factorization, 311, 675
spectral
  density, 719–720
  enrichment, 715, 731
  factorization, 634, 682
spectrogram
  for bell tone, 616, 709
  generation of, 702–704
  for Risset's glissando, 726
  for shaped noise, 721, 723
  for Twinkle, Twinkle, 703
spectroscopy, 21
spectrum of arginine, 22
standard deviation σ, 756
standard normal probability distribution, A32
Stockham's autosort FFT, 344, 366
Strang, G., *xii*, A1
structure of this book, *xi*
suitably regular, 3
sum of independent random variables, 765
  and central limit thoerem, 771–779
  mean, variance for, 767
  probability density for, 766
summation rule, 204, 209, 258, 262
support-limited function, 426
support-limited wavelets, 609
synthesis
  using bandlimited functions, 75–76, 84, 489, 491, 497, 499, 503
  using cas, 249
  using complex exponentials, 3, 5, 7, 10, 430, 440
  using sin, cos, *ix*, 67, 247
  using solutions of PDE, 15, 524, 582
  using wavelets, 594, 597, 684

symmetry
  for boundary conditions, 535, 545, 550–552, 569, 585
  for deriving the Maxwell density, 791
  via Fourier transforms, 62, 64, 66–67, 477
  via operators, 239, 254–256, 277
  for solutions of PDEs, 575, 584, 590

$\mathbb{T}_p$ **for circle**, 4
tag notation, 251–252, 267, 286, 607
tapered box, 445–447, 488
Tartaglia formula for roots of cubic, 19, 72
Taylor's formula, 122, 624, 685, 786
temperature, 15–16, 540–553
test functions, 451
thick coin, 779
timbre of tone, 707
transformation of music theme, 724–725
transformation shift rule (*see* modulation rule)
translation invariant, 17, 282 (*see also* LTI)
translation rule, 136, 140, 177, 199, 413
trapazoid rule
  and Euler-Maclaurin formula, 213
  and Fourier coefficients, 234
tree diagram for FFT, 298, 302
triangle function, 144
truncated exponential, 131, 143, 165, 470
truncated power function, 382

uncertainty relation, 736, 761, 792
unification of Fourier analysis
  via Fourier-Poisson cube, 36
  via generalized functions, 448–550
unit gaussian function, 132
units for s, x, k, n, 68
universal constant β, 244
upsampling (*see* zero packing)

validity of Fourier's representation
  for generalized functions, 413, 440–441, 450
  impact on mathematics, A1–A2
  for ordinary functions, 37–58, 77, 81–82, 85
validity of wavelet representation, 600, 684
variance σ^2, 756, 759
  for sum of random variables, 766–767
vector operation for FFT, 343
velocity of traveling wave
  group, 589
  phase, 566, 589
  on string, 528
  on water, 591
vibrating string, 523
  bowed, plucked, struck, 537, 582
  equation of motion, 526
  frequency, 536
  normal vibrational modes, 536
  overtones, 578

vibrating string (*cont.*)
  send message, 526
  shake to rest, 577
  with stiffness, 581
  tone synthesis, 539, 710, 731
Vieta's formula, 570

water waves, 554, 591
wave equation, 523, 527
  with boundary conditions, 532, 535
  conservation of energy, 531
  derivation, 526
  Fourier synthesis, 524, 532
  initial conditions, 527, 531, 587
  kernel, 528, 532, 575, 576
  polynomial solutions, 574
  symmetry, 575
  traveling solution, 529–530, 534, 575–576
  velocity, 592
wavelet, 594
  analysis equation, 600, 640
  coefficients, 609
  continuity, 683
  Daubechies, 610, 614
  via dilation equation, 609
  frame-detail, 598–604
  Haar's prototype, 594
  mother-father, 594, 597
  music score, 597, 600, 674
  samples, 687
  scaling function, 597, 609
  having smoothness, 616, 683
  support-limited, 594, 616, 677
  synthesis equation, 594, 684
  vs wave, 593
weak limit, 428
  for central limit theorem, 772
  for "continuity", 431
  for derivative, 430
  for Fourier series, 440–441, 450
  for initial conditions, 587
  for partial derivative, 438
  for sampling theorem, 489
  for solving dilation equation, 614
  for solving PDEs, 587
  transformations of, 434
Weierstrass theorem, 26, 29, 76, 78
Weierstrass tone, 731
Weyl's equidistribution theorem, 195
Whittaker-Robinson Ωowchart for harmonic analysis, 349, A22
Wiener's series for Fourier analysis, 167
window for DFT, 229, 705
Wirtinger's inequality, 226, 231

Young's double slit, 560, 588

$\mathbb{Z}$ **for integers**, 5
zero packing (upsampling), 187, 201, 232, 257, 261, 656
Zipper identity, 314
  for FFT, 312, 327
  for FHT, 323, 359
  as Kronecker product, 344